Radio Frequency Integrated Circuits and Systems

Second Edition

This updated and expanded new edition equips students with a thorough understanding of the state of the art in RF design and the practical knowledge and skills needed in industry. Introductory and advanced topics are covered in depth, with clear, step-by-step explanations, including core topics such as RF components, signals and systems, two-ports, noise, distortion, low-noise amplifiers, power amplifiers, and transceiver architectures. New material has been added on wave propagation, skin effect, antennas, mixers and oscillators, and digital PAs and transmitters. Two new chapters detail the analysis and design of RF and IF filters (including SAW and FBAR duplexers and N-path filters), phase-locked loops, frequency synthesizers, digital PLLs, and frequency dividers. Theory is linked to practice through real-world applications, practical design examples, and exploration of the pros and cons of various topologies. Over 250 homework problems are included, with solutions and lecture slides for instructors available online. With its uniquely practical and intuitive approach, this is an essential text for graduate courses on RFICs and a useful reference for practicing engineers.

Hooman Darabi is a Fellow of Broadcom, California, and a lecturer at the University of California, Los Angeles. His research interests include analog and RF IC design for wireless communications.

Radio Frequency Integrated Circuits and Systems

Second Edition

HOOMAN DARABI

Broadcom Inc., Irvine

Shaftesbury Road, Cambridge CB2 8EA, United Kingdom

One Liberty Plaza, 20th Floor, New York, NY 10006, USA

477 Williamstown Road, Port Melbourne, VIC 3207, Australia

314–321, 3rd Floor, Plot 3, Splendor Forum, Jasola District Centre, New Delhi – 110025, India

103 Penang Road, #05–06/07, Visioncrest Commercial, Singapore 238467

Cambridge University Press is part of Cambridge University Press & Assessment, a department of the University of Cambridge.

We share the University's mission to contribute to society through the pursuit of education, learning and research at the highest international levels of excellence.

www.cambridge.org
Information on this title: www.cambridge.org/9781107194755

DOI: 10.1017/9781108163644

© Cambridge University Press & Assessment 2020

This publication is in copyright. Subject to statutory exception and to the provisions of relevant collective licensing agreements, no reproduction of any part may take place without the written permission of Cambridge University Press & Assessment.

First published 2015
Second edition 2020

A catalogue record for this publication is available from the British Library

Library of Congress Cataloging-in-Publication data
Names: Darabi, Hooman, 1972– author.
Title: Radio frequency integrated circuits and systems / Hooman Darabi Broadcom Inc., Irvine.
Description: Second edition. | Cambridge, United Kingdom ; New York, NY, USA : Cambridge University Press, 2020.
Identifiers: LCCN 2019021297 | ISBN 9781107194755 (hardback)
Subjects: LCSH: Radio frequency integrated circuits–Design and construction.
Classification: LCC TK7874.78 .D37 2020 | DDC 621.3841/2–dc23
LC record available at https://lccn.loc.gov/2019021297

ISBN   978-1-107-19475-5   Hardback

Additional resources for this publication at www.cambridge.org/darabi2

Cambridge University Press & Assessment has no responsibility for the persistence or accuracy of URLs for external or third-party internet websites referred to in this publication and does not guarantee that any content on such websites is, or will remain, accurate or appropriate.

To my family

CONTENTS

Preface to the Second Edition — page xiii
Preface to the First Edition — xv
Glossary — xviii

1. RF Components — 1
1.1 Electric Fields and Capacitance — 2
1.2 Magnetic Fields and Inductance — 5
1.3 Time-Varying Fields and Maxwell Equations — 9
1.4 Circuit Representation of Capacitors and Inductors — 11
1.5 Distributed and Lumped Circuits — 12
1.6 Energy and Power — 15
1.7 LC and RLC Circuits — 17
1.8 The Uniform Plane Wave — 26
1.9 Antennas — 36
1.10 Integrated Capacitors — 43
1.11 Integrated Inductors — 47
1.12 Summary — 70
1.13 Problems — 70
1.14 References — 77

2. RF Signals and Systems — 79
2.1 Fourier Transform and Fourier Series — 80
2.2 Impulses — 83
2.3 Fourier Transform of Periodic Signals — 85
2.4 Impulse Response — 86
2.5 Network Functions — 88
2.6 Hilbert Transform and Quadrature Signals — 93
2.7 Stochastic Processes — 95
2.8 Analog Linear Modulation — 109
2.9 Analog Nonlinear Modulation — 115
2.10 Modern Radio Modulation Scheme — 119
2.11 Single-Sideband Receivers — 121
2.12 Summary — 123
2.13 Problems — 123
2.14 References — 129

3. RF Networks — 130
- 3.1 Introduction to Two-Ports — 130
- 3.2 Available Power — 139
- 3.3 Impedance Transformation — 149
- 3.4 Lossless Transmission Lines — 166
- 3.5 Low-Loss Transmission Lines — 173
- 3.6 Receive–Transmit Antennas as Two-Port Circuits — 176
- 3.7 Smith Chart — 179
- 3.8 Scattering Parameters — 185
- 3.9 Differential Two-Ports — 196
- 3.10 Summary — 197
- 3.11 Problems — 197
- 3.12 References — 205

4. RF and IF Filters — 206
- 4.1 Ideal Filters — 207
- 4.2 Doubly Terminated LC Filters — 208
- 4.3 Active Filters — 238
- 4.4 Surface and Bulk Acoustic Wave Filters — 248
- 4.5 Duplexers — 253
- 4.6 N-Path Filters — 255
- 4.7 Quadrature Filters — 260
- 4.8 Summary — 270
- 4.9 Problems — 270
- 4.10 References — 274

5. Noise — 278
- 5.1 Types of Noise — 279
- 5.2 Two-Port Equivalent Noise — 296
- 5.3 Noise Figure — 299
- 5.4 Minimum NF — 303
- 5.5 Impact of Feedback on Noise Figure — 309
- 5.6 Noise Figure of Cascade of Stages — 312
- 5.7 Phase Noise — 316
- 5.8 Sensitivity — 317
- 5.9 Noise Figure Measurements — 322
- 5.10 Summary — 325
- 5.11 Problems — 325
- 5.12 References — 330

6. Distortion — 331
- 6.1 Blockers in Wireless Systems — 332
- 6.2 Full-Duplex Systems and Coexistence — 335

6.3	Small Signal Nonlinearity	336
6.4	Large Signal Nonlinearity	356
6.5	Reciprocal Mixing	359
6.6	Harmonic Mixing	363
6.7	Transmitter Nonlinearity Concerns	364
6.8	Summary	382
6.9	Problems	382
6.10	References	385

7. Low-Noise Amplifiers — 386

7.1	Matching Requirements	387
7.2	RF Tuned Amplifiers	392
7.3	Common-Source and Common-Gate LNAs	397
7.4	Shunt Feedback LNAs	401
7.5	Series Feedback LNAs	405
7.6	Feedforward LNAs	410
7.7	LNA Practical Concerns	413
7.8	LNA Power–Noise Optimization	421
7.9	Signal and Power Integrity	425
7.10	LNA Design Case Study	431
7.11	Summary	433
7.12	Problems	433
7.13	References	436

8. Mixers — 437

8.1	Mixers Fundamentals	437
8.2	Evolution of Mixers	442
8.3	Active Mixers	445
8.4	Passive Current-Mode Mixers	460
8.5	Passive Voltage-Mode Mixers	484
8.6	Transmitter Mixers	486
8.7	Harmonic Folding in Transmitter Mixers	491
8.8	LNA/Mixer Case Study	494
8.9	Summary	502
8.10	Problems	503
8.11	References	507

9. Oscillators — 510

9.1	The Linear LC Oscillator	511
9.2	The Nonlinear LC Oscillator	517
9.3	Phase Noise Analysis of the Nonlinear LC Oscillator	521
9.4	LC Oscillator Topologies	540
9.5	Q-Degradation	551

9.6	Frequency Modulation Effects	554
9.7	More LC Oscillator Topologies	562
9.8	Ring Oscillators	566
9.9	Quadrature Oscillators	577
9.10	Crystal and FBAR Oscillators	581
9.11	Summary	588
9.12	Problems	589
9.13	References	592

10. PLLs and Synthesizers 595

10.1	Phase-Locked Loops Basics	596
10.2	Type I PLLs	598
10.3	Type II PLLs	601
10.4	Integer-N Frequency Synthesizers	611
10.5	Fractional-N Frequency Synthesizers	618
10.6	Frequency Dividers	630
10.7	Introduction to Digital PLLs	640
10.8	Summary	647
10.9	Problems	647
10.10	References	649

11. Power Amplifiers 651

11.1	General Considerations	652
11.2	Class A PAs	654
11.3	Class B PAs	656
11.4	Class C PAs	660
11.5	Class D PAs	662
11.6	Class D Digital PAs	665
11.7	Class E PAs	669
11.8	Class F PAs	672
11.9	PA Linearization Techniques	673
11.10	Summary	685
11.11	Problems	685
11.12	References	688

12. Transceiver Architectures 690

12.1	General Considerations	691
12.2	Receiver Architectures	692
12.3	Blocker-Tolerant Receivers	707
12.4	Receiver Filtering and ADC Design	715
12.5	Receiver Gain Control	718
12.6	Transmitter Architectures	719

12.7	Transceiver Practical Design Concerns	734
12.8	Summary	751
12.9	Problems	751
12.10	References	754

Index 757

PREFACE TO THE SECOND EDITION

While the first edition was written to be equally used as a textbook in graduate-level RF courses in academia, as well as among RF professional engineers, it was felt that it may be more suitable to practicing RF engineers in industry. Personal teaching experience in the past few years and feedback from some colleagues and friends inspired me to work on a new edition to address the needs of entry-level graduate and senior undergraduate students more properly. To that extent, without altering the overall organization of the book significantly and retaining the bulk of material from the first edition, the book has been rewritten to address the fundamental concepts more deeply, give more step-by-step design explanations, and provide many more examples and clarifying points. The examples particularly are designed to cover both fundamental theoretical concepts as well as hands-on circuit-level design and simulations and have been organized more distinctly to both complement the basic concepts and provide guidelines for more advanced discussions. It has been the author's intention to familiarize the readers early on with two of RF designers' best friends, Spectre-RF and EMX, through many of these examples covering circuit-level analysis followed by verification through actual simulations. Furthermore, over 50 new additional problems, with guidance and hints, have been added throughout the 12 chapters and may be assigned as homework problems or exam questions or simply used as means of practicing the fundamental concepts.

Apart from this main motivation, there were several important concepts crucial in the RF design that were missing from the first edition or at least needed more elaboration. One of the main additions of this edition is the inclusion of two new chapters, one to cover in-depth analysis and design of RF and IF filters, including SAW, FBAR, duplexers, and N-path filters, as well as a dedicated chapter to cover phase-locked loops, frequency synthesizers, digital PLLs, and frequency dividers. Additionally, new discussions on wave propagation, skin effect, and antennas were added to Chapter 1. This is introductory material but is deemed vital for RF engineers as they are an essential part of RF design. Among other new additions are more expanded discussion on RF two-ports and reciprocal networks, additional discussions of mixers and oscillators, and a section on digital power amplifiers. One of the main innovations of this edition is inclusion of a section to address signal and power integrity, a critical part of the RF design. This was added to the low-noise amplifiers chapter, as being the most susceptible block, though is applicable in general to any part of the radio. It covers various topics such as electric and magnetic coupling, shielding, and power rails transient noise.

Neither in the first edition nor in the current one does a dedicated chapter on the subject of the wireless standards exist. The wireless standards are evolving, and as such a chapter examining them would be drastically different today versus five or ten years in the future. It is more important for RF designers to attain the ability to translate high-level system specifications into circuit-level requirements rather than memorizing the standard itself. This is fully addressed in Chapter 5 and especially Chapter 6, along with numerous references to the existing mainstream RF standards.

I have been very fortunate to enjoy the company of many talented individuals at Broadcom, from whom I have learned tremendously in my 20-year tenure there. I am thankful to all of them. I would like to specifically thank the following Broadcom colleagues and friends who helped proofread various chapters of the new edition: Wei-Liat Chan, Yuyu Chang, Saeed Chehrazi, Matteo Conta, Milad Darvishi, Valentina Della Torra, Dale Douglas, Mohyee Mikhemar, David Murphy, Bevin Perumana, and Hao Wu. I am also thankful to Dr. Ed Chien, who helped on the PLL chapter in general with many useful discussions, and particularly for his guidance on the digital PLL section. Likewise, thanks to Dr. Mikhemar, who co-wrote the digital PA section, and Dr. Rich Ruby, who provided valuable feedback on the SAW and FBAR section. I would also like to thank my teammates in Broadcom, Saeed Chehrazi, Milad Darvishi, Daivd Murphy, and Hao Wu, with whom I have had many useful discussions on various topics, directly or indirectly related to the book.

I am grateful to my Ph.D. advisor with whom I have kept close contact throughout the years, Prof. Asad Abidi from UCLA, who has been a great inspiration. I have benefitted tremendously from his insights and innovative teaching methods in general, and particularly in many sections of the book. I would also like to acknowledge my undergraduate instructor from Sharif University, Prof. Masood Jahanbegloo, a UCLA alumnus himself, for his dedicated teaching, and to whom I owe my fundamentals of basic circuit theory and electronics.

Lastly, I would like to thank my wife, Dr. Shahrzad Tadjpour, for her patience and technical feedback, and my family for their support.

PREFACE TO THE FIRST EDITION

In the past 20 years, radio frequency (RF) integrated circuits in CMOS have evolved dramatically, and matured. Started as a pure research topic in mid-1990 at several universities, they have made their way into complex systems-on-a-chip (SoCs) for wireless connectivity and mobile applications. The reason for this dramatic evolution comes primarily from two main factors: the rapid improvement of CMOS technology and innovative circuits and architectures. In contrary to the common belief that RF and analog circuits do not improve much with technology, a faster CMOS process has enabled a number of topologies that have led to substantially lower cost and power consumption. In fact to some extent, the recent inventions may not have been possible if not for better and faster technology. This rapid change has caused the modern RF design to be somewhat industry-based, and consequently it is timely, and perhaps necessary, to provide an industry perspective. One main goal of this book has been to cover possibly fewer topics but in a much deeper fashion. Even for RF engineers working on routine products in industry, a deep understanding of fundamental concepts of RF IC design is very critical, and it is the intention of this work to break this gap. During the course of writing the book, I have tried to address the topics that I would have wanted as a *wish list* for my fellow future colleagues. Our main focus then has been to elaborate the basic definitions and fundamental concepts, while an interested designer with a strong background can explore other variations on his or her own.

The contents of this book stem largely from the RF courses taught at the University of California, Los Angeles and Irvine, as well as many years of product experience at Broadcom. Accordingly, the book is intended to be useful both as a text for students and as a reference book for practicing RF circuit and system engineers. Each chapter includes several worked examples that illustrate the practical applications of the material discussed, with examples of real-life products and a problem set at the end to complement that.

RF circuit design is a *multidisciplinary* field where a deep knowledge of analog integrated circuits, as well as communication theory, signal processing, electromagnetics, and microwave engineering is crucial. Consequently, the first three chapters as well as parts of Chapter 4 cover selected topics from the aforementioned fields, but customized and shaped properly to fit the principles of RF design. It is, however, necessary for the interested students or RF engineers to have already taken appropriate senior-level undergraduate courses.

An outline of each chapter is given below along with suggestions for the material to be covered if the book is to be used for a 20-lecture quarter-based course. Furthermore, in the beginning of each chapter a list of specific items to be covered as well as more detailed suggestion of which sections to include for the class use are outlined. For beginner and intermediate practicing engineers we recommend following the selected topics suggested for the class use, while more advanced readers may focus on the other topics assigned for reading.

Chapter 1 contains a review of basic electromagnetic concepts and particularly the inductors and capacitors. Among many students and RF engineers, often the basic definition of capacitors and inductors is neglected, despite using them very regularly. A short reminder is deemed necessary. Furthermore, some basic understanding of Maxwell's equations is needed to understand transmission lines, electromagnetic waves, the antenna concept, and scattering parameters. These are discussed in Chapter 3. The chapter also gives an overview of integrated inductors and capacitors in modern CMOS. Two lectures are expected to be needed to cover the basic concepts.

Chapter 2 deals with basic communication and signal processing concepts, which are a crucial part of RF design. The majority of the material is gathered to provide a reminder and may be left to be studied outside the class, depending on the students' background. However, we cannot emphasize enough the importance of them. Spending a total of two or three lectures on the stochastic processes, modulation section, as well as a brief general reminder of passive filters and Hilbert transform may be helpful.

Chapter 3 is concerned with several key concepts in RF design such as available power, matching topologies, transmission lines, as well as scattering parameters and complements Chapter 1. Two lectures may be dedicated to cover the first three sections, while the more advanced material on transmission lines, the Smith chart, and scattering parameters may be very briefly touched or omitted altogether depending on the students' background.

In Chapter 4 we discuss, noise, noise figure, sensitivity, and an introduction to phase noise. The introductory part on types of noise may be assigned as reading, but the noise figure definition, minimum noise figure, and sensitivity sections must be covered in full. A total of two or three lectures suffices.

Chapter 5 covers the distortion and blockers. A large portion of this chapter (as well as Chapter 10) may be left for a more advanced course, and one lecture should suffice to cover only the basic concepts. However, the material may be very appealing to RF circuit and system engineers who work in industry. A thorough knowledge of this chapter is crucial to understand Chapter 10.

Chapters 6 to 9 deal with the RF circuit design. Chapter 6 is mostly built upon the concepts covered in Chapters 3 and 4 and deals with low-noise amplifiers. Three lectures may be dedicated to cover most of the topics presented in this chapter.

Chapter 7 provides a detailed discussion on receiver and transmitter mixers. Roughly two lectures may be dedicated to this chapter to cover basic active and passive topologies with some limited discussion on noise. The majority of the material on M-phase and upconversion mixers may be assigned as reading.

Chapter 8 discusses oscillators, including LC, ring, crystal oscillators, and an introduction to phase-locked loops. The chapter is long, and the latter three topics may be assigned as reading, while two lectures could be dedicated to LC oscillators, and a brief introduction to phase noise. A detailed discussion of phase noise is very math intensive, and may be beyond the scope of an introductory RF course. Thus, it may be sufficient to focus mostly on the premises of an abstract linear oscillator, and summarize Bank's general results to provide a more practical perspective.

Power amplifiers are discussed in Chapter 9. Basic classes are presented in the first few sections, followed by efficiency improvement and linearization techniques. Most of the material on the latter subject may be skipped, and one or two lectures may be assigned to cover a few examples of classes (perhaps only classes A, B, and F), as well as the introductory material on the general concerns and the trade-offs.

Finally, in Chapter 10 transceiver architectures are presented. This is one of the longest chapters of the book, and much of the material can be assigned as reading. The last section covers some practical aspects of the design, such as packaging and production issues. It presents a few case studies as well. The topics may be appealing for practicing RF engineers, but the entire section may be skipped for class use. A maximum of two lectures is sufficient to cover selected key transceiver architectures.

I have been very fortunate to be working with many talented RF designers and instructors throughout my carrier at UCLA, and subsequently at Broadcom. They have had an impact on this book one way or another. However, I wish to acknowledge the following individuals who have directly contributed to the book: Dr. David Murphy from Broadcom who co-wrote most of Chapter 8, and provided very helpful insight on Chapter 6, particularly the LNA topologies; Dr. Ahmad Mirzaei from Broadcom as well, who helped on the write up of sections of Chapters 9 and 10, and proofread the entire book painstakingly. They both have been major contributors to this book beyond the chapters mentioned. I would also like to thank Dr. Hwan Yoon from Broadcom with whom I had numerous helpful discussions on Chapter 1 material, and particularly the integrated inductors. My sincere thanks go to Prof. Eric Klumperink of the University of Twente, who proofread most of the book diligently, and provided valuable insight on various topics. I would also like to acknowledge my sister Hannah, who helped design the book cover. Lastly, I wish to thank my wife, Shahrzad Tadjpour, not only for her technical feedback on the book, but for her general support throughout all these years.

GLOSSARY

ACLR	alternate adjacent channel leakage ratio
ADC	analog-to-digital converter
AM	amplitude modulation
BALUN	balanced-unbalanced
BJT	bipolar junction transistor
BNC	Bayonet Neill–Concelman
BPF	bandpass filter
BT	Bluetooth
BW	bandwidth
CG	common-gate
CMOS	complementary metal-oxide semiconductor
CP	charge pump
CS	common-source
DAC	digital-to-analog converter
dBc	decibels relative to the carrier
DSB	double sideband
DSP	digital signal processor
DUT	device under test
EDGE	enhanced data rate for GSM evolution
EMF	electromotive force
ENR	excess noise ratio
EVM	error vector magnitude
FBGA	fine-pitch ball grid array
FDD	frequency division duplexing
FDMA	frequency division multiple access
FET	filed effect transistor
FM	frequency modulation
FSK	frequency shift keying
GSM	global system for mobile communications
HD	harmonic distortion
HPF	highpass filter
IF	intermediate frequency
IMN	Nth-order intermodulation
IO	input/output
IPN	Nth-order intermodulation product

I/Q	in/quadrature phase
ISM	industrial–scientific–medical
LNA	low-noise amplifier
LO	local oscillator
LPF	lowpass filter
LTE	long-term evolution
MATLAB	matrix laboratory
NB	narrowband
NF	noise figure
OFDM	orthogonal frequency division multiplexing
OOB	out of band
P1dB	1 dB compression point
PA	power amplifier
PAPR	peak-to-average power ratio
PCB	printed circuit board
PD	phase detector
PDF	probability distribution function
PFD	phase-frequency detector
PGA	programmable gain amplifier
PLL	phase-locked loop
PM	phase modulation
PPM	parts per million
PTAT	proportional-to-absolute temperature
QAM	quadrature amplitude modulation
RDL	redistribution layer
RMS	root mean square
RSSI	received signal strength indicator
RX	receiver
SAW	surface acoustic wave
SNR	signal-to-noise ratio
SSB	single sideband
TDD	time division duplexing
TDMA	time division multiple access
TIA	transimpedance amplifier
TR	transmit–receive
TX	transmitter
VCO	voltage-controlled oscillator
WB	wideband
WCDMA	wideband code division multiple access
Wi-Fi	wireless fidelity
WLAN	wireless local area network
XO	crystal oscillator

1 RF Components

In this chapter basic components used in RF design are discussed. Detailed modeling and analysis of MOS transistors at high frequency can be found already in many analog books [1], [2]. Although mainly offered for analog and high-speed circuits, the model is good enough for most RF applications operating at several GHz and beyond, especially for nanometer CMOS processes used today. Thus, instead we will have a more detailed look at inductors, capacitors, and LC resonators in this chapter. We will also briefly discuss the fundamental operation of distributed circuits and transmission lines and follow up with more in Chapter 3. In Chapters 5 and 7 we will discuss some of the RF aspects of the transistors, including a more detailed noise analysis as well as substrate and gate resistance. New to this edition are Sections 1.8 and 1.9, which cover the fundamentals of wave propagation and antennas.

LC circuits are widely used in RF design, with applications ranging from tuned amplifiers to matching circuits and LC oscillators. Inspired by superior noise and linearity compared to transistors, historically radios have heavily relied on inductors and capacitors, with large portions of the RF blocks occupied by them. Although this dependence has been reduced in modern radios mostly for cost concerns, RF designers deal with integrated inductors and capacitors quite often.

We start the chapter with a brief introduction to electromagnetic fields and take a closer look at capacitors and inductors from a field perspective. We then discuss capacitors, inductors, and LC resonators from the circuit point of view. We conclude the chapter by presenting the principles and design trade-offs of integrated inductors and capacitors. Throughout this section several examples of inductor and transformer design are presented using theory, as much as possible, validated by EMX simulation, with the goal to familiarize students and junior RF designers with the tool and its applications.

The specific topics covered in this chapter are:

- Capacitance and inductance electromagnetic and circuit definitions
- Maxwell's equations
- Distributed elements and introduction to transmission lines
- Energy, power, and quality factor
- Wave propagation and antennas
- Integrated capacitors and inductors

For class teaching, we recommend focusing on Sections 1.7, 1.10, and 1.11, while Sections 1.1–1.6 as well as 1.8 and 1.9 may be assigned for reading or only a brief summary presented if deemed necessary.

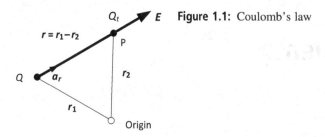

Figure 1.1: Coulomb's law

1.1 ELECTRIC FIELDS AND CAPACITANCE

Let us start with a brief overview of electric fields and electric potential. We shall define the concept of capacitance accordingly.

Published first in 1875 by Charles Coulomb, the French army officer, Coulomb's law states that the force between two point charges separated in a vacuum or free space by a distance is proportional to each charge and inversely proportional to the square of the distance between them (Figure 1.1). It bears a great similarity to Newton's gravitational law, discovered about a hundred years earlier. Writing the force (F_t) as a force per unit charge gives the electric field intensity, E, measured in V/m (or volt per meter) as follows:

$$E = \frac{F_t}{Q_t} = \frac{Q}{4\pi\epsilon_0 r^2} a_r,$$

where the **bold** notation indicates a vector in 3D space, $\epsilon_0 = \frac{1}{36\pi} \times 10^{-9}$ F/m (or farad per meter) is the permittivity in free space, and Q[1] is the charge in C (or coulomb).[2] a_r is the unit vector pointing to the direction of the field, which is in the same direction as the vector r connecting the charge Q to the point of interest P in space (see Figure 1.1). Q_t is a test charge to which the force (or field) created by Q is applied.

In many cases the electric field can be calculated more easily by applying Gauss's law instead,[3] expressing that the electric flux density $D = \epsilon_0 E$[4] (measured in C/m^2) passing through any closed surface is equal to the total charge enclosed by that surface,[5] and mathematically expressed as

$$\oint_S D \cdot dS = Q,$$

where $\oint_S D \cdot dS$ indicates the integral over a closed surface. The *dot* product indicates the product of the magnitude and the cosine of the smaller angle. The charge Q could be the sum of

---

[1] Not to be confused with Q, used as quality factor later in this chapter.
[2] Like many units used in electronics throughout this book (farad, henry, tesla, weber, watt, etc.), coulomb is not one of the seven base SI (international system) units. The SI unit for electric charge is s.A, or second times ampere.
[3] Johann Carl Friedrich Gauss (1777–1855) was a German mathematician and physicist who made significant contributions to many mathematics, physics, and engineering fields.
[4] Only in free space.
[5] The expression itself is a result of Michael Faraday's experiment. Gauss's contribution is providing the mathematical tools to formulate it.

several charge points, that is $Q = \sum Q_i$, a volume charge distribution $Q = \int_V \rho_V dV$, a surface distribution, and so forth. The nature of the surface integral implies that only the normal component of $\boldsymbol{D}$ at surface contributes to the charge, whereas the tangential component leads to $\boldsymbol{D} \cdot d\boldsymbol{S}$ equal to zero.

Example: Consider a long coaxial cable[6] with the inner radius of a, and an outer radius of b, carrying a uniform charge distribution of ρ_S per area on the outer surface of the inner conductor (and $-\frac{b}{a}\rho_S$ on the inner surface of the outer conductor) as shown in Figure 1.2. For convenience, let us use the cylindrical coordinates [3].

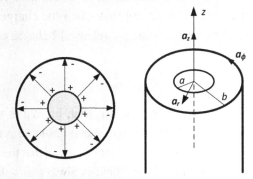

Figure 1.2: Electric flux in a coaxial cable

The flux will have components in the $\boldsymbol{a_r}$ direction, normal to the surface. For an arbitrary length of L in the z-axis direction, we can write:

$$\int_{z=0}^{L} \int_{\phi=0}^{2\pi} D_r (r d\phi dz) = Q = \rho_S (2\pi a L),$$

Thus inside the cable, that is for $a < r < b$:

$$\boldsymbol{D} = \frac{\rho_S a}{r} \boldsymbol{a_r}.$$

The electric field and flux density are both zero outside the cable as the net charge is equal to zero.

Based on the electric energy definition,[7] the potential difference between points A and B (V_{AB}) is defined as

$$V_{AB} = \frac{W}{Q} = -\int_B^A \boldsymbol{E} \cdot d\boldsymbol{L},$$

where W is the energy in J (or joule), and the right side is the *line integral* of the electric field. The physical interpretation of energy or potential is such that moving a charge Q along with the electric field from point B to A results in energy reduction (the charge

---

[6] Coaxial cable was invented by English engineer and mathematician Oliver Heaviside, who patented the design in 1880.
[7] We shall discuss the electric energy shortly.

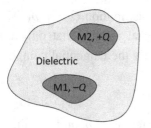

Figure 1.3: Capacitance definition

releases energy), and accordingly we expect point A to be at a *lower potential*. By definition of the line integral, one can see that the sum of *static* potentials in a closed path must be equal to zero, that is, $\oint E \cdot dL = 0$ which is a general representation of *Kirchhoff's voltage law* or KVL. This is physically understood by noting that when the charge is moved around a closed path, the total energy received and the energy released balance each other, thus no *net* work is done.

Example: An interesting property of a charged piece of metal is that, no matter what its shape is and if current is zero, the electric field inside the piece of metal has to be zero. Free charges in the metal go to the surfaces of the metal and arrange themselves so that the electric field is zero everywhere inside the metal. In fact the electric field is always in a direction normal to the surface. Consequently, a closed metal surface, no matter what its shape, screens out external sources of electric field, often referred to as a *Faraday cage*,[8] which has applications in RF shielding (see Chapter 7 for more details on shields and signal integrity).

We close this section by defining capacitance. Suppose we have two oppositely charged (each with a charge of Q) conductors M1 and M2 within a given dielectric with permittivity[9] of $\epsilon = \epsilon_r \epsilon_0$ (Figure 1.3). Assuming a potential difference of V_0 between the conductors, we define capacitance C measured in farad as

$$C = \frac{Q}{V_0}.$$

Alternatively, one can rewrite C as

$$C = \epsilon \frac{\oint_S E \cdot dS}{-\int E \cdot dL},$$

which indicates that capacitance is independent of the charge or potential, as E (or D) linearly depends on Q according to Gauss's law.

Physically capacitance indicates the capability of energy or equivalently electric flux storage in electrical systems, analogous to inductors that store magnetic flux.

---

[8] Michael Faraday (1791–1867) was a British scientist.   [9] $\epsilon_r = 1$ for free space.

> **Example:** Returning to our previous example of the coaxial cable, the potential between the inner and outer conductors is calculated by taking the line integral of the $E = D/\epsilon$, where D was obtained previously. This yields
>
> $$V_0 = -\frac{1}{\epsilon}\int_b^a \frac{a\rho_S}{r}dr = \frac{a\rho_S \ln\frac{b}{a}}{\epsilon}.$$
>
> Thus the capacitance per unit length is equal to
>
> $$C = \frac{2\pi\epsilon}{\ln\frac{b}{a}},$$
>
> which is clearly only a function of the coaxial cable radii and the dielectric permittivity.

1.2 MAGNETIC FIELDS AND INDUCTANCE

A steady magnetic field can be created in one of three ways: through a permanent magnet, through a linear time-varying electric field, or simply due to a direct current. The permanent magnet has several applications in RF and microwave devices, such as passive gyrators used in a lossless circulator, which is a *passive,* but *nonreciprocal* circuit [4], [5]. However, we will mostly focus on the latter two methods of creating magnetic fields, and defer the gyrator implementation details to [4].

In 1820, the law of Biot–Savart[10] was proposed as follows which associates magnetic field intensity H (expressed in A/m) at a given point P to a current of I flowing in a differential vector length dL of an ideal filament:

$$dH = \frac{IdL \times a_r}{4\pi r^2} = \frac{IdL \times r}{4\pi r^3}.$$

The *cross* product ($\times$) indicates the product of the magnitude and the sine of the smaller angle. The magnetic field will then be perpendicular to the plain containing the current filament and vector r, and whose direction is determined based on the right-hand rule. The law states that the magnetic field intensity is directly proportional to the current (I), but inversely proportional to the square of the distance between P and differential length (r), and also proportional to the magnitude of the differential element times the sinus of the angle θ shown in Figure 1.4.

A more familiar law describing the magnetic field was proposed by Ampere[11] shortly afterward in 1823, widely known as Ampere's circuital law,[12] and is mathematically expressed as

$$\oint H \cdot dL = I,$$

---

[10] Named after Jean-Baptiste Biot and Félix Savart, who discovered this relationship in 1820.
[11] André-Marie Ampère (1775–1836) was a French mathematician and physicist who is considered the father of electrodynamics. Ampere, the current unit, is one of the seven SI base units.
[12] Ampere's law may be derived from Biot–Savart's law.

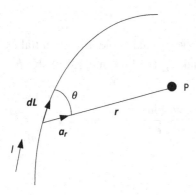

Figure 1.4: Biot–Savart's law expression

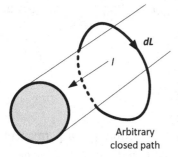

Figure 1.5: Ampere's law

indicating that the line integral of the magnetic field (**H**) about any *closed path* is exactly equal to the current enclosed by that path (Figure 1.5). This law proves to be more useful as it allows us to calculate the field more easily as long as it is properly determined which components of the field are present, and that the symmetry is invoked appropriately. By comparison, Ampere's circuital law is more analogous to Gauss's law, whereas the law of Biot–Savart could be considered similar to Coulomb's law.

Example: Consider a long coaxial cable carrying a uniform current of I in the center conductor, and $-I$ in the outer one shown in Figure 1.6. Clearly the field cannot have any component in the z direction, as it must be normal to the current direction. Moreover, the symmetry shows that **H** cannot be a function of ϕ or z, and thus could be expressed as a general form of $\mathbf{H} = H_r \mathbf{a}_\phi$. Inside the coaxial cable, that is $a < r < b$, applying the line integral then leads to

$$H = \frac{I}{2\pi r} \mathbf{a}_\phi.$$

Moreover, similar to the electric field, the magnetic field is also zero outside the cable as the net current flow is zero, showing the concept of *shielding* provided by the coaxial cable. Note that inside the cable, the magnetic field consists of closed lines circling around the current, as opposed to the electric field lines that start on a positive charge and end on a negative one.

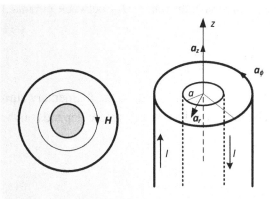

Figure 1.6: Magnetic field in a coaxial cable

In free space, magnetic flux density B (measured in weber/m^2 or tesla), is defined as

$$B = \mu_0 H,$$

where $\mu_0 = 4\pi \times 10^{-7}$ H/m (or henry per meter) in free space and is the *permeability*. The magnetic flux, ϕ, is then the flux passing through a designated area S, measured in weber, and is defined as

$$\phi = \int_S B \cdot dS.$$

Generally the magnetic flux is a linear function of the current (I), that is, $\phi = LI$, where the proportionality constant, L, is known as the *inductance*, and is measured in henry. We can thus say

$$L = \mu_0 \frac{\int_S H \cdot dS}{\oint H \cdot dL},$$

and since H is a linear function of I, as established by Ampere's (or Biot–Savart's) law, the inductance is a function of the conductor geometry and the distribution of the current, but not the current itself.

Example: By calculating the total flux inside the coaxial cable of the previous example, one can simply show that the cable inductance per unit length is

$$L = \frac{\mu_0}{2\pi} \ln \frac{b}{a},$$

whereas the capacitance per unit length of the same coaxial cable was calculated before by applying Gauss's law, equal to

$$C = \frac{2\pi\epsilon}{\ln \frac{b}{a}}.$$

Clearly

$$LC = \mu_0 \epsilon.$$

We conclude this section by defining mutual inductance M_{12} between circuits 1 and 2 in terms of their flux linkage:

$$M_{12} = \frac{N_2 \phi_{12}}{I_1},$$

where ϕ_{12} signifies the flux produced by I_1 that links the path of the filamentary current I_2, and N_2 is the number of turns in circuit 2. The mutual inductance therefore depends on the magnetic interaction between the two currents.

Example: Consider an N-turn solenoid with a finite length of d, consisting of N closely wound filaments that carry a current of I shown in Figure 1.7. We assume the solenoid is long enough with respect to its diameter.

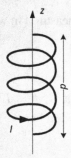

Figure 1.7: An N-turn solenoid

The magnetic field is in the $\boldsymbol{a_z}$ direction, as the current is in the $\boldsymbol{a_\phi}$ direction, and Ampere's law readily shows that within the solenoid

$$\boldsymbol{H} = \frac{NI}{d} \boldsymbol{a_z}.$$

If the radius is r, corresponding to an area of $A = \pi r^2$, the self-inductance is

$$L = \frac{N\phi}{I} = \mu_0 N^2 \frac{A}{d}.$$

Example: Now consider two coaxial solenoids, with radius r_1, and $r_0 < r_1$, carrying currents of I_1 and I_0, and with different number of turns N_1 and N_0, respectively. The top view is shown in Figure 1.8.

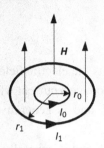

Figure 1.8: Two coaxial solenoids top view

To find the mutual inductance M_{01}, we can write

$$\phi_{01} = \mu_0 A_0 H_0,$$

where $H_0 = \frac{N_0 I_0}{d}$ is the magnetic field intensity created by the smaller solenoid, and ϕ_{01} is the magnetic flux created in the larger solenoid by the smaller one. Note that H_0 is zero outside the radius of the smaller solenoid. By definition we have

$$M_{01} = \frac{N_1}{I_0} \mu_0 A_0 H_0 = \mu_0 N_0 N_1 \frac{A_0}{d}.$$

A similar procedure leads to $M_{10} = \frac{N_0 \phi_{10}}{I_1} = \frac{N_0}{I_1} \mu_0 A_0 H_1$, which comes out to be equal to M_{01}. This is in agreement with reciprocity of course.

1.3 TIME-VARYING FIELDS AND MAXWELL EQUATIONS

As described earlier, time-varying fields could also be a source of electric or magnetic field creation. In 1831, Faraday published his findings from the following experiment where he proved that a time-varying magnetic field does indeed result in a current. He wound two separate coils on an iron toroid and placed a galvanometer in one and a battery and switch in the other (Figure 1.9). Upon closing the switch, he realized that the galvanometer was momentarily deflected. He observed the same deflection but in an opposite direction when the battery was disconnected. In terms of fields, we can say that a time-varying magnetic field (or flux) produces an *electromotive force* (emf, measured in volts) that may establish a current in a closed circuit. A time-varying magnetic field may be a result of a time-varying current, or relative motion of a steady flux and a closed path, or a combination of the two.

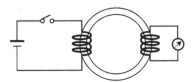

Figure 1.9: Faraday's experiment

Faraday's law as stated above is customarily formulated as

$$\text{emf} = \oint E \cdot dL = -\frac{d\phi}{dt},$$

where the line integral comes from basic definition of voltage (E is the electric field intensity). The minus sign indicates that the emf is in such a direction as to produce a current whose flux, if added to the original one, would reduce the magnitude of emf, and is generally known as Lenz's law.[13]

Similarly, a time-varying electric flux results in a magnetic field, and is generally formulated by modifying the Ampere's circuital law as follows:

$$\oint H \cdot dL = I + \int_S \frac{\partial D}{\partial t} \cdot dS,$$

where D is the electric flux density, and $\int_S \frac{\partial D}{\partial t} \cdot dS$ is termed as the *displacement current* by Maxwell.[14] To summarize, we can state the four Maxwell's equations in the integral form as follows:

$$\oint E \cdot dL = -\int_S \frac{\partial B}{\partial t} \cdot dS$$

$$\oint H \cdot dL = I + \int_S \frac{\partial D}{\partial t} \cdot dS$$

$$\oint_S D \cdot dS = \int_V \rho_V dV$$

$$\oint_S B \cdot dS = 0.$$

The third equation is Gauss's law as we discussed earlier. The fourth equation[15] states that unlike the electric fields that begin and terminate on positive and negative charges, the magnetic field forms *concentric circles*. In other words the magnetic flux lines are closed and do not terminate on a magnetic charge[16] (Figure 1.10). Therefore, the closed surface integral of a magnetic field (or magnetic flux density) is zero.

In free space where the medium is sourceless, I (or ρ_V) is equal to zero. The first two of Maxwell equations, when combined, lead to a differential equation relating the second-order derivative of E (or H) versus space, to its second order derivative versus time, describing the *wave propagation* in free space. For example, if $E = E_x a_x$, or if the electric field is *polarized* only in the x direction, with some straightforward math [6], and using Maxwell's equations in differential form, it can be shown that[17]

---

[13] Emil Lenz (1804–1865) was a Russian physicist.
[14] James Clark Maxwell was a Scottish scientist in the 19th century whose most notable achievement was to formulate the theory of electromagnetic radiation. Maxwell's equations are often described as the second great unification in physics, after the first one realized by Newton.
[15] The fourth equation is often known as Gauss's law for magnetism.
[16] Magnetic charges or monopoles have not been found in nature, although the magnetic monopole is used in physics as a *hypothetical* elementary particle.
[17] The more general form of the wave equation is $\nabla^2 E = \mu_0 \epsilon_0 \frac{\partial^2 E}{\partial t^2}$.

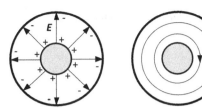

Figure 1.10: Lines of electric and magnetic fields in a coaxial cable

$$\frac{\partial^2 E_x}{\partial z^2} = \mu_0 \epsilon_0 \frac{\partial^2 E_x}{\partial t^2},$$

and the propagation is in the z direction, whose velocity is defined as

$$v = \frac{1}{\sqrt{\mu_0 \epsilon_0}} = c,$$

where $c = 3 \times 10^8$ m/s is the speed of light in free space. More on this will be covered in Section 1.8.

1.4 CIRCUIT REPRESENTATION OF CAPACITORS AND INDUCTORS

From a circuit point of view, a capacitor is symbolically represented as shown in Figure 1.11, whose voltage and current ($v(t)$ and $i(t)$) as shown satisfy the following relations [7]:

$$i(t) = \frac{dq}{dt},$$

where q is the charge stored in the capacitor. The above expression is widely known as the *continuity equation*. For the case of a linear and time-invariant capacitor, since $q = Cv$, we can write the well-known expression for the capacitor:

$$i(t) = C \frac{dv}{dt}.$$

Note that the continuity equation as expressed in most physics books is $i(t) = -\frac{dq}{dt}$, indicating that the *outward flow* of the positive charge must be balanced by a *decrease* of the charge within the closed surface (that is q). The minus sign is omitted since in Figure 1.11 as the current flow is associated into one terminal of the capacitor with the time rate of increase of charge on that terminal, and not the *outward* current.

An inductor is symbolically represented as shown in Figure 1.11, whose voltage and current ($v(t)$ and $i(t)$) as shown satisfy the following relations:

$$v(t) = \frac{d\phi}{dt},$$

where ϕ is the magnetic flux linkage. The above equation is a direct result of Faraday's law, and since $\phi = Li$, we arrive at the well-known expression:

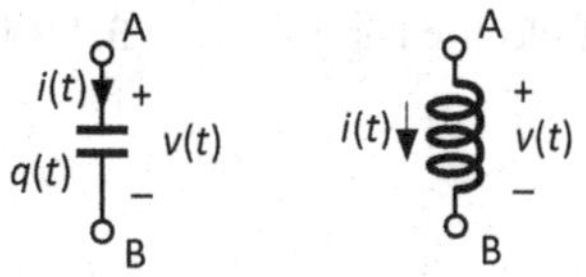

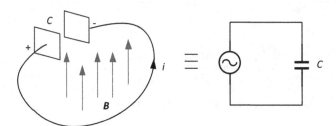

Figure 1.11: Capacitor and inductor circuit representation

Figure 1.12: Description of Faraday's law in a closed path

$$v(t) = L\frac{di}{dt}.$$

Note that the minus sign is again omitted from the inductor i–v equation, so let us verify if it agrees with Lenz's law. Suppose the current $i(t)$ increases, that is $\frac{di}{dt} > 0$. This indicates that the magnetic field must also increase, hence $d\phi/dt > 0$, which follows $v(t) > 0$, that is, the potential in node A is greater than node B. This is precisely the polarity required to oppose a further increase in current, as required by Lenz's law.

1.5 DISTRIBUTED AND LUMPED CIRCUITS

Kirchhoff's voltage law or KVL states that the sum of electric potentials in a closed path is equal to zero, that is, $\oint E \cdot dL = 0$, whereas Maxwell's first equation (or Faraday's law as described before) says otherwise. The time-varying term in Maxwell's second equation, that is, the displacement current, is similarly in violation of KCL. To clarify further, let us study the simple circuit shown in Figure 1.12, consisting of an ideal (zero inductance and resistance) piece of wire attached to a parallel-plate capacitor, forming a loop around it.

Assume that within the loop an external magnetic field is applied, varying sinusoidally with time. Thus, an emf $= V_0 \cos \omega_0 t$ across the capacitor is produced, as predicted by Faraday's law. On the other hand, if the wire is ideal, KVL indicates that the *shorted* capacitor must have zero voltage across it. Interestingly, the voltage across the capacitor creates a current i in the wire:

$$i = -\omega_0 C V_0 \sin \omega_0 t = -\omega_0 \frac{\epsilon A}{d} V_0 \sin \omega_0 t,$$

where ϵ, A, and d are parallel plate capacitor parameters. In any closed path the Ampere's circuital law gives us the magnetic field as a result of this current. Particularly, for a specific closed path, which passes between the capacitor plates, we can determine the displacement current. Within the capacitor

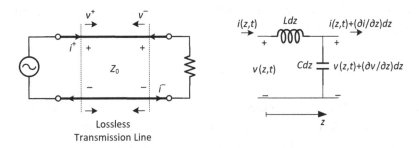

Figure 1.13: Lossless transmission line and its lumped differential equivalent

$$D = \epsilon E = \epsilon \left(\frac{V_0}{d} \cos \omega_0 t \right),$$

and according to Maxwell's second equation, the displacement current is

$$i_D = \frac{\partial D}{\partial t} A = -\omega_0 \frac{\epsilon A}{d} V_0 \sin \omega_0 t,$$

which is equal to the earlier result obtained for the current in the loop.

This brings us to a general discussion about *lumped* and *distributed* circuits. We can say that the basic elements in a circuit, and the connections between them, are considered *lumped* (and thus KVL or KCL are applicable) if the time delay in traversing the elements is negligible, and hence they can be treated as static. If the components are large enough, or the frequency is high enough (or equivalently the delays are small enough), one must use *distributed* elements. This means that the resistive, capacitive, or inductive characteristics must be evaluated on a per unit distance basis. Common examples of the latter are transmission lines or waveguides, which are intended to deliver electric energy from one point to the other, and naturally are separated by long (relative to the wavelength) distances. To illustrate an example of how to deal with such circuits, consider a lossless line, as shown in Figure 1.13, connecting a generator to a load. We can construct a model for this *transmission line* using lumped capacitors and inductors. An equivalent circuit of a differential section of the line (where the length dz is approaching zero) with no loss is shown in Figure 1.13. Since each section corresponds to a very small portion of the line, that is dz is approaching zero, KVL and KCL are valid for that section, despite the distributed nature of the line.

Writing KVL leads to

$$v(z,t) = (Ldz)\frac{\partial i(z,t)}{\partial t} + v(z,t) + \frac{\partial v(z,t)}{\partial z} dz,$$

which results in

$$L\frac{\partial i}{\partial t} = -\frac{\partial v}{\partial z}.$$

Similarly by writing KCL we obtain

$$\frac{\partial i}{\partial z} = -C\frac{\partial v}{\partial t}.$$

Taking derivative of one of the two equations above versus space (or z), and the other versus time (or t), the current component, i, could be eliminated, arriving at

$$\frac{\partial^2 v}{\partial z^2} = LC \frac{\partial^2 v}{\partial t^2}.$$

Note the similarity of this differential equation with the one given for the wave propagation in the previous section, where electric field is replaced by voltage, and the propagation velocity (as we will define shortly) is now equal to $\frac{1}{\sqrt{LC}}$. Since L and C are the inductance and capacitance per unit length, they have the same units as μ and ϵ in the wave equation, that is H/m and F/m.

The solution of this differential equation is of the form

$$v(z,t) = f_1\left(t - \frac{z}{v}\right) + f_2\left(t + \frac{z}{v}\right) = v^+ + v^-,$$

which could be verified by replacing $v(z,t)$ above in the original differential equation describing the distributed wave propagation. The functions f_1 and f_2 could be anything as long as they are differentiable twice, and take arguments $t \pm \frac{z}{v}$. The arguments of f_1 and f_2 indicate, respectively, travel of the functions in the *forward and backward z directions*, and thus we assign symbols v^+ and v^- to them. To understand this better, suppose we would like to keep the argument of f_1 constant at zero. As time increases (as it should), z has to also increase with a rate of $v \times t$ (hence we call v the velocity). Therefore the function f_1 needs to move forward or in positive z direction. On the other hand, for f_2, z has to decrease, indicating a backward motion. The forward moving signal is illustrated in Figure 1.14, where we assume $f_{1/2}$ are sinusoidal. This in fact will be the case, for the sinusoidal steady state solution, as we show in Chapter 3.

The propagation velocity is obtained by replacing the solution in the original differential equation. This yields

$$v = \frac{1}{\sqrt{LC}}.$$

A similar procedure results in the following solution for the current:

$$i(z,t) = \frac{1}{Z_0} f_1\left(t - \frac{z}{v}\right) - \frac{1}{Z_0} f_2\left(t + \frac{z}{v}\right) = i^+ + i^-,$$

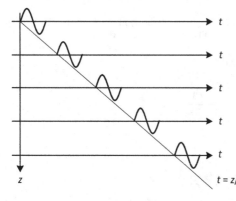

Figure 1.14: Wave propagating in a transmission line

where Z_0 is defined as the *characteristic impedance* of the line measured in ohm and is equal to

$$Z_0 = \sqrt{\frac{L}{C}}.$$

Even though Z_0 is measured in Ω, it is not a physical resistor as we already assumed that the line is lossless. It simply relates the forward and backward voltages and currents in the line as follows (Figure 1.13):

$$v^+ = Z_0 i^+$$
$$v^- = -Z_0 i^-.$$

Example: Returning to our previous coaxial cable example, since the values of L and C were obtained already, the characteristic impedance is readily given by

$$Z_0 = \sqrt{\frac{\mu}{\epsilon}} \ln \frac{b}{a} \; \Omega.$$

Typical values of a, b, and ϵ result in a characteristic impedance of several tens of ohm, commonly set to 50Ω.

1.6 ENERGY AND POWER

From electromagnetic field perspective, we can define electrostatic and magnetic energy stored as follows [6]:

$$W_E = \frac{1}{2} \int_V \mathbf{D} \cdot \mathbf{E} \, dV = \frac{1}{2} \epsilon \int_V |\mathbf{E}|^2 dV$$

$$W_H = \frac{1}{2} \int_V \mathbf{B} \cdot \mathbf{H} \, dV = \frac{1}{2} \mu \int_V |\mathbf{H}|^2 dV,$$

where W_E and W_H denote electric and magnetic energy respectively in joules and the integrals are performed over volume.[18]

From circuit perspective, let us consider Figure 1.15, where a generator is connected to a one-port, with current $i(t)$ entering the one-port, and a voltage $v(t)$ across it.

The instantaneous power *delivered* to the one-port by the generator is defined as

$$p(t) = v(t)i(t),$$

and the energy produced by the generator from initial time t_0 to t is

$$W(t_0, t) = \int_{t_0}^{t} p(\theta) d\theta = \int_{t_0}^{t} v(\theta) i(\theta) d\theta.$$

---

[18] Both equations may be proven from basic definitions.

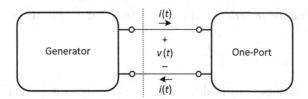

Figure 1.15: Instantaneous power and energy concept

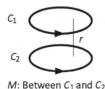

Figure 1.16: Self and mutual inductances as only functions of geometry

For an ideal capacitor with initial zero voltage or charge, that is $q(t_0)=0$, we have

$$W(t) = \int_{t_0}^{t} v(\theta)i(\theta)d\theta = \int_{0}^{q(t)} \frac{q(\theta)}{C}\frac{dq}{d\theta}d\theta = \frac{q(t)^2}{2C} = \frac{1}{2}Cv(t)^2,$$

where v and i are replaced with their q equivalents inside the integral. Similarly for an inductor

$$W(t) = \frac{1}{2}Li(t)^2.$$

Again we can identify the resemblance between μ and ϵ to L and C, and E and H to V and I in the energy equations. Since E and ϵ have units per distance (/m), the field energy integral is performed over volume. Note that sometimes it is more convenient to calculate the inductance or capacitance for a given geometric structure from energy definition, for instance to use

$$L = \frac{2W_H}{I^2}$$

for inductance calculation (as opposed to $L = \phi/I$ presented earlier). W_H is obtained from its basic definition as a function of B and H. Expressing W_H and I based on H, it can be shown that[19]

$$L = \frac{\mu}{4\pi}\oint\left(\oint\frac{d\mathbf{L}}{r}\right)\cdot d\mathbf{L} = \frac{\mu}{4\pi}\oint\oint\frac{d\mathbf{L}_1 \cdot d\mathbf{L}_2}{r},$$

which indicates that only the inductance is a function of geometry, and not the current. A similar expression for the mutual inductance exists as well, where the integral is defined between two circuits carrying current:

$$M = \frac{\mu}{4\pi}\oint\oint\frac{d\mathbf{L}_1 \cdot d\mathbf{L}_2}{r}.$$

---

[19] The proof requires the use of vector magnetic potential ($\mathbf{A}$), defined as $\mathbf{B} = \nabla \times \mathbf{A}$, analogous to electric potential as defined before, where $\mathbf{E} = -\nabla V$.

It is perhaps worthwhile to summarize the following similarities between electric and magnetic fields, and to voltages and currents:

$$C(F) \leftrightarrow \epsilon(F/m)$$
$$L(H) \leftrightarrow m(H/m)$$
$$V(V) \leftrightarrow E(V/m)$$
$$I(A) \leftrightarrow H(A/m)$$
$$E \leftrightarrow H$$
$$D \leftrightarrow B$$

Also note the similarity (or duality) between Gauss's and Ampere's laws, as well as Coulomb's and Biot–Savart's laws.

1.7 LC AND RLC CIRCUITS

With the background presented, we have now the right tools to analyze LC circuits. We start off with ideal LC resonators and extend the analysis to more a practical, lossy LC circuit.

1.7.1 Lossless LC Resonator

An ideal (lossless) LC circuit is shown in Figure 1.17 (left side).

Let us assume that the capacitor is charged initially to a voltage of V_0. From a circuit point of view, the initial charge may be a result of an impulse current source of $i(t) = V_0 \delta(t)$ appearing in parallel with the circuit. If the impulse has a magnitude of V_0 coulomb, the capacitor initial voltage will be $v_C(0^+) = V_0$.

Taking the capacitor voltage $v_C(t)$ as the variable of interest, we can write

$$\frac{\partial^2 v_C}{\partial t^2} + \frac{1}{LC} v_C = 0.$$

The differential equation is solved by taking the Laplace transform of the two sides. This yields [7], [8]

$$s^2 + \frac{1}{LC} = 0.$$

This results in the poles of the circuit at $s_{1,2} = \pm \frac{j}{\sqrt{LC}} = \pm j\omega_0$. The final solution is

$$v_C(t) = V_0 \cos \omega_0 t.$$

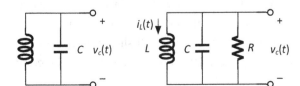

Figure 1.17: Ideal and lossy LC circuits

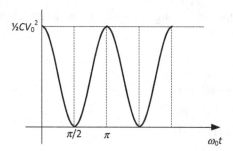

Figure 1.18: Energy of capacitor in a lossless LC circuit

Next we will calculate the energy stored in the capacitor and the inductor. From the previous section

$$W_C(t) = \frac{1}{2}Cv_C(t)^2 = \frac{1}{2}CV_0^2 \cos \omega_0 t^2.$$

Similarly solving for the inductor current yields

$$i_L(t) = \frac{1}{L}\int v_C(t)dt = \frac{V_0}{L\omega_0} \sin \omega_0 t,$$

and since $C = \frac{1}{L\omega_0^2}$, we have

$$W_L(t) = \frac{1}{2}CV_0^2 \sin \omega_0 t^2.$$

Therefore the total energy at any point of the time is $W_T(t) = W_c(t) + W_L(t) = \frac{1}{2}CV_0^2$, which is constant and equal to the initial energy stored in the capacitor. This is expected, as the LC circuit is lossless. The energy is exchanged only between the inductor and the capacitor as shown in Figure 1.18, indicating a steady *oscillation*.

Note the rate of the energy exchange is twice as fast as that of the capacitor voltage or the inductor current.

1.7.2 Practical LC Resonator

In practice, both the capacitor and the inductor are lossy, and for now let us model the total loss as a parallel resistor as shown in Figure 1.17 (right side). We assume the loss is moderate, that is the value of R is large compared to the impedances of L or C at the frequency of interest. The new differential equation is

$$\frac{\partial^2 v_C}{\partial t^2} + \frac{1}{RC}\frac{\partial v_C}{\partial t} + \frac{1}{LC}v_C = 0,$$

resulting in the complex poles $s_{1,2} = -\frac{\omega_0}{2Q} \pm j\omega_0\sqrt{1 - \frac{1}{4Q^2}} = -\alpha \pm j\omega_d$, where, for the moment, we define the quality factor or Q as $Q = \frac{R}{L\omega_0} = RC\omega_0$. In order to have complex poles, Q must be greater than ½, which is consistent with our assumption of moderate loss. This definition is only a mathematical expression for Q, and we will shortly attach a more physical meaning to it based on the energy concept. Figure 1.19 shows the location of the complex poles on the s-plane.

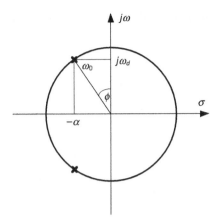

Figure 1.19: Pole locations of an RLC circuit in $j\omega$ plane

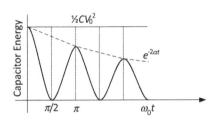

Figure 1.20: Energy in a lossy LC circuit

Defining $\phi = \cos^{-1}\frac{\omega_d}{\omega_0}$, the capacitor voltage $v_C(t)$ and the inductor current $i_L(t)$ are given by

$$v_C(t) = V_0 \frac{\omega_0}{\omega_d} e^{-\alpha t} \cos(\omega_0 t + \phi)$$

$$i_L(t) = \frac{V_0}{L\omega_d} e^{-\alpha t} \sin \omega_0 t.$$

Assuming $Q \gg 1$, then the total energy stored in the LC tank is approximately

$$W_T(t) = W_c(t) + W_L(t) \approx \frac{1}{2} C V_0^2 e^{-2\alpha t}.$$

Indicating an initial energy of $\frac{1}{2}CV_0^2$ decaying exponentially as shown in Figure 1.20.

Similar to the ideal tank, the capacitor and the inductor energy move back and forth between the two, but at a rate of ω_d, slightly lower than ω_0, but eventually decays to zero. The total energy decay rate, or equivalently the power dissipated in the resistor is

$$p = -\frac{dW_T}{dt} = 2\alpha W_T = \frac{\omega_0}{Q} W_T.$$

Rearranging the equation above, we arrive at a more physical and perhaps a more fundamental definition for the quality factor:

$$Q = \omega_0 \frac{W_T}{p} = \omega_0 \frac{\text{Total energy stored}}{\text{Average power dissipated}}.$$

Note that as shown in Figure 1.21 the normalized decay rate versus normalized time (that is the number of cycles) is a constant and is equal to Q.

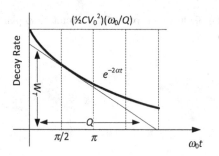

Figure 1.21: Power dissipated in a resistor of an RLC circuit vs. cycles

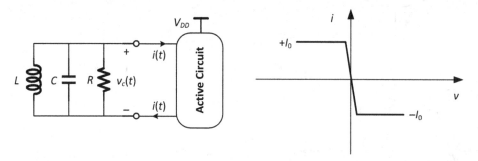

Figure 1.22: Active circuit to compensate for the loss of RLC circuit, and its $i-v$ characteristics

To sustain a steady oscillation, the power dissipated in the resistor must be compensated. Since the passive circuits are incapable of generating energy, this must be done by an active circuit as shown in Figure 1.22. The power *creation* requires the $v(t) \times i(t)$ ($v(t) = v_C(t)$) product to be negative, so we expect the $i-v$ characteristics of the active circuit to have a negative slope as shown in Figure 1.22, effectively acting as a *negative resistance* compensating the loss associated with the *positive resistance*. The sharper the slope, the more efficient this would be. Without being distracted with details of the active circuit, we could ideally model that as a one-port whose $i-v$ curve is shown in Figure 1.22, illustrating a sharp transition in the current around origin. The conservation of energy requires the energy created to originate from somewhere, and that is usually from the DC power supplying the active circuit (V_{DD} in Figure 1.22). So it is reasonable to assume that as the voltage or the current increases, they eventually reach a plateau where they can no longer increase, as denoted by I_0 in Figure 1.22.

Now at steady state when the power dissipated and the power created are balanced, the capacitor voltage is equal to $v_C(t) = V_0 \cos \omega_0 t$ as in an ideal tank, and the total energy stored in the tank must be $\frac{1}{2}CV_0^2$.

Since this voltage appears across the active circuit, we expect the current supplied by that to the lossy LC tank ($i(t)$) to be as shown in Figure 1.23, as we assumed a sharp transition in the active element $i-v$ characteristic.

Therefore, the average power dissipated in the tank (which is equal to power created in the active one-port) is given by (Figure 1.23):

$$p = -\frac{1}{T}\int_T (V_0 \cos \omega_0 t) i(t) dt = \frac{2V_0 I_0}{\pi}.$$

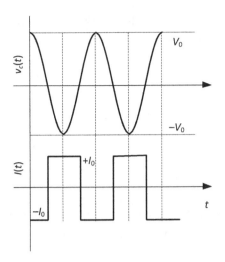

Figure 1.23: Voltage and current waveforms in the case of steady oscillation

Hence, according to the Q definition

$$Q = \omega_0 \frac{W_T}{p} = \omega_0 \frac{\frac{1}{2}CV_0^2}{\underbrace{\frac{2V_0I_0}{\pi}}} = \frac{\pi}{4}\omega_0 C \frac{V_0}{I_0},$$

and since $Q = RC\omega_0$ for a parallel RLC tank, then

$$V_0 = \frac{4}{\pi} R I_0.$$

This shows that the steady oscillation amplitude, V_0, is a function only of the active one-port saturation current, I_0, and the amount of loss. Since regardless of its voltage the one-port always drains a steady current of $2I_0$ from the supply (to produce the waveform shown in Figure 1.22), then the efficiency is equal to

$$\eta = \frac{\frac{2V_0I_0}{\pi}}{V_{DD}2I_0} = \frac{V_0}{\pi V_{DD}} = \frac{2}{\pi},$$

assuming V_0 can reach a maximum swing of $2V_{DD}$. This outcome can be intuitively explained by realizing that the assumed high-Q nature of the tank produces a sinusoidal waveform, whereas the current delivered by the active circuit is square-wave in nature (Figure 1.23), given the sharp slope of its i–v curve. Taking only the fundamental then leaves a *loss* factor of $2/\pi$. We will arrive at a similar result for hard-switching mixers, as we will discuss in Chapter 8.

We have not offered much insight into how to realize the active one-port necessary for sustaining the oscillation. As our focus in this chapter has been LC circuits, we will leave it at that, and present various circuit topologies that implement this active one-port in Chapter 9.

Contrary to our model, in practice inductors experience an ohmic loss due to the finite wire conductivity, physically modeled as a small *series* resistance as shown in Figure 1.24. While we will systematically prove it in Chapter 3, it can be easily shown that if the Q is large, the two circuits shown in Figure 1.24 are equivalent. The quality factor (Q) is

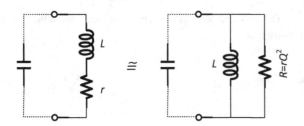

Figure 1.24: Inductor with series resistor, and its equivalent model

$$Q = \frac{R}{L\omega_0} = \frac{L\omega_0}{r}.$$

Furthermore, assuming the elements are high-Q, if the capacitor has also a parallel resistive loss, we can derive the equivalent parallel RLC circuit whose quality factor, Q_T, is

$$\frac{1}{Q_T} = \frac{1}{Q_L} + \frac{1}{Q_C},$$

where Q_L and Q_C are the inductor and the capacitor quality factors (see Problem 15).

Example: A 1nH inductor has a series resistance of 2Ω, and resonates with a capacitance whose Q is 25 at 5.5GHz. We would like to find the overall LC tank quality factor, and the equivalent shunt resistance at 5.5GHz.
The inductor Q is

$$Q_L = \frac{L\omega_0}{r} = 17.3.$$

The total Q is then

$$Q_T = \frac{Q_L Q_C}{Q_L + Q_C} = 10.2.$$

The total shunt resistance is

$$R_T = L\omega_0 Q_T = 352.5\Omega.$$

Note that the required capacitance to achieve resonance at 5.5GHz is about 837fF. The corresponding shunt resistance due to the capacitor loss only is then $R_C = \frac{Q_C}{C\omega_0} = 864\Omega$, whereas that of the inductor is $R_L = L\omega_0 Q_L = 597.8\Omega$. These two resistors in parallel will lead to the same $R_T = 352.5\Omega$.

1.7.3 Resonator Analysis Based on Energy Conservation

Using energy conservation, the analysis offered in the previous section (Figure 1.22) may be extended to any general nonlinear active circuit employed to compensate the resonator loss.[20]

---

[20] Since the waveforms are periodic, a more rigorous analysis is offered in Chapter 9 based on Fourier series.

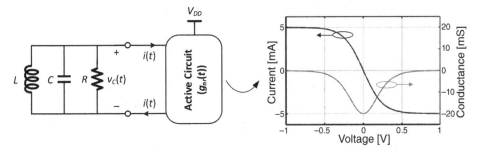

Figure 1.25: Arbitrary active circuit acting as a negative resistance to compensate the LC resonator loss

Depicted in Figure 1.25 is an arbitrary nonlinear active device whose i–v characteristics are shown on the right side. The slope is negative, and thus it pumps energy to the lossy RLC circuit, as desired. We assume the active element has a conductance of $g_{nr}(t)$ (also shown in the figure), defined as follows:

$$g_{nr}(t) = \frac{di_{nr}(t)}{dv_{nr}(t)} = \frac{di_{nr}(t)/dt}{dv_{nr}(t)/dt},$$

where $i_{nr}(t)$ and $v_{nr}(t)$ are the current and voltage of the active element.

Assuming the voltage across the resonator is mostly sinusoidal, that is $v_C(t) = V_0 \cos \omega_0 t$, at steady state, we can write

$$\frac{V_0^2}{2R} + \frac{1}{T}\int_T (V_0 \cos \omega_0 t) i(t) dt = 0.$$

The left side of the equation is the power dissipated in the resistor, whereas the right integral is the *average power* created by the active circuit over one cycle, which must balance each other at steady state.

The sinusoidal resonator *voltage* implies a high quality factor, which is a practical assumption. Nonetheless, the current produced by the active element may be very nonlinear, e.g., a square-wave for the example of Figure 1.23.

Using the identity $\int x \, dy = xy - \int y \, dx$, the integral is simplified as

$$\int_T (V_0 \cos \omega_0 t) i(t) dt = \left[\left(\frac{V_0}{\omega_0} \sin \omega_0 t\right) i(t)\right]\bigg|_0^T - \int \frac{V_0}{\omega_0} \sin \omega_0 t \, di = -\int \frac{V_0}{\omega_0} \sin \omega_0 t \, di.$$

Since

$$di = g_{nr}(t) dv_C = g_{nr}(t)(-V_0 \omega_0 \sin \omega_0 t \, dt),$$

we can further simplify

$$\int_T (V_0 \cos \omega_0 t) i(t) dt = V_0^2 \int_T \sin \omega_0 t^2 g_{nr}(t) dt,$$

which results in

$$\frac{1}{2R} + \frac{1}{T}\int_T \sin^2 \omega_0 t \, g_{nr}(t) dt = 0.$$

Equivalently, for steady oscillation, the resonator conductance, $G = \frac{-1}{R}$ must be

$$G = \frac{-2}{T} \int_T \sin^2 \omega_0 t \, g_{nr}(t) dt.$$

As the RLC circuit quality factor, and hence the conductance is often given, the equation above sets a constraint on the *oscillation amplitude* by which the shape of $g_{nr}(t)$ is defined.

The right-hand side integral deserves some more explanation. Using the simple trigonometric identity we may further write

$$\frac{2}{T} \int_T \sin^2 \omega_0 t \, g_{nr}(t) dt = \frac{1}{T} \int_T (1 - \cos 2\omega_0 t) g_{nr}(t) dt = \frac{1}{T} \int_T g_{nr}(t) dt - \frac{1}{T} \int_T \cos 2\omega_0 t \, g_{nr}(t) dt.$$

Hence

$$G = G_{nr}[2] - G_{nr}[0],$$

where $G_{nr}[0] = \frac{1}{T} \int_T g_{nr}(t) dt$ is the DC average of the periodic waveform $g_{nr}(t)$, and $G_{nr}[2] = \frac{1}{T} \int_T g_{nr}(t) \cos 2\omega_0 t \, dt$ is the average normalized to $\cos 2\omega_0 t$.

Example: For the previous case of Figure 1.22 (sharp $i - v$ transition), the corresponding voltage and current are re-depicted in Figure 1.26, along with the waveform of $g_{nr}(t)$.

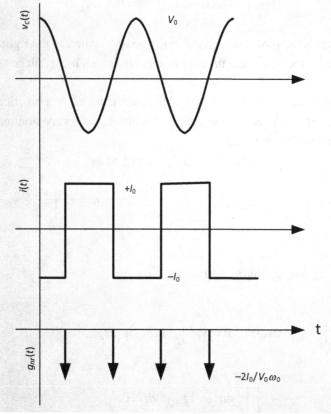

Figure 1.26: Waveforms corresponding to square-wave-like current produced by the active element

Since the active element current is assumed to be an ideal square-wave, we have

$$g_{nr}(t) = \frac{di_{nr}(t)/dt}{dv_{nr}(t)/dt} = \frac{di_{nr}(t)/dt}{-V_0\omega_0 \sin\omega_0 t}dt = \frac{-2I_0}{V_0\omega_0}\sum_{k=-\infty}^{+\infty}\delta\left(t - \frac{T}{4} - k\frac{T}{2}\right).$$

Clearly, the oscillation amplitude directly impacts the shape of $g_{nr}(t)$; the higher the amplitude, the smaller the conductance.

The value of the integral is easily found to be

$$\frac{-2}{T}\int_T \sin^2\omega_0 t\, g_{nr}(t)dt = \frac{8I_0}{V_0\omega_0 T} = \frac{4}{\pi}\frac{I_0}{V_0},$$

or $V_0 = \frac{4}{\pi}RI_0$, as found previously. The efficiency may be calculated as

$$\eta = \frac{\frac{V_0^2}{2R}}{V_{DD}2I_0},$$

and is maximized for $V_0 = 2V_{DD}$ to

$$\eta_{max} = \frac{V_0}{2RI_0} = \frac{2}{\pi}.$$

The efficiency may be further improved if the active element injects currents only at the peaks or valleys of the resonator voltage. The current will then consist of narrow pulses of width Δt, whose height is now $I_0 T/2\Delta t$ (Figure 1.27). In the limit case then, the current resembles a train of impulses with the height of $I_0 T$, appearing at every maximum or minimum of the resonator voltage. Compared to the previous case, if the current injected by the active element is square-wave, it results in the first harmonic of current (the rest are filtered by the resonator) to be $\frac{4}{\pi}I_0$, whereas the DC average is $2I_0$, leading to an efficiency of $\frac{2}{\pi}$. On the contrary, in the idealistic case of impulse current, the DC and all harmonics are of the same value, and hence we expect a perfect efficiency.

From the figure it readily follows

$$g_{nr}(t) = \frac{-I_0 T}{2\omega_0 \Delta t V_0 \sin\left(\frac{\omega_0 \Delta t}{2}\right)}\left[\sum_{k=-\infty}^{+\infty}\delta\left(t - \frac{\Delta t}{2} - k\frac{T}{2}\right) + \sum_{k=-\infty}^{+\infty}\delta\left(t + \frac{\Delta t}{2} - k\frac{T}{2}\right)\right]$$

and

$$\frac{-2}{T}\int_T \sin^2\omega_0 t\, g_{nr}(t)dt = \frac{2I_0}{V_0}\frac{\sin\left(\frac{\omega_0 \Delta t}{2}\right)}{\frac{\omega_0 \Delta t}{2}}.$$

Thus

$$V_0 = 2RI_0\frac{\sin\left(\frac{\omega_0 \Delta t}{2}\right)}{\frac{\omega_0 \Delta t}{2}}.$$

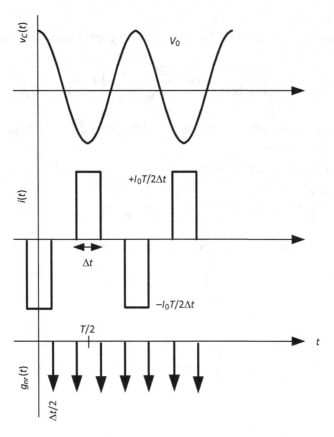

Figure 1.27: Waveforms corresponding to impulse current generated by the active element

The maximum efficiency is now

$$\eta_{max} = \frac{V_0}{2RI_0} = \frac{\sin\left(\frac{\omega_0 \Delta t}{2}\right)}{\frac{\omega_0 \Delta t}{2}}.$$

For the previous case of square-wave current, $\Delta t = \frac{T}{2}$, and hence $V_0 = \frac{4}{\pi} R I_0$, leading to a maximum efficiency of $\frac{2}{\pi}$. However, as Δt approaches zero, the amplitude reaches to a maximum of $2RI_0$, and the efficiency approaches 100%.

Although ideally a perfect efficiency is promised when the active element generates impulse current, realizing such a device proves to be a challenge in practice. Whereas the square-wave current is simply generated through a properly sized differential pair, producing impulses requires biasing the transistors to be on only at the peaks of the voltage, which is not as practical. Nonetheless, this type of circuitry known as class C has been successfully demonstrated in the context of both efficient oscillators and power amplifiers (see Chapters 9 and 11 for details of circuit realization and the challenges associated with the design).

1.8 THE UNIFORM PLANE WAVE

In this section as well as the next, we discuss the principles of wave propagation and provide a brief introduction to antennas [5], [6]. A thorough understanding of electromagnetic waves and antennas perhaps requires several chapters or an entire book [9], [10]. The goal here is to offer just an overview of basic RF design.

The uniform plane wave represents one of the simplest applications of Maxwell's equations and is the basic entity by which the energy is propagated. We focus on the free space first, and explore the solutions for a good (low-loss) conductor subsequently. The latter is important to understand the *skin effect* and loss mechanisms associated with it when discussing integrated inductors later.

1.8.1 Wave Propagation in Free Space

Earlier we presented the four Maxwell equations in integral form as follows:

$$\oint H \cdot dL = I + \int_S \frac{\partial D}{\partial t} \cdot dS$$

$$\oint E \cdot dL = -\int_S \frac{\partial B}{\partial t} \cdot dS$$

$$\oint_S D \cdot dS = \int_V \rho_V dV$$

$$\oint_S B \cdot dS = 0.$$

Using the divergence and Stokes's theorems,[21] we can express Maxwell's equations in differential form, which proves to be more useful when deriving the wave equations,

$$\nabla \times H = J + \epsilon \frac{\partial E}{\partial t}$$

$$\nabla \times E = -\mu \frac{\partial H}{\partial t}$$

$$\nabla \cdot E = \rho_V$$

$$\nabla \cdot H = 0,$$

where J is the current density, that is, $I = \int_S J \cdot dS$.

In free space where the medium is sourceless, thus J and ρ_V are equal to zero, and the four equations simplify to

$$\nabla \times H = \epsilon_0 \frac{\partial E}{\partial t}$$

$$\nabla \times E = -\mu_0 \frac{\partial H}{\partial t}$$

$$\nabla \cdot E = 0$$

$$\nabla \cdot H = 0.$$

---

[21] According to divergence theorem, for an arbitrary vector A, we have $\oint_S A \cdot dS = \int_V \nabla \cdot A \, dV$, where V is the volume enclosed by the closed surface S. Stokes's theorem states that: $\oint A \cdot dL = \int_S \nabla \times A \cdot dS$, where S is the area enclosed by the closed line. Both theorems can be readily proven using basic calculus principles.

Qualitatively speaking, we can infer the wave motion by considering the first two equations above. The first equation states that if E is changing with time at some point, then H has a curl at that point, and thus can be considered as forming a small closed loop linking the changing E field. A similar deduction can be made considering the second equation. Furthermore, the changing field is a small distance away from the disturbance point, and we will show (as it may be already guessed) that the velocity by which the effect moves is the velocity of light.

To solve the wave equation, we use the identity $\nabla \times \nabla \times A = \nabla(\nabla \cdot A) - \nabla^2 A$, and take a curl of the two sides of the first equation:

$$\nabla \times \nabla \times H = \nabla(\nabla \cdot H) - \nabla^2 H = \epsilon_0 \frac{\partial}{\partial t}(\nabla \times E).$$

Note that the $\nabla \times$ operator is performed over space. Since $\nabla \cdot H = 0$ and $\nabla \times E = -\mu_0 \frac{\partial H}{\partial t}$, this leads to

$$\nabla^2 H = \mu_0 \epsilon_0 \frac{\partial^2 H}{\partial t^2}.$$

Similarly,

$$\nabla^2 E = \mu_0 \epsilon_0 \frac{\partial^2 E}{\partial t^2}.$$

The above equation is widely known as the Helmholtz[22] equation for the plane wave, named after the German physicist.

Expanding the ∇^2 (Laplacian) operator is quite formidable, and does not lead to a closed solution. To gain insight, we postulate uniform wave in which both fields are in the *transverse* plane, that is the plane whose normal is the direction of propagation. For this, such a wave is sometimes called a transverse electromagnetic (TEM) wave. Consequently, we may assume that for example $E = E_x a_x$, that is, the electric field is *polarized* only in the x direction. To simplify matters further, we assume both the electric and magnetic fields vary only in the z direction, that is the direction of wave propagation. With those assumptions we have

$$\nabla \times E = \frac{\partial E_x}{\partial z} a_y = -\mu_0 \frac{\partial H}{\partial t} = -\mu_0 \frac{\partial H_y}{\partial t} a_y.$$

Note that while E is assumed to be in the x direction, H has a component in the y direction only, both normal to the wave that propagates in the z direction. Similarly,

$$\nabla \times H = -\frac{\partial H_y}{\partial z} a_x = \epsilon_0 \frac{\partial E}{\partial t} = \epsilon_0 \frac{\partial E_x}{\partial t} a_x.$$

The two equations above lead to

$$\frac{\partial^2 E_x}{\partial z^2} = \mu_0 \epsilon_0 \frac{\partial^2 E_x}{\partial t^2}.$$

---

[22] Hermann Helmholtz was a German physicist in the 19th century (1821–1894). Helmholtz's equation is the general partial differential equation of the form: $\nabla^2 A + k^2 A = 0$.

This is very similar to the transmission line equation derived earlier, whose solution was shown to be of the form

$$E_x(z,t) = f_1\left(t - \frac{z}{v}\right) + f_2\left(t + \frac{z}{v}\right) = E_x^+ + E_x^-.$$

The propagation is in the z direction, as stated earlier, and has velocity

$$v = \frac{1}{\sqrt{\mu_0 \epsilon_0}} = c,$$

where $c = 3 \times 10^8$ m/s is the speed of light in free space.

If the wave is sinusoidal, as it generally is, the functions f_1 and f_2 assume sinusoidal forms, and we may write

$$E_x(z,t) = E_{01} \cos(\omega t - k_0 z + \varphi_1) + E_{02} \cos(\omega t + k_0 z + \varphi_2),$$

where $k_0 \equiv \frac{\omega}{c}$ is the *wave number*.

In a manner consistent with our analysis of transmission lines, we can identify the forward and backward wave propagations from the wave solution above. Furthermore, we define the *wavelength* (λ) in free space as the distance over which the spatial phase shifts by 2π, assuming fixed time:

$$k_0 \lambda = 2\pi.$$

Thus

$$\lambda = \frac{2\pi}{k_0} = \frac{c}{f}.$$

Now, consider an arbitrary point in the cosine of the first term of the wave equation ($E_{01} \cos(\omega t - k_0 z + \varphi_1)$). To keep track of the chosen point, we then require the argument of the cosine to be an integer multiple of 2π:

$$\omega t - k_0 z = 2\pi m.$$

As time increases (as it should), the position z must also increase to satisfy the above condition. As such, the entire wave moves in the *forward* z direction with a velocity of $c = \frac{\omega}{k_0}$. For the second portion of the wave equation ($E_{02} \cos(\omega t + k_0 z + \varphi_2)$), we can identify a motion in the negative z direction, or backward propagation.

In the common case of sinusoidal propagation in time, it is more convenient to use complex notations and *phasor*. This leads to the following for the first two of Maxwell's equations:

$$\nabla \times \mathbf{H} = j\omega \epsilon_0 \mathbf{E}$$
$$\nabla \times \mathbf{E} = -j\omega \mu_0 \mathbf{H},$$

where $\mathbf{E}$ and $\mathbf{H}$ represent *phasor* vectors, and ω is the angular frequency of propagation. Following a procedure similar to the one in the time domain, we arrive at

$$\nabla^2 \mathbf{H} = -\omega^2 \mu_0 \epsilon_0 \mathbf{H} = -k_0^2 \mathbf{H},$$

and

$$\nabla^2 \mathbf{E} = -k_0^2 \mathbf{E},$$

where $k_0 = \omega\sqrt{\mu_0\epsilon_0} = \frac{\omega}{c}$, as defined earlier.

If for example the electric field is polarized in the x direction, and assuming both magnetic and electric fields vary only with z, in the phasor domain we have

$$E_x(z) = E_{01}e^{-jk_0z+j\varphi_1} + E_{02}e^{+jk_0z+j\varphi_2},$$

which is the solution to the phasor domain wave equation derived from the general equation ($\nabla^2 \mathbf{E} = -k_0^2 \mathbf{E}$) expressed below:

$$\frac{\partial^2 E_x}{\partial z^2} = -k_0^2 E_x.$$

Taking the real part of the phasor solution multiplied by $e^{j\omega t}$ results in the same time domain solution as we had before:

$$E_x(z,t) = \text{Re}\left[(E_{01}e^{-jk_0z+j\varphi_1} + E_{02}e^{+jk_0z+j\varphi_2})e^{j\omega t}\right].$$

Given the complex notation, we may conveniently arrive at a solution for the magnetic field as well. Since

$$\nabla \times \mathbf{E} = -j\omega\mu_0\mathbf{H},$$

and given the TEM assumptions earlier, we have

$$\frac{dE_x}{dz} = -j\omega\mu_0 H_y.$$

Hence

$$H_y = \frac{1}{-j\omega\mu_0}\left[(-jk_0)E_{01}e^{-jk_0z+j\varphi_1} + ((jk_0))E_{02}e^{+jk_0z+j\varphi_2}\right],$$

which could be rearranged as follows:

$$H_y = H_{01}e^{-jk_0z+j\varphi_1} + H_{02}e^{+jk_0z+j\varphi_2} = E_{01}\sqrt{\frac{\epsilon_0}{\mu_0}}e^{-jk_0z+j\varphi_1} - E_{02}\sqrt{\frac{\epsilon_0}{\mu_0}}e^{+jk_0z+j\varphi_2}.$$

and in the time domain

$$H_y(z,t) = E_{01}\sqrt{\frac{\epsilon_0}{\mu_0}}\cos(\omega t - k_0z + \varphi_1) - E_{02}\sqrt{\frac{\epsilon_0}{\mu_0}}\cos(\omega t + k_0z + \varphi_2).$$

Two key observations can be made: First, the forward and backward electric and magnetic fields amplitudes are related by:

$$E_{x01} = \eta_0 H_{y01}$$
$$E_{x02} = -\eta_0 H_{y02},$$

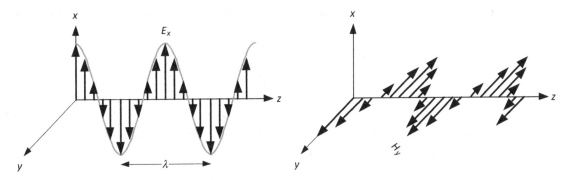

Figure 1.28: Instantaneous values of the electric and magnetic fields for $t=0$

where

$$\eta_0 = \sqrt{\frac{\mu_0}{\epsilon_0}} = 120\pi = 377\Omega.$$

The dimension of η_0 is in ohms, evident from the fact that it relates electric filed intensity (measured in V/m) to the magnetic field intensity (measured in A/m). It is a direct analogy to the transmission line characteristic impedance $Z_0 = \sqrt{\frac{L}{C}}$, which was defined as the ratio of the voltage to the current of a traveling wave.

Second, we notice the minus sign between the forward and backward components of the two fields. This is again consistent with our transmission line analysis where there was a minus sign between forward and backward voltages and currents.

To have a feeling for the way the fields behave, plotted in Figure 1.28 are the electric and magnetic fields where they vary over space (versus z), but are shown at $t=0$.

In reality, a uniform plane wave cannot exist physically, as it extends to infinity and requires an infinite amount of energy. Nonetheless, the distant field of a transmitting antenna is basically a uniform wave in some limited region. We apply our findings here to obtain insight into antennas discussed in the next section. Before that, let us discuss the power as well as wave propagation in conductors (as opposed to free space) first.

1.8.2 Wave Propagation in a Good Conductor: Skin Effect

An important discussion related to wave propagation pertains the behavior of a good conductor when a uniform plane wave is established in it. We will show that the primary transmission of energy must take place in the region outside the conductor (or on the surface), as all the time-varying fields attenuate very quickly within a good conductor.

Whereas in free space we assumed the current density J is zero, in the case of conductive materials, in which currents are formed by the motion of free electrons (or holes) under the influence of an electric field, the governing relation is $J = \sigma E$. With finite conductivity (σ), the wave loses power through resistive heating of the material.

Consequently, Maxwell's first equation in sinusoidal steady state may be written as

$$\nabla \times H = J + j\omega\epsilon E = \sigma E + j\omega\epsilon E = (\sigma + j\omega\epsilon)E = j\omega\left[\epsilon\left(1 - j\frac{\sigma}{\omega\epsilon}\right)\right]E.$$

The solution then may be found exactly as before, but with $k_0 = \omega\sqrt{\mu_0\epsilon_0}$ replaced with k defined as

$$jk = j\omega\sqrt{\mu\left[\epsilon\left(1 - j\frac{\sigma}{\omega\epsilon}\right)\right]} = \alpha + j\beta,$$

which is directly resulted from ϵ being replaced by $\epsilon\left(1 - j\frac{\sigma}{\omega\epsilon}\right)$. The new k is a complex number, and the values of α and β may be obtained by expanding the term under square root. The general solution for the electric field phasor is

$$E_x(z) = E_0 e^{-jkz} = E_0 e^{-\alpha z} e^{-j\beta z}.$$

While $e^{-j\beta z}$ results in the same time domain expression as before, the presence of $e^{-\alpha z}$ indicates an exponential decay of the magnitude over space, resulting from losses in the conductor.

In case of a good conductor, where the conduction current σE as expressed by Ohm's law is much larger than the displacement current (that is to say σ is large), we may approximate

$$jk = j\omega\sqrt{\mu\epsilon}\sqrt{1 - j\frac{\sigma}{\omega\epsilon}} \cong j\omega\sqrt{\mu\epsilon}\sqrt{-j\frac{\sigma}{\omega\epsilon}} = j\sqrt{-j\omega\mu\sigma}.$$

Accordingly, as $\sqrt{-j} = \frac{1-j}{\sqrt{2}}$, we can readily obtain

$$\alpha = \beta = \sqrt{\frac{\omega\mu\sigma}{2}}.$$

The electric field will be

$$E_x(z,t) = E_0 e^{-\alpha z} \cos(\omega t - \beta z),$$

and the current density in the conductor is

$$J_x = \sigma E_x = \sigma E_0 e^{-\alpha z} \cos(\omega t - \beta z),$$

which shows an exponential decay rate of

$$\delta = \frac{1}{\alpha} = \left(\frac{1}{2}\omega_0\mu_0\sigma\right)^{-1/2}.$$

The parameter δ measured by the meter is known as *skin depth*, and denotes the depth by which the electric field (or the current) decays by e^{-1}. In typical CMOS processes, δ is several μm at GHz frequency range.

If the skin depth is comparable to the width of the conductor, the effective resistance of the conductor increases, as the current tends to stay only at the surface (Figure 1.29). Clearly, the impact exacerbates as the frequency increases.

The solution to the magnetic field is readily obtained, considering

$$\eta = \sqrt{\frac{\mu}{\epsilon + \frac{\sigma}{j\omega\epsilon}}} = \sqrt{\frac{j\omega\mu\epsilon}{\sigma + j\omega\mu\epsilon}},$$

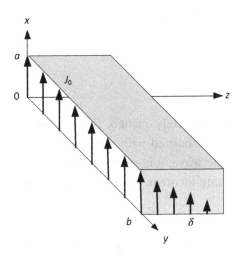

Figure 1.29: The current density in a good conductor

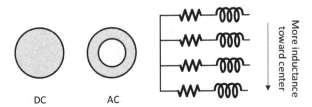

Figure 1.30: Physical interpretation of the skin effect in a wire

which in the case of a good conductor may be approximated as

$$\eta \cong \sqrt{\frac{j\omega\mu\epsilon}{\sigma}} = \frac{1+j}{\sigma\delta}.$$

Similar to k, η is a complex number, which results in an additional phase shift (of 45°) between magnetic and electric fields. Thus

$$H_y(z,t) = \frac{\sigma\delta E_0}{\sqrt{2}} e^{-\alpha z} \cos\left(\omega t - \beta z - \frac{\pi}{4}\right).$$

At a given point, the maximum amplitude of the magnetic field happens one-eighth of a cycle later with respect to that of the electric field.

Example: In 16nm CMOS, the top metal layer has a thickness of about 2.8μm, and a conductivity of 3.4×10^7 S/m. The skin depth at 2.4GHz is

$$\delta = \sqrt{\frac{1}{\pi f \mu \sigma}} = \sqrt{\frac{1}{\pi f \mu_0 \sigma}} = 1.8\mu m.$$

We see that the skin depth is quite comparable with the metal thickness at this frequency.

Physically, the skin effect can be interpreted considering that a unit filament of current at the center links more flux, and thus has more inductance. Since the current tends to flow into the path of *least impedance* (not necessarily the least resistance), it clings toward the outside of the wire (Figure 1.30).

Interestingly, the current sees a smaller resistance inside, but then larger inductance, and overall takes an exponential distribution as we showed earlier.

1.8.3 Power Considerations

When dealing with waves and antennas, what ultimately matters is the amount of power transmitted or received. We presented the power associated with static electric and magnetic fields earlier. To find the power in a uniform plane wave, it is necessary to describe a theorem for the electromagnetic field known as the Poynting theorem, originally developed by English physicist John Poynting in 1884.

Using the vector identity

$$\nabla \cdot (\boldsymbol{E} \times \boldsymbol{H}) = -\boldsymbol{E} \cdot \nabla \times \boldsymbol{H} + \boldsymbol{H} \cdot \nabla \times \boldsymbol{E},$$

and some straightforward algebraic steps performed on Maxwell's equations, one can show

$$-\nabla \cdot (\boldsymbol{E} \times \boldsymbol{H}) = \boldsymbol{J} \cdot \boldsymbol{E} + \frac{\partial}{\partial t}\left(\frac{\epsilon E^2}{2} + \frac{\mu H^2}{2}\right),$$

where $E = |\boldsymbol{E}|$, $H = |\boldsymbol{H}|$. Taking a volume integral, and using the divergence theorem, we have

$$-\oint_S (\boldsymbol{E} \times \boldsymbol{H}) \cdot d\boldsymbol{S} = \int_V (\boldsymbol{J} \cdot \boldsymbol{E}) dV + \frac{\partial}{\partial t} \int_V \left(\frac{\epsilon E^2}{2} + \frac{\mu H^2}{2}\right) dV.$$

The first integral on the right is the total instantaneous ohmic power dissipated within the volume (assuming a sourceless medium). The second integral on the right is the total energy stored in the electric and magnetic fields, and the partial derivatives with respect to time lead to the instantaneous power going to increase the stored energy within this volume. Thus, the right side of the equation must be the total power flowing into this volume. Consequently, on the left side the term $\oint_S (\boldsymbol{E} \times \boldsymbol{H}) \cdot d\boldsymbol{S}$ indicates the total power flowing out of the volume. The integral is taken over the closed surface surrounding the volume. The product $\boldsymbol{E} \times \boldsymbol{H}$ is known as the Poynting vector,

$$\boldsymbol{\mathcal{P}} = \boldsymbol{E} \times \boldsymbol{H},$$

and may be interpreted as the instantaneous power density measured in W/m². Considering the cross product, it is clear that the Poynting vector must be normal to electric and magnetic fields. This is consistent with our earlier example of a wave propagating in the z direction with electric and magnetic fields having components at the x and y directions. Considering that in free space

$$E_x(z,t) = E_0 \cos(\omega t - k_0 z)$$

$$H_y(z,t) = \frac{E_0}{\eta_0} \cos(\omega t - k_0 z)$$

we have

$$\mathcal{P}_x(z,t) = \frac{E_0^2}{\eta_0} \cos^2(\omega t - k_0 z),$$

and the average power density is

$$\mathcal{P}_{x,avg} = \frac{1}{T}\int_0^T P_x(z,t)dt = \frac{E_0^2}{2\eta_0},$$

which greatly resembles the average power dissipated in a lossy passive circuit. Note that the equation above denotes the average power *density*, which is evident from the fact that the electric field is measured in V/m.

Example: For a good conductor discussed in the previous section (Figure 1.29), the average power density becomes

$$\mathcal{P}_{x,avg} = \frac{\sigma\delta E_0^2}{2\sqrt{2}}e^{-2z/\delta}\cos\left(\frac{\pi}{4}\right) = \frac{\sigma\delta E_0^2}{4}e^{-2z/\delta}.$$

Note the energy decay of $e^{-2z/\delta}$. In terms of current density, since $J_x = \sigma E_x$, the above equation can be rewritten as

$$\mathcal{P}_{x,avg} = \frac{\delta J_0^2}{4\sigma}e^{-2z/\delta},$$

where J_0 is the current density on the surface (but decays exponentially over the z direction). The total power loss at the surface of the conductor is

$$P_{avg} = \frac{\delta J_0^2}{4\sigma}ab.$$

Example: As a thought experiment, let us find the effective resistance of the conductor of Figure 1.29. To do so, we shall find the total current first. We showed in the previous section

$$J_x = \sigma E_x = J_0 e^{-z/\delta}\cos(\omega t - \beta z).$$

The current is moving in the x direction (same as the electric field). Thus

$$I = \int_0^\infty \int_0^b J_0 e^{-z/\delta}\cos(\omega t - \beta z)dydz = \frac{J_0 b\delta}{\sqrt{2}}\cos\left(\omega t - \frac{\pi}{4}\right).$$

Continued

Assuming an effective resistance of R_{eff}, the total power dissipated is

$$P_L = \frac{1}{2} R_{eff} |I|^2 = \frac{(J_0 b \delta)^2}{4} R_{eff}.$$

When compared to the earlier expression for the average power ($P_{avg} = \frac{\delta J_0^2}{4\sigma} ab$), we conclude

$$R_{eff} = \frac{1}{\sigma} \frac{a}{b\delta} = \frac{1}{\sigma} \frac{a}{A},$$

which is a familiar expression for the resistance of a piece of conductor with the length a, and the total area of $A = b\delta$.

If it were not for the skin effect (say at low frequency), the resistance would have been zero as the conductor is assumed to have an infinite width (in the z direction), despite its finite conductivity. Interestingly, the presence of the skin effect reduces the conductor area effectively to $A = b\delta$, as if it has a finite width of equal to skin depth. This holds well for a conductor with a circular cross section as well (e.g., a coaxial cable), if the skin depth is small compared to the actual radius. We shall use this outcome in Chapter 3 to arrive at an expression for a coaxial cable loss.

1.9 ANTENNAS

An antenna is any device that radiates electromagnetic fields into space. The fields originate from a source that feeds the antenna through (typically) a transmission line (say a micro strip line printed on the board). The antenna thus serves as an interface between the radio and space when used as a transmitter, or between space and the radio when used as a receiver. As we will discuss in Chapter 3, reciprocity results in the receive or transmit antenna characteristics to be the same, however.

As stated earlier, a complete coverage of antennas is well beyond the scope of this book. However, as they are an essential part of any wireless system, we shall have a brief discussion of antenna basics and associated field radiation in this section. Further reading on the subject may be found in [9], [10], [11].

1.9.1 Antenna Basic Principles

Consider an ideal current filament of infinitesimally small cross section with a length of l, placed in a lossless medium (e.g., free space), as shown in Figure 1.31. The structure is widely known as a Hertzian[23] dipole, and is one of the earliest and most popular realizations of antennas.

---

[23] Heinrich Hertz was the German physicist in the 19th century to whom the invention of the dipole antenna is assigned. He was the first to experimentally verify Maxwell's equations and one of the pioneers of the invention of the radio in its current form today. The SI unit of frequency, hertz, is named in his honor.

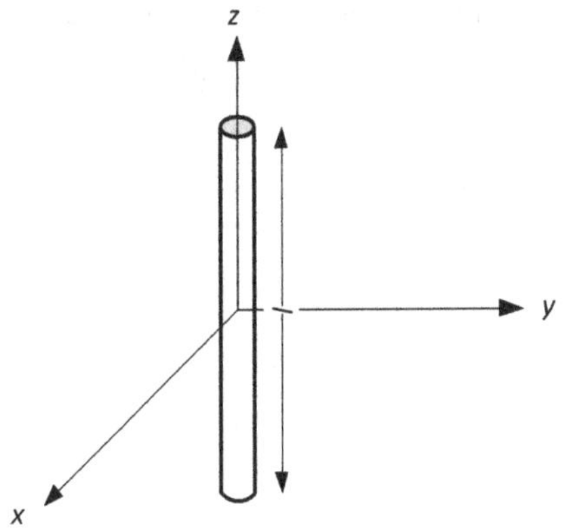

Figure 1.31: A Hertzian dipole carrying a uniform sinusoidal current

We assume that the filament is carrying a uniform (with respect to z) sinusoidal current (of $I_0 \cos \omega t$), and that it is short, particularly with respect to the wavelength of the signal it is carrying. For the moment, let us not concern ourselves with the source of this current, and the apparent discontinuities at each end.

To obtain the electric and magnetic fields caused by the differential length, we use the vector magnetic potential ($\boldsymbol{A}$), briefly discussed earlier, whose curl leads to magnetic flux density,

$$\boldsymbol{B} = \nabla \times \boldsymbol{A}.$$

From Biot–Savart's law, it readily follows that

$$\boldsymbol{A} = \oint \frac{\mu I d\boldsymbol{L}}{4\pi R},$$

where R is the distance in space from the differential length of $d\boldsymbol{L}$, carrying a current of I. For the case of the short filament of Figure 1.31, since l is small, the integration is no longer necessary, and thus

$$\boldsymbol{A} = \frac{\mu I\left(t - \frac{R}{v}\right) l}{4\pi R} \boldsymbol{a}_z.$$

Note that the current is represented by $I\left(t - \frac{R}{v}\right)$, and that $\boldsymbol{A}$ is in the z direction only, as is the filament itself (Figure 1.31). $v = \frac{\omega}{k}$ is the velocity, and $k = \omega \sqrt{\mu \epsilon}$ is as defined earlier. Using phasor notation, we have

$$A_z = \frac{\mu I_0 l}{4\pi R} e^{-jkR}.$$

It is more convenient to express the results in the spherical coordinates (Figure 1.32), which will readily lead to

$$A_r = \frac{\mu I_0 l}{4\pi r} e^{-jkr} \cos \theta$$

$$A_\theta = -\frac{\mu I_0 l}{4\pi r} e^{-jkr} \sin \theta,$$

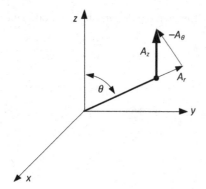

Figure 1.32: Expressing the magnetic vector potential of a dipole in spherical coordinates

where θ is the angle between the z axis and the vector $\mathbf{R}$, connecting the origin to the point of interest in the space (whose coordinates are (r, θ, φ)).

From the definition of $\mathbf{A}$, it follows that

$$\mathbf{H} = \frac{1}{\mu} \nabla \times \mathbf{A},$$

which leads to

$$H_\varphi = \frac{I_0 l \sin\theta}{4\pi} e^{-jkr} \left(\frac{jk}{r} + \frac{1}{r^2} \right).$$

Note that the magnetic field has only a component in φ direction, that is, it circles around the filament. The interested reader may prove that if the filament current were static (see also the problem sets), the magnetic potential would have been

$$\mathbf{A} = \frac{\mu I_0 l}{4\pi R} \mathbf{a}_z,$$

leading to the familiar solution of the magnetic field as expressed by Biot–Savart's law:

$$\mathbf{H} = \frac{I_0 l \sin\theta}{4\pi R^2} \mathbf{a}_\varphi = \frac{I_0 \mathbf{R} \times d\mathbf{L}}{4\pi R^3}.$$

Having the magnetic field already, the electric field is obtained by

$$\nabla \times \mathbf{H} = j\omega\epsilon \mathbf{E},$$

leading to electric field components in the r and θ directions:

$$E_r = \frac{I_0 l \cos\theta}{2\pi j\omega\epsilon} e^{-jkr} \left(\frac{jk}{r^2} + \frac{1}{r^3} \right)$$

$$E_\theta = \frac{I_0 l \sin\theta}{4\pi j\omega\epsilon} e^{-jkr} \left(\frac{-k^2}{r} + \frac{jk}{r^2} + \frac{1}{r^3} \right).$$

In the case of a far field, that is, if $kr \gg 1$, only $\frac{1}{r}$ terms remain, and consequently

$$H_\varphi \cong jk \frac{I_0 l \sin\theta}{4\pi r} e^{-jkr}$$

$$E_\theta \cong jk\eta \frac{I_0 l \sin\theta}{4\pi r} e^{-jkr} = \eta H_\varphi,$$

and $\eta = \sqrt{\mu/\epsilon}$, as defined before. Note that the E_r terms goes away, and just like uniform plane wave, the electric and magnetic fields are now related by $E_\theta = \eta H_\varphi$. In time domain, we can now write

$$H_\varphi = -\frac{I_0 l k \sin\theta}{4\pi r} \sin(\omega t - kr) = \frac{E_\theta}{\eta}.$$

The power density vector is in r direction and whose magnitude is equal to

$$\mathcal{P}_r = E_\theta H_\varphi = \eta \left(\frac{I_0 l k \sin\theta}{4\pi r}\right)^2 \sin^2(\omega t - kr).$$

Note that the power density drops by $\frac{1}{r^2}$. The total instantaneous power crossing an arbitrary sphere of radius r is then obtained by

$$P = \int_0^{2\pi} \int_0^\pi \mathcal{P}_r r^2 \sin\theta d\theta d\varphi,$$

and the average power is readily calculated to be

$$P_{avg} = 10(I_0 l k)^2 = 40\pi^2 \left(\frac{I_0 l}{\lambda}\right)^2.$$

Compared to a simple resistive circuit, we can deduce the effective *radiation resistance* of the dipole as

$$R_{radiation} = 80\pi^2 \left(\frac{l}{\lambda}\right)^2.$$

For the average radiated power to be meaningfully large, the dipole length needs to be comparable to the wavelength (due to the term $\left(\frac{l}{\lambda}\right)^2$), as is often noted in the literature. Equivalently, one must ensure that the effective radiation resistance is large with respect to the ohmic loss of the wire and other parasitic effects.

It is possible to extend the analysis of the differential current filament to a short dipole. If the dipole is short, the current distribution is generally not uniform, and one expects it to be zero at the two ends and maximal at the center, as depicted in Figure 1.33. Also shown is a practical current feed to the antenna through a transmission line. The antenna has identical currents at the two halves, and the gap at the center point is typically small and has a negligible impact.

A linear current distribution results in an average current of $I_0/2$, and thus one-fourth smaller radiation resistance. In practice the current distribution may be somewhat different, perhaps closer to sinusoidal. Furthermore, retardation effects may cause the signal arriving at any field point from the two ends of the antenna not to be in phase, and thus may result in some cancellation.

For a half-wave dipole ($l = \frac{\lambda}{2}$), which is arguably one of the world's most popular antennas, we can show that the actual radiation resistance in about 73Ω, whereas the simple equation

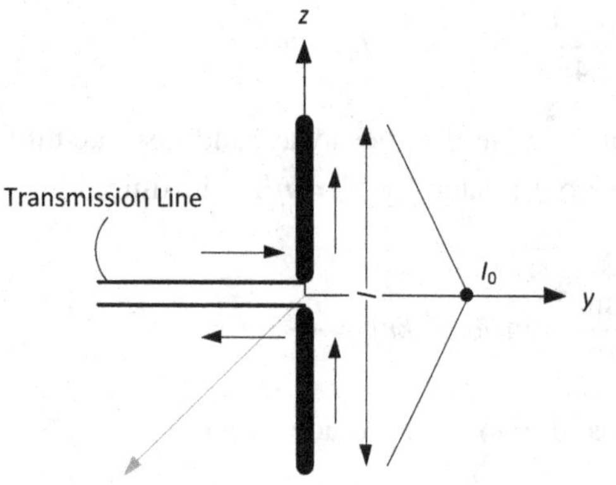

Figure 1.33: A dipole with proper transmission line feed

developed earlier underestimates a resistance of 49Ω (if an average current of $I_0/2$ assumed). If the current distribution were assumed to be sinusoidal, the average current would have been $\frac{2}{\pi}I_0$, leading to a closer estimate of 80Ω for the resistance.

1.9.2 Antenna Characteristics

To be able to fully describe and quantify the radiation of an antenna, several key parameters are typically defined.

For *any* antenna, at far-field, we can show that the electric field has the general phasor of the form

$$\boldsymbol{E} = E_0\left[F(\theta,\varphi)\boldsymbol{a_\theta} + G(\theta,\varphi)\boldsymbol{a_\varphi}\right]\frac{e^{-jkr}}{r},$$

where $F(\theta,\varphi)$ and $G(\theta,\varphi)$ are normalized functions of θ and φ only. The solution for the magnetic field is

$$\boldsymbol{H} = \frac{E_0}{\eta}\left[-G(\theta,\varphi)\boldsymbol{a_\theta} + F(\theta,\varphi)\boldsymbol{a_\varphi}\right]\frac{e^{-jkr}}{r}.$$

Accordingly, from our discussion in Section 1.8.3, the *average* power density is

$$\boldsymbol{\mathcal{P}}_{avg} = \frac{1}{2}\text{Re}[\boldsymbol{E}\times\boldsymbol{H}^*] = \frac{|E_0|^2}{2\eta r^2}\left[|F(\theta,\varphi)|^2 + |G(\theta,\varphi)|^2\right]\boldsymbol{a_r}.$$

For a current filament, for example,

$$F(\theta,\varphi) = \sin\theta$$

$$E_0 = j\eta kl\frac{I_0}{4\pi},$$

while the electric field has no component in the φ direction, and hence $G(\theta,\varphi) = 0$.

For longer length dipoles, one can consider the antenna as made up of a stack of filaments of infinitesimally short length, with a sinusoidal current distribution. If each half of the antenna (Figure 1.33) has an arbitrary length of $\frac{l}{2}$ (thus a total length of l), then by integration we obtain

$$F(\theta, \varphi) = \frac{\cos\left(\frac{kl\cos\theta}{2}\right) - \cos\left(\frac{kl}{2}\right)}{\sin\theta},$$

and

$$E_0 = j\eta \frac{I_0}{2\pi}.$$

For more details, see Problem 24.

Example: For a half-wave dipole, $l = \frac{\lambda}{2}$, thus

$$F(\theta, \varphi) = F(\theta) = \frac{\cos\left(\frac{\pi}{2}\cos\theta\right)}{\sin\theta}.$$

This yields

$$P_{avg} = \int_0^{2\pi}\int_0^\pi \frac{|E_0|^2}{2\eta}|F(\theta)|^2 \sin\theta d\theta d\varphi = 30I_0^2 \int_0^\pi |F(\theta)|^2 \sin\theta d\theta,$$

and the radiation resistance will be

$$R_{radiation} = 30\int_0^\pi |F(\theta)|^2 \sin\theta d\theta.$$

The integral may be evaluated numerically, leading to a radiation resistance of 73Ω.

With this background, let us take a look at a few key requirements.

- **Radiated power:** In general,

$$P_{avg} = \int_0^{2\pi}\int_0^\pi \frac{|E_0|^2}{2\eta}\left[|F(\theta,\varphi)|^2 + |G(\theta,\varphi)|^2\right]\sin\theta d\theta d\varphi$$

$$R_{radiation} = \frac{2P_{avg}}{I_0^2}.$$

Furthermore, the average radiated power density (in W/m²) may be expressed by

$$\mathcal{P}_{avg} = \frac{1}{2}\text{Re}[E \times H^*],$$

while the radiated power is in the r direction clearly.

For the dipole, since $F(\theta,\varphi) = \sin\theta$, we showed that

$$P_{avg} = 40\pi^2\left(\frac{I_0 l}{\lambda}\right)^2$$

$$R_{radiation} = 80\pi^2\left(\frac{l}{\lambda}\right)^2.$$

- **Antenna directivity:** The total radiated power was shown to be

$$P_{avg} = \int_0^{2\pi} \int_0^{\pi} \mathcal{P}_{r,avg} r^2 \sin\theta d\theta d\varphi,$$

defining the differential angle as

$$d\Omega = \sin\theta d\theta d\varphi,$$

and the *radiation intensity* as

$$K(\theta,\varphi) = r^2 \mathcal{P}_{r,avg}.$$

Since in general $\mathcal{P}_{r,avg} = \frac{|E_0|^2}{2\eta r^2}\left[|F(\theta,\varphi)|^2 + |G(\theta,\varphi)|^2\right]$, then

$$K(\theta,\varphi) = \frac{|E_0|^2}{2\eta}\left[|F(\theta,\varphi)|^2 + |G(\theta,\varphi)|^2\right].$$

The radiated power is

$$P_{avg} = \int_0^{2\pi} \int_0^{\pi} K d\Omega.$$

Interestingly, for all antennas at the far-field (say ten wavelengths or more away), $\mathcal{P}_{r,avg}$ exhibits a $1/r^2$ dependence. This is best understood from the conservation of energy; in a lossless medium, the total power at an arbitrary space of r is evaluated by integration at the sphere surface, which is equal to $4\pi r^2$. The radiation intensity thus comes out to be independent of r, and only a function of θ and φ.

For the special case of an *isotropic* radiator, that is to say, if the antenna radiation intensity is constant ($K(\theta,\varphi) = K_0$), then

$$P_{avg} = \int_0^{2\pi} \int_0^{\pi} K d\Omega = 4\pi K_0.$$

Consequently, we define the antenna *directivity* as

$$D(\theta,\varphi) = \frac{K(\theta,\varphi)}{K_0} = 4\pi \frac{K(\theta,\varphi)}{\oint_S K d\Omega}.$$

The significance of directivity is that it indicates whether the antenna intensity is more prominent in certain directions than in others.

Example: For a Hertzian dipole, we have

$$K(\theta,\varphi) = K(\theta) = \frac{1}{2}\eta\left(\frac{I_0 l k}{4\pi}\right)^2 \sin^2\theta.$$

Thus

$$D(\theta,\varphi) = 2\pi \frac{\eta\left(kl\frac{I_0}{4\pi}\right)^2 \sin^2\theta}{10(I_0 lk)^2} = \frac{\eta}{80\pi}\sin^2\theta = \frac{3}{2}\sin^2\theta.$$

Hence, the maximum directivity of a Hertzian dipole is $\frac{3}{2}$ or 1.76dB.

By contrast, for a half-wavelength antenna, by solving the integral numerically, we can show that the maximum directivity of 1.6dB.

- **Antenna gain and efficiency:** Given the resistive losses within the antenna, the radiated power (P_r), worked out earlier, is less than the total power delivered to the antenna (P_{in}). To quantify, we define the radiation efficiency as

$$\eta_r = \frac{P_r}{P_{in}},$$

which is clearly always less than one.

Furthermore, assuming a lossy antenna with an isotropic radiation, we have

$$P_{in} = 4\pi K_0.$$

Accordingly, the antenna gain is defined as

$$G(\theta,\varphi) = \frac{K(\theta,\varphi)}{K_0} = \eta_r D(\theta,\varphi).$$

For a lossless antenna hence, the gain and directivity are the same.

The interested reader as an exercise may work out the above parameters for a general dipole of length l, based on the $F(\theta,\varphi) = \frac{\cos\left(\frac{kl\cos\theta}{2}\right) - \cos\left(\frac{kl}{2}\right)}{\sin\theta}$ function calculated earlier.

1.10 INTEGRATED CAPACITORS

It is often very desirable to tune the resonance frequency of LC tanks. For instance, in a tuned amplifier a discrete tuning may be required to extend the bandwidth, whereas in an oscillator locked in a phase-locked loop, both discrete and continuous tuning are needed. Due to their physical structure, inductors are usually not viable options, although several structures have been proposed that vary the inductance through using multi-tap switched segments [12]. This usually comes at the cost of performance, and at best offers only a discrete tuning. Capacitors, on the other hand, are very well suited to provide the tuning. In this section we discuss a few schemes that are commonly used in RF integrated circuits. Before that, let us first briefly take a look at the common fixed capacitors available in integrated circuits.

The gate capacitance of MOS transistors may be exploited to realize high density but nonlinear capacitors. Shown in Figure 1.34 (the gray curve) is the simulated capacitance versus gate voltage of a 40nm regular NMOS capacitor. Depending on the gate voltage, the device

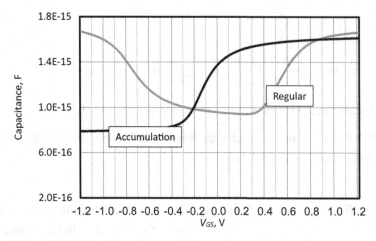

Figure 1.34: Regular and accumulation-mode MOS capacitance

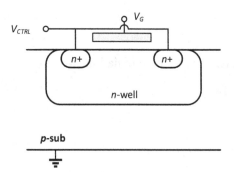

Figure 1.35: Accumulation-mode MOS capacitor

operates in accumulation (approximately for $V_{GS} < 0$), depletion ($0 < V_{GS} < V_{TH}$), or inversion ($V_{GS} > V_{TH}$) [2]. The device threshold voltage (V_{TH}) is estimated to be around 400mV. In inversion or accumulation the capacitance reaches a maximum, close to gate-oxide capacitance, C_{OX}. To achieve a reasonably linear response, the device should be biased at voltages well above threshold (say greater than 500mV in our example below), which makes it unsuitable for a low supply voltage application. Moreover, MOS capacitors have usually a very large gate leakage, which may be problematic. For that reason, a *thick-oxide* device that is less dense but has substantially less leakage may be chosen.

To avoid inversion, the NMOS may be placed inside an *n*-well [13] at no extra cost, known as accumulation-mode MOS capacitor (Figure 1.35). The naming has to do with the fact that the transistor must stay either in depletion or in accumulation.

This leads to the *C–V* characteristics also shown in Figure 1.34 (the black curve). Even though the capacitance is still quite nonlinear, the accumulation starts right around 0V rather than the threshold voltage. Consequently, it is easier to bias the device at low voltages and benefit from an almost flat region where the gate capacitance has reached C_{OX}.

In many cases, for instance when the capacitance is placed in the feedback path of an op-amp, it may not be practical to apply a relatively large bias voltage. An alternative is to use a linear capacitor formed by the fringe fields that are quite strong in most modern CMOS processes due to the close proximity of metal lines. While problematic for routing signals

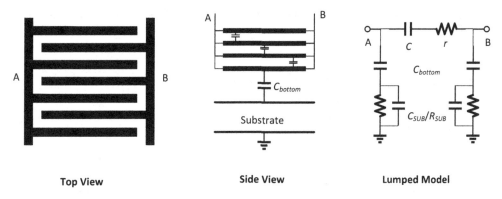

Figure 1.36: Fringe capacitors

and connecting blocks, this along with the large number of metal layers available could be taken advantage of in building linear capacitors. An example is shown in Figure 1.36. To maximize density, minimum width metal lines are placed at minimum spacing allowed by technology, and the two terminals form a comb-like structure. Moreover, several layers of metals connected on each end may be placed on top of each other to improve density further.

It is common for CMOS processes to provide a thick or ultra-thick top metal layer used for clock tree routing or inductors. Since the minimum spacing allowed is usually too large (for example 1.8μm for the very top layer known as the AP layer in 16nm CMOS), the top thick metal may not be used. Additionally, there is a concern over how large the bottom plate parasitic capacitance would be. Therefore, it may be advantageous to drop one or two of the bottom metal layers as well (particularly the poly and metal one due to large sheet resistance), to place the capacitor further away from the substrate and reduce the bottom plate capacitance. However, if fewer metal layers are used, for the same value of capacitance the structures need to be bigger, and hence the bottom plate parasitic increases accordingly. In a 40nm CMOS process with a thick M6 option, the best compromise appears to be using M3–M5, leading to a density of about 2fF/μm^2. The bottom (or top) plate parasitic is generally very small, about 1–2%. For a given structure, the exact amount of capacitance is very difficult to calculate using closed formulas, and it is often best to use extraction tools (such as EMX)[24] to predict the capacitance.

A lumped model of the capacitance is presented in Figure 1.36. The bottom and top plate parasitics that are generally symmetrical due to the physical structure of the capacitor are connected to substrate. The substrate is lossy and is typically modeled by a parallel RC circuit. Since the bottom plate parasitic capacitance (C_{bottom}) is small, and R_{SUB} is usually large for bulk processes, this loss is negligible for frequencies up to several GHz. Additionally, there is the metal series resistance forming the comb lines. If for a given capacitance the structure is built to consist of N smaller units in parallel, this resistance is then reduced by N^2, typically leading to a very high-Q capacitor for well-designed structures.

The fringe capacitors scale with technology to a good extent, as the physical spacing between metal lines generally improves. Shown in Figure 1.37 is multi-finger fringe capacitance density in fF/μm^2 for several recent generations of standard CMOS processes. Note that unlike MIM

---

[24] EMX is a planar 3D integral equation solver that uses an accurate representation of Maxwell's equations.

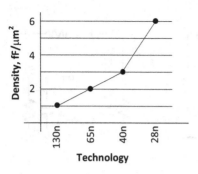

Figure 1.37: Fringe capacitance density in various processes

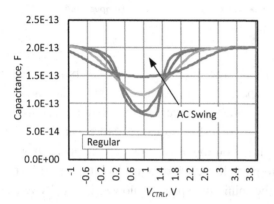
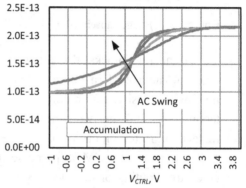

Figure 1.38: Large signal MOS capacitor simulation

(metal–insulator–metal) capacitors, fringe capacitors do not require any additional process steps, and thus incur no extra costs. By comparison, a thin-oxide MOS capacitor in 28nm is 23fF/μm² (corresponding to an oxide thickness of about 1.25nm), whereas a thick oxide one is 10fF/μm². Note that if MOS capacitors are used, it is still possible to fill fringe capacitors on top to further boost the density.

Continuous tuning may be achieved by using one of the two MOS structures discussed earlier. Particularly NMOS in n-well is desirable as it provides a greater tuning range. For a regular NMOS, if the voltage swing across the tank were large (as is the case for most CMOS oscillators) regardless of the DC bias applied, the effective capacitance would be mostly C_{OX}. That is because the depletion capacitance corresponds to a relatively narrow region of the C–V curve (Figure 1.34). To illustrate this important point further, Figure 1.38 shows a large signal simulation of a 28nm MOS capacitor. The effective capacitance is plotted versus the control voltage, for four different values of signal swing across it, 0, 0.5V, 1V, and 1.8V. The *effective capacitance* here is defined as the magnitude of the fundamental component of the current, divided by the voltage swing across the capacitance, normalized to the angular frequency.

In the case of a regular MOS capacitor, the ratio of maximum to minimum capacitance (that is, approximately C_{OX} to $C_{OX}\|C_{DEP}$) is around 2.5, but diminishes to less than 1.4 as the signal swing increases. The accumulation mode varactor, on the other hand, has a maximum to minimum capacitance ratio of about 2, which is maintained reasonably well for signal swings of as high as 1.8V. The two varactors are identical in size and use a thick oxide NMOS with a channel length of 0.75μm. Although a shorter channel results in a better Q, it has a worse tuning range and handles less swing due to reliability.

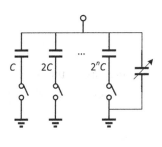 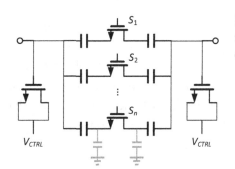

Figure 1.39: Discrete tuning using switched capacitors

Discrete Tuning Concept **Differential Design**

Since the Q of continuously tuned capacitors may not be very high, a wider tuning range without compromising Q can be achieved by incorporating discrete tuning using switched linear capacitors along with MOS varactors, as illustrated in Figure 1.39 [14]. This also results in less VCO gain, and thus less sensitivity to noise and interference at the VCO control voltage. The MOS varactor needs to provide only enough range to cover the worst case discrete step size.

A larger switch results in lower resistance and thus a better quality factor; however, the switch parasitic capacitance in the off mode limits the tuning range. If designed differentially, the same tuning range is achieved but with twice as much Q, as the on resistance is half. A differential design in 28nm CMOS as an example is shown in Figure 1.39. It consists of 32 units of 40fF linear capacitors, with the total capacitance varying from 430fF to 1.36pF (about $3\times$), in steps of 29fF. The Q at 3.5GHz varies from a maximum of 80 to 45 when all the capacitors are turned on. The switches are $11\times1/0.1\mu$m.

In the differential design, it may often be necessary to bias the switches' drain and source, which can easily be done by placing a large resistance there.

1.11 INTEGRATED INDUCTORS

First introduced into silicon in 1990 [15], [16], monolithic inductors have since been widely used in RF and mm-wave applications. Due to fabrication limitations, on-chip inductors are typically realized as metal spirals. To achieve a lower loss and thus better Q, it is common to use the top metal layer, which is typically thick or ultra-thick.

Applying Biot–Savart's law to calculate the magnetic field, one can show that the self-inductance of a piece of wire with a length of l, and a rectangular cross section at moderate frequencies (several GHz) is [17]

$$L \approx \frac{\mu_0}{2\pi}l\left(\ln\frac{2l}{W+t}+0.5\right),$$

where t is the thickness of the metal, and is fixed for a given technology, while W is the metal width and is a design parameter. All units are metric, and we assume that the length is much larger than the width. To gain some insight, the series resistance of the same line at low frequency is given by

$$r = R_\square \frac{l}{W},$$

where $R_\square$ is the metal sheet resistance (about 10.4mΩ/□ for ultra-thick top layer for instance). If the only concern is to maximize Q for a given inductance, then increasing the width helps, but it must be noted that a larger W needs a larger length to keep the inductance constant, leading to some increase in resistance and of course area.

Example: For a typical CMOS process considering an ultra-thick metal layer with $R_\square = 10.4$ mΩ/□ and $t = 2.8$μm, the corresponding theoretical inductance suggested by the equations above and Q have been plotted in Figure 1.40 at 4GHz, assuming a 1mm trace.

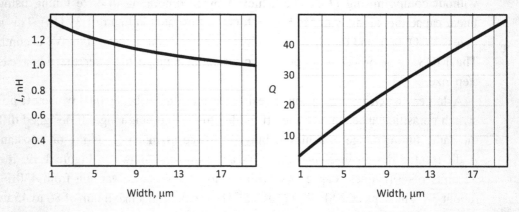

Figure 1.40: Low-frequency theoretical inductance and quality factor for a 1mm trace versus its width

The quality factor predicted in Figure 1.40 applies only to low frequencies, as it assumes the low-frequency series resistance is the only loss mechanism. This is certainly not true, as we already know that at higher frequency the *skin effect* results in a higher resistance, and hence lower value of Q as suggested in the example. As we discussed in Section 1.8.2, in a conducting medium where the conduction current, $\sigma \mathbf{E}$, as expressed by Ohm's law is much larger than the displacement current, the solution for the electric field is given by

$$\mathbf{E} = E_0 \mathbf{a}_x e^{-jkz} = E_0 \mathbf{a}_x e^{-\alpha z} e^{-\beta z},$$

where

$$\alpha = \beta = \sqrt{\frac{\omega \mu \sigma}{2}}$$

and $\delta = \left(\frac{1}{2} \omega_0 \mu_0 \sigma\right)^{-1/2}$ is the *skin depth*. As pointed out, this suggests that in the conductor, the field decays by an amount of e^{-1} in a distance of one skin depth δ, which could be comparable to metal width or its thickness at high frequencies. For instance, the skin depth for the top metal layer used in the previous example is about 1.4μm at 4GHz. As a result, effectively the current tends to flow more at the surface, and the resistance will increase. Accordingly, a modified expression for the metal resistance to include the skin effect is

$$r = R_\square \frac{l}{W} \frac{t/\delta}{1 - e^{-t/\delta}},$$

where t is the metal thickness. At low frequency, δ is large and the equation simplifies to the original expression for the resistance. However, at very high frequencies, the exponential term approaches zero and thus

$$r = R_\square \frac{l}{W} \frac{t}{\delta} = \frac{1}{\sigma} \frac{l}{\delta W}.$$

That is, t is replaced by δ at high frequency, which could lead to considerably higher resistance.

Example: A more thorough study of the inductance of a straight line is shown in Figure 1.41. This not only is beneficial for understanding the spiral inductors, but also gives a good perspective on the implications of the long routing often needed for signals or power rails in RF. Shown in Figure 1.41 is the simulated inductance and quality factor of a 1mm piece of metal in 16nm CMOS using EMX. Three different widths are simulated using the top ultra-thick metal layer (sheet resistance of 10.4mΩ/□ and about 2.8μm thickness), along with a fourth case of 20μm wide trace, but with the top metal as well as the next two top metal layers (sheet resistance of 16.7mΩ/□ and about 12μm thickness) all shorted together.

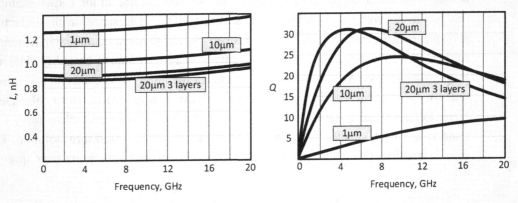

Figure 1.41: Simulated inductance and quality factor of a 1mm piece of metal in 16nm CMOS

For the case of 10μm wide metal, the calculated low-frequency inductance is about 1.1nH, and the quality factor is 0.6 at 100MHz (Figure 1.40), both matching reasonably well with the simulations. As do the other three cases. For instance, doubling the width to 20μm raises the Q to about 1.1, and shorting all the top three metals brings it up to 2.1. However, evidently the quality factors do not rise linearly with frequency as explained earlier. In fact, in addition to the skin effect, which is comparable to the metal widths used, the substrate loss discussed shortly is another important contributor for integrated inductors or transmission lines, especially at higher frequencies. At 4GHz, for instance, the quality factor is simulated to be around 18.3 for 10μm wide metal, and increases only to 27.4 if the metal width is doubled. Keeping the width at 20μm but shorting the top three metals leads to only a small rise of about 10% to 30.8, primarily limited by the skin effect.

1.11.1 Spiral Inductors

Using a long piece of wire as described above is clearly not a viable option, as apart from the area, connecting it to any circuit is impractical. It is more common to form spirals to make it more compact and practical to use. The most natural choice would be a circle. However, circles are not physically possible to be laid out in an integrated circuit. The same length could be wound to a square spiral as shown in Figure 1.42. According to Biot–Savart's law, the magnetic fields of all four legs add up in the center, with a direction normal to this page of the book, although they will not add quite constructively at the edges. A better approximation of the circle may be made by using a hexagon or an octagon, as 45° angles are allowed. This results in a more constructive addition of the magnetic fields (Figure 1.42), typically leading to a higher Q.

The inductance of the spiral could be given with the general expression as follows [17], [18]:

$$L_T = \sum L + \sum M^+ - \sum M^-,$$

where the first term indicates the sum of the self-inductances of each leg, and the second and third terms indicate the mutual inductances between *parallel* legs that carry the *same* or *opposite* direction currents. In the case of a single-turn square spiral shown in Figure 1.42, the second term does not exist as the parallel legs carry only opposite currents. Ignoring the mutual inductances for the moment (the third term) and assuming a width of 10μm and a length of roughly 1mm / 4 = 250μm for each leg, the total inductance is 4 × 0.21nH = 0.84nH, which is somewhat less than the 1.1nH obtained for the straight line in the earlier example. Clearly, this has to do with the logarithmic dependence of the inductance on the inductor length as shown earlier. In reality the inductance is even less as the negative mutual inductance reduces it further. Since the series resistance is roughly the same, the upper bound Q obtained earlier is proportionally less. A closed form expression for the mutual inductance between two pieces of wires with identical length l, separated by a space d, is quite tedious to calculate [18], [19]. Here, to gain some insight, we provide only the expression for two filamentary lines (that is, t and W are much smaller than l and d), which has a simple closed form solution. The equation may be readily obtained by applying Biot–Savart's law and integrating over space:

$$M = \frac{\mu_0}{2\pi} l \left[\ln\left(\frac{l}{d} + \sqrt{1 + \frac{l^2}{d^2}}\right) - \sqrt{1 + \frac{d^2}{l^2}} + \frac{d}{l} \right].$$

For our previous example, if $\frac{l}{d} \cong 1$, which is a good approximation for a square single-turn inductor, then the mutual inductance between two legs is roughly 12% of the inductance of each leg.

Figure 1.42: Spiral inductor in integrated circuits

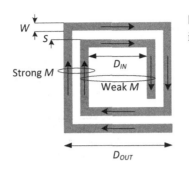

Figure 1.43: Multi-turn versus single-turn inductor

The inductance may increase if another turn is added (Figure 1.43), assuming that the inner diameter (D_{IN}) is still large, that is, the inner and outer diameters (D_{OUT}) are comparable. The increase is due to the large contribution of positive mutual inductance between the adjacent legs that carry the same current, while the negative mutual inductance of the opposite legs is still small as they are far apart. Therefore, when designing spiral inductors, it is common to have *hollow* structures, with the adjacent legs drawn as close as possible. Keeping the inner diameter large, however, limits the number of turns allowed.

Multi-turn inductors typically enjoy a more compact design but somewhat larger capacitance. If the inductance needed is small (say a few tenths of henry), then a single-turn inductor with reasonably large diameter remains the best choice.

For a given structure in general, the inductance can be fully expressed by knowing the metal width (W), the spacing between adjacent legs (S), the number of turns, and either the inner diameter, the outer diameter, or the total length (Figure 1.43). For most RF applications, practical values of integrated inductances range from a few tenths of nH to several nH. Apart from very simple structures, having closed form expressions for the inductance value is not possible. Numerous attempts have been made to calculate approximate closed form expressions for the spiral inductors with somewhat limited accuracy [18], [19], [20], [21]. Given their efficiency and precision, it is best to utilize common 3D electromagnetic simulators such as EMX or HFSS,[25] widely used among many RF designers.

1.11.2 Second-Order Effects

Apart from the ohmic loss, there are several other contributors that limit the inductor performance. Shown in Figure 1.44, the metal strips inevitably have a nonzero capacitance to the substrate.

This capacitance limits the maximum allowable frequency that the inductor can be used at. This is typically expressed as the *self-resonance* frequency, that is, the frequency that the total parasitic capacitance resonates with the inductance. To function properly, the self-resonance frequency must be obviously well above the maximum frequency that the inductor is intended to be used. Moreover, this capacitance is connected to the lossy silicon substrate which could degrade the quality factor at higher frequencies. There is also a capacitance between various legs of the inductor branches. This capacitance may be ignored for a single-turn structure. However, for multi-turn inductors it would be relatively important as the adjacent legs are laid out very close

---

[25] HFSS is a commercial finite element method solver for electromagnetic structures.

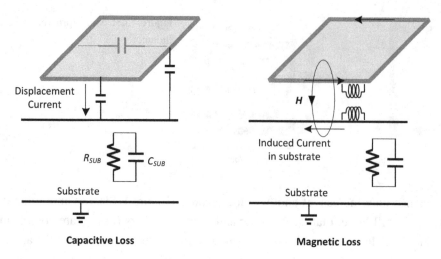

Figure 1.44: Substrate loss in on-chip inductors

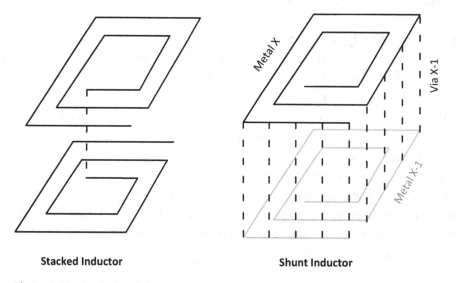

Figure 1.45: Stacked and shunt inductors

to one another to maximize positive mutual inductance. This capacitance is substantially higher in *multi-stacked* structures (Figure 1.45) where several inductors of similar structure are placed in *series* using lower metal layers to increase the inductance without increasing the area. Since this structure leads to Q degradation as well, given much larger sheet resistance of lower level metals, it is not very common unless very large values of inductance are needed.

An alternative structure uses several inductors designed with lower metal layers connected to each other in *parallel* instead (Figure 1.45). While this will not lead to an increase in the inductance, it does improve the ohmic loss to some extent. However, that comes at the expense of lower self-resonance frequency, as lower metals have higher capacitance to substrate. The Q improvement as a result of this may not be very significant for two reasons: First, the lower metal levels typically have worse sheet resistance. Second, since the parasitic capacitance to substrate is increased, Q degradation due to capacitive coupling is worse. At lower frequencies, say 1GHz or less, this structure may be still helpful. At higher frequencies, say 2GHz or above, apart from the two reasons mentioned, the skin effect becomes an issue as well, and shunting the metal layers may not improve the Q at all.

As demonstrated in Figure 1.44, there is also a magnetic loss created at high frequencies. The magnetic field created due to the AC current flowing in the inductor branches results in a magnetic flux that is varying with time. Faraday's law suggests that an electric field E_{Si} is induced in the substrate. This electric field leads to a current density flow of $J = \sigma_{Si} E_{Si}$ according to Ohm's law, where σ_{Si} is the silicon substrate conductivity. It is as if there is a transformer reflecting the substrate resistance (or loss) in parallel with the inductor (Figure 1.44, right side). A higher substrate resistance (lower σ_{Si}) is preferred to lower this loss. Fortuitously, most modern CMOS processes use *bulk substrate*, where the resistivity is relatively high. Unlike capacitive loss, where adding a metal shield could help, in the case of the magnetic loss the shield is generally not helpful, as it effectively shorts out the inductor. The impact of the shield is going to be further discussed in the context of an example shortly.

In addition to the substrate factor, there is another mechanism degrading the inductor quality factor in the multi-turn spirals known as the *proximity effect* or *current crowding* [22], [23], [24]. In the case of a single-turn inductor, the current and magnetic field distributions are fairly uniform except at the corners, which is expected. On the other hand, in multi-turn structures, the assumption of uniform current distribution is no longer valid, especially for the metal strips close to the center of the inductor, as the current as well as the magnetic field distributions tend to become higher near the inner edge. This non-uniform current distribution means that the resistance increases because the majority of the current flows through a smaller area, leading to Q degradation because of the higher resistance.

Example: Shown in Figure 1.46 is a 3.6nH inductor designed at 2GHz for LTE applications. The inductor uses the AP (or RDL) layer, as well as the top two metal layers all

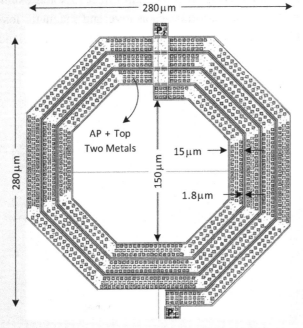

Figure 1.46: A 3.6nH inductor in 16nm CMOS

Continued

shorted to improve the ohmic losses (the AP and the top two metals have a sheet resistance of 10.4mΩ/□, and 16.7mΩ/□, respectively, in the 16nm CMOS process used here). The inductor uses a relatively large metal width of 15µm, and consequently is rather large (280µm in dimension).

Shown in Figure 1.47 is the EMX simulated inductance and quality factor over frequency (that is, the black solid curve labeled "no shield"; we shall discuss the others shortly). The self-resonance frequency is about 9.6GHz, which is reasonable given the relatively large inductance and is well above the 2GHz intended frequency of operation. The quality factor is about 9.4 at 2GHz.

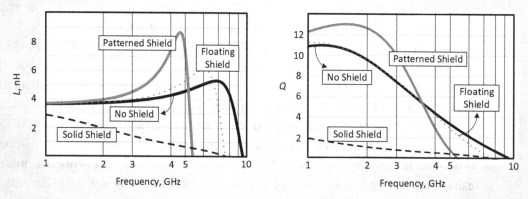

Figure 1.47: The simulated performance of the 3.6nH inductance designed for LTE applications

Shorting the top two metals along with AP layer results in a net sheet resistance of 4.6mΩ/□, which is quite low, and we suspect that along with the skin effect to some extent, the main reason for a relatively low Q of 9.4 is the substrate capacitive and magnetic losses. Hence, inserting a shield as shown in Figure 1.48 may seem to be helpful at the first glance.

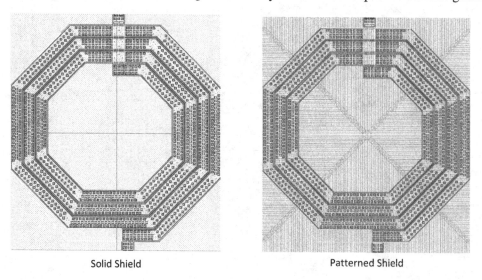

Figure 1.48: Two variations of the inductor of Figure 1.46 with a metal one-layer shield added

On the left shown is a solid metal one-layer shield that once appropriately connected to an ideal ground could potentially eliminate the capacitive losses to the substrate. This kind of shielding is commonly done for the IO pads or RF capacitors and to eliminate the capacitive coupling to the lossy substrate [25], [26], as if in the simple model of Figure 1.44, R_{SUB}/C_{SUB} is bypassed by the low-resistance metal shield. In the case of an inductor, in contrast, while it potentially eliminates the capacitive loss, the solid ground shield also disturbs the inductor's magnetic field. According to Lenz's law, an image current will be induced in the solid ground shield by the magnetic field of the spiral inductor. The image current in the solid ground shield will flow in a direction opposite to that of the current in the spiral. The resulting negative mutual coupling between the currents reduces the magnetic field, and thus the overall inductance. This can be intuitively explained by placing a very small resistance in the secondary of the transformer in the simple model of Figure 1.44. Problem 31 shows that assuming a small shield resistance, the effective inductance is $L_{eff} \approx L_1(1-k^2)$, where L_1 is the intended inductance, and k is the coupling coefficient between the inductance and the shield. As the losses stay the same, not only the inductance is reduced, but also the quality factor could significantly degrade.

A *patterned shield* realized by poly or metal one with slots orthogonal to the spiral as illustrated in Figure 1.48 helps, on the other hand, as it increases the resistance of the image current [27]. The slots effectively act as an open circuit to cut off the path of the induced loop current. The technique is mostly helpful in singled-ended inductors though, and it naturally tends to degrade the self-resonance frequency. The implications of the shield in the differential inductors are explained in the next section. Figure 1.47 illustrates three additional simulated cases for a patterned shield, a solid shield, and a floating patterned shield added to the same inductor of the previous example. The shield results in a reduction of the self-resonance frequency, as expected. For this reason, the effective inductance at 2GHz increases to about 4nH for the patterned shield. The solid shield results in lower inductance and much worse Q, as anticipated, while the patterned shielded inductor Q improves to about 12.4. Even though the self-resonance frequency drops to 5.4GHz, it is still acceptable, as a positive trade-off. It must be emphasized that the shield must be connected to a good ground with low inductance. As shown in the figure, the extreme case of a floating shield results in no net improvement to Q, but a little worse self-resonance frequency due to the extra metal layer.

Another interesting and related phenomenon is the lack or presence of the *native layer* underneath the inductor. To illustrate this further, shown in Figure 1.49 is the simulated Q of the 3.6nH inductor of the previous example with and without the native layer. The lack of a native layer results in the excess substrate implant to increase the threshold voltage of regular NMOS devices.[26] Consequently, the channel conductivity increases from 10mS/m to about 50mS/m. Interestingly, the removal of the native layer, despite a lower substrate resistance, leads to somewhat improved Q for both the patterned shield and the no shield design, and a Q of about 14 is achievable now.

One may argue that the reason for this behavior has to do with the fact that at the frequency of interest the capacitive loss of the substrate is a dominant factor. While the shield is

---

[26] Native NMOS transistors have a threshold voltage of close to zero.

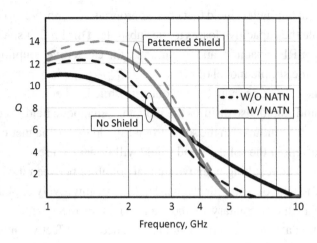

Figure 1.49: Simulated Q of the inductor of Figure 1.13 with and without the native layer

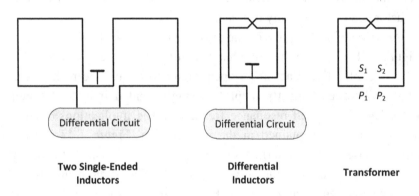

Figure 1.50: Differential inductors

expected to help, it will not entirely eliminate that due to the nonzero resistance of the metal 1 layer (in this process, the sheet resistance of metal 1 is quite high, about $1.1\Omega/\square$). Thus, reducing the substrate resistance by removal of the native layer still helps both the unshielded and the shielded inductor. At higher frequencies (3–4GHz and above), it is clear that removal of the NATN layer hurts the Q, suggesting that the substrate magnetic loss is a more dominant factor. This implies that improving the shield resistance, e.g., using the metal 2 layer shorted to the original metal 1 layer, could help, a task we will defer to the interested reader to experiment with.

1.11.3 Differential Inductors

If two identical inductors are used in a differential circuit, they may be replaced by a *differential* inductor, where the two single inductors are combined (Figure 1.50) [28]. This naturally leads to a more compact design. Furthermore, a smaller area means less substrate loss and capacitance, which is important at high frequencies. This of course assumes that the size of the differential inductor to realize an inductance twice as big as each single-ended one remains the same. This is not exactly the case, and the area tends to grow some, but still there is substantial saving.

Even though the capacitance to substrate is expected to decrease (ideally halve), the main drawback of the differential topology is lower self-resonance frequency. This is caused by higher capacitance between the adjacent legs, compared to two single-spaced legs far apart, as shown in Figure 1.50. Another disadvantage of differential inductors is the fact that any unwanted coupling to the inductor (through parasitic capacitive and particularly magnetic sources) appears as an undesirable differential signal at the two terminals. For two single-ended inductors, however, if the parasitic source is far enough, it appears as common-mode noise at the outputs.

Example: A 2-turn 1nH differential inductor using the ultra-thick AP layer only in 16nm CMOS has been laid out and simulated in EMX, as shown in Figure 1.51. The inductor has a Q of 16.2 at 5.5GHz.

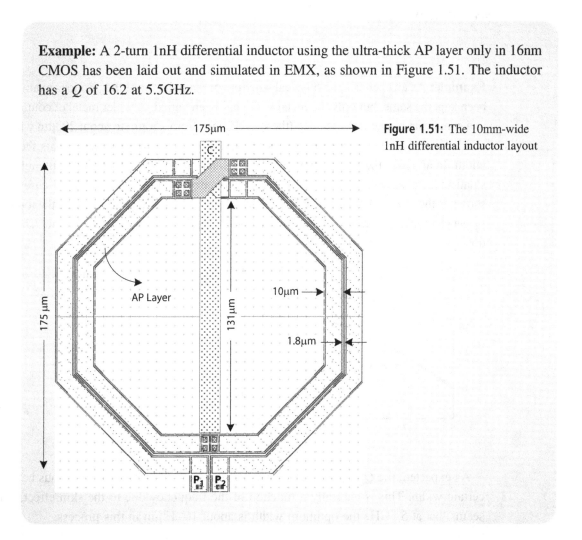

Figure 1.51: The 10mm-wide 1nH differential inductor layout

The inductor simulated performance over frequency is shown in Figure 1.52. In the case of differential inductors, the presence of the patterned shield proves to be less helpful, as the electric fields do not penetrate deeply into the substrate, rather they stay on surface between the adjacent differential legs of the spiral. In practice, the degradation of the self-resonance frequency is more of a concern, and often the shield is not used in the differential inductors.

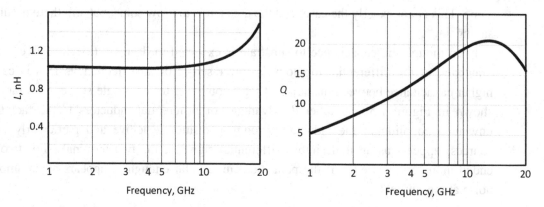

Figure 1.52: EMX simulation results of the inductor of Figure 1.51

Example: As an exercise, the physical structure of the inductor of the previous example has been kept the same, but only the metal width has been varied. A wider metal of course leads to a somewhat bigger inductor size (the size changes from 150μm to about 200μm when the width varies). Shown in Figure 1.53 is the simulated Q of the structure versus the metal width. In all cases the differential inductance is kept constant at 1nH. Given the relatively small inductance value, the self-resonance frequency is above 20GHz in all cases. Also shown is the Q for 10μm width, but two additional cases: AP layer changed to the top metal (sheet resistance of about 16.7mΩ/□), and both AP and top metal shorted together (shown as dots).

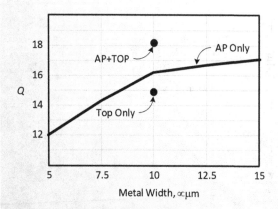

Figure 1.53: Q versus the metal width of a 1nH 2-turn differential inductor

As expected, the Q improves for a wider metal width, but the effect plateaus beyond a certain width. This is naturally a function of the frequency due to the skin effect, and it seems that at 5.5GHz the optimum width is about 10–12μm in this process.

1.11.4 Transformers

An ideal transformer is realized by winding two coils with turn ratios of n_1 and n_2 respectively on a magnetic core, as shown in Figure 1.54.

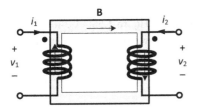

Figure 1.54: An ideal transformer

If the core permeability is large (or ideally infinite), the magnetic flux is contained, and the flux linkages for each coil will be $\phi_1 = n_1\phi$ and $\phi_2 = n_2\phi$. From Faraday's law, as $v_1 = \frac{d\phi_1}{dt}$ and $v_2 = \frac{d\phi_2}{dt}$, we have

$$\frac{v_1(t)}{v_2(t)} = \frac{n_1}{n_2}.$$

To find the relation between currents, we notice that analogous to Ohm's law and concept of *resistance* for electric fields, we can define *magnetic reluctance*, $\mathcal{R}_m$, relating the current and magnetic flux as follows (see [6] for more details):

$$n_1 i_1 + n_2 i_2 = \mathcal{R}_m \phi,$$

and since $\mathcal{R}_m = 0$ for an ideal core, this results in

$$\frac{i_1(t)}{i_2(t)} = -\frac{n_2}{n_1}.$$

From this it follows that $v_1(t)i_1(t) + v_2(t)i_2(t) = 0$, for all t, stating that an ideal transformer is lossless without *energy storage capability* (unlike inductors or capacitors). Moreover, the energy conservation tells us that in an ideal transformer the self-inductance of the two coils must be infinite, with a coupling coefficient of one.

Since a magnetic core with high permeability is not available in integrated circuits, practical transformers behave more like coupled inductors with reasonably high coupling factor if designed properly [29]. A differential inductor is in fact a transformer whose secondary ports are shorted together and connected to a common voltage, as was shown in Figure 1.50. Therefore, we face more or less similar trade-offs when designing transformers. Obviously it is key to lay out the primary and secondary lines as close as possible to maximize the coupling coefficient (k). A k factor as high as 0.8 is achievable with a proper design.

Example: A 2-turn transformer using the ultra-thick AP layer is shown in Figure 1.55. The secondary has a center tap (marked as C in the figure) as is usually commonly used in many applications (e.g., RF amplifiers) for biasing purposes. The center tap has been brought out with a lower metal (the dotted stripe). Since the length of this metal is short, the impact on the quality factor is usually small. The primary and secondary parts are labeled P_1, P_2, S_1, S_2.

Continued

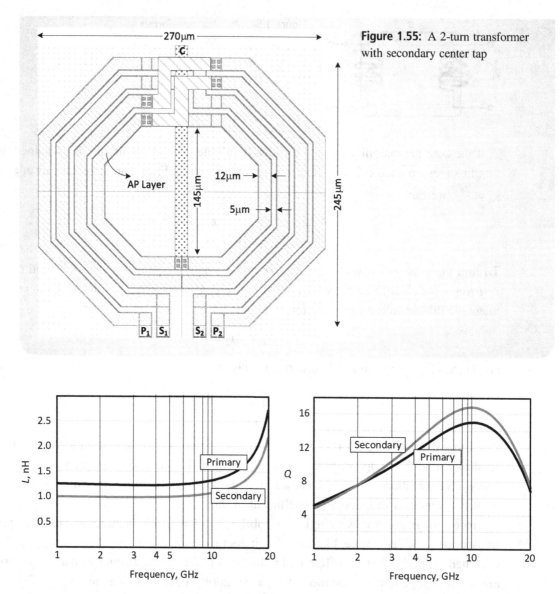

Figure 1.55: A 2-turn transformer with secondary center tap

Figure 1.56: EMX simulation results of the transformer of Figure 1.55

The primary has a simulated inductance of 1.25nH and a Q of 13.4 at 5.5GHz, and the secondary inductance is 1nH with a Q of 15 at 5.5GHz. The coupling factor is about 0.7 at 5.5GHz and remains relatively flat over frequency. More or less a similar trend as differential inductors exists here in terms of the metal thickness, the type of metal layer used, etc. The EMX simulation results of the transformer are shown in Figure 1.56.

Example: Shown in Figure 1.57 is another design, where primary is single turn, while the secondary is kept at two turns with no center tap. The transformer uses both the AP and the top metal layer shorted to improve the loss, but that comes at the expense of worse self-resonance frequency (about 18 GHz). The primary has a simulated inductance of 0.44nH and a Q of 7 at 5.5GHz, and the secondary inductance is 1.14nH with a Q of 12.5 at 5.5GHz. The coupling factor is about 0.8 at 5.5GHz.

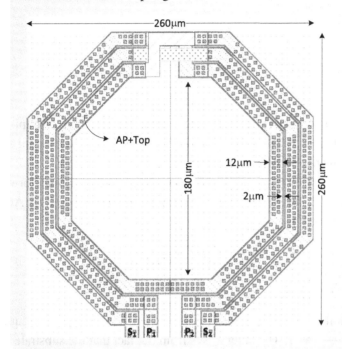

Figure 1.57: A 1-turn primary, 2-turn secondary transformer with no center tap

The primary reason for worse Q compared to the previous transformer is the large difference between the primary and secondary inductance. In some applications such as impedance transformation used in RF matching networks (see Chapter 3), different primary to secondary turn ratios are needed, and the transformer of Figure 1.57 may become very handy. Rather than specifying the quality factor of either the primary or the secondary, it may be more meaningful to describe the insertion loss of the transformer. We will take a closer look at this topic in Chapter 3.

1.11.5 Inductor Lumped Circuit Model

Although they are distributed in nature due to their relatively large size, it is convenient to model inductors by simple lumped elements. The most common circuit is shown in Figure 1.58. It consists of the low-frequency inductance (L), the series ohmic resistance (r), the oxide capacitance to substrates (C_{OX}), the substrate model (R_{SUB} and C_{SUB}), and C_F, which models the capacitance between adjacent legs. The main advantage of the model is that all its elements are physical, while it produces a reasonable approximation valid over a wide range of frequency. It is therefore very common among RF designers to model the inductor as such. Since the substrate characteristics are not very well known, R_{SUB} and C_{SUB} are usually fitting parameters. The combination of C_{OX} and R_{SUB}/C_{SUB} is typically sufficient to account for both magnetic and capacitive loss of substrate. While the model could be fit to represent the inductor at least at one exact frequency, and possibly over a reasonable range, if one is interested in a

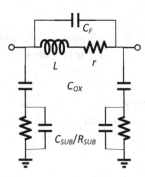

Figure 1.58: Inductor lumped model

true wideband model, S-parameters generated by the simulator (EMX for example) may be used. To speed up the simulations, EMX may be used to produce a lumped element equivalent circuit comprising RLC element and dependent sources.[27]

Without loss of generality, let us take a closer look at the single-ended inductor equivalent circuit model of Figure 1.58 where we assume one terminal is connected to an AC ground. Ignoring C_F, the input impedance is

$$Z_{IN} = \frac{(r+jL\omega)\left(1 + \frac{C_{SUB}}{C_{OX}} + \frac{1}{jR_{SUB}C_{OX}\omega}\right)}{1 + \frac{C_{SUB}}{C_{OX}} + \frac{r}{R_{SUB}} + j\omega\left(\frac{L}{R_{SUB}} + rC_{SUB}\right) - LC_{Si}\omega^2 + \frac{1}{jR_{SUB}C_{OX}\omega}}.$$

To simplify, we recognize that generally $r \ll R_{SUB}$, $C_{SUB} \ll C_{OX}$, and at frequencies of interest and beyond $\left|\frac{1}{jR_{SUB}C_{OX}\omega}\right| \ll 1$. The latter arises from the fact that the substrate resistivity is large, while at higher frequencies the impedance of C_{OX} is relatively small.

This leads to

$$Z_{IN} \cong \frac{r + jL\omega}{(1 - LC_{SUB}\omega^2) + j\omega\left(\frac{L}{R_{SUB}} + rC_{SUB}\right)}.$$

The impedance is bandpass, although at very low frequencies it does not approach zero. Rather it becomes equal to the low-frequency series resistance, r, which is expected. The self-resonance frequency is $\omega_{SRF} = 1/\sqrt{LC_{SUB}}$, where the magnitude of impedance peaks to a value of $\frac{L}{\frac{L}{R_{SUB}} + rC_{SUB}} \cong R_{SUB}$. We made the assumption that at higher frequencies $r \ll L\omega$, which is true if the Q is reasonably large.

The inductance is naturally frequency dependent and by definition is equal to $L(\omega) = \frac{|Im[Z_{IN}]|}{\omega}$. This leads to

$$L(\omega) \equiv L\frac{1 - LC_{SUB}\omega^2}{(1 - LC_{SUB}\omega^2)^2 + \left[\left(\frac{L}{R_{SUB}} + rC_{SUB}\right)\omega\right]^2}.$$

Evidently, the low-frequency inductance is equal to L, while the inductance eventually approaches zero at self-resonance frequency. This makes sense, as at the self-resonance

---

[27] The equivalent circuit is not physical and is curve fit to merely replace S-parameters to improve convergence and simulation speed.

frequency the input impedance has a phase of zero (and a peak magnitude of roughly R_{SUB}, as discussed before). Beyond the self-resonance frequency Z_{IN} is capacitive. Taking the derivative of $L(\omega)$ versus ω, we can show that the inductance is expected to peak right before approaching the self-resonance frequency. Defining a unitless parameter, $\eta = \frac{1}{2R_{SUB}}\sqrt{\frac{L}{C_{SUB}}}$, then the frequency where the inductance peaks is roughly $(1-\eta)\omega_{SRF}$, and the value of the inductance at that frequency is $\frac{L}{4\eta}$. For typical values, note that $\eta \ll 1$. The reason that the inductance peaks before self-resonance is due to the fact that the term $(1-LC_{SUB}\omega^2)^2$ in the denominator starts approaching zero faster than the numerator due to the square function.

Example: The modeled and simulated inductance of a differential inductor designed in 28nm CMOS is shown in Figure 1.59. The inductance is designed to be 1nH at 4GHz, where the EMX and lumped equivalent are plotted. The measured characteristics of the inductor are very close to the one predicated by EMX. The lumped model values are $L = 1$nH, $r = 0.44\Omega$, $C_{OX} = 5.57$pF, $C_{SUB} = 61$fF, and $R_{SUB} = 872\Omega$. The inductor is a 2-turn design using the top metal layer only. The total length is about 1.1mm, and the width is 22μm. The DC resistance is then calculated to about 0.5Ω. If it were designed as a long piece of wire, the DC inductance would be 1.1nH.

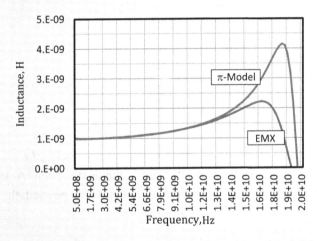

Figure 1.59: Simulated inductance for a 1nH inductor

The self-resonance frequency is about 20GHz, very close to what is predicted by our analysis, where the inductance reaches zero. Moreover, EMX and lumped model match fairly well for a wide range of frequencies. According to our derivations, $\eta = 0.074$, and we expect the inductance to peak to 3.4nH, at 18.5GHz.

Next, we shall present quantitative description of Q based on the lumped model input impedance derived earlier. Defining Q as

$$Q = \frac{|\text{Im}[Z_{IN}]|}{\text{Re}[Z_{IN}]}$$

we will arrive at a simplified expression for the quality factor of the inductor. Given the approximate expression for Z_{IN} derived earlier,

$$Z_{IN} \cong \frac{r+jL\omega}{\left(1+\frac{r}{R_{SUB}}-LC_{SUB}\omega^2\right)+j\omega\left(\frac{L}{R_{SUB}}+rC_{SUB}\right)} = \frac{r+jL\omega}{A+jB},$$

we can write

$$\text{Re}[Z_{IN}] = \frac{r+\frac{r^2}{R_{SUB}}+\frac{(L\omega)^2}{R_{SUB}}}{A^2+B^2}$$

and

$$\text{Im}[Z_{IN}] = \frac{L\omega\left[1-LC_{SUB}\omega^2-\frac{r^2 C_{SUB}}{L}\right]}{A^2+B^2}.$$

Consequently, we have

$$Q = \frac{L\omega\left|1-LC_{SUB}\omega^2-\frac{r^2 C_{SUB}}{L}\right|}{r+\frac{r^2}{R_{SUB}}+\frac{(L\omega)^2}{R_{SUB}}}.$$

Since $r \ll R_{SUB}$, and given that $\frac{r^2 C_{SUB}}{L} \ll 1$, we may rewrite

$$Q \cong \frac{L\omega\left|1-LC_{SUB}\omega^2\right|}{r+\frac{(L\omega)^2}{R_{SUB}}}.$$

Interestingly, the Q approaches zero at self-resonance frequency, $\omega_{SRF} = 1/\sqrt{LC_{SUB}}$, as Z_{IN} is purely real. This is of course erroneous and will be clarified in the next section. At frequencies well below the self-resonance, where the inductor is typically intended to be used, the Q may be expressed as

$$Q \cong \frac{L\omega}{r+\frac{(L\omega)^2}{R_{SUB}}}.$$

Two mechanisms contribute to the loss: At low frequencies, the second branch of the π-model is not effective, and thus $Q = \frac{L\omega}{r}$, which rises linearly with frequency but flattens somewhat due to the skin effect (although not included in the simple π-model). On the other hand, at higher frequencies, C_{OX} becomes a short circuit, and the model simplifies to a parallel RLC equivalent circuit, consisting of L, R_{SUB}, and C_{SUB} (note that r may be ignored compared to $L\omega$ if the frequency is sufficiently high). Thus $Q = \frac{R_{SUB}}{L\omega}$, which linearly falls with frequency. Moreover, the high- and low-frequency quality factors become equal to each other at a frequency of approximately $\omega_{Q_{opt}} = \frac{1}{L}\sqrt{rR_{SUB}}$, where we expect the Q to peak. This suggests that balancing the low- and high-frequency losses optimizes the inductor quality factor at a given frequency. The optimum Q, as such, will be equal to

1.11 Integrated Inductors

$$Q_{opt} = \frac{1}{2}\sqrt{\frac{R_{SUB}}{r}},$$

which is only a function of the low-frequency ohmic loss, and the substrate resistance.

Example: The simulated Q for the same inductor of previous example is shown in Figure 1.60. The quality factor due to only series resistance is 57 at 4GHz, whereas the substrate resistance loss yields a Q of 35. Thus, the combined quality factor at 4GHz sums up to be 21.7, which is slightly overestimating EMX results. The Q is expected to peak at 3.2GHz, which is close to EMX simulation results.

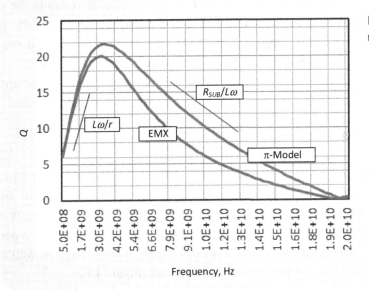

Figure 1.60: Simulated Q of the 1nH inductor

As one final note on this topic, shown in Figure 1.61 is the lumped model of a differential inductor created by EMX. It is substantially more complex than the simple model of Figure 1.58.

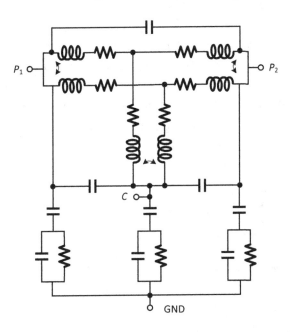

Figure 1.61: An example of a differential inductor lumped model created by EMX

1.11.6 Fundamental versus Inductor Q Definitions

In the previous example, the quality factor approaching zero may have sounded non-intuitive and perhaps deserves some clarification here. The definition of quality factor according to $Q = \frac{|\text{Im}[Z_{IN}]|}{\text{Re}[Z_{IN}]}$ is based on the assumption that the circuitry of interest is inherently an inductor along with some small parasitic capacitances or resistances. This is certainly the case if the inductor is to be utilized at frequencies well below self-resonance, as it typically is. For this reason, this definition is widely adopted in both measurement equipment and the literature. Close to the self-resonance frequency and beyond, however, it clearly falls apart.

To get around this shortcoming, one may suggest using the *fundamental* definition of the Q presented earlier based on the energy stored and power dissipated at resonance:

$$Q = \omega_0 \frac{\text{Total energy stored}}{\text{Average power dissipated}}.$$

The problem with this definition is that it is mainly defined for a second-order RLC circuit at resonance, which is not directly applicable to an inductor especially at lower frequencies. Nonetheless, if the inductor is well behaved, that is to say if the quality factor is reasonably high, and parasitic capacitances are small, then one may effectively model it as a parallel RLC circuit, where the parallel capacitance comprises that of the inductor itself as well as an extra added part to create resonance at the frequency of interest (Figure 1.62).

Once expressed in terms of the parallel components (R_p, L_p, and C_p in the figure), at resonance ($\omega = \frac{1}{\sqrt{L_p C_p}}$), the fundamental Q as derived earlier is

$$Q = \frac{R_p}{L_p \omega} = \frac{|B_{pL}|}{G_p},$$

where G_p is the effective parallel conductance, and B_{pL} is the inductive part of the parallel susceptance at the frequency of interest. Note that G_p and B_{pL} are frequency dependent, and that B_{pL} is negative. Furthermore, the assumption made here is that C_p consists of enough external capacitance (C_{ext} in the figure) to establish resonance at the frequency of interest.

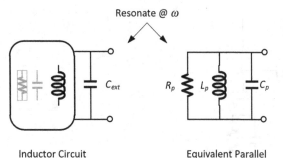

Figure 1.62: Inductor circuit with parallel external resonance

1.11 Integrated Inductors

To gain insight, let us derive the fundamental Q expression for the π-circuit developed earlier as an exercise. The π-circuit input admittance was found to be

$$Y_{IN} = \frac{\left[r + \dfrac{r^2}{R_{SUB}} + \dfrac{(L\omega)^2}{R_{SUB}}\right] - jL\omega\left[1 - \left(LC_{SUB}\omega^2 + \dfrac{r^2 C_{SUB}}{L}\right)\right]}{r^2 + (L\omega)^2}.$$

Thus, the parallel effective conductance and the *inductive* susceptance are

$$G_p = \frac{r + \dfrac{r^2}{R_{SUB}} + \dfrac{(L\omega)^2}{R_{SUB}}}{r^2 + (L\omega)^2}$$

$$B_{pL} = \frac{-L\omega}{r^2 + (L\omega)^2}.$$

Note the term $\dfrac{L\omega\left(LC_{SUB}\omega^2 + \frac{r^2 C_{SUB}}{L}\right)}{r^2 + (L\omega)^2}$ is the *capacitive* part of the susceptance, and is always smaller than the inductive part unless at the self-resonance frequency that they equate. The Q definition, however, assumes that enough external capacitance is added to create the resonance at any arbitrary frequency below self-resonance.

Consequently, the fundamental Q is

$$Q = \frac{L\omega}{r + \dfrac{r^2}{R_{SUB}} + \dfrac{(L\omega)^2}{R_{SUB}}} \cong \frac{L\omega}{r + \dfrac{(L\omega)^2}{R_{SUB}}}.$$

Interestingly (but not surprisingly), the Q does not depend on the parasitic capacitance anymore.

Furthermore, compared to the previous definition, the two expressions come out to be very similar, except that the fundamental Q does not have the term $\left|1 - LC_{SUB}\omega^2 - \dfrac{r^2 C_{SUB}}{L}\right|$, and hence never approaches zero (unless at $\omega \to \infty$).

Example: Compared in Figure 1.63 are the previous example inductor quality factors simulated in EMX according to both definitions.

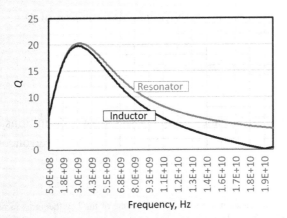

Figure 1.63: Quality factor of 1nH inductor based on two Q definitions

Continued

Clearly, the two curves match very well at frequencies well below the self-resonance, where the inductor is typically intended to be used. Whereas the inductor definition predicts a Q of zero at self-resonance, the fundamental definition based on the resonance circuitry leads to a Q of about 5, which is a more meaningful value.

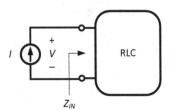

Figure 1.64: An arbitrary RLC one-port depicting the relation of its input impedance to energy dissipated and/or stored

The results above are not a coincidence, nor are limited to the simple π model of Figure 1.58. Using Tellegen's theorem,[28] one can show (see [7] and Problems 26, 27 and 28) that for an arbitrary RLC one-port (Figure 1.64) we have

$$Z_{IN} = \frac{2P_{avg} + 4j\omega(W_L - W_C)}{|I|^2},$$

where P_{avg} is the total power dissipated in all the resistors, W_L is the total energy stored in all the inductors, and W_C is the total energy stored in all the capacitors. They are naturally all real and positive quantities.

The inductor definition of Q will then yield

$$Q_{IND} = \frac{|\text{Im}[Z_{IN}]|}{\text{Re}[Z_{IN}]} = \omega \frac{2|W_L - W_C|}{P_{avg}}.$$

Given a well-designed inductor considered as the one-port, at the frequencies well below the resonance, the inductive part is dominant, which is to say, $W_C \ll W_L$ and thus

$$Q_{IND} \approx \omega \frac{2W_L}{P_{avg}}.$$

On the other hand, according to the resonator (or fundamental) definition of Q,

$$Q_{FUND} = \omega \frac{W_L + W_C}{P_{avg}} = \omega \frac{2W_L}{P_{avg}},$$

since at resonance $W_L = W_C$.

As expected, well below the self-resonance frequency, the two definitions are nearly identical. Around resonance, since $W_L \approx W_C$, Q_{IND} diminishes (erroneously), but Q_{FUND} holds.

---

[28] Published by Bernard Tellegen, the Dutch electrical engineer in 1952, it is one of the key theorems in network theory. Notably, Tellegen was also the inventor of the pentode, a commonly used type of vacuum tube, and the gyrator.

1.11.7 Transformer Modeling

A similar modeling may be developed for the transformers as well. Shown in Figure 1.65 is an integrated transformer model, which consists of two inductors with the equivalent π-model presented before, but also coupled to each other.

To gain some insight, ignoring the losses and capacitances, consider the circuits in Figure 1.66.

For the coupled inductor on the left, we have

$$\begin{cases} \phi_1 = L_1 i_1 + M i_2 \\ \phi_2 = M i_1 + L_2 i_2 \end{cases}$$

or equivalently, $[\phi] = [L][i]$, where $[L] = \begin{bmatrix} L_1 & M \\ M & L_2 \end{bmatrix}$ is the *inductance matrix* [7]. The coupling factor by definition is $k = \frac{|M|}{\sqrt{L_1 L_2}}$. Note that M could be positive or negative, but k is always positive.

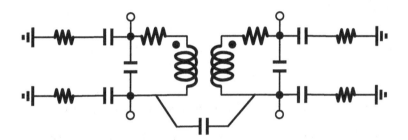

Figure 1.65: A simplified model of an integrated transformer

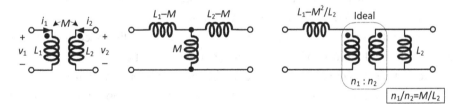

Figure 1.66: Lossless integrated transformer equivalent models

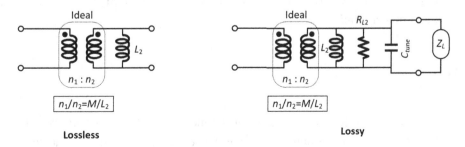

Figure 1.67: Integrated transformer equivalent model when $k \approx 1$

Moreover, from an energy point of view, we can show that $k \leq 1$, otherwise the overall energy of the coupled inductor could become negative. One can derive the L matrix for the other two circuits on the right and show that all three circuits have the same inductance matrix, and hence they are equivalent. If the coupled inductor has a coupling factor close to one, then $L_1 - \frac{M^2}{L_2} \approx 0$, and the transformer can be represented by the equivalent models shown in Figure 1.67.

If the transformer circuit model is available from simulations, say by EMX, it follows that

$$L_1 = \frac{|z_{11}(j\omega)|}{\omega}$$
$$L_2 = \frac{|z_{22}(j\omega)|}{\omega},$$

and the mutual inductance is

$$M = \frac{|z_{12}(j\omega)|}{\omega} = \frac{|z_{21}(j\omega)|}{\omega},$$

where z_{11}, z_{21} ($=z_{12}$), and z_{22}, are the transformer open impedance parameters. The coupling factor is thus $k = \frac{|z_{12}(j\omega)|}{\sqrt{z_{11}(j\omega)z_{22}(j\omega)}}$.

1.12 Summary

This chapter has dealt with basic RF elements and the concepts associated with them.

- Sections 1.1, 1.2, and 1.4 dealt with the electromagnetic fields and the basic definition of capacitors and inductors.
- Time-varying fields and Maxwell's equations were discussed in Section 1.3.
- An introduction to distributed circuits was presented in Section 1.5, and is going to be followed up on in Chapter 3 in the context of transmission lines.
- Resonance circuit properties and the energy dissipation were discussed in Sections 1.6 and 1.7, along with the definition of quality factor and its properties.
- A brief introduction to electromagnetic waves and antennas was presented in Sections 1.8 and 1.9.
- Sections 1.11 and 1.12 offered an in-depth discussion of properties and the design of integrated capacitors, single-ended and differential inductors, and integrated capacitors.

Much of the discussion of the integrated inductors and tuned LC circuits in this chapter will be used in Chapters 7, 9, and 11, as these elements are widely used in low-noise amplifiers, oscillators, and power amplifiers.

1.13 Problems

1. Using spherical coordinates, find the capacitance formed by two concentric spherical conducting shells of radii a and b. What is the capacitance of a metallic marble with a diameter of 1cm in free space? **Hint:** let $b \rightarrow \infty$, thus, $C = 4\pi\epsilon_0 a = 0.55$pF.

2. Consider the parallel plate capacitor containing two different dielectrics. Find the total capacitance as a function of the parameters shown in the figure.

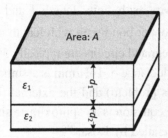

3. What would be the capacitance of the structure in Problem 2 if there were a third conductor with zero thickness at the interface of the dielectrics? How would the electric field lines look? How does the capacitance change if the spacing between the top and bottom plates is kept the same, but the conductor thickness is not zero?

4. Repeat Problem 2 if the dielectric boundary were placed normal to the two conducting plates as shown below.

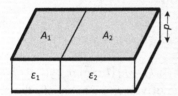

5. Analogs to the capacitance, using Ohm's law, show that the leakage conductance of an almost perfect conductor with a non-infinite conductivity of σ is given by $G = \sigma \dfrac{\int_S E \cdot dS}{-\int E \cdot dL}$.

Calculate the leakage conductance of a coaxial cable with radii a and b as was used throughout the chapter.

6. Consider a very long hollow charge-free superconductor cylindrical shell with inner and outer radii of a and b, respectively. A wire with a current I is placed at the center of the cylinder. Calculate the magnetic field inside and outside considering that the magnetic field inside the shell would have to be zero. If the current I is moved away from the center but inside the shell, how would the magnetic fields inside and outside alter?

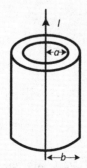

7. What is the internal inductance (per length) of a long straight wire with a circular cross section of radius a (use energy definition)? **Answer:** $\frac{\mu_0}{8\pi}$.

8. Show that the DC inductance of a piece of wire with finite length l and radius r is $L = \frac{\mu_0 l}{2\pi}\left(\ln\frac{2l}{r} - \frac{3}{4}\right)$. What is the inductance of a copper bond-wire with length of 2mm and a diameter of 25μm (practical bonding pads in integrated circuits are typically $50\times 50\mu m^2$)? Argue why traditionally, as a rule of thumb, an inductance of 1nH/mm is assumed for bond-wires. **Hint:** Calculate both the internal (previous problem) and the external (using Biot–Savart's law and flux definition) inductances. The equation is an approximate simplified for the practical case of $r \ll l$. The calculations are detailed by Rosa.[29]

9. In Faraday's experiment, assume the switch has a resistance of R, and the two coils are identical with an inductance of L. The battery voltage is V_{BAT}. Find the time-varying current in the coil. Assuming the iron toroid has a large permeability, find the magnetic flux in the second coil and estimate the emf read by the galvanometer.

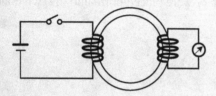

10. Show that the solution of the general form $V(z,t) = f_1\left(t - \frac{z}{v}\right) + f_2\left(t + \frac{z}{v}\right)$ satisfies the transmission line equation. Find the proper value of velocity v to satisfy the equation.

11. Consider the series RLC circuit below where the inductor has an initial current of I_0. Solve the circuit differential equation and find the inductor current. What are the total energies stored in the inductor, and dissipated in the resistor over time?

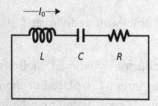

12. For the circuit below, the inductor L_1 has an initial stored current of I_0. The switch is closed at $t=0$. Find the final current of the inductors at $t=\infty$. **Answer:** $i_{L1}(\infty) = -i_{L2}(\infty) = \frac{L_1}{L_1+L_2}I_0$

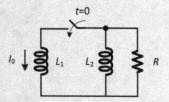

---

[29] Edward B. Rosa, "On the Self-Inductance of Circles," *Bulletin of the Bureau of Standards*, 4, no. 2, 301–305, 1907.

13. For the problem above argue intuitively how the final currents of the two inductors look with respect to each other. Find the total energy dissipated in the resistor, and from that find the final energy and the currents of the two inductors. Is there a condition that leads to the initial energy of the inductor L_1 completely dissipated, leading to zero final current?

    Answer: Resistor energy: $E_R = \int_0^\infty \frac{\left(-RI_0 e^{-t/\tau}\right)^2}{R} dt = \frac{1}{2}\left(\frac{L_1 L_2}{L_1 + L_2}\right) I_0^2$.

14. Suppose an LC tank used in a voltage-controlled oscillator (VCO) consists of a switchable capacitance C_F and a varactor with nominal capacitance $C(v)$. We define the VCO gain as $K_{VCO} = \frac{\partial \omega}{\partial V}$, where V is the varactor voltage. Show that K_{VCO} varies with frequency cubed (ω_0^3) as the switchable capacitance changes the nominal frequency of oscillation, ω_0.

15. Find the Q of parallel RC, RL circuit shown below. Show that the overall Q can be expressed by $\frac{1}{Q} = \frac{1}{Q_L} + \frac{1}{Q_C}$, assuming the inductor and capacitor are high-Q.

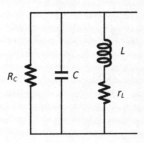

16. Find the electric potential difference at distances r_1 and r_2, associated with a point charge $+Q$ located at the origin. Assuming the potential reference ($V=0$) is associated with $r_2 \to \infty$, find the potential of the point charge at an arbitrary distance from the origin.

17. An electric dipole is defined as two point charges with the same magnitude, but opposite polarity separated by a small distance (l), as shown below. Find the electric potential at a distance of r from the origin. Assuming $r \gg l$, find a simplified expression for the potential, and from that calculate the electric field associated with the dipole. **Answer:** $E = \frac{Q}{4\pi\epsilon r^3}(2\cos\theta \mathbf{a}_r + \sin\theta \mathbf{a}_\theta)$.

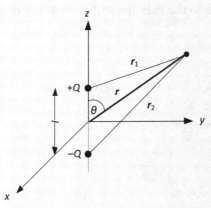

18. Find the magnetic field and magnetic vector potential of the ideal current filament (shown below) carrying a static current of I_0.

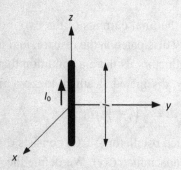

19. Consider an ideal circular current loop carrying a total current of I_0 as shown below. Find the vector potential and the magnetic field along the z axis. **Answer:** $H = \dfrac{a^2 I_0}{2(z^2 + a^2)^{3/2}} a_z$.

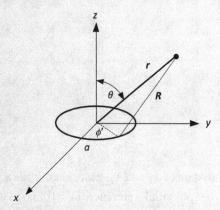

20. For the previous problem, find the vector potential and the magnetic field off the symmetry axis. **Hint:** The solution for an arbitrary point in space leads to elliptic integrals that can be solved only numerically. For far-field ($r \gg a$), the integral may be approximated by Taylor expansion and ignoring higher order terms, leading to $H = \dfrac{I_0(\pi a^2)}{4\pi r^3}(2\cos\theta a_r + \sin\theta a_\theta)$.

21. Show that the magnetic field lines of the circular current loop are as depicted below. Argue whether or not there is any resemblance to the field lines of a small bar magnet.

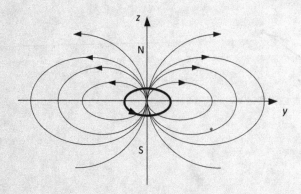

22. Argue the similarity of the magnetic and electric fields between the electric dipole and the circular current loop. How do the charge and current relate?

23. A *magnetic* dipole consists of a circular current loop of radius a, carrying a sinusoidal current of $I_0 \cos \omega t$, as shown below. It is closely related to the Hertzian dipole discussed earlier. To solve for the fields, we may take an indirect but easier approach considering the *duality* between the E and H fields in Maxwell's equations. Show that for a sourceless medium, Maxwell's equations remain intact if E is replaced with H, H with $-E$, and ϵ with μ. Invoking duality between the electric and magnetic dipoles (previous problem), the symmetry of Maxwell's equations, and that $I = \frac{dQ}{dt} = j\omega Q$, show that the Hertzian dipole solution could be used to obtain the magnetic dipole fields, but with l replaced by $j\omega\epsilon\pi a^2$. **Answer:** For far-field, $H_\varphi \cong (\omega\mu\pi a^2)\frac{j}{\eta}\frac{I_0\beta \sin\theta}{4\pi r}e^{-j\beta r}$, and $E_\theta \cong -\eta H_\varphi$.

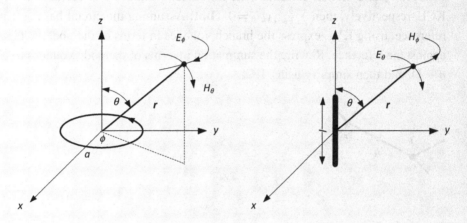

24. Show that for a general dipole of length l,

$$F(\theta, \varphi) = \frac{\cos\left(\frac{kl\cos\theta}{2}\right) - \cos\left(\frac{kl}{2}\right)}{\sin\theta}.$$

Assume that the current distribution is $I(z) = I_0 \sin(k(\frac{l}{2} - |z|))$. **Hint:** Using the figure below, show that for far-field, the differential electric field is $dE_\theta = j\eta k\frac{I(z)dz}{4\pi r'}\sin\theta' e^{-jkr'} \cong j\eta k\frac{I(z)dz}{4\pi r}\sin\theta e^{-jk(r-z\cos\theta)}$. The total electric field is obtained by $E_\theta(r,\theta) = \int_{-l/2}^{l/2} dE_\theta$.

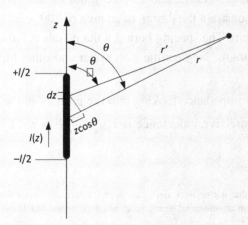

25. For an RLC one-port with an input impedance of $Z(j\omega)$, the quality factor is sometimes defined as

$$Q = \frac{|\text{Im}[Z]|}{\text{Re}[Z]}.$$

Using the energy definition of the Q and the concept of complex power, justify this definition.

26. Find a similar equation based on the admittance of the one-port: $Y(j\omega) = \frac{1}{Z(j\omega)}$. Discuss how this definition works for a series (or parallel) RLC circuit.

27. (*Tellegen's theorem*) Consider an arbitrary network with b branches. We assign branch voltages v_k and branch currents i_k. If all branch voltages and currents satisfy KVL and KCL, respectively, then $\sum_{k=1}^{b} v_k i_k = 0$. **Hint:** Assuming the circuit has $n - 1$ nodes plus a reference, using KVL express the branch voltages in terms of the node voltages compared against the reference. Rewrite the summation in terms of the node voltages (e_l, $l = 1, 2, \ldots, n-1$), and then simplify using KCL.

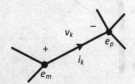

28. For the one-port of Figure 1.64, Tellegen's theorem states $\sum_k V_k I_k^* = 0$, where V_k and I_k are the branch voltages and currents phasors in sinusoidal steady state.[30] Operator $*$ denotes the conjugate.
    a. Show that $Z_{IN}|I|^2 = \sum_k (R_k I_k) I_k^* + \sum_m (j\omega L_m I_m) I_m^* + \sum_n V_n (j\omega C_n V_n)^*$, where the summation is broken into groups of resistors, inductors, and capacitors.
    b. Using the results in part a, show that $Z_{IN} = \frac{2P_{avg} + 4j\omega(W_L - W_C)}{|I|^2}$.

29. Design a 4nH single-layer spiral inductor assuming the inductance may be approximated by $L = \mu_0 N^2 r$, where N is the number of turns, and r is the spiral radius. Assume a metal sheet resistance of 10mΩ/□, and constrain the design to an area of $200 \times 200 \mu m^2$, with an inner diameter of greater than 150μm. The spacing between the metals is 5μm. The goal is to maximize the Q given the constrains. Neglect the skin effect and other high-frequency factors, and find the optimum Q.

30. For the figure below, find the input impedance looking into the primary of the transformer, Z_1. Show that if R_2 is small, the effective inductance is $L_{eff} = \frac{\text{Im}[Z_1]}{\omega} \approx L_1(1 - k^2)$, where $k = \frac{|M|}{\sqrt{L_1 L_2}}$ is the coupling factor.

---

[30] The only constraint on Tellegen's theorem is that the network must be lumped. Given the conclusions drawn in the problem, the theorem is stated in the special form of sinusoidal steady state, which is limited to LTI networks only.

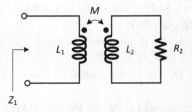

31. For the circuit shown below,
    a. Show that equivalent inductance is $L_{eq} = \frac{L_1 L_2 + L_1 L_3 + L_2 L_3 - M^2 - 2ML_3}{L_2 + L_3}$.
    b. Show that L_{eq} is always positive. **Hint:** Use the fact that the coupling factor $k = \frac{|M|}{\sqrt{L_1 L_2}}$ is equal to one at maximum. That sets an upper bound for M.

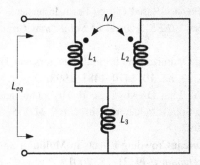

32. Using EMX, try an experiment where the inductor of Figure 1.15 is shielded with a slotted plane consisting of metal 1 and 2 layers shorted together. Use the sheet resistance values provided in the text considering that metal 1 and metal 2 are the same. Assume NATN layer is present. Find the inductance, quality factor, and self-resonance frequency. **Answer:** Q improves from 12.4 to 12.8 at 2GHz.

1.14 References

[1] P. R. Gray and R. G. Meyer, *Analysis and Design of Analog Integrated Circuits*, John Wiley, 1990.
[2] Y. Tsividis and C. McAndrew, *Operation and Modeling of the MOS Transistor*, vol. 2, Oxford University Press, 1999.
[3] R. Ellis and D. Gulick, *Calculus, with Analytic Geometry*, Saunders, 1994.
[4] R. E. Colline, *Foundation for Microwave Engineering*, McGraw-Hill, 1992.
[5] D. M. Pozar, *Microwave Engineering*, John Wiley, 2009.
[6] W. H. Hayt and J. A. Buck, *Engineering Electromagnetics*, vol. 73104639, McGraw-Hill, 2001.
[7] C. A. Desoer and E. S. Kuh, *Basic Circuit Theory*, McGraw-Hill, 2009.
[8] W. E. Boyce, R. C. DiPrima, and C. W. Haines, *Elementary Differential Equations and Boundary Value Problems*, vol. 9, John Wiley, 1992.
[9] R. E. Collin and F. J. Zucker, *Antenna Theory*, McGraw-Hill, 1969.
[10] C. A. Balanis, *Antenna Theory: Analysis and Design*, vol. 1, John Wiley, 2005.
[11] W. L. Stutzman and G. A. Thiele, *Antenna Theory and Design*, John Wiley, 2012.
[12] M. Zargari, M. Terrovitis, S.-M. Jen, B. J. Kaczynski, M. Lee, M. P. Mack, S. S. Mehta, S. Mendis, K. Onodera, H. Samavati, et al., "A Single-Chip Dual-Band Tri-Mode CMOS Transceiver for IEEE 802.11 a/b/g Wireless LAN," *IEEE Journal of Solid-State Circuits*, 39, no. 12, 2239–2249, 2004.

[13] F. Svelto, S. Deantoni, and R. Castello, "A 1.3GHz Low-Phase Noise Fully Tunable CMOS LC VCO," *IEEE Journal of Solid-State Circuits*, 35, no. 3, 356–361, 2000.

[14] A. Kral, F. Behbahani, and A. Abidi, "RF-CMOS Oscillators with Switched Tuning," in *Custom Integrated Circuits Conference, 1998. Proceedings of the IEEE*, 1998.

[15] N. Nguyen and R. Meyer, "A 1.8-GHz Monolithic LC Voltage-Controlled Oscillator," *IEEE Journal of Solid-State Circuits*, 27, no. 3, 444–450, 1992.

[16] N. Nguyen and R. Meyer, "Si IC-Compatible Inductors and LC Passive Filters," *IEEE Journal of Solid-State Circuits*, 25, no. 4, 1028–1031, 1990.

[17] F. Grover, *Inductance Calculations*, Dover, 1946.

[18] H. Greenhouse, "Design of Planar Rectangular Microelectronic Inductors," *IEEE Transactions on Parts, Hybrids, and Packaging*, 10, no. 2, 101–109, 1974.

[19] S. S. Mohan, M. del Mar Hershenson, S. P. Boyd, and T. H. Lee, "Simple Accurate Expressions for Planar Spiral Inductances," *IEEE Journal of Solid-State Circuits*, 34, no. 10, 1419–1424, 1999.

[20] S. Jenei, B. K. Nauwelaers, and S. Decoutere, "Physics-Based Closed-Form Inductance Expression for Compact Modeling of Integrated Spiral Inductors," *IEEE Journal of Solid-State Circuits*, 37, no. 1, 77–80, 2002.

[21] A. Niknejad and R. Meyer, "Analysis, Design, and Optimization of Spiral Inductors and Transformers for Si RF ICs," *IEEE Journal of Solid-State Circuits*, 33, no. 10, 1470–1481, 1998.

[22] H.-S. Tsai, J. Lin, R. C. Frye, K. L. Tai, M. Y. Lau, D. Kossives, F. Hrycenko, and Y.-K. Chen, "Investigation of Current Crowding Effect on Spiral Inductors," in *IEEE MTT-S Symposium on Technologies for Wireless Applications Digest*, 1997.

[23] W. B. Kuhn and N. M. Ibrahim, "Analysis of Current Crowding Effects in Multiturn Spiral Inductors," *IEEE Transactions on Microwave Theory and Techniques*, 49, 31–38, 2001.

[24] W. B. Kuhn and N. M. Ibrahim, "Approximate Analytical Modeling of Current Crowding Effects in Multi-turn Spiral Inductors," in *IEEE MTT-S International Microwave Symposium Digest*, 2000.

[25] A. Rofougaran, J. Y. Chang, M. Rofougaran, and A. A. Abidi, "A 1 GHz CMOS RF Front-End IC for a Direct-Conversion Wireless Receiver," *IEEE Journal of Solid-State Circuits*, 31, 880–889, 1996.

[26] T. Tsukahara and M. Ishikawa, "A 2 GHz 60 dB Dynamic-Range Si Logarithmic/Limiting Amplifier with Low Phase Deviations," in *IEEE International Conference on Solid-State Circuits*, 1997.

[27] C. Yue and S. Wong, "On-Chip Spiral Inductors with Patterned Ground Shields for Si-Based RF ICs," *IEEE Journal of Solid-State Circuits*, 33, no. 5, 743–752, 1998.

[28] M. Danesh, J. R. Long, R. Hadaway, and D. Harame, "A Q-Factor Enhancement Technique for MMIC Inductors," in *IEEE MTT-S International Microwave Symposium Digest*, 1998.

[29] J. R. Long, "Monolithic Transformers for Silicon RF IC Design," *IEEE Journal of Solid-State Circuits*, 35, 1368–1382, 2000.

2 RF Signals and Systems

In this chapter we review some of the basic concepts in communication systems. We start with a brief summary of Fourier and Hilbert transforms, both of which serve as great tools for analyzing RF circuits and systems. We also present an overview of network functions and the significance of poles and zeros in circuits and systems. To establish a foundation for the noise analysis presented in Chapter 5, we also provide a brief summary of stochastic processes and random variables. We conclude this chapter by briefly describing the fundamentals of analog modulation schemes and analog modulators.

A majority of the material presented in this chapter is a review of various concepts that exist in signal processing, communication systems, and basic circuit theory. However, we feel it is important to present a reminder as well as a summary as throughout this book they will be referred to continuously.

The specific topics covered in this chapter are:

- Fourier series and Fourier transform
- Impulse response, poles, zeros, and network functions
- Hilbert transform and quadrature signals
- Random processes, Gaussian signals, and stationary and cyclostationary processes
- Amplitude, phase, and frequency modulation
- Narrowband FM and Bessel functions
- Introduction to modern digital communication

For class teaching, we recommend focusing on selected topics from Sections 2.1, 2.8, 2.9, and 2.10, while Sections 2.1 through 2.5 and 2.7 may be comfortably assigned as reading.

The stochastic processes (Section 2.7) are crucial to attain a thorough understanding of noise in RF circuits, and particularly oscillator phase noise. For an introductory RF course, however, proper coverage of noise and phase noise may not be feasible. Consequently, we defer Section 2.7 and the majority of the phase noise discussion (in Chapter 9) to a more advanced course, and certainly to more astute readers. Section 2.7 offers a summary of selected topics that are most relevant to RF design, which will greatly complement our discussions on mixer and oscillator noise.

2.1 FOURIER TRANSFORM AND FOURIER SERIES

Electrical communication signals are analog time-varying quantities such as voltages or currents. Although a signal physically exists in the time domain, we can also represent it in the frequency domain, where we can view it as various sinusoidal components at different frequencies, known as *spectrum*. We will mostly focus on nonperiodic signals, concentrated over relatively short periods, and briefly overview *Fourier transform*, which is used to represent such signals in the frequency domain.

For a given nonperiodic signal, $v(t)$, whether strictly time limited (such as a pulse) or asymptotically time limited in the sense that it eventually approaches zero (such as an exponential function over time), we can define the signal energy over one period as

$$E = \int_{-\infty}^{\infty} |v(t)|^2 dt,$$

consistent with our previous definition of energy.[1] If the integral yields $E < \infty$, then the signal has a well-defined energy, and accordingly the Fourier[2] transform of $v(t)$, symbolized as $V(f)$, is defined as

$$V(f) = \mathcal{F}[v(t)] = \int_{-\infty}^{\infty} v(t) e^{-j2\pi ft} dt.$$

Similarly, an inverse Fourier transform performed on $V(f)$ yields the time domain signal back:

$$v(t) = \mathcal{F}^{-1}[V(f)] = \int_{-\infty}^{\infty} V(f) e^{+j2\pi ft} df.$$

The Fourier transform is evidently a complex function. If $v(t)$ is real, then $V(-f) = V^*(f)$, where * denotes the conjugate operator.

Example: One can easily show that for a rectangular pulse with a pulse width of τ, defined as

$$\Pi_\tau(t) = \begin{cases} 1 & |t| < \frac{\tau}{2} \\ 0 & |t| > \frac{\tau}{2} \end{cases}.$$

The Fourier transform is a sinc function,[3] whose magnitude and phase are shown in Figure 2.1:

$$V(f) = \frac{1}{\pi f} \sin \pi f \tau = \tau \operatorname{sinc} f \tau$$

---

[1] For instance, it represents the voltage energy $v(t)$ delivered to a 1Ω resistor.
[2] Though initially presented in 1807 but rejected (largely because of his former advisor Lagrange), Joseph Fourier's findings on Fourier series were first published in a book titled *The Analytic Theory of Heat* in 1822. The Fourier transform as well as Fourier laws have also been named in his honor.
[3] Our definition of sinc here includes a factor π in the argument, that is, $\operatorname{sinc}(x) = \sin \pi x / \pi x$, consistent with information theory and signal processing definition. In mathematics, sinc is defined as $\operatorname{sinc}(x) = \sin x / x$.

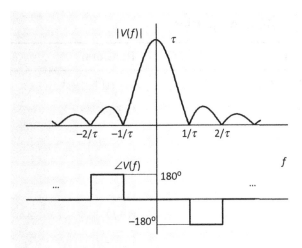

Figure 2.1: A rectangular pulse Fourier transform

The basic definition of the integral allows us to express the energy in the frequency domain as well, that is,

$$E = \int_{-\infty}^{\infty} |v(t)|^2 dt = \int_{-\infty}^{\infty} |V(f)|^2 df,$$

which is known as *Parseval's energy theorem*.

Example: Suppose the function of interest ($v(t)$) is multiplied by $e^{j2\pi f_c t}$, thus the new function is

$$v'(t) = v(t)e^{j2\pi f_c t}.$$

The Fourier transform of the new function is

$$V'(f) = \int_{-\infty}^{\infty} v(t)e^{j2\pi f_c t}e^{-j2\pi ft} dt = V(f - f_c).$$

Thus multiplying by $e^{j2\pi f_c t}$ results in a *shift in the frequency domain*. The outcome, though very trivial, is the basis of how mixers and image-reject systems operate, as we shall discuss later in Chapter 8.

In Table 2.1, a summary of basic properties of Fourier transform is listed. Refer to [1], [2] for actual proof and more details. All the relations can be easily proven based on the integral definition.

If $v(t)$ is a *periodic* function with period $T = \frac{1}{f_0}$, it may be represented by its Fourier series instead,

$$v(t) = \sum_{k=-\infty}^{\infty} a_k e^{j2\pi k f_0 t},$$

where $a_k = \frac{1}{T}\int_T v(t)e^{-j2\pi k f_0 t} dt$ is the Fourier coefficient.[4] The signal energy is

$$E = \int_T |v(t)|^2 dt,$$

---

[4] The Fourier coefficients of a variable $v(t)$ may also be denoted by $V[k]$.

Table 2.1: **Summary of Fourier transform properties**

Operation	Function	Transform
Superposition	$\alpha_1 v_1(t) + \alpha_2 v_2(t)$	$\alpha_1 V_1(f) + \alpha_2 V_2(f)$
Time delay	$v(t - \tau)$	$V(f)e^{-j2\pi f \tau}$
Scaling	$v(\alpha t)$	$\dfrac{1}{|\alpha|} V\left(\dfrac{f}{\alpha}\right)$
Conjugation	$v^*(t)$	$V^*(-f)$
Duality	$V(t)$	$v(-f)$
Frequency translation	$v(t)e^{j2\pi f_c t}$	$V(f - f_c)$
Modulation	$v(t)\cos(\omega_c t + \phi)$	$\dfrac{1}{2}\left[V(f-f_c)e^{j\phi} + V(f+f_c)e^{-j\phi}\right]$
Differentiation	$\dfrac{dv(t)}{dt}$	$j2\pi f V(f)$
Integration	$\int_{-\infty}^{t} v(\theta)d\theta$	$\dfrac{1}{j2\pi f} V(f)$
Convolution	$v * w(t) = \int_{-\infty}^{\infty} v(\tau)w(t-\tau)d\tau$	$V(f)W(f)$
Multiplication	$v(t)w(t)$	$V * W(f)$

which must be finite so that the Fourier coefficients are finite. Parseval's energy equation becomes

$$\frac{1}{T}\int_T |v(t)|^2 dt = \sum_{k=-\infty}^{\infty} |a_k|^2.$$

If $v(t)$ is real, then $a_{-k} = a^*_k$, and it is possible to represent the Fourier series as the sum of cosines [1]. Similar properties as the ones shown in Table 2.1 exist for the Fourier series as well.

Example: As a very useful case study, consider Figure 2.2, where a slow-varying signal $x(t)$ is compared against a high-frequency sawtooth signal varying between ± 1. The resulting waveform $x_P(t)$ has a constant amplitude of 1, but its width varies *linearly* with the input signal amplitude at the time location t_k where $x(t)$ intersects with the sawtooth waveform. We would like to express $x_P(t)$ in terms of its Fourier series.

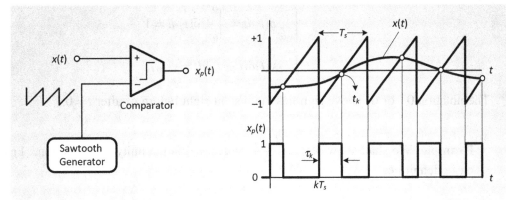

Figure 2.2: A slow-varying signal compared against a sawtooth waveform

The duration of each pulse, τ_k, may be defined as

$$\tau_k = \frac{T_s}{2}(1+x(t)),$$

where T_s is the sawtooth signal period. To prevent missing pulses or negative durations, let us assume that $|x(t)| < 1$. Since the rate of the input signal variation is assumed to be much less than the sampling frequency, we may assume uniform sampling, which is to say τ_k could be treated nearly as constant. Thus, we can write

$$a_n = \frac{1}{T_s}\int_{T_s} x_P(t)e^{-j2\pi n f_s t}dt = \frac{1}{T_s}\int_{-\tau_k/2}^{\tau_k/2} e^{-j2\pi n f_s t}dt = \frac{1}{\pi n}\sin\left(\frac{n\pi}{2}(1+x(t))\right),$$

which leads to

$$x_P(t) = \frac{1}{2}(1+x(t)) + \sum_{n=1}^{\infty}\frac{2}{n\pi}\sin n\phi(t)\cos n\omega_s t,$$

where $\phi(t) = \frac{\pi}{2}(1+x(t))$.

This kind of waveform is known as a pulse-width modulated signal and is commonly used in class D power amplifiers, as we will discuss in Chapter 11.

2.2 IMPULSES

The impulse has no mathematical or physical meaning, unless it appears under the operation of integration. Nevertheless, it proves to have a valuable role when analyzing linear networks and systems. Particularly when dealing with spectrum, an impulse in the frequency domain could represent a discrete frequency component.

In time domain, an impulse is represented as

$$\delta(t) = \lim_{\Delta\to 0}\delta_\Delta(t) = \lim_{\Delta\to 0}\frac{1}{\Delta}\Pi_\Delta(t),$$

where $\Pi_\Delta(t)$ represents a rectangular pulse defined in the previous section. By virtue of the definition above, it follows

$$\int_{-\infty}^{\infty} \delta(t)dt = \int_{0^-}^{0^+} \delta(t)dt = 1$$

$$\int_{-\infty}^{\infty} v(t)\delta(t-t_0)dt = v(t_0).$$

The notation 0^- or 0^+ above signifies the instant right before or after $t=0$.

Example: We shall show $\delta(t) = \dfrac{du(t)}{dt}$, where $u(t)$ is the unity step function. From our basic definition,

$$\delta(t) = \lim_{\Delta \to 0} \frac{u\left(t+\dfrac{\Delta}{2}\right) - u\left(t-\dfrac{\Delta}{2}\right)}{\Delta},$$

which is the very definition of $\dfrac{du(t)}{dt}$.

Closely related to the unit impulse function is the unit *doublet* $\delta'(t)$, which is defined by

$$\delta'(t) = \begin{cases} singular & t=0 \\ 0 & t \neq 0 \end{cases},$$

where the singularity at $t=0$ is defined such that $\delta(t) = \int_{-\infty}^{t} \delta'(\theta)d\theta$. The symbols of a unit impulse and doublet are shown in Figure 2.3.

In the frequency domain, the impulse represents a phasor or a constant. In particular, let $v(t)=1$ be a constant for all time. Although this signal has infinite energy, we can still obtain the Fourier transform in a limiting case considering that

$$v(t) = \lim_{W \to 0} \text{sinc}(2Wt) = 1,$$

and since the Fourier transform of a sinc is already shown to be a pulse, then

$$V(f) = \lim_{W \to 0} \frac{1}{2W} \Pi_{2W}\left(\frac{f}{2W}\right) = \delta(f),$$

based on basic properties of Fourier transform summarized in the previous section. We can generalize

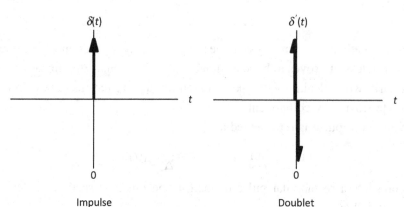

Figure 2.3: Symbols of impulse and doublet functions

$$Ae^{j2\pi f_c t} \leftrightarrow A\delta(f-f_c).$$

Similarly, according to duality, an impulse in the time domain will have a flat response in the frequency domain. This is very important, as it suggests that to characterize a system in the frequency domain, one can analyze the impulse response where all the *natural frequencies* of the system are excited.

2.3 FOURIER TRANSFORM OF PERIODIC SIGNALS

Fourier transform may be developed for periodic signals as well. Consider a signal $v(t)$, which is periodic and represented by its Fourier series as

$$v(t) = \sum_{k=-\infty}^{\infty} a_k e^{j2\pi k f_0 t}.$$

Since we already showed that the Fourier transform of $e^{j2\pi k f_0 t}$ is an impulse at kf_0, then, the Fourier transform of $v(t)$ may be expressed as a linear combination of impulses equally spaced in frequency:

$$V(f) = \sum_{k=-\infty}^{\infty} a_k \delta(f - kf_0).$$

Example: If $v(t) = \cos 2\pi f_0 t$, since $a_1 = a_{-1} = \frac{1}{2}$ with the rest of coefficients zero, its Fourier transform consists of a pair of impulses at $\pm f_0$ with a height of $\frac{1}{2}$. Similarly, the Fourier transform of a sine is a pair of impulses at $\pm f_0$ with a height of $\frac{\pm j}{2}$.

Example: Consider a train of impulses $v(t) = \sum_{k=-\infty}^{\infty} \delta(t - kT)$. This signal is periodic and its Fourier coefficients are

$$a_k = \frac{1}{T} \int_T \left(\sum_{n=-\infty}^{\infty} \delta(t - nT) \right) e^{-j2\pi k f_0 t} dt = \frac{1}{T} \int_{-T/2}^{T/2} \delta(t) e^{-j2\pi k f_0 t} dt = \frac{1}{T}.$$

Thus, the Fourier transform is also a train of impulses in frequency:

$$V(f) = \frac{1}{T} \sum_{k=-\infty}^{\infty} \delta(f - kf_0).$$

Table 2.2 shows the Fourier transform of some of the well-known signals. If periodic, also given is their Fourier series coefficients. They can all be easily verified by the reader given the basic definition.

Table 2.2: **Basic Fourier transform pairs**

Signal	Fourier transform	Fourier series coefficient
$\sum_{k=-\infty}^{\infty} a_k e^{j2\pi k f_0 t}$	$\sum_{k=-\infty}^{\infty} a_k \delta(f - kf_0)$	a_k
$A\cos(2\pi f_0 t + \phi)$	$\frac{A}{2}\left[e^{j\phi}\delta(f-f_0) + e^{-j\phi}\delta(f+f_0)\right]$	$a_{\pm 1} = \frac{A}{2} e^{\pm j\phi}$, ϕ otherwise
1	$\delta(f)$	$a_0 = 1$, ϕ otherwise
$\sum_{k=-\infty}^{\infty} \delta(t - kT)$	$\frac{1}{T}\sum_{k=-\infty}^{\infty} \delta(f - kf_0)$	$\frac{1}{T}$
Periodic squarewave: $\Pi(t) = \begin{cases} 1 & \|t\| < \frac{\tau}{2} \\ 0 & \|t\| > \frac{\tau}{2} \end{cases}$ $\Pi(t+T) = \Pi(t)$	$\sum_{k=-\infty}^{\infty} \frac{\sin \pi k \tau / T}{\pi k} \delta(f - kf_0)$	$\frac{\sin \pi k \tau / T}{\pi k}$
$\delta(t - t_0)$	$e^{j2\pi f t_0}$	None
$u(t)$	$\frac{1}{j2\pi f} + \frac{\delta(f)}{2}$	None
$e^{-\alpha t} u(t)$	$\frac{1}{\alpha + j2\pi f}$	None
$\Pi_\tau(t) = \begin{cases} 1 & \|t\| < \frac{\tau}{2} \\ 0 & \|t\| > \frac{\tau}{2} \end{cases}$	$\frac{1}{\pi f} \sin \pi f \tau = \tau \operatorname{sinc} f\tau$	None

2.4 IMPULSE RESPONSE

It is well known that if a linear, time-invariant system's impulse response is $h(t)$, then the response of the output $y(t)$ to an arbitrary input $x(t)$ is

$$y(t) = h * x(t) = \int_{-\infty}^{\infty} h(\tau) x(t - \tau) d\tau,$$

and in the frequency domain,

$$Y(f) = H(f) X(f),$$

where $H(f)$ is the system transfer function. It is important to point out that determining $H(f)$ does not necessarily involve $h(t)$. In fact, if one knows the differential equations for a lumped system, $H(f)$ can be directly expressed as a ratio of two polynomials:

$$H(f) = \frac{b_0 + b_1(j2\pi f) + \cdots + b_m(j2\pi f)^m}{a_0 + a_1(j2\pi f) + \cdots + a_n(j2\pi f)^n}.$$

Example: Consider the parallel resonance circuit shown in Figure 2.4 left, where the response is the inductor current $i_L(t)$.

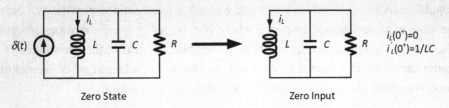

Figure 2.4: A parallel resonance circuit driven by an impulse current

By definition, the impulse response is a *zero-state* response, that is, the response of a circuit with an input excitation and no initial conditions (or state). Thus, the corresponding differential equation describing the circuit is

$$i_L'' + \frac{1}{RC}i_L' + \frac{1}{LC}i_L = \frac{1}{LC}\delta(t)$$
$$i_L(0^-) = 0$$
$$i_L'(0^-) = 0.$$

The second initial condition, $i_L'(0^-) = 0$, implies that the capacitor has no initial charge. Perhaps the most convenient way of solving the equation is by applying Laplace transform, but instead we try to directly solve it in the time domain to gain some insight. To do so, we integrate the two sides of the differential equation from 0^- to 0^+, attempting to find the initial conditions at 0^+. Thus,

$$\int_{0^-}^{0^+} i_L'' dt + \frac{1}{RC}\int_{0^-}^{0^+} i_L' dt + \frac{1}{LC}\int_{0^-}^{0^+} i_L dt = \frac{1}{LC}\int_{0^-}^{0^+} \delta(t) dt.$$

The last two terms on the left side are zero, as the only way for them not to be zero is if i_L' or i_L contain δ. If they did, then i_L'' will contain δ' or δ'', and that will violate the original differential equation. Thus,

$$i_L'(0^+) - i_L'(0^-) = \frac{1}{LC},$$

leading to $i_L'(0^+) = \frac{1}{LC}$. Obviously $i_L(0^+)$ remains zero, or otherwise, as stated, i_L' must contain δ, which is not possible. Now for $t > 0$ we can then write

$$i_L'' + \frac{1}{RC}i_L' + \frac{1}{LC}i_L = 0$$
$$i_L(0^+) = 0$$
$$i_L'(0^+) = \frac{1}{LC}.$$

Continued

This is a simple zero-input circuit as shown on the right of Figure 2.4, and the response is the *homogenous* solution of the differential equation,

$$i_L(t) = \frac{\omega_0^2}{2\sqrt{\alpha^2 - \omega_0^2}} (e^{s_1 t} - e^{s_2 t}) u(t),$$

where $s_{1,2} = -\alpha \pm \sqrt{\alpha^2 - \omega_0^2}$, $u(t)$ is the unit step function, signifying the fact that the impulse response is valid for $t > 0$, and ω_0 and α were defined previously.[5] Note that in the case of underdamped response where $\alpha < \omega_0$, $\sqrt{\alpha^2 - \omega_0^2}$ becomes imaginary, but also $s_{1,2}$ are complex and the response is still real. By taking the derivatives of $i_L(t)$ the reader can show that indeed i_L' does not contain any δ, whereas i_L'' does, making up for the $\delta(t)$ term on the right-hand side of the differential equation.

These results can be intuitively explained as follows: At $t = 0$, the source puts out an infinite current. The capacitor acts like a short (while the inductor is open) to absorb the current. Effectively, the source wants to deliver 1 coulomb of charge instantly to the capacitor, which results in the capacitor voltage to rise to $v_c(0^+) = \frac{1}{c} \int_{0^-}^{0^+} \delta(t) dt + v_c(0^-) = \frac{1}{c}$. Consequently, $i_L'(0^+) = \frac{v_c(0^+)}{L} = \frac{1}{LC}$. The inductor current does not change at 0^+. If it did, it would require an infinite voltage in the capacitor, which leads to a doublet current.

2.5 NETWORK FUNCTIONS

In general, for a linear and time-invariant network (or system), we can define a *network function*, $H(s)$ as follows,

$$\text{Network function} = \frac{L[0 - \text{state response}]}{L[\text{input}]} = \frac{B(s)}{A(s)},$$

where L denotes the Laplace transform [3], [4], and the zero-state response corresponds to the response of the network to a given input at no initial conditions. The variable $s = \sigma + j\omega = \sigma + j2\pi f$ is the complex frequency. From basic circuit theory [4], it can be shown that the network functions are rational functions of the complex frequency, with real coefficients, that is, a_i and b_i are real numbers. Thus, we can write in general

$$H(s) = \frac{b_0 + b_1 s + \cdots + b_m s^m}{a_0 + a_1 s + \cdots + a_n s^n} = K \frac{\prod_{i=1}^{m}(s - z_i)}{\prod_{j=1}^{n}(s - p_j)},$$

where z_i and p_i are called *zeros and poles* of the network.[6] Since the network coefficients are real, the poles and zeros of the system are either real or complex conjugate pairs. Moreover, if

---

[5] As we showed in Chapter 1: $\alpha = \frac{\omega_0}{2Q} = \frac{1}{RC}$ and $\omega_0 = \frac{1}{\sqrt{LC}}$.

[6] In the previous section we used the symbol Π to denote the pulse function. Here it has been used to indicate the product, analogous to Σ showing the sum.

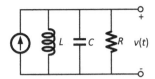

Figure 2.5: Parallel RLC circuit driven by a current source

one replaces s with $j\omega = j2\pi f$, then one can obtain the *frequency response* of the system, which is commonly expressed in terms of its magnitude and phase: $H(j\omega) = |H(j\omega)|e^{\angle H(j\omega)}$.

Finally, as was the case for the Fourier transform, by applying inverse Laplace transform, the corresponding impulse response of the system is obtained:

$$h(t) = L^{-1}[H(s)].$$

To arrive at a physical interpretation of the poles and zeros, let us consider the example of a parallel RLC circuit driven by a current source discussed earlier. Let us take the voltage across the circuit to be the response. The network function is

$$H(s) = \frac{1}{C} \frac{s}{s^2 + \frac{s}{RC} + \frac{1}{LC}} = \frac{1}{C} \frac{s}{s^2 + 2\alpha s + \omega_0^2}.$$

For the underdamped case where $Q > \frac{1}{2}$, we have

$$h(t) = \frac{1}{C} \frac{\omega_0}{\omega_d} e^{-\alpha t} \cos(\omega_d t + \phi) \qquad t > 0.$$

The pole locations of this circuit were presented before in Chapter 1. Two examples of the frequency response as well as the corresponding step responses are sketched in Figure 2.6. For both cases, the resistance R is kept constant along with $\omega_d = 1$, and C, and L are modified accordingly. The value of α is 0.3 in one case, and 0.1 in the other case. For convenience, the time domain plots are normalized to have the same initial magnitude (which is $\frac{1}{C}$).

Figure 2.6 shows clearly that the distance between the pole and the $j\omega$ axis determines completely the rate of decay, as well as the sharpness of the frequency response. In fact, if the poles were on the $j\omega$ axis, there would be no decay. Moreover, the ordinate of the pole, ω_d, determines the distance between successive zero crossings of the impulse response.

In general, we can say that isolated poles close to the $j\omega$ axis tend to produce sharper peaks in the magnitude curve. The distance of the pole to the $j\omega$ axis may be estimated by the fact that $2\alpha \cong$ 3dB bandwidth (in rad/s).

Moreover, any pole of the network function is the natural frequency of the corresponding output.[7] Suppose the input is an impulse. The Laplace transform of the output of interest will be equal to $H(s)$, which based on partial fraction expansion can be expressed as

$$V(s) = H(s) = \sum_{i=1}^{n} \frac{K_i}{s - p_i},$$

---

[7] Although the opposite is not necessarily true: any natural frequency of a network variable *need not* be a pole of a given network function that has this variable as the output. See Problem 6 for an example.

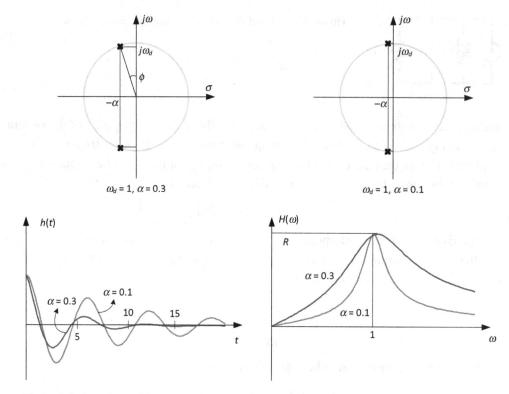

Figure 2.6: Impulse and frequency response of the parallel RLC circuit

where K_i is the residue of the pole p_i, and $v(t)$ is the output of interest. Thus

$$v(t) = \sum_{i=1}^{n} K_i e^{p_i t},$$

and since for $t > 0$ the input is zero, the equation above may be considered as the zero-input response, and thus p_i is a natural frequency of the system.

Example: Consider the second-order circuit in Figure 2.7. It is a phase shift network commonly used in *Wien bridge oscillators*.[8] For simplicity, assume $RC = 1$. We shall first determine the impulse response of the circuit.

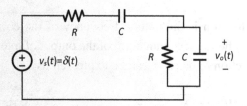

Figure 2.7: Second-order RC circuit used in Wien bridge oscillators

---

[8] Wien bridge oscillators were developed by Max Wien, the German physicist, in 1891.

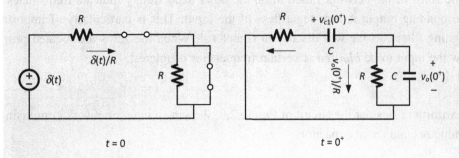

Figure 2.8: Circuit to determine the initial conditions of the circuit of Figure 2.7 at $v(t)$

The circuit differential equation is

$$v_o'' + \frac{3}{RC}v_o' + \frac{1}{(RC)^2}v_o = \frac{1}{RC}\delta'(t).$$

To obtain the initial conditions at $t=0^+$, we use the circuit shown on the left side of Figure 2.8, considering that the capacitors are short at $t=0$ when the infinite current is passing through. That creates a current of $\frac{\delta(t)}{R}$ going through both capacitors, raising their voltage to $\frac{1}{RC}$ at $t=0^+$.

At $t=0^+$, with the aid of the equivalent circuit on the right, it can be seen that a current of $\frac{v_o(0^+)}{R} = \frac{1}{R^2C}$ passes through the parallel resistor, and the series resistor current will be $\frac{v_o(0^+)+v_{c1}(0^+)}{R} = \frac{2}{R^2C}$. Thus, a net current of $\frac{-3}{R^2C}$ passes through the shunt capacitor, and as a result we have

$$v_o(0^+) = \frac{1}{RC} = 1$$
$$v_o'(0^+) = \frac{-3}{(RC)^2} = -3.$$

The circuit natural frequencies are directly obtained from the differential equation and are equal to

$$s_{1,2} = \frac{-3 \pm \sqrt{5}}{2RC} = \frac{-3 \pm \sqrt{5}}{2},$$

indicating that the circuit is overdamped, as any second-order RC network should. Accordingly, the impulse response is

$$v_o(t) = \left(\frac{-3+\sqrt{5}}{2\sqrt{5}}e^{s_1 t} + \frac{3+\sqrt{5}}{2\sqrt{5}}e^{s_2 t}\right)u(t).$$

As for the network function, it is easy to show that

$$\frac{V_o}{V_s}(s) = \frac{\frac{s}{RC}}{s^2 + \frac{3}{RC}s + \frac{1}{(RC)^2}}.$$

The two poles are on the real axis on the left plane and are equal to the natural frequencies obtained earlier, and there is one zero at the origin.

The zeros of the network function on the other hand simply indicate frequencies where the corresponding output is zero regardless of the input. This is particularly of importance when designing filters, as we will discuss in Chapter 4. When the zeros are located properly, they allow the input to be *blocked* at certain frequencies of interest.

Example: Consider the circuit of Figure 2.9, showing a *ladder* network comprising series inductors and shunt capacitors.

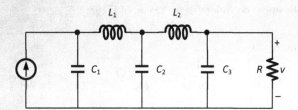

Figure 2.9: A lowpass filter with zeros at infinity

At very low frequencies, all the inductors are short, whereas all the capacitors are open, and thus the input current appears entirely at the output. On the other hand, at very high frequencies the capacitors are short, while the inductors are open, and input current does not appear at the output resistor. Thus the circuit is known to be *lowpass* and comprises only *zeros at infinity*. In fact, for such a lowpass ladder in general, we expect the transfer function to be of the form

$$H(s) = \frac{K}{s^n + a_{n-1}s^{n-1} + \cdots + a_0},$$

where n is the total number of reactive components, and K is the product of all capacitors and inductors. To prove the latter intuitively, consider a given section, say one consisting of C_i and L_i, at very high frequencies. We expect the ratio of the output to the input current to be $\frac{\frac{1}{C_i s}}{L_i s + \frac{1}{C_i s}} \cong \frac{1}{L_i C_i s^2}$. Thus, by considering $|H(\omega)|$ at very large frequencies, and realizing how the current is divided accordingly, we expect

$$|H(\omega \to \infty)| = \frac{1}{L_1 C_1 \omega^2} \times \frac{1}{L_2 C_2 \omega^2} \times \cdots \times \frac{1}{L_m C_m \omega^2} = \frac{K}{\omega^n},$$

assuming we have an equal number of Ls and Cs, that is, n is even and equal to $2 \times m$. Moreover, the lowpass ladder consists of n zeros at infinity.

Example: In the previous example of Figure 2.7, the zero at DC is justifiable considering that the series capacitance is open, and hence there is no voltage at the output at DC.

As for the usage of the network in the context of an oscillator, one can see that at the frequency of $\omega = \frac{1}{RC}$, we have

$$\frac{V_o}{V_s}\left(\frac{j}{RC}\right) = \frac{1}{3}.$$

If the output is fed back to the input with zero phase shift and a gain of higher than 3, a positive feedback is established and the circuit is going to oscillate. The interested reader may see Problem 10 for more details.

2.6 HILBERT TRANSFORM AND QUADRATURE SIGNALS

Quadrature signals and filters are widely used in RF receivers and transmitters. A quadrature filter is an *allpass* network that merely shifts the phase of the positive frequency components by $-90°$, and negative frequency components by $+90°$. Since $\pm 90°$ phase shift is equivalent by multiplying by $e^{\pm j90} = \pm j$, the transfer function can be written as

$$H_Q(f) = \begin{cases} -j & f > 0 \\ +j & f < 0 \end{cases},$$

which is plotted in Figure 2.10.

By duality (see Problem 12), we obtain the time domain representation of the sign function in the frequency domain expressed above to be $h_Q(t) = \frac{1}{\pi t}$. For an arbitrary input $x(t)$, passing through such quadrature filter results in an output $\hat{x}(t)$ defined as the *Hilbert transform* of the input:

$$\hat{x}(t) = x(t) * \frac{1}{\pi t} = \frac{1}{\pi}\int_{-\infty}^{\infty}\frac{x(\tau)}{t-\tau}d\tau.$$

From the impulse response derived, it is clear that Hilbert transform, or in general a 90° phase shift is physically not realizable as $h_Q(t)$ is noncausal, although its behavior can be well *approximated* over a finite frequency range using a real network.

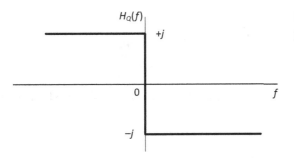

Figure 2.10: Transfer function of a quadrature phase shifter

Example: To illustrate further, consider a rectangular pulse with a width of t_0, as shown in Figure 2.11.

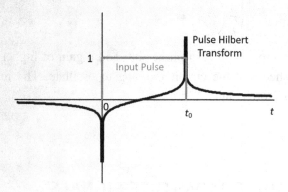

Figure 2.11: Hilbert transform of a rectangular pulse

The Hilbert transform, that is the pulse shifted by 90° is given by solving the convolution integral and is plotted in Figure 2.11 as well:

$$\hat{x}(t) = \frac{1}{\pi} \ln \left| \frac{t}{t - t_0} \right|,$$

which is clearly noncausal. Specifically, the output approaches infinity at $t = 0$ and $t = t_0$, where there are sharp transitions in the input.

Example: We wish to find the Hilbert transform of the function $x(t) = \frac{\sin t}{t}$, which is basically a sinc. In the frequency domain, the Fourier transform of $x(t)$ is a pulse with a height of π and a width of $\frac{1}{\pi}$:

$$X(f) = \begin{cases} \pi & |f| < \frac{1}{2\pi} \\ 0 & |f| > \frac{1}{2\pi} \end{cases}.$$

The Hilbert transform in the frequency domain will then consist of two pulses with half the width shifted by $\pm \frac{1}{4\pi}$,

$$\hat{X}(f) = -jX_{\frac{1}{2}}\left(f - \frac{1}{4\pi}\right) + jX_{\frac{1}{2}}\left(f + \frac{1}{4\pi}\right),$$

where $X_{\frac{1}{2}}(f) = \begin{cases} \pi & |f| < \frac{1}{4\pi} \\ 0 & |f| > \frac{1}{4\pi} \end{cases}$. In the time domain each pulse is a sinc itself. Therefore,

$$\hat{x}(t) = -j\left[\frac{\sin\frac{t}{2}}{t}e^{j\frac{1}{4\pi}2\pi t} - \frac{\sin\frac{t}{2}}{t}e^{-j\frac{1}{4\pi}2\pi t}\right] = \frac{2\sin^2\frac{t}{2}}{t} = \frac{1-\cos t}{t}.$$

Finally, if $x(t) = A\cos(\omega_0 t + \varphi)$, then $\hat{x}(t) = A\sin(\omega_0 t + \varphi)$. Moreover, any signal that consists of a sum of sinusoids follows the above accordingly.

2.7 STOCHASTIC PROCESSES

All meaningful communication signals are unpredictable or random, as viewed from the receiver end. Otherwise there would be little value in transmitting a signal whose behavior is known beforehand. Furthermore, the noise that is a great part of our discussion here and affects communication systems in a number of ways is random in nature. Therefore, we dedicate this section to discussing *random or stochastic processes* briefly. Such signals are random in nature, but also a function of time. Therefore, unlike the deterministic signals that we are mostly used to, they have to be dealt with *statistically*, as their exact value at a given point of time is not predictable, or known.

A random variable, $X(s)$, maps the outcomes of a chance experiments into numbers along the real line [5], [6]. For instance, the random variable $X(s)$ may be mapping the outcome of the random experiment, rolling the dice, into a discrete number 1, 2, ... 6. The *probability density function* (PDF), $f_X(x)$, is then discrete, and consists of points with equal probability of ⅙, as is shown in Figure 2.12.

By definition, for the case of a discrete process such as the one above, *Probability* $\{X \leq x_k\} = \sum_{i=1}^{k} f_X(x_i)$. It follows that $\sum_{i=1}^{n} f_X(x_i) = 1$. In the case of a continuous PDF the $\sum$ is replaced by $\int$.

A *random (or stochastic) process*, on the other hand, maps the outcomes into real functions of *time* [5]. The collection of time functions is called an *ensemble*, represented by $v(t, s)$, where each member is called a *sample*. For example, consider this experiment: pick a resistor at random out of a large number of identical resistors, and observe the voltage across its terminals. The random electron motion produces a random voltage waveform for a given resistor, which differs from the other ones, and hence is not predictable. Figure 2.13 depicts some of the waveforms from the ensemble $v(t, s)$.

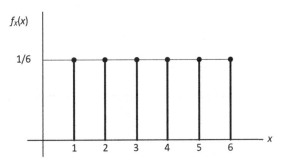

Figure 2.12: Probability density function of experiment rolling the dice

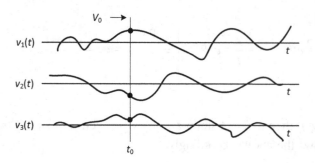

Figure 2.13: Waveforms in an ensemble $v(t,s)$

At a given point of time, say t_0, $v(t_0, s)$ is now a random variable. For instance, the noise voltage of a given resistor at a given point of time could assume any random value with a given probability density function. This we can represent by simply V_0, with a certain probability density function of $f_{V_0}(v_0)$. So the probability of this event such that $v(t_0) < a$ can be then found by performing the integration: $\int_{-\infty}^{a} f_{V_0}(v_0) dv_0$. For the case of the resistor example, that is essentially the probability of a given selected resistor noise voltage to be less than a. Naturally, for the random process $v(t)$, the probability density function is a function of time and may be generally expressed as $f_v(v, t)$.

For a given random process $v(t)$ with a probability density function $f_v(v, t)$, at an arbitrary time t, the *expected value* of $v(t)$ is defined as

$$\overline{v(t)} = E[v(t)] \triangleq \int_{-\infty}^{\infty} v(t) f_v(v, t) dv.$$

Likewise, the autocorrelation function between two random variables, $v(t_1)$ and $v(t_2)$, representing the random process at two points of time, t_1 and t_2, is the expected value of the product $v(t_1) v^*(t_2)$:

$$R_v(t_1, t_2) \triangleq E[v(t_1) v^*(t_2)] = \int_{-\infty}^{\infty} \int_{-\infty}^{\infty} v_1 v_2^* f_v(v_1, v_2, t_1, t_2) dv_1 dv_2,$$

which simply signifies how the two variables measured at two different times are correlated.

For the remainder of this section, we assume that the random signals we deal with are real, and thus drop the conjugate notation, that is,

$$R_v(t_1, t_2) = E[v(t_1) v(t_2)]$$

assuming $v(t)$ is real.

Example: A randomly phased sinusoid is a random process defined as

$$v(t) = A \cos(\omega_0 t + \phi),$$

where A and ω_0 are constants while ϕ is a random variable uniformly distributed between 0 and 2π. The underlying experiment might be picking an oscillator at random over a large collection of oscillators with constant amplitude and frequency, but with no phase synchronization. For the phase, the probability density function is $f_\phi(\phi) = \frac{1}{2\pi}$, for $0 \leq \phi \leq 2\pi$. The expected value is

$$E[v(t)] = \overline{v(t)} = \int_0^{2\pi} A\cos(\omega_0 t + \phi) f_\phi(\phi) d\phi = 0.$$

The autocorrelation is

$$R_v(t_1, t_2) = \int_0^{2\pi} A\cos(\omega_0 t_1 + \phi) A\cos(\omega_0 t_2 + \phi) f_\phi(\phi) d\phi,$$

and after expansion, the ϕ dependent term averages to zero, and we are left with

$$R_v(t_1, t_2) = \frac{A^2}{2} \int_0^{2\pi} \cos(\omega_0(t_1 - t_2)) P_\phi(\phi) d\phi = \frac{A^2}{2} \cos(\omega_0(t_1 - t_2)).$$

Interestingly, the autocorrelation is only a function of the time difference, $t_1 - t_2$. Moreover, for $t_1 = t_2 = t$,

$$\overline{v^2(t)} = R_v(t, t) = \frac{A^2}{2},$$

indicating the notion of the power.

2.7.1 Stationary Processes and Ergodicity

For the example above, we notice that the expected value is a constant, and the autocorrelation is a function of the time difference: $t_1 - t_2$. This type of process is known as a *stationary* process. To generalize, if the statistical characteristics of a random process remain invariant over time (of course not necessarily the process itself), we call the process *stationary* (in the strict sense). Two important consequences of stationarity are:

The mean value must be independent of time: $\eta(t) = E[v(t)] = \eta_0$.
The autocorrelation must depend only on the time difference, that is, $R_v(t_1, t_2) = E[v(t_1) v(t_2)] = R_v(t_1 - t_2) = R_v(\tau)$, where $z = t_1 - t_2$.

If only these two conditions hold, the process is said to be at least *stationary in the wide sense*. From Condition 2, it follows that the mean square value[9] (that is, $E[v(t)^2]$) and the variance of a stationary process are constant. Furthermore, we have $E[v^2(t)] = R_v(0)$, analogous to the autocorrelation of a real deterministic power signal.

Properties of $R_v(\tau)$ that follow from the second condition are:

- $R_v(0) = \overline{v^2}$

---

[9] Also known as the *second moment*.

- $R_v(-\tau) = R_v(\tau)$[10]
- $R_v(\tau) \leq R_v(0)$

The latter can be proven easily considering that

$$E[|v(t_1) - v(t_2)|^2] = E[(v(t_1))^2 - 2v(t_1)v(t_2) + (v(t_2))^2] \geq 0,$$

which leads to

$$R_v(0) \geq R_v(\tau).$$

For the previous example of randomly phased oscillator, we further notice that the ensemble average equals the corresponding *time averages* of an arbitrary sample function. Specifically, if we take an arbitrary sample of $v(t) = A\cos(\omega_0 t + \phi_k)$, and apply the *time average*,

$$<v(t)> = \lim_{T \to \infty} \frac{1}{T} \int_{-T/2}^{T/2} v(t) dt,$$

we will find that $<v(t)> = E[v(t)]$, and $<v(t)v(t-\tau)> = E[v(t)v(t-\tau)] = R_v(\tau)$.

When all *ensemble averages* equal to the corresponding *time averages*, the process is said to be *ergodic*.[11] An ergodic process must be strictly stationary, but an arbitrary stationary process is not necessarily ergodic. There is no ergodic simple test for a given process. If a process is stationary, and we can reasonably argue that a typical long enough sample function captures all statistical properties of the process, we can assume ergodicity. Fortuitously, many communication processes that we deal with, including noise, happen to be ergodic, which results in great flexibility in characterizing them.

2.7.2 Gaussian Processes

A random process $v(t)$ is called a Gaussian (or normal) process if its probability density function (f_v) as well as all higher order joint PDFs are Gaussian for all values of t. The probability density function of a Gaussian process is

$$f_v(v,t) = \frac{1}{\sqrt{2\pi}\sigma} e^{-\frac{(v-\mu)^2}{2\sigma^2}},$$

where μ and σ are the mean and standard deviation, respectively. Although its proof is well beyond the scope of this book,[12] and it can be found in [5], the *central limit theorem* (CLT) may be invoked to determine if a given process is Gaussian. The theorem states that, given

---

[10] If the random signal is not real, then $R_v(-\tau) = R^*_v(\tau)$.

[11] The word *ergodic* is Greek and was chosen by Boltzmann while he was working on a problem in statistical mechanics.

[12] It can be proven by establishing the characteristics function of a random vector. It follows that the density function of the sum of independent random variables equals the convolution of their densities.

certain conditions, the arithmetic mean of a sufficiently large number of iterates of independent random variables, each with a well-defined expected value and well-defined variance, will be approximately normally distributed. Since the random variables are independent, the mean and variance of the sum is equal to the sum of the mean and variance of each variable.

A good example is a resistor thermal noise, which is produced by the random (*Brownian*) motion of many electrons in a microscopic fashion. This results in a normal distribution of the noise itself at a macroscopic level.

It can be shown that if $v(t)$ is Gaussian, then:

1. The process is completely described by $\overline{v(t)}$ and $R_v(t_1, t_2)$, leading to the values of μ and σ.
2. If $v(t)$ satisfies the conditions for wide-sense stationarity, then the process is strictly stationary and ergodic.
3. Any linear operation on $v(t)$ produces another Gaussian process.

Example: Consider a normal process X with zero mean passing through a squarer with the output $Y = X^2$. We wish to find the output probability density function.

If $y < 0$, clearly $y = x^2$ has no solution, and thus, $f_Y(y) = 0$. For $y \geq 0$, there are two solutions, $x = \pm \sqrt{y}$. By definition,

$$f_Y(y)dy = P(y \leq Y \leq y + dy),$$

where $P(.)$ denotes the probability. Referring to Figure 2.14, and considering the symmetry of a normal process:

$$P(y \leq Y \leq y + dy) = 2P(x \leq X \leq x + dx),$$

where the values of x is obtained from $y = x^2$ relationship or $x = \pm \sqrt{y}$ as depicted in the figure. Shown as well is the probability of $x \leq X \leq x + dx$ highlighted by the shaded area on the variable X normal density curve.

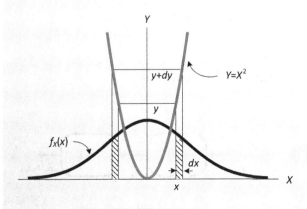

Figure 2.14: A normal random variable passing through a squarer block

Continued

Thus

$$f_Y(y)dy = 2f_X(\sqrt{y})dx = \frac{f_X(\sqrt{y})}{\sqrt{y}}dy,$$

owing to the fact that $dy = 2xdx$ or $dx = \frac{dy}{2x} = \frac{dy}{2\sqrt{y}}$. Therefore

$$f_Y(y) = \frac{1}{\sqrt{2\pi\sigma y}} e^{-\frac{y}{2\sigma^2}} u(y),$$

where $u(.)$ is the unity step function.

2.7.3 Systems with Stochastic Inputs

In most communication systems, it is common to pass a certain random signal (or noise) through a filter (or any type of linear and time-invariant system), as shown in Figure 2.15.

The resulting output is

$$y(t) = L[x(t)] = \int_{-\infty}^{\infty} h(\tau)x(t-\tau)d\tau,$$

where L signifies the linear operation that the system imposes on $x(t)$, and $h(t)$ is the impulse response. Since convolution is a linear operation, we expect a Gaussian input produces a Gaussian output whose properties will be completely described by its mean and variance. We assume $x(t)$ is stationary and real, and that $h(t)$ is real too. We expect then $y(t)$ to be real and stationary as well.

From convolution integral basic definition, it is easy to show that

$$E\{L[x(t)]\} = L\{E[x(t)]\}.$$

In other words, the mean of output equals to the response of the system to the mean of the input. Furthermore, the autocorrelation of the output, $R_y(t_1, t_2)$, can be obtained by taking a two-step approach as follows. For a given time t_1, $x(t_1)$ is a constant, and thus

$$x(t_1)y(t) = x(t_1)L[x(t)] = L[x(t_1)x(t)],$$

which yields

$$E[x(t_1)y(t)] = L\{E[x(t_1)x(t)]\}.$$

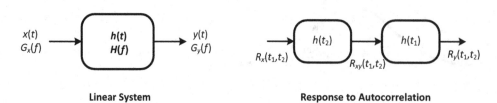

Figure 2.15: A stationary input passing through a linear time-invariant system

Similarly,

$$E[y(t)y(t_2)] = L\{E[x(t)y(t_2)]\}.$$

Letting $t = t_2$ in either one of the equations above, leads to auto- or cross-correlation functions, as depicted in Figure 2.15.

For a stationary process, the input output correlations are convolutions with the impulse response as proved earlier, with $\tau = t_2 - t_1$, which follows:

$$R_{yx}(\tau) = h(\tau) * R_x(\tau)$$
$$R_y(\tau) = h(-\tau) R_{yx}(\tau) = h(-\tau) * h(\tau) * R_x(\tau).$$

Note that the last result could have been directly obtained from the integral itself:

$$R_y(\tau) = E[y(t+\tau)y(t)] = E\left[\int h(\alpha)x(t+\tau-\alpha)d\alpha \int h(\beta)x(t-\beta)d\beta\right].$$

Or

$$R_y(\tau) = \iint h(\alpha)h(\beta)E[x(t+\tau-\alpha)x(t-\beta)]d\alpha d\beta = \iint h(\alpha)h(\beta)R_x(\tau-\alpha+\beta)d\alpha d\beta,$$

which could be rewritten as

$$R_y(\tau) = \int h(-\beta)(\int h(\alpha)R_x(\tau-\alpha-\beta)d\alpha)d\beta,$$

leading to $R_y(\tau) = h(-\tau) * h(\tau) * R_x(\tau)$.

Example: A normal stationary random signal $X(t)$ with zero mean and autocorrelation $R_X(\tau)$ passes through a squarer with an output $Y(t) = X^2(t)$. We have

$$E[Y(t)] = E[X^2(t)] = R_X(0).$$

Furthermore, it can be shown ([6] and also see Problem 23) that if X and Y are jointly normal random processes (or variables) with zero mean, then

$$E[X^2Y^2] = E[X^2]E[Y^2] + 2E^2[XY].$$

Therefore

$$R_Y(\tau) = E[X^2(t+\tau)X^2(t)] = E[X^2(t+\tau)]E[X^2(t)] + 2E^2[X(t+\tau)X(t)],$$

resulting in

$$R_Y(\tau) = R_X^2(0) + 2R_X^2(\tau).$$

The density functions as calculated in the previous example are shown in Figure 2.16.

Continued

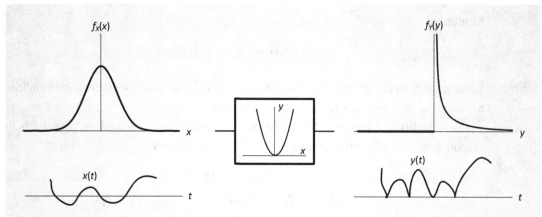

Figure 2.16: A normal random process passing through a squarer

Notably,

$$E[Y(t)^2] = R_Y(0) = 3R_X^2(0).$$

Example: A hard limiter is a memoryless system creating ± 1 output as shown in Figure 2.17:

$$y = \begin{cases} -1 & \text{if } x < 0 \\ +1 & \text{if } x > 0 \end{cases}.$$

The output expected value is

$$E[y(t)] = 1 \cdot P(x > 0) - 1 \cdot P(x < 0) = (1 - F_x(0)) - F_x(0) = 1 - 2F_x(0),$$

where $F_x(x)$ is the distribution function of random signal x (the integral of the probability density function, that is, $f_x(x) = \frac{\partial F_x(x)}{\partial x}$).

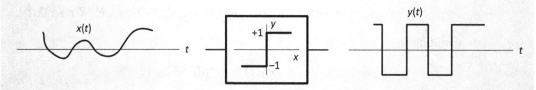

Figure 2.17: Random signal $x(t)$ passing through a limiter

There is not a simple form for the autocorrelation of $y(t)$ in general, though interestingly, if $x(t)$ is a stationary normal process, it can be shown that (see Problem 21 for the proof) the autocorrelation of the limiter output is $R_y(\tau) = \frac{2}{\pi}\sin^{-1}\left(\frac{R_x(\tau)}{R_x(0)}\right)$.

2.7.4 Power Spectral Density

It is often more insightful to characterize the stochastic processes in the frequency domain. Applying Fourier transform to the time domain signal directly is not meaningful. Instead, we need to resort to statistical characteristics of the random signal.

Let $v(t)$ be a wide-sense stationary random signal.[13] Let us define

$$P_T = \frac{1}{T}\int_{-T/2}^{T/2} v^2(t)dt,$$

which is a *random variable* itself. We define the average power of the random signal $v(t)$ to be

$$\bar{P} = \lim_{T\to\infty} E[P_T] = \lim_{T\to\infty} \frac{1}{T}\int_{-T/2}^{T/2} E[v^2(t)]dt = \; <E[v^2(t)]>,$$

which involves time averaging ($<.>$) after ensemble averaging ($E[\cdot]$). Since $v(t)$ is stationary, we expect $E[v^2(t)] = \overline{v^2}$ to be a constant. Consequently, $<\overline{v^2}> \; = \overline{v^2} = \bar{P}$. If the source is ergodic as well as stationary, then $\overline{v^2}$ and $\bar{v}$ may be obtained from simply time domain averages.

The *power spectral density* or simply *power spectrum* $S_v(f)$ tells us how the power is distributed in the frequency domain. The autocorrelation function of a wide-sense stationary random process has a spectral decomposition given by the power spectrum of that process, which is defined as

$$S_v(f) = \mathcal{F}[R_v(\tau)] = \int_{-\infty}^{\infty} R_v(\tau)e^{-j2\pi f\tau}d\tau,$$

and conversely,

$$R_v(\tau) = \mathcal{F}^{-1}[S_v(f)] = \int_{-\infty}^{\infty} S_v(f)e^{j2\pi f\tau}df.$$

Just as in the case of deterministic signals, the autocorrelation and spectral density constitute a Fourier transform. Since $R_v(\tau) = R_v^*(-\tau)$, the power spectrum is always real. Furthermore, if $v(t)$ is real, then since $R_v(\tau)$ becomes even, so will be $S_v(f)$, and it may be expressed as

$$S_v(f) = 2\int_0^{\infty} R_v(\tau)\cos\omega\tau d\tau.$$

---

[13] There is generally little value in defining the power spectral density for nonstationary processes.

From the basic definition it follows

$$\int_{-\infty}^{\infty} S_v(f)df = R_v(0) = \overline{v^2} = \bar{P},$$

which shows that the area of the power spectrum of any process is positive. Moreover, according to the *Wiener–Khinchin theorem*,[14] $S_v(f) \geq 0$ for every f, which will be proven in the next section, after discussing the power spectrum of filtered random signals. Conversely, if $S_v(f) \geq 0$ for every f, then a process $v(t)$ can be found that has the power spectrum $S_v(f)$. The proof is as follows: Consider the process $v(t) = e^{j(2\pi Ft + \phi)}$, where F is a random variable with the density $f_F(f)$, and ϕ is a random variable independent of F with a uniform density in the interval $0 \leq \phi \leq 2\pi$. This process is stationary in wide-sense as the mean is zero, and the autocorrelation is

$$R_v(\tau) = E\left[e^{j(2\pi F(t+\tau)+\phi)}e^{-j(2\pi Ft+\phi)}\right] = E\left[e^{j2\pi F\tau}\right] = \int_{-\infty}^{\infty} e^{j2\pi f\tau} f_F(f)df.$$

By comparison, since $R_v(\tau) = \int_{-\infty}^{\infty} S_v(f)e^{j2\pi f\tau}df$, then $S_v(f) = f_F(f)$. Note that the density function $f_F(f)$ is obviously always positive.

In summary, the function $S_v(f)$ is a power spectrum if and only if it is positive.

Example: For the case of the randomly phased sinusoid, we calculated

$$R_v(\tau) = \frac{A^2}{2}\cos(2\pi f_0 \tau),$$

so

$$S_v(f) = \frac{A^2}{4}\delta(f - f_0) + \frac{A^2}{4}\delta(f + f_0),$$

which is plotted in Figure 2.18.

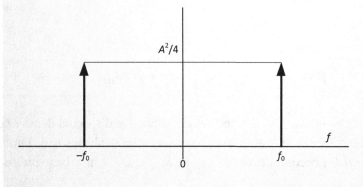

Figure 2.18: Power spectrum of a randomly phased sinusoid

---

[14] Norbert Wiener proved this theorem for the case of a deterministic function in 1930. Aleksandr Khinchin later formulated a similar result for stationary stochastic processes. Albert Einstein explained, without proof, the idea in a memo in 1914.

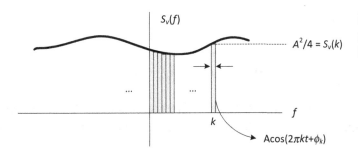

Figure 2.19: An arbitrary stationary process modeled as an infinite sum of impulses in the frequency domain

The results derived above could be generalized to model an arbitrary stationary process. Consider the stationary process $v(t)$, modeled as an infinite sum of sinusoids that are uncorrelated in phase and separated in frequency by 1Hz,[15]

$$v(t) = \sum_{k=0}^{\infty} \sqrt{4S_v(k)} \cos(2\pi kt + \phi_k),$$

where $S_v(k)$ is the spectral density (per unit hertz) of $v(t)$ at frequency of k hertz, and ϕ_k is the uncorrelated random phase offset of each sinusoid with a uniform distribution. The spectral density of each cosine is an impulse with an amplitude of $S_v(k)$, and frequency of k, as shown in Figure 2.19.

In Chapter 6 we shall exploit this to calculate the spectral density of band-limited white noise squared. We also use this in Chapter 9 to discuss phase and frequency noise.

Example: Consider the *modulation* operation defined as

$$z(t) = v(t)\cos(2\pi f_c t + \phi),$$

where $v(t)$ is a stationary random signal, and ϕ is a random angle independent of $v(t)$ uniformly distributed over 2π radians. If we didn't include ϕ, $z(t)$ would not be stationary. Inclusion of ϕ simply recognizes arbitrary choice of the time with respect to $v(t)$ and $\cos 2\pi f_c t$ as two independent functions,

$$R_z(\tau) = E[v(t+\tau)v(t)\cos(2\pi f_c(t+\tau)+\phi)\cos(2\pi f_c t+\phi)],$$

and after trigonometric expansion, and realizing that the ϕ dependent term averages to zero, we obtain

$$R_z(\tau) = 1/2 R_v(\tau)\cos 2\pi f_c \tau,$$

Continued

---

[15] The choice of 1Hz is somewhat inaccurate; rather the frequency separation is in general Δf, with Δf approaching zero, assuming there is an infinite number of infinitesimal frequency bins. If the frequencies of interest are well above 1Hz, which is usually the case, 1Hz is a reasonable approximation.

and the power spectrum is

$$S_z(f) = 1/4[S_v(f-f_c) + S_v(f+f_c)].$$

So the modulation translates the power spectrum of $v(t)$ up and down by f_c. Modulation can be generalized to the case of the product of two independent and stationary processes $z(t) = v(t)w(t)$:

$$R_z(\tau) = R_v(\tau)R_w(\tau)$$

and

$$S_z(f) = S_v(f) * S_w(f),$$

similar to the multiplication property of Fourier transform of deterministic signals.

2.7.5 Filtered Random Processes

We showed previously that if a stationary random process $x(t)$ passes through an LTI system with an impulse response of $h(t)$, the output $y(t)$ autocorrelation is

$$R_y(\tau) = h(-\tau) * h(\tau) * R_x(\tau).$$

Therefore in terms of power spectral density

$$S_y(f) = |H(f)|^2 S_x(f),$$

from which

$$\overline{y^2} = R_y(0) = \int_{-\infty}^{\infty} |H(f)|^2 S_x(f) df.$$

Also, we can easily show

$$E[y(t)] = \left[\int_{-\infty}^{\infty} h(t)dt\right] E[x(t)] = H(0)E[x(t)],$$

where $H(0)$ equals the system DC gain.

One important outcome of ergodicity is the basic principle that a spectrum analyzer operates upon. Although random signals such as noise require statistical data collection to be characterized, as long as they are ergodic, a simple time domain measurement is sufficient. Consider Figure 2.20, illustrating a simplified block diagram of a typical spectrum analyzer, where $x(t)$ is a random signal under test.

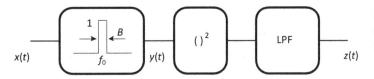

Figure 2.20: An example of a simplified spectrum analyzer function

The input is first passed through a narrow filter tuned at a desired frequency of f_0. Sweeping frequency (that is, varying f_0) extracts the signal spectrum over the range of interest. Squaring and lowpass filtering are equivalent to time averaging, that is,

$$z(t) = <y(t)^2>,$$

and from ergodicity,

$$z(t) = \overline{y^2} = R_y(0).$$

On the other hand, assuming the filter is sufficiently narrow with a passband gain of unity, we can write

$$R_y(0) = \int_{-\infty}^{\infty} |H(f)|^2 S_x(f) df \cong \int_{f_0-B/2}^{f_0+B/2} S_x(f) df \cong S_x(f_0),$$

which proves that the output signal $z(t)$ is the power spectrum of the input at f_0. It is important to point out that the time average set by the lowpass filter must be sufficiently long to ensure that the time and ensemble averages are identical. This is generally controlled by adjusting the sweep time, whereas the bandpass filter bandwidth, B, is set by adjusting the resolution bandwidth.

Since $S_x(f_0)$ is equal to the output power of filtered signal ($S_x(f_0) = \overline{y^2}$), it must be positive. This indicates that for an arbitrary stationary process, always $S_x(f) \geq 0$ for all f, which serves as proof of the Wiener–Khinchin theorem, stated in Section 2.7.4.

Example: Let us consider the Hilbert transform $\hat{x}(t) = x(t) * \frac{1}{\pi t} = \frac{1}{\pi} \int_{-\infty}^{\infty} \frac{x(\tau)}{t-\tau} d\tau$ as we defined earlier, where $x(t)$ (which is itself a random signal) is a random process. Since $H_Q(f) = -j \operatorname{sgn} f$, then $|H_Q(f)|^2 = 1$, and thus the power spectral density of a Hilbert transformed signal is identical to that of the input signal. It also follows that the autocorrelations are the same, indicating that 90° phase shift does not alter the random signal statistics. The complex process,

$$z(t) = x(t) + j\hat{x}(t),$$

is known as the *analytic signal* associated with $x(t)$. Then the response of such system is $1 + j(-j \operatorname{sgn} f) = 2u(f)$, and $S_z(f) = 4S_x(f)u(f)$, where u is the unity step function.

Example: Consider the random process $y(t) = \frac{1}{2T}\int_{t-T}^{t+T} x(\theta)d\theta$. The process $y(t)$ is obtained by passing $x(t)$ (which itself is a random signal) through a system whose impulse response is a rectangular pulse with a height of $\frac{1}{2T}$ and width of $\pm T$ around zero. Clearly

$$H(f) = \frac{1}{2T}\int_{-T}^{T} e^{-j2\pi f\tau} d\tau = \frac{\sin 2\pi fT}{2\pi fT}.$$

Thus

$$S_y(f) = \left(\frac{\sin 2\pi fT}{2\pi fT}\right)^2 S_x(f).$$

Also we can show that

$$R_y(\tau) = \frac{1}{2T}\int_{-2T}^{2T}\left(1 - \frac{|\theta|}{2T}\right) R_X(\tau - \theta)d\theta.$$

Effectively, $y(t)$ is the moving average of $x(t)$, causing the signal energy to be mostly concentrated around DC.

2.7.6 Cyclostationary Processes

In some RF circuits such as oscillators or mixers, the spectral density of the noise sources vary periodically due to changing of the operating points of the transistors. This type of process is called a *cyclostationary* process. By definition, a process $v(t)$ is called strict-sense cyclostationary with period T if its statistical properties are invariant to a shift of the origin by integer multiples of T [6], [7]. Similarly, $v(t)$ is cyclostationary in the wide-sense if the following two conditions are met:

$$E[v(t+mT)] = E[v(t)] = \eta(t)$$
$$R_v(t_1 + mT, t_2 + mT) = R_v(t_1, t_2),$$

for every integer m.

Example: A commonly used practical example of a cyclostationary process is the *pulse amplitude modulated* signal,

$$v(t) = \sum_{n=-\infty}^{\infty} a_n h(t - nT),$$

where a_n is the random amplitude sequence, and $h(t)$ is a deterministic time-sampling function. If a_n is stationary, clearly $v(t)$ satisfies the above two conditions for mean and autocorrelation.

System analysts have, for the most part, treated cyclostationary signals as if they were stationary [7]. This is simply done by time averaging the statistical parameters over one cycle:

$$\bar{\eta} = \frac{1}{T}\int_0^T \eta(t)dt$$

$$\overline{R_v}(\tau) = \frac{1}{T}\int_0^T R_v(t+\tau,t)dt.$$

This, in fact, is exactly what is viewed if a cyclostationary process is displayed over a spectrum analyzer.

Example: As a constructive case study, let us define the *shifted* process as follows:

$$v_s(t) = v(t-\theta),$$

where θ is a random variable uniform in the interval $(0,T)$, and is independent of $v(t)$. If $v(t)$ is cyclostationary in wide-sense, then the shifted process is stationary in wide-sense. To prove that, let us first find the expected value of the shifted process. Since $v(t)$ and θ are independent,

$$E[v_s(t)] = \int_{-\infty}^{\infty}\int_0^T v(t-\theta)f_V(v)\frac{1}{T}dvd\theta = \frac{1}{T}\int_0^T \eta(t-\theta)dt,$$

where $\eta(t) = E[v(t)]$. Since $\eta(t)$ is periodic,

$$\eta_s = E[v_s(t)] = \frac{1}{T}\int_0^T \eta(t-\theta)d\theta = \frac{1}{T}\int_0^T \eta(t)dt.$$

Similarly, for the autocorrelation, we can show

$$R_{v_s}(\tau) = \frac{1}{T}\int_0^T R_v(t+\tau,t)dt.$$

We will follow up with our discussion on cyclostationary processes in Chapter 5 in the context of periodically varying noise in active devices.

2.8 ANALOG LINEAR MODULATION

Effective communication over appreciable distance usually requires a high-frequency sinusoidal carrier. Consequently, most practical transmission signals are *bandpass*. A general block diagram of such a system is shown in Figure 2.21. The input signal contains the desired *information* intended to be transmitted, for example voice or data, whose spectrum is typically

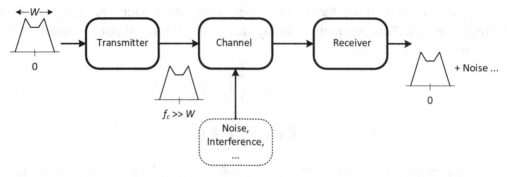

Figure 2.21: A generalized communication system

located around DC, hence known as the *baseband signal*. Among various functionalities, one important role of the transmitter is to translate the baseband signal to a known and desirable high-frequency carrier, resulting in a *modulated spectrum* at carrier frequency (f_c). The receiver is expected to perform the reverse functionality, and thus extract the original baseband, information-bearing signal from the modulated spectrum. This reverse process is known as *demodulation*. The transmission channel is air in the case of wireless systems. In general, we expect an attenuated version of the transmitter output to appear at the receiver input, but with some contamination, such as noise, and interference added to it.

Among the many reasons for which modulation is required are the following:

1. Accommodating a large number of users. Imagine that if all the FM radios were to transmit information at baseband, the stations would interfere with each other. The concept of one *tuning* to one's desired station would not exist.
2. Efficient transmission where the frequency and thus efficiency of the corresponding traveling electromagnetic waves are chosen appropriately [10], [11].
3. Practical design of antennas.
4. Compatible frequency assignment for several wireless standards operating concurrently.
5. Wideband noise or interference reduction by exploiting a narrowband bandpass receiver.

We will describe transmitter and receiver functionality in detail in Chapter 12. In this section we will briefly describe the fundamentals of analog linear modulation. This is followed by a nonlinear modulation description in the next section. More details may be found in [10], [11] as here we only recast a summary to establish the basic concepts needed in RF circuit design. Furthermore, the analog and digital modulation schemes may be found in [12], [13] in details.

Let us start with amplitude modulation. There are two types of double-sideband amplitude modulation: standard amplitude modulation (AM) and suppressed-carrier double-sideband modulation (DSB). Before defining those, let us review some conventions regarding modulating signals.

The analog modulation is performed on an arbitrary baseband message waveform $x(t)$, which is a stochastic process, and could stand for a sample function from the ensemble of possible messages produced by the information source. It must have a reasonably well defined bandwidth, W, so there is negligible energy for $|f| > W$, as shown in Figure 2.22. Also for mathematical convenience, we will normalize all messages to have a magnitude not exceeding

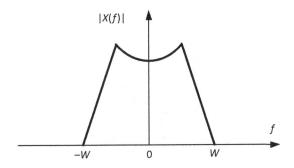

Figure 2.22: Message spectrum with a bandwidth of W

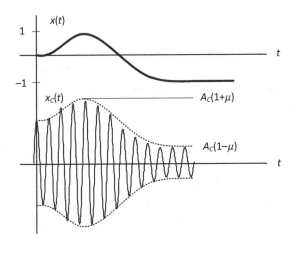
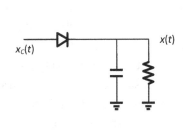

Figure 2.23: Typical AM waveform

unity, so $|x(t)| \leq 1$. Furthermore, even though $x(t)$ is a stochastic process, for convenience we may often express it in the frequency domain directly by applying Fourier transform.

For an ergodic message source (or a deterministic one), this normalization puts an upper limit on the average message power: $<x^2(t)> \leq 1$. Finally, sometimes for the ease of analysis, $x(t)$ may be expressed as a single tone, that is, $x(t) = A_m \cos \omega_m t$ where $A_m \leq 1$, and $f_m < W$.

The unique property of AM is that the envelope of the modulated carrier has the same shape as the message. Thus for a modulated envelope of $A(t)$ with an unmodulated amplitude of A_c, we can write

$$A(t) = A_c[1 + \mu x(t)],$$

where μ is a positive constant called the *modulation index*. The complete amplitude modulated signal $x_c(t)$ is then

$$x_c(t) = A_c[1 + \mu x(t)] \cos(\omega_c t).$$

An example of a typical message and its corresponding AM signal is shown in Figure 2.23. We have assumed $\mu < 1$, and that the message bandwidth W is much smaller than the carrier

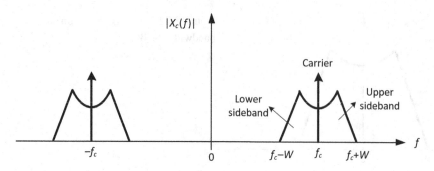

Figure 2.24: AM spectrum

frequency f_c. With these conditions satisfied, a simple envelope detector may be used to extract $x(t)$ as shown in Figure 2.23. With $\mu = 1$, the envelop $A_c[1+\mu x(t)]$ varies between zero and a maximum of $2A_c$. If $\mu > 1$ however, the envelope could become negative, resulting in phase reversal and ultimately envelope distortion [10]. This is known as overmodulation, and is typically avoided in AM.

In the frequency domain, applying Fourier transform to $x_c(t)$ results in an impulse created due to the cosine term from the carrier itself, as well as the message signal shifted to $\pm f_c$, as we established before in our example for modulation. This is shown in Figure 2.24.

The AM spectrum consists of symmetrical lower and upper sidebands, and thus the name double-sideband AM. Evidently, AM requires twice the bandwidth compared to unmodulated baseband signals.

Expanding $x_c(t)^2$, and assuming that the baseband signal carrier has no DC content, we have

$$<x_c(t)^2> = \frac{1}{2}A_c^2\left(1+\mu^2 <x(t)^2>\right).$$

Statistical averaging yields a similar result for $E[x_c(t)]$. The term $\frac{1}{2}A_c^2$ represents the unmodulated carrier power, whereas $\frac{1}{4}A_c^2\mu^2 <x(t)^2>$ is the power per sideband (half of $\frac{1}{2}A_c^2\mu^2 <x(t)^2>$ at each sideband). Since $|\mu x(t)| \leq 1$, at least 50% of the total transmitted power resides in a carrier that conveys no information.

The wasted carrier power can be eliminated by suppressing the unmodulated term and setting $\mu = 1$, leading to

$$x_c(t) = A_c x(t)\cos(\omega_c t),$$

known as *double-sideband suppressed-carrier* or simply DSB. The spectrum is the same as regular AM but without unmodulated carrier impulses. However, unlike AM, as the envelope takes the shape of $|x(t)|$ rather than $x(t)$, the modulated signal undergoes a phase reversal whenever $x(t)$ crosses zero. That is clear by considering the envelope and phase of the DSB signal as given below:

$$A(t) = A_c|x(t)| \quad \text{and} \quad \phi(t) = \begin{cases} 0 & x(t) > 0 \\ 180° & x(t) < 0 \end{cases}.$$

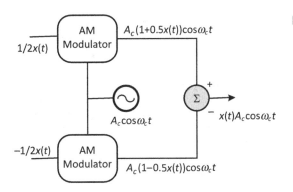

Figure 2.25: Balanced modulator

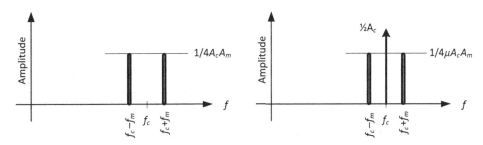

Figure 2.26: AM and DSB spectrum for tone modulation

This implies that suppressed-carrier DSB, though more power efficient, does come at a cost, as a simple envelope detector can no longer be used. In fact, detection of DSB calls for a much more sophisticated demodulation process [10], [11]. As for DSB generation, perhaps the most practical way is to use a *balanced modulator* shown in Figure 2.25, where DSB is created by using two identical AM modulators, save for the reversed sign.

For phase noise analysis of oscillators in Chapter 9, it is beneficial to calculate the spectra of AM and DSB signals in the simple case of tone modulation. Setting $x(t) = A_m \cos \omega_m t$ we obtain for DSB

$$x_c(t) = A_c A_m \cos \omega_m t \cos \omega_c t = \frac{A_c A_m}{2} \cos(\omega_c - \omega_m)t + \frac{A_c A_m}{2} \cos(\omega_c + \omega_m)t.$$

For AM we have an extra term for the unmodulated carrier. The results are shown in Figure 2.26.

The upper and lower bands of DSB are related due to symmetry, so either one contains all the information. Thus, having both sidebands seems unnecessary. Therefore, suppressing either sideband can cut the transmission bandwidth. This also results in reducing the transmission power by half. A conceptual approach to produce single-sideband AM is shown in Figure 2.27, where a sideband filter is used to suppress one sideband. Also shown is the modulated signal spectrum in the case of upper sideband.

Producing such a sharp sideband filter may be a challenge in practice. Only if the baseband signal has no or little energy around DC might the filter be realizable. A more practical approach may be considered if one obtains the SSB signal in the time domain. Although the frequency

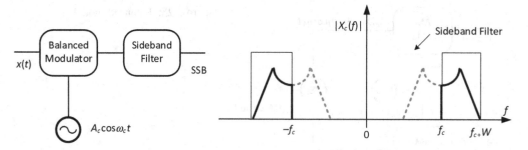

Figure 2.27: Single-sideband AM generation and its spectrum

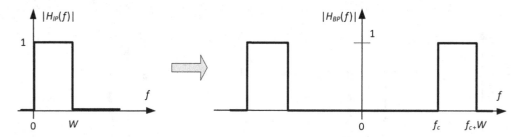

Figure 2.28: Sideband filter lowpass equivalent

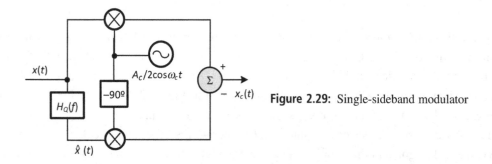

Figure 2.29: Single-sideband modulator

domain spectrum is very easy to depict, the time domain signal requires a few steps to be calculated. We show only the final result here, while more details may be found in [10], [11]:

$$x_c(t) = 1/2 A_c [x(t)\cos\omega_c t - \hat{x}(t)\sin\omega_c t].$$

Appearance of the Hilbert transform may seem surprising at the first glance, but we can observe that the sideband filter, say in the case of upper sideband, is created by shifting the lowpass equivalent as shown in Figure 2.28 to $\pm f_c$. The lowpass equivalent frequency response is therefore $1/2[1 + \text{sgn}(f)]$ for $|f| < W$. The function sgn(f) is the signum function and is represented by Hilbert transform in the time domain, as we showed in the earlier section. The remaining steps follow fairly easily.

The time domain signal representation leads to a very interesting system level realization that has been widely adopted in almost all RF transmitters. As shown in Figure 2.29, the appropriate sideband may be selected through a 90° phase shift of the carrier, and applying Hilbert transform to the baseband signal. Although our example considered the upper sideband, lower sideband is realized by adding the two terms, rather than subtracting.

The quadrature phase shifter, $H_Q(f)$, is not realizable as we showed that Hilbert transform leads to a noncausal impulse response. However, it may be approximated usually with the help of additional but identical phase networks in both branches. It still creates phase distortion at very low frequencies, and thus it works best if the baseband signal has low energy around DC. In most modern transmitters today, this is done in the digital domain and the analog baseband signal is produced by using a data converter. Note that the carrier should be also shifted by 90°.

2.9 ANALOG NONLINEAR MODULATION

Two fundamental features of linear modulation described in the previous section are that the modulated signal is a direct translation of the baseband spectrum and the transmission bandwidth is at most twice that of the baseband. Moreover, it can be shown [10] that the final signal-to-noise ratio is no better than that of the baseband signal, and can be improved only by increasing the power.

In contrast, in nonlinear modulation, such as phase (PM) or frequency (FM) modulation, the transmitted signal is not related to the baseband signal in a simple fashion. Consequently, the modulated signal bandwidth may very well exceed twice the message bandwidth. However, it can provide a better signal-to-noise ratio without increasing the transmitter power. We will momentarily show that the phase modulation is very similar to frequency modulation, so in this section we will mostly explore the properties of FM only.

Consider the CW signal with a constant envelope but time-varying phase:

$$x_c(t) = A_c \cos(\omega_c t + \phi(t)).$$

The instantaneous angle is defined as

$$\theta_c(t) = \omega_c t + \phi(t),$$

while the instantaneous frequency is the derivative of the angle,

$$f(t) = \frac{1}{2\pi} \frac{d}{dt} \theta_c(t) = f_c + \frac{1}{2\pi} \frac{d}{dt} \phi(t).$$

We can see that the phase and frequency modulations are simply related by an integral. In the case of FM, we can represent the instantaneous frequency in terms of the baseband message signal, $x(t)$, where, as before, we assume $|x(t)| \leq 1$,

$$f(t) = f_c + f_\Delta x(t),$$

where $f_\Delta < f_c$ is called the *frequency deviation*. The upper bound of f_Δ ensures that $f(t)$ is always positive. The FM waveform then can be expressed as

$$x_c(t) = A_c \cos\left(\omega_c t + 2\pi f_\Delta \int x(\tau) d\tau\right).$$

We assume that the message has no DC component. If it did, it would simply result in a constant carrier frequency shift.

The PM waveform, on the other hand, is

$$x_c(t) = A_c \cos(\omega_c t + \phi_\Delta x(t)),$$

where ϕ_Δ represents the maximum phase shift produced by $x(t)$. By comparison, it is clear that there is little difference between FM and PM.

One very important aspect of both PM and FM is that unlike linear modulation, they are constant envelope and the transmission power is always $1/2A_c^2$. In contrast, the zero crossings are not periodic and change according to the baseband signal. Figure 2.30 shows a comparison between AM and FM for a simple step-like baseband signal.

Despite similarities between FM and PM, frequency modulation turns out to have superior noise related properties. For example, for a given noise level, since instantaneous frequency is $f(t) = f_c + f_\Delta x(t)$, the signal can be raised by increasing frequency deviation f_Δ, without raising the transmitted power. This, however, comes at the expense of a wider modulated bandwidth. Ironically frequency modulation was first conceived as a means of reducing the bandwidth. The argument was that if instead of modulating the amplitude we modulate the frequency by swinging it over a certain range, say 100kHz, then the transmission bandwidth is always twice that, 200kHz, regardless of the baseband signal bandwidth. The major flaw with this argument is that there is a fundamental difference between the *instantaneous frequency* and the actual *FM bandwidth*.

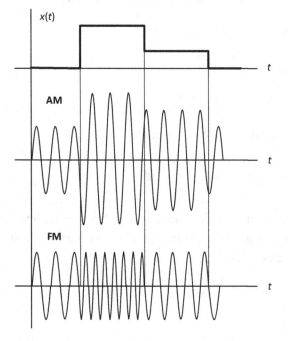

Figure 2.30: A comparison between AM and FM

2.9 Analog Nonlinear Modulation

To be able to calculate the FM bandwidth, let us start with a special case where we assume the phase variations are small: $|\phi(t)| \ll 1$. The modulated signal is

$$x_c(t) = A_c \cos(\omega_c t + \phi(t)) = A_c[\cos \omega_c t \cos \phi(t) - \sin \omega_c t \sin \phi(t)],$$

which under the above assumption simplifies to

$$x_c(t) \approx A_c[\cos \omega_c t - \phi(t) \sin \omega_c t],$$

and in the frequency domain for positive f,

$$X_c(f) = \frac{1}{2} A_c \delta(f - f_c) + \frac{j}{2} A_c \Phi(f - f_c).$$

In the case of FM, $\Phi(f) = -jf_\Delta X(f)/f$, given that frequency is the derivative of phase. Thus if $x(t)$ has a bandwidth of W, the FM signal has a bandwidth of $2W$, assuming $|\phi(t)| \ll 1$. This special case is known as the *narrowband FM*, but proves to be very useful in analyzing various circuits, including phase noise of oscillators.

Next let us consider the tone modulation, where $x(t) = A_m \cos \omega_m t$. Thus $\phi(t) = \beta \sin \omega_m t$, where $\beta = (A_m/f_m)f_\Delta$, and is known as the *modulation index*. Narrowband tone modulation requires $\beta \ll 1$, and thus we will have

$$x_c(t) \approx A_c \cos \omega_c t - \frac{A_c \beta}{2} \cos(\omega_c - \omega_m)t + \frac{A_c \beta}{2} \cos(\omega_c + \omega_m)t.$$

The corresponding spectrum and phasor diagram are shown in Figure 2.31. The spectrum is a lot like that of AM shown in Figure 2.26, with one fundamental difference: there is a phase reversal of the lower sideband, which produces a component perpendicular or quadrature to the carrier phasor as shown in the phasor diagram. In the case of AM, it is in line with that carrier.

For an arbitrary modulation index where the narrowband approximation will not hold, we can write

$$x_c(t) = A_c[\cos(\beta \sin \omega_m t) \cos \omega_c t - \sin(\beta \sin \omega_m t) \sin \omega_c t],$$

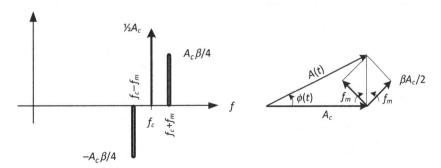

Figure 2.31: Narrowband tone modulation spectrum

despite the fact that $x_c(t)$ itself is not periodic, the terms $\cos(\beta \sin \omega_m t)$ and $\sin(\beta \sin \omega_m t)$ are, and can be expanded using *Bessel functions* as follows:

$$\cos(\beta \sin \omega_m t) = J_0(\beta) + \sum_{\substack{n=2 \\ n \text{ even}}}^{\infty} 2J_n(\beta) \cos(n\omega_m t)$$

$$\sin(\beta \sin \omega_m t) = \sum_{\substack{n=1 \\ n \text{ odd}}}^{\infty} 2J_n(\beta) \sin(n\omega_m t).$$

There is no closed form of expressing $J_n(\beta)$, and one must use numerical methods or available tables to calculate the values. Nevertheless, to gain some insight, knowing that $J_{-n}(\beta) = (-1)^n J_n(\beta)$, we arrive at

$$x_c(t) = A_c \sum_{n=-\infty}^{\infty} J_n(\beta) \cos(\omega_c t + n\omega_m)t.$$

The equation above shows that the FM spectrum consists of a carrier at f_c, as well as an infinite number of sidebands at $f_c \pm nf_m$ as shown in Figure 2.32 (for simplicity the ½ factors are dropped from the spectrum). The sidebands are equally spaced, but odd-order lower sidebands are reversed in phase. Note that as n increases, $|J_n(\beta)|$ tends to reduce, so the sidebands get smaller. Also to avoid the far negative sidebands from folding back to positive frequency $\beta f_m \ll f_c$, which is usually the case.

The relative magnitude of the carrier $J_0(\beta)$ varies with the modulation index (β), and thus depends only partly on the modulating signal. Interestingly, $J_0(\beta)$ could be zero for certain values of β, such as 2.4, 5.5, and so on. For very small values of modulation index, only J_0 and J_1 are significant, and moreover $J_0 \approx 1$, which is the case of narrowband FM. On the other hand, for larger values of β, the higher order terms are still significant, and thus the FM bandwidth will extend to well beyond twice that of the modulation signal.

From the spectrum shown in Figure 2.32 we can see that the FM has theoretically an infinite bandwidth. However, practical FM systems are band-limited and thus inevitably some frequency distortion could occur. The actual FM bandwidth depends on how fast $J_n(\beta)$ decays, and what kind of distortion is acceptable. This can be done by estimating $J_n(\beta)$ behavior, and is discussed in [10].

Another interesting observation is that from Figure 2.31, the magnitude of the modulated signal is not exactly equal to A_c. The difference may be justified by the even-order sidebands that were previously ignored for the narrowband approximation. This is graphically shown in

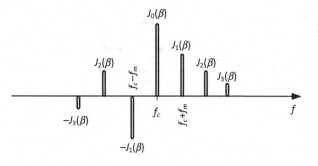

Figure 2.32: FM spectrum for tone modulation

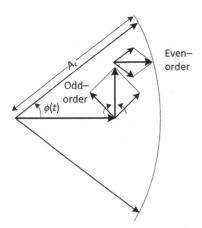

Figure 2.33: FM phasor diagram for arbitrary β

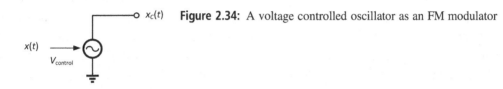

Figure 2.34: A voltage controlled oscillator as an FM modulator

Figure 2.33, where all spectral lines are included. The odd-order pairs are in quadrature with the carrier, and provide the desired frequency modulation, plus *unwanted* amplitude modulation. The resultant even-order sidebands, collinear with the carrier, correct for the amplitude variations.

As for realizing analog FM modulators, there are a number of different methods that mostly rely on fundamental property of a voltage controlled oscillator (VCO), shown in Figure 2.34.

For a baseband signal $x(t)$ applied to the VCO control voltage, the VCO output is

$$x_c(t) = A_c \cos\left(2\pi f_c t + K_{VCO} \int x(\tau)d\tau\right),$$

where K_{VCO} is the VCO gain, A_c is the VCO oscillation amplitude, and f_c is the free running frequency, all design parameters. This is precisely what is needed to produce an FM signal. In practice, since the oscillator free running frequency that sets the carrier frequency could vary, it is common to place it inside a phase-locked loop [14], [15]. More details on analog FM modulators and FM detectors can be found in [10], [16].

2.10 MODERN RADIO MODULATION SCHEME

Most modern radios use more complex modulation schemes where both phase (or frequency) and amplitude vary with time. The RF signal can be expressed in the general form of

$$x_c(t) = A(t)\cos(\omega_c t + \phi(t)).$$

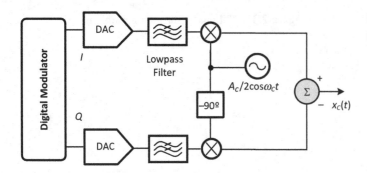

Figure 2.35: Generic transmitter using single-sideband modulation

Moreover, the actual modulation scheme and the modulator itself are digital and the analog equivalent as described above is created through a data converter. A generic transmitter is accordingly shown in Figure 2.35, which is evolved from the single-sideband modulation concept depicted in Figure 2.29. The digital modulator produces two components that are multiplied with signals that are 90° out of phase. For that reason, the two branches are commonly known as I (in-phase) and Q (quadrature phase). Although in a single-sideband AM modulator the two inputs are created by using Hilbert transform, and thus one can be obtained from the other, in most modern transmitters the I and Q inputs to the modulator are uncorrelated. This is necessary to ensure that the baseband bandwidth is half of the bandwidth of the final RF output. If the two signals were Hilbert transform of each other, producing an RF signal with a bandwidth of W, they would have needed a baseband signal spanning from $-W$ to $+W$. This is clear from Figure 2.27. The required bandwidth can be halved if the architecture presented in Figure 2.35 is employed. Each of the baseband signals in the I and Q channels has a bandwidth equal to $\frac{W}{2}$. The information in the I and Q channels is images of the trajectory of the complex baseband signal on the constellation diagram when mapped to the horizontal and vertical axes. This explains why the two baseband signals are statistically independent.

Now, assume that $X_{BB,I}(f)$ and $X_{BB,Q}(f)$ are Fourier transforms of the baseband signals in Figure 2.35. For a give frequency offset f_m, Fourier transforms of the RF signal at $f_c + f_m$ and $f_c - f_m$ are proportional to $X_{BB,I}(+f_m) + j \times X_{BB,Q}(+f_m)$ and $X_{BB,I}(-f_m) + j \times X_{BB,Q}(-f_m)$. Since $X_{BB,I}(-f_m) + j \times X_{BB,Q}(-f_m) \neq \{X_{BB,I}(f_m) + j \times X_{BB,Q}(f_m)\}^*$, the up-converted components at $f_c + f_m$ and $f_c - f_m$ are not complex conjugate of each other. In fact, assuming a normal distribution, independency of the baseband signals at $+f_m$ leads to the independency of the frequency components at $f_c + f_m$ and $f_c - f_m$.

The dual of the above discussion can be also applied to a wireless receiver. Assume that the received channel is centered at f_c with a channel bandwidth equal to W. Considering only the desired signal, the down-converted baseband signals each occupy a channel bandwidth equal to $\frac{W}{2}$ and are statistically independent. Outputs of the two channels are treated as a single complex signal ($x_{BB,I} + j \times x_{BB,Q}$). After passing through a fixed rotation, the resulting complex signal can be directly mapped to the nearest point on the constellation diagram.

The transmitter in Figure 2.35 is known as a *Cartesian modulator*, where the baseband signals are produced as quadrature (or I and Q) components, as opposed to a *polar modulator*, which

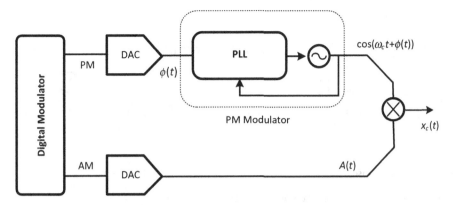

Figure 2.36: A conceptual polar transmitter

takes phase and amplitude directly, as shown in Figure 2.36. The polar modulator then consists of a frequency modulator, or to be precise a PM modulator, which could be realized as a VCO inside a phase-locked loop, as well as a second path that takes the envelope (or the AM) information. Both the AM and PM information may be created in digital domain and converted to analog through a data converter, similar to the Cartesian transmitter. Although a polar modulator seems a more natural way of realizing the intended modulated signal, there are several implementation issues that make them less suitable for especially application that the baseband signal is wideband. Various types of transmitters and their properties will be presented in Chapter 12.

2.11 SINGLE-SIDEBAND RECEIVERS

While the transceiver architectures will be extensively discussed in Chapter 12, as an exercise and also a dual discussion to single-sideband transmitters presented earlier (Figure 2.29), we shall say a few words here regarding the *single-sideband receivers*, which perform the reverse function of a single-sideband transmitter.

A single-sideband transmitter as introduced earlier in Section 2.8 is shown in Figure 2.37 on the left, in which the baseband signal ($x(t)$) is shifted up in frequency by means of quadrature multiplication and Hilbert transform around the carrier frequency (f_c). A single-sideband receiver shown on the right could be thought of as the dual of the transmitter, where the modulated signal ($x_c(t)$) at *only one side of the carrier* is shifted down in frequency. Similar to the transmitter, the single-sideband receiver relies on quadrature multiplication and Hilbert transform.

We will illustrate the functionality of the single-sideband receiver through an example shown in Figure 2.38. Suppose the receiver input ($x_c(t)$ in Figure 2.37) consists of two modulated spectrums one at the upper side of the carrier (the triangles), and one at the lower side (the rectangles), as shown in Figure 2.38.

The input is first multiplied by a cosine at the carrier frequency on the I side, and a sine on the Q side. The Fourier transforms of the cosine and sine are also shown, which consist of a pair of impulses at the carrier frequency. Since multiplication in the time domain is equivalent to

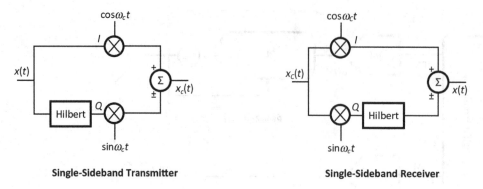

Figure 2.37: Single-sideband receiver compared against a single-sideband transmitter

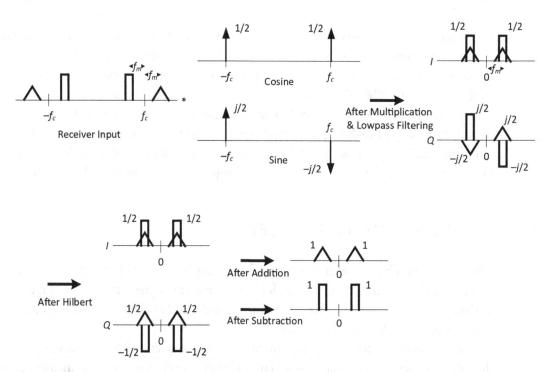

Figure 2.38: Signals at various points of the single-sideband receiver of Figure 2.37

convolving in the frequency domain, the input signal is shifted up and down in frequency. The up shift of the positive spectrum (or the down shift of the negative spectrum) results in a signal around twice the carrier frequency, which is a very high frequency, and is typically lowpass filtered at the multiplier output. The remaining signals in the I and Q channels will now be at baseband, as desired. However, the upper and lower sideband signals are not distinguishable, as they both appear at the same frequency. Since their phases are different though, Hilbert transform could be used to extract one. As shown in Figure 2.38 once the Hilbert transform is applied to the Q channel (with the I signal remaining intact), the upper and lower sidebands are aligned, and depending on whether the I and Q outputs are added or subtracted at this point, only one sideband remains as desired.

The single-sideband receiver, also known as the *image-reject receiver*, is the basis of most modern radio receivers. The Hilbert transform may be performed entirely in the digital domain once the *I* and *Q* channels are digitized after multiplication, or could be approximated by a polyphase filter (see Chapter 4) in the analog domain.

2.12 Summary

This chapter covered basic concepts of circuit theory and communication systems, and presented a summary of several key topics that will be revisited throughout the book.

We started with a brief summary of Fourier and Hilbert transforms, both of which serve as great tools to analyze RF circuits and systems. To establish some foundation for noise analysis presented in Chapter 5, a brief summary of stochastic processes and random variables was provided. The majority of the material presented in this chapter was a review of various concepts that exist in signal processing, communication systems, and basic circuit theory. However, we felt it was important to remind the reader, especially a less expert one, of those basic concepts.

- Sections 2.1 and 2.3 presented a brief summary of Fourier transform and Fourier series and their properties.
- Fundamental properties of impulse response, natural frequencies, and poles and zeros were reviewed in Sections 2.2, 2.4, and 2.5.
- Hilbert transform was discussed in Section 2.6, and will be followed up in Chapter 4 in the context of quadrature filter design.
- Section 2.7 discussed stochastic processes and their properties. Many of the signals we deal with in radio design, such as noise, are random in nature. A good understanding of random signals is hence very critical.
- Fundamentals of analog modulation schemes and analog modulators were presented in Sections 2.8 and 2.9.
- A brief introduction to modern radio modulation schemes was presented in Section 2.10.

Of special importance is Section 2.7, where random signals were discussed; it is important to understand noise in time-varying circuits such as mixers (Chapter 8) and oscillators (Chapter 9).

2.13 Problems

1. Using Fourier transform basic definition, prove Parseval's energy theorem:

$$\int_{-\infty}^{\infty} |x(t)|^2 dt = \int_{-\infty}^{\infty} |X(f)|^2 df.$$

2. Show that multiplying the Fourier transform by $e^{-j2\pi f\tau}$ results in a shift in the time domain.

3. Show that if $x(t)$ is real, then $X(-f) = X^*(f)$.
4. Prove Parseval's energy theorem for the Fourier series:

$$\frac{1}{T}\int_T |v(t)|^2 dt = \sum_{k=-\infty}^{\infty} |a_k|^2.$$

5. Show that the Fourier transform of $x(t) = \frac{1}{1+t^2}$ is $X(f) = \pi e^{-2\pi|f|}$. **Hint:** Use the duality property.

6. This exercise shows that while every pole of a system is a natural frequency, the opposite is not necessarily true. Consider the circuit below.
   a. If the input is an impulse $i_s(t) = \delta(t)$, find the initial capacitor voltage and inductor initial current at $t = 0^+$.
   b. Find the impulse response of the circuit $v(t)$. Determine the natural frequencies.
   c. Find the transfer impedance $\frac{V(s)}{I_s(s)}$. What are the circuit poles?

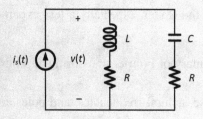

7. In the ladder structure shown below, show that the transfer function is of the form

$$\frac{v_{OUT}(s)}{i_{IN}(s)} = \frac{1}{a_n s^n + a_{n-1} s^{n-1} + \cdots + a_1 s + a_0},$$

where n is the number of reactive components, and a_n is their product.

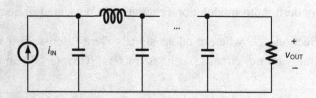

8. In the circuit below, find the impulse response in three different ways (1) Laplace transform, (2) solving the time domain differential equation by integrating from 0^- to 0^+, and (3) by intuitively finding the circuit initial condition. The response is the inductor current $i(t)$.

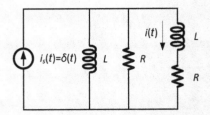

9. Shown below is the circuit of a Wien bridge oscillator. The RC phase shift network operates as explained in the example shown in Figure 2.7. The resistors R_f and R_b along with the opamp create the required gain from the point A to the output.
   a. Determine the frequency of oscillation and explain the circuit operation.
   b. What is the minimum $\frac{R_f}{R_b}$ required for the circuit to oscillate?

   Hint: Find the loop gain by following the output to point A through the RC phase shift circuit, and from there back to the output through the amplifier. Determine the magnitude and phase of the loop gain.

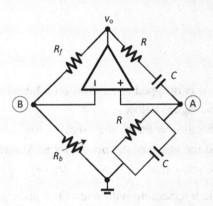

10. Find the Fourier transform of $x(t) = \begin{cases} e^{-at} & t > 0 \\ -e^{at} & t < 0 \end{cases}$.

11. Prove that the impulse response of the Hilbert transform is $h_Q(t) = \frac{1}{\pi t}$. **Hint:** In the previous problem set $\alpha = 0$, and use duality.

12. Show that the Hilbert transform of $x(t) = \frac{1}{1+t^2}$ is $\hat{x}(t) = \frac{t}{1+t^2}$.

13. A transfer function $H(f)$ is plotted below, which has a unity magnitude and phase of $-\alpha$ and $+\alpha$ for positive and negative frequencies, respectively. Prove that the impulse response is given by the following expression: $h(t) = \frac{\sin \alpha}{\pi t} + \cos \alpha \times \delta(t)$.

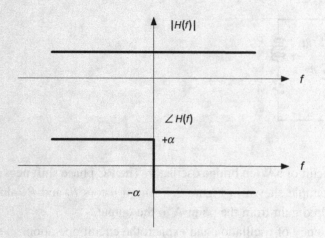

14. Consider the following comb filter, where the input $x(t)$ is a stationary random signal. Prove that the autocorrelations of the input and output are related according to the following equation: $R_y(\tau) = 2R_x(\tau) - R_x(\tau - T) - R_x(\tau + T)$.

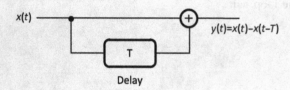

15. Buffon's needle: A fine needle of length a is dropped at random on a board covered with parallel line distance $b > a$ apart given the figure below.
  a. Assume that the distance from the needle center to the edge of the adjacent line is denoted by the random variable x, and the angle of the needle by θ. Assuming uniform distribution for both random variables, show that $f(x,\theta) = f_X(x)f_\Theta(\theta) = \frac{1}{b}\frac{1}{\pi}$.
  b. Show that the probability of the needle intersecting with one of the lines is $\frac{2a}{\pi b}$. As such, this can be used to experimentally determine the number π.

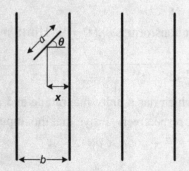

16. For the process $x(t) = r\cos(\omega t + \phi)$, we assume that random variables r and ϕ are independent, and ϕ is uniform in the interval $(-\pi, \pi)$. Find the mean and autocorrelation.

17. For the process $x(t) = a\cos(\omega t + \phi)$, the random variable ω has the probability density function of $f(\omega)$, and the random variable ϕ is uniform in the interval $(-\pi, \pi)$, and independent of ω. Find the mean and autocorrelation of $x(t)$.

18. Prove the equation $E[L\{x(t)\}] = L\{E[x(t)]\}$, where L denotes the linear operation imposed by convolution integral.

19. Prove the *Schwarz* inequality: $(E\{xy\})^2 \leq E\{x^2\}E\{y^2\}$. **Hint:** Consider that the quadratic $E\{(ax-y)^2\} = a^2 E\{x^2\} - 2aE\{xy\} + E\{y^2\}$ is always positive for any a, where a is an arbitrary constant.

20. Suppose $x(t)$ is a stationary process with zero mean, and autocorrelation $R_x(\tau)$, and random variable s is defined as $s = \int_{-T}^{T} x(t)dt$. Find the variance of s (σ_s^2) in terms of $R_x(\tau)$. **Answer:** $\sigma_s^2 = \int_{-2T}^{2T}(2T - |\tau|)R_x(\tau)d\tau$.

21. Consider the random variables X and Y. We form a new random variable $Z = X/Y$.
    a. Show that the density function of Z is $f_Z(z) = \int_{-\infty}^{\infty} |y|f(zy, y)dy$, where $f(x, y)$ is the joint density function of X and Y.
    b. If X and Y are jointly normal with zero mean, their spectral density is given by $f(x,y) = \frac{1}{2\pi\sigma_1\sigma_2\sqrt{1-r^2}} \exp\left(-\frac{1}{2(1-r^2)}\left(\frac{x^2}{\sigma_1^2} - 2r\frac{xy}{\sigma_1\sigma_2} + \frac{y^2}{\sigma_2^2}\right)\right)$, where σ_1 and σ_2 are the standard deviations of X and Y respectively, and $r = \frac{E[XY]}{\sigma_1\sigma_2}$ is the correlation coefficient. Show that $f_Z(z) = \frac{\sigma_1\sigma_2\sqrt{1-r^2}}{\pi\sigma_2^2\left(z - r\frac{\sigma_1}{\sigma_2}\right)^2 + \pi\sigma_1^2(1-r^2)}$, known as a *Cauchy* density.
    c. Prove that $|r| \leq 1$.

22. If the random variables X and Y are jointly normal with zero mean, show that $E[X^2 Y^2] = E[X^2]E[Y^2] + 2E^2[XY]$.

23. For the limiter of Figure 2.17, assume $x(t)$ is a normal stationary process.
    a. Show that $R_y(\tau) = P(x(t+\tau)x(t) > 0) - P(x(t+\tau)x(t) < 0) = 1 - 2P(x(t+\tau)x(t) < 0)$.
    b. Given that the processes $x(t+\tau)$ and $x(t)$ are jointly normal with zero mean, show that their variance is $R_x(0)$, and the correlation coefficient $\left(r = \frac{E[XY]}{\sigma_1\sigma_2}\right)$ is $\frac{R_x(\tau)}{R_x(0)}$.
    c. Using parts a and b, and Problem 9b, show that $R_y(\tau) = \frac{2}{\pi}\sin^{-1}\left(\frac{R_x(\tau)}{R_x(0)}\right)$. This result is known as the arcsine law.

24. For the pulse amplitude modulated process $v(t) = \sum_{n=-\infty}^{\infty} a_n h(t - nT)$, we assume a_n is a stationary sequence, with autocorrelation $R_a(n) = E[a_{n+m}a_m]$, and spectral density $S_a(f) = \sum_{n=-\infty}^{\infty} R_a(n)e^{-j2\pi fn}$. We form the impulse train $w(t) = \sum_{n=-\infty}^{\infty} a_n \delta(t - nT)$. Show that for the impulse train process the autocorrelation of the shifted process $w_s(t) = w(t - \theta)$ is $R_{w_s}(\tau) = \frac{1}{T}\sum_{n=-\infty}^{\infty} R_a(n)\delta(\tau - nT)$. Show that the spectral density for the shifted process $v_s(t) = v(t - \theta)$ is $S_{v_s}(f) = \frac{1}{T}S_a(f)|H(f)|^2$. **Hint:** $v(t)$ is the output of a linear system with input $w(t)$. Thus $v(t) = h * w(t)$ and $v_s(t) = h * w_s(t)$.

25. In the previous problem, suppose $h(t)$ is a pulse with width T, and is a_n is white noise taking the values ± 1 with equal probability. The resulting process is called *binary transmission*. It is a cyclostationary process taking the values of ± 1 in every interval T. Show that $S_{v_s}(\omega) = \frac{4\sin^2(\omega T/2)}{\omega^2 T}$.

26. Assume that x is a random variable with mean μ and standard deviation σ.
   a. Prove that the following inequality is held for any arbitrary values of a and b (b is positive): $p\{|x-a| \geq b\} \leq \frac{E\{(x-a)^2\}}{b^2}$. Hint: $p\{|x-a| \geq b\} = \int_{-\infty}^{a-b} f_X(x)dx + \int_{a+b}^{+\infty} f_X(x)dx \leq \int_{-\infty}^{+\infty} \frac{(x-a)^2}{b^2} f_X(x)dx = \frac{E[(x-a)^2]}{b^2}$, since $\frac{(x-a)^2}{b^2} > 1$.
   b. Use the above inequality to prove the following extension of Chebyshev's inequality: $p\{k_1 < x < k_2\} \geq \frac{4\{(\mu-k_1)(k_2-\mu)-\sigma^2\}}{(k_2-k_1)^2}$. Chebyshev's inequality is a special case of the above inequality where $k_1 = \mu - k\sigma$ and $k_2 = \mu + k\sigma$.

27. A random telegraph signal assumes only two values, 0 and A, which happen with equal probabilities. The signal makes independent random shifts between these two values. The number of shifts per unit time is governed by the Poisson distribution, with μ being the average shift rate $\left(p\{x=k, \text{over time } \tau\} = \frac{\mu^k}{k!}e^{-\mu\tau}\right)$. (a) Prove that the autocorrelation function of the signal is found to be $R_v(\tau) = \frac{A^2}{4}\left(1 + e^{-2\mu|\tau|}\right)$. (b) Use the above autocorrelation to find the power spectral density.

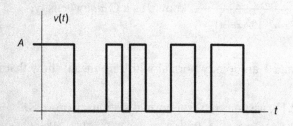

28. Show that the FM signal, $v_c(t) = A_c \cos(\omega_c t + 2\pi f_\Delta \int x(\tau)d\tau)$, satisfies the following equation, known as the *FM differential equation*,

$$v_c(t) - \frac{v_c'(t)\omega_i'(t)}{\omega_i(t)^3} + \frac{v_c''(t)}{\omega_i(t)^2} = 0,$$

where $\omega_i(t)$ is the instantaneous frequency.

29. Show that the FM signal also satisfies the following integro-differential equation:

$$v_c(t) + \int_0^t \omega_i(\theta)\left(\int_0^\theta \omega_i(\tau)v_c(\tau)d\tau\right)d\theta.$$

Accordingly, devise an FM modulator by a circuit comprising multipliers and integrators (known as *analog computer*) that satisfies the equation.

30. Show that the following circuit may be employed as an FM demodulator. Propose a proper circuit to perform the differentiation suitable for high frequencies.

Hint: Use the circuit below, known as a *balanced slope demodulator*.

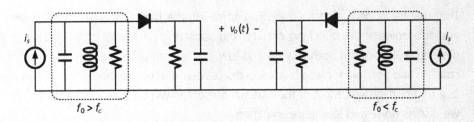

2.14 References

[1] A. V. Oppenheim, A. Willsky, and I. Young, *Signals and Systems*, Prentice Hall, 1983.
[2] A. V. Oppenheim, R. W. Schafer, J. R. Buck, et al., *Discrete-Time Signal Processing*, Prentice Hall, 1989.
[3] W. E. Boyce, R. C. DiPrima, and C. W. Haines, *Elementary Differential Equations and Boundary Value Problems*, John Wiley, 1992.
[4] C. A. Desoer and E. S. Kuh, *Basic Circuit Theory*, McGraw-Hill, 2009.
[5] A. Papoulis, *Stochastic Processes*, McGraw-Hill, 1996.
[6] A. Papoulis and S. U. Pillai, *Probability, Random Variables, and Stochastic Processes*, McGraw-Hill, 2002.
[7] W. Gardner and L. Franks, "Characterization of Cyclostationary Random Signal Processes," *IEEE Transactions on Information Theory*, 21, no. 1, 4–14, 1975.
[8] R. E. Colline, *Foundation for Microwave Engineering*, McGraw-Hill, 1992.
[9] D. M. Pozar, *Microwave Engineering*, John Wiley, 2009.
[10] A. B. Carlson and P. B. Crilly, *Communication Systems: An Introduction to Signals and Noise in Electrical Communication*, vol. 1221, McGraw-Hill, 1975.
[11] H. Taub and D. L. Schilling, *Principles of Communication Systems*, McGraw-Hill, 1986.
[12] J. G. Proakis, *Digital Communications*, McGraw-Hill, 1995.
[13] K. S. Shanmugam, *Digital and Analog Communication Systems*, John Wiley, 1979.
[14] F. M. Gardner, *Phaselock Techniques*, John Wiley, 2005.
[15] D. H. Wolaver, *Phase-Locked Loop Circuit Design*, Prentice Hall, 1991.
[16] K. K. Clarke and D. T. Hess, *Communication Circuits: Analysis and Design*, Krieger, 1994.

3 RF Networks

This chapter is dedicated to reviewing some of the basic concepts in RF design such as available power gain, matching circuits, and scattering parameters. We also present a more detailed discussion on both lossless and low-loss transmission lines and introduce the Smith chart. In addition, we recast a follow-up discussion on a receive–transmit antenna pair viewed as a two-port system. Most of the material presented will be used in Chapters 5 and 7, when we discuss noise and low-noise amplifiers.

The majority of our discussions in this chapter, including topics on antennas and transmission lines, are closely tied with the material presented in Chapter 1.

The specific topics covered in this chapter are:

- Reciprocal and lossless networks
- Available power and matching
- Wideband and narrowband transformers
- Series–parallel transformation
- Antennas circuit model
- Lossless and low-loss transmission lines
- Smith chart
- Scattering parameters

For class teaching, we recommend focusing on Sections 3.2 and 3.3, which are crucial for the LNA design. The transmission lines, Smith chart, and S parameters, while important for RF design, could be skipped for the class. However, the material is relatively easy to follow, and the interested student or engineer may study individually.

3.1 INTRODUCTION TO TWO-PORTS

3.1.1 Two-Port Definition

A two-port is simply a network inside a black box that has only two pairs of accessible terminals, usually one representing the input, the other one the output. Let us consider a one-port network first shown in Figure 3.1.

Given a linear network and a pair of terminals, we obtain a one-port network, which could be fully characterized by its Thevenin equivalent shown on the right. N_0 is the same network with

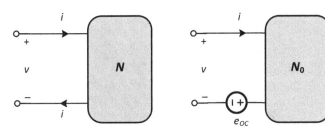

Figure 3.1: A linear one-port and its Thevenin equivalent

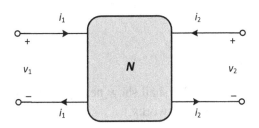

Figure 3.2: A two-port network

all independent source turned off (voltage sources shorted, and current sources opened), known as the relaxed network, and e_{OC} is the Thevenin open circuit voltage across the one-port.

A two-port is simply an extension of the one-port concept, and is shown in Figure 3.2, which is essentially a four-terminal network with two pairs of accessible terminals. What distinguishes a two-port from any arbitrary four-terminal network is the additional constraint that the current going into a given port must come out as shown in the figure.

From basic circuit theory [1], since there are four unknown elements attached to the two-port (i_1, v_1, i_2, and v_2), the two-port can impose only two linear constraints on these four variables. Since there are only six ways of picking two elements out of four, there are six ways of characterizing a linear and time-invariant two-port, namely impedance, admittance, two hybrid, and two transmission matrices. Knowing one, any other one of the five remaining matrices could be obtained. For instance, admittance matrix is inverse of the impedance matrix. We will later show that a two-port could be also represented by scattering parameters, which is a more convenient way of characterizing it at microwave frequencies. The scattering matrix could be also converted into any of the six aforementioned forms matrices.

Example: Let us find the Z matrix of the simple resistive network shown in Figure 3.3.

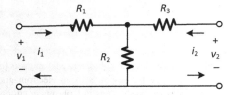

Figure 3.3: A resistive T two-port

From the basic definition, Z_{11} (or Z_{22}) may be found by calculating the impedance of the corresponding port, with the other port open-circuited. Hence

Continued

$$Z_{11} = R_1 + R_2$$
$$Z_{22} = R_3 + R_2.$$

Also,

$$Z_{12} = \left.\frac{V_1}{I_2}\right|_{I_1=0} = R_2$$

$$Z_{21} = \left.\frac{V_2}{I_1}\right|_{I_2=0} = R_2.$$

In this particular example, it happens that $Z_{12} = Z_{21}$. We shall show next that this condition is met for any *reciprocal* circuit, such as linear resistive networks.

3.1.2 Reciprocal Two-Ports

Consider the network of Figure 3.4, comprising only linear and time-invariant passive elements (no active circuits, plasmas,[1] and ferrites).[2] As such, the network may be constructed by linear resistors, capacitors, self, and mutual inductors (RLCM). Such network is known to be *reciprocal*, possessing several useful properties, some of which we will briefly discus here, particularly the ones related to our discussion of available power gain.

First, from the electromagnetic point of view, of the several theorems related to the reciprocal networks, perhaps the most common one is the one proposed by Hendrik Lorentz,[3] over a century ago, as follows:

Suppose the network of Figure 3.4 contains two current sources with densities of $\mathbf{J}_1$, and $\mathbf{J}_2$, producing the corresponding electric and magnetic fields of $\mathbf{E}_1$, $\mathbf{H}_1$, $\mathbf{E}_2$, and $\mathbf{H}_2$.

Then the Lorentz theorem states that for an arbitrary surface S enclosing a volume V, we have

$$\int_V (\mathbf{J}_1 \cdot \mathbf{E}_2 - \mathbf{J}_2 \cdot \mathbf{E}_1) dV = \oint_S (\mathbf{E}_1 \times \mathbf{H}_2 - \mathbf{E}_2 \times \mathbf{H}_1) \cdot d\mathbf{S}.$$

A common situation arises when the volume V entirely encloses the current sources $\mathbf{J}_1$ and $\mathbf{J}_2$, and that there are no incoming waves from far away. Under such conditions, the right-hand side surface integral above diminishes to zero, leaving

---

[1] Plasma is one of the four fundamental states of matter (the other three being solid, liquid, and gas) and was first detected by William Crookes in 1879. Like gas, it does not have a defined shape but responds to an electromagnetic field very differently. It is abundant in stars, such as the sun. By studying Maxwell's equations in a plasma environment, it can be shown that when a magnetic field is present, the plasma becomes *anisotropic* and hence in general leads to nonreciprocity.

[2] Some of the most practical anisotropic microwave materials are *ferrimagnetic* compounds, or simply ferrites. In contrast to ferromagnetic materials (such as iron), they have a significant amount of anisotropy, induced by applying a DC magnetic bias. Consequently, a microwave field may interact very strongly with the resultant magnetic dipole in certain directions rather than others.

[3] Hendrik Lorentz was a Dutch physicist who was awarded the Nobel Prize in 1902.

3.1 Introduction to Two-Ports

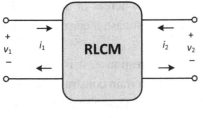

Figure 3.4: A general reciprocal network comprising resistors, capacitors, self, and coupled inductors

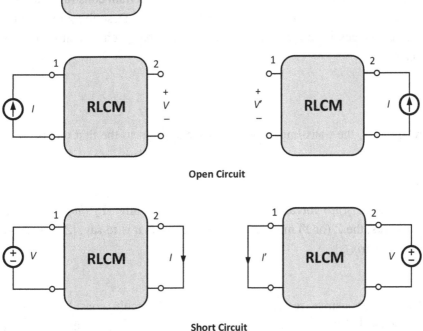

Figure 3.5: Two common criteria to test reciprocity in a given network

$$\int_V \boldsymbol{J}_1 \cdot \boldsymbol{E}_2 dV = \int_V \boldsymbol{J}_2 \cdot \boldsymbol{E}_1 dV.$$

From the latter equation, it may be deduced that for any reciprocal circuit, the voltage at position 1 from a current at 2 is identical to the voltage at 2 from the same current at 1 (this is shown in Figure 3.5 top in the context of a two-port with positions 1 and 2 being the input/output ports). Equivalently, one can say that the current at position 1 from a voltage at 2 is identical to the current at 2 from the same voltage at 1 (Figure 3.5 bottom).

Strictly speaking from the circuit point of view, we can make the following statements for the *zero-state response* of the circuit:[4]

1. Connect a current source $i(t)$ to the port labeled 1, and measure the voltage $v(t)$ across the second port. Next reverse the operation, that is, apply the same current source $i(t)$ to the second port, and observe the *open-circuit* voltage across the first terminal, $v'(t)$. The reciprocity asserts that for all t: $v(t) = v'(t)$. This is described in Figure 3.5 top.
2. Connect a voltage source $v(t)$ to the port labeled 1, and measure the current response $i(t)$ at the second port. Next, apply the same voltage source $v(t)$ to the second port, and observe the *short-circuit* current at the first terminal, $i'(t)$. The reciprocity asserts that for all t: $i(t) = i'(t)$. This is described in Figure 3.5 bottom.

---

[4] The zero-state response is the circuit response with an initial state of zero.

Note that in the first case we observe open-circuit voltages, whereas in the second case short-circuit currents are observed. Consistent with that, in the first case a current source is applied, while in the second case a voltage source is used.

Since the reciprocity theorem deals with the zero-state response, it is often convenient to describe it in terms of network functions. This could lead to certain constraints in the impedance (or other network) matrices. Consider Figure 3.5 for instance. Describing the input current and the open circuit voltages in terms of their steady state phasor equivalents ($I(j\omega)$ and $V(j\omega)$), clearly we have

$$Z_{21}(j\omega) = \frac{V(j\omega)}{I(j\omega)}.$$

But, from reciprocity, the trans-impedance from second port to the first one is

$$Z_{12}(j\omega) = \frac{V'(j\omega)}{I(j\omega)}.$$

Hence, $Z_{21}(j\omega) = Z_{12}(j\omega)$ for all ω. Similarly, we may infer $Y_{21}(j\omega) = Y_{12}(j\omega)$. Thus, for reciprocal networks, the Z (or Y) matrices are symmetric, that is to say, $[Z] = [Z]^t$ (superscript t denotes traverse matrix).

Example: To demonstrate reciprocity in a practical circuit, consider the T-inductive network depicted below. This circuit is often used to model a transformer or a coupled inductor, in general (see Section 1.11.7). We shall try the case of voltage source/short-circuit current to determine the reciprocity.

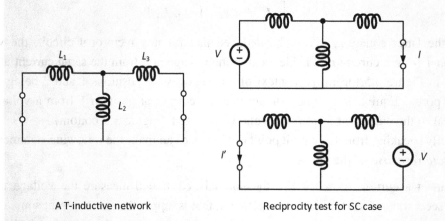

A T-inductive network   Reciprocity test for SC case

Figure 3.6: A T-inductive circuit used to demonstrate reciprocity

For the circuit shown on the top right of Figure 3.6, we have

$$I(j\omega) = \frac{V(j\omega)}{jL_1\omega + (jL_2\omega \| jL_3\omega)} \frac{L_2}{L_2 + L_3} = \frac{L_2}{j\omega(L_1L_2 + L_2L_3 + L_3L_1)} V(j\omega).$$

For the bottom right circuit, if the roles are reversed, only $L_1 \leftrightarrow L_3$, and clearly the observed short-circuit current remains the same.

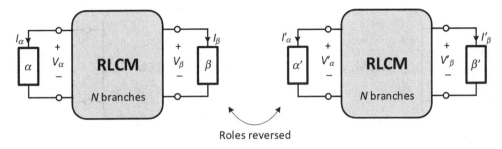

Figure 3.7: Proof of reciprocity using Tellegen's theorem

The above statements for reciprocity may be readily proven using Tellegen's theorem [1].[5] Suppose the network of interest consists of N branches, in addition to the input and output ports labeled as α and β. Then the associated voltage and current phasors are V_α, V_β, V_1, V_2, ..., V_N and I_α, I_β, I_1, i_2, ..., I_N. When the roles are reversed, we assume the corresponding branch voltages and currents are V'_α, V'_β, V'_1, V'_2, ..., V'_N and I'_α, I'_β, I'_1, i'_2, ..., I'_N (Figure 3.7). The boxes labeled α, β, α', and β' represent voltage or current sources on one side, and short or open circuits on the other side. For instance, in the short circuit experiment of Figure 3.5, α and β' are the voltage sources, whereas β and α' and are the short circuits.

Since all four sets of voltage and current phasors satisfy KVL/KCL, we can write

$$V_\alpha I'_\alpha + V_\beta I'_\beta + \sum_{k=1}^{N} V_k I'_K = 0$$

$$V'_\alpha I_\alpha + V'_\beta I_\beta + \sum_{k=1}^{N} V'_k I_K = 0.$$

The internal branches of the network are composed of linear and time-invariant resistors, capacitors, or inductors,[6] and as such

$$V_k = Z_k I_k,$$

where Z_k is the corresponding branch impedance. Consequently,

$$\sum_{k=1}^{N} V_k I'_K = \sum_{k=1}^{N} (Z_k I_k) I'_K = \sum_{k=1}^{N} (Z_k I'_K) I_k = \sum_{k=1}^{N} V'_k I_K.$$

---

[5] Tellegen's theorem states that in an arbitrary *lumped* network comprising b branches of voltage v_k and current i_k, we have $\sum_{k=1}^{b} v_k i_k = 0$, as long as the branch voltages and branch currents satisfy KVL and KCL. See Chapter 1 for more details and a proof. See also [1] for a detailed proof and other relevant discussions.

[6] The case of coupled inductors is handled taking the two corresponding branches and including the mutual inductance.

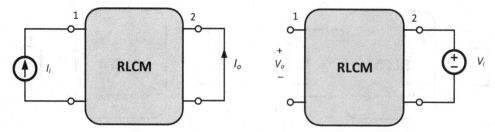

Figure 3.8: Third criteria for reciprocity

Thus,

$$V_\alpha I'_\alpha + V_\beta I'_\beta = V'_\alpha I_\alpha + V'_\beta I_\beta.$$

Now consider the first statement for instance (Figure 3.5 top). Clearly $I_\beta = I'_\alpha = 0$, given the open-circuit condition. Thus if $I_\alpha = I'_\beta$, that is, if in either condition the network is excited by the same current, then $V_\beta = V'_\alpha$, which implies same open-circuit voltages observed at the other terminal.

A third statement dealing with reciprocity theorem is demonstrated in Figure 3.8. If for all time the waveforms $i_i(t)$, and $v_i(t)$ are identical, the reciprocity asserts that, for all t,

$$v_o(t) = i_o(t).$$

The proof readily follows from the Tellegen's theorem and the equation above (see also Problem 19).

As a final remark, we shall show that in general *reciprocity* and *passivity* are not equivalent in the context of the following example.

Example: A common example of a passive nonreciprocal circuit is a gyrator [2] whose circuit symbol is shown in Figure 3.9.[7] In practice passive gyrators are made of ferrites, and have applications in many microwave circuits such as circulators or isolators [3], [4]. We will show in Chapter 4 that an active implementation of the gyrator is widely used in active filter design.

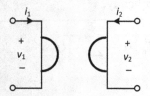

Figure 3.9: Circuit of a gyrator. The symbol has been proposed by Tellegen.

---

[7] The gyrator idea and its circuit symbol were proposed by Tellegen.

An ideal gyrator is described by the equations

$$v_1(t) = i_2(t)$$
$$v_2(t) = -i_1(t),$$

or in terms of the impedance matrix,[8]

$$[Z] = \begin{bmatrix} 0 & 1 \\ -1 & 0 \end{bmatrix},$$

Clearly the circuit is linear and time-invariant. Furthermore, the power delivered by the external world to the gyrator is zero for all t, as

$$v_1(t)i_1(t) + v_2(t)i_2(t) = 0,$$

and hence, it is passive. However, it is not reciprocal as one can easily verify by applying the same currents to each terminal, and observing that the open-circuit voltages will be opposite in polarity.

3.1.2.1 Reciprocity in Nonlinear and Time-Variant Networks

A critical step in proof of reciprocity came from substituting the branch voltage, V_k by $Z_k I_k$. Clearly, this can apply only to a linear and time-invariant element. Here, through a couple of examples we shall show heuristically that reciprocity does not hold in general in either *nonlinear* or *time-variant* networks.

Example: To demonstrate nonreciprocity in nonlinear circuits, consider the common-source amplifier depicted in Figure 3.10.

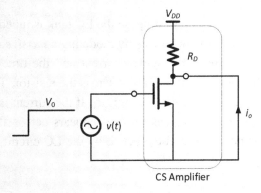

Figure 3.10: A common-source amplifier as an example of a nonlinear and nonreciprocal circuit. A step voltage at gate results in a step current at drain, whereas the same voltage applied to drain leads to no current at gate.

Suppose a step voltage of $v(t) = V_0 u(t)$ is applied to the gate. As long as V_0 is greater than the transistor threshold voltage, a steady short-circuit current is produced at the drain. If the step voltage is applied to the drain, however, clearly no current is created at the gate.

---

[8] In general $[Z] = \begin{bmatrix} 0 & \alpha \\ -\alpha & 0 \end{bmatrix}$, where α is called the *gyration ratio*.

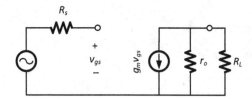

Figure 3.11: Small signal model of a common-source amplifier at low frequency. The presence of the dependent current source leads to nonreciprocity

It is important to note that even if the amplifier is linearized around some operating point, it would remain nonreciprocal, despite consisting of linear and time-invariant elements (Figure 3.11). This is due to the dependent current source modeling the transconductance gain from gate to drain. This shows that when *dependent sources* are present, reciprocity does not hold in general.

As an exercise, the interested reader may consider modeling the gyrator of Figure 3.9 through dependent sources, and examine reciprocity in the equivalent circuit.

Example: Consider the network of Figure 3.12, comprising a time-varying resistor. We shall show that the circuit is not reciprocal.

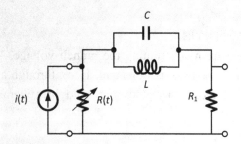

Figure 3.12: Network comprising a time-variant linear resistor

Assume the time-varying resistor is described by

$$R(t) = R_0 \sin \omega_0 t,$$

and the input current is $i(t) = \cos \omega_0 t$. Furthermore, suppose the LC tank is tuned to $2\omega_0$, that is, $\omega_0 = \frac{1}{2\sqrt{LC}}$. The time-varying resistor may be loosely modeling a *passive mixer* as we shall discuss later on. Given the nature of the source current and the time-varying resistor, we expect a voltage at the frequency of $2\omega_0$ present across the resistor. The ideal tank blocks this component to appear across the other resistor, R_1. If the current is applied to the other side, however, a voltage at the frequency of ω_0 appears across the time-invariant resistor, R_1. This voltage is not going to be blocked by the LC circuit, and we expect some kind of voltage at $2\omega_0$ present across $R(t)$.

In conclusion, in general, time-variant networks are not reciprocal. An exception to this is a network comprising linear but *time-varying resistors only*. If the summation derived earlier according to Tellegen's theorem is expressed in time domain, we will have

$$v_\alpha(t)i'_\alpha(t) + v_\beta(t)i'_\beta(t) + \sum_{k=1}^{N} v_k(t)i'_k(t) = 0.$$

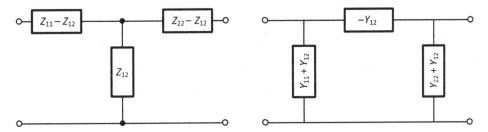

Figure 3.13: Equivalent circuits of any arbitrary reciprocal two-port

In a resistive network (time-variant or not) $v_k(t) = R_k(t)i_k(t)$, and the proof readily follows from this (see also Problem 20).

We shall use this important outcome when describing *passive mixers* in Chapter 8. Since passive mixers are commonly realized by time-varying linear switches, they fall into the category of linear resistive networks, and thus could be treated as reciprocal (at least if the frequency is low enough where the capacitances are ignored).

As a final remark, one can easily show that any reciprocal network may be represented by its π or T equivalent circuits shown in Figure 3.13.

It is interesting to note that the equivalent circuits do not include any dependent sources, while a general two-port represented by its Z or Y matrices does.

3.2 AVAILABLE POWER

An important concept that would be repeatedly used throughout noise discussion and low-noise amplifier design is the available power, and the available power gain. In this section we examine this concept and discuss its key properties in the context of amplifiers as well as passive reciprocal networks such as filters.

3.2.1 Basic Concept

Consider the simple resistive circuit of Figure 3.14. Assume the source is sinusoidal with a peak voltage of $|V_s|$. Shown also is the load line (the solid line representing $V = R_L I$) as well as the source line (the dashed line, representing $V = V_s - R_s I$) on the V–I plane.

The point that the load and source lines intercept is naturally the operation point that corresponds to $(V, I) = \left(\frac{R_L}{R_s + R_L} V_s, \frac{V_s}{R_s + R_L} \right)$.

Shown also on the figure is the power hyperbola representing the $\frac{1}{2} VI$ constant, which for a given load of R_L corresponds to $\frac{1}{2} VI = \frac{R_L}{2(R_s + R_L)^2} |V_s|^2$. The power hyperbola intersects the source line in general at two points $\left((V, I) = \left(\frac{R_L V_s}{R_s + R_L}, \frac{V_s}{R_s + R_L} \right) \text{ and } (V, I) = \left(\frac{R_s V_s}{R_s + R_L}, \frac{\frac{R_L}{R_s} V_s}{R_s + R_L} \right) \right)$,

one of which is naturally the operating point. As the load varies, the operating point moves on the source line, and there is one point where the power hyperbola is tangent to the source line, in which case the hyperbola is at its farthest right. This corresponds to the maximum power at the

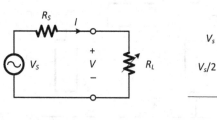
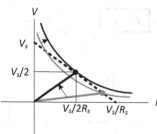

Figure 3.14: Simple circuit to demonstrate the available power concept

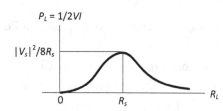

Figure 3.15: Power delivered to the load of the circuit of Figure 3.14 as a function of the load impedance

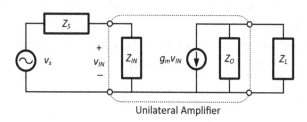

Figure 3.16: A simplified unilateral amplifier model

load and happens when $R_L = R_s$, that is, when the operating point voltage and current are in the middle. The power delivered to the load in this condition is

$$P_a = \frac{|V_s|^2}{8R_s},$$

which is only a function of the source, and not the load. This power is known as the source *available power*, and is the maximum power that the load can extract from the source. In other words, regardless of what the load is, the source cannot deliver any more power, and hence, the name available power. This is graphically illustrated in Figure 3.15.

For very small or large values of load, the power delivered is small, as either the load voltage or its current is very small, and naturally there exists some optimum value somewhere in the middle.

3.2.2 Unilateral Two-Ports

Consider the amplifier shown in Figure 3.16, where the source and load are represented with a more general complex impedances. The amplifier can be expressed in any of the matrices mentioned before, but for the moment we assume it is unilateral, as is the case for most well-designed open-loop amplifiers. We shall represent it here by its input impedance, transconductance gain, and output impedance. A voltage amplifier may be realized by simply using a Thevenin equivalent circuit at the output.

From basic circuit theory [1], the *complex power* delivered from the source to the amplifier input is

$$P = \frac{1}{2}V_{IN}I_{IN}^* = \frac{1}{2}\frac{Z_{IN}|V_s|^2}{|Z_s+Z_{IN}|^2},$$

where V_{IN} and I_{IN} represent the *peak* voltage and current to the amplifier input. Defining $Z_{IN} = R_{IN} + jX_{IN}$, and $Z_s = R_s + jX_s$, the *average power* delivered is

$$P_{avg} = \mathrm{Re}[P] = \frac{1}{2}|V_s|^2\frac{R_{IN}}{(R_s+R_{IN})^2+(X_s+X_{IN})^2},$$

which is maximized when $R_{IN} = R_s$, and $X_{IN} = -X_s$, or when $Z_{IN} = Z_s^*$. This condition is known as power matching, or source *conjugate matching*. The power delivered under such condition, which is the source available power, is

$$P_{a,IN} = \frac{|V_s|^2}{8R_s}.$$

The total power generated by the source is easily shown to be

$$P_s = \frac{|V_s|^2}{4R_s},$$

and thus, under source conjugate matching condition, half of the source average power is delivered to the amplifier, resulting in a power efficiency of 50%.

For cases such as a radar receiver, the input is conjugate matched, otherwise the incoming electromagnetic energy will be lost if not absorbed entirely at the input. On the other hand, if in a certain application power efficiency is of a more importance, conjugate matching is not a desirable condition as half the power produced will be lost. As we will see in Chapter 11, this becomes very important when designing power amplifiers.

Similarly, using Thevenin equivalent at the output, we can define the available output power as

$$P_{a,OUT} = \frac{|g_m Z_o v_{IN}|^2}{8R_o},$$

where $R_o = \mathrm{Re}[Z_o]$ The amplifier *available power gain* is defined as the ratio of the available powers at the output to the input:

$$G_a = \frac{P_{a,OUT}}{P_{a,IN}} = \frac{\frac{|g_m Z_o v_{IN}|^2}{8R_o}}{\frac{|v_s|^2}{8R_s}} = \frac{R_s}{R_o}\left|\frac{Z_{IN}}{Z_s+Z_{IN}}\right|^2|g_m Z_o|^2.$$

For a resistive source, and assuming the amplifier is matched at its input and output ports, $Z_o = Z_{IN} = Z_s^*$, then

$$G_a = \frac{|g_m Z_o|^2}{4}.$$

3.2.3 General Two-Port Available Power Gain

The analysis performed on unilateral amplifiers can be readily extended to any two-port. Consider Figure 3.17, showing a general two-port described by its Z matrix. Problems 9–11 deal with the case of a two-port represented by its Y matrix.

The input impedance of the amplifier is (see Problem 9)

$$Z_{IN} = Z_{11} - \frac{Z_{12}Z_{21}}{Z_{22} + Z_L},$$

which simplifies to Z_{11} if the amplifier is unilateral ($Z_{12} = 0$). The available input power is

$$P_{a,IN} = \frac{|V_s|^2}{8R_s},$$

which is delivered to the two-port input under condition $Z_s = Z_{IN*}$. To find the available power at the output, consider the Thevenin equivalent of the output port depicted in Figure 3.18.

To find the Thevenin impedance, we set the source voltage to zero, and obtain

$$Z_{TH} = Z_{OUT} = Z_{22} - \frac{Z_{12}Z_{21}}{Z_s + Z_{11}}.$$

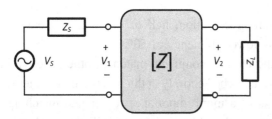

Figure 3.17: A general two-port represented by its Z matrix

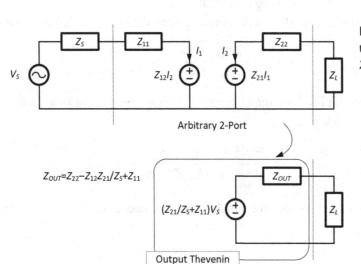

Figure 3.18: Thevenin equivalent of the two-port output based on the Z matrix

The Thevenin voltage is

$$V_{TH} = Z_{21} \frac{V_s}{Z_S + Z_{11}}.$$

Now considering the Thevenin equivalent shown at the bottom of Figure 3.18, the available power delivered to the load impedance Z_L is

$$P_{a,OUT} = \frac{|V_{TH}|^2}{8R_{OUT}} = \left|\frac{Z_{21}}{Z_s + Z_{11}}\right|^2 \frac{|V_s|^2}{8R_{OUT}},$$

where $R_{OUT} = \text{Re}[Z_{OUT}]$.

The available power gain is then

$$G_a = \frac{R_s}{R_{OUT}} \left|\frac{Z_{21}}{Z_s + Z_{11}}\right|^2.$$

Note that the available power gain *is not* a function of the load.

If the amplifier is unilateral,

$$G_a = \frac{R_s}{R_{OUT}} \left|\frac{Z_{21}}{Z_s + Z_{IN}}\right|^2,$$

which is the result obtained previously.

3.2.4 Reciprocal Networks

3.2.4.1 Available Power Gain of Reciprocal Networks

With this background, let us now derive a general expression for the available power gain of reciprocal networks. Consider the RLCM circuit shown in Figure 3.19.

Since the available power gain does not depend on the load, the inclusion of the load impedance will not affect the final result, and thus for simplicity is left out.

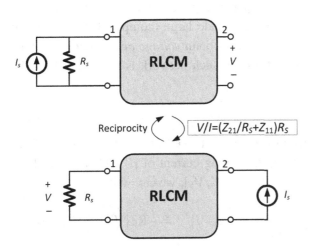

Figure 3.19: The available power gain of a reciprocal network

With the source represented by its Norton equivalent, the open-circuit voltage at the output is

$$V = \frac{R_s}{R_s + Z_{11}} Z_{21} I_s.$$

If the output is excited by the same current I_s, then according to reciprocity the open-circuit voltage observed at the other end must be the same, that is, $V = \frac{R_s}{R_s + Z_{11}} Z_{21} I_s$, as shown in the bottom of Figure 3.19.

Considering Figure 3.19 bottom side, the total power delivered to the output port by the exciting current source I_s is

$$\frac{1}{2}\text{Re}[Z_{OUT}]|I_s|^2 = \frac{1}{2} R_{OUT}|I_s|^2,$$

where $R_{OUT} = \text{Re}[Z_{OUT}]$. This power must be greater than that dissipated in R_s, as the network does not contain any active elements. Hence,

$$\frac{1}{2} R_{OUT}|I_s|^2 \geq \frac{1}{2}\frac{|V|^2}{R_s},$$

but $V = \frac{R_s}{R_s + Z_{11}} Z_{21} I_s$, and consequently

$$R_s \left|\frac{Z_{21}}{R_s + Z_{11}}\right|^2 \leq R_{OUT}.$$

Recall that for an arbitrary two-port in general, the available power gain was shown to be

$$G_a = \frac{R_s}{R_{OUT}} \left|\frac{Z_{21}}{Z_s + Z_{11}}\right|^2.$$

Thus, the available power gain of a reciprocal network is always less than one. Note, this outcome may be derived through an alternative and more general approach presented in Problem 11.

An important observation that may be made is the contrast between *unilateral* and *reciprocal* networks. A unilateral network, and particularly a unilateral amplifier, is highly non-reciprocal, as $Z_{12} = 0$, while Z_{21} is typically made large (compared to the amplifier R_{OUT}) given the gain requirements. Whereas one may obtain *voltage or current gain* in a reciprocal network (say a step-up or -down transformer), such network is never capable of producing *power gain*.

3.2.4.2 Lossless Reciprocal Networks

In the special case of a reciprocal *lossless* N-port, the net average power delivered to the network must be zero. As such, assuming exciting current of $[I] = [I_1, I_2, \ldots, I_N]$ at the ports, and the corresponding voltage of $[V] = [V_1, V_2, \ldots, V_N]$, we can write

$$P_{avg} = \frac{1}{2}\text{Re}\{[V]^t[I]^*\} = \frac{1}{2}\text{Re}\{([Z][I])^t[I]^*\} = \frac{1}{2}\text{Re}\{[I]^t[Z][I]^*\} = 0,$$

where, given reciprocity, we have assumed $[Z] = [Z]^t$. Hence

$$P_{avg} = \frac{1}{2}\text{Re}\left[\sum_{m=1}^{N}\sum_{n=1}^{N} Z_{nm} I_n I_m^*\right] = 0.$$

Since the currents are independent, by setting all the currents except the one at the nth port to zero, we must have the real part of each term inside the summation to be zero, that is, $\text{Re}[Z_{nn} I_n I_n^*] = |I_n|^2 \text{Re}[Z_{nn}] = 0$. Thus,

$$\text{Re}[Z_{nn}] = 0.$$

Similarly, by setting all port currents except for I_n and I_m ($n \neq m$) to be zero, we arrive at

$$\text{Re}[I_n I_m^* Z_{nm} + I_m I_n^* Z_{mn}] = \text{Re}[(I_n I_m^* + I_m I_n^*) Z_{mn}] = 0.$$

However, generally speaking, $(I_n I_m^* + I_m I_n^*)$ is a real quantity that cannot be necessarily zero, and hence

$$\text{Re}[Z_{mn}] = 0.$$

Thus, in a lossless reciprocal network, the Z (or Y) matrixes are symmetric, and imaginary. Evidently, such network may be only constructed by capacitors, as well as self and coupled inductors (or transformers).

From the argument presented in the previous section, it is obvious that a lossless reciprocal network always has an available power gain of one.

3.2.5 Stability of Two-Port Amplifiers

Consider the two-port shown below (Figure 3.20) expressed by its Y parameters. We wish to find a condition that the two-port is unconditionally stable.

The two-port will be *potentially unstable* if the admittance (or impedance) of either port has a negative conductance for a passive termination on the other port. This means that once the source (or load) impedance is added, the real part of the combined impedances could become negative, in which case the amplifier will oscillate.

Consider the expression for the admittance of the input port,

$$Y_{IN} = Y_{11} - \frac{Y_{12} Y_{21}}{Y_{22} + Y_L},$$

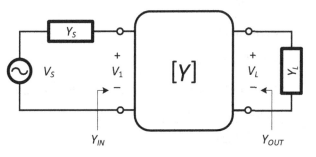

Figure 3.20: Two-port amplifier driven by a source and terminated by a load

where Y_L is the load impedance. Defining

$$Y_{11} = G_{11} + jB_{11}$$
$$Y_{22} = G_{22} + jB_{22}$$
$$Y_L = G_L + jB_L$$
$$Y_{12}Y_{21} = P + jQ,$$

we can rearrange the input admittance expression as

$$Y_{IN} = G_{11} + jB_{11} - \frac{P + jQ}{G_{22} + jB_{22} + G_L + jB_L}.$$

With some manipulation we obtain

$$\text{Re}[Y_{IN}] = \frac{(G_{22} + G_L)^2 + (B_{22} + B_L)^2 - \frac{P}{G_{11}}(G_{22} + G_L) - \frac{Q}{G_{11}}(B_{22} + B_L)}{\frac{(G_{22} + G_L)^2 + (B_{22} + B_L)^2}{G_{11}}}.$$

For the amplifier to be stable, G_{11} must be always positive, otherwise for a sufficiently large Y_L (say shorting the output port), Re $[Y_{IN}]$ becomes negative. A similar reasoning can be made for $G_{22} > 0$. With $G_{11} > 0$, we note that the denominator of Re $[Y_{IN}]$ is always positive, and thus we need to seek the condition where the numerator is always positive. To do so, we rearrange the numerator as

$$(G_{22} + G_L)^2 + (B_{22} + B_L)^2 - \frac{P}{G_{11}}(G_{22} + G_L) - \frac{Q}{G_{11}}(B_{22} + B_L)$$
$$= \left(G_L + \left(G_{22} - \frac{P}{2G_{11}}\right)\right)^2 + \left(B_L + \left(B_{22} - \frac{Q}{2B_{11}}\right)\right)^2 - \frac{P^2 + Q^2}{4G_{11}^2}.$$

The numerator is minimum for the load admittance of $G_L = 0$ and $B_L = -\left(B_{22} - \frac{Q}{2B_{11}}\right)$ (a reactive load). Under this condition, we must have

$$\left(G_{22} - \frac{P}{2G_{11}}\right)^2 > \frac{P^2 + Q^2}{4G_{11}^2},$$

or

$$|2G_{11}G_{22} - P| > \sqrt{P^2 + Q^2}.$$

Defining

$$C = \frac{\sqrt{P^2 + Q^2}}{2G_{11}G_{22} - P},$$

known as the *Linvil stability factor*, we must then have

$$0 < C < 1.$$

(Note that the condition $-1 < C < 0$ cannot be met. Why?) Alternatively, we can write

$$K = \frac{2\text{Re}(Y_{11})\text{Re}(Y_{22}) - \text{Re}(Y_{12}Y_{21})}{|Y_{12}Y_{21}|} > 1.$$

Since the above expression will not change if the input and output ports are switched, it is often used as a general criterion for the two-port *unconditional stability*.

Example: For a given amplifier, $Y = \begin{bmatrix} (2+j2) \times 10^{-3} & (-2-j20) \times 10^{-6} \\ (20-j3) \times 10^{-3} & (20+j60) \times 10^{-6} \end{bmatrix}$. We wish to evaluate the stability of the amplifier.

We form

$$Y_{12}Y_{21} = (-2 - j20) \times 10^{-6} \times (20 - j3) \times 10^{-3} = (-100 - j394) \times 10^{-9}$$
$$= 406 \times 10^{-9} e^{j256}.$$

Thus

$$K = \frac{2\text{Re}(Y_{11})\text{Re}(Y_{22}) - \text{Re}(Y_{12}Y_{21})}{|Y_{12}Y_{21}|} = 0.44 < 1.$$

So the amplifier is potentially unstable.

3.2.6 Maximum Power Gain

Considering the two-port of Figure 3.20, we shall define the power gain as the ratio of the power delivered to the load, to the power appearing at the two-port input,

$$G_p = \frac{P_L}{P_{IN}},$$

where $P_L = \frac{1}{2}\text{Re}[Y_L]|V_L|^2$, and $P_{IN} = \frac{1}{2}\text{Re}[Y_{IN}]|V_1|^2$. Since $V_L = \frac{-Y_{21}}{Y_{22}+Y_L}V_1$, we have

$$G_p = \frac{\text{Re}[Y_L]}{\text{Re}[Y_{IN}]}\left|\frac{V_L}{V_1}\right|^2 = \frac{\text{Re}[Y_L]|Y_{21}|^2}{\text{Re}\left[Y_{11} - \frac{Y_{12}Y_{21}}{Y_{22}+Y_L}\right]|Y_{22}+Y_L|^2}.$$

Clearly the power gain does not depend on the source impedance, and is only a function of the two-port and the load.

Assuming the two-port is unconditionally stable ($K > 1$), we shall find the load admittance for which the power gain is maximized. To do so we must naturally solve for $\frac{\partial G_p}{\partial G_L} = 0$, and $\frac{\partial G_p}{\partial B_L} = 0$.

This leads to

$$G_{L,opt} = \frac{1}{2G_{11}}\sqrt{(2G_{11}G_{22} - P)^2 - (P^2 + Q^2)}$$

$$B_{L,opt} = \frac{Q}{2G_{11}} - B_{22}$$

$$G_{p,MAX} = \frac{|Y_{21}|^2}{2G_{11}G_{22} - P + \sqrt{(2G_{11}G_{22} - P)^2 - (P^2 + Q^2)}}.$$

The optimum load admittance obtained above does not necessarily mean the power delivered to the load is maximized; rather it only guarantees that for a given power delivered to the input (P_{IN}), the power absorbed by the load is maximized. To maximize the power at the load for a given source, we must also have

$$Y_s = Y_{IN}^*,$$

where $Y_{IN} = Y_{11} - \frac{Y_{12}Y_{21}}{Y_{22} + Y_{L,opt}}$ is a function of the load admittance. This leads to the following for the optimum source impedance:

$$G_{s,opt} = \frac{1}{2G_{22}} \sqrt{(2G_{11}G_{22} - P)^2 - (P^2 + Q^2)}$$

$$B_{s,opt} = \frac{Q}{2G_{22}} - B_{11}.$$

The above equations may be recast as the following more concise formulas (see Problems 9–11 as well as [5] for more details):

$$Y_{s,opt} = \frac{Y_{12}Y_{21} + |Y_{12}Y_{21}|\left(K + \sqrt{K^2 - 1}\right)}{2\text{Re}(Y_{22})} - Y_{11}$$

$$Y_{L,opt} = \frac{Y_{12}Y_{21} + |Y_{12}Y_{21}|\left(K + \sqrt{K^2 - 1}\right)}{2\text{Re}(Y_{11})} - Y_{22}$$

$$G_{p,MAX} = \left|\frac{Y_{21}}{Y_{12}}\right| \left(K - \sqrt{K^2 - 1}\right),$$

where K is the stability factor obtained in the previous section.

Under the optimum source and load impedances, the two-port is known to be *biconjugate matched*, that is, the source available power is delivered to the input, and simultaneously the output available power is delivered to the load as well. In other words, if $\begin{cases} Y_s = Y_{s,opt} \\ Y_L = Y_{L,opt} \end{cases}$, then $\begin{cases} Y_s = Y_{IN}^* \\ Y_L = Y_{OUT}^* \end{cases}$. In this case the power gain and the available power gains are equal and are both maximum.

Clearly for the conditions above to hold we must have $K > 1$. If the two-port is potentially unstable, $G_{p,MAX}$ is meaningless, as for a certain load admittance, the power delivered to the input ($P_{IN} = \frac{1}{2}\text{Re}[Y_{IN}]|V_1|^2$) will be zero (since Re $[Y_{IN}]$ can become negative, for some Y_L it must be zero). Therefore, the power gain will become infinite.

Example: The Y parameters of the transistor 2N3783 at 200MHz are as follows:

$$Y = \begin{bmatrix} (20 + j13) \times 10^{-3} & (-0.015 - j0.502) \times 10^{-3} \\ (41.5 - j64) \times 10^{-3} & (0.25 + j1.9) \times 10^{-3} \end{bmatrix}.$$

Thus,

$$Y_{12}Y_{21} = P + jQ = (-32.75 - j19.84) \times 10^{-3}.$$

The stability factor is then found to be

$$K = \frac{2\text{Re}(Y_{11})\text{Re}(Y_{22}) - \text{Re}(Y_{12}Y_{21})}{|Y_{12}Y_{21}|} = 1.116,$$

so it is unconditionally stable. The optimum source and load admittances are

$$Y_{s,opt} = (37.6 - j52.7) \times 10^{-3}$$
$$Y_{L,opt} = (0.47 - j2.41) \times 10^{-3}.$$

We see that $B_{s,opt}$ and $B_{L,opt}$ are negative. This makes sense as the internal capacitances of the transistor must resonate with the optimum source and load inductances to create conjugate matching.

3.2.6.1 Reciprocal and Unilateral Two-Ports

If the two-port is reciprocal, $Y_{21} = Y_{12}$. Thus,

$$G_{p,MAX} = \left|\frac{Y_{21}}{Y_{12}}\right|\left(K - \sqrt{K^2 - 1}\right) = K - \sqrt{K^2 - 1} < 1.$$

In a unilateral two-port, however, $Y_{12} = 0$. Therefore, as long as $\text{Re}(Y_{11})$ and $\text{Re}(Y_{22})$ are positive, the two-port is unconditionally stable. Additionally, with $\begin{cases} Y_s = Y_{11}{}^* \\ Y_L = Y_{22}{}^* \end{cases}$, the power gain is maximized, and

$$G_{p,MAX} = \frac{|Y_{21}|^2}{4G_{11}G_{22}}.$$

3.3 IMPEDANCE TRANSFORMATION

In RF circuits, and particularly RF amplifiers, it is often necessary to transform the input impedance to some desirable value. There are several reasons for this, mainly arising from the fact that the input impedance of the amplifier, often constrained by its performance parameters such as gain or power consumption, may not match what is required say by maximum power transfer or other considerations such as minimum noise figure.[9] As shown in Figure 3.21, there is often a need for an intermediate network, known as the *matching network*, to transform the amplifier input impedance to the optimum value.

---

[9] We will discuss minimum noise figure in Chapter 5.

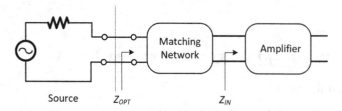

Figure 3.21: Role of matching networks in RF amplifiers

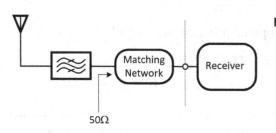

Figure 3.22: External SAW filter connected to a receiver

Another common situation is a receiver often preceded by high-quality external filters that are required to be terminated to 50Ω at their inputs and outputs (Figure 3.22). Typical example of such filters include SAW (surface acoustic wave) filters, that are *electromechanical* devices constructed of a piezoelectric crystal or ceramic (see Chapter 4 for more details). A non-50Ω termination often degrades the filter passband loss, and reduces its stopband attenuation. On the other hand, as we will discuss in Chapter 7, due to noise and power consumption trade-offs, it is desirable to design the amplifier with a different, and often much higher input impedance than the 50Ω needed by the filter. Thus a matching network, so long as it does not contribute much to the cost, becomes very handy.

Matching networks are not unique to the receivers, and are common in transmitter as well, especially the power amplifiers, for similar reasons.

Since the matching network is at the very input of the amplifier and often the entire radio, its performance becomes very critical. For that reason, it is typically built of very low-loss passive components such as high-Q inductors and capacitors. If the quality factor of integrated components is not high enough, it is not uncommon to realize the matching network, or part thereof, using external, higher Q elements. Given its importance, we will spend this section on discussing a few common topologies. Shown in Figure 3.23 is various commonly used matching networks in radios. Their functionality and their properties will be discussed in the next few sections.

3.3.1 Lossless Matching Network Basic Properties

Before we get to the matching network implementation, we will discuss some general properties first.

Consider the circuit in Figure 3.24, consisting of a lossless (LCM) network (representing the matching circuit) terminated to a source and a load. If the source or load have any reactive part, it may be lumped into the LCM network without the loss of generality. V_S represents the RMS value of the source voltage.

3.3 Impedance Transformation

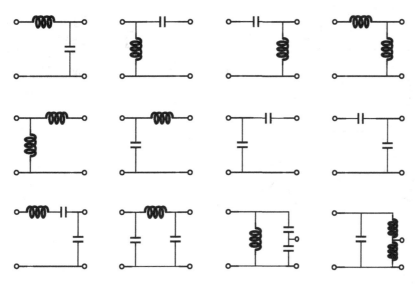

Figure 3.23: Various types of matching circuits commonly used in radios

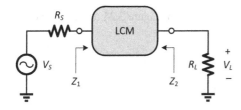

Figure 3.24: A terminated lossless matching network

Assume the impedances looking into the right and left of the lossless circuit are Z_1 and Z_2, respectively, as shown. Excluding the load, the circuit may be represented by its Thevenin equivalent illustrated in Figure 3.25.

To find the open circuit Thevenin voltage V_{OC}, reciprocity dictates that if the current source I_S is applied to the right side of LCM network, it will induce the same open circuit voltage, V_{OC} on the resistor R_S (bottom figure). Given the LCM circuit is lossless, the power conservation demands

$$\text{Re}[Z_2]|I_S|^2 = \frac{|V_{OC}|^2}{R_S},$$

and hence,

$$|V_{OC}|^2 = R_S \text{Re}[Z_2]|I_S|^2 = \frac{\text{Re}[Z_2]}{R_S}|V_S|^2.$$

Therefore we can replace the circuit of Figure 3.24 with the equivalent circuit as follows.

The power delivered to the load, P_L is

$$P_L = \left|\frac{V_{OC}}{R_L + Z_2}\right|^2 R_L,$$

which must be equal to the power delivered to the input of the LCM network of Figure 3.24, P_1, where

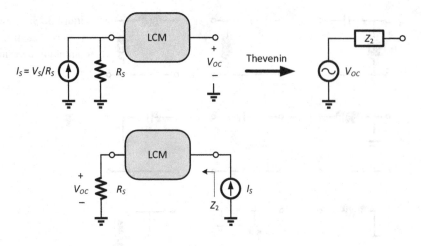

Figure 3.25: The Thevenin equivalent of the source and LCM circuit of Figure 3.24

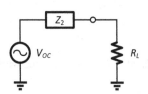

Figure 3.26: The equivalent circuit of Figure 3.24 using the Thevenin circuit found in Figure 3.25

$$P_1 = \left|\frac{V_S}{R_S+Z_1}\right|^2 \operatorname{Re}[Z_1].$$

Thus,

$$\frac{\operatorname{Re}[Z_1]R_S}{|R_S+Z_1|^2} = \frac{\operatorname{Re}[Z_2]R_L}{|R_L+Z_2|^2}.$$

With some simple algebra we arrive at

$$\left|\frac{Z_1-R_S}{Z_1+R_S}\right| = \left|\frac{Z_2-R_L}{Z_2+R_L}\right|.$$

By definition, $\rho_1 \triangleq \frac{Z_1-R_S}{Z_1+R_S}$ is the input *reflection coefficient* (or reflection factor), and $\rho_2 \triangleq \frac{Z_2-R_L}{Z_2+R_L}$ is the output reflection coefficient.[10] Thus, reciprocity, and the lossless nature of the LCM network demand equal magnitude for the input and output reflection coefficients. Furthermore, if one port is matched, say $Z_1 = R_S$, the other port is automatically matched too, that is, Z_2 must be equal to R_L. The latter can be physically explained as follows. If the input is matched, the source available power $\left(\frac{|V_S|^2}{4R_S}\right)$ is absorbed by the lossless LCM network, which is then entirely delivered to the load. Hence, the output must be matched as well.

---

[10] In some books reflection factor is defined as $\rho_1 = \frac{R_S-z_1}{R_S+z_1}$. We will use this definition in the next chapter when we discuss filters.

3.3 Impedance Transformation

For the network of Figure 3.24, we can define the voltage gain as follows:

$$A_V = \frac{V_L}{V_S},$$

which may be greater than one. We may further define

$$\frac{P_a}{P_L} = \frac{\frac{|V_S|^2}{4R_S}}{\frac{|V_L|^2}{R_L}} = \left|\frac{1}{2}\sqrt{\frac{R_L}{R_S}}\frac{1}{A_V}\right|^2,$$

which is always greater or equal to one. $P_a = \frac{|V_S|^2}{4R_S}$ is the source available power. Consequently, the *transducer factor* is defined as

$$H(s) \triangleq \frac{1}{2}\sqrt{\frac{R_L}{R_S}}\frac{1}{A_V(s)}.$$

Given that the matching network is lossless, the power delivered to its input is always equal to the power delivered to the load: $P_1 = P_L$. On the other hand, $P_1 \leq P_a$. The difference,

$$P_r = P_a - P_1,$$

is known as the *reflected power*, that is, the amount of power not delivered to the network when it is not matched. It is easy to show that

$$P_r = |\rho_1|^2 P_a,$$

which can be physically explained knowing that $P_1 = P_a$ only when the input is matched, that is when $|\rho_1| = 0$.

Example: Consider the matching network shown in Figure 3.27 designed to match a load $R_L > R_s$ to the source R_s. The matching circuit is used for impedance downconversion, and the details will be discussed in Section 3.3.3. For now, we would like to find the available power gain and the reflection coefficients.

The LC network Z parameters are

$$Z_{11}(s) = Ls + \frac{1}{Cs}$$
$$Z_{12}(s) = Z_{21}(s) = \frac{1}{Cs}$$
$$Z_{22}(s) = \frac{1}{Cs},$$

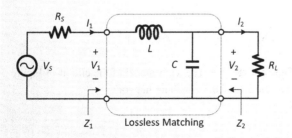

Figure 3.27: LC circuit to match an amplifier to the source

Continued

and the impedances looking into the matching network are

$$Z_1(s) = \frac{R_L + Ls + R_L LCs^2}{1 + R_L Cs}$$

$$Z_2(s) = \frac{R_s + Ls}{1 + R_s Cs + LCs^2}.$$

From this, the reflection coefficients are found to be

$$\rho_1(s) = \frac{Z_1 - R_S}{Z_1 + R_S} = \frac{R_L - R_S + (L - R_S R_L Cs)s + R_L LCs^2}{R_L + R_S + (L + R_S R_L Cs)s + R_L LCs^2}$$

$$\rho_2(s) = \frac{Z_2 - R_L}{Z_2 + R_L} = \frac{R_S - R_L + (L - R_S R_L Cs)s - R_L LCs^2}{R_S + R_L + (L + R_S R_L Cs)s + R_L LCs^2}.$$

Clearly, $|\rho_1(j\omega)| = |\rho_2(j\omega)|$.

Finally, the available power gain is

$$G_a = \frac{R_s}{\text{Re}[Z_2(j\omega)]} \left| \frac{Z_{21}(j\omega)}{R_s + Z_{11}(j\omega)} \right|^2 = \frac{R_s}{\frac{R_s}{(1-LC\omega^2)^2 + (R_s C\omega)^2}} \left| \frac{\frac{1}{jC\omega}}{R_s + jL\omega + \frac{1}{jC\omega}} \right|^2 = 1,$$

as expected.

Although we proved in Section 3.2.4.1 that the available power gain of any LCM network is 1, the results above is worth some further discussion. To arrive at $G_a = 1$ in our example, there was no assumption made on whether the circuit is matched or not. If not matched, only a portion of the source available power ($P_{a,IN}$) reaches the LC circuit, as some ($P_r = |\rho_1|^2 P_{a,IN}$) is reflected back. The astute reader may question then how despite this the available gain is still 1. To answer that, let us consider the Thevenin circuit of the source and the LC network as shown in Figure 3.28.

By definition, the output available power is

$$P_{a,OUT} = \frac{|V_{TH}|^2}{8\text{Re}[Z_2]},$$

and the power delivered to the load is

$$P_L = \frac{|V_2|^2}{2R_L} = \frac{1}{2R_L} \left| \frac{R_L}{R_L + Z_2} V_{TH} \right|^2.$$

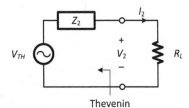

Figure 3.28: The Thevenin model of the circuit of Figure 3.27 looking into the left of the matching network

Since the LC circuit is lossless, the power must be equal to the one delivered to it, that is,

$$P_L = P_1 = (1 - |\rho_1|^2)P_{a,IN}.$$

From this we can write

$$P_{a,OUT} = \frac{|V_{TH}|^2}{8\text{Re}[Z_2]} = \frac{(1 - |\rho_1|^2)P_{a,IN}R_L}{8\text{Re}[Z_2]}\left|\frac{R_L + Z_2}{R_L}\right|^2.$$

With some algebraic steps, we arrive at

$$P_{a,OUT} = P_{a,IN}\frac{1 - |\rho_1|^2}{1 - |\rho_2|^2} = P_{a,IN},$$

since $|\rho_1| = |\rho_2|$ for a lossless circuit. Equivalently,

$$(1 - |\rho_1|^2)P_{a,IN} = (1 - |\rho_2|^2)P_{a,OUT}.$$

While a mismatch at the input leads to less power absorbed by the LC network, and ultimately reaching to the load, the input and output available powers always track since $|\rho_1| = |\rho_2|$.

Example: We shall find the values of L and C in the previous example to match the load to the source.

To do so, $Z_1 = R_S$, or $\rho_1(s) = 0$. This leads to

$$\text{Re}[Z_1] = \frac{R_L}{1 + (R_L C\omega)^2} = R_s$$

$$\text{Im}[Z_1] = L\omega - \frac{R_L^2 C\omega}{1 + (R_L C\omega)^2} = 0,$$

which results in $C = \frac{1}{R_L\omega}\sqrt{\frac{R_L}{R_S} - 1}$, and $L = \frac{R_S}{\omega}\sqrt{\frac{R_L}{R_S} - 1}$. These are frequency dependent, meaning that the matching works only at one specific frequency or is narrowband in general. Also, this type of matching network can only reduce or *downconvert* the load impedance. More details to follow in Section 3.3.3.

Our previous definition of the reflection coefficient assumes a real reference impedance (e.g. $\rho_1 = \frac{Z_1 - R_S}{Z_1 + R_S}$, where R_S is real). As we will discuss in Section 3.4, it is also consistent with the transmission line reflection coefficient definition, as most practical (i.e., lossless or low-loss) transmission lines have a real characteristic impedance. It is however worthwhile discussing the situation where the source impedance is complex.

Consider Figure 3.29, where a source with a complex impedance of Z_S is attached to a two-port with an input impedance of Z_1.

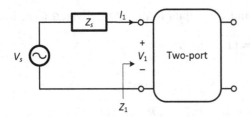

Figure 3.29: Two-port attached to an arbitrary complex source impedance

The average power delivered to the two-port is

$$P_1 = \text{Re}\left[\frac{1}{2}V_1 I_1^*\right] = \frac{1}{2}\text{Re}[Z_1]\frac{|V_s|^2}{|Z_S+Z_1|^2}.$$

The source available power is

$$P_a = \frac{|V_s|^2}{8\text{Re}[Z_s]}.$$

So the reflected power is

$$P_r = P_a - P_1 = \frac{|V_s|^2}{8\text{Re}[Z_s]}\left(1 - \frac{4\text{Re}[Z_s]\text{Re}[Z_1]}{|Z_S+Z_1|^2}\right) = P_a \frac{|Z_1 - Z_s^*|^2}{|Z_1+Z_s|^2}.$$

Consistent with our definition of the reflection coefficient, one may say

$$\rho_1 = \frac{Z_1 - Z_s^*}{Z_1 + Z_S}.$$

This is also consistent with the fact that the source available power is absorbed entirely by the two-port only if $Z_1 = Z_s^*$.

3.3.2 Wideband Transformers

A transformer may be used to provide wideband impedance transformation. For an ideal transformer, the i–v relation,

$$\frac{v_1(t)}{v_2(t)} = \frac{n_1}{n_2},$$

and

$$\frac{i_1(t)}{i_2(t)} = -\frac{n_2}{n_1},$$

may be exploited for impedance conversion. Shown in Figure 3.30, it is clear that $Z_{IN}(j\omega) = \left(\frac{n_1}{n_2}\right)^2 Z_L(j\omega)$.

As we showed in Chapter 1, in a practical transformer, if the coupling factor is close to one, then the transformer can be represented by the equivalent model shown on the left of

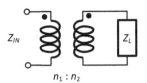

Figure 3.30: Impedance transformation using an ideal transformer

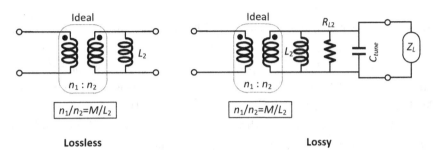

Figure 3.31: Integrated transformer equivalent model when $K \approx 1$

Figure 3.31. The transformer loss may be also modeled by a parallel resistor at the secondary as shown on the right side of Figure 3.31.

Thus, a practical transformer can still upconvert or downconvert the impedance depending on the primary to secondary turn ratios, but the self-inductances of the coils appear as well. If needed, they may be tuned out as shown in Figure 3.31, but this reduces the bandwidth. The capacitances that were ignored certainly lead to additional unwanted reactive components.

Example: As a related useful case study, let us consider the circuit of Figure 3.32, known as a *double-tuned circuit*, consisting of two coupled identical RLC circuits.

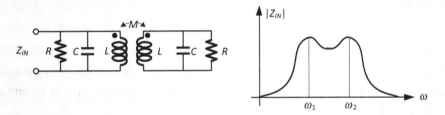

Figure 3.32: A double-tuned circuit based on coupled inductors

By performing a simple node analysis, we can show that

$$Z_{IN} = \frac{R}{2}\left[\frac{1}{1+jQ_1\left(\frac{\omega}{\omega_1}-\frac{\omega_1}{\omega}\right)} + \frac{1}{1+jQ_2\left(\frac{\omega}{\omega_2}-\frac{\omega_2}{\omega}\right)}\right],$$

where $\omega_{1/2} = \frac{1}{\sqrt{LC(1\pm k)}}$, $Q_{1/2} = RC\omega_{1/2}$, and $k = \frac{M}{L}$ is the coupling factor. Thus, the circuit effectively consists of two parallel RLC circuit in series, one slightly high-tuned, and the

Continued

other one slightly low-tuned, depending on the value of k. A plot of the magnitude of Z_{IN} is also shown in Figure 3.32. For more details, see Problem 4.

Example: We wish to match a resistance of 250Ω modeling the input resistance of an amplifier to the 50Ω source using the 1-2 transformer of Figure 1.57. The primary had a simulated inductance of 0.44nH and a Q of 7 at 5.5GHz, and the secondary inductance was 1.14nH with a Q of 12.5 at 5.5GHz. The coupling factor was found to be about 0.8 at 5.5GHz leading to a mutual inductance of about $M = 0.627$nH. A simplified model of the transformer as developed previously is shown in Figure 3.33. The effective inductance at primary, $L_1 - \frac{M^2}{L_2} = L_1(1-k^2)$, comes out to be around 0.158nH, which we have ignored. Additionally, the resistive part at primary, which is $\frac{L_1 \omega_0}{Q_1} = 2.2\Omega$, contributing about a tenth of a dB loss, is ignored as well, Finally, $R_L = 250\Omega$ represents the amplifier input impedance, and R_2 captures the transformer loss.

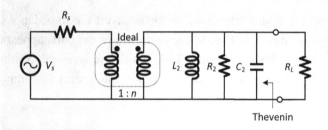

Figure 3.33: Simplified model of the transformer along with parasitic elements

The transformer ratio, n, is

$$n = \frac{L_2}{M} = \sqrt{5},$$

which is suitable to match $R_L = 250\Omega$ to $R_s = 50\Omega$. The capacitance C_2 is a combination of the transformer parasitic, potentially the amplifier input capacitive part, and enough additional part to resonate with L_2 at 5.5GHz.

Shown in Figure 3.34 is the simulated input reflection as well as the available power gain to the amplifier input. Judging from the reflection coefficient, clearly the transformer provides a good match to the source. The resonance created by C_2 makes the matching somewhat narrow, though still a reflection factor of less than −10dB over 1GHz is covered.

To calculate the available gain, we find the Thevenin equivalent to the left of the amplifier:

$$R_{TH} = n^2 R_s \| R_2$$

$$V_{TH} = n \frac{\frac{R_2}{n^2}}{R_s + \frac{R_2}{n^2}} V_s = \frac{nR_2}{R_2 + n^2 R_s} V_s.$$

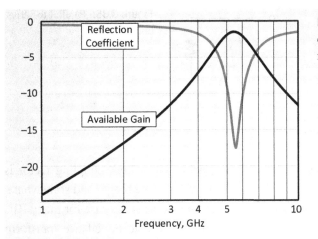

Figure 3.34: Insertion loss and reflection coefficient of the transformer in Chapter 1 matching a 250Ω load

Therefore, the available power gain is found to be

$$G_a = \frac{\frac{|V_{TH}|^2}{4R_{TH}}}{\frac{|V_s|^2}{4R_s}} = \frac{1}{1+\frac{n^2 R_s}{R_2}}.$$

Or equivalently the *insertion loss* is

$$IL = \frac{1}{G_a} = 1 + \frac{n^2 R_s}{R_2} = 1 + \frac{R_L}{R_2}.$$

This is rather expected, as the source power delivered to the right of the transformer is basically divided between the amplifier (R_L), and the undesirable resistor (R_2) modeling the transformer loss.

Two important conclusions can be drawn: First, with no transformer loss, $R_2 = \infty$ and the insertion loss approaches 0dB, which is rather obvious. Second, and more importantly, for a given transformer loss, the higher the impedance transformation ratio, the larger the insertion loss. This is has very important implications in design of matching networks for the low-noise or power amplifier amplifiers.

The resistance R_2 is estimated to be

$$R_2 = L_2 \omega_0 Q_2 = 604\Omega,$$

which leads to an insertion loss of about 1.5dB at the resonance frequency of 5.5GHz and agrees well with the simulation. As mentioned, the loss of primary, which was ignored, adds about another 0.1–0.2dB to that.

Evidently, the matching network provides some moderate bandpass filtering as well, which is usually very desirable to attenuate the unwanted signals.

3.3.3 Parallel–Series Circuits

Although a transformer can ideally provide a wideband impedance transformation, practical integrated transformers are somewhat narrowband due to finite self-inductance, and parasitic

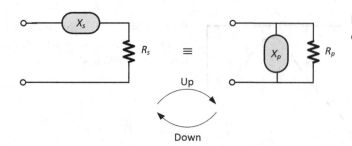

Figure 3.35: Parallel to series impedance conversion

capacitances associated with them. Additionally, achieving a high coupling factor is typically challenging. Finally, practical integrated transformers tend to be bigger and somewhat more lossy than the integrated inductors. For narrowband applications, as is the case for many RF standards, one could use lumped inductors and capacitors to approximate impedance transformation at a single frequency, and possibly over a reasonable bandwidth around that. A convenient approach to do so is through parallel to series impedance transformation, as shown in Figure 3.35.

Let us find the input impedance of the parallel circuit on the right

$$Z_{IN} = \frac{R_p(jX_p)}{R_p + jX_p} = R_p \frac{X_p^2}{R_p^2 + X_p^2} + jX_p \frac{R_p^2}{R_p^2 + X_p^2},$$

which is of the form of a resistance in series with a reactance. Since X_p is frequency dependent, we expect the series resistance and reactance be frequency dependent as well. However, as long as we are looking at only one frequency or a narrow band, we can represent them by the series equivalent circuit shown on the left. To satisfy equivalency, we must have

$$R_s = R_p \frac{X_p^2}{R_p^2 + X_p^2} \quad \text{and} \quad X_s = X_p \frac{R_p^2}{R_p^2 + X_p^2}.$$

Alternatively, we can express the parallel circuit in terms of the series one:

$$R_p = \frac{R_s^2 + X_s^2}{R_s} \quad \text{and} \quad X_p = \frac{R_s^2 + X_s^2}{X_s}.$$

We can also show that the following identity must be met:

$$R_s R_p = X_s X_p.$$

The series RL circuit that was discussed in Chapter 1 to model the induction loss is a special case where $R_s = r$, and $X_s = L\omega$. Since $Q = \frac{L\omega}{r}$, we have

$$R_p = r(1 + Q^2) \quad \text{and} \quad X_p = L\omega\left(1 + \frac{1}{Q^2}\right) \approx L\omega,$$

which we had shown previously in Chapter 1.

The property of series–parallel circuits just shown could be exploited to change the real part of an amplifier input impedance. If the amplifier real part turns out to be larger than what is needed for example, then placing a shunt reactance will reduce that to the desirable value, or vice versa. The remaining reactive part can be always tuned out by placing the proper reactive component with the opposite polarity as long as we are operating at a specific frequency or around a narrow bandwidth.

Example: Consider Figure 3.36, where we wish to match the input impedance of an amplifier with $R_{IN} > R_s$, to the source resistance, R_s at a given frequency of ω_0.

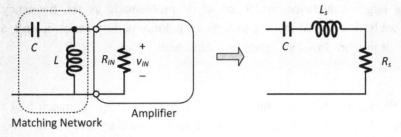

Amplifier
Matching Network

Figure 3.36: An LC circuit to match $R_{IN} > 50\Omega$ to 50Ω

Since we like to reduce the resistance, we need to insert a parallel reactance, which could be either an inductor, or a capacitor. We choose to place an inductor with the value of L in parallel with R_{IN}. Since the equivalent series network will be inductive, we have to naturally place a capacitor C in series to absorb the inductance. Thus our matching network consists of a parallel L and a series C, as shown in Figure 3.36. The parallel network consisting of R_{IN} and L could be converted to a series one, where the new resistance must be equal to the source resistance R_s, and the series inductance L_s to be tuned out by C. Thus

$$R_s = R_{IN} \frac{(L\omega_0)^2}{R_{IN}^2 + (L\omega_0)^2},$$

which leads to $L = \frac{R_{IN}}{\omega_0}\sqrt{\frac{R_s}{R_{IN}-R_s}}$. For this to work, R_{IN} must be obviously greater than R_s. If smaller, we should have chosen a series L and a shunt C for instance. The new series inductance can be also calculated easily:

$$L_s = L\frac{R_{IN}^2}{R_{IN}^2 + (L\omega_0)^2} = \frac{\sqrt{R_s(R_{IN}-R_s)}}{\omega_0},$$

which must resonate with C at ω_0. Thus

$$C = \frac{1}{L_s\omega_0^2} = \frac{1}{\omega_0\sqrt{R_s(R_{IN}-R_s)}}.$$

The matching network components are clearly a function of ω_0, and thus frequency dependent. To have a sense of how much the frequency can deviate from ω_0, let us calculate the quality factor of the series RLC circuit:

$$Q = \frac{1}{R_sC\omega_0} = \sqrt{\frac{R_{IN}}{R_s}-1}.$$

Since as we showed before Q is an indication of the 3dB bandwidth for an RLC circuit, we have

$$\omega_{3dB} = \frac{\omega_0}{Q} = \omega_0\sqrt{\frac{R_s}{R_{IN}-R_s}}.$$

As a rule of thumb, we can say that the circuit provides a reasonable match as long as we are within the bandwidth given above. An important conclusion as a direct outcome of the equation above is an upper limit for how big of an input impedance can be realistically matched. The higher the R_{IN}, the larger the Q, and the narrower the matching network becomes. Also, a larger R_{IN} implies a larger inductance, which could be problematic as the frequency increases. Furthermore, we have assumed that the inductance is lossless. In practice, it isn't, and a larger R_{IN} would result in more loss for a given inductor quality factor.

> **Example:** For $R_{IN} = 250\Omega$, we obtain $L = 10\text{nH}$ and $C = 0.8\text{pF}$ to match 50Ω at 2GHz. The corresponding Q is equal to 2. Using the same matching network but operating at 2.5GHz (the edge of the 3dB bandwidth), the impedance obtained is $70 + j33\Omega$, which may be marginally acceptable for many applications.

If the input impedance of the amplifier had some reactive component in addition to R_{IN}, then all we had to do was to modify the shunt L to absorb that reactance. The remaining steps would have been the same.

The matching network of Figure 3.36 has several other features beyond the impedance transformation. First, even though a lossless network as the one discussed here does not impact the power, it is indeed capable of providing either voltage or current gain. Let us first find the voltage at the input of the amplifier across the resistor R_{IN},

$$\frac{v_{IN}}{v_s} = -\frac{\frac{R_{IN}}{R_{IN}-R_s}\left(\frac{\omega}{\omega_0}\right)^2}{1 - \frac{R_{IN}+R_s}{R_{IN}-R_s}\left(\frac{\omega}{\omega_0}\right)^2 + 2j\sqrt{\frac{R_s}{R_{IN}-R_s}}\frac{\omega}{\omega_0}},$$

which is a highpass function, and whose magnitude peaks to $\frac{1}{2}\sqrt{\frac{R_{IN}}{R_s}}$ at $\omega = \omega_0$, the center frequency where the matching network components were originally chosen. Thus with $R_{IN} > R_s$, the *available voltage gain* from the source to the amplifier input is $\sqrt{\frac{R_{IN}}{R_s}} \approx Q > 1$. The higher the Q, the narrower the response, and the higher the voltage gain. This is important, as it helps reduce the amplifier power consumption for a given input referred noise allowed. However, the presence of the voltage gain from the source to the input implies that the unwanted signals will be also amplified along with the desired input, making the design more susceptible to nonlinearity and distortion.

Shown in Figure 3.37 is the available voltage gain from the source to the input versus the frequency for $R_{IN} = 250\Omega$ and $f_0 = 2\text{GHz}$, for which the components were calculated before. For frequencies lower than f_0 and outside the network bandwidth, the matching circuit acts like a filter, thus attenuating the unwanted signals that fall outside the bandwidth. The presence of a series capacitor and a shunt inductor makes the matching network highpass. Thus, the attenuation at frequencies above the center is not much as the transfer function flattens out to $\frac{2R_{IN}}{R_{IN}+R_s}$ (the extra factor of two is to account for *available* voltage gain). If needed, however, one could choose other types of matching components that behave as lowpass or bandpass.

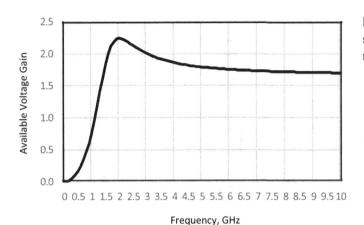

Figure 3.37: Transfer function from the source to the input for the matching network of Figure 3.36

Example: We wish to find the appropriate biconjugate matching network for the example of Section 3.2.6. The complete schematic of the amplifier along with the matching networks details are shown in Figure 3.38. Resistors R_{B1} and R_{B2} are for biasing purposes, and are large enough not to affect the transistor Y parameters. Similarly, C_E is assumed to create a short at the frequency of interest (200MHz).

We had already found

$$Y_{s,opt} = (37.6 - j52.7) \times 10^{-3}$$

$$Y_{L,opt} = (0.47 - j2.41) \times 10^{-3}.$$

Assuming $R_s = 50\Omega$ for the source, at the input we insert a shunt inductance (L_1) to reduce the source resistance to the desired value of 9Ω (note that $Z_{s,opt} = 9 + j12.6\Omega$). The resultant reactive part is too inductive, and capacitor C_1 is used to reduce to the required reactance of 12.6Ω. From the previous example,

$$X_{L1} = R_s \sqrt{\frac{R_{s,opt}}{R_s - R_{s,opt}}} = 23.4\Omega,$$

which leads to $L_1 = 18.6\text{nH}$. The corresponding reactive part will be $\sqrt{R_{s,opt}(R_s - R_{s,opt})} = 19.2\Omega$. Since what is required is 12.6Ω, we find

$$C_1 = \frac{1}{2\pi \times 200 \times 10^6 \times (19.2 - 12.6)} = 120\text{pF}.$$

Note that if the required $X_{s,opt}$ were greater than 19.2Ω, we could not have achieved the required matching with a series capacitance, and instead we had to use a series inductance along with the original shunt inductance (L_1).

Continued

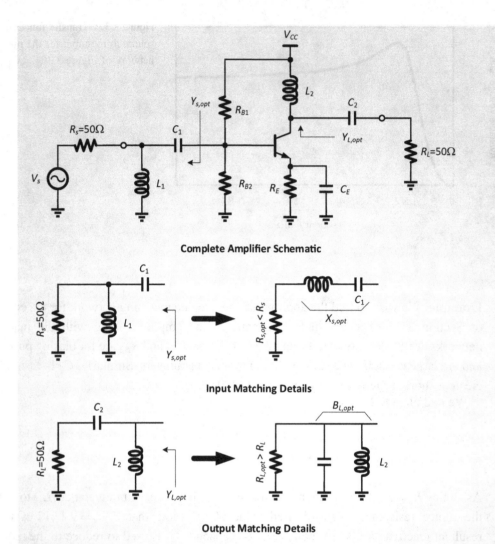

Figure 3.38: 2N3783 amplifier with source and load matching circuits

The input matching network effective quality factor is

$$Q = \left|\frac{B_{s,opt}}{2G_{s,opt}}\right| = 0.7,$$

leading to a bandwidth of about 280MHz at the input.

For the output matching, we choose a series capacitance C_2 to raise the load resistance (which is 50Ω) to the desired value $\frac{1}{0.47 \times 10^{-3} S} = 2.13\text{k}\Omega$ (note that $Z_{L,opt}$ consists of 2.13kΩ resistance in parallel with a 415Ω reactance). The inductance L_2 then creates the required reactive part. The capacitance will be given by

$$C_1 = \frac{1}{\omega\sqrt{R_L(R_{L,opt} - R_L)}} = \frac{1}{2\pi \times 200 \times 10^6 \times 323} = 2.45\text{pF}.$$

The reader can verify that $L_2 = 144\text{nH}$. The output quality factor is found to be 2.56, and the effective bandwidth at the output would be 78MHz.

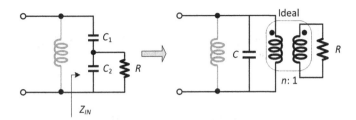

Figure 3.39: Narrowband transformer-like matching circuit

3.3.4 Narrowband Transformers

There is yet a third common method of transforming impedances that approximates an ideal transformer with only capacitors or inductors, shown in Figure 3.39. Consequently, like the previous circuit it is inherently narrowband. Let us find the impedance looking into the circuit shown on the left, ignoring the inductance for the moment.

We have

$$Y_{IN} = \frac{jC_1\omega(1+jRC_2\omega)}{1+jR(C_1+C_2)\omega}.$$

Assuming the capacitive loss due the parallel resistor R is small, or equivalently $|RC_{1/2}\omega| \gg 1$ at the frequency of interest, then we can write

$$Y_{IN} \approx \frac{1}{(R(C_1+C_2)\omega)^2}jC_1\omega(1+jRC_2\omega)(1-jR(C_1+C_2)\omega),$$

which simplifies to

$$Y_{IN} \approx \frac{1}{R\left(1+\frac{C_2}{C_1}\right)^2} + j\frac{C_1C_2}{C_1+C_2}\omega\left[1+\frac{1}{C_2(C_1+C_2)(R\omega)^2}\right].$$

Defining $n = 1 + \frac{C_2}{C_1}$, and $C = \frac{C_1C_2}{C_1+C_2}$, and ignoring the last term assuming the loss is moderate or small, we have

$$Y_{IN} \approx \frac{1}{n^2R} + j\frac{C_1C_2}{C_1+C_2}\omega,$$

and thus it simplifies to the circuit shown on the right side, consisting of an ideal transformer with a coil ratio of n and a net shunt C. The inductor may be used to tune out the equivalent capacitance C, and provide a resistive component.

The transformer may have been realized by two series inductances L_1 and L_2, a shunt capacitance to provide the resonance [6]. However, since it requires two inductors, it is not as common as the one already shown. The arrangement shown in Figure 3.39 is commonly used in Colpitts oscillators (Chapter 9).[11]

---

[11] Edwin Colpitts (1872–1949) was an American engineer.

Example: We shall show that the two circuits below (Figure 3.40) are equivalent, where the circuit on the left illustrates an inductive narrowband transformer.

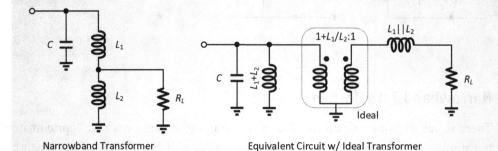

Figure 3.40: Inductive narrowband transformer and its equivalent circuit with an ideal transformer

The steps are self-explanatory, and shown in Figure 3.41 (see also Problems 2 and 3). If $R_L \gg (L_1 \| L_2)\omega$, then the ideal transformer upconverts R_L to the proper value with the transformation ratio of $n^2 = \left(1 + \frac{L_1}{L_2}\right)^2$. $L_1 + L_2$ at the input must resonate with C.

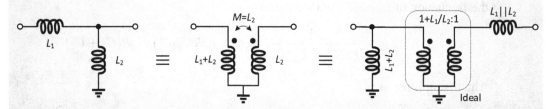

Figure 3.41: Steps to simplify the narrowband transformer of Figure 3.40

3.4 LOSSLESS TRANSMISSION LINES

In addition to lumped LC elements, transmission lines may be incorporated to provide matching. In Chapter 1 we showed that in a lossless transmission line the general solution appears to be of the form

$$v(z,t) = f_1\left(t - \frac{z}{v}\right) + f_2\left(t + \frac{z}{v}\right) = v^+ + v^-,$$

where f_1 and f_2 are arbitrary functions, signifying the forward and backward propagations. Now suppose we are interested only in the sinusoidal steady state solution, where a signal with a specific frequency of $f = \omega/2\pi$ is of our interest. We expect the solution to be sinusoidal as well, that is,

$$v_{b/f}(z,t) = |V_0| \cos(\omega t \pm \beta z + \phi),$$

where $\beta = \frac{\omega}{v}$ is the *phase constant* (in rad/m), and v is the *phase velocity* (in m/s). Same as before, we expect the $+$ to show the backward propagation, whereas the $-$ to correspond to the forward propagation of the signal, denoted by the indexes b and f, respectively. Choosing $\phi = 0$ for now, if we were to fix the time at $t = 0$, the signal becomes

$$v_{b/f}(z,t) = |V_0|\cos(\beta z).$$

Evidently, β signifies the *spatial* frequency. Defining wavelength $\lambda = \frac{2\pi}{\beta} = \frac{v}{f}$, we note that the function above repeats every integer increment of λ. In fact, for the forward propagating waveform, setting the condition

$$\omega t - \beta z = \omega(t - z/v) = 2\pi m,$$

the waveform at a given point of time is kept constant. With increasing time, z must also increase in the positive direction, and at a rate of v. A similar argument can be made for the backward wave, although z in this case needs to decrease.

For any sinusoidal steady state condition we could use phasor to describe the forward and backward signals. For the lossless transmission line, the original wave equation derived in Chapter 1,

$$\frac{\partial^2 v}{\partial z^2} = LC\frac{\partial^2 v}{\partial t^2},$$

then becomes

$$\frac{\partial^2 V}{\partial z^2} = -\omega^2 LCV,$$

in phasor form in the case of steady state sinusoidal, where V denotes the complex phasor voltage, and has a solution in the form of

$$V(z) = V_0^+ e^{-j\beta z} + V_0^- e^{+j\beta z},$$

consistent with our previous results. The time dependence is dropped, as we know the waves will always be in the form of a cosine with a frequency of ω. Note that the line is assumed to be lossless. If not, $j\beta$ must be replaced by $\gamma = \alpha + j\beta$, where α captures the loss of the line [3], [7]. Similarly,

$$I(z) = I_0^+ e^{-j\beta z} + I_0^- e^{+j\beta z}.$$

We shall use the above two general equations for the voltage and current when dealing with the transmission lines. By virtue of the wave differential equation, the two equations below always hold as well:

$$I_0^+ = \frac{V_0^+}{Z_0}$$

$$I_0^- = -\frac{V_0^-}{Z_0},$$

where Z_0 is the line characteristic impedance.

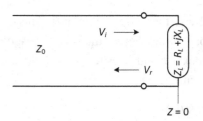

Figure 3.42: Voltage reflected from a complex load

3.4.1 Terminated Transmission Lines

Any transmission line is inevitably terminated to some load. The need to satisfy all voltages and currents *boundary conditions* at discontinuities, e.g., a load, results in *reflected* waves. The basic reflection problem is shown in Figure 3.42.

For convenience, let us assume that the load is located at $z=0$, thus the rest of the line will be at $z<0$. Suppose we have an incident voltage in the phasor format:

$$V_i(z) = V_0^+ e^{-j\beta z}.$$

We expect when the wave reaches the load, it creates a reflected wave propagating backward:

$$V_r(z) = V_0^- e^{+j\beta z}.$$

At the load where $z=0$, we have

$$V_L = V_0^+ + V_0^-,$$

where V_L is the load voltage. The load current is

$$I_L = \frac{1}{Z_0}(V_0^+ - V_0^-) = \frac{V_L}{Z_L} = \frac{V_0^+ + V_0^-}{Z_L}.$$

We can solve for V_0^+ and V_0^-, and more importantly, their ratio, known as the *reflection coefficient*:

$$\Gamma = \frac{V_0^-}{V_0^+} = \frac{Z_L - Z_0}{Z_L + Z_0}.$$

The reflection coefficient is a complex number in general, and has a similar format compared to the one derived in Section 3.3.1 for lumped circuits. Knowing the incident and reflected voltages and currents, we can evaluate the power associated with each as well. We can show that the ratio of reflected to incident power is

$$\frac{P_r}{P_i} = \Gamma\Gamma^* = |\Gamma|^2,$$

again consistent with our derivation earlier.

3.4.2 Voltage Standing Wave Ratio

It is instructive to monitor the signal in a terminated transmission line at different points. In practice, this may be accomplished by inserting a voltage probe in a *slotted* transmission line to measure the magnitude of the voltage at a desired point. We have already expressed the voltage phasor in a transmission line in the general form of

$$V(z) = V_0^+ e^{-j\beta z} + V_0^- e^{+j\beta z} = V_0 e^{-j\beta z} + \Gamma V_0 e^{+j\beta z},$$

where $\Gamma = |\Gamma| e^{j\phi}$ is the reflection coefficient we just obtained as function of the load impedance. After a few algebraic steps, the equation above can be expanded, and expressed as follows:

$$V(z) = V_0(1 - |\Gamma|) e^{-j\beta z} + 2V_0|\Gamma| e^{j\phi/2} \cos(\beta z + \phi/2).$$

Converting from phasor format above to a time domain signal,

$$v(z,t) = \text{Re}\,[V(z)e^{j\omega t}] = V_0(1 - |\Gamma|)\cos(\omega t - \beta z) + 2V_0|\Gamma|\cos(\beta z + \phi/2)\cos(\omega t + \phi/2).$$

The first term has the form of $\cos(\omega t - \beta z)$, and we expect it to propagate in the forward z direction. Hence, we call it the *traveling wave*, with an amplitude of $(1 - |\Gamma|)V_0$. On the other hand, the second term is known as *standing wave*, whose amplitude is $2V_0|\Gamma|$. As we move along the line in the z direction, we expect the two terms to add or subtract, and hence result in different probe readings. Evidently, the maximum amplitude observed in the line is $(1 + |\Gamma|)V_0$, when the standing and traveling waves add constructively. Finding the minimum is not as obvious. Let us first make the observation that since $V(z) = V_0 e^{-j\beta z} + V_0|\Gamma|e^{j\phi}e^{+j\beta z}$, the minimum occurs when the two terms have 180° of phase shift, that is, when $z = -\frac{1}{2\beta}(\phi + (2n+1)\pi)$. In that case the minimum amplitude would be $(1 - |\Gamma|)V_0$. Although we already calculated the maximum amplitude, through a similar deduction we can find the location that it occurs to be $z = -\frac{1}{2\beta}(\phi + 2n\pi)$. The results are depicted in Figure 3.43.

The distance between each consecutive peak is $\lambda/2$, whereas the distance between a peak and the neighboring valley is $\lambda/4$. The ratio of the maximum to minimum voltage observed in the line is called the *voltage standing wave ratio*, and is known as VSWR in short. From our analysis then we have

$$VSWR = \frac{1 + |\Gamma|}{1 - |\Gamma|}.$$

If the load is matched, there is no reflection and VSWR is one. This implies that there is no standing wave in the line. On the other extreme, if the load is a short or an open circuit, $|\Gamma| = 1$, and VSWR is infinite.

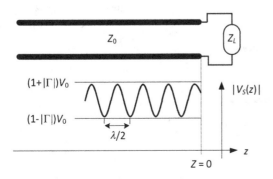

Figure 3.43: Magnitude of the voltage in the transmission line

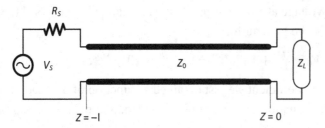

Figure 3.44: Finite-length transmission line

3.4.3 Transmission Line Input Impedance

Let us consider Figure 3.44, which shows a transmission line of finite length of l. We would like to find the impedance at a given point of the line knowing that it is terminated at $z = 0$ with a complex impedance of Z_L.

This can be done simply by finding the voltage and current phasors in the line, that is,

$$V(z) = V_0^+ e^{-j\beta z} + V_0^- e^{+j\beta z}$$

$$I(z) = I_0^+ e^{-j\beta z} + I_0^- e^{+j\beta z} = \frac{1}{Z_0}\left(V_0^+ e^{-j\beta z} - V_0^- e^{+j\beta z}\right).$$

Considering that $V_0^- = \Gamma V_0^+$, we have the impedance at an arbitrary point:

$$Z(z) = Z_0 \frac{e^{-j\beta z} + \Gamma e^{+j\beta z}}{e^{-j\beta z} - \Gamma e^{+j\beta z}} = Z_0 \frac{Z_L \cos(\beta z) - jZ_0 \sin(\beta z)}{Z_0 \cos(\beta z) - jZ_L \sin(\beta z)}.$$

As a sanity check, we can see that at $z = 0$, $Z(0) = Z_L$. At the source side of the line, where $z = -l$, the input impedance looking into the line is

$$Z_{IN} = Z_0 \frac{Z_L \cos(\beta l) + jZ_0 \sin(\beta l)}{Z_0 \cos(\beta l) + jZ_L \sin(\beta l)}.$$

There are interesting properties associated with the equation above. For instance, if the line length is equal to half the wavelength or any integer multiple of that, the input impedance is always equal to the load impedance. If the length is a quarter wavelength, however, $Z_{IN} = \frac{Z_0^2}{Z_L}$ instead. Hence, a short appears as an open circuit on the other end, and vice versa.

Example: Consider the lossless transmission line in Figure 3.45 with $Z_0 = 50\Omega$, terminated to two equal loads of $R_L = 50\Omega$. The source is a 2GHz sinewave with a

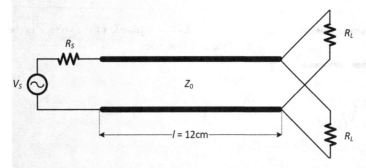

Figure 3.45: A 50Ω transmission line connected to two equal 50Ω loads

magnitude of 1V RMS, and impedance of $R_s = 50\Omega$. Assuming $v = 3 \times 10^8$m/s, then $\lambda = 15$cm, and $\beta l = 2\pi \frac{l}{\lambda} = 1.6\pi$.

Since the load impedance is effectively 25Ω, the reflection coefficient is $\Gamma = -\frac{1}{3}$, and the VSWR is 2. The line input impedance is

$$Z_{IN} = 50 \frac{25\cos(1.6\pi) + j50\sin(1.6\pi)}{50\cos(1.6\pi) + j25\sin(1.6\pi)} = 85e^{-j24°}\Omega,$$

which is capacitive. Physically, this means that the line stores more energy in its electric field than its magnetic field. The current going into the line is $\frac{V_s}{Z_{IN}+R_s} = 7.6e^{j15°}$mA, and the power delivered to the line is $\text{Re}[Z_{IN}]\left|\frac{V_s}{Z_{IN}+R_s}\right|^2 = 4.4$mW. Since the line is lossless, this power is split between the two load resistors, so each one absorbs 2.2mW, corresponding to a load voltage of $\frac{1}{3}$V.

3.4.4 Transmission Lines Transient Response

So far, we have mostly paid attention to the sinusoidal steady state behavior of the transmission lines, which is of course of great value. However, it is often constructive to study the transient behavior of the forward and backward signals in the transmission lines as well, as it allows us to study how the line could be used to store and release energy.

Consider the transmission line of Figure 3.46 of a length l. As a simple case study first, let us assume the source impedance is zero (ideal voltage source), and that the load is matched to the line characteristic impedance Z_0.

Now assuming the input is a step function of value V_0 as shown in Figure 3.47, the incident voltage at the input of the line is $v^+ = V_0$ right after the step signal is enforced. This voltage propagates in the line, and reaches the load after $\frac{l}{v}$ seconds, at which point is entirely absorbed by the matched load, as shown in the figure. Obviously if the length of the line is short, which is to say if the circuit is lumped, this delay is negligible, and we see the output rising to V_0 immediately.

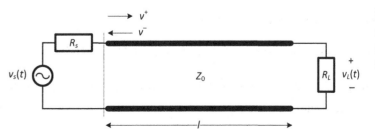

Figure 3.46: A lossless transmission line attached to arbitrary source and load impedances

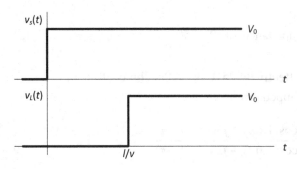

Figure 3.47: Step response of the transmission line of Figure 3.46 with matched termination

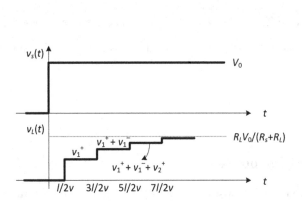
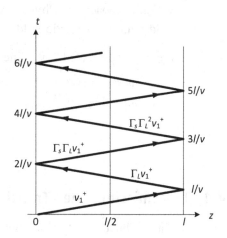

Figure 3.48: Transient response of the transmission line of Figure 3.46 with unmatched load and source

Next, let us assume the source and load are two arbitrary impedances. Upon applying the step voltage, the forward voltage at the line input is

$$v_1^+ = Z_0 i^+ = Z_0 \frac{V_0 - v_1^+}{R_s},$$

or $v_1^+ = \frac{Z_0}{Z_0 + R_s} V_0$, which is a simple voltage division. Once reaching the load after $\frac{l}{v}$ seconds, a reflected voltage (and current) of $v_1^- = v_1^+ \Gamma_L$ is created, where $\Gamma_L = \frac{R_L - Z_0}{R_L + Z_0}$ is the load reflection coefficient. Now this reflected signal makes it to the source side at $t = 2\frac{l}{v}$, at which point a new forward voltage of $v_2^+ = v_1^+ \Gamma_L \Gamma_s$ is created, where $\Gamma_s = \frac{R_s - Z_0}{R_s + Z_0}$ is the source reflection coefficient. Voltage v_1^+ exists everywhere ahead of the v_1^- wave until it reaches the battery, whereupon the entire line now is charged to voltage $v_1^+ + v_1^-$. Now, the new forward voltage $v_1^+ \Gamma_L \Gamma_s$ propagates to the load, and the process repeats. Shown in Figure 3.48 is the graphical representation of the forward and backward voltages, as well as the voltage in the line at the halfway point $z = \frac{l}{2}$.

The voltage gradually builds up to a final values of

$$v_1^+ + v_1^- + v_2^+ + v_2^- + \cdots = \frac{Z_0}{Z_0 + R_s} V_0 (1 + \Gamma_L + \Gamma_L \Gamma_s + \Gamma_L^2 \Gamma_s + \cdots).$$

Replacing Γ_L and Γ_s with their values expressed in terms of source and load impedances, and considering that both are less than one, the infinite series converges to $\frac{R_L}{R_L + R_s} V_0$ at steady state as expected.

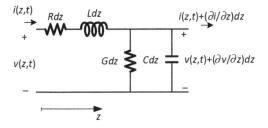

Figure 3.49: Low-loss transmission line lumped model

3.5 LOW-LOSS TRANSMISSION LINES

Although we dedicated most of our transmission line analysis to the lossless propagation, we mentioned that in the general case of a lossy transmission line, the voltage phasor is

$$V(z) = V_0^+ e^{-\gamma z} + V_0^- e^{+\gamma z},$$

where $\gamma = \alpha + j\beta$, and parameter α captures the loss of the line [3], [4], [7]. These results can be obtained by modifying the transmission line lumped equivalent circuit as shown in Figure 3.49, where R and G signify the transmission line losses and, as with L and C, are per unit length. The new differential equation describing the line is obtained as

$$\frac{\partial^2 v}{\partial z^2} = LC\frac{\partial^2 v}{\partial t^2} + (LG + RC)\frac{\partial v}{\partial t} + RGv,$$

which simplifies to our original differential equation if $R = G = 0$. Moreover, in steady-state conditions, the phasor solution is of the form $V(z) = V_0^+ e^{-\gamma z} + V_0^- e^{+\gamma z}$, as stated before, and γ is easily shown to be [7]

$$\gamma = \alpha + j\beta = \sqrt{(R + jL\omega)(G + jC\omega)}.$$

For low-loss propagation, α and β are obtained as follows:

$$\alpha = \frac{1}{2}\left(\frac{R}{Z_0} + GZ_0\right)$$

$$\beta = \omega\sqrt{LC}\left(1 + \frac{1}{8}\left(\frac{G}{C\omega} - \frac{R}{L\omega}\right)^2\right),$$

where $Z_0 = \sqrt{\frac{L}{C}}$ is the characteristic impedance. We can further show that when the waves propagate in such a low-loss line, the power at a given point in the line decays as $P(z) = P_0 e^{-2\alpha z}$, very much like our RLC analysis performed in Chapter 1.

Example: A better model for a practical transmission line segment is shown in Figure 3.50, where there is no assumption of an ideal ground (or return) plane. Note that the return path is not assumed to be necessarily identical to the other path.

Continued

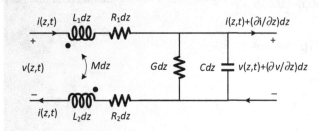

Figure 3.50: A more proper model of a segment of a lossy transmission line

The reader can show that the new line equation is

$$\frac{\partial^2 v}{\partial z^2} = L_{eq} C \frac{\partial^2 v}{\partial t^2} + (L_{eq} G + R_{eq} C) \frac{\partial v}{\partial t} + R_{eq} G v,$$

where $L_{eq} = L_1 + L_2 + 2M$, and $R_{eq} = R_1 + R_2$. So the overall behavior of the line will be similar to the one described by the simpler model of Figure 3.49.

3.5.1 Reasons for Adopting 50Ω

Historically, most transmission lines are designed to be 50Ω or 75Ω. As a very constructive case study, we shall discuss the reasons as to why this common practice has been adopted. Let us consider the coaxial transmission line shown below (Figure 3.51).

Upon our analysis in Chapter 1, we obtained $C = \frac{2\pi\epsilon}{\ln\frac{b}{a}}$, and $L = \frac{\mu_0}{2\pi} \ln\frac{b}{a}$, leading to $Z_0 = \sqrt{\frac{\mu_0}{\epsilon}} \ln\frac{b}{a} = \frac{60}{\sqrt{\epsilon_r}} \ln\frac{b}{a}$. Moreover, we showed that $\boldsymbol{D} = \frac{\rho_s a}{r} \boldsymbol{a}_r$, which may be expressed in terms of the voltage as $\boldsymbol{D} = \frac{\epsilon V_0}{\ln\frac{b}{a}} \frac{\boldsymbol{a}_r}{r}$. We can further find the values of G and R for the low-loss coaxial line (Figure 3.49). The dielectric leakage, G, is found very much like the capacitance, all needs to be done is to note that the current density $J = \sigma E$, where σ is the dielectric conductivity. Consequently, this leads to a very similar expression (see Chapter 1 problem sets for the proof):

$$G = \frac{I}{V} = \sigma \frac{\int_S \boldsymbol{E} \cdot d\boldsymbol{S}}{-\int \boldsymbol{E} \cdot d\boldsymbol{L}} = \frac{2\pi\sigma}{\ln\frac{b}{a}}.$$

It is reasonable to assume that the dielectric leakage, say in the case of an air coaxial line, is very small, as conductivity is small, and thus $G \approx 0$. To find the conductor loss, R, we assume

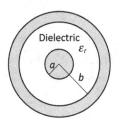

Figure 3.51: Cross section of a lossy transmission line

that the frequency is high enough that the current is flowing only into the skin depth, δ. Moreover, let us assume that the current is distributed uniformly, and that the conductor has a conductivity of σ_c. Thus, given that the area that the current flows is roughly $2\pi\delta a$ for the inner conductor, and $2\pi\delta b$ for the outer conductor, we have

$$R \approx \frac{1}{2\pi\delta\sigma_c}\left(\frac{1}{a}+\frac{1}{b}\right),$$

where we have added the resistance of the inner and outer conductors, as they appear effectively in series. For an air transmission line, ignoring G, the loss factor is then

$$\alpha = \frac{1}{240\pi\delta\sigma_c} \frac{\left(\frac{1}{a}+\frac{1}{b}\right)}{\ln\frac{b}{a}}.$$

Now by taking the derivative of α versus $\frac{b}{a}$, the loss factor is minimized if $\frac{b}{a} = 3.6$, leading to a characteristic impedance of 77Ω.

Moreover, since

$$E = \frac{V_0}{\ln\frac{b}{a}}\frac{a_r}{r},$$

the strongest value of the electric field occurs at $r = a$, leading to $E_{max} = \frac{V_0}{a\ln\frac{b}{a}}$. Thus the power delivered to the load is

$$P_L = \frac{V_0^2}{2Z_0} = \frac{E_{max}^2}{120}a^2\ln\frac{b}{a}.$$

For a given acceptable maximum filed, E_{max}, the power may be maximized if $\frac{b}{a} = \sqrt{e}$, leading to $Z_0 = 30\Omega$.

Historically, the best coaxial cable impedances in high-power and low-attenuation applications were experimentally determined at Bell Laboratories in 1929 to be 30Ω, and 77Ω, respectively, agreeing with our analysis above. It appears that a compromise between the two conditions, that is maximum power delivery (for a certain maximum field), and the minimum power dissipated, sets the characteristic impedance to about 50Ω, as commonly used. On the other hand, the approximate impedance required to match a center-fed dipole antenna in free space is 73Ω, so 75Ω coax was commonly used for connecting shortwave antennas to receivers. These typically involve such low levels of RF power that power-handling and high-voltage breakdown characteristics are unimportant when compared to attenuation.[12]

There is obviously no need to consider a 50Ω interface in modern radios, depending on the chip size and frequency. However, the majority of external components such as the RF SAW filters or the antennas are traditionally designed to be 50Ω. As mentioned at the beginning, to interface such elements it is often desirable to present a well-defined and close to 50Ω impedance at the boundaries of RF IC and the outside world. On the other hand, in an ideal

---

[12] The interested reader may refer to the following online articles for more details: "Why 50 Ohms?," *Microwaves 101*, January 13, 2009; "Coax Power Handling," *Microwaves 101*, September 14, 2008.

world, with a custom designed radio, there is no need to set the characteristic impedance to 50Ω. This number is arbitrary and is simply a legacy design parameter.

3.6 RECEIVE–TRANSMIT ANTENNAS AS TWO-PORT CIRCUITS

Throughout our discussion of antennas in Chapter 1, we treated them as single transmitting devices, producing electromagnetic waves propagating in the air. In this section, we turn to other fundamental purpose of an antenna, which is a means to detect (or receive) radiation originated from a distant source, namely a transmit antenna. We will approach this as a two-port network, comprising a receive and a transmit antenna, as well as their supporting circuitry.

Shown in Figure 3.52 is a simple example of a receive–transmit antenna arrangement, where the two coupled antennas establish a linear two-port network.

The voltage and the current of the first antenna affect the voltage and current of the second one, and vice versa. Quantifying this coupling through trans-impedance parameters (Z_{12} and Z_{21}), we may write

$$V_1 = Z_{11}I_1 + Z_{12}I_2$$
$$V_2 = Z_{21}I_1 + Z_{22}I_2.$$

The impedances Z_{11} and Z_{22} are each antenna input impedance when the other one is isolated, or equivalently, located very far away. In practice, they consist of the radiation resistance as shown in Chapter 1, any associated ohmic loss, and possibly some reactance depending on the antenna physical structure, and design parameters. On the other hand, trans-impedances Z_{12} and Z_{21} depend on the distance and relative position of the two antenna. Regardless of the absolute values of trans-impedances, given the reciprocity, we can always say

$$Z_{12} = Z_{21}.$$

One important conclusion drawn from the above is that an antenna's *radiation and reception patterns* are identical. In other words, the extent to which the receiving antenna accepts power is determined by its radiation pattern.

Now consider Figure 3.53, where the second antenna is terminated by some load impedance Z_L. This could model the receiver input impedance for instance. The two-port equivalent is also depicted on the bottom of the figure. An important but realistic assumption here is that the antennas are far enough that only the forward coupling is appreciable, that is, $Z_{12}I_2 \approx 0$. Consequently, we assume the induced current I_2 by the first antenna is much less than I_1. So the current induced back in the first antenna as a result of I_2 is negligible compared to I_1.

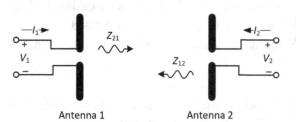

Figure 3.52: A pair of receive and transmit antennas coupled to each other

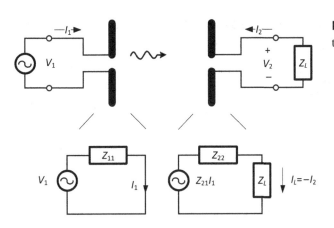

Figure 3.53: Loaded coupled antennas with the equivalent circuit model

We can then write

$$I_L = -I_2 = \frac{Z_{21} I_1}{Z_{22} + Z_L}.$$

The average power dissipated in the load is

$$P_L = \frac{1}{2}\text{Re}[V_L I_L^*] = \frac{1}{2} R_L |I_1|^2 \left|\frac{Z_{21}}{Z_{22} + Z_L}\right|^2,$$

under maximum power transfer condition, $Z_L = Z_{22}^*$, and thus

$$P_L = \frac{|I_1|^2 |Z_{21}|^2}{8 R_{22}},$$

where $R_{22} = \text{Re}[Z_{22}]$, and is equal to the receiving antenna radiation resistance if ohmic losses ignored. The average power transmitted by the first antenna is

$$P_r = \frac{1}{2} R_{11} |I_1|^2,$$

where R_{11} is the radiation resistance of the transmitting antenna. Consequently,

$$\frac{P_L}{P_r} = \frac{|Z_{21}|^2}{4 R_{11} R_{22}},$$

whereas R_{11} and R_{22} are equal to each antenna's radiation resistance (if resistive losses are negligible), the trans-impedance Z_{21} (or Z_{12}) is a function of each antenna characteristics, as well as their relative spacing and orientation.

3.6.1 Antenna Effective Area

To have a better understanding of Z_{21}, consider Figure 3.54 as an example of a pair of dipole antennas separated by a radial distance r, and a relative orientation angle of θ.

A common and convenient way of expressing the received power in an antenna is through its *effective area*, expressed in m^2. In Chapter 1 we showed that the power density radiated by an antenna is given by

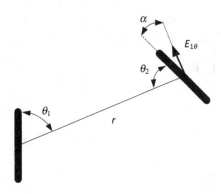

Figure 3.54: A pair of receive–transmit antenna with some arbitrary orientation

$$\mathcal{P}_1(r,\theta_1,\varphi_1) = \frac{P_r}{4\pi r^2} D_1(\theta_1,\varphi_1),$$

where $D_1(\theta_1,\varphi_1)$ is the directivity (of the first antenna), P_r is the average power radiated, and $\mathcal{P}_1(r,\theta_1,\varphi_1)$ is the antenna average power density. The latter quantity, that is, $\mathcal{P}_1(r,\theta_1,\varphi_1)$, produced by the first antenna, induces a certain power on the receiver antenna, whose value is a function of the antennas positioning and spacing. As such, we define the effective area of the second (receiving antenna) as

$$P_L = \mathcal{P}_1(r,\theta_1,\varphi_1) \times A_2(\theta_2,\varphi_2),$$

where P_L is the power delivered to the load of the second antenna, and $A_2(\theta_2,\varphi_2)$ is defined as the second antenna *effective area*. Note the subscript 1 (i.e., θ_1, φ_1) signifies the power density created by the *first* antenna ($\mathcal{P}_1(r,\theta_1,\varphi_1)$), which is a function of its positing, whereas the subscript 2 (i.e., θ_2, φ_2) denotes the relative positioning of the second antenna. Combining the two equations above, we can write

$$\frac{P_L}{P_r} = \frac{D_1(\theta_1,\varphi_1)A_2(\theta_2,\varphi_2)}{4\pi r^2}.$$

However, since we showed in the previous section that $\frac{P_L}{P_r} = \frac{|Z_{21}|^2}{4R_{11}R_{22}}$, then

$$|Z_{21}|^2 = \frac{R_{11}R_{22}D_1(\theta_1,\varphi_1)A_2(\theta_2,\varphi_2)}{\pi r^2}.$$

As expected, Z_{21} depends not only on the individual characteristics of each antenna (R_{11} and R_{22}), but also on the spacing, and the relative positioning of the two, specifically the directivity of the transmitting antenna, as well as the effective area of the receiving antenna.

We next note that if the roles of the antennas are reversed, that is, the second antenna transmits to the first, we must have

$$|Z_{12}|^2 = \frac{R_{11}R_{22}D_2(\theta_2,\varphi_2)A_1(\theta_1,\varphi_1)}{\pi r^2}.$$

From reciprocity it follows

$$\frac{D_1(\theta_1,\varphi_1)}{A_1(\theta_1,\varphi_1)} = \frac{D_2(\theta_2,\varphi_2)}{A_2(\theta_2,\varphi_2)}.$$

That is, the ratio of the directivity to effective area of any antenna is a constant.

3.6.2 Friis Transmission Formula

As we already established that the ratio $\frac{D(\theta_1,\varphi_1)}{A(\theta_1,\varphi_1)}$ is a constant, one may attempt to calculate it for a known simple antenna, for instance, the Hertzian dipole. To do so, consider again Figure 3.54, showing a pair of Hertzian dipoles. We showed in Chapter 1 that the electric field created by the first antenna at far field is

$$E_{1\theta}(r,\theta,\varphi) = jk\eta \frac{I_0 l \sin\theta}{4\pi r} e^{-jkr} = \eta H_{1\varphi}.$$

Note that in a short dipole, given the symmetry the fields are not a function of φ. Consequently, the electric field of the transmitting antenna, when projected on the receiving antenna will be $E_{1\theta}\cos\alpha$, where $\alpha = 90° - \theta_2$ as shown in the figure. Thus, the voltage induced on the second antenna given the electric field of the first antenna is

$$V_2 = (E_{1\theta}\cos\alpha)l = E_{1\theta}l\sin\theta_2.$$

From this one can calculate the power delivered to the second antenna under the matched load condition, and the resultant effective area of

$$A_2(\theta_2,\varphi_2) = \frac{3}{8\pi}\lambda^2 \sin^2\theta_2.$$

Since the directivity of a short dipole was shown to be

$$D_2(\theta_2,\varphi_2) = \frac{3}{2}\sin^2\theta_2,$$

we conclude then

$$\frac{D(\theta,\varphi)}{A(\theta,\varphi)} = \frac{4\pi}{\lambda^2}$$

for any antenna.

Interestingly, the directivity of an antenna, which is a transmit characteristic, is related to its effective area, which is a receive characteristic by $\frac{4\pi}{\lambda^2}$.

Using the above relation, we may further write

$$\frac{P_L}{P_r} = \frac{A_1(\theta_1,\varphi_1)A_2(\theta_2,\varphi_2)}{\lambda^2 r^2} = \left(\frac{\lambda}{4\pi r}\right)^2 A_1(\theta_1,\varphi_1)A_2(\theta_2,\varphi_2),$$

which is known as the *Friis transmission formula*. It simply suggests that the ratio of the power delivered to the receive antenna to the power radiated from the transmit antenna is directly proportional to the product of their directivity, whereas inversely proportional to the square of the distance between the two antennas.

3.7 SMITH CHART

Transmission line problems often require manipulations with complex numbers. The work involved could be greatly reduced without impacting accuracy much by using graphical methods, perhaps the most common one being the Smith chart [8]. The basic principle upon which the chart is constructed is the reflection coefficient equation

$$\Gamma = \frac{Z_L - Z_0}{Z_L + Z_0}.$$

Since Γ is a complex number, it may expressed as $\Gamma = |\Gamma|e^{j\phi} = \Gamma_r + j\Gamma_i$. In addition, as $|\Gamma| \leq 1$, for any Z_L, the information always falls inside a unity circle depicting Γ in the complex plane. It is customary to normalize the load impedance to the characteristics impedance of the line, and express that as a complex number as follows:

$$z_L = \frac{Z_L}{Z_0} = r + jx.$$

Consequently,

$$\Gamma = \frac{z_L - 1}{z_L + 1}.$$

Alternatively, if we were to define the normalized admittance $y_L = \frac{Y_L}{Y_0} = Z_0 Y_L$, where $Y_L = \frac{1}{Z_L}$, then the reflection coefficient may be expressed as

$$\Gamma = -\frac{y_L - 1}{y_L + 1}.$$

This indicates that using the normalized admittance instead leads to the same magnitude of the reflection coefficient, and a 180° shift for the phase. We shall use this outcome to interchangeably work with both admittances and impedances on the chart, whichever more convenient. Clearly, $y_L = \frac{1}{z_L}$.

Using normalized impedance, we have

$$\Gamma = \frac{z_L - 1}{z_L + 1} = \frac{r + jx - 1}{r + jx + 1} = \Gamma_r + j\Gamma_i.$$

With some simple steps of algebra, we arrive at the following set of equations expressing the real and imaginary parts of Γ as functions of r and x:

$$\begin{cases} \left(\Gamma_r - \frac{r}{1+r}\right)^2 + \Gamma_i^2 = \left(\frac{1}{1+r}\right)^2 \\ (\Gamma_r - 1)^2 + \left(\Gamma_i - \frac{1}{x}\right)^2 = \left(\frac{1}{x}\right)^2 \end{cases}.$$

Either equation describes a family of circles where each circle is associated with a specific value of r (or x) and is shown in Figure 3.55.

The real part of impedance is expected to be positive, so r is always greater than zero, whereas x assumes positive (inductive) or negative (capacitive) values. The circle $r = 0$ corresponds to the unity circle, as $|\Gamma| = 1$. The two families of both circles appear on the Smith chart, where one can obtain the magnitude and phase of Γ for a given load impedance. For instance, if $Z_L = 100 + j25\Omega$, $r = 2$, and $x = 0.5$, corresponding to the point shown in the simplified Smith chart of Figure 3.56.

Accordingly, by virtue of measuring on the chart, we obtain: $\Gamma \cong 0.37 \angle 23°$. Even in this very simple example, using the original format of Γ equation, one has to take several steps of $\tan^{-1}$ and magnitude calculations to obtain the same results. The Smith chart proves to be

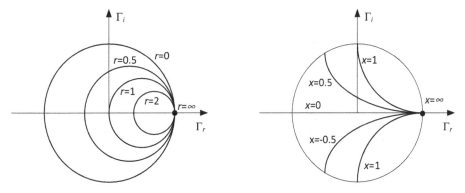

Figure 3.55: Constant-r and -x circles

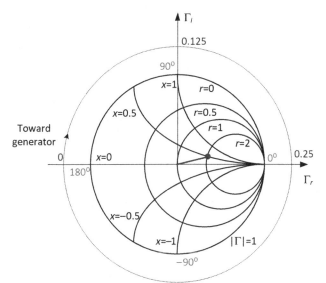

Figure 3.56: The Smith chart consisting of constant-x and -r circles

considerably more powerful however when the impedance moving away from the load across the transmission line needs to be calculated. We showed before at a given point z on the line

$$Z(z) = Z_0 \frac{e^{-j\beta z} + \Gamma e^{+j\beta z}}{e^{-j\beta z} - \Gamma e^{+j\beta z}}.$$

Thus, the *normalized input impedance* at $z = -l$, that is at a distance of l away from the load, is

$$z_{IN} = \frac{1 + \Gamma e^{-2j\beta l}}{1 - \Gamma e^{-2j\beta l}},$$

which shows that once Γ is obtained at the load position, that is for $l = 0$, the corresponding impedance at a distance of $-l$ from the load is found by keeping $|\Gamma|$ constant, but rotating its phase *clockwise* by $2\beta l = \frac{4\pi}{\lambda} l$. A half-circle rotation then corresponds to traveling a quarter wavelength, which will rotate the phase by 180°. What is shown in an actual Smith chart is not the angle of rotation, but the distance traveling toward generator (that is rotating clockwise) normalized to half-wavelength (or 360°). This additional scale is shown in the simplified chart of Figure 3.56 as well in dashed line. If impedances are known on the chart, the corresponding admittance is obtained as the mirror image of the impedance. This, we established before, knowing that

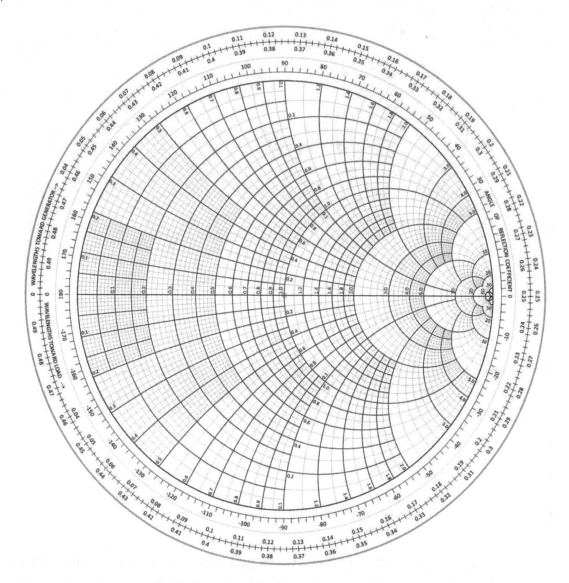

Figure 3.57: A commercially available Smith chart

$$\Gamma = \frac{Z_L - Z_0}{Z_L + Z_0} = -\frac{Y_L - Y_0}{Y_L + Y_0},$$

showing that $|\Gamma|$ must remain the same, but its phase is $180°$ (or a quarter of a wavelength) different.

Shown in Figure 3.57 is an actual commercially available Smith chart, widely used among RF designers. Note that the calculations may be performed using online software, easing the use of the chart.

Example: Let us consider a 50Ω transmission line shown in Figure 3.58, terminated with a load impedance of $Z_L = 250\Omega$, representing an amplifier input. We showed how this impedance may be matched to 50Ω using a lumped LC network or a transformer. The goal is to match that to 50Ω using transmission lines. This is commonly done by inserting a short circuited stub with a constant length of d_S at a distance d from the load.

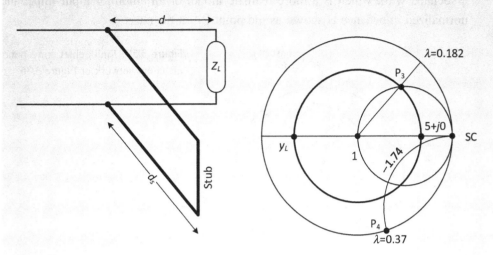

Figure 3.58: A transmission line with a short-circuited stub

We first note that the shorted circuit stub, regardless of its length is always reactive, and it will be placed in parallel with the line impedance at the point $z = -d$. Since it is easier to add admittances, we will perform everything using admittances instead. The normalized load impedance is $5 + j0$ as shown on the Smith chart in Figure 3.58. To convert to admittance, we add a quarter wavelength, as the impedance is transferred to Z_0^2/Z_L, or alternatively z_L is changed to y_L. Clearly y_L corresponds to the $r = 0.2$ circle. That is shown as point y_L in the chart which is now at 0λ. Next, to match to 50Ω, we need to reside on the $r = 1$ circle. Since moving across the transmission line only changes the angle of Γ, then the desired point will be the intercept point of the $r = 1$ circle with constant $|\Gamma|$ circle for the point y_L. This would be point P_3 on the chart, which has a reading of 0.182 wavelength. As a result, we have moved the load by a net length of 0.182 wavelength, or $d = 0.182\lambda$. Point P_3 has a real part equal to 1, but it is capacitive (note that we are dealing with admittances). Reading the x family circles, we obtain the normalized imaginary part is equal to 1.74. If the stub presents an imaginary part of −1.74, then once added to the line, it will result in a pure real normalized admittance of 1 (or 50Ω). The appropriate length of stub is found by intercepting the $x = -1.74$ circle with $|\Gamma| = 1$ circle, labeled as P_4, and whose reading is 0.37λ. Since the short circuit is at 0.25λ, then the length $d_S = 0.12\lambda$.

It is clear that without using the Smith chart, the calculations would have been very difficult. If the load has some reactive component, the calculations would have been very similar.

Example: To show that the applications of the Smith chart go beyond transmission lines and distributed elements, let us redo our previous example of matching the input impedance of an amplifier to 50Ω, through a lumped LC network (Figure 3.36), but this time

Continued

using the Smith chart instead. Since R_{IN} is greater than 50Ω, it will reside on the right-half side of the circle (Figure 3.59). We also assume that it has a capacitive component associated with, which is a more realistic model of an amplifier input impedance. The normalized impedance is shown as the point P_1 on the chart.

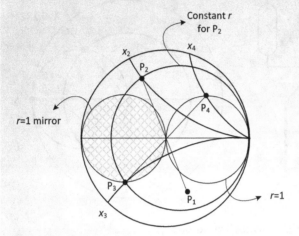

Figure 3.59: Smith chart corresponding to matching network of Figure 3.36

Since first a shunt inductance is placed, we convert z_{in} to y_{in} by adding a quarter wavelength to arrive at point P_2. Now to obtain a real part of eventually 50Ω, we would like the matching inductance to move the point P_2 to somewhere on the mirrored image of the $r = 1$ circle (the shaded circle). If so, when converted back to impedance (to conveniently add the series capacitance) it must reside on the $r = 1$ circle. This is obtained by intercepting the constant r circle where P_2 resides with the mirrored $r = 1$ circle. There are two solutions, but only the point P_3 is valid. That is because the other point when converted back to impedance will reside in the lower half of the circle, corresponding to a capacitive component. This point then can be matched to 50Ω only through a series inductor, whereas our matching network comprises a series capacitor. The value of the inductance needed is obtained by considering the original susceptance, and the new one corresponding to point P_3. Point P_3 is converted back to impedance, resulting in point P_4, which now resides on the $r = 1$ circle, and whose reactance is x_4. This point is inductive, and all that is needed is to add a series capacitance C such that $x_4 = \frac{\frac{1}{C\omega_0}}{50\Omega}$. It is clear by inspection that if R_{IN} is less than 50Ω, or in general for any point residing inside the shaded circle, there will be no solution for this specific matching network, an outcome that is not so obvious by using the series–parallel calculations. Back to our numerical example, if $R_{IN} = 250\Omega$, then $y_{in} = 0.2 + j0$. After intercepting with the mirrored $r = 1$ circle, x_3 is read to be –0.4. Since at this point, we are still dealing with normalized admittances, $0.4 = \frac{50\Omega}{L\omega_0}$, or $L\omega_0 = 125\Omega$. This leads to $L = 10$nH at 2GHz. Point P_4 reading on the chart is $1 + j2$, which is now impedance. Thus $2 = \frac{\frac{1}{C\omega_0}}{50\Omega}$, leading to $C = 0.8$pF at 2GHz. One other nice feature of the Smith chart is that what is obtained is reactance or susceptance, and is not frequency dependent. It becomes a function of the frequency only when converted to inductance or capacitance.

3.8 SCATTERING PARAMETERS

In addition to the six ways of characterizing a linear and time-invariant N-port described earlier in Section 3.1.1 (impedance, admittance, two hybrid, and two transmission matrices), there is yet another common method widely used in RF and especially microwave applications, known as scattering parameters. Their basic properties and application will be presented in this section. For a more detailed discussion, see [3], [4].

3.8.1 Basic Properties of Scattering Parameters

Describing microwave circuits with impedance (or admittance) matrixes may not be practical, since voltages, currents, and impedances cannot be measured directly at microwave frequency. Obtaining these matrices requires terminating the two-port by ideal short or open circuits, which prove to be challenging at high frequencies. The quantities that are directly measurable by means of a small probe used to sample the relative field strength are the standing wave ratio and power, which directly lead to reflection coefficient. Also directly measurable is a relative relation of the amplitude and phase of the transmitted signal as compared to those of the incident (by using a directional coupler for instance). In other words, directly measurable quantities are the relative (to the incident wave) amplitude and phases of the waves reflected, or *scattered*. The matrix describing such relationship is called a scattering matrix, or S matrix. Adopted from microwave circuits, it is not uncommon to use scattering parameters for RF circuits as well, especially when dealing with the interface of the RF circuit and outside world, i.e., receiver input or transmitter output.

Consider the N-port shown in Figure 3.60. If a wave with an associated equivalent voltage V_1^+ is incident at terminal 1, the reflected wave is $V_1^- = S_{11}V_1^+$ in that terminal, where S_{11} is the reflection coefficient. Moreover, it is natural to assume that waves will be also scattered out of the other ports, expressed as $V_n^- = S_{n1}V_n^+$, $n = 2, 3, \ldots, N$.

When waves are incident at all ports, we can write in general

$$\begin{bmatrix} V_1^- \\ V_2^- \\ \ldots \\ V_N^- \end{bmatrix} = \begin{bmatrix} S_{11} & S_{12} & \cdots & S_{1N} \\ S_{21} & S_{21} & \cdots & S_{2N} \\ \ldots & \ldots & \ldots & \ldots \\ S_{N1} & S_{N2} & \cdots & S_{NN} \end{bmatrix} \begin{bmatrix} V_1^+ \\ V_2^+ \\ \ldots \\ V_N^+ \end{bmatrix},$$

or $[V^-] = [S][V^+]$, where $[S]$ is the scattering matrix.

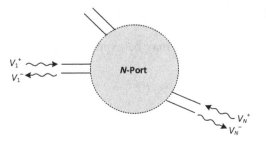

Figure 3.60: An N-port showing scattered waves

We assume that all the ports use the same characteristic impedance of Z_0, and thus the same as before

$$V^+ = Z_0 I^+$$
$$V^- = -Z_0 I^-,$$

and

$$V = V^+ + V^-$$
$$I = I^+ + I^- = \frac{1}{Z_0}(V^+ - V^-).$$

Combining the equations above, one can show that the S matrix could be expressed in terms of impedance or admittance matrixes. For example, in matrix format we have

$$[V] = [V^+] + [V^-] = [Z][I] = [\bar{Z}][V^+] - [\bar{Z}][V^-],$$

where $[\bar{Z}] = \frac{1}{Z_0}[Z]$ is the N-port *normalized impedance* matrix. It follows that

$$[V^-] = ([\bar{Z}] + [U])^{-1}([\bar{Z}] - [U])[V^+],$$

where $[U] = \begin{bmatrix} 1 & 0 & \cdots & 0 \\ 0 & 1 & \cdots & 0 \\ \cdots & \cdots & \cdots & \cdots \\ 0 & 0 & \cdots & 1 \end{bmatrix}$ is the *unit* matrix. According to the definition of the S matrix then,

$$[S] = ([\bar{Z}] + [U])^{-1}([\bar{Z}] - [U]).$$

This results to two important observations:

1. First, even though our discussion so far has involved incident and reflected waves, S parameters are not solely limited to distributed elements. Any N-port, lumped or distributed, may be represented by the S matrix. However, using S parameters proves to be a necessity at microwave frequencies due to the measurement limitation pointed out earlier. We will have a more detailed discussion on this topic at the end of this section.
2. For any reciprocal N-port, the known symmetry of the impedance matrix readily results into the symmetry of the S matrix. That is, $[S] = [S]^t$, where the superscript t denotes the S matrix *transpose*. This can be proven easily based on the basic definition.

In addition to the symmetry of the S matrix for a reciprocal N-port, the power conservation leads to more simplification if the circuit is lossless as well. Since the total power leaving the lossless N-port must be equal the total incident power, we have

$$\sum_{n=1}^{N} |V_n^-|^2 = \sum_{n=1}^{N} |V_n^+|^2.$$

Replacing $V_n^- = \sum_{i=1}^{N} S_{ni} V_i^+$, the power conservation may be expressed as

$$\sum_{n=1}^{N} \left| \sum_{i=1}^{N} S_{ni} V_i^+ \right|^2 = \sum_{n=1}^{N} |V_n^+|^2.$$

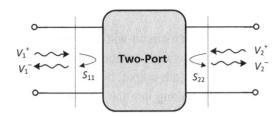

Figure 3.61: A two-port circuit

The V_n^+ are all independent incident voltages, if we choose all to be zero except for one of them, say, V_i^+ we have

$$\sum_{n=1}^{N} |S_{ni} V_i^+|^2 = |V_i^+|^2,$$

which leads to

$$\sum_{n=1}^{N} |S_{ni}|^2 = \sum_{n=1}^{N} S_{ni} S^*_{ni} = 1,$$

where the index i is arbitrary. Similarly, we may choose for all V_n^+ to be zero, except for V_s^+ and V_r^+ ($s \neq r$). This leads to (see Problem 27 for proof)

$$\sum_{n=1}^{N} S_{ns} S^*_{nr} = 0.$$

The above two conditions are sufficient to restrict the scattering matrix to $\frac{1}{2}N(N+1)$, as opposed to N^2. Such a matrix is known as the *unitary* matrix.

Since many common RF circuits are two-ports, let us focus on the scattering matrix of a two-port shown in Figure 3.61.

The incident and scattered waves are related as

$$V_1^- = S_{11} V_1^+ + S_{12} V_2^+$$
$$V_2^- = S_{21} V_1^+ + S_{22} V_2^+.$$

If the output is terminated with a matched load, $V_2^+ = 0$, and thus S_{11} will represent the reflection coefficient. However, if the output is terminated with an arbitrary load of Z_L, the ratio of V_2^+/V_2^- must be equal to the reflection coefficient of the load (as V_2^- is incident to the load), thus

$$\frac{V_2^+}{V_2^-} = \frac{Z_L - Z_0}{Z_L + Z_0} = \Gamma_L.$$

Moreover, solving for V_1^+ and V_1^- we obtain

$$\frac{V_1^-}{V_1^+} = S_{11} - \frac{S_{12} S_{21} \Gamma_L}{S_{22} \Gamma_L - 1},$$

which is the modified input reflection coefficient.

Example: Consider the circuit of Figure 3.62, where a circuit with an arbitrary reflection coefficient ($\Gamma_{IN} = \frac{Z_{IN}-Z_0}{Z_{IN}+Z_0}$) is connected to the source with an attenuator in middle. We further assume that the circuit is either terminated at the output, or is unilateral such that its input S_{11} is equal to Γ_{IN}. We wish to find S_{11} looking into the attenuator.

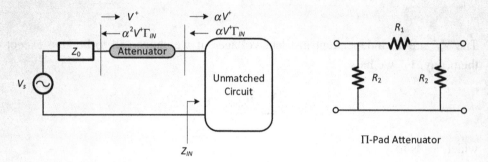

Figure 3.62: Unmatched load preceded by an attenuator

The attenuator is typically realized by a Π resistive network as shown on the right, often known as the Π-pad attenuator.[13] Assuming the attenuator is matched at both ends, and has an attenuation of $L = -20\log\alpha$, ($\alpha < 1$), then we can show that (see also Problem 32 for more details)

$$R_1 = Z_0 \frac{1-\alpha^2}{2\alpha}$$

$$R_2 = Z_0 \frac{1+\alpha}{1-\alpha},$$

where Z_0 is the reference (termination) impedance.

With reference to the figure, the incident voltage of V^+ is attenuated by α coming into the circuit, and the reflected signal of $\alpha V^+\Gamma_{IN}$ is created. The reflected signal travels to the left, and once passed by the attenuator, a total reflected voltage of $\alpha^2 V^+\Gamma_{IN}$ is received back by the source. Thus the reflection coefficient at source is $\alpha^2\Gamma_{IN}$, or in dB, it improves by two times the loss of the pad or $2L$.

It is not uncommon in RF to insert a pad to improve the interface in terms of matching for sensitive measurements. Even a small attenuation of 3dB effectively boosts the S_{11} by 6dB, which could be quite helpful. The loss of the attenuator must be naturally de-embedded from the system. Note that the attenuator loss is defined for matched ports, and thus the poor S_{11} of the circuit under the test may affect it. Hence, that must also be included. See the example at the end of the section.

For a reciprocal two-port, $S_{12} = S_{21}$. If the two-port is lossless, according to power conservation we have

$$|S_{11}| = |S_{22}|,$$

---

[13] Attenuators are sometimes referred to as pads due to their effect of padding down a signal by analogy with acoustics.

which tells us that the reflection coefficients of the input and output ports are equal in magnitude. Moreover,

$$|S_{12}| = \sqrt{1 - |S_{11}|^2}.$$

Example: Consider the shunt susceptance jB, connected across a transmission line with a characteristics impedance of $Z_0 = \frac{1}{Y_0}$, as shown in Figure 3.63.

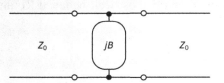

Figure 3.63: Shunt element in a transmission line

The S_{11} is the input reflection coefficient for a matched load, and thus

$$S_{11} = \frac{Y_0 - Y_{in}}{Y_0 + Y_{in}} = \frac{Y_0 - (Y_0 + jB)}{Y_0 + (Y_0 + jB)} = \frac{-jB}{2Y_0 + jB},$$

which must be equal to S_{22} due to symmetry. To find S_{21}, let us terminate the output, forcing V_2^+ to zero. For a shunt element, we must have $V_1^+ + V_1^- = V_2^-$, which leads to

$$S_{21} = S_{21} = 1 + S_{11} = \frac{2Y_0}{2Y_0 + jB}.$$

As an exercise, one can show that the obtained S parameters for the example of Figure 3.63, indeed satisfy the two restraints set by the energy conservation.

Hypothetically, if the right side transmission line had a different characteristics impedance of $Z_0' = \frac{1}{Y_0'}$ instead, we can show

$$S_{11} = \frac{Y_0 - Y_0' - jB}{Y_0 + Y_0' + jB},$$

and S_{22} and S_{21} could be obtained similarly.

Example: Consider the lumped amplifier shown in Figure 3.64. We further assume that the source and load are connected very close to the amplifier, so no distributed element is involved. Since the amplifier is lumped, the incident and reflected waves may not be meaningful. However, we shall show that they still give us some insight on the circuit properties, particularly if one considers the available power concept.

Continued

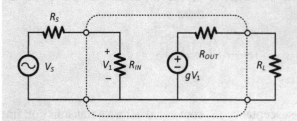

Figure 3.64: A lumped amplifier

We set the scattering matrix reference impedance to be equal to that of the source, that is, R_S. Since the circuit is unilateral, as is the case for most well-behaved RF amplifiers, $S_{12} = 0$. It follows that

$$S_{11} = \frac{R_{IN} - R_S}{R_{IN} + R_S},$$

which is equal to the reflection coefficient, regardless of the output termination. Although the circuit is lumped, we could still calculate V_1^+ and V_1^- based on the basic definition presented earlier. Since

$$V_1 = V_1^+ + V_1^-$$

$$I_1 = \frac{V_1^+ - V_1^-}{R_S},$$

we have

$$V_1^+ = \frac{V_1 + R_S I_1}{2} = \frac{1}{2} V_S$$

$$V_1^- = S_{11} V_1^+.$$

Now let us define the power associated with V_1^+ and V_1^-, which we will call P_1^+ and P_1^-:

$$P_1^+ = \frac{1}{2R_S} |V_1^+|^2 = \frac{|V_S|^2}{8R_S}$$

$$P_1^- = |S_{11}|^2 P_1^+.$$

We note that P_1^+ is in fact the circuit *available power* (P_a) as defined before. Moreover, the total power delivered to the amplifier input, P_{IN}, is found to be

$$P_{IN} = \frac{|V_1|^2}{2R_{IN}} = P_1^+ - P_1^- = \left(1 - |S_{11}|^2\right) P_a.$$

This shows that P_1^+ and P_1^- signify similar concepts as incident and scattered waves. The total power delivered to the circuit is the difference between the two, as if P_1^+ is the power incident, while P_1^- is the power reflected back to the source. For a matched input, $S_{11} = 0$, and thus the power reflected is zero, meaning that the source available power is entirely delivered to the amplifier.

If we set the reference impedance of the output terminal to be R_{OUT}, then
$$S_{22} = \frac{R_L - R_{OUT}}{R_L + R_{OUT}}.$$
If the reference impedance Z_0 is chosen such that $Z_0 = R_S = R_{OUT}$, then the available power gain is equal to
$$G_a = |S_{21}|^2.$$

The general case of arbitrary source and load impedances will be discussed next.

Example: Consider the two-port of Figure 3.65 with arbitrary source and load impedances. We would like to find the available power gain in terms of the scattering parameters that are directly measurable at high frequencies.

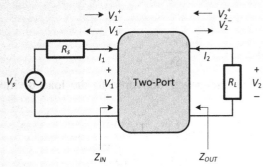

Figure 3.65: A two-port with arbitrary source and load terminations

We define the two-port input and output reflection coefficients as
$$\Gamma_{IN} = \frac{Z_{IN} - Z_0}{Z_{IN} + Z_0}$$
$$\Gamma_{OUT} = \frac{Z_{OUT} - Z_0}{Z_{OUT} + Z_0},$$
where Z_0 is the reference impedance. As we pointed out, $\Gamma_{IN} = S_{11}$, and $\Gamma_{OUT} = S_{22}$ if the other port is terminated by Z_0. Likewise, we define the reflection coefficients for the source and load, which are also directly measurable as
$$\Gamma_s = \frac{Z_s - Z_0}{Z_s + Z_0}$$
$$\Gamma_L = \frac{Z_L - Z_0}{Z_L + Z_0}.$$

We can further express the current going into the two-port in terms of the incident and reflected input voltages:
$$Z_0 I_1 = V_1^+ - V_1^-$$
$$Z_{IN} I_1 = V_1^+ + V_1^-.$$

Continued

Also the power delivered to the two-port is

$$P_{IN} = \text{Re}\,[Z_{IN}]|I_1|^2.$$

Combining the three equations above and expressing Z_{IN} in terms of Γ_{IN} leads to

$$P_{IN} = \frac{|V_1^+|^2}{2Z_0}\left(1 - |\Gamma_{IN}|^2\right).$$

Similarly, the power delivered to the load would be

$$P_L = \frac{|V_2^-|^2}{2Z_0}\left(1 - |\Gamma_L|^2\right).$$

Next, we attempt to express V_2^-, the voltage absorbed by the load, versus V_1^+, the voltage delivered to the two-port in terms of the two-port scattering parameters. That is easily done using the basic definition of scattering parameters:

$$V_2^- = S_{21}V_1^+ + S_{22}V_2^+ = S_{21}V_1^+ + S_{22}\Gamma_L V_2^-.$$

Hence

$$V_2^- = \frac{S_{21}}{1 - \Gamma_L S_{22}} V_1^+.$$

Therefore, the power gain, defined as the power delivered to the load divided by the power delivered to the input, is

$$G_P = \frac{P_L}{P_{IN}} = \frac{|S_{21}|^2\left(1 - |\Gamma_L|^2\right)}{|1 - \Gamma_L S_{22}|^2\left(1 - |\Gamma_{IN}|^2\right)}.$$

To find the available power gain, we note that the power delivered to the input becomes the available power if the two-port input is conjugate matched to the source, that is, if $\Gamma_{IN} = \Gamma_s^*$. Likewise, a similar statement can be made for the output available power and P_L if we have $\Gamma_L = \Gamma_{OUT}^*$. With a few steps of algebra (see problem 40), the available power gain would be

$$G_a = \frac{|S_{21}|^2\left(1 - |\Gamma_s|^2\right)}{|1 - \Gamma_s S_{11}|^2\left(1 - |\Gamma_{OUT}|^2\right)}.$$

In conclusion, in general the available power gain is not equal to $|S_{21}|^2$. If the two-port output impedance is matched to Z_0, then $G_a = \frac{|S_{21}|^2}{(1-|\Gamma_s|^2)}$, and only if its input is also matched to Z_0, then $G_a = |S_{21}|^2$, a conclusion we had already reached for unilateral amplifiers.

Example: Shown in Figure 3.66 is a circulator circuit realized based on a gyrator introduced earlier (Figure 3.9). Considering the left figure, we have

$$\begin{bmatrix} V_1 \\ V_3 \end{bmatrix} = \begin{bmatrix} 0 & -1 \\ 1 & 0 \end{bmatrix} \begin{bmatrix} I_a \\ I_b \end{bmatrix},$$

where $\begin{bmatrix} 0 & -1 \\ 1 & 0 \end{bmatrix}$ is the gyrator impedance matrix.

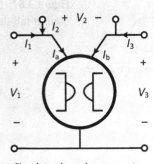 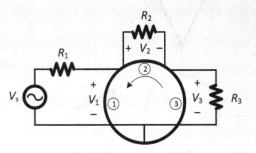

Circulator based on a gyrator | Typical circulator configuration

Figure 3.66: Circulator circuit realized based on a gyrator

We can write

$$I_a = I_1 + I_2$$
$$I_b = I_3 - I_2,$$

and

$$V_1 = -I_b = I_2 - I_3$$
$$V_3 = -I_a = I_1 + I_2$$
$$V_2 = V_1 - V_3 = -I_1 - I_3.$$

Thus, the corresponding open impedance matrix is

$$Z = \begin{bmatrix} 0 & 1 & -1 \\ -1 & 0 & -1 \\ 1 & 1 & 0 \end{bmatrix}.$$

Assuming a reference impedance of $Z_0 = 1\Omega$, then the circulator scattering matrix is found to be

$$S = (Z+U)^{-1}(Z-U) = \begin{bmatrix} 0 & 1 & 0 \\ 0 & 0 & -1 \\ 1 & 0 & 0 \end{bmatrix},$$

where $U = \begin{bmatrix} 1 & 0 & 0 \\ 0 & 1 & 0 \\ 0 & 0 & 1 \end{bmatrix}$ is the unit matrix as introduced earlier in this section.[14]

A typical circuit level configuration of the circulator is shown in the right side of Figure 3.66, where all the ports are terminated. For the moment, let us assume that all the ports are matched, that is, $R_1 = R_2 = R_3 = Z_0 = 1\Omega$. Given that, we can write

$$\frac{\text{Power reflected from port 1}}{\text{Source available power}} = |S_{11}|^2 = 0$$

$$\frac{\text{Power delivered to port 2}}{\text{Source available power}} = |S_{21}|^2 = 0$$

$$\frac{\text{Power delivered to port 3}}{\text{Source available power}} = |S_{31}|^2 = 1.$$

---

[14] For the general case of a $\begin{bmatrix} 0 & a \\ -a & 0 \end{bmatrix}$ gyrator, the S matrix is $S = \frac{1}{3a^2+1} \begin{bmatrix} a^2-1 & 2a(a-1) & 2a(a+1) \\ 2a(a+1) & a^2-1 & -2a(a-1) \\ 2a(a-1) & -a(a+1) & a^2-1 \end{bmatrix}$. Clearly, for either case of $a = \pm 1$, an ideal circulator with either 1-2-3 or 3-2-1 port rotation is achieved.

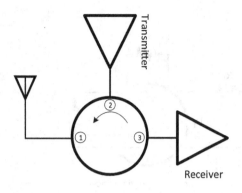
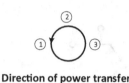

Figure 3.67: Circulator used in full-duplex radios

Consequently, the source available power that is absorbed by port 1 is entirely delivered to port 3, with no signal appearing at port 2. Moving the source to the other two ports, we can see that the circulator achieves isolation between its adjacent ports in the clockwise direction, while the source available power is delivered entirely to its adjacent port in the counterclockwise direction (the direction of the arrow shown in the figure).

As mentioned earlier, the gyrators and thus the circulators, while passive, are nonreciprocal. One of their applications is in full duplex transceivers as the receiver and the transmitter, which are operating concurrently, need to be isolated (Figure 3.67). This way the receiver is protected from the large transmitter output, while the antenna signal still shows in the receiver input. Full duplex radios will be discussed in Chapter 6.

It must be noted that the perfect isolation is obtained only if all the ports are matched. For instance, if port 3 is not matched ($\Gamma_3 = \frac{R_3-1}{R_3+1} \neq 0$), then the power delivered to the port 2 resistor is $P_2 = |\Gamma_3|^2 P_a$, and the power delivered to the port 3 resistor is $P_3 = \frac{|1+\Gamma_3|^2}{R_3} P_a$, where $P_a = \frac{|V_s|^2}{8}$ is the source available power. The reader can verify that $P_2 + P_3 = P_a$, which agrees with power conservation. Accordingly, the isolation from port 1 to port 2 is limited to how good of a match can be achieved at port 3. We can physically justify this considering that if port 3 is not matched, not all the source power would be delivered to R_3, and the portion reflected must inevitably leak to the second port. For more details and the general case, please see Problem 34 at the end of this chapter.

3.8.2 Two-Port Stability Using S Parameters

The two-port stability through admittance (or impedance) parameters was already discussed earlier in Section 3.2.5. We can also derive stability criteria in terms of the input reflection coefficient and S parameters. Consider the two-port of Figure 3.68.

We showed earlier that the input reflection coefficient can be expressed in terms of the two-port S parameters

$$\Gamma_{IN} = S_{11} - \frac{S_{12}S_{21}\Gamma_L}{S_{22}\Gamma_L - 1},$$

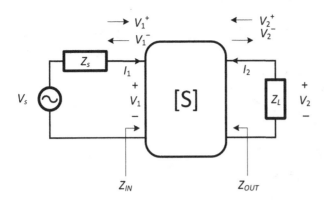

Figure 3.68: Two-port described by its S parameters

where $\Gamma_L = \frac{Z_L - Z_0}{Z_L + Z_0}$ is the load reflection coefficient. Since

$$\Gamma_{IN} = \frac{Z_{IN} - Z_0}{Z_{IN} + Z_0},$$

then for Re[Z_{IN}] to be positive (i.e., for the two-port to be unconditionally stable),

$$|\Gamma_{IN}| = \left| \frac{S_{11} - (S_{12}S_{21} - S_{11}S_{22})\Gamma_L}{S_{22}\Gamma_L - 1} \right| < 1.$$

To simplify, we define $\Delta = S_{12}S_{21} - S_{11}S_{22}$, which is the S matrix determinant. Hence,

$$\left| \frac{S_{11} - \Delta\Gamma_L}{S_{22}\Gamma_L - 1} \right| < 1 \text{ for all } \Gamma_L.$$

To find the boundary between stability/instability, we have to set $|\Gamma_{IN}| = 1$. With some algebraic manipulations, we arrive at

$$\left| \Gamma_L - \frac{S_{21}^* - \Delta^* S_{11}}{|S_{22}|^2 - |\Delta|^2} \right| = \frac{|S_{12}S_{21}|}{|S_{22}|^2 - |\Delta|^2},$$

which is a circle with the center $C = \frac{S_{21}^* - \Delta^* S_{11}}{|S_{22}|^2 - |\Delta|^2}$, and radius $R = \frac{|S_{12}S_{21}|}{|S_{22}|^2 - |\Delta|^2}$.

Example: The region of stability/instability for a given amplifier is demonstrated by a circle on the Smith chart as shown in Figure 3.69.

Continued

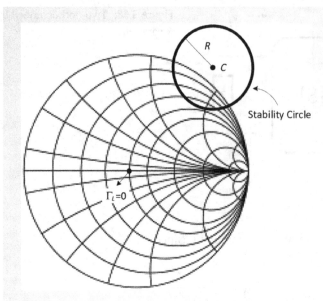

Figure 3.69: Stability circle on the Smith chart for a given amplifier

The boundary line between stable and unstable operation at the input is when $|\Gamma_{IN}| = 1$. The contour of values for Γ_L that produces a unity magnitude reflection coefficient at the input is then the input stability circle. Note that for $\Gamma_L = 0$ (the Smith chart origin), $|\Gamma_{IN}| = |S_{11}|$. Thus, since $|S_{11}|$ must be less than one (or else $\text{Re}[Y_{IN}] < 0$), the stable region is outside the stability circle on the Smith chart.

3.9 DIFFERENTIAL TWO-PORTS

Differential two-ports in general and differential amplifiers in particular are widely used in RF design for obvious well-known reasons. This includes better supply rejection, lower second-order nonlinearity, and higher swings among other advantages. On the other hand, most signal generators and network analyzers available to characterize our circuits are single-ended. Typically external single-ended to differential hybrids are exploited to measure such circuits.

Consider Figure 3.70, illustrating a differential amplifier with a differential input impedance of $2R_{IN} = 2R_s$ and an available voltage gain of g.

A $1{:}\sqrt{2}$ ideal transformer at the input presents a net single-ended input impedance of R_s to the source as intended. Since the transformer is ideal the power is expected to split in half at each end of the secondary, with voltages $\sqrt{2}$ times smaller and 180° out of phase. Each voltage is

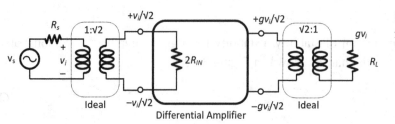

Figure 3.70: A differential matched amplifier

amplified by a gain of g to the output. Once combined with a similar ideal transformer with a turn ratio of $\sqrt{2}$:1 this time, a total available voltage gain of g is measured, as it would have been the case for an identical single-ended design. While most receivers are matched at the input, it is common to have a high impedance output, as the signal is residing at low frequencies. In practice, the transformers used to make such measurements are realized with either on-board narrowband Baluns or wideband hybrid circuits, whose loss must be characterized beforehand, and be de-embedded from the measurement.

Although we are expected to discuss noise in the next section, one can show that the noise performance of the circuit would be characterized similarly, as if there were two identical single-ended amplifiers, but driven differentially [9].

3.10 Summary

In this chapter we have covered a broad range of topics essential to RF and microwave engineering, such as available power gain, matching circuits, and scattering parameters.

- Section 3.1 presented a summary of two-ports and their circuit properties. Of great importance in RF design are the reciprocal two-ports that consist of RLCM elements.
- The concepts of available power and available power gain were discussed in Section 3.2. The available power of unilateral networks, reciprocal networks, and lossless reciprocal networks was specifically derived and discussed.
- Impedance transformation and matching were discussed in Section 3.3. The concept of matching is critical to the design of low-noise and power amplifiers. Several ways of matching an amplifier, such as lumped LC circuits, and transformers were discussed.
- Lossless and low-loss transmission lines were discussed in Sections 3.4 and 3.5. These are examples of distributed networks that were introduced in Chapter 1, and are widely used in RF and microwave design.
- Section 3.6 dealt with some additional properties of the antennas.
- In Section 3.7 the Smith chart was introduced and its applications in impedance transformation and matching the amplifiers were discussed.
- Finally, scattering parameters as another common way of expressing an LTI two-port was presented in Section 3.8.

3.11 Problems

1. Find the L matrix for the following circuits:

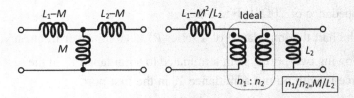

2. Show that the two circuits below are equivalent.

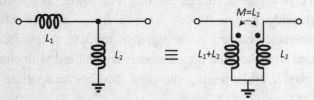

3. Show that the two circuits below are equivalent, where $k = \frac{|M|}{\sqrt{L_1 L_2}}$ is the coupling factor and $n = k\sqrt{\frac{L_1}{L_2}}$.

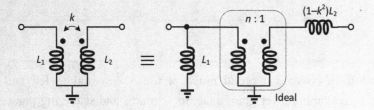

4. Find the transfer function and input impedance of the double-tuned circuit shown below.

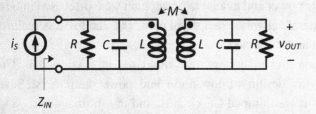

5. Using parallel–series transformation, find the values of L and C to match the 250Ω amplifier input resistance to 50Ω.

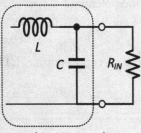

Matching Network

6. Repeat the problem by using the Smith chart.

7. Using parallel–series transformation, design an LC matching network to match an amplifier with an input impedance of 20Ω||1pF to 50Ω.

8. Plot on the Smith chart the impedance of a series RLC circuit over frequency.

9. Consider the following two-port system terminated to admittance Y_L in the second port and to a voltage source v_s with output admittance Y_s in the first port.

a. Prove that

$$Y_{IN} = Y_{11} - \frac{Y_{12}Y_{21}}{Y_{22} + Y_L}$$

$$Y_{OUT} = Y_{22} - \frac{Y_{12}Y_{21}}{Y_{11} + Y_S}$$

$$A_v = \frac{V_L}{V_s} = -\frac{Y_S Y_{21}}{(Y_S + Y_{11})(Y_L + Y_{22}) - Y_{12}Y_{21}}.$$

b. Repeat part a and calculate Z_{IN}, Z_{OUT}, and A_v in terms of Z parameters.

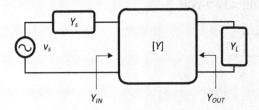

10. Using the circuit from Problem 7, prove that the condition of $Y_s + Y_{IN} = 0$ or $Y_L + Y_{OUT} = 0$ for the boundary between stable and unstable regions both lead to $(Y_s + Y_{11})(Y_L + Y_{22}) - Y_{12}Y_{21} = 0$, which is equivalent to the condition that A_v approaches infinity.

11. Using the circuit from Problem 7, when the input and output are simultaneously conjugate matched the power gain becomes maximum. This would happen if Y_s and Y_L are found such that the following two equations are met:

$$Y_{IN} = Y_{11} - \frac{Y_{12}Y_{21}}{Y_{22} + Y_L} = Y_s^*$$

$$Y_{OUT} = Y_{22} - \frac{Y_{12}Y_{21}}{Y_{11} + Y_s} = Y_L^*.$$

Such Y_s and Y_L are called $Y_{s,opt}$ and $Y_{L,opt}$.

a. Prove that $Y_{s,opt}$ and $Y_{L,opt}$ are given by

$$Y_{s,opt} = \frac{Y_{12}Y_{21} + |Y_{12}Y_{21}|\left(K + \sqrt{K^2 - 1}\right)}{2\text{Re}(Y_{22})} - Y_{11}$$

$$Y_{L,opt} = \frac{Y_{12}Y_{21} + |Y_{12}Y_{21}|\left(K + \sqrt{K^2 - 1}\right)}{2\text{Re}(Y_{11})} - Y_{22},$$

where K is equal to $K = \frac{2\text{Re}(Y_{11})\text{Re}(Y_{22}) - \text{Re}(Y_{12}Y_{21})}{|Y_{12}Y_{21}|}$. This is a more compact equivalent to the results obtained in Section 3.2.6.

b. Prove that under the above optimum condition the power gain is equal to $G_p = \left|\frac{Y_{21}}{Y_{12}}\right|\left(K - \sqrt{K^2 - 1}\right)$, which also proves that for a reciprocal network ($Y_{21} = Y_{12}$)

the power gain is less than unity. As a special case, an LTI passive network cannot amplify the power.

12. Common in transistor modeling, a two-port (refer to Figure 3.2) may be expressed by *hybrid* parameters as follows:

$$V_1 = h_{11}I_1 + h_{12}V_2$$
$$I_2 = h_{21}I_1 + h_{22}V_2.$$

Express the hybrid matrix $H = \begin{bmatrix} h_{11} & h_{12} \\ h_{21} & h_{22} \end{bmatrix}$ in terms of the two-port admittance matrix.

13. Find the Z matrix for the following π two-port.

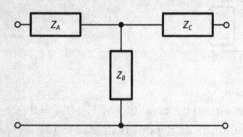

14. A student makes the following argument as to why an ideal transformer is not reciprocal: A voltage $v_0(t)$ in primary results in a voltage of $\frac{n_2}{n_1}v_0(t)$ in the secondary, while the same voltage of $v_0(t)$ in the secondary results in a different voltage of $\frac{n_1}{n_2}v_0(t)$ in the primary, as in general $n_1 \neq n_2$. What is the flaw of the argument?

15. In the circuit shown below, the source is inductive, and is connected to a 50Ω load through a lossless inductive network. Find $\rho_1 = \frac{Z_1 - Z_S^*}{Z_1 + Z_S^*}$ and $\rho_2 = \frac{Z_2 - Z_L^*}{Z_2 + Z_L^*}$. Is $|\rho_1| = |\rho_2|$?

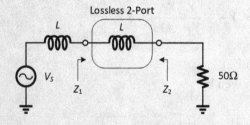

16. The gyrator shown below is described by the Z matrix $[Z] = \begin{bmatrix} 0 & \alpha \\ -\alpha & 0 \end{bmatrix}$, and is loaded with an ideal capacitance C. By finding the relation between the input current and voltage, show that the circuit resembles an inductor. What is the input impedance?

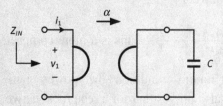

17. For the circuit below, calculate and plot the open-circuit voltages for the input current $i(t) = +I_0$ and $i(t) = -I_0$. Is the circuit reciprocal? Why?

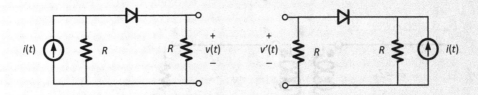

18. Repeat the previous problem for a step input current: $i(t) = u(t)$.
19. Prove the third reciprocity statement shown in Figure 3.8.
20. Show that the reciprocity theorem would be still valid to a network comprising linear but time-varying resistors only.
21. For the pair of Hertzian dipole antennas shown below, work out the effective area of the receiving antenna. Assume the load is matched ($Z_L = Z_{22}^*$ in Figure 3.53). **Hint:** Show that the load current is $I_L = \frac{(E_{1\theta} l \sin\theta_2)^2}{8R_{22}}$, and thus the power delivered is $P_L = \frac{1}{2}R_{22}|I_L|^2$. Also use $\mathcal{P}_1(r,\theta_1,\varphi_1) = \frac{E_{1\theta}^2}{2\eta}$.

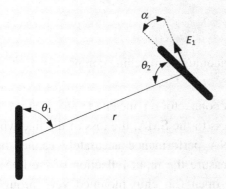

22. Calculate the differential equation of a low-loss transmission line and the value of γ.
23. Consider a transmission line with distributed series impedance of Z_s and parallel admittance of Y_P, both per unit length. (a) Prove that in the sinusoidal steady state case the differential equation for the phasor of line voltage (or current) is given by this expression: $\frac{\partial^2 V(z)}{\partial z^2} = Z_s Y_P \times V(z)$. (b) Show that the solution for this differential equation would be of the form $V^+ e^{-\gamma z} + V^- e^{+\gamma z}$ where $\gamma = \sqrt{Z_s Y_P}$. (c) Calculate the real and imaginary parts of γ for a lossy transmission line where $Z_s = j\omega L_s + R_s$ and $Y_P = j\omega C_P + G_P$. The real part of γ is the decay rate of the wave magnitude.
24. Consider a loss-less distributed transmission line with a series inductance of L per unit length and the distributed parallel capacitance of $C(v)$ per unit length. The distributed capacitance is a varactor that is a function of voltage. (a) Prove that any traveling wave in the line must satisfy the following modified wave equation: $LC(v)\frac{\partial^2 v}{\partial t^2} + L\frac{dC(v)}{dv}\left(\frac{\partial v}{\partial t}\right)^2 = \frac{\partial^2 v}{\partial z^2}$. (b) Assume that $C(v) = C_0 + \alpha v$, $\alpha \neq 0$. Assuming that a solution could be of the form

$f(\omega t - \beta z)$ where $\beta = \omega\sqrt{LC_0}$ and replacing it in the derived wave equation, prove that the solution could be of the form $v(z,t) = k_1\sqrt{\omega t - \beta z} + k_2$.

25. For the circuit below, find the input reflection coefficient.

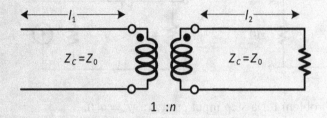

26. For the circuit shown below, find the input reflection coefficient and the VSWR. What value of X and R_L minimizes the VSWR?

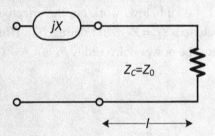

27. Using the Smith chart, recalculate the biconjugate matching components for the example of Figure 3.38.

28. A 50Ω line on PCB connects the SMA connector on one end to an on-chip LNA input on the other end. Since there is only access to the SMA, the loss of the line which must be de-embedded to characterize the LNA performance accurately cannot be measured directly. A common practice is to measure the input reflection of the line on the bare board, where the LNA input is left open (no chip mounted yet) through the SMA connector. Show that for a low-loss transmission line, the insertion loss of the line is half this value in dB.

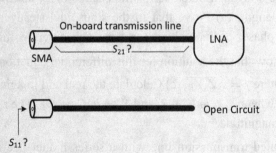

29. Prove the equation $\sum_{n=1}^{N} S_{ns} S^*_{nr} = 0$, where $s \neq r$ for an arbitrary N-port.

30. Find the S-matrix of a series reactance on a transmission line with two different characteristics impedances as shown below. **Answer (partial):** $S_{21} = \frac{2Z_{02}}{jX + Z_{02} + Z_{01}}$.

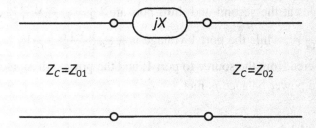

31. Convert the S matrix of a two-port to a Y matrix and vice versa. **Answer:** $[S] = ([U]+[\bar{Y}])^{-1}([U]-[\bar{Y}])$.

32. For the Π-pad attenuator shown below,
   a. Show that for the two ports to be matched to Z_0, we must have $\frac{1}{Z_0} = \sqrt{\frac{1}{R_2^2} + \frac{2}{R_1 R_2}}$.
   b. Show that the available power gain of the circuit if the matching condition (part a) is satisfied is $\dfrac{1}{\left(1+\frac{R_1}{R_2}+\frac{R_1}{Z_0}\right)^2} = \alpha^2$, assuming $R_1 = Z_0 \frac{1-\alpha^2}{2\alpha}$ and $R_2 = Z_0 \frac{1+\alpha}{1-\alpha}$.

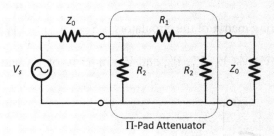

Π-Pad Attenuator

Hint: You could take the advantage of the symmetry, and use the *image impedance* theorem, suggesting that the attenuator may be split in half and be terminated by itself.

33. Calculate the S matrix of the T network shown below by:
   a. Direct analysis of forward and backward signals.
   b. Finding the Y matrix first and converting to the S matrix.

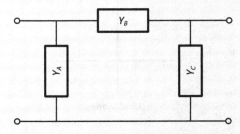

34. In the circulator below, the scattering matrix is $S = \begin{bmatrix} 0 & 1 & 0 \\ 0 & 0 & -1 \\ 1 & 0 & 0 \end{bmatrix}$, and $Z_0 = 1\,\Omega$. For arbitrary port resistances:
   a. Show that port 1 reflection coefficient $\Gamma_1 = \frac{Z_1-1}{Z_1+1} = -\Gamma_2 \Gamma_3$, where $\Gamma_2 = \frac{R_2-1}{R_2+1}$ is the port 2 reflection coefficient, and $\Gamma_3 = \frac{R_3-1}{R_3+1}$ is the port 3 reflection coefficient. Thus, if either one of the second or third ports are matched, the first port is matched too.

b. Show that the voltages at the second and third port are $V_2 = \frac{-(1+\Gamma_2)\Gamma_3}{(1+R_1)-(1-R_1)\Gamma_2\Gamma_3} V_s$ and $V_3 = \frac{1+\Gamma_3}{(1+R_1)-(1-R_1)\Gamma_2\Gamma_3} V_s$, while the port 1 voltage is $\frac{1-\Gamma_2\Gamma_3}{(1+R_1)-(1-R_1)\Gamma_2\Gamma_3} V_s$.

c. Find the power delivered from the source to port 1, and the powers dissipated in R_2 and R_3, and show that the power balance is met.

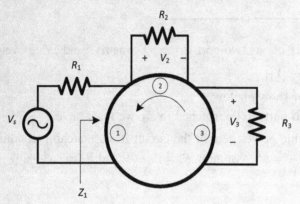

35. In the circuit below, the scattering matrix of the circulators is $S = \begin{bmatrix} 0 & 1 & 0 \\ 0 & 0 & -1 \\ 1 & 0 & 0 \end{bmatrix}$, and $Z_0 = 1\Omega$.

Find all the voltages labeled (V_1, V_2, V_3, V'_1, V'_2, and V'_3) in terms of V_{s1} and V_{s2}.

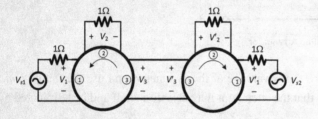

36. Discuss the stability criteria for a unilateral amplifier ($S_{12} = 0$) on the Smith chart.

37. Consider the following AM detector driven by a sinusoidal source, where $v_s(t) = A\cos\omega_0 t$.

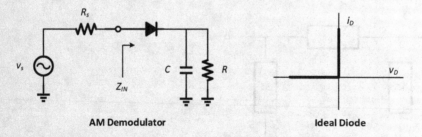

AM Demodulator   Ideal Diode

We assume for simplicity that the diode whose i–v characteristics are shown on the right is ideal. Sketch the capacitor voltage and current, and from that find the input current. Calculate the instantaneous and average power delivered by the source to the detector. What is the instance of the diode turning on? Knowing the average power, estimate the average input resistance of the detector. Assume: $R \gg R_s$, $RC\omega_0 \gg 1$. (**Answer:** $t_{on} = T - \frac{1}{RC\omega_0^2}(\sqrt{1+4\pi RC\omega_0} - 1)$, $R_{IN} \approx \frac{R}{2}$).

38. Propose an LC circuit to match the nonlinear input resistance of the AM detector to source. How does the *instantaneous* reflection coefficient look?

39. Aside from full duplex transceivers, a circulator could be utilized to build an *isolator*. Shown below is a realization, where the circulator second port is connected to the reference impedance Z_0, while the input and output (ports 1 and 3) are terminated with arbitrary resistances R_s and R_L.
   a. Find the transfer function from port 1 to port 3, and vice versa.
   b. Explain why the circuit works as an isolator.

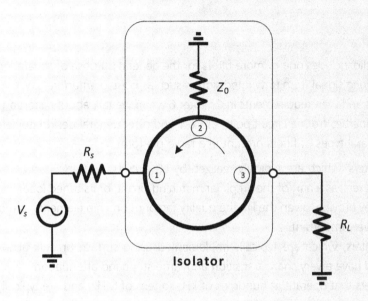

40. In this problem, we will show the steps to derive the available power gain of a two-port.
   a. Knowing that $\Gamma_{IN} = S_{11} + \frac{S_{12}S_{21}\Gamma_L}{1-S_{22}\Gamma_L}$, show that in a similar fashion, $\Gamma_{OUT} = S_{22} + \frac{S_{12}S_{21}\Gamma_s}{1-S_{11}\Gamma_s}$.
   b. Using the two equations in part a, eliminating $S_{12}S_{21}$ term, show that: $\frac{1-\Gamma_{IN}\Gamma_s}{1-\Gamma_L\Gamma_{OUT}} = \frac{1-S_{11}\Gamma_s}{1-S_{22}\Gamma_L}$.
   c. Using $P_{IN} = \frac{|V_1^+|^2}{2Z_0}\left(1-|\Gamma_{IN}|^2\right)$ and $P_L = \frac{|V_1^+|^2}{2Z_0}\left|\frac{S_{21}}{1-\Gamma_L S_{22}}\right|^2\left(1-|\Gamma_L|^2\right)$ derived earlier, and given that for available power gain we must have $\Gamma_{IN} = \Gamma_s^*$ and $\Gamma_L = \Gamma_{OUT}^*$, calculate the expression for the available power given in section 3.8.1.

3.12 References

[1] C. A. Desoer and E. S. Kuh, *Basic Circuit Theory*, McGraw-Hill, 2009.
[2] B. D. H. Tellegen, "The Gyrator, a New Electric Network Element," *Philips Research Report*, 3, 81–101, 1948.
[3] R. E. Colline, *Foundation for Microwave Engineering*, McGraw-Hill, 1992.
[4] D. M. Pozar, *Microwave Engineering*, John Wiley, 2009.
[5] R. Carson, *High-Frequency Amplifiers*, John Wiley, 1975.
[6] K. K. Clarke and D. T. Hess, *Communication Circuits: Analysis and Design*, Krieger, 1994.
[7] W. H. Hayt and J. A. Buck, *Engineering Electromagnetics*, 6th edn, McGraw-Hill, 2001.
[8] P. H. Smith, "Transmission Line Calculator," *Electronics*, 12, no. 1, 29–31, 1939.
[9] A. Abidi and J. Leete, "De-embedding the Noise Figure of Differential Amplifiers," *IEEE Journal of Solid-State Circuits*, 34, no. 6, 882–885, 1999.

4 RF and IF Filters

Almost every radio includes one or more filters for the general purpose of separating a *useful information-bearing* signal from unwanted signals such as *noise or interferers*. We will discuss the role of filters and their requirements in Chapter 6 when we talk about distortion. Here we discuss their properties from a circuit point of view and offer general design guidelines.

There are several types of filters present in a radio platform:

- RF on-chip filters, which are typically realized by LC circuits, and are used as standalone filters or duplexers, as a part of the amplifier matching circuit, or its tuned load. The amount of filtering they provide, given the limited quality factor of on-chip elements, is modest but, nonetheless, very important.
- RF external filters, which are typically implemented using surface or bulk acoustic wave elements, and have a very sharp transition and large stopband attenuation.
- IF on-chip filters that operate at hundreds of kHz to tens of MHz, and are typically realized by transistors, resistors, and capacitors. They usually provide partial or full channel selection in the receiver, as well as anti-aliasing for the receiver ADC, or are utilized to suppress the DAC noise or image in the transmitter.

In this chapter we start by reviewing some of the basic concepts of filter theory and design. We turn our attention to LC filters first. While LC filters, given the inductor size concerns, are used only at RF, the concepts are applicable to both active and SAW/FBAR filters that will be discussed later in the chapter. We also present a detailed discussion of quadrature filters and quadrature signal generation, and discuss a brief overview of N-path filters. The N-path filters will be discussed more thoroughly in Chapter 8, as they effectively operate like mixers.

Filter design and circuit synthesis is at least a book of its own, and the purpose of this chapter is to highlight what is necessary as far as RF design is concerned. More details may be found in [1], [2], [3], [4].

The specific topics covered in this chapter are:

- Passive LC filters
- Active opamp-RC and g_m-C filters
- SAW and FBAR filters
- Duplexers
- N-path filters
- Polyphase filters and quadrature generation

For class teaching, we recommend focusing on selected topics from Sections 4.1, 4.3, and 4.7, while Sections 4.2, 4.4, 4.5, and 4.6 may be assigned as reading.

4.1 IDEAL FILTERS

The transfer function of an ideal bandpass filter (BPF) as an example is shown in Figure 4.1, and is described below:

$$H(f) = \begin{cases} e^{-j2\pi f t_d} & f_l \leq |f| \leq f_h \\ 0 & \text{elsewhere} \end{cases}.$$

The filter has a bandwidth of $B = f_h - f_l$ with a passband gain of unity, and a constant delay of t_d. f_h and f_l are the cutoff frequencies. Similarly we can define lowpass (LPF) and highpass filters (HPF) by either setting $f_l = 0$, or $f_h = \infty$.

Such an ideal filter, however, is physically not realizable, as its characteristics cannot be achieved with a *finite number* of elements. We will defer the mathematical proof[1] to Problem 3, and instead provide a qualitative perspective here.

Consider an ideal LPF as an example with a bandwidth of B, whose transfer function is shown in Figure 4.2, and is equal to

$$H(f) = e^{-j2\pi f t_d} \Pi\left(\frac{f}{2B}\right).$$

The impulse response will be the inverse Fourier of $H(f)$,

$$h(t) = 2B \operatorname{sinc}[2B(t - t_d)],$$

also sketched in Figure 4.2.

Since $h(t)$ is the response to $\delta(t)$, and has nonzero values for $t < 0$, it implies that the output appears before the input is applied. Such a system is said to be *noncausal*, and is physically impossible to realize. This has to do with infinitely sharp transition in the frequency domain, which results in a time domain response with infinite duration.

Shown in Figure 4.3 is the magnitude of the frequency response of a *practical* BPF. The cutoff frequencies f_h and f_l are typically defined as where the gain drops by $\sqrt{2}$ or 3dB, and $B = f_h - f_l$ is the 3dB bandwidth. More importantly, between the passband and stopband are *transition regions*, where the filter is expected to neither pass nor reject signals. This lies on the premises that no unwanted signals are expected to exist at such a region. For instance, for a GSM radio operating at band V, the receiver signal lies between 869–894MHz, whereas the transmitter (which could potentially act as an interferer) is at 824–849MHz, and thus there is a 20MHz guard-band allowed between the edges of receiver and transmitter frequencies. This makes the receiver filter realizable, but comes at the expense of an *unoccupied portion of the spectrum* left unused.

---

[1] It mathematically may be proven in the frequency domain using the Paley–Wiener criterion, stating that for a causal function $h(t)$ with finite energy, $\int_{-\infty}^{\infty} \frac{Ln|H(\omega)|}{1+\omega^2} d\omega$ must exist and be finite. See Problem 3 for more details.

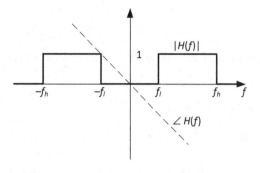

Figure 4.1: Frequency response of an ideal BPF

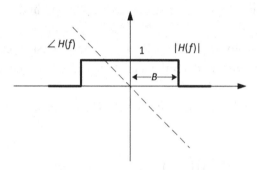
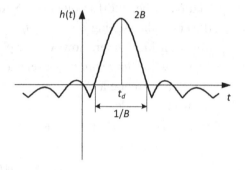

Figure 4.2: An ideal LPF and its impulse response

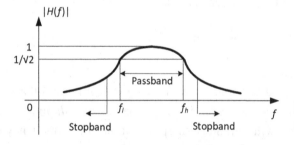

Figure 4.3: Practical bandpass filter frequency response

4.2 DOUBLY TERMINATED LC FILTERS

Traditionally, LC filters have been realized as *reactance* two-ports, which are resistively terminated at both ports (Figure 4.4).

There are several important reasons for this:

- In practice, all physical generators have a nonzero internal impedance, and in most cases the loads are at least partially resistive.
- Doubly terminated reactive two-ports offer the possibility of simultaneously matching both the generator and the load to the rest of the system as was shown in the previous chapter.
- They offer the least sensitivity to the components variation [5], [6], [7].

The last point here deserves more explanation. If the two-port is matched to the source, that is, the impedance looking into the two-port Z_1 is made equal to that of the source, then the source

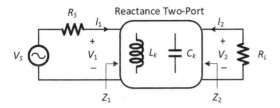

Figure 4.4: A doubly terminated reactive two-port

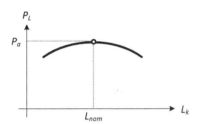

Figure 4.5: Power delivered to the output as a function of the LC two-port inductance

available power, $P_a = \frac{|V_S|^2}{4R_S}$ is entirely absorbed by the two-port. Since the two-port is lossless, that power is delivered to the load, that is, $P_L = P_a$. This is usually the case around the filter passband as is desirable to minimize the loss. Assume now that any element of the two-port, say the inductor L_k, is slightly varied from its nominal value L_{nom}. Whether the inductor increases or decreases, the power delivered to the load is going to decrease. This situation is graphically illustrated in Figure 4.5.

The important conclusion made here is that in the immediate vicinity of its nominal value, L_k has little impact on the power delivered to the load, or equivalently the filter transfer function. More explicitly, the sensitivity of P_L to changes in L_k is zero, or

$$\frac{\partial P_L}{\partial L_k} = 0.$$

This argument can be of course made for any element of the reactive two-port.

There are several reasons as to why the two-port itself is often chosen to be reactive (or LC):

- A reactance two-port can process signal without dissipating any power of its own, keeping the passband loss ideally zero.
- A reactance two-port exhibits a loss versus frequency characteristic, which can vary very rapidly with frequency. This is of course crucial in the filter design.

While the first point is rather obvious, the second point above may be clarified by considering the frequency response of a doubly terminated RC two-port compared to that of an LC circuit as illustrated in Figure 4.6.

Without the loss of generality, we assume in each case the system has three poles, all *real* poles for the case of RC two-port, whereas the LC two-port has one real pole, and a pair of complex conjugate poles.[2]

---

[2] It can be shown that any RC network can have only real poles, whereas RLC networks can have real or complex poles.

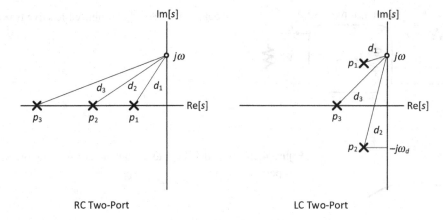

Figure 4.6: Magnitude of transfer function of RC and LC two-ports based on their pole location

The voltage transfer function may be written as

$$A_V(s) = \frac{A_0}{(s-p_1)(s-p_2)(s-p_3)},$$

and the magnitude of the transfer function is

$$|A_V(j\omega)| = \frac{|A_0|}{|(j\omega-p_1)||(j\omega-p_2)||(j\omega-p_3)|} = \frac{|A_0|}{d_1 d_2 d_3},$$

where d_i is the distance from the pole p_i to an arbitrary point on the $j\omega$ axis. For the RC network, the magnitude of the transfer function is maximum at DC ($\omega=0$), as the length of all the three vectors d_1, d_2, and d_3 is minimum. As we move up along the $j\omega$ axis, that is to say, as the frequency increases, $|A_V(j\omega)|$ monotonically reduces. The LC circuit has a local maximum at DC, as well as another local maximum around the complex poles ordinate, $\pm\omega_d$, as the length of d_1 (or d_2) is minimum. In either case, the transfer function magnitude has a local maxima around the pole frequency, but in the case of the RC circuit, since all the poles are real, this happens only at DC. Furthermore, around the pole there is a sharp variation in the frequency response, and clearly the LC two-port is advantageous from this point of view, as complex poles with ordinate around the passband create sharp transition to stopband. This also gives us a useful recipe as to how design the filter: place poles with ordinates around the passband edge close to $j\omega$ axis to create a sharp transition.

Example: Shown in Figure 4.7 is the frequency response of the following two transfer functions:

$$H_1(s) = \frac{1}{(s+1)(s^2+s+1)}$$

$$H_2(s) = \frac{6}{(s+1)(s+2)(s+3)}.$$

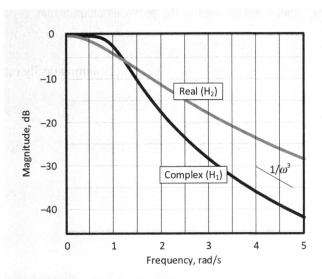

Figure 4.7: Magnitude of two transfer functions, one with complex poles, the other with all real poles

They both have a real pole at -1, whereas H_1 has a pair of complex conjugate poles at $\frac{-1\pm j\sqrt{3}}{2}$, while H_2 has two additional real poles at -2, and -3. The transfer function are normalized to have a magnitude of 1 at DC. The two transfer functions both drop at a rate proportional to $\frac{1}{\omega^3}$ at high frequencies, but clearly H_1 has a flatter passband with sharper stopband transition.

In summary, since RC two-ports have all of their poles on the σ axis, they show large passband droop with modest stopband sharpness. The LC two-port on the other hand can have complex poles arbitrarily close to the $j\omega$ axis. A similar argument may be made for the RL networks as well.

4.2.1 Transducer Parameters

Consider the doubly terminated LC circuit shown in Figure 4.4, consisting of a lossless (LC) network (representing the filter) terminated to a source and a load. V_S represents the RMS value of the source voltage.

Assuming the impedances looking into the right and left of the lossless circuit are Z_1 and Z_2 respectively, we define the input and output reflection factors as

$$\rho_1 \triangleq \frac{R_S - Z_1}{R_S + Z_1}$$

$$\rho_2 \triangleq \frac{R_L - Z_2}{R_L + Z_2},$$

which is slightly different from the one we introduced in the previous chapter (ρ_1 was defined as $\frac{Z_1 - R_S}{Z_1 + R_S}$, which is negative of the one above, and so on), but it is more consistent with filter design textbooks.

Using the power balance argument, we showed in the previous chapter that

$$|\rho_1| = |\rho_2|,$$

which is to say, if one port is matched, say $Z_1 = R_S$, the other port is automatically matched too, that is, Z_2 must be equal to R_L.

We also defined the voltage gain as

$$A_V \triangleq \frac{V_2}{V_S},$$

and the transducer factor as

$$H(s) \triangleq \frac{1}{2}\sqrt{\frac{R_L}{R_S}} \frac{1}{A_V(s)}.$$

Given that the source available power is $P_a = \frac{|V_S|^2}{4R_S}$, and the power delivered to the load is $P_L = \frac{|V_2|^2}{R_L}$, then

$$\frac{P_a}{P_L} = \frac{\frac{|V_S|^2}{4R_S}}{\frac{|V_2|^2}{R_L}} = \left|\frac{1}{2}\sqrt{\frac{R_L}{R_S}}\frac{1}{A_V}\right|^2 = |H(j\omega)|^2 \geq 1.$$

Since the LC network is lossless, the power delivered it to its input is always equal to the power delivered to the load, that is, $P_1 = P_L$. On the other hand, $P_1 \leq P_a$. The difference,

$$P_r = P_a - P_1,$$

is known as the *reflected power*, that is, the amount of power not delivered to the network when it is not matched. This is graphically shown in Figure 4.8.

To find the relation between the available power and the reflected power, we can write

$$P_r = P_a - P_1 = \frac{|V_S|^2}{4R_S} - \text{Re}[Z_1]\left|\frac{V_S}{R_S+Z_1}\right|^2 = \frac{|V_S|^2}{4R_S}\frac{|R_S+Z_1|^2 - 4R_S\text{Re}[Z_1]}{|R_S+Z_1|^2} = \frac{|V_S|^2}{4R_S}\left|\frac{R_S-Z_1}{R_S+Z_1}\right|^2$$

or

$$P_r = |\rho_1|^2 P_a.$$

This can be physically explained knowing that $P_1 = P_a$ only when the input is matched, that is, when $|\rho_1| = 0$.

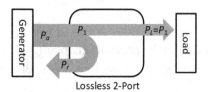

Figure 4.8: Power flow from the source to load in a lossless two-port

Next we introduce the *characteristic function*, $K(s)$, of the terminated two-port

$$K(s) \triangleq \rho_1(s)H(s),$$

or in the $j\omega$ domain

$$|K|^2 = |\rho_1|^2|H|^2 = \frac{P_r}{P_a}\frac{P_a}{P_L} = \frac{P_r}{P_L} = \frac{P_a - P_L}{P_L} = |H|^2 - 1,$$

which leads to the *Feldtkeller equality*:

$$|H|^2 = 1 + |K|^2.$$

Since $H(s)$ and $K(s)$ are real rational functions of s, $|H(j\omega)|^2 = H(j\omega)H(-j\omega)$, and $|K(j\omega)|^2 = K(j\omega)K(-j\omega)$. Replacing $j\omega$ by s, we can rewrite the Feldtkeller equation in a more familiar form:

$$H(s)H(-s) = 1 + K(s)K(-s).$$

Example: Consider the circuit shown in Figure 4.9, representing a 3rd-order Butterworth filter.

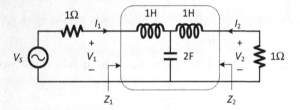

Figure 4.9: A 3rd-order Butterworth filter

Using a mesh current analysis, we can write

$$\begin{bmatrix} 1+s+\dfrac{1}{2s} & \dfrac{-1}{2s} \\ \dfrac{-1}{2s} & 1+s+\dfrac{1}{2s} \end{bmatrix} \begin{bmatrix} I_1 \\ -I_2 \end{bmatrix} = \begin{bmatrix} V_S \\ 0 \end{bmatrix},$$

and

$$A_V = \frac{V_2}{V_S} = \frac{-I_2}{V_S} = \frac{\Delta_{12}}{\Delta} = \frac{\frac{1}{2}}{(s+1)(s^2+s+1)},$$

where Δ_{12} and Δ are the mesh matrix cofactor and determinant [8]. Thus, by definition

$$H(s) = \frac{1}{2}\sqrt{\frac{R_L}{R_S}}\frac{1}{A_V(s)} = (s+1)(s^2+s+1) = s^3 + 2s^2 + 2s + 1.$$

Clearly, $|H|^2 = \omega^6 + 1 \geq 1$. In fact $|H|$ equals one only at DC, where the lowpass LC circuit is a short, and the load and source are matched.

Continued

Similarly, the admittance looking to the right of the source is

$$Y = \frac{I_1}{V_S} = \frac{\Delta_{11}}{\Delta} = \frac{2s^2 + 2s + 1}{2(s+1)(s^2+s+1)}.$$

Thus,

$$Z_1 = \frac{1}{Y} - 1 = \frac{2s^3 + 2s^2 + 2s + 1}{2s^2 + 2s + 1}.$$

The reflection factor is then readily calculated to be

$$\rho_1(s) = \frac{1 - Z_1}{1 + Z_1} = \frac{-s^3}{s^3 + 2s^2 + 2s + 1}.$$

As expected, $|\rho_1| = 0$ at DC. Also note that the denominator of $\rho_1(s)$ is the same as the numerator of $H(s)$. This will be proven for a general case in the next section.

Finally, the characteristic function is

$$K(s) = \rho_1(s)H(s) = -s^3.$$

Now assuming $|V_S| = 2$V RMS, the source available power would be

$$P_a = \frac{|V_S|^2}{4R_S} = 1\text{W}.$$

To find the reflected power and the power delivered to the load at $\omega = 1$rad/s, we have

$$|\rho_1(j\omega)|^2 = \frac{\omega^6}{\omega^6 + 1},$$

which gives $P_r = |\rho_1(j1)|^2 P_a = \frac{1}{2}$W, and $P_L = P_a - P_r = \frac{1}{2}$W. Alternatively, $|H(j1)|^2 = 2$, which yields the same $P_L = \frac{P_a}{|H(j1)|^2} = \frac{1}{2}$W. We will shortly see that $\omega = 1$rad/s is the filter 3dB passband edge, which justifies why at this frequency only half of the power is absorbed at the output.

4.2.2 Relation between Transducer and Immittance[3] Parameters

It is common to express any two-port through its impedance or admittance parameters, as typically that's how the two-port and ultimately the filter are synthesized. With reference to Figure 4.10, using the impedance (or z) parameters for instance, there are two constraints set on the two-port input/output voltages/currents, and two additional constraints set by the source and load as the following:

$$V_1 = z_{11}I_1 + z_{12}I_2$$
$$V_2 = z_{21}I_1 + z_{22}I_2$$
$$V_s = V_1 + R_s I_1$$
$$V_2 = -R_L I_2,$$

---

[3] "Immittance" is a term coined by Bode describing either an impedance or an admittance.

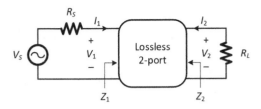

Figure 4.10: Double-terminated lossless two-port

where z_{11}, z_{12}, z_{21}, and z_{22} are the two-port z parameters. It must be noted that z_{11} and z_{22} are the *driving point input and output impedances* when the other port is left open. Furthermore, since the two-port is reciprocal, $z_{12} = z_{21}$.

It follows that (see also Problem 16)

$$H(s) = \frac{(z_{11} + R_s)(z_{22} + R_L) - z_{12}^2}{2\sqrt{R_s R_L} z_{12}}$$

and

$$K(s) = \frac{(R_s - z_{11})(R_L + z_{22}) + z_{12}^2}{2\sqrt{R_s R_L} z_{12}}.$$

Now with $H(s)$ and $K(s)$ known based on certain filter characteristics (passband loss, stopband rejection, etc.), one can determine the two-port z parameters. At first, this may seem an impossible task, as it involves solving two equation ($H(s)$ and $K(s)$) for three unknowns (z_{11}, z_{12}, and z_{22}). However, since z_{ij}s specify the impedance parameters of a *reactance* two-port, they are all *odd* rational functions of s. While this can be mathematically proven [3], [9] by Tellegen's theorem [8], it can be physically justified considering that a reactance network is lossless, and thus neither dissipates nor generates power. This requires $\text{Re}[z_{ij}(j\omega)] = 0$ for all ω, which means $z_{ij}(s)$ must be an odd function of s. With that, and by identifying the even and odd parts of $H(s)$, and $K(s)$ (H_e, H_o, and K_e, K_o respectively), it follows

$$z_{11} = R_s \frac{H_e - K_e}{H_o + K_o}$$

$$z_{22} = R_L \frac{H_e + K_e}{H_o + K_o}$$

$$z_{12} = \frac{\sqrt{R_s R_L}}{H_o + K_o}.$$

Note that there are now four equations in terms H_e, H_o, K_e, and K_o, but the same three unknowns. However, the four functions H_e, H_o, K_e, and K_o are related given the Feldtkeller equality.

A similar expression can be developed in terms of the two-port y or ABCD (chain or transmission) parameters [3]:

Example: For the previous filter of Figure 4.9, it is easy to show that for the π-LC network (Figure 4.11):

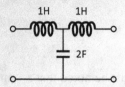

Figure 4.11: The LC section of the example of Figure 4.9

$$z_{11} = z_{22} = s + \frac{1}{2s} = \frac{2s^2 + 1}{2s}$$

$$z_{12} = \frac{1}{2s}.$$

On the other hand, since we already have the transducer and characteristic functions from the previous section, we can write:

$$H_e = 2s^2 + 1$$
$$H_o = s^3 + 2s$$
$$K_e = 0$$
$$K_o = -s^3.$$

Consequently,

$$z_{11} = R_s \frac{H_e - K_e}{H_o + K_o} = \frac{2s^2 + 1}{2s}$$

$$z_{12} = \frac{\sqrt{R_s R_L}}{H_o + K_o} = \frac{1}{2s}.$$

These results of course match the ones calculated directly from the LC two-port. The difference is that we do not have the actual LC circuit to begin with; rather what is known is the characteristic or transducer functions for a given filter requirements. Once the z parameters are obtained as shown above, then one can synthesize the corresponding LC circuit, and not the other way around.

4.2.3 Transducer Parameters Properties

Next, we will discuss the properties of the transducer and characteristic functions from physical point of view. This is important, since what is given in the course of the filter design is typically the characteristic function. On the other hand, once $K(s)$ is determined given a certain filter characteristics (we will deal with this in the next section), one must ensure that it is realizable, and if so, how it can be implemented.

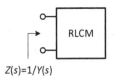

Figure 4.12: An RLCM one-port

$Z(s)=1/Y(s)$

4.2.3.1 Positive Real Property

Before we get to the properties of the transducer parameters, it must be emphasized that for minimum z_{11} and z_{22} representing the open port driving impedances of the lossless two-port (as well as Z_1 and Z_2, the terminated impedances) must be *realizable* impedances. The realizability conditions for an RLCM one-port impedance (or admittance) was first given by Otto Brune[4] in his classical 1931 paper [10] as follows:

An impedance function $Z(s)$ is realizable using lumped RLCM elements (Figure 4.12) if and only if it is a positive real (PR) function, that is,

1. $Z(s)$ is a real rational function of s
2. $\text{Re}[Z(s)] \geq 0$ if $\text{Re}[s] \geq 0$

The proof is through stability argument [9], [10], or by Tellegen's theorem [3], and will be skipped here. Note that these are both necessary and sufficient conditions. The sufficiency follows from a synthesis algorithm developed by Brune that can successfully realize any PR impedance using positive RLCM elements.

The conditions above, though elegant and concise, are somewhat useless from a practicality point of view. There are several equivalent conditions, the most notable one being

1. $Z(s)$ is a real rational function of s.
2. $\text{Re}[Z(j\omega)] \geq 0$ for all ω.
3. All poles of $Z(s)$ are in the closed[5] left half-plane (LHP) of the s-plane. All the $j\omega$ axis poles (including zero and infinity) must be simple with positive real residues.[6]

The second condition is simply justified considering that the power dissipated in the RLCM one-port (assuming positive elements of course) must be positive or zero, and is easily deduced from the original PR second condition. While the proof of the third condition is beyond the scope of our RF book, one can tie that to the stability requirement of the RLCM one-port: Poles in the RHP of the s-plane lead to terms like $e^{\alpha t}$ ($\alpha > 0$) in the impulse response of the one-port, while multiple $j\omega$ axis poles lead to terms like $t^n \cos(\omega t + \varphi)$ ($n \geq 1$), both of which grow indefinitely as $t \to \infty$.

The simplicity of the poles of $Z(s)$ (or $Y(s) = \frac{1}{Z(s)}$) at infinity necessitates that the degree of the numerator and denominator of $Z(s)$ differ by no more than one, a condition that again may be physically justified by examining the RLCM impedances at very high frequencies (see Problem 9).

---

[4] Otto Brune completed his Ph.D. at MIT in 1929. His Ph.D. work, later published in [10], is one of the key contributions to modern network synthesis.

[5] Inside the LHP or on the $j\omega$ axis, including zero and infinity.

[6] Residues of partial fraction of $Z(s)$. If $Z(s) = \frac{N(s)}{D(s)} = \frac{N(s)}{(s-p_1)(s-p_2)\ldots} = \frac{K_1}{(s-p_1)} + \frac{K_2}{(s-p_2)} + \cdots$, then K_1 is a residue of p_1. If p_1 lies on $j\omega$ axis, then K_1 must be real and positive to satisfy the third equivalent condition.

In the case of z_{11} and z_{22} representing the driving point impedances of a *reactive* circuit, it is easy to show that all the poles and zeros lie on the $j\omega$ axis (in conjugate pairs), and of course are simple with positive real residues. As stated earlier, $z_{11}(s)$ and $z_{22}(s)$ are odd rational functions of s, or $z_{11}(j\omega)$ and $z_{22}(j\omega)$ are purely imaginary.

Example: $Z(s) = \frac{2s^2+1}{2s^3+2s}$ is a PR function, and since it is odd, it is a reactance function. If expanded by partial fraction,

$$Z(s) = \frac{2s^2+1}{2s^3+2s} = \frac{\frac{1}{2}}{s} + \frac{\frac{1}{2}s}{s^2+1} = \frac{\frac{1}{2}}{s} + \frac{\frac{1}{4}}{s+j} + \frac{\frac{1}{4}}{s-j}.$$

Thus, all the poles are simple on the $j\omega$ axis, with positive real residues. Of course, $\mathrm{Re}[Z(j\omega)] = 0$.

From the partial fraction expansion (generally known as Foster's method of synthesis),[7] since,

$$Z(s) = \frac{\frac{1}{2}}{s} + \frac{\frac{1}{2}s}{s^2+1} = \frac{\frac{1}{2}}{s} + \frac{1}{2s + \frac{2}{s}},$$

then the circuit representing $Z(s)$ consists of a capacitor in series with an LC parallel circuit, as shown in Figure 4.13 left. It is clear that if the residue condition is not met, it would lead to complex or negative values of the inductors or capacitors.

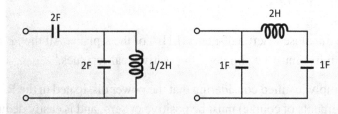

Figure 4.13: Two possible realizations of the impedance $Z(s) = \frac{2s^2+1}{2s^3+2s}$

Partial Fraction (Foster)   Pole Removal (Cauer)

While the topic of synthesis is well beyond the scope of this chapter, one can show that by using Cauer's method[8] (removing poles of $Y(s) = \frac{1}{Z(s)}$ at infinity) the LC circuit on the right is obtained, which represents a 3rd-order Butterworth ladder filter as will be discussed shortly. Clearly, unlike analysis, the synthesis does not lead to a unique answer.

It is interesting to point out that for any reactance function, with $Z(j\omega) = jX(j\omega)$, we can show $\frac{d}{d\omega}X > 0$ for all ω (see Problem 11 for the proof). Consequently, the poles and zeros are located

---

[7] Ronald Martin Foster was a Bell Labs engineer who made significant contributions to the field of circuit synthesis and filter design.

[8] Wilhelm Cauer was a German mathematician and scientist who sadly was executed by Soviet soldiers in 1945 at the end of World War II during the fall of Berlin. He was the co-supervisor of Brune (along with Ernst Guillemin), and is most noted for his significant contributions to circuit synthesis and filter theory.

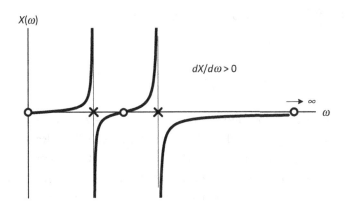

Figure 4.14: Poles and zeros of a reactance function $\left(Z(s) = \frac{s(s^2+2)}{(s^2+1)(s^2+4)}\right)$ interlaced on the $j\omega$ axis

interlaced on the $j\omega$ axis [11], as shown in Figure 4.14. This is certainly true for the previous example where there are two zeros at $\frac{\sqrt{2}}{2}$, and ∞, and two poles at 0 and 1.

With this background, let us derive other properties of the transducer function of the double-terminated LC filter.

4.2.3.2 Realizability Conditions for Transducer Parameters

While z_{11} and z_{22} must be PR impedances, there are additional requirements for the two-port impedance, as well as its transducer parameters to be realizable.

Listed below are the conditions under which the transducer function, $H(s) = \frac{E(s)}{P(s)}$, is realizable. A detailed proof may be found in [3], and we will just offer a physical justification here.

1. The numerator ($E(s)$) and the denominator ($P(s)$) are polynomials with real coefficients.
2. The numerator ($E(s)$) is a strictly Hurwitz polynomial,[9] that is all of its roots are on the left-half of the s-plane (LHP). This is justified from a stability point of view.
3. $|H(j\omega)| \geq 1$ for all ω.
4. $P(s)$ is either an even or an odd polynomial, unless common factors of $P(s)$ and $E(s)$ cancel. This follows from $H(s) = \frac{(z_{11}+R_s)(z_{22}+R_L)-z_{12}^2}{2\sqrt{R_s R_L} z_{12}}$, and that z_{ij}s are odd functions of s. Multiplying by the common denominator of z_{11}, z_{12}, and z_{22} leaves an either odd or even denominator for $H(s)$.

The degree of $E(s)$ must be greater or equal than that of $P(s)$. If not, for sufficiently large ω, $|H(j\omega)|$ will become less than one, violating the third condition.

Now let us consider $K(s) = \frac{F(s)}{P(s)}$. The expressions derived earlier for $K(s)$ and $H(s)$ in terms of the two-port z parameters have both z_{12} in their denominator, and we therefore suspect $K(s)$ and $H(s)$ have the same denominator, $P(s)$. Furthermore, from Feldtkeller equality,

$$E(s)E(-s) = F(s)F(-s) + P(s)P(-s).$$

Since the reflection factor, $\rho_1(s) = \frac{K(s)}{H(s)}$, then $\rho_1(s) = \frac{F(s)}{E(s)}$. Also as $|\rho_1(j\omega)| \leq 1$, then the degree of $E(s)$ must be greater than or equal to that of $F(s)$ as well.

---

[9] In a strictly Hurwitz polynomial, all the zeros (or roots) happen either on the negative σ axis or in conjugate pairs inside LHP. Thus, all the coefficients of the polynomial must be positive, with no missing terms. This is, however, only a *necessary* condition. See Problem 10 for an example.

4.2.3.3 Physical Interpretation of the Poles and Zeros of Transducer Function

On the second condition, since

$$|H|^2 = \frac{|Z_1 + R_s|^2}{4R_s R_L},$$

the only way for $H(s)$ and consequently $E(s)$ to be zero is if $Z_1(s) = -R_s$. Now since $Z_1(s)$ is a realizable impedance, it must be positive real, and can be negative only if $\text{Re}[s] < 0$, which implies $E(s)$ is strictly Hurwitz. Additionally, since $|H|^2 = \frac{P_a}{P_L}$, $E(s) = 0$ requires the simultaneous physical condition of a nonzero output with a zero source voltage. Therefore the zeros of $E(s)$ are the nonzero *natural frequencies* of the network. It is now clear why these zeros must be in open LHP.

The poles of $K(s)$, which are also the poles of $H(s)$, are the zeros $P(s)$. At these frequencies, $|H|$ becomes infinite, which implies no power is delivered to the output ($|H|^2 = \frac{P_a}{P_L}$, thus $P_L = 0$), or the loss becomes infinite. Therefore, the zeros of $P(s)$ are often called *transmission zeros* or *loss poles*, and are a critical part of the filter design. If $E(s)$ is of higher degree than that of $P(s)$, then $\omega \to \infty$ is a loss pole.

There is no restriction for the zeros $P(s)$ to be in the LHP, but since $P(s)$ have real coefficients, they must either be real or appear in complex conjugate pairs.

Example: Consider the same network of the example of Figure 4.9, where $K(s) = -s^3$ was obtained. As we will see in the next section, this represents the characteristic function of a 3rd-order maximally flat (or Butterworth) filter. We wish to find the transducer function not having any knowledge of the actual circuit. Using the Feldtkeller equation,

$$H(s)H(-s) = 1 + K(s)K(-s) = -s^6 + 1,$$

the function $H(s)H(-s) = -s^6 + 1$ has six zeros grouped on the unity circle as shown in Figure 4.15.

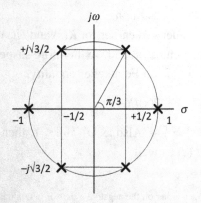

Figure 4.15: Zero location of the function $-s^6 + 1$

Now, knowing that the zeros of $H(s)$ must all be on the LHP, then for $H(s)$ we choose only the three zeros on the LHP, and assign the other three on the RHP to $H(-s)$. Thus,

$$H(s) = (s+1)\left(s + \frac{1+j\sqrt{3}}{2}\right)\left(s + \frac{1-j\sqrt{3}}{2}\right) = (s+1)(s^2+s+1),$$

which is the same result previously obtained.

4.2.4 Specifying the Filter

We showed earlier that an ideal filter cannot be implemented with a finite number of elements, and hence given a certain requirement, the filter characteristic must be *approximated* with realizable transducer parameters.

In most cases (unless the phase or time response of the filter is prescribed)[10] the filter is defined through its *transducer loss* α, as follows:

$$\alpha(\omega) = 10\log_{10}(|H(j\omega)|^2) = 10\log_{10}(1 + |K(j\omega)|^2).$$

The filter is expected to have minimum loss in its passband, and hence $|H(j\omega)|^2 = \frac{P_a}{P_L} \approx 1$ (the source available power is mostly absorbed by the load), whereas at stopband, it blocks the source power reaching the load, and hence $|H(j\omega)|^2 \to \infty$, or the loss is infinite. A typical lowpass filter transducer loss is illustrated in Figure 4.16 as an example.

The transducer loss is expected to be small (ideally zero) in the passband, and very large in the stopband. For a practical filter, there is of course a transition region in between. Generally the smaller the transition region, the sharper the filter, and the closer we become to an ideal filter.

For example, a lowpass filter for WLAN applications may be specified as follows:

$$\begin{cases} \alpha \leq 0.5\text{dB} & \text{for} \quad 0 \leq f \leq 8.5\text{MHz} \\ \alpha \geq 40\text{dB} & \text{for} \quad 35\text{MHz} \leq f \leq \infty, \end{cases}$$

or equivalently, $\alpha_p = 0.5\text{dB}$, $\omega_p = 2\pi \times 8.5\text{MHz}$, $\alpha_s = 40\text{dB}$, and $\omega_s = 2\pi \times 35\text{MHz}$.

The next step would be to find the suitable transducer parameters. The branch of circuit theory that deals with this is known as *approximation theory*, and is briefly discussed next.

Knowing the transducer loss, it is possible to determine either $|H(j\omega)|^2$ or $|K(j\omega)|^2$. It is however usually advantageous to find $|K(j\omega)|^2$, due to the similarity between transducer loss, and characteristic function frequency behavior: in passband, $\alpha(\omega) \approx 0$, and so is $|K(j\omega)|^2$, while in stopband both $\alpha(\omega) \to \infty$ and $|K(j\omega)|^2 \to \infty$. This allows us to readily determine the pole and zero locations of $K(s)$: put zeros in $K(s)$ within the passband, and poles within the stopband.

---

[10] If we are interested in phase or group delay, $T_{phase}(\omega) = \frac{\beta(\omega)}{\omega} = \frac{1}{\omega}\tan^{-1}\frac{\text{Im}[H(j\omega)]}{\text{Re}[H(j\omega)]}$, and $T_{delay}(\omega) = \frac{d\beta(\omega)}{d\omega}$ are commonly used.

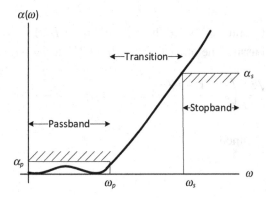

Figure 4.16: Transducer loss of a typical lowpass filter

Example: Suppose we would like to have a lowpass filter with a passband of $0 \leq \omega \leq 1$, and a stopband of $\omega \geq 3$. Then one may choose

$$K(s) = K_0 \frac{(s-z_1)(s-z_2)(s-z_3)(s-z_4)}{(s-p_1)(s-p_2)(s-p_3)(s-p_4)}.$$

To satisfy the passband and stopband values, we further select $z_1 = z_2{}^* = j0.25$, $z_3 = z_4{}^* = j0.75$, $p_1 = p_2{}^* = j3.5$, and $p_3 = p_4{}^* = j5.5$. Setting K_0 to 200, we have

$$K(j\omega) = 200 \frac{(-\omega^2 + 0.25^2)(-\omega^2 + 0.75^2)}{(-\omega^2 + 3.5^2)(-\omega^2 + 5.5^2)}.$$

Figure 4.17 shows the plot of the transducer loss versus frequency, which is well within the requirements, and confirms the effectiveness of the aforementioned pole/zero distribution.

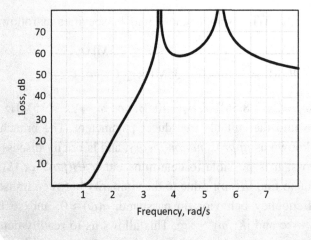

Figure 4.17: Loss response of a filter with

$$K(j\omega) = 200 \frac{(-\omega^2 + 0.25^2)(-\omega^2 + 0.75^2)}{(-\omega^2 + 3.5^2)(-\omega^2 + 5.5^2)}$$

There are generally two ways of approximating the filter loss with what is ideally needed:

1. To compare the two functions and their first $n-1$ derivatives at one specific point of the independent variable (in the case of filter that would be frequency).

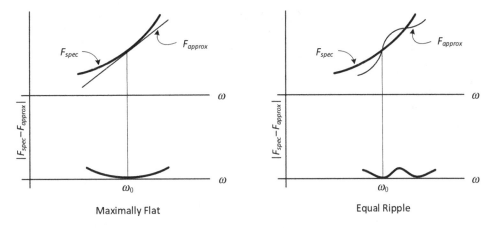

Figure 4.18: Illustration of maximally flat and equal ripple approximation

2. To evaluate the maximum deviation between the two functions in a range of the independent variable.

In the first criterion, known as the *maximally flat* approximation [12], at certain frequency $\omega = \omega_0$ we must have

$$F_{spec}(\omega_0) = F_{approx}(\omega_0)$$

$$\frac{dF_{spec}(\omega_0)}{d\omega} = \frac{F_{approx}(\omega_0)}{d\omega}$$

$$\vdots$$

$$\frac{d^{n-1}F_{spec}(\omega_0)}{d\omega^{n-1}} = \frac{d^{n-1}F_{approx}(\omega_0)}{d\omega^{n-1}}.$$

We take up to the $n-1$ derivatives since an nth-order filter gives us n degrees of freedom. The second criterion, known as the *equal ripple* approximation, suggests that the maximum absolute error

$$Error = max|F_{spec}(\omega) - F_{approx}(\omega)|$$

be minimized in a range of $\omega_1 \leq \omega \leq \omega_2$. Assuming again an nth-order filter, this usually happens if the error function has $n+1$ equal alternating extrema (minima and maxima) in the range of $\omega_1 \leq \omega \leq \omega_2$. The two criteria are compared against in Figure 4.18.

As a case study, we will discuss the maximally flat criterion next in some details, and briefly go over the equal ripple case as well.

4.2.4.1 Maximally Flat Approximation

A simple yet effective way of finding a realizable function is to carry out the maximally flat (or Butterworth)[11] approximation at $\omega = 0$. We first observe that wherever $\frac{d^k \alpha}{d\omega^k}$ is zero, so is $\frac{d^k |K|^2}{d\omega^{2k}}$

---

[11] Stephen Butterworth (1885–1958) was a British physicist and mathematician who invented Butterworth filters.

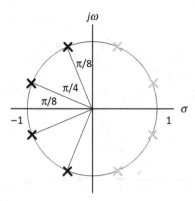

Figure 4.19: Natural frequencies and their mirror image of a 4th-order Butterworth filter

($k = 0, 1, \ldots, n-1$). If one chooses $|K|^2$ to be a polynomial rather than a rational function, that is to say

$$|K|^2 = C_n \omega^{2n} + C_{n-1} \omega^{2n-2} + \cdots + C_0,$$

then maximally flat condition requires all the coefficients to be zero except for C_n. Thus,

$$|K|^2 = C_n \omega^{2n}$$
$$K(s) = \pm \sqrt{C_n} s^n.$$

From the Feldtkeller equation,

$$H(s)H(-s) = 1 + (-1)^n C_n s^{2n}.$$

The zeros of $H(s)$ that are the natural frequencies of the network thus lie on a unity circle in LHP as shown in Figure 4.19.

The transducer loss is

$$\alpha(\omega) = 10 \log(1 + |K(j\omega)|^2) = 10 \log(1 + C_n \omega^{2n}),$$

which is plotted in Figure 4.20 for various values of n, with $C_n = 1$.

A higher order not only gives a higher stopband loss but also exhibits more passband flatness. The order of the filter is naturally a function of the required passband and stopband loss, and must be such that (see Figure 4.16)

$$10 \log(1 + C_n \omega_p^{2n}) \leq \alpha_p$$
$$10 \log(1 + C_n \omega_s^{2n}) \geq \alpha_s.$$

Combining the two equations, with a few simple algebraic steps we have

$$n \geq \frac{\log \frac{1}{k_1}}{\log \frac{1}{k}} = \frac{\log \sqrt{\frac{10^{\alpha_s/10} - 1}{10^{\alpha_p/10} - 1}}}{\log \frac{\omega_s}{\omega_p}},$$

where $k_1 \triangleq \sqrt{\frac{10^{\alpha_p/10} - 1}{10^{\alpha_s/10} - 1}}$ is the discrimination parameter, and $k \triangleq \frac{\omega_p}{\omega_s}$ is the selectivity factor.

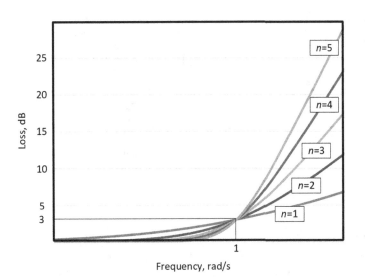

Figure 4.20: Loss response of 1st- to 5th-order Butterworth filter

Alternatively, one can find the filter order through filter design handbooks [13], [14], [15], MATLAB,[12] or filter design software, many of which are available online.

Example: For the WLAN filter example earlier, we had $\alpha_p = 0.5$dB, $\omega_p = 2\pi \times 8.5$MHz, $\alpha_s = 40$dB, and $\omega_s = 2\pi \times 35$MHz. Thus $n \geq \dfrac{\log\sqrt{\dfrac{10^{\frac{40}{10}} - 1}{10^{\frac{0.5}{10}} - 1}}}{\log\frac{35}{8.5}} = 3.99$. So a 4th-order Butterworth filter meets the specifications.

Once the filter order is known, so are the loci of the $H(s)$ zeros, which are the network natural frequencies (Figure 4.19). For instance, for a 2nd-order maximally flat filter (with 1Ω termination), with $K(s) = \pm s^2$, the zeros of $H(s)H(-s) = 1 + s^4$ will be at $s_k = e^{j\pi(n-1+2k)/2n}$, where $n = 4$ and $k = 0, 1, 2, 3$. Selecting the two zeros that lie on the LHP,

$$H(s) = \left(s + \frac{\sqrt{2} + j\sqrt{2}}{2}\right)\left(s + \frac{\sqrt{2} - j\sqrt{2}}{2}\right) = s^2 + \sqrt{2}s + 1.$$

Consequently,

$$z_{11} = R_s \frac{H_e - K_e}{H_o + K_o} = \frac{1}{\sqrt{2}s}$$

$$z_{22} = R_L \frac{H_e + K_e}{H_o + K_o} = \frac{2s^2 + 1}{\sqrt{2}s}$$

$$z_{12} = \frac{\sqrt{R_s R_L}}{H_o + K_o} = \frac{1}{\sqrt{2}s}.$$

---

[12] MATLAB (Matrix Laboratory) is a numerical computing environment developed by MathWorks. Among its many capabilities, it has an extensive filter design library.

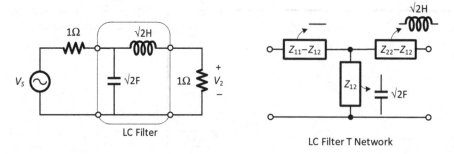

Figure 4.21: Second-order Butterworth filter schematic

The filter circuit is illustrated in Figure 4.21. We will show in the next section how the LC filter is obtained knowing its impedance matrix. However, in this simple case of a 2nd-order, one may realize the filter using the T equivalent of the LC two-port as shown on the right.

The reader can verify that the z parameters above correspond to the LC circuit of Figure 4.21. The filter is normalized, e.g., its 3dB bandwidth is 1rad/s, with impractical values of inductors and capacitors. The filter scaling subject will be covered in Section 4.2.6.

4.2.4.2 Equal Ripple Approximation

In the case of equal ripple approximation, we are interested to minimize the absolute error in a certain range of say $0 \leq \omega \leq \omega_p$, i.e., the filter passband. For convenience, we may choose $\omega_p = 1$ rad/s, and subsequently scale the filter to match our needs (Section 4.2.6). The desired oscillatory behavior in Figure 4.18 suggests a squared and horizontally compressed trigonometric function, for instance, something like

$$|K|^2 = k_p^2 \cos^2 nu(\omega),$$

where

$$u(\omega) = \cos^{-1}\omega.$$

Upon expansion (see Problem 14), it can be shown that $|K|^2$ is in fact a polynomial in ω^2, which oscillates between 0 and k_p^2 for $-1 \leq \omega \leq +1$, taking on the value 0 n times, and the value of k_p^2 $n+1$ times. Furthermore, $|K|^2$ tends to monotonically increase with frequency, approaching $k_p^2 2^{2n-2} \omega^{2n}$ as $\omega \to \infty$ as desired. In fact, for $\omega \geq 1$,

$$|K|^2 = \frac{k_p^2}{4}\left[\left(\omega + \sqrt{\omega^2 - 1}\right)^n + \left(\omega + \sqrt{\omega^2 - 1}\right)^{-n}\right]^2.$$

Given these properties, the filter prescribed as above satisfies our requirements. Besides the filter degree, unlike the Butterworth filter, there is an additional degree of freedom set by the k_p that determines the maximum *passband ripple* as follows:

$$\alpha_p = 10\log(1 + k_p^2).$$

The loss function of a 4th- and a 5th-order filter is plotted in Figure 4.22 for $\alpha_p = 1$dB.

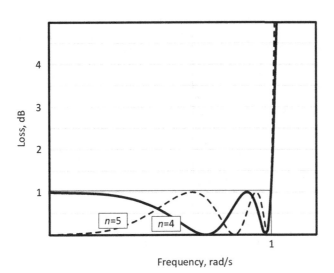

Figure 4.22: Passband transducer loss of a normalized Chebyshev filter

The filters that have $|K|^2 = k_p^2 \cos^2(n \cos^{-1} \omega)$ are known as Chebyshev filters after the mathematician first analyzing the properties of the polynomials $\cos(n \cos^{-1} x)$, known as Chebyshev polynomials.[13]

At high frequencies, for a Butterworth filter, the loss may be approximated as

$$\alpha \approx 10 \log C_n \omega^{2n} = 10 \log C_n + 20n \log \omega.$$

For an nth-order Chebyshev filter, assuming $k_p^2 = C_n$ to have the same passband loss at $\omega = 1$, then

$$\alpha \approx 10 \log k_p^2 2^{2n-2} \omega^{2n} = 10 \log C_n + 20n \log \omega + 6.02(n-1).$$

The first two terms are the same, but there is an additional term, $6.02(n-1)$, that accounts for more stopband loss with the same degree for the Chebyshev filter. This suggests that the Chebyshev implementation is a lot more efficient than Butterworth, which is very desirable in many applications. This comes at a cost though, as there is more phase variation associated with the Chebyshev filters. A comparison between various degrees of Chebyshev (labeled as C) filter with a 3rd-order Butterworth (labeled as B3) is shown Figure 4.23.

With a similar procedure as before, we can show that the order of the Chebyshev filter follows from

$$n \geq \frac{\cosh^{-1} \frac{1}{k_1}}{\cosh^{-1} \frac{1}{k}} = \frac{\cosh^{-1} \sqrt{\frac{10^{\alpha_s/10} - 1}{10^{\alpha_p/10} - 1}}}{\cosh^{-1} \frac{\omega_s}{\omega_p}}.$$

---

[13] Pafnuty Lvovich Chebyshev (1821–1894) was a Russian mathematician.

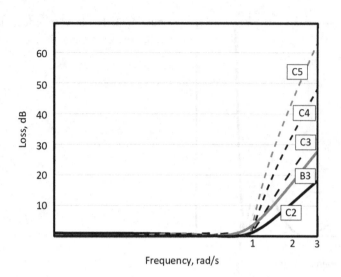

Figure 4.23: Comparison between Chebyshev filters of different order and a 3rd-order Butterworth filter

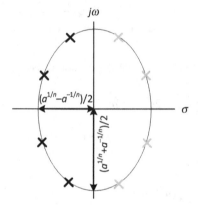

Figure 4.24: Natural frequencies of a 4th-order Chebyshev filter

Example: Back to the WLAN filter example, $\frac{1}{k} = \frac{35}{8.5} = 4.12$, $\frac{1}{k_1} = \sqrt{\frac{10^{40/10}-1}{10^{0.5/10}-1}} = 286.26$, which leads to $n \geq 3.03$. So still a 4th-order filter is needed, but $n = 3$ marginally works as well!

Finally, it can be shown that [1], [3], [9] the zeros of $H(s)$ lie on an ellipse, with half axes $\frac{a^{1/n}+a^{-1/n}}{2}$ and $\frac{a^{1/n}-a^{-1/n}}{2}$, where $a = \frac{1}{k_p} + \sqrt{\frac{1}{k_p^2}+1}$, as shown in Figure 4.24.

Once the zeros are found, then $H(s)$, and ultimately the LC filter z parameters are determined. The hand calculations may be tedious, and one can readily use the aforementioned tables or filter software.

A closely related filter known as *inverse Chebyshev* may be obtained by replacing ω with ω^{-1} and $|K|^2$ with $|K|^{-2}$ in the original $|K|^2 = k_p^2 \cos^2(n \cos^{-1} \omega)$ equation. Thus

$$|K|^2 = \frac{k_s^2}{\cos^2\left(n \cos^{-1} \frac{1}{\omega}\right)}.$$

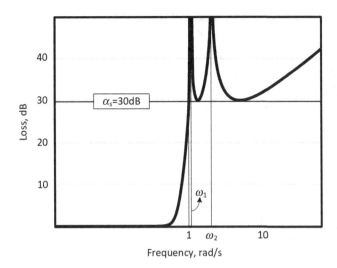

Figure 4.25: A 5th-order inverse Chebyshev filter with $\alpha_s = 30$dB

What is given now is the minimum stopband loss k_s (or $\alpha_s = 10\log(1+k_s^2)$), and the unit frequency is the stopband limit frequency (Figure 4.25).

The filter has the same degree as the Chebyshev for a given passband/stopband loss, but has transmission zeros (or loss poles) at finite frequencies (ω_1 and ω_2 in Figure 4.25). This makes the filter synthesis somewhat more difficult, but provides flexibility and advantages in many applications.

4.2.4.3 General Stopband Filters

Important in many applications, a generalization of Chebyshev filters is provided by a class of filters with equal ripple passband loss, and *finite* loss poles. The mathematical details are beyond the scope of this chapter, and the interested reader may find more details in [1]. A special case is obtained when the loss poles are in such a manner that the stopband as well as the passband are equal ripple. Since such filter $K(s)$ can be analytically constructed in terms of *elliptic* functions [16], they are called *elliptic filters*. Shown in Figure 4.26 is an example of a 3rd-order elliptic filter with 0.2dB passband ripple.

The filter has a loss pole at $\omega = 1.57$rad/s.

4.2.5 LC Filter Design

Let us summarize what we have discussed so far:

- Based on the filter requirements, transducer parameters are determined. This involves approximating the desired filter characteristic with one of the well-known forms (maximally flat, Chebyshev, etc.). The transducer parameters must be circuit realizable. Section 4.2.3.2 highlights the conditions required to guarantee that.
- Transducer parameters lead to a unique value of the LC two-port impedance (or admittance, chain, etc.) parameters.

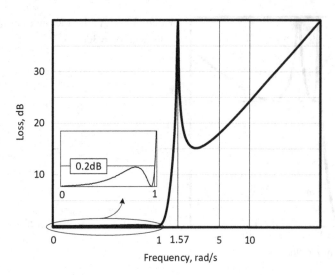

Figure 4.26: Loss function of a 3rd-order elliptic filter with 0.2dB passband ripple

- An LC circuit (typically ladder) is synthesized accordingly. The solution is usually not unique.

The last part is what we have not discussed yet and requires the knowledge of circuit synthesis.

Synthesizing double-terminated LC two-ports is a rather difficult task, since the transducer parameters involve all three of z_{11}, z_{12}, and z_{22} to be satisfied simultaneously. In general, synthesizing double-terminated two-ports starts with realizing either z_{11} or z_{22} as one-port impedances by removing poles such that the transmission zeros, which are typically the zeros of z_{12}, are properly realized. In most cases all three of z_{11}, z_{12}, and z_{22} have the same poles, and often z_{11} and z_{22} are the same except for a private lumped impedance that can be realized separately.

Both Butterworth and Chebyshev polynomials belong to a class of functions known as *all-pole* transfer functions. All the transmission zeros or loss poles (zeros of z_{12} or poles of $K(s)$) are at infinity, and can be physically realized using a ladder structure consisting of series inductors and shunt capacitors as shown in Figure 4.27. That is simply due to the fact that a series inductor or a shunt capacitor blocks the signal from reaching the output at very large frequencies, hence creating a zero at infinity in the transfer function (or infinite loss).

With the transmission zeros figured out, the driving point open impedance of the LC ladder (z_{11} or z_{22}) can be expanded by Stieltjes continued fraction,[14] which will be explained in the context of an example shortly:

$$z_{11} = \alpha_1 s + \cfrac{1}{\alpha_2 s + \cfrac{1}{\alpha_3 s + \cfrac{1}{\ddots + \cfrac{1}{\alpha_n s}}}}$$

Since z_{11} is a PR reactance function, it can be shown that all the α_i factors are real and positive, describing the values of the corresponding series inductors and shunt capacitors. Furthermore, the continued fraction will not end prematurely. Thus, for an nth-order all-pole lowpass filter,

---

[14] Thomas Joannes Stieltjes (1856–1894) was a Dutch mathematician. He was a pioneer in the field of moment problems and contributed to the study of continued fractions.

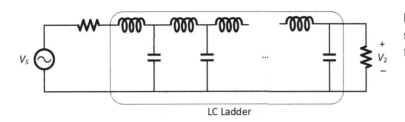

Figure 4.27: LC ladder structure for all-pole lowpass filters

there is a total of n LC elements. As can be seen, such realization requires the minimum number of elements, and the realized circuit is often called *canonical*.

Example: Assume z_{11} and z_{22} of a 4th-order Butterworth filter are given as below:

$$z_{11} = \frac{\frac{L_2}{C_1}s^2 + \frac{1}{C_1 C_3}}{L_2 s^3 + \left(\frac{1}{C_1} + \frac{1}{C_3}\right)s}$$

$$z_{22} = \frac{L_2 L_4 s^4 + \left(\frac{L_2}{C_3} + \frac{L_4}{C_1} + \frac{L_4}{C_3}\right)s^2 + \frac{1}{C_1 C_3}}{L_2 s^3 + \left(\frac{1}{C_1} + \frac{1}{C_3}\right)s}.$$

It will be clear momentarily as to why they are represented in this format. Both functions are odd, indicating reactance functions, and it can be shown that they are PR (for instance by partial fraction). This in turn has to do with the fact that the transducer parameters for the filter satisfy the realizability criteria pointed out in Section 4.2.3.2. Also, despite the fact that z_{11} and z_{22} are not identical, they have the same poles.

Let us start with z_{22}. Since it has a pole at infinity, we can first try to remove that pole, corresponding to a series inductor in the ladder. This is simply done by dividing the numerator by the denominator as follows:

$$\frac{L_2 L_4 s^4 + \left(\frac{L_2}{C_3} + \frac{L_4}{C_1} + \frac{L_4}{C_3}\right)s^2 + \frac{1}{C_1 C_3}}{L_2 s^3 + \left(\frac{1}{C_1} + \frac{1}{C_3}\right)s} = L_4 s + \frac{\frac{L_2}{C_3}s^2 + \frac{1}{C_1 C_3}}{L_2 s^3 + \left(\frac{1}{C_1} + \frac{1}{C_3}\right)s}.$$

The remaining impedance, $\frac{\frac{L_2}{C_3}s^2 + \frac{1}{C_1 C_3}}{L_2 s^3 + \left(\frac{1}{C_1} + \frac{1}{C_3}\right)s}$, is one degree simpler, and now has a zero at infinity. We can therefore remove a pole at infinity from the admittance function, that is, divide the denominator by the numerator. This will correspond to a shunt admittance in the ladder, as it should be.

Continued

$$\frac{L_2 s^3 + \left(\frac{1}{C_1} + \frac{1}{C_3}\right)s}{\frac{L_2}{C_3}s^2 + \frac{1}{C_1 C_3}} = C_3 s + \frac{\frac{1}{C_3}s}{\frac{L_2}{C_3}s^2 + \frac{1}{C_1 C_3}}$$

The procedure can be carried out until we are left with a single capacitor (in the shunt branch) or an inductor (in the series branch). This leads to the following for z_{22}:

$$z_{22} = L_4 s + \cfrac{1}{C_3 s + \cfrac{1}{L_2 s + \cfrac{1}{C_1 s}}}.$$

Carrying out the continued fraction for z_{11} results in

$$z_{11} = \cfrac{1}{C_1 s + \cfrac{1}{L_2 s + \cfrac{1}{C_3 s}}}.$$

Note that since $\frac{1}{z_{11}}$ has a pole at infinity, we start with $\frac{1}{z_{11}}$. The pole removal at infinity by alternating between the impedance and admittance functions resulting in a ladder structure is known as Cauer 1 realization.[15]

The realized filter is shown in Figure 4.28. For the Butterworth realization, the elements are $L_4 = 0.7654$, $C_3 = 1.8478$, $L_2 = 1.8478$, $C_1 = 0.7654$, which will be directly obtained from transducer parameters, and subsequently the corresponding z_{11} and z_{22}.

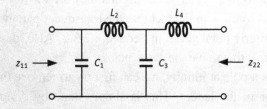

Figure 4.28: A 4th-order Butterworth filter specified by

$$z_{22} = \frac{L_2 L_4 s^4 + \left(\frac{L_2}{C_3} + \frac{L_4}{C_1} + \frac{L_4}{C_3}\right)s^2 + \frac{1}{C_1 C_3}}{L_2 s^3 + \left(\frac{1}{C_1} + \frac{1}{C_3}\right)s}$$

Other types of filters such as elliptic approximate more complex polynomials that are not all-pole. The finite transmission zeros are often realized by either a parallel resonance circuit in the series path or a series resonance circuit in the shunt path, as shown in Figure 4.29. The resonance frequency of the series or parallel circuits is naturally at the transmission zero(s).

The values of the element are often readily obtained through filter design tables [13], [14], [15] or simple software programs, though the circuit can be directly synthesized as well [3], [9]. The synthesis often involves (partial) removal of poles at the transmission zeros frequency. Shown in Figure 4.30 is an example of a 3rd-order elliptic filter with 0.2dB passband ripple (see Figure 4.26 for filter response).

---

[15] Cauer 2 realization involves removing poles at zero from the immittance function.

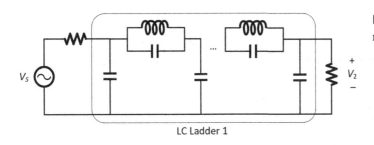

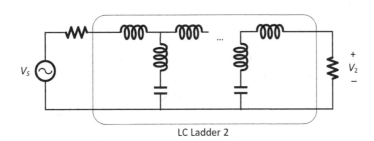

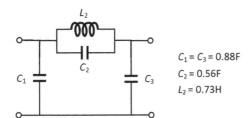

$C_1 = C_3 = 0.88F$
$C_2 = 0.56F$
$L_2 = 0.73H$

Figure 4.29: LC ladder structures to realize finite transmission zeros

Figure 4.30: A 3rd-order elliptic filter and its corresponding elements

4.2.6 Scaling Filters

In the WLAN filter example, we found $n=4$. The degree of filter is independent of C_n. However, since $10\log(1+C_n\omega_p^{2n}) \leq \alpha_p$ for $\alpha_p = 0.5$dB, and $\omega_p = 2\pi \times 8.5$MHz, we obtain, $C_n = \frac{10^{\alpha_p/10}-1}{\omega_p^{2n}} = 1.84 \times 10^{-63}$. The reader can verify that the other equation ($10\log(1+C_n\omega_s^{2n}) \geq \alpha_s$) is going to be automatically met. Such an extremely small value for C_n, which will make the subsequent design very tedious, results from carrying out the calculations without normalization.

To avoid this we start with a normalized filter often with $\omega_p = 1$rad/s. This usually leads to element values on the order of 1H or 1F. The termination resistor is also left at typically 1Ω, or on that order. There are two types of scaling that can be made afterward:

1. Frequency scaling: Since the impedance of an inductor is $L\omega$, and that of a capacitor is $\frac{1}{C\omega}$, and that the resistors are frequency independent, then the filter frequency is scaled by a factor α if all the capacitors and inductors are divided by α. This scaling shrinks or expands only the frequency axis, and the overall shape or rejection of the filter remains intact.
2. Impedance scaling: By a similar argument, if all the resistors and inductors are multiplied by α, and all the capacitors are divided by α, the filter response remains unchanged. Increasing or reducing the impedance levels has only noise implications. This however allows us to achieve more practical values for the components.

Example: Looking back at the WLAN filter example, let us start with a normalized 4th-order Butterworth filter with 1Ω termination at both sides as shown in Figure 4.31.

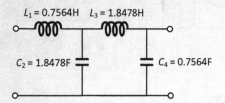

Figure 4.31: Normalized 4th-order Butterworth filter

For the normalized filter, at $\omega = 1\,\text{rad/s}$ the loss increases by 3dB. Since the requirement is 0.5dB loss at 8.5MHz, the scaling factor is $\frac{2\pi \times 8.5}{0.77} = 6.97 \times 10^7$, where $\omega = 0.77\,\text{rad/s}$ is the normalized frequency of 0.5dB passband loss. Furthermore, we would like to design the filter for 50Ω termination, which results in $L_1 = \frac{0.7564 \times 50}{6.97 \times 10^7} = 0.542\,\mu\text{H}$, $C_2 = \frac{1.8478}{50 \times 0.53\,\text{nF}\,(6.97 \times 10^7)} = 530\,\text{pF}$, $L_3 = \frac{1.8478 \times 50}{6.97 \times 10^7} = 1.33\,\mu\text{H}$, $C_4 = \frac{0.7564}{50 \times 0.53\,\text{nF}\,(6.97 \times 10^7)} = 217\,\text{pF}$. These values are a lot more practical for such filter at this frequency range.

The scaled filter simulated response is plotted in Figure 4.32 along with the new values.

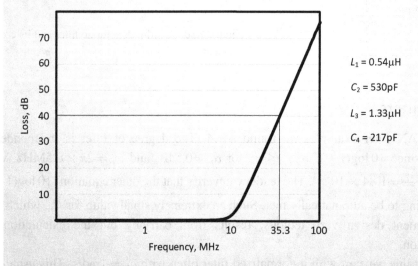

Figure 4.32: Scaled filter response along with the element values for the 4th-order Butterworth filter of Figure 4.31

The response follows that of Figure 4.20, except the frequency axis is scaled to the WLAN filter requirements.

4.2.7 Bandpass LC Filters

All the filter types discussed thus far approximate an ideal *lowpass* filter. To create other types of filters (e.g., highpass, bandpass, etc.) there are common impedance transformations that may

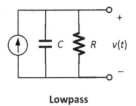
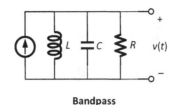

Figure 4.33: A 1st-order RC lowpass circuit compared to a 2nd-order bandpass circuit

be used [17]. Given its importance, we will mostly discuss the bandpass transformation in this section. Highpass or bandstop filters follow a very similar procedure.

To create a bandpass filter, let us first take another look at the simple parallel RLC circuit discussed earlier, and compare that to a 1st-order RC circuit also shown in Figure 4.33.

The frequency response of each circuit is

$$H_{LP}(j\omega) = \frac{R}{1+jRC\omega}$$

$$H_{BP}(j\omega) = \frac{R}{1+jRC\dfrac{\omega^2-\omega_0^2}{\omega}}.$$

Comparing the two transfer functions, one can realize that the lowpass response can be converted to bandpass if one shifts the frequency from $\omega \leftrightarrow \frac{\omega^2-\omega_0^2}{\omega}$, or in the s domain, from $s \leftrightarrow \frac{s^2+\omega_0^2}{s}$.

To generalize, if the *normalized* lowpass transfer function is given by $H_{LPn}(s)$, the bandpass response is obtained by $H_{BP}(s) = H_{LPn}\left(\frac{s^2+\omega_0^2}{sB}\right)$, where B is the bandpass filter bandwidth, and ω_0 is the center frequency. The appearance of B (bandwidth) in the denominator is due to the fact that what is used is the normalized lowpass transfer function. In the simple circuit of Figure 4.33, for instance, the bandwidth is $\frac{1}{RC}$. Given that the normalized lowpass response is $H_{LPn}(s) = \frac{1}{1+s}$, the bandpass response is readily obtained by applying the transformation.

From a circuit point of view, this is equivalent to replacing each shunt capacitor with a parallel LC circuit whose resonance frequency is set by ω_0, as the lowpass capacitor admittance needs to change from $Cs \leftrightarrow \left(\frac{C}{B}\right)s + \frac{1}{\left(\frac{B}{C\omega_0^2}\right)s}$. This is also evident from Figure 4.33. Moreover, one can show that the transformation also requires replacing each series inductor with a series LC circuit resonating at ω_0 as well. As an example, our previous 3rd-order lowpass filter could be converted to bandpass through the equivalent ladder circuit shown in Figure 4.34. For the same order, the number of reactive components doubles. At the center frequency ω_0, clearly all the series LC branches are short, whereas all the parallel LC legs are open, thus the circuit is expected to pass. Moreover, at very high or very low frequencies the input signal is blocked, and hence the bandpass response is achieved.

Assuming a lowpass frequency of Ω, and a bandpass frequency of ω, the transformation may be expressed as

$$\omega^2 - \omega\Omega B - \omega_0^2 = 0,$$

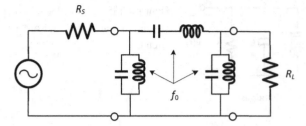

Figure 4.34: A 3rd-order bandpass ladder filter

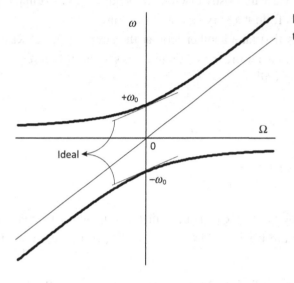

Figure 4.35: Plot of lowpass to bandpass frequency transformation

which has the following two roots:

$$\omega = \frac{B}{2}\Omega + \sqrt{\frac{B^2}{4}\Omega^2 + \omega_0^2}, \omega > 0$$

$$\omega = \frac{B}{2}\Omega - \sqrt{\frac{B^2}{4}\Omega^2 + \omega_0^2}, \omega < 0.$$

The equation above shows that while the center frequency shifts as intended, the lowpass frequency response is not preserved entirely. Only if $\Omega \ll \frac{2\omega_0}{B}$, that is, if the filter is narrow enough, are its characteristics shifted intact. This is shown in Figure 4.35, indicating that the transformation is not exactly linear. Also shown in the dashed line is the ideal curve. For small deviations from the center, we have

$$\omega \approx \omega_0 \pm \frac{B}{2}\Omega.$$

However, if Ω is large,

$$\omega \approx B\Omega,$$

which clearly shows a large deviation from the ideal response.

4.2 Doubly Terminated LC Filters

Example: The corresponding transfer functions of a 3rd-order Butterworth design is shown in Figure 4.36.

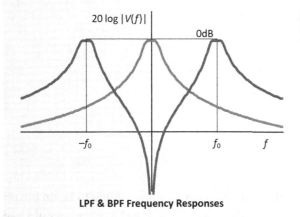

Figure 4.36: Lowpass and bandpass Butterworth transfer functions

As a final note, let us examine the filter response in the time domain to a pulse or step. Since a step function contains abrupt transitions in the time domain, we expect it to contain very high-frequency components that will be attenuated by the filter. For a given lowpass filter as an example, the output $y(t)$ corresponding to a step input, that is, $x(t) = u(t)$, is

$$y(t) = \int_{-\infty}^{\infty} h(\tau)u(t-\tau)d\tau = \int_{-\infty}^{t} h(\tau)d\tau.$$

To gain some insight, consider the extreme case of an ideal LPF with a bandwidth of B, where $h(t)$ is a sinc. The integral then becomes

$$y(t) \int_{-\infty}^{t} 2B\,\text{sinc}(2B\tau)d\tau = \frac{1}{2} + \frac{1}{\pi}Si(2\pi Bt),$$

where $Si(t) = \int_0^t \frac{\sin \alpha}{\alpha} d\alpha$ is the sine integral function. The filter step response is shown in Figure 4.37, and compared to the response of a 1st-order RC filter.

As expected, the step response of the ideal filter has components at $t < 0$, signifying the fact that such filter is not physically realizable. Nevertheless, it gives us some insight: for both cases the step response approaches unity, and the rise time (t_r)[16] is $0.35/B$ for a 1st-order design, whereas the ideal filter rise time is $0.44/B$. Since the rise time appears not to be a strong function of the filter order, as a general rule of thumb we approximate

$$t_r \cong \frac{1}{2B}.$$

---

[16] Rise time is typically defined as the time it takes to reach from 10% to 90% of the final value.

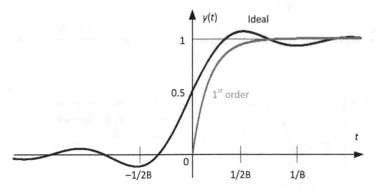

Figure 4.37: Step response of an ideal and a 1st-order filter

4.3 ACTIVE FILTERS

In the previous example of WLAN filter, the inductances were found to be on the order of μH, and capacitances on the order of 100s of pF. These values are large, and would become even larger for lower frequency filters (applications such as Bluetooth, GPS, or WCDMA need filters on the order of a few MHz or less). Impedance scaling is not helpful, for instance, if at the expense of higher noise the termination resistors are scaled by 20 times to 1kΩ, only the capacitances reduce, while the inductances increase further. This certainly rules out LC filters, despite all their good properties, as suitable candidates for IF filters as such large values of inductor are impractical in integrated circuits. In this section we briefly discuss the fundamental properties of the active filters, an alternative to the LC ladder that can be realized on chip with practical element values. More detailed discussion of active filters may be found in [1], [3], [4], [18].

4.3.1 Active Filters Ladder Design

Let us start with a 3rd-order Chebyshev filter realized as a passive ladder structure (Figure 4.38). Also shown is the transducer loss in dB versus the frequency. As pointed out through the example, at lower frequencies, say several hundreds of kHz to a few tens of MHz, the size of the inductor L_2 will be too big to be realized on chip.

An alternative realization may be conceived if one considers the *signal flow graph* of the passive filter. For that, we assign a voltage to every capacitor, and a current to every inductor, as shown in Figure 4.38, and ensure that the KVL and KCL are satisfied at every loop and node of the ladder filter. For instance, writing a KCL at V_1 node yields

$$\frac{V_1 - V_S}{R_S} + sC_1 V_1 + I_2 = 0,$$

which can be rearranged as

$$-V_1 = \frac{-1}{sC_1}\left(\frac{V_S}{R_S} - \frac{V_1}{R_S} - I_2\right).$$

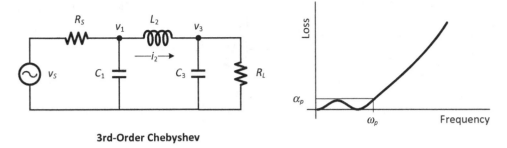

Figure 4.38: A 3rd-Order Chebyshev ladder filter

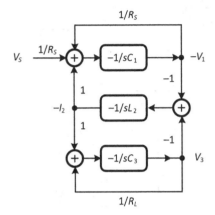

Figure 4.39: Signal flow graph of the Chebyshev filter

Similarly, a KVL for I_2 and a KCL for V_3 after rearranging lead to the following:

$$-I_2 = \frac{-1}{sL_2}(V_1 - V_3)$$

$$V_3 = \frac{-1}{sC_3}\left(\frac{V_3}{R_L} - I_2\right).$$

The corresponding flow graph to realize the transfer function of the 3rd-order filter according to the three equations above is shown in Figure 4.39.

By inspection, evidently the filter may be realized by analog building blocks: integrators, adders, and multipliers. There are as many integrators as the reactive elements. The integrator may be realized by an active-RC circuit, a g_m-C circuit involving transconductors, or a switched capacitor circuit.[17] Examples of a g_m-C and active-RC integrators are shown in Figure 4.40.

Feeding multiple currents into a node realizes the addition, and multiplication is done by scaling values of resistors or transconductance. A g_m-C realization of the 3rd-order filter is shown in Figure 4.41 as an example. Negative gain is easily realized in a differential scheme.

Similarly, the filter may be realized through opamp-RC integrators and is shown in Figure 4.42. The addition and multiplication functions are performed by feeding various current with appropriate impedance levels into the virtual ground of a given opamp.

---

[17] A switched current integrator is also an option, although is seldom used in practice.

RF and IF Filters

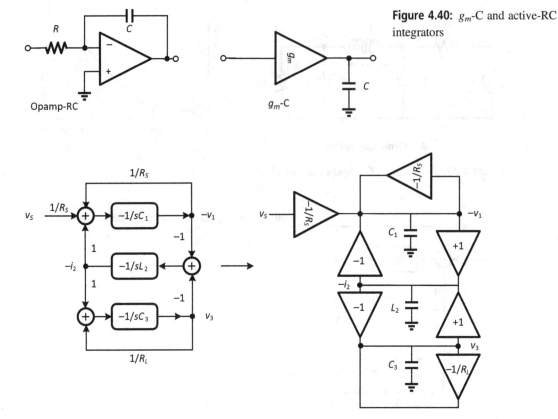

Figure 4.40: g_m-C and active-RC integrators

Figure 4.41: g_m-C realization of the 3rd-order filter of Figure 4.38

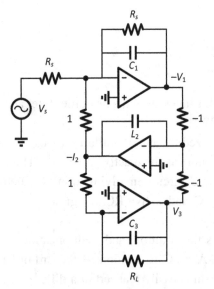

Figure 4.42: Active-RC realization of the 3rd-order filter of Figure 4.38

The negative resistors are readily realized in fully differential designs through cross coupling the differential outputs of the corresponding stage. If a single-ended design is to be used, one has no choice but to insert an inverting unity gain buffer in front of the negative resistors to flip their sign.

Example: We wish to design a 3rd-order active-RC Chebyshev filter with 1dB passband ripple and 10MHz passband frequency. The element values for the normalized filter ($\omega_p = 1\text{rad/s}$) are

$$C_1 = C_3 = 2F, \ L_2 = 1H, \ R_s = R_L = 1\Omega.$$

To scale to the desired frequency, all the capacitors must be multiplied by $\frac{1}{2\pi \times 10^7}$. To further achieve more practical component values, we raise all the resistance to 4kΩ, leading to the capacitance values of $\frac{\frac{2}{2\pi \times 10^7}}{4000} = 8\text{pF}$ for the first and last lossy integrators, and $\frac{\frac{1}{2\pi \times 10^7}}{4000} = 4\text{pF}$ for the middle integrator. Furthermore, to achieve a passband gain of 1, as opposed to $\frac{1}{2}$, which is inherent in the double-terminated ladder, we reduce the source resistance to 2kΩ. This completes the design, and the active-RC filter simulated frequency response with the aforementioned elements is shown in Figure 4.43.

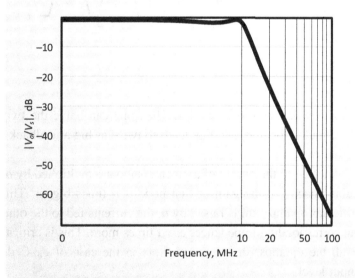

Figure 4.43: A 10MHz 3rd-order Chebyshev filter

In the previous example, the capacitances on the order of pF and resistances on the order of kΩ were obtained, which are very suitable values for integrated circuits. The resistance scaling to 4kΩ was quite arbitrary, and is primarily driven by the noise constraints; e.g., at the expense of doubling the noise, the filter capacitances may be halved, while the resistances may be doubled. This generally imposes a trade-off between the filter size (cost), noise, and opamps driving capability and consequently their power consumption.

The filter may be further scaled considering the following:

- If all the impedances connected to the input of a given opamp are multiplied by α, the filter response remains intact. This is simply because the currents fed to that opamp input are all α times less, but the overall gain of that opamp remains the same as the feedback impedance is also α times bigger. This allows us to locally optimize the elements around each opamp to further minimize the components *spread* if needed. For instance, it may be desirable to have

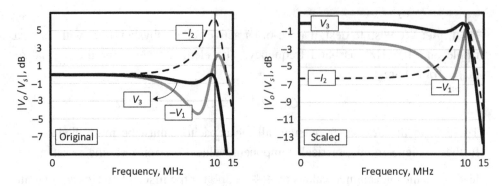

Figure 4.44: Scaling the filter of Figure 4.43 for the optimum dynamic range

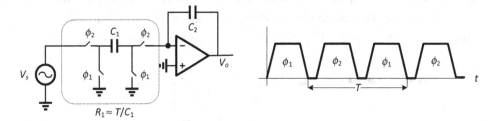

Figure 4.45: A switched capacitor integrator

equal feedback capacitors for all the three opamps. Then one can halve the two negative resistors connected to the middle opamp, and that leads to twice as big a feedback capacitor, now the same as the other two.

- If all the impedances connected to the output of a given opamp are multiplied by α, the filter overall response remains intact, but the gain of that node is α times bigger. This again is obvious; while the particular opamp gain is raised by α, the currents fed to the other opamps remain the same as the corresponding impedances are α times more. This is critical in active filters, to ensure that all the opamps (or transconductors in the case of g_m-C design) are clipped at the same input level.

To clarify the previous point further, shown in Figure 4.44 is the transfer gain of the filter of Figure 4.43 for each of the internal nodes, as well as that of the output. Evidently, the middle opamp is clipped as much as 6.4dB earlier than the output. This can be remedied by reducing all the impedances connected to that opamps by 2.09×. A similar scaling must be done for the first opamp as well, resulting in the scaled filter response shown on the right, where all the three stages are clipped at the same input level.

While this technique raises the filter noise [4], it overall leads to a more optimum *dynamic range*.[18]

Another common realization of active filters is through switched-capacitor integrators [4], the details of which we will defer to the reader (Figure 4.45). It is often possible to start with an

---

[18] Dynamic range and noise will be extensively discussed in Chapters 5 and 6. For a filter, the dynamic range is usually defined as the ratio of the filter maximum input for some acceptable level of distortion to its integrated noise.

active-RC equivalent, and simply replace each resistor with a switched capacitor equivalent (as shown in the dashed box).

More elaborate designs using continuous- to discrete-domain mapping techniques, for instance *bilinear* mapping, are also possible [19].

4.3.2 Active Filters Cascaded Design

The ladder realization of the filter leads to the least sensitivity to the elements variation as pointed out in Section 4.2. However, the design is somewhat tedious and not quite straightforward. A simpler approach is to obtain the overall filter transfer function, through MATLAB or by looking up (or calculating) the natural frequencies of the filter (recall those are the zeros of $H(s)$). Then the filter may be synthesized by breaking the transfer function into a cascade of second-order (or biquadratic) sections, known as biquads. If the filter order is odd, one first-order stage is also needed.

Shown in Figure 4.46 is a general realization of a 1st-order stage. It is easy to show that the transfer function is

$$\frac{V_o}{V_s} = \frac{K_1 s + K_0}{s + \omega_0}.$$

Figure 4.47 shows a general 2nd-order or biquad stage. The biquad transfer function can be shown to be

$$\frac{V_o}{V_s} = \frac{K_2 s^2 + K_1 s + K_0}{s^2 + \frac{\omega_0}{Q} s + \omega_0^2}.$$

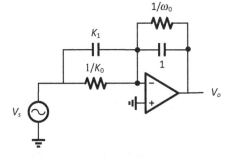

Figure 4.46: Active-RC realization of a generic 1st-order stage

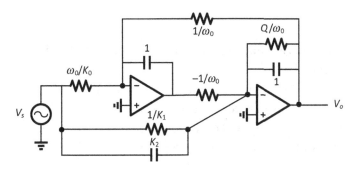

Figure 4.47: A generic active-RC biquad stage

This biquad is mostly suitable for the cases where the required quality factor Q is low. An example of a high-Q biquad is discussed in Problem 16.

Example: For the same 3rd-order Chebyshev filter of previous example, from filter tables or MATLAB the normalized natural frequencies can be found to be

$$s_1 = -0.5$$
$$s_{2,3} = -0.25 \pm j0.97.$$

Thus, the filter transfer function may be expressed as

$$\frac{V_o}{V_s} = \frac{\frac{1}{4}}{\left(s+\frac{1}{2}\right)\left(s^2+\frac{1}{2}s+1\right)} = \frac{\frac{1}{2}}{\left(s+\frac{1}{2}\right)} \frac{\frac{1}{2}}{\left(s^2+\frac{1}{2}s+1\right)}.$$

The $\frac{1}{4}$ in the numerator is arbitrarily spilt between the two stages. We raise all the resistances to 4kΩ, and apply the same scaling factor of $\frac{2\pi \times 10^7}{4000}$ for the capacitors as in the ladder example, which leads to the filter schematic of Figure 4.48. The input resistor is halved to eliminate the 6dB inherent loss of the double-terminated filter. The biquad does not need any additional scaling as both nodes (V_2 and V_o) have the same peak gain. However, the 1st-order stage output (V_1) has 6dB higher gain, and thus the impedances at the output of the 1st opamp are scaled down by a factor of 2, as shown in the figure. Note that the 1st-order stage is usually preferred to go first, and is often built in to the previous stage (for example, the receiver downconversion mixer)

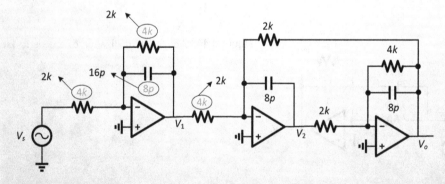

Figure 4.48: Third-order Chebyshev filter with 10MHz passband using a cascade of a 1st- and a 2nd-order stage

The filter has the same overall transfer function as the ladder design, and the final scaled filter characteristics is shown in Figure 4.49.

As mentioned earlier, the cascaded design doesn't necessarily have the good property of low sensitivity to element variations like the ladder design. However, in general, in integrated circuits good matching between the resistors and capacitors is enjoyed. The key is to design the resistors and the capacitors out of unit elements. For the previous case, for instance, one may

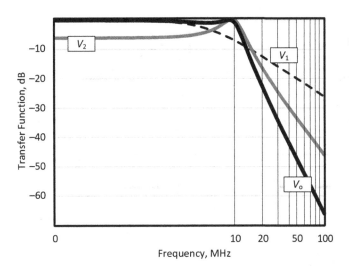

Figure 4.49: Transfer function of the cascade design of Figure 4.48

choose a unit of $2k$ for the resistors, and a unit of $8p$ for the capacitors. In many cases the element values need to be rounded, which could cause some sensitivity.

Example: Shown in Figure 4.50 is another realization of an active-RC biquad known as the Sallen–Key lowpass biquad stage [20].

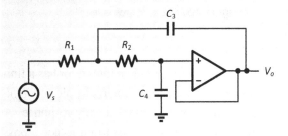

Figure 4.50: Sallen–Key lowpass filter

The reader can verify that

$$\frac{V_o}{V_s} = \frac{\omega_0^2}{s^2 + \frac{\omega_0}{Q}s + \omega_0^2},$$

where $\omega_0 = \frac{1}{\sqrt{R_1 R_2 C_1 C_2}}$, and $Q = \frac{\sqrt{R_1 R_2 C_1 C_2}}{C_2(R_1+R_2)}$.

A more generic realization of Sallen–Key topology is discussed in Problem 26. The Sallen–Key biquad requires only one opamp and thus half the power consumption, but generally suffers from more sensitivity to element mismatches and rounding. Nonetheless, given the more superior power consumption it is commonly used.

4.3.3 Nonideal Effects in Active Filters

There are two fundamental limitations implementing high-frequency active filters, regardless of the type of the integrator used. Let us consider a g_m-C realization, though the conclusions may

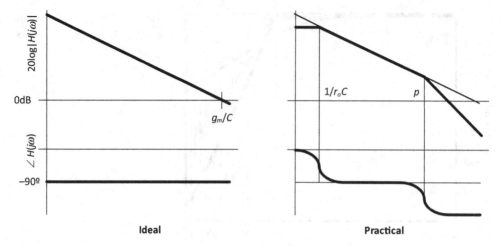

Figure 4.51: Integrator phase and frequency response

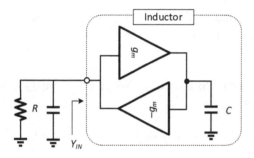

Figure 4.52: A g_m-C resonator

be extended to opamp-RC or other types as well. First, a practical integrator has a finite DC gain (of $a_0 = g_m \times r_o$ for the transconductor for instance), and an internal pole (of ω_p). Shown in Figure 4.51 is the ideal versus practical integrator phase and frequency response. A finite DC gain introduces a phase lead in the transfer function, which tends to introduce a passband loss. On the other hand, a finite pole frequency creates a phase lag, which in turn results in peaking and phase distortion at the edges of passband [21]. Both impacts must be carefully avoided especially as the filter bandwidth becomes narrower.

To quantify these impacts, let us consider a 2nd-order resonator built out of a two back-to-back nonideal integrators, as shown in Figure 4.52. Once loaded with a resistor R, a parallel RLC network is built whose bandwidth is set by the resistor. The cross-coupled transconductors create what is known as an *active gyrator* [22], [23]. When loaded by a capacitor (Figure 4.52), the gyrator results into an inductor ($L = \frac{C}{g_m^2}$) when looking into the other end:[19]

$$Y_{IN} = g_m^2 Z_L = \frac{g_m^2}{jC\omega}.$$

We shall first find the impact of finite output resistance of the integrators.

---

[19] As a side note, Orchard shows in [68], [69] that active gyrators lead to bandpass filters with comparable sensitivity to the element variations to that of double-terminated LC ladders if every inductor is replaced by the gyrator-C equivalent.

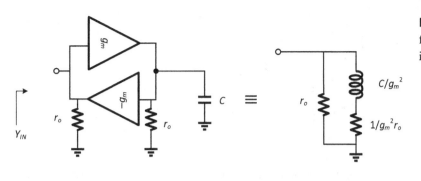

Figure 4.53: Impact of finite DC gain of the integrator

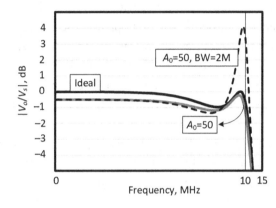

Figure 4.54: Impact of finite gain and bandwidth in the filter of example of Figure 4.43

Shown in Figure 4.53, one integrator resistor appears directly at the input. The other one, after transformation by the gyrator, appears as a small resistance in series with the inductor. Once converted to parallel RL circuit using the high-Q approximation showed in Chapter 1, it results in a net resistance of $r_o/2$. Consequently, a finite Q is observed even for the unloaded resonator.

Regarding the finite pole, let us assume that each transconductor has a transfer function of

$$\frac{g_m}{1+\frac{s}{p}}.$$

Looking into the other end of the gyrator, the input admittance is

$$Y_{IN} = \frac{g_m^2}{Cs\left(1+\frac{s}{p}\right)^2} \approx \frac{g_m^2}{Cs}\left(1-\frac{2s}{p}\right) = \frac{g_m^2}{Cs} - \frac{2g_m^2}{Cp}.$$

Thus, the finite pole results in a parallel negative resistance of $\frac{-Cp}{2g_m^2}$ at the input. Similar expressions may be derived for the opamp-RC integrators.

Shown in Figure 4.54 is the ladder filter of the previous example (Figure 4.43) simulated with ideal opamps, along with two additional cases of a DC gain of 50 (or 34dB) but infinite bandwidth, and the same 34dB DC gain but a 3dB bandwidth of 2MHz (or a gain bandwidth product of 1GHz) for all the opamps.

The finite DC gain leads to a passband loss of 0.5dB, and some rounding of the passband edge, whereas the finite bandwidth leads to severe peaking around the passband edge as expected.

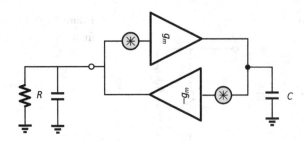

Figure 4.55: g_m-C resonator noise

These two factors cause a degradation of the resonator quality factor. As a consequence, the overall Q of the actual filter, comprising several of these resonators, is limited. Considering typical values of gain-bandwidth product of transconductor or opamp-based integrators, the active filters are typically limited to lowpass structures up to several tens of MHz. If implemented bandpass, their center frequency is limited to a few MHz with fairly wide bandwidth.

Next let us consider the noise properties of the active filters.[20] Consider the g_m-C resonator shown in Figure 4.55, where we model each transconductor with an input referred noise voltage whose spectral density is $\frac{4KTF}{g_m}$.

If the transconductor is built of only one transistor, $F \approx 1$, whereas in practice F is somewhat larger due to the other contributors. We can show that if the loaded resonator has a quality factor of Q, set by the resistor R (Figure 4.55), its output noise is [24]

$$\overline{v_n^2} = \frac{KT}{C}FQ,$$

which indicates that the noise is exacerbated by a factor of Q. Similar to finite gain and bandwidth issues, the excess noise sets how narrow the bandwidth could be for a certain center frequency and the filter area (mostly dominated by the total capacitance).

We will see in Chapter 12 that these shortcomings fundamentally limit the choices of the receiver architecture.

4.4 SURFACE AND BULK ACOUSTIC WAVE FILTERS

In many wireless applications such as mobile phones, very aggressive front-end filtering is required for the reasons that will be discussed in Chapter 6. The modest quality factor of the integrated resonators along with IC variations often precludes the use of on-chip filters. Consequently, in such platforms, RF filtering is often dominated by surface acoustic wave (SAW) and bulk acoustic wave (BAW) filters (also known as FBAR or thin film bulk acoustic resonator filters), which offer tremendous advantages in performance, cost, and size. Given their importance, we shall recast a brief summary of their physical structure and properties in this section. More general details may be found in [25], [26], [27], [28]. Also, for further reading on FBAR resonators and filters, see [29], [30], [31], [32].

---

[20] We will discuss noise in Chapter 5, but assume the reader already has basic knowledge of noise from analog circuit design courses.

4.4.1 Filter Structure

Surface acoustic waves (SAWs) were first explained in 1885 by Rayleigh, who described the surface acoustic mode of propagation and predicted its properties [33]. However, it was not until the early 1980s that FBAR devices appeared in the literature [34], [35]. SAW or FBAR resonators and filters are essentially *electromechanical* devices where electrical signals are converted to a mechanical wave in a device constructed of a piezoelectric crystal or ceramic; this wave is delayed as it propagates across the device, before being converted back to an electrical signal by further electrodes.

Given the desirable properties described in Section 4.2, the most common filter topology for both FBAR and SAW filters is a doubly terminated ladder structure shown in Figure 4.56.

The same as crystals (see Chapter 9), either resonator may be modeled by an equivalent RLC circuit shown in Figure 4.57.

The top branch consisting of R_M, L_M, and C_M model the motional (or acoustic) components, whereas C_O is the plate capacitance, and R_O is used to model the loss associated with it. Typically C_M is much smaller than C_O, resulting in good stability (see the crystal stability discussion in Chapter 9 for more details on this). Furthermore, the quality factor of the resonator is very high, often on the order of several thousands.

The resonator is a second-order circuit, which has a pole and zero, and whose magnitude of its admittance is shown in Figure 4.58. The zero of admittance (pole of the impedance) defines the parallel resonance, whereas the pole of the admittance define the series resonance.

As such, from a circuit point of view the filter may be treated the same as a well-understood LC ladder, with the exception that the resonators demonstrate substantially higher quality factor and better stability.

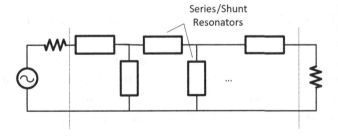

Figure 4.56: A SAW (or BAW) filter based on a doubly terminated ladder structure comprising series and shunt resonators

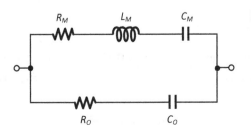

Figure 4.57: A SAW (or BAW) resonator circuit model

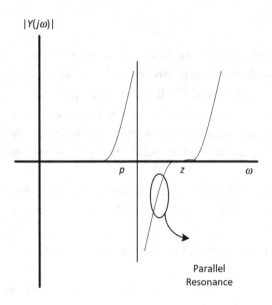

Figure 4.58: The BAW (or SAW) resonator pole and zero define parallel and series resonance in its admittance

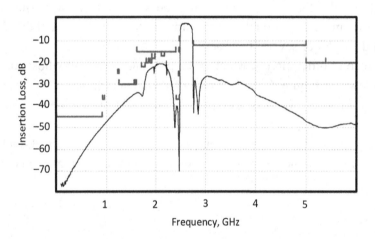

Figure 4.59: Frequency response of a Murata SAW filter intended for use in cellular platforms

As an example, shown in Figure 4.59 is the frequency response of a commercial SAW filter designed by Murata.[21]

The filter is intended to be used for cellular applications, and is tuned to the LTE (long-term evolution) band 41 (2496~2690MHz). The filter has a stopband rejection of about 40dB at the 2.4GHz ISM band,[22] which spans from 2402~2480MHz. This kind of steep stopband rejection is not possible to be accomplished using integrated inductors and capacitors with modest quality factors of 10–20.

4.4.2 Resonator Physical Implementation

The core element of the filter is the resonator. What differentiates BAW and SAW technologies from the other technologies is their substantially higher Q as the acoustic materials have low

---

[21] Murata, PN: SAFRE2G59MA0F0A.   [22] ISM stands for industrial, scientific, and medical.

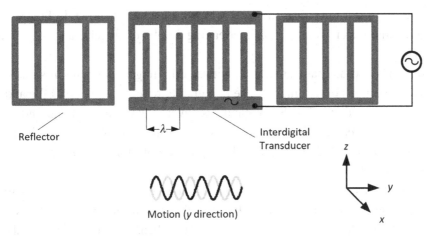

Figure 4.60: SAW resonator physical structure

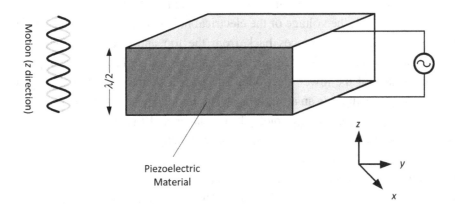

Figure 4.61: FBAR resonator structure and the propagation of the acoustic wave

propagation loss. Moreover, the acoustic resonators and filters are much smaller than the other technologies because the acoustic wave speed (v) is several orders of magnitude less than the speed of the electromagnetic waves (about 300m/s, compared to the speed of light). Hence, the corresponding acoustic wavelength $\lambda = \frac{v}{f}$ is much shorter than that of the electromagnetic wave. Finally, the resonators and the filters are often mass produced through semiconductor technologies such as photolithography or thin-film deposition, which makes them cheaper.

Although the impedance responses of FBAR and SAW resonators look the same (Figure 4.57), the resonances are realized quite differently. As their name suggests, a SAW is an acoustic wave that travels along the surface of a crystal substrate in the y direction (Figure 4.60), whereas FBAR travels inside solid material and in the z direction (Figure 4.61). In a SAW, the amplitude of the motion decays rapidly with depth, and loses most of its acoustic energy with one wavelength.

The key to making SAW and FBAR devices is the use of piezoelectric materials as the medium, and the use of transducers to convert between electrical signals and acoustic waves. In a piezoelectric material the application of an electric field produces mechanical stress or force. There is also an inverse effect where the imposition of a stress on the material produces an electrical charge or electrical field.

A basic SAW resonator operating on a piezoelectric substrate is shown in Figure 4.60. The IDT (interdigital transducer) is a pair of interleaving combs of fingers, and it generates and receives SAWs.

The electric fields and mechanical stresses between the fingers alternate in sign because of the alternating connections of the fingers. The two grating reflectors reflect SAWs and form an acoustic cavity between them. The SAW cavity is several wavelengths long, and a standing wave is formed in this cavity. The frequency at which the resonance occurs is determined mainly by the pitch of the IDT. The IDT and the grating reflectors are fabricated of thin-film metal, usually aluminum.

The basic FBAR resonators consist of a thin film layer of an aluminum nitride (AlN) piezoelectric material sandwiched between two metal thin film electrodes [30], as shown in Figure 4.61. The FBAR filter itself is manufactured on a silicon substrate, which makes it easy to be implemented in a CMOS fab, thus eliminating much of the wafer cost. Furthermore, an all-silicon package may be developed that eliminates much of the back-end costs.

Referring to Figure 4.61, the voltage or the electrical field between the two electrodes excites the acoustic wave. The wave bounces back from the top and bottom surfaces of the two electrodes, and an acoustic cavity is formed between the top surface of the upper electrode and the bottom surface of the lower electrode. The perfect boundary for totally reflecting the acoustic wave into the medium is an air interface. The frequency at which the resonance occurs is determined by the thickness of the piezoelectric layer and the thickness and mass of the electrodes. There is only one-half of an acoustic wavelength in this cavity at fundamental resonance.

It is noteworthy to point out that FBAR resonators have also been used in the context of oscillators and reference generation [36], [37]. Their operation is very similar to crystal oscillators, which will be discussed in Chapter 9 and thus will be skipped here.

4.4.3 Comparison between FBAR and SAW Filters

While both structures are commonly used in today's radios, there are several advantages/disadvantages of one versus the other [27], [28], [38]:

- FBAR resonators have demonstrated a higher quality factor than SAW resonators at RF frequencies. Consequently, BAW filters can have lower insertion loss, and better selectivity.
- FBAR filters have better power handling (up to 36dBm), since BAW is based on a parallel plate capacitor geometry, not the long, narrow, and thin interdigital fingers as used in the SAW filter (Figure 4.60). This is obviously critical in transmitters.
- The temperature coefficient of an FBAR resonator appears to be somewhat better than SAW, but not as good as ceramic filters. For mobile applications though, ceramic filters are substantially larger, which precludes them from being used.[23] See [39] for more on power handling and temperature coefficient studies in FBAR duplexers.

---

[23] For example, a ceramic filter might be $5 \times 5 \times 10 \text{mm}^3$, whereas a hermetically sealed FBAR filter is $0.5 \times 0.5 \times 0.2 \text{mm}^3$, about $5{,}000\times$ smaller.

- SAW filters are easier to manufacture, and hence typically cheaper. FBAR has many more masks, and thus a longer manufacturing time.
- FBAR resonators generally have a higher figure of merit (FOM). A commonly used FOM is defined as [40]

$$\text{FOM} = k_{teff}^2 \times Q,$$

where Q is the unloaded quality factor, and k_{teff}^2 is the coupling coefficient, which is a function of the electrode material thickness ratio between electrode and piezoelectric materials. Since the discovery of Scandium in films, the coupling coefficient can be as high as 20% in development and about 10% in production. The unloaded quality factor hovers between 2000 and 3000. Thus, the corresponding FOM is around 200 for FBAR resonators, which is superior to that of a SAW resonator.

Despite all these, both technologies have been evolving rapidly, and both are commonly used. Particularly, SAW filters appear to be more suitable for lower frequency range (say around 1GHz), while BAW filters are more popular at higher frequencies (2GHz and above).

4.5 DUPLEXERS

The duplexer is a network with three ports to which are connected, respectively, the antenna (ANT), the transmitter output (TX), and the receiver input (RX) as shown in Figure 4.62. In certain applications that the RX and TX operate concurrently, the duplexer is needed to ideally isolate the receiver from the transmitter; convey the available output power from the TX to the antenna; and transfer the voltage induced on the antenna to the receiver input with no attenuation. Given their generally very stringent requirements (see Chapter 6 for detailed discussion of concurrent or *full-duplex* transceivers), they are typically realized through SAW or FBAR technologies [30], [38], [41].

SAW (or FBAR) duplexers operate in well-defined nonoverlapping narrow bands of frequency for TX and RX, while providing an approximately constant resistance at the TX port. They consist of RX and TX SAW bandpass filters with a sharp interband transition, so that the RX filter presents a small input reactance across the TX sub-band. An integrated $\frac{\lambda}{4}$ line then transforms this to a large reactance at the TX filter output where the antenna is also connected.

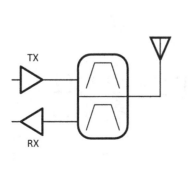
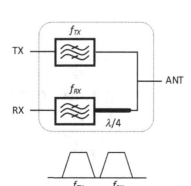

Figure 4.62: SAW/FBAR duplexer schematic and its realization

Thus, across the TX band, the SAW filter connected to the PA output is terminated essentially only by the antenna impedance. The physical length of the $\frac{\lambda}{4}$ line is small, as it is set by the very short acoustic wavelength.

It is obviously very desirable to integrate the duplexer functionality on chip, given the size and cost concerns, especially in multiband applications where several of these duplexers, one for each band of operation, are needed. On the other hand, mimicking the exact same design of the SAW or FBAR duplexer on chip by using integrated inductors and capacitors is very challenging, due to the reasons pointed out in the previous section, namely the very modest quality factor and the variation of on chip passives.

To overcome this fundamental limitation, there have been recent efforts to integrate a programmable duplexer on chip using the concept of electrical balance based on the hybrid transformers [42], [43], [44], [45], [46]. The hybrid transformer's roots stretch back to the earliest years of telephony [47], [48]. In the pre-electronic telephone handset it served to isolate the microphone from the earpiece, enabling signals on a pair of wires at each transducer on a two-wire loop to the central office, while suppressing crosstalk from microphone to the headset. This concept could be revived to ideally isolate the RX and TX rather than the microphone and headset through an integrate hybrid transformer. Three variations of such transformer are shown in Figure 4.63.

The three circuits can be analyzed very similarly, and we for now will focus on the one on the far right, which is perhaps the most suited for an RF electrical balance duplexer. The hybrid transformer primary is symmetric, and is connected to the antenna on one end, and to a balancing impedance (Z_{BAL}) that is ideally exactly equal to the antenna impedance. Given the symmetry of the circuit then, the TX output that is applied in the center tap of the primary is split equally on the antenna and balancing impedance, and delivers half its power to the antenna, with the other half *wasted* in the balancing network. On the other hand, assuming perfect symmetry, since the signals at the two ends of the primary appear common mode, no signal will be induced on the secondary, which will be driving a differential RX. Thus the receiver is ideally isolated from the transmitter. The circuit thus has the added advantage that a differential input is created for the receiver. If the symmetry is preserved, the circuit is capable of providing large isolation over a wide bandwidth. Particularly, at balance there is no net differential voltage across the autotransformer, and so zero self-inductance current flows. The bandwidth is ultimately limited by the asymmetric parasitic elements in the transformer or the surrounding circuitry. In contrast, the left circuit, which works conceptually very similarly, is not as wideband as the self-inductance of the autotransformer

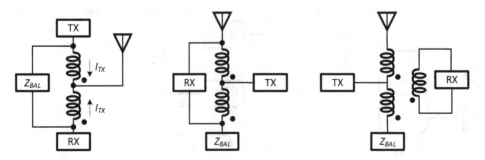

Figure 4.63: Different realizations of electrical balance duplexer

introduces phase shift and limits the isolation bandwidth substantially. Finally, the middle circuit has the drawback of the RX connected directly to antenna, and thus it can be heavily compressed, while the isolation is merely achieved through the common mode rejection of the receiver. A more complete analysis may be found in [42].

There are, however, several important drawbacks associated with the electrical balance duplexer: First, half the TX signal is wasted in the balancing network, and when the implementation loss of the transformer due to finite Q is added, could lead up to 4dB of signal loss, which is quite significant. In contrast, an external duplexer using SAW technology has a typical insertion loss of about 1.5–2dB. Second, the isolation created in the RX port depends on how well the balancing network follows the antenna impedance. This will require a complicated programmable network, which should tolerate a signal as large as the one at the antenna output, and thus will be hard to design. Nonetheless, tremendous effort has been done in the recent literature to overcome these issues [43], [44], [45], [46].

4.6 N-PATH FILTERS

Aside from the SAW or FBAR filters, another method of creating accurately controlled and sharp filters at RF is through N-path filtering concept. Unlike acoustic wave filters, N-path filters are programmable, which is a substantial advantage, but they have their own limitations and disadvantages. The N-path filters are best understood in the context of passive mixers, and will be analyzed in detail in Chapter 8. Given the relevance however, we will provide an overview in this section, and defer a more analytical discussion to the passive mixer section. For further reading aside from our mixer chapter, see [49], [50], [51], [52], [53]. Also [54] gives a summary and some historical perspective.

Probably the first instance of an N-path filter, also referred to as a Barber filter, was described by Barber in 1947 [55], and is conceptually depicted in Figure 4.64.

Suppose to the filter input a desired signal at f_0 along with a close-by unwanted signal at f_1 are applied. For instance, a desired WLAN signal in the band of 2.4–2.48GHz may be accompanied by an undesirable band-41 LTE signal at 2.51GHz. A traditional RF filter must then have a passband of 2.4 to 2.48GHz, that is 80MHz, centered at 2.44GHz. The ratio of the center

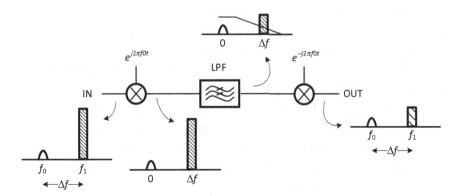

Figure 4.64: N-path filter as described in [55]

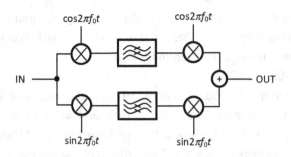

Figure 4.65: A 2-path filter to realize the frequency shift in practice

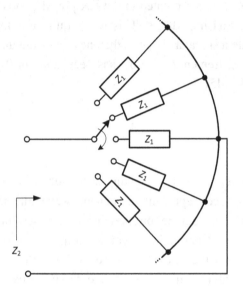

Figure 4.66: N-path filter proposed by Smith [56]

frequency to pass band is about 30. For the filter to provide enough rejection at the close-by frequency offset of 30MHz away where the LTE interferer is, the Q of the filter elements must be much larger than 30, say few hundred or thousands, a task that may be accomplished only by a SAW or FBAR resonator. On the other hand, as proposed in Figure 4.64, one may *frequency shift* (that is to say *downconvert*) the signals at the filter input using complex multiplication at f_0. Now the desired signal is at DC, and the unwanted interferer at $f_1 - f_0 = \Delta f$ is easily removed by a lowpass filter. For instance, a 3rd-order Butterworth filter with a 3dB passband edge of 10MHz provides over 28dB of rejection at 30MHz where the interferer is. The filtered signal is then shifted back up to the original desired location of f_0 through another set of quadrature multipliers. Effectively, by means of frequency shifting of the signals, one can frequency shift a lowpass filter to a narrow bandpass filter at a known and precise frequency of f_0. The lowpass filter on the other hand is easy to implement using integrated elements very efficiently as described earlier.

The actual complex multiplication may be accomplished using the arrangement shown in Figure 4.65, representing a 2-path filter. Note that around where the desired signal is, the LPF has a gain of one, and the signal is effectively multiplied by $\cos^2 2\pi f_0 t + \sin^2 2\pi f_0 t = 1$ as desired. In most cases the multiplier and the signals are differential, effectively realizing a 4-path filter, which is the most common implementation of an N-path filter.

Shown in Figure 4.66 is the realization of an N-path filter as a one-port proposed by Smith [56]. Apart from the fact that the one in Figure 4.65 is a two-port implementation, the ideas are

4.6 N-Path Filters

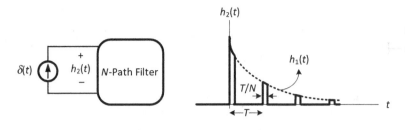

Figure 4.67: Impulse response of Z_1 and that of the N-path filter

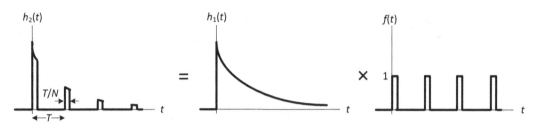

Figure 4.68: Representation of the impulse response $h_2(t)$ as the product of $h_1(t)$ and a sampling function $f(t)$

very similar and both involve frequency shifting. The one-port N-path filter comprises a rotary switch connected to one terminal of each of N identical impedances Z_1. The switch is assumed to be rotating at a constant frequency, $\omega_0 = \frac{2\pi}{T}$, making contact with each of the networks for a short period of $\frac{T}{N}$ once each revolution. We wish to find the impedance Z_2 looking into the one-port.

Although [56] makes the simplifying assumption of treating the network time-invariant, the results are reasonably accurate and give us some perspective, and thus we will recast a summary here.

Since Z_1 is an LTI network, we can associate some typical impulse response ($h_1(t)$) to it which is the dashed line in Figure 4.67. Next, we try to find the impulse response of the N-path network, that is, the voltage produced across it if one coulomb of charge were to be dumped into the network instantaneously. The current impulse passes through the commutating switch to one of the networks and causes a voltage response, $h_1(t)$, in that one network. At some short time later the commutator leaves that network and thereafter samples the voltage across it for a short time, $\frac{T}{N}$, once each revolution. Thus the voltage on the switch arm, which is the impulse response $h_2(t)$, would be as shown in the figure, which can be written as

$$h_2(t) = \begin{cases} h_1(t) & nT < t < (n+1)T \\ 0 & \text{elsewhere} \end{cases},$$

where n is an integer greater or equal than 0.

The impulse response $h_2(t)$ can be expressed as follows, as graphically shown in Figure 4.68,

$$h_2(t) = h_1(t).f(t) = h_1(t) \sum_k a_k e^{jk\omega_0 t},$$

where $a_k = \frac{1}{N} e^{-\frac{jk\pi}{N}} \operatorname{sinc}\left(\frac{k}{N}\right)$ is the Fourier series coefficients of $f(t)$.

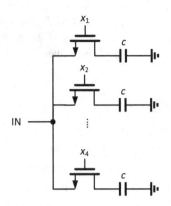 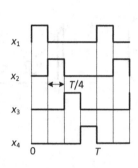

Figure 4.69: CMOS implementation of a 4-path filter and the corresponding waveforms of the switches

Thus, in the frequency domain

$$Z_2(\omega) = F[h_2(t)] = \sum_k a_k Z_1(\omega - k\omega_0),$$

which clearly shows the impedance Z_1 is shifted to ω_0 (the switch rotation frequency) and its harmonics with a sinc function envelope. Around the main harmonic

$$Z_2(\omega) \approx a_1 Z_1(\omega - \omega_0) + a_1^* Z_1(\omega + \omega_0).$$

The analysis in [56], as shown already, as well as the one in [57], [58] involve sampled data approximation, and hence the terminology sampled data bandpass filter has been commonly used. A more rigorous analysis performed on the passive mixers in Chapter 8 shows that for $\omega = 4$, around the fundamental

$$Z_2(\omega) \approx R_{SW} + \frac{2}{\pi^2}[Z_1(\omega - \omega_0) + Z_1(\omega + \omega_0)],$$

where R_{SW} is one switch on resistance. Since at any point of time one and only one switch is on, this simply comes at the series with the shifted impedance.

The N-path filters were quite popular in the 1970s and 1980s for lower frequency applications in the context of switched capacitor filters [4], but did not make it into RF until very recently [59], [60], [61], [62] mainly due to the advent of *good* RF switches offered in nanometer CMOS technology. An example of a 4-path filter in 16nm CMOS is shown in Figure 4.69. Impedance Z_1 is realized using a simple capacitor, which is advantageous as it is inherently very linear and noiseless.

Example: The simulated reactance of the 4-path network of Figure 4.69 (driven by an ideal current source) with the switches commutating at a frequency of 2GHz, and a capacitance value of 20pF is shown in Figure 4.70. Evidently, the capacitance is shifted to $\pm\omega_0$ (and its harmonics), and it effectively behaves like a parallel resonance circuit at the vicinity of $\pm\omega_0$. Even with ideal switches, the tank is still lossy given the time variant nature of the circuit and the aliasing of higher frequency harmonics back to the frequency of interest [62]. Nonetheless, it still achieves superior quality factor compared to integrated LC tanks, and more importantly, its resonance frequency is tightly controlled, and is easily programmable.

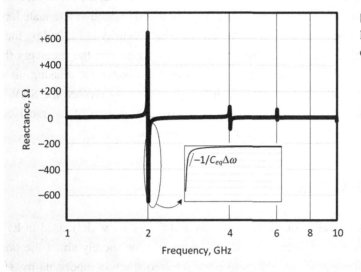

Figure 4.70: The reactance looking into the input of the circuit of Figure 4.69

With the N-path resonator of Figure 4.69 one can now build high-Q bandpass filters. A simple 2nd-order realization is shown in Figure 4.71. At 2GHz, the N-path circuit is ideally an open circuit and the signal passes. However, given the finite Q of the N-path network an insertion loss of about 2.4dB is simulated. Besides the finite Q, the switches' parasitic capacitance at 2GHz also adds to the loss. From [50], the passband insertion loss due to the resistive part of a 4-path filter is $\frac{8}{\pi^2}$ or 1.8dB, and apparently the remaining 0.6dB is due to the switches' capacitance (estimated to about 100fF).

The filter rejection at far-out frequencies is mostly limited by the nonzero switch resistance. In this case, the switches have a resistance of about 20Ω, limiting the rejection to about 11dB or so.

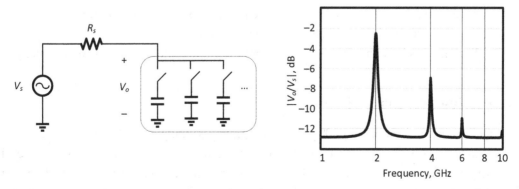

Figure 4.71: A simple second-order bandpass filter based on the N-path resonator of Figure 4.69 and its frequency response

Higher order filtering is possible through the use of active circuits (e.g., transconductors) to stagger tune and couple the resonators [52], [53].

N-path networks with a simple capacitor being commutated are generally low-noise. However, the noise of the clock signal controlling the switches ultimately decides the maximum allowable interferer through reciprocal mixing (see Chapter 6). That means in most cases these filters cannot be used for noise reduction; rather they are mostly suited to attenuate large signals and relax the amplifier compression. Furthermore, the filter linearity is limited by the switches, which makes it difficult to be used in transmitters' very output. In the receivers the filter is mostly limited to a few-dBm signals. They also suffer from harmonic aliasing, as is evident from Figure 4.71. Given all these, the N-path filters cannot really replace the SAW or FBAR filters, but certainly are very helpful to relax the radio and especially the receiver linearity requirements, and are widely used in modern radio products.

4.7 QUADRATURE FILTERS

As we pointed out in Chapter 2, quadrature signals and filters are widely used in RF receivers and transmitters. A quadrature filter is an *allpass* network that merely shifts the phase of the positive frequency components by −90°, and negative frequency components by +90°. Since ±90° phase shift is equivalent by multiplying by $e^{\pm j90} = \pm j$, the transfer function can be written as

$$H_Q(f) = \begin{cases} -j & f > 0 \\ +j & f < 0 \end{cases}.$$

We also showed that for an arbitrary input $x(t)$ passing through such quadrature filter (with $h_Q(t) = 1/\pi t$ being impulse response) results in an output $\hat{x}(t)$, defined as the *Hilbert transform* of the input:

$$\hat{x}(t) = x(t) * \frac{1}{\pi t} = \frac{1}{\pi} \int_{-\infty}^{\infty} \frac{x(\tau)}{t - \tau} d\tau.$$

The Hilbert transform or in general 90° phase shift is physically not realizable as $h_Q(t)$ is noncausal, although its behavior can be well *approximated* over a finite frequency range using a real network. If $x(t) = A\cos(\omega_0 t + \varphi)$, then $\hat{x}(t) = A\sin(\omega_0 t + \varphi)$. Moreover, any signal that consists of a sum of sinusoids follows the above accordingly.

With this background, let us take a closer look at quadrature filters and quadrature generation.

4.7.1 Passive Polyphase Filters

Quadrature generation and 90° phase shift may be accomplished by using a class of networks known as polyphase filters [63]. Although their implementation is symmetric, they have an asymmetric frequency response, that is, the filter responds to positive and negative frequencies differently. Consider a 4-phase network in Figure 4.72, where two sets of four identical vectors each 90° phase shifted with respect to one another is applied. The difference between each set is that in one case the vectors are lagging when rotating counterclockwise, whereas in the other

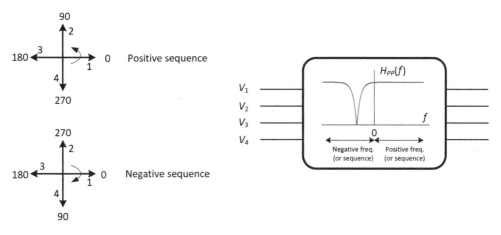

Figure 4.72: A 4-phase network with inputs having negative and positive sequences

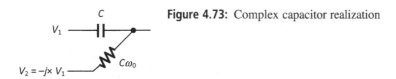

Figure 4.73: Complex capacitor realization

case they are leading. We expect $H_{pp}(f) \neq H_{pp}(-f)$, which at first glance implies $H_{pp}(f)$ is not real, although we will see this is not the case.

Although in practice real RF signals have only positive frequencies, in complex domain the set of vectors could be viewed as having identical magnitudes, but opposite frequencies as they rotate in opposite directions. The polyphase filter is expected to respond to each set differently. The underlying principle behind creating such asymmetry in the frequency response may be understood by examining the LP to BP frequency transformation. For a *real* BPF, such as the ones presented earlier, the transformation requires the frequency to change from $\omega \leftrightarrow \frac{\omega^2 - \omega_0^2}{\omega}$. This will accordingly preserve the symmetry of the negative and positive frequencies, as is evident from Figure 4.36. On the other hand, if one transforms the frequency by applying: $\omega \leftrightarrow \omega - \omega_0$, the symmetry will not be preserved. Such transformation, however, cannot be achieved using real components. For example, the capacitor admittance needs to change from $jC\omega \leftrightarrow jC\omega - jC\omega_0$. This results in the original capacitor in parallel with a frequency independent element (say a resistor with a value of $-jC\omega_0$), whose value is complex. This problem may be resolved by considering that polyphase networks take multiphase inputs, for example a quadrature sequence for a 4-phase realization. Thus the factor j for the resistor may be achieved by connecting that resistor to the second input which is identical in magnitude, but 90° phase shifted. The resulting complex capacitor is shown in Figure 4.73.

A 1st-order 4-phase RC filter may be realized using this concept as shown in Figure 4.74. The circuit is symmetric, and takes a differential quadrature signal as an input. Since a real capacitor in series with the signal creates a highpass function, that is, produces a null at DC, we expect the polyphase filter to have the same characteristics but shifted to $\omega_0 = \frac{-1}{RC}$. This is shown in Figure 4.74. Alternatively the polyphase transfer function may be understood based on the fact that, in the case of the positive sequence (or positive frequency) input, after experiencing phase shift from each RC branch, the signals are added *constructively*, thus

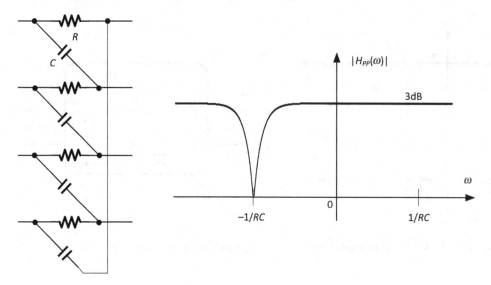

Figure 4.74: A 1st-order RC polyphase filter

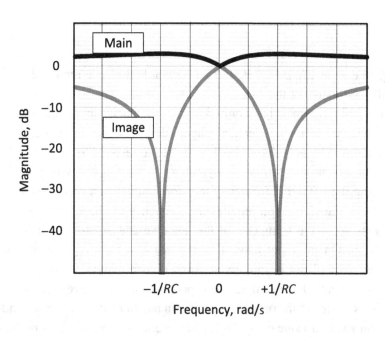

Figure 4.75: Frequency response of a 1st-order polyphase filter to two input sequences

providing a flat passband with a gain of $\sqrt{2}$ at or around $\omega = \frac{1}{RC}$. On the other hand, for the negative sequence, the inputs are subtracted after experiencing the phase shift. In fact at the exact frequency of $-1/RC$ for the negative sequence, each of the four signals that are $-90°$ apart, experiences exactly another $+45°$ of phase shift, and they cancel entirely, creating a null in the transfer function.

Note that the filter response plotted in Figure 4.74 is not exactly accurate. The polyphase filter is indeed composed of real components (resistors and capacitors), and driven by real signals, and hence its frequency response is symmetric along the y-axis. What the plot ignores is the fact that at negative frequencies, the phase sequence of the input changes (Figure 4.72). In other words, the filter will have a symmetric flat passband for one sequence (labeled as main), and has the null for the other sequence (labeled as image), as shown in Figure 4.75. The

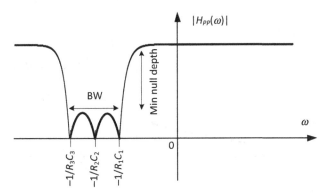

Figure 4.76: Frequency response of a 3rd-order passive polyphase filter

reasoning for calling the sequences main versus image will be clear in Section 4.7.4 in the context of single-sideband or image-reject receivers. The plot of Figure 4.74 is simply a convenient representation of the filter characteristics, combining the response of the two sequences and assigning one to the negative frequency, and the other to the positive frequencies.

Note that for the main sequence, the passband is not quite flat. The passband has a gain of 1 (or 0dB) at very low and high frequencies, and peaks to $\sqrt{2}$ (or 3dB) at exactly $\omega = \pm\frac{1}{RC}$. The transfer function of the 1st-order polyphase filter (for *positive frequencies* only) is given below for the main and image sequences:

$$H_{MAIN}(j\omega) = \frac{1 + RC\omega}{1 + jRC\omega}$$

$$H_{IMAGE}(j\omega) = \frac{1 - RC\omega}{1 + jRC\omega}.$$

The null created by a 1st-order polyphase filter may be too narrow for many applications. To extend the null width, as is common in any filter design, several staggered tuned stages may be cascaded. The corresponding transfer function of three stages cascaded is shown in Figure 4.76 (with main and image responses combined), where the stages are identical but only tuned to $1/R_1C_1 < 1/R_2C_2 < 1/R_3C_3$. In addition to the number of stages, the extended bandwidth is also a function of the minimum stopband null depth. Unlike an unloaded stage that has a passband gain of 3dB, when two identical stages are connected to each other, they result in a passband loss of 3dB, and adding more stages rapidly increases the loss further.

Practical concerns when using polyphase filters include passband loss, loading presented to the previous stages, noise, as well as null depth limitation due to finite matching between components. A more comprehensive study may be found in [64].

4.7.2 Active Polyphase Filters

Using a similar technique shown in Figure 4.73, active polyphase filters combined may be realized [65]. An example of a biquad is shown in Figure 4.77 with the corresponding transfer function on the right. As expected, the response of a first order active lowpass filter (each branch on top or bottom) is simply shifted in frequency. The cross-coupled resistors are connected to create complex capacitors as illustrated in Figure 4.73 already. A negative resistor is simply realizable

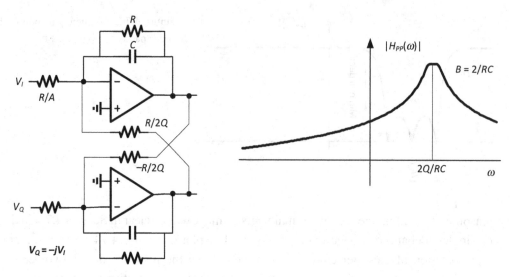

Figure 4.77: An active polyphase biquad

in a differential structure, although for convenience the filter is drawn single-ended. The filter has a passband gain of A, a 3dB bandwidth of $\frac{2}{RC}$, and a center frequency of $\omega_0 = QB = 2Q/RC$. Though not common due to their complexity, other variations of polyphase filters such as ladder structures are also available based on the same concept shown in Figure 4.73 [66].

Example: Shown in Figure 4.78 is a single-stage active polyphase filter intended for Bluetooth applications. Also shown is the filter transfer function for the I output. The Q output has the same magnitude, but is 90° out of phase. The filter is intended to have a 3dB bandwidth of 2MHz, centered at 2MHz. The input resistor is adjusted for a passband voltage gain of 20dB.

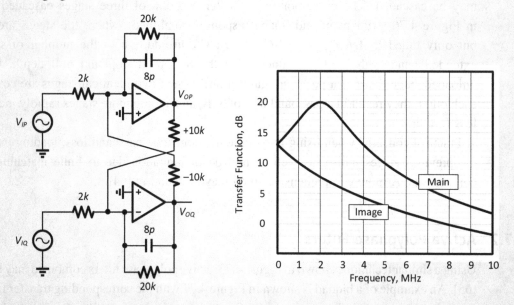

Figure 4.78: Bluetooth biquad and the simulated filter transfer function

The two transfer functions shown belong to the two cases of I leading Q, and I lagging. The latter, labeled as image, could be thought of as negative frequencies, so when flipped and attached to the first transfer function, it will resemble the composite plot of Figure 4.77. The amount of image rejection for this first order design is fairly small, and about 13dB. Like other types of filter, several stages must be stagger-tuned to provide sharper rejection.

Example: To improve the rejection of the filter of previous example we would like to design a 2nd-order Butterworth filter with the same 2MHz bandwidth and center frequency.

The characteristic function of a 2nd-order Butterworth filter is $K(s) = s^2$, leading to $H(s) = s^2 + \sqrt{2}s + 1$. So there are two complex poles (or natural frequencies) at $\frac{\sqrt{2}}{2} \pm j\frac{\sqrt{2}}{2}$. A single-stage active polyphase filter (Figure 4.77) realizes a transfer function of

$$\frac{\frac{-A}{RC}}{j\omega + \frac{1}{RC} - \frac{j2Q}{RC}}.$$

Assuming normalized lowpass poles of $\alpha \pm j\beta$, and a scaling factor of k, shifting to ω_0 will result in

$$\frac{1}{RC} - \frac{j2Q}{RC} = k\alpha \pm jk\beta - j\omega_0.$$

Thus,

$$\frac{1}{RC} = k\alpha$$
$$Q = \frac{\omega_0 \pm k\beta}{2k\alpha}.$$

Making the arbitrary choice of setting all the feedback resistors to $R = 20k$, since $\alpha = \frac{\sqrt{2}}{2}$, and $k = 2\pi \times 10^6$ (note that the equivalent lowpass bandwidth is 1MHz), we have $C = 11.3p$. With $\omega_0 = 2\pi \times 2 \times 10^6$, for the two stages we obtain $Q_1 = 0.93$, and $Q_2 = 1.92$. We intentionally assign the lower Q to the first stage, as it will be lower tuned, which results in more stopband filtering at the output of the first stage.

Shown in Figure 4.79 is the complete 2nd-order filter schematic. The gain of the first stage is adjusted to 20dB, but since it is tuned at 1.41MHz, it has about 6dB of droop at 2MHz, and thus to get a net gain of 20dB, the input resistor of the 2nd stage is scaled to have a net passband gain of 20dB.

Continued

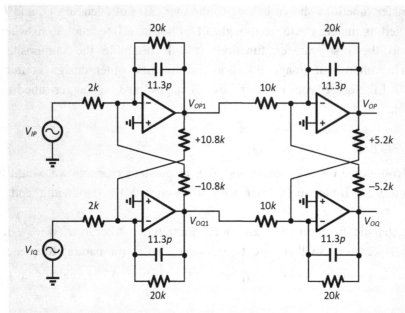

Figure 4.79: A 2nd-order active polyphase filter

The filter frequency response is plotted in Figure 4.80. Also shown is the filter response at the output of its first stage, which is tuned at 1.41MHz. The image rejection at 2MHz is now 25dB.

Unlike their passive counterparts, the active polyphase stages provide gain, and also filtering in general, whereas the passive polyphase stages only reject the image, while not affecting the other frequencies much. This of course comes at the expense of more noise and power consumption for active filters.

4.7.3 Quadrature Generation

The polyphase filter to function properly requires a polyphase signal. For example, a quadrature input is needed in the case of 4-phase networks shown earlier. There are a number of ways of

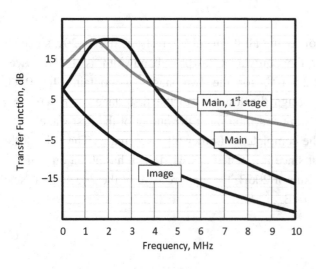

Figure 4.80: Frequency response of the 2nd-order filter of Figure 4.80

4.7 Quadrature Filters

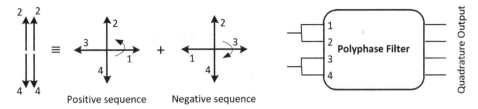

Figure 4.81: Quadrature generation using polyphase filters

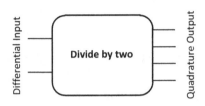

Figure 4.82: A divide by two used to produce quadrature outputs

creating quadrature signals; perhaps the most natural one is to use the same polyphase network shown earlier. Consider Figure 4.81, where a differential (2-phase) input is applied to a polyphase filter whose adjacent inputs are shorted. The 2-phase input can be broken into two 4-phase signals with opposite sequences.

One sequence will be rejected by the polyphase filter, assuming its frequency is selected appropriately to lie in the stopband of the filter. The resultant quadrature output has 6dB less magnitude compared to the input, but since a balanced unloaded polyphase stage has a passband gain of 3dB, the overall passband loss is 3dB. To achieve a reasonable bandwidth, and to cover RC process variations, it is likely needed to cascade several stages, leading to further loss. Thus using a polyphase filter may not be the most viable option to generate quadrature signals.

A more common way is to use a frequency divider with a divide ratio of 2^n, for example, a divide by two. This results in an accurate quadrature generation with much wider bandwidth as long as the divider is functional (Figure 4.82). The drawback is the need to run the master oscillator at higher frequency, at least twice as much. There are however other advantages of running the main oscillator at a different frequency from the carrier, which we will discuss in Chapter 12. For these reasons, division by 2 (or 4) seems the most appealing method of producing quadrature signals in most modern radios. The dividers and their limitations will be extensively covered in Chapter 10.

Example: Consider a two-stage polyphase filter shown in Figure 4.83 used for quadrature generation. The values of R and C to resonate at 5.5GHz would be $R = 1\text{k}\Omega$, and $C = 29\text{fF}$. Each of the stages are shifted in frequency by $\pm 10\%$ to create as much flatness as possible around 5.5GHz center frequency, as shown in the figure.

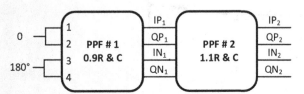

Figure 4.83: A two-stage polyphase quadrature generator

Continued

In Problem 24 we show that if the two stages are identical, the transfer function to the I output of the first state is

$$\frac{1+jRC\omega}{1+4jRC\omega-(RC\omega)^2},$$

and to the I output of the second stage is

$$\frac{1+RC\omega}{1+4jRC\omega-(RC\omega)^2}.$$

Thus, the first stage has a passband loss of $\frac{|1+j|}{|4j|} = 9\text{dB}$ (to be precise at $\omega = \frac{1}{RC}$), whereas the total loss is about $\frac{|1+1|}{|4j|} = 6\text{dB}$ as the second stage has a gain of 3dB. The simulated losses at 5.5GHz are 8.1dB and 5.6dB, respectively, for each stage. The small difference is due to the approximation of assuming identical stages.

Similarly, the transfer function to the Q output of the first state is

$$\frac{jRC\omega(1+jRC\omega)}{1+4jRC\omega-(RC\omega)^2},$$

and to the Q output of the second stage is

$$\frac{jRC\omega(1+RC\omega)}{1+4jRC\omega-(RC\omega)^2}.$$

Thus, the *magnitude* of gain imbalance between the I and Q outputs is

$$\frac{|1+RC\omega||1-RC\omega|}{\sqrt{\left(1-(RC\omega)^2\right)^2+(4RC\omega)^2}}.$$

The simulated magnitude of I/Q gain imbalance for each of the stages is shown in Figure 4.84.

Note that all the outputs have a perfect phase difference at all frequencies given the symmetric nature of the filter (as long as there are no mismatches), whereas the gain imbalance

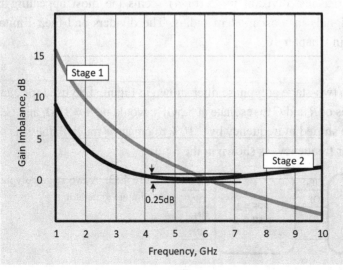

Figure 4.84: Simulated gain imbalance of the polyphase filter of Figure 4.83

is frequency dependent, as shown in Figure 4.84, and predicted by the equation. The addition of a second stage extends the gain flatness quite a bit, e.g., the gain imbalance remains below 0.25dB from 4.2GHz to 7.2GHz, roughly 3GHz.

4.7.4 Polyphase Filters Application in Single-Sideband Receivers

A common use of polyphase filters is in *single-sideband* or *image-reject* receivers introduced in Chapter 2. There we showed that by exploiting quadrature multiplication along with the Hilbert transform, the spectrum around one sideband of the carrier can be rejected (Figure 4.85 left block diagram). As shown in the right block diagram of Figure 4.85, the Hilbert transform may be realized in analog domain through a polyphase filter. In many cases the multiplier output is differential, and hence the quadrature differential multiplication creates a 4-phase sequence labeled by $I+$, $I-$, $Q+$, and $Q-$ as needed in a polyphase filter.

To understand how the receiver functions, let us assume the input to the receiver is a tone at the upper side of the carrier, that is,

$$x_C(t) = \cos(\omega_C + \omega_m)t.$$

After multiplication, assuming the multipliers lowpass filter the high-frequency component around $2\omega_C$, the following signals appear at the polyphase filer inputs: $\pm \cos \omega_m t$ at the differential I input, and $\mp \sin \omega_m t = \pm \cos(\omega_m t + 90°)$ at the differential Q input. So the Q input is *leading* the I input by 90°. Now if the input lies at the lower sideband, that is,

$$x_C(t) = \cos(\omega_C - \omega_m)t,$$

then the differential Q input changes to $\pm \cos(\omega_m t - 90°)$, while the differential I input remains intact, which is to say the Q input is now *lagging* the I input. Thus, the polyphase input sees two different sequences for the signals at main and image (or upper and lower) sidebands, and consequently one is rejected if the filter is tuned to ω_m. The choice of upper or lower sidebands is of course arbitrary, and depends only on how the polyphase filter inputs are connected to the differential I and Q multipliers.

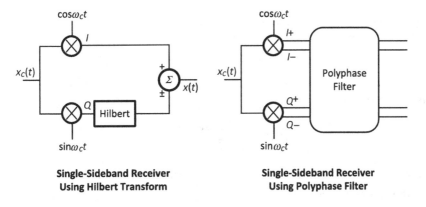

Figure 4.85: Single-sideband receiver employing a polyphase filter

4.8 Summary

This chapter deals with filter topologies used in RF and IF stages of radios.

- Ideal filters and their causality were discussed in Section 4.1.
- Section 4.2 discussed the general principles of filter specification and realizability, and presented the design procedure of double-terminated LC ladder filters.
- Following the LC filters, opamp-RC and g_m-C active filters were presented in Section 4.3. These filters are widely used at the IF of both transmitters and receivers in most modern radios.
- Surface and bulk acoustic wave filters were discussed in Section 4.4. Both filters are commonly used at the RF input of transceivers.
- A brief recast of SAW/FBAR duplexers was presented in Section 4.5. Additionally, an introduction to electrical balance duplexers as an alternative for external duplexers was offered in that section.
- Section 4.6 dealt with N-path filters, which have become recently popular in mostly receivers.
- Finally in Section 4.7 we discussed the principles of quadrature signals, polyphase filters, and quadrature generation.

4.9 Problems

1. The input impedance of the circuit shown below is expressed as

$$Z(s) = \frac{N(s)}{D(s)} = \frac{a_n s^n + a_{n-1} s^{n-1} + \cdots + a_1 s + a_0}{b_m s^m + b_{m-1} s^{m-1} + \cdots + b_1 s},$$

where $|n - m| \leq 1$.

Show that the value of the series input capacitor is

$$C = \frac{\frac{\partial}{\partial s} D(s)|_{s=0}}{N(0)}.$$

2. A one-port has the input impedance

$$Z(s) = \frac{2s^3 + 2s + 1}{s^2 + 1}.$$

Using a similar approach as Problem 1, find the value of the series input inductance, and synthesize the rest of the circuit.

3. For a causal function $h(t)$ with finite energy ($\int_{-\infty}^{\infty} |H(\omega)|^2 d\omega < \infty$), the Paley–Wiener theorem states that $\int_{-\infty}^{\infty} \frac{Ln|H(\omega)|}{1+\omega^2} d\omega$ must exist and be finite. Show, as such, an ideal filter like the one shown in Figure 4.1 cannot exist.

4. As an alternative to the proof presented in the previous problem, in this problem we show that an ideal filter requires infinite number of elements, and hence is not realizable.
   a. Show that no rational function
   $$|K|^2 = \frac{a_0 + a_1\omega^2 + \cdots + a_n\omega^{2n}}{b_0 + b_1\omega^2 + \cdots + b_m\omega^{2m}}$$
   cannot be zero at all points of a frequency range $0 \leq \omega \leq \omega_p$, unless $a_i = 0$ for all i, and hence $|K|^2 = 0$.
   b. Using the same reasoning as part a, show that $\frac{1}{|K|^2}$ cannot be zero at every point of a frequency range. What are the implications?

5. Consider the 3rd-order maximally flat filter below.
   a. Calculate the transducer parameters directly based on the definition.
   b. Find the LC two-port impedance parameters.
   c. Redo part a indirectly by using the two-port z parameters.
   d. Assuming $|V_S| = 1$ V RMS, find the reflected power and the power delivered to the load at $\omega = 0$, $\omega = 1$, and $\omega = 10$ rad/s.

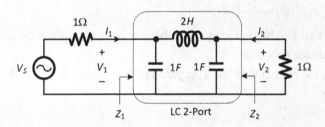

6. While the transducer factor ($H(s)$) can be obtained directly based the z parameters of the two-port, a simpler yet more elegant approach is described below.
   a. Show that the two-ports shown below are equivalent.
   b. For the two-port on the right, show that $\frac{I_L}{V_s} = y_{21}$.
   c. Knowing that $[Y] = [Z]^{-1}$, find an expression for $\frac{V_L}{V_s}$, and subsequently $H(s)$.

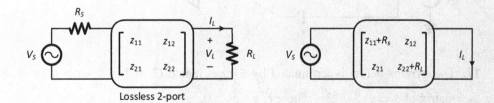

7. Prove the Feldtkeller equation using the $H(s)$ and $K(s)$ derived based on the two-port impedance parameters.

8. Show that $\Delta_z = z_{11}z_{22} - z_{12}^2 = R_s R_L \frac{H_o - K_o}{H_o + K_o}$. **Hint:** Use the Feldtkeller equation.

9. Consider $Z(s)$, representing the impedance of an RLCM one-port (Figure 4.12). Arguing that at infinity $Z(s)$ is dominated by either its capacitors or inductors, show that the degrees of the numerator and the denominator can differ by no more than one.

10. Show that the polynomial $P(s) = s^5 + s^4 + 6s^3 + 6s^2 + 25s + 25$ is not Hurwitz, despite having positive coefficients with no missing terms.

11. Assume $Z(j\omega) = jX(j\omega)$ describes an LC one-port impedance. Considering that the LC impedance has only simple $j\omega$ axis poles, using partial fraction of the impedance prove that $\frac{d}{d\omega}X > 0$ for all ω.

12. Show that the zeros of $H(s)H(-s)$ for a normalized maximally flat filter are at

$$s_k = e^{j\pi(n-1+2k)/2n} \text{ for } k = 1, 2, \ldots, 2n.$$

13. Prove that for a Butterworth filter, the filter degree is $n \geq \dfrac{\log \sqrt{\frac{10^{\alpha_s/10}-1}{10^{\alpha_p/10}-1}}}{\log \frac{\omega_s}{\omega_p}}$.

14. For a Chebyshev filter with $|K|^2 = k_p^2 \cos^2 nu = k_p^2 \cos^2(n \cos^{-1}\omega)$, show that upon expansion $|K|^2$ is indeed a polynomial in ω^2. **Hint:** Use the identity $\cos nu = \text{Re}\,[e^{jnu}] = \text{Re}\,[(\cos u + j \sin u)^n]$, and expand.

15. Prove that in a Chebyshev filter, for $\omega \geq 1$: $|K|^2 = \frac{k_p^2}{4}\left[\left(\omega + \sqrt{\omega^2-1}\right)^n + \left(\omega + \sqrt{\omega^2-1}\right)^{-n}\right]^2$. **Hint:** Use the identity $\cos(nu) = \frac{e^{jnu}+e^{-jnu}}{2}$.

16. Shown below is an active-RC biquad stage, suitable for high-Q applications.
    a. Find the biquad transfer function.
    b. Show the spread of the elements is less compared to the biquad of Figure 4.47 when the required Q is large.

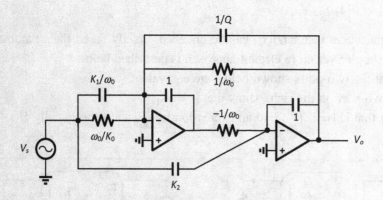

17. The gyrator below is terminated by a capacitance C, and is described by an admittance matrix of $Y = \begin{bmatrix} 0 & g_1 \\ -g_2 & \epsilon \end{bmatrix}$, where $\epsilon, g_1, g_2 > 0$.

    a. With $\epsilon = 0$, find the impedance Z_{IN} looking into the gyrator.
    b. Explain how ϵ affects Z_{IN} and construct an equivalent circuit model.

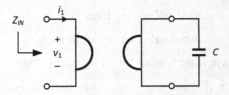

18. An active-RC gyrator to simulate a *grounded* inductor known as the Riordan gyrator [67] is shown below.
    a. Show that if the opamps are ideal, the effective inductance is given by $L = \frac{R_1 R_3 R_4}{R_2} C$.
    b. If the opamps have a finite DC gain of A_0, show that the inductance quality factor is equal to $Q = \frac{A_0}{\frac{R_1 R_3 C \omega}{R_2} + \frac{1}{R_3 C \omega}}$.

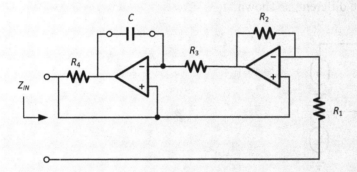

19. For the following active gyrator, the capacitor C is voltage-dependent given by $C = f(v)$, prove that the effective inductor is current-dependent with value given by $L = \frac{1}{gm^2} f\left(\frac{1}{gm} i_{in}\right)$.

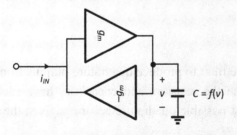

20. A generic Sallen–Key topology is illustrated below. Show that the filter transfer function is
$$\frac{V_o}{V_s} = \frac{Z_3 Z_4}{Z_1 Z_2 + Z_3(Z_1 + Z_2) + Z_3 Z_4}.$$

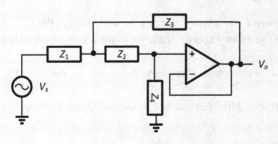

RF and IF Filters

21. Show that the input impedance of an unloaded polyphase filter is $\frac{R}{2} + \frac{1}{j2C\omega}$.
22. Find the input impedance of a polyphase filter loaded by an arbitrary load Z_L.
23. Find the input impedance of a polyphase filter loaded by an identical stage.
24. Consider a two-stage polyphase filter (Figure 4.83) with identical resistors of R and capacitors of C used for quadrature generation. Show that the transfer function to the I output of the first state is $\frac{1+jRC\omega}{1+4jRC\omega-(RC\omega)^2}$, and to the I output of the second stage is $\frac{1+RC\omega}{1+4jRC\omega-(RC\omega)^2}$. **Hint:** Use the findings of Problem 21 to load the first stage properly.
25. Consider the following one-stage polyphase filter, where all the resistors and capacitors are identical, loaded by nonidentical loads. Calculate the transfer function. What is the passband loss and image rejection if the loads are identical? Repeat for the case the I and Q loads are different, as shown.

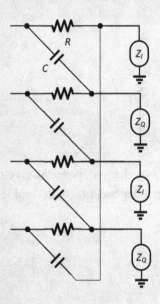

26. Design a two-stage passive polyphase filter to produce quadrature outputs from a differential clock signal at 1.6GHz. Assume the resistors and capacitors have each a process variation of 10%. Design for the best possible quadrature accuracy given the variation of the components.

4.10 References

[1] G. C. Temes and S. K. Mitra, *Modern Filter Theory and Design*, John Wiley, 1973.
[2] G. Temes, "The Present and Future of Filter Theory," *IEEE Transactions on Circuit Theory*, 15, no. 12, 302 1968.
[3] G. C. Temes and J. W. Lapatra, *Introduction to Circuit Synthesis and Design*, vol. 15, McGraw-Hill, 1977.
[4] R. Gregorian and G. C. Temes, *Analog MOS Integrated Circuits for Signal Processing*, vol. 1, Wiley-Interscience, 1986.

[5] H. J. Orchard, G. C. Temes, and T. Cataltepe, "General Sensitivity Formulas for Lossless Two-Ports," *Electronics Letters*, 19, no. 7, 576–578, 1983.

[6] H. Orchard, G. Temes, and T. Cataltepe, "Sensitivity Formulas for Terminated Lossless Two-Ports," *IEEE Transactions on Circuits and Systems*, 32, no. 5, 459–466, 1985.

[7] G. Temes and H. Orchard, "First-Order Sensitivity and Worst Case Analysis of Doubly Terminated Reactance Two-Ports," *IEEE Transactions on Circuit Theory*, vol. 20, no. 9, 567–571, 1973.

[8] C. A. Desoer and E. S. Kuh, *Basic Circuit Theory*, McGraw-Hill Education, 2009.

[9] M. E. Van Valkenburg, *Introduction to Modern Network Synthesis*, John Wiley, 1965.

[10] O. Brune, "Synthesis of a Finite Two-Terminal Network Whose Driving-Point Impedance is a Prescribed Function of Frequency," *Journal of Mathematics and Physics*, 10, 191–236, 1931.

[11] R. M. Foster, "A Reactance Theorem," *Bell System Technical Journal*, 3, no. 4, 259–267, 1924.

[12] H. J. Orchard and G. C. Temes, "Maximally Flat Approximation Techniques," *Proceedings of the IEEE*, 56, no. 1, 65–66, 1968.

[13] A. I. Zverev, et al., *Handbook of Filter Synthesis*, vol. 47, John Wiley, 1967.

[14] L. Weinberg, *Network Analysis and Synthesis*, RE Krieger, 1975.

[15] E. Christian and E. Eisenmann, *Filter Design Tables and Graphs*, John Wiley, 1966.

[16] A. J. Grossman, "Synthesis of Tchebycheff Parameter Symmetrical Filters," *Proceedings of the IRE*, 45, 454–473, 1957.

[17] H. Orchard and G. Temes, "Filter Design Using Transformed Variables," *IEEE Transactions on Circuit Theory*, 15, no. 12, 385–408, 1968.

[18] Y. Tsividis and J. Voorman, *Integrated Continuous-Time Filters: Principles, Design, and Applications*, IEEE Press, 1993.

[19] G. Temes, H. Orchard, and M. Jahanbegloo, "Switched-Capacitor Filter Design Using the Bilinear z-Transform," *IEEE Transactions on Circuits and Systems*, 25, no. 12, 1039–1044, 1978.

[20] R. P. Sallen and E. L. Key, "A Practical Method of Designing RC Active Filters," *IRE Transactions on Circuit Theory*, 2, no. 3, 74–85, 1955.

[21] H. Khorramabadi and P. Gray, "High-Frequency CMOS Continuous-Time Filters," *IEEE Journal of Solid-State Circuits*, 19, no. 6, 939–948, 1984.

[22] B. D. H. Tellegen, "The Gyrator, a New Electric Network Element," *Philips Research Reports*, 3, 81–101, 1948.

[23] A. G. J. Holt and J. Taylor, "Method of Replacing Ungrounded Inductors by Grounded Gyrators," *Electronics Letters*, 1, no. 6, 105, 1965.

[24] Y.-T. Wang and A. Abidi, "CMOS Active Filter Design at Very High Frequencies," *IEEE Journal of Solid-State Circuits*, 25, no. 6, 1562–1574, 1990.

[25] K. M. Lakin, J. R. Belsick, J. P. McDonald, K. T. McCarron, and C. W. Andrus, "Bulk Acoustic Wave Resonators and Filters for Applications above 2GHZ," in *IEEE MTT-S Microwave Symposium Digest*, 2002.

[26] J. Tsutsumi, S. Inoue, Y. Iwamoto, T. Matsuda, M. Miura, Y. Satoh, M. Ueda, and O. Ikata, "Extremely Low-Loss SAW Filter and Its Application to Antenna Duplexer for the 1.9GHZ PCS Full-Band," in *IEEE Proceedings of Frequency Control Symposium*, 2003.

[27] F. Z. Bi and B. P. Barber, "Bulk Acoustic Wave RF Technology," *IEEE Microwave Magazine*, 9, 65–80, 2008.

[28] R. Ruby, "FBAR—From Technology Development to Production," in *Proceedings of the 2nd International Symposium on Acoustic Wave Devices for Future Mobile Communication Systems*, Chiba, Japan, 2004.

[29] R. Ruby and P. Merchant, "Micromachined Thin Film Bulk Acoustic Resonators," in *Proceedings of the IEEE 48th Annual Symposium on Frequency Control*, 1994.

[30] R. Ruby, P. Bradley, J. D. Larson, and Y. Oshmyansky, "PCS 1900 MHz Duplexer Using Thin Film Bulk Acoustic Resonators (FBARs)," *Electronics Letters*, 35, no. 5, 794–795, 1999.

[31] R. Ruby, P. Bradley, J. Larson, Y. Oshmyansky, and D. Figueredo, "Ultra-Miniature High-Q Filters and Duplexers Using FBAR Technology," in *Proceedings of the IEEE International Solid-State Circuits Conference Digest of Technical Papers*, 2001.

[32] K. M. Lakin, G. R. Kline, and K. T. McCarron, "Thin Film Bulk Acoustic Wave Filters for GPS," in *Proceedings of the IEEE 1992 Ultrasonics Symposium*, 1992.

[33] L. Rayleigh, "On Waves Propagated Along the Plane Surface of an Elastic Solid," *Proceedings of the London Mathematical Society*, 17, 4–11, 1885.

[34] T. W. Grudkowski, J. F. Black, T. M. Reeder, D. E. Cullen, and R. A. Wagner, "Fundamental Mode UHF/VHF Miniature Resonators and Filters," *Applied Physics Letters*, 39, 993–995, 1980.

[35] K. M. Lakin and J. S. Wang, "Acoustic Bulk Wave Composite Resonators," *Applied Physics Letters*, 38, 125–127, 1981.

[36] K. A. Sankaragomathi, J. Koo, R. Ruby, and B. P. Otis, "25.9 A ±3ppm 1.1mW FBAR Frequency Reference with 750MHz Output and 750mV Supply," in *Proceedings of the IEEE International Solid-State Circuits Conference – (ISSCC) Digest of Technical Papers*, 2015.

[37] W. Pang, R. C. Ruby, R. Parker, P. W. Fisher, M. A. Unkrich, and J. D. Larson, "A Temperature-Stable Film Bulk Acoustic Wave Oscillator," *IEEE Electron Device Letters*, 29, no. 4, 315–318, 2008.

[38] R. Ruby, P. Bradley, D. Clark, D. Feld, T. Jamneala, and K. Wang, "Acoustic FBAR for Filters, Duplexers and Front End Modules," in *Proceedings of the IEEE MTT-S International Microwave Symposium Digest*, 2004.

[39] J. D. Larson, J. D. Ruby, R. C. Bradley, J. Wen, S.-L. Kok, and A. Chien, "Power Handling and Temperature Coefficient Studies in FBAR Duplexers for the 1900 MHz PCS Band," in *Proceedings of the IEEE Ultrasonics Symposium*, 2000.

[40] Y. Wang, C. Feng, T. Lamers, D. Feld, P. Bradley, and R. Ruby, "FBAR Resonator Figure of Merit Improvements," in *Proceedings of the IEEE International Ultrasonics Symposium*, 2010.

[41] N. Kamogawa, S. Dokai, N. Shibagaki, M. Hikita, T. Shiba, S. Ogawa, S. Wakamori, K. Sakiyama, T. Ide, and N. Hosaka, "Miniature SAW Duplexers with High Power Capability," in *Proceedings of the IEEE Ultrasonics Symposium*, 1998.

[42] M. Mikhemar, H. Darabi, and A. A. Abidi, "A Multiband RF Antenna Duplexer on CMOS: Design and Performance," *IEEE Journal of Solid-State Circuits*, 48, no. 9, 2067–2077, 2013.

[43] B. Debaillie, D. Broek, C. Lavín, B. Liempd, E. A. M. Klumperink, C. Palacios, J. Craninckx, B. Nauta, and A. Pärssinen, "Analog/RF Solutions Enabling Compact Full-Duplex Radios," *IEEE Journal on Selected Areas in Communications*, 32, no. 9, 1662–1673, 2014.

[44] M. Elkholy, M. Mikhemar, H. Darabi, and K. Entesari, "Low-Loss Integrated Passive CMOS Electrical Balance Duplexers with Single-Ended LNA," *IEEE Transactions on Microwave Theory and Techniques*, 64, no. 5, 1544–1559, 2016.

[45] B. Hershberg, B. Liempd, X. Zhang, P. Wambacq, and J. Craninckx, "20.8 A Dual-Frequency 0.7-to-1GHz Balance Network for Electrical Balance Duplexers," in *Proceedings of the IEEE International Solid-State Circuits Conference*, 2016.

[46] G. Qi, B. Liempd, P. Mak, R. P. Martins, and J. Craninckx, "A SAW-Less Tunable RF Front End for FDD and IBFD Combining an Electrical-Balance Duplexer and a Switched-LCN-Path LNA," *IEEE Journal of Solid-State Circuits*, 53, no. 5, 1431–1442, 2018.

[47] G. A. Campbell and R. M. Foster, "Maximum Output Networks for Telephone Substation and Repeater Circuits," *Transactions of the American Institute of Electrical Engineers*, 39, no. 1, 231–290, 1920.

[48] E. Sartori, "Hybrid Transformers," *Materials and Packaging IEEE Transactions on Parts*, 4, no. 9, 59–66, 1968.

[49] H. Darabi and A. Mirzaei, *Integration of Passive RF Front End Components in SoCs*, Cambridge University Press, 2013.

[50] A. Ghaffari, E. A. M. Klumperink, M. C. M. Soer, and B. Nauta, "Tunable High-Q N-Path Band-Pass Filters: Modeling and Verification," *IEEE Journal of Solid-State Circuits*, 46, 998–1010, 2011.

[51] A. Ghaffari, E. A. M. Klumperink, and B. Nauta, "Tunable N-Path Notch Filters for Blocker Suppression: Modeling and Verification," *IEEE Journal of Solid-State Circuits*, 48, no. 6, 1370–1382, 2013.

[52] M. Darvishi, R. Zee, and B. Nauta, "Design of Active N-Path Filters," *IEEE Journal of Solid-State Circuits*, 48, no. 12, 2962–2976, 2013.

[53] M. Darvishi, R. Zee, E. A. M. Klumperink, and B. Nauta, "Widely Tunable 4th Order Switched G_m-C Band-Pass Filter Based on N-Path Filters," *IEEE Journal of Solid-State Circuits*, 47, no. 12, 3105–3119, 2012.

[54] E. A. M. Klumperink, H. J. Westerveld, and B. Nauta, "N-Path Filters and Mixer-First Receivers: A Review," in *Proceedings of the IEEE Custom Integrated Circuits Conference*, 2017.

[55] N. Barber, "Narrow Band-Pass Filter Using Modulation," *Wireless Engineer*, 24, 132–134, 1947.

[56] B. D. Smith, "Analysis of Commutated Networks," *Transactions of the IRE Professional Group on Aeronautical and Navigational Electronics*, PGAE-10, 21–26, 1953.

[57] L. E. Franks and I. W. Sandberg, "An Alternative Approach to the Realization of Network Transfer Functions: The N-Path Filter," *IEEE RFIC Virtual Journal*, 39, 1321–1350, 1960.

[58] L. Franks and F. Witt, "Solid-State Sampled-Data Bandpass Filters," in *Proceedings of the IEEE International Solid-State Circuits Conference Digest of Technical Papers*, 1960.

[59] A. Oualkadi, M. Kaamouchi, J. Paillot, D. Vanhoenacker-Janvier, and D. Flandre, "Fully Integrated High-Q Switched Capacitor Bandpass Filter with Center Frequency and Bandwidth Tuning," in *Proceedings of the IEEE Radio Frequency Integrated Circuits (RFIC) Symposium*, 2007.

[60] A. Mirzaei, H. Darabi, and J. Leete, *Frequency Translated Filter*, US patent 8,301,101, filed May 22, 2009, issued October 13, 2012.

[61] A. Mirzaei, H. Darabi, A. Yazdi, Z. Zhou, E. Chang, and P. Suri, "A 65 nm CMOS Quad-Band SAW-Less Receiver SoC for GSM/GPRS/EDGE," *IEEE Journal of Solid-State Circuits*, 46, no. 4, 950–964, 2011.

[62] A. Ghaffari, E. A. M. Klumperink, and B. Nauta, "A Differential 4-Path Highly Linear Widely Tunable On-Chip Band-Pass Filter," in *Proceedings of the IEEE Radio Frequency Integrated Circuits Symposium*, 2010.

[63] M. Gingell, "Single Sideband Modulation Using Sequence Asymmetric Polyphase Networks," *Electrical Communication*, 48, nos. 1–2, 21–25, 1973.

[64] F. Behbahani, Y. Kishigami, J. Leete, and A. Abidi, "CMOS Mixers and Polyphase Filters for Large Image Rejection," *IEEE Journal of Solid-State Circuits*, 36, no. 6, 873–887, 2001.

[65] J. Crols and M. Steyaert, "An Analog Integrated Polyphase Filter for a High Performance Low-IF Receiver," in *Digest of Technical Papers: 1995 Symposium on VLSI Circuits*, 1995.

[66] J. Haine, "New Active Quadrature Phase-Shift Network," *Electronics Letters*, 13, no. 7, 216–218, 1977.

[67] R. H. S. Riordan, "Simulated Inductors Using Differential Amplifiers," *Electronics Letters*, 3, no. 2, 50–51, 1967.

[68] H. J. Orchard, "Inductorless Filters," *Electronics Letters*, 2, no. 6, 224–225, 1966.

[69] H. J. Orchard and D. F. Sheahan, "Inductorless Bandpass Filters," *IEEE Journal of Solid-State Circuits*, 5, no. 6, 108–118, 1970.

5 Noise

An unavoidable cause of electrical noise is the random thermal motion of electrons in conducting media such as wires or resistors. Active circuits comprising MOS transistors suffer from similar random motion of electrons leading to noise. As long as communication systems are constructed from such devices, the noise will be with us. Amplifying arbitrarily weak signals to make them detectable is unrealistic, as the presence of noise sets a lower limit to minimum detectable signal.[1] This ultimately limits the range where the transmitter and receiver can communicate, yet maintaining a minimum signal-to-noise ratio at the detector output.

This chapter starts with a review of noise sources present in RF components, including how they are modeled and dealt with in the context of two-ports. We then define the noise figure as a universal figure of merit describing the noise properties of a circuit, and link that to the minimum detectable signal achievable in a given radio. Finally we present the concept of minimum noise figure, and a systematical approach to optimize the signal-to-noise in amplifiers. Most of this chapter deals with linear and time-invariant circuits such as RF amplifiers. We will also present the cyclostationary noise to set the foundation for the mixer and oscillator noise analysis. A detailed study of noise in time-variant or nonlinear circuits such as mixers or oscillators is presented in Chapters 8, and 9, respectively, although many basic concepts introduced in this chapter are still applicable.

The specific topics covered in this chapter are:

- Thermal and white noise, cyclostationary noise, as well as FET thermal and flicker noise
- Noise of lossy passive circuits
- Noise figure
- Minimum noise figure, and noise optimization
- Introduction to phase noise
- Sensitivity
- Noise measurement techniques

For class teaching, we recommend presenting a brief summary of Sections 5.1 and 5.2, while fully covering Sections 5.3, 5.4, 5.6, and 5.8. Particularly, much of the material on noise figure definition and noise optimization (Sections 5.3 and 5.4) will be directly used in Chapter 6. Sections 5.7 and 5.9 may be assigned as reading.

---

[1] The upper limit is set by the distortion that will be discussed in the next chapter.

5.1 TYPES OF NOISE

In this section we briefly discuss the types of noise present in RF components, namely resistors, capacitors, inductors, as well as transistors.

5.1.1 Thermal Noise

Thermal noise is the noise due to the random motion of charged particles, and particularly the electrons in a conducting device. In 1906 Einstein predicted that the *Brownian* motion of charged particles results in an electromotive force (emf) across the two open terminals of a resistor. The effect was first observed by Johnson in 1928, and was mathematically formulated by Nyquist.[2] For that reason the thermal noise is sometimes designated as Johnson or Nyquist noise as well. The results appeared in the July 1928 issue of a physical review magazine as side-by-side articles [11], [12].

For a resistor R at a temperature T, the random electron motion produces a noise voltage $v(t)$ across the two open terminals of the resistor. From the *equipartition law* in thermodynamics [17], for each degree of freedom there corresponds an average energy equal to $\frac{1}{2}KT$, where T is the absolute temperature (measured in Kelvin), and $K = 1.37 \times 10^{-23}$ J/K is the Boltzmann constant.

Now consider two of such resistors connected with a lossless line of length l (Figure 5.1). According to the second law of thermodynamics, at a temperature T where the thermal equilibrium is reached, the total energy as a result of thermal agitation created in one resistor and delivered to the other is equal to the one produced by the second resistor and delivered to the first one.

With the help of this circuit it is proven in [11] that $\overline{v^2}$, the average voltage squared across the resistor given a 1Hz integration bandwidth, is[3]

$$\overline{v^2} = 2KTR.$$

To present the proof of above equation in a more intuitive way [6], let us consider a parallel RC circuit (Figure 5.2). We shall first find the average energy stored on the capacitor as a result of the resistor noise, and consequently, find the noise spectral density.

Figure 5.1: Two resistors connected with a lossless line

---

[2] Both Nyquist and Johnson were born in Sweden, immigrated to the United States, and worked at Bell Labs.
[3] This noise is of course specified in a 1Hz bandwidth. The correct notation in general is $\overline{v^2} = 2KTR\Delta f$ for an arbitrary bandwidth. Throughout this chapter, and most of the book, we take the simplifying assumption that $\Delta f = 1$Hz whenever appropriate.

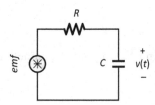

Figure 5.2: Parallel RC circuit with a thermal noise emf associated with the resistor

As a result of the random thermal agitation of the electrons in the resistor, the capacitor will be charged and discharged at random. The average energy stored in the capacitor will be

$$\frac{1}{2}C\overline{v^2},$$

where $\overline{v^2}$ is the mean-square value of the voltage fluctuation impressed across the capacitor. Taken from thermodynamics, if a system has a temperature T, Maxwell–Boltzmann statistics[4] states that the probability that it has an energy E is proportional to $e^{\left(\frac{-E}{KT}\right)}$. Since in the RC circuit the energy stored in the capacitor C is $\frac{1}{2}Cv^2$, then the capacitor voltage has a probability density function of

$$f_V(v) = K_0 e^{\left(\frac{-Cv^2}{2KT}\right)}.$$

The proportionality factor, K_0, is obtained by considering that

$$\int_{-\infty}^{\infty} f_V(v)dv = 1,$$

which is to say that the probability of the capacitor voltage to be between $-\infty$ and $+\infty$ is one. Since $\int_{-\infty}^{\infty} e^{-x^2}dx = \sqrt{\pi}$, with some simple algebraic steps we find

$$K_0 = \sqrt{\frac{C}{2\pi KT}}.$$

Now we shall find the mean-squared voltage $\overline{v^2}$ (the second moment):

$$\overline{v^2} = \int_{-\infty}^{\infty} v^2 f_V(v)dv = \sqrt{\frac{C}{2\pi KT}} \int_{-\infty}^{\infty} v^2 e^{\left(\frac{-Cv^2}{2KT}\right)} dv,$$

and since $\int_{-\infty}^{\infty} x^2 e^{-x^2} dx = \frac{\sqrt{\pi}}{2}$, the integral yields

$$\overline{v^2} = \frac{KT}{C}.$$

Now we will prove that $\overline{v^2} = 2KTR$ with the help of a parallel RC circuit. We can model the thermal noise in the resistor as an *electromotive force* (emf) in series with a noiseless resistor R, as shown in Figure 5.2.

---

[4] Maxwell–Boltzmann statistics states that the gas atom density is proportional to $\exp(-E/KT)$, where E is the potential energy [18]. As a result, as the height increases, the gas is expected to be less dense due to higher potential energy. Maxwell–Boltzmann as well as Fermi–Dirac statistics are widely used in quantum mechanics to describe electron and hole density in semiconductors.

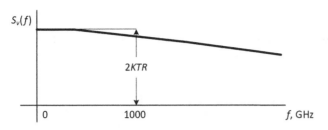

Figure 5.3: Thermal noise power spectral density

If the emf source is white and has a spectral density of $S_{emf}(f)$, considering that the system transfer function is $H(f) = \frac{1}{1+j2\pi RCf}$, the RMS voltage across the capacitor is

$$\overline{v^2} = S_{emf}(0) \int_{-\infty}^{\infty} |H(f)|^2 df = 2S_{emf}(0) \int_0^{\infty} \frac{df}{1+(2\pi RCf)^2} = \frac{S_{emf}(0)}{2RC},$$

which results in a spectral density of $2KTR$ for the resistor.

As the thermal noise spectrum is symmetric, it is common to consider only positive frequencies. If so, an extra factor of two appears in the equation to account for the other half of the energy residing at negative frequencies. Consequently, the thermal noise may be expressed as $4KTR$, and the energy integration will be performed from 0 to ∞ instead. We shall use both notations interchangeably throughput the book, whichever convenient. In the case of linear amplifiers, we will mostly use $4KT$. On the other hand, during our analysis of nonlinear circuits such as oscillators where aliasing is involved, we will use $2KT$ more often.

The equipartition law assigns a total energy of $\frac{1}{2}KT$ per degree of freedom, regardless of the frequency. This in fact is not quite accurate. Quantum mechanics shows [18] that the average energy per degree of freedom is[5]

$$\frac{1}{2}\frac{hf}{e^{hf/KT}-1},$$

where f is the frequency, and $h = 6.6 \times 10^{-34}$ J.s is the Planck constant. Clearly, it simplifies to $\frac{1}{2}KT$ if the frequency is low enough.

Consequently, the power spectral density of noise is modified to

$$S_v(f) = \frac{2Rhf}{e^{hf/KT}-1},$$

expressed in V²/Hz, and plotted in Figure 5.3.

Consistent with central limit theorem, $v(t)$ has a Gaussian distribution, with zero mean. The zero mean has to do with the fact that the noise voltage is a result of purely random motion of many electrons. Hence, when observed over a large period of time the opposite direction movements cancel each other.

---

[5] This is derived from Planck's law, proposed in 1900. Planck postulated that any physical entity can possess only total energies E in discrete levels (i.e. $E = nhf$, where n is an integer).

Example: We shall find the thermal noise variance, calculated by finding $\overline{v^2} = E[v^2(t)]$. By definition, we have

$$\overline{v^2} = \int_{-\infty}^{\infty} S_v(f)df = \int_0^{\infty} \frac{4Rhf}{e^{hf/KT}-1}df = 4Rh\int_0^{\infty} \frac{fe^{-hf/KT}}{1-e^{-hf/KT}}df.$$

The integral may be solved using the following expansion (for $x > 0$):

$$\frac{1}{1-e^{-x}} = \sum_{n=0}^{\infty} e^{-nx}.$$

This leads to

$$\overline{v^2} = \frac{4R}{h}(KT)^2 \sum_{n=0}^{\infty} \frac{1}{(n+1)^2}.$$

Since $\sum_{n=0}^{\infty} \frac{1}{(n+1)^2} = \frac{\pi^2}{6}$, the variance is found to be

$$\overline{v^2} = \sigma_v^2 = \frac{2(\pi KT)^2}{3h}R.$$

The noise variance calculated above has been experimentally proven as well. The noise of a 50Ω resistor for instance has a standard deviation of $\sigma_v = 4.66$mV.

As depicted in Figure 5.3, thermal noise power spectrum is not flat, although it is typically assumed so. In fact if it were, it would have erroneously predicted $\overline{v^2} = \infty$ when integrated over all f.[6] However, it is instructive to have a sense of h/KT as it appears in the exponential term in the denominator of the power spectral density equation. For standard room temperature of 25°C, we obtain $KT \approx 4\times 10^{-21}$J, leading to $\frac{KT}{h} \approx 10^{13}$Hz, which falls into *infrared* portion of the spectrum, clearly well above our frequencies of interest. Thus, for all practical purposes (at least for now), we can safely assume that

$$S_v(f) = 2KTR.$$

It is common to construct the Thevenin equivalent model of a thermal resistor as shown in Figure 5.4, where only positive frequencies are considered. Similarly a Norton equivalent can be obtained considering that $S_i(f) = \frac{S_v(f)}{R^2}$. The resistors in Figure 5.4 are therefore noiseless. Moreover, we have included an extra factor of 2 to assign the spectral density to only positive frequencies where $S_v(f) = 4KTR$.

As a useful example, a 50Ω resistor has an RMS noise voltage of 0.91nV/√Hz.

---

[6] This primarily arises from the fact that an integral over all possible energy from classical mechanics is replaced by a summation over discrete energy levels in quantum mechanics.

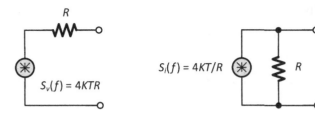

Figure 5.4: Resistor noise Thevenin and Norton equivalents

We could also describe the noise based on its *available power*. As we defined in Chapter 3, for a given source whose resistance is $Z_s = R_s + jX_s$, when driving a certain load with an impedance Z_L, the maximum power transfer then requires the load to be matched to the source: $Z_L = Z_s^*$. The noise of the source, $v_s(t)$, appears as $v_s/2$ on the load, and the available power is

$$P_a = \frac{<v_s(t)^2>}{4R_s}.$$

Using the resistor Thevenin model, we can extend this concept, arriving at *available spectral density* at the load resistor

$$S_a(f) = \frac{S_v(f)}{4R} = KT,$$

which depends only on the temperature, and is independent of the value of R.

5.1.2 White Noise and Noise Bandwidth

Noise of thermal resistors as well as many other types of noise sources are Gaussian and have a flat power spectral density over practical range of frequencies of interest. Since such a spectrum has all frequency components in equal proportion, the noise is said to be *white*, analogous to white light. The power spectral density can be written in general as

$$S(f) = \frac{\eta}{2}.$$

Hence

$$R(\tau) = \frac{\eta}{2}\delta(\tau).$$

The autocorrelation is zero for $\tau \neq 0$, so any two different samples of Gaussian white noise signal are *uncorrelated* and hence statistically independent. As a result, if we display white noise on oscilloscope, successive sweeps are always different from each other, yet the waveform looks the same, since all rates of time variation (or equivalently frequency components) are contained in equal proportion. Similarly, if white noise drives a loudspeaker, it always sounds the same, although it is not!

Now suppose a Gaussian white noise with the power spectral density of η is applied to an LTI system with a transfer function $H(f)$. The output, $y(t)$, will be Gaussian but colored, that is, the power spectral density of the filtered white noise takes the shape $|H(f)|^2$. Note that the integration is from 0 as opposed to $-\infty$, which we used in Chapter 2. This takes into account the fact that we are only considering positive frequencies, and thus assigned twice as much power to the white noise spectral density.

$$S_y(f) = \frac{\eta}{2}|H(f)|^2$$

$$\overline{y^2} = \eta \int_0^\infty |H(f)|^2 df$$

Example: Figure 5.5 shows a white noise passed through an ideal lowpass filter. Evidently, $R_y(\tau) = B\eta \,\text{sinc}\, 2B\tau$, and $\overline{y^2} = \eta B$. Thus, the output power is directly proportional to the filter bandwidth as expected.

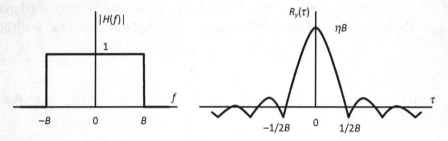

Figure 5.5: White noise passed by an ideal lowpass filter

Example: Consider an RC circuit shown in Figure 5.6, where the capacitor is ideal, but the resistor is noisy.

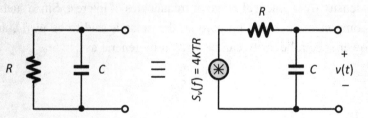

Figure 5.6: RC circuit with noisy resistor

Using the Thevenin equivalent, we arrive at the circuit on the right. We have $S_x(f) = 4KTR$, and $|H(f)|^2 = \frac{1}{1+\left(\frac{f}{B}\right)^2}$, where B is the 1st-order lowpass bandwidth equal to $1/2\pi RC$. Therefore for the output

$$S_y(f) = \frac{4KTR}{1+\left(\frac{f}{B}\right)^2},$$

and the inverse transform yields

$$R_y(\tau) = \frac{KT}{C} e^{-|\tau|/RC}.$$

The output power is then $\overline{y^2} = \frac{KT}{C}$, independent of the resistor value (this was already proven earlier in the context of resistor thermal noise spectral density).

Given the example above, we deduce that the 1st-order RC filter could have been replaced by an ideal lowpass filter whose bandwidth is $\frac{1}{4RC}$, producing the same output power of $\frac{KT}{C}$. This is known as the *equivalent noise bandwidth*. We can show that for a 2nd-order RLC circuit, the equivalent noise bandwidth is also equal to $\frac{1}{4RC}$. For the general case of a Butterworth nth-order filter as we defined in Chapter 4, one can prove (see Problem 3) that the equivalent noise bandwidth (B_N) is related to the 3dB bandwidth (B) by

$$B_N = \frac{\pi B}{2n \sin\left(\frac{\pi}{2n}\right)},$$

and hence $B_N \to B$ as $n \to \infty$.

5.1.3 Inductors and Capacitors Noise

The inductors, capacitors, and transformers are generally noiseless. If there is a loss associated with them, say an inductor wiring resistance, then that resistor is noisy whose power spectral density was given earlier. If high-Q, this noise is negligible, unless the loss occurs at the very input stage of the radio. For example, the loss of the matching network of a receiver comprising Ls and Cs could lead to nonnegligible noise degradation in the radio.

5.1.4 Passive Lossy Network Noise

Consider Figure 5.7 showing a passive lossy circuit comprising resistors, capacitors, inductors, or possibly coupled inductors. Since the circuit consists of passive RLCM elements, it is reciprocal.

Before we analyze the circuit, let us consider the network on the right first. The *Nyquist theorem* states that the power spectrum of the output voltage $v(t)$ is

$$S_v(f) = 2KT \, \text{Re}\,[Z(f)],$$

where $Z(s)$ is the impedance associated with that terminal.

We shall present the proof of the Nyquist theorem in a two-step approach. First, we assume the circuit consists of only one resistor, along with the rest of inductors and capacitors. The resistor is presented by a noiseless resistance, R, in parallel with its equivalent noise current, and

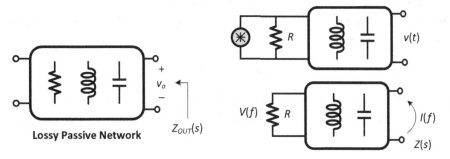

Figure 5.7: Passive lossy circuit

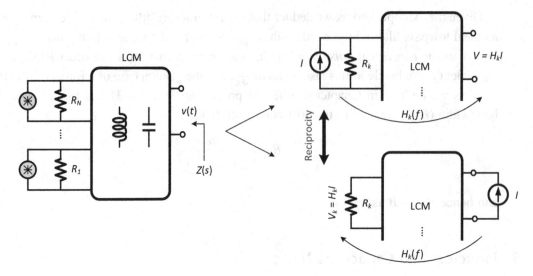

Figure 5.8: RLCM one-port

the remaining network contains only reactive elements. Therefore, $v(t)$ is the voltage resulted from the noise current associated with the resistor in response to the system function, $H(f)$. From reciprocity, $H(f) = \frac{V(f)}{I(f)}$, where $I(f)$ is the current injected to the network as shown in Figure 5.7 on the bottom, and $V(f)$ is the amplitude of the voltage across resistor. The input power is $|I(f)|^2 \operatorname{Re}[Z(f)]$, and the power delivered to the resistor is $\frac{|V(f)|^2}{R}$. Since the connecting network is only reactive and thus lossless, the two must be equal, which yields

$$|H(f)|^2 = \frac{|V(f)|^2}{|I(f)|^2} = R \times \operatorname{Re}[Z(f)].$$

This results in

$$S_v(f) = \frac{4KT}{R}|H(f)|^2 = 2KT\operatorname{Re}[Z(f)].$$

The general case where the circuit consists of more than one resistor may be proven similarly, considering the *independence of resistor noise* sources. Consider Figure 5.8, where the one-port now consists of N arbitrary resistors each with a noise current in parallel, and the rest of the network is reactive.

Assume that the transfer function from the noise source of resistor R_k to the output is $H_k(f)$, indicating that if the resistor is excited with a current I in parallel (Figure 5.8, right), the voltage appearing at the output port is

$$V(f) = H_k(f)I(f).$$

According to *reciprocity* then, if we excite the output port with a current I, the voltage appearing across the resistor R_k (v_k) must also obey that same transfer function, that is,

$$V_k(f) = H_k(f)I(f).$$

Now, since the rest of the network consists of only reactive components (ideal inductors, capacitors, coupled inductors or transformers), then *energy conservation*[7] entails that power delivered to the one-port by the exciting current I must be equal to the sum of the powers dissipated in the resistors,

$$\text{Re}[Z(f)]|I(f)|^2 = \sum_{k=1}^{N} \frac{|V_k(f)|^2}{R_k} = \sum_{k=1}^{N} \frac{|H_k(f)|^2|I(f)|^2}{R_k},$$

which results in

$$\text{Re}[Z(f)] = \sum_{k=1}^{N} \frac{|H_k(f)|^2}{R_k}.$$

Since the noise current experiences a transfer function of $H_k(f)$ from the corresponding resistor R_k to the output, the spectral density of the output considering that the noise sources are uncorrelated is

$$S_v(f) = \sum_{k=1}^{N} \frac{2KT}{R_k} |H_k(f)|^2 = 2KT\text{Re}[Z(f)],$$

which concludes our general proof of the Nyquist theorem. As a dual to the Nyquist theorem, we can readily show that

$$S_i(f) = 2KT\text{Re}[Y(f)],$$

where $S_i(f)$ is the *noise current* spectral density and $Y(f)$ is the output *admittance* of the one-port.

There are several other interesting results arisen from the Nyquist theorem. First, the autocorrelation of the noise voltage for the circuit at the left side of Figure 5.8 is

$$R_v(\tau) = KT(z(\tau) + z(-\tau)),$$

where $z(t)$ represents $Z(s)$ in time domain. The proof is readily obtained, considering that for an arbitrary function $x(t)$ with a Fourier transform of $X(f)$, we have

$$\mathcal{F}^{-1}(\text{Re}[X(f)]) = \frac{x(t) + x(-t)}{2}.$$

Since for the output impedance of any causal system, $z(t) = 0$, $t < 0$,

$$R_v(\tau) = KTz(\tau) \quad \tau > 0.$$

Second, defining the network equivalent output capacitance, C, as

$$\frac{1}{C} = \lim_{\omega \to \infty} j\omega Z(j\omega),$$

---

[7] The energy conservation, though it appears obvious, is a result of Tellegen's theorem [13], which states that in an arbitrary lumped network comprising b branches of voltage v_k and current i_k, we have $\sum_{k=1}^{b} v_k i_k = 0$, as long as the branch voltages and branch currents satisfy KVL and KCL. It follows that the sum of the complex powers in each branch is zero, and since the reactive branches have imaginary impedances associated with them, the energy conservation is resulted. See also Chapter 1 for the proof and more details.

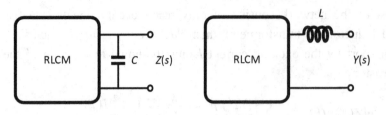

Figure 5.9: Equivalent capacitor or inductor in a passive RLCM network

we have

$$E\left[v(t)^2\right] = \frac{KT}{C}.$$

The proof immediately follows from *initial value theorem* stating that $z(0^+) = \lim_{s \to \infty} sZ(z)$. Similarly, duality suggests that

$$L = \lim_{\omega \to \infty} j\omega Y(j\omega),$$

where Y is the output admittance (Figure 5.9).

Example: Now let us reconsider our previous examples of RC and RLC equivalent noise bandwidth shown in Figure 5.6. Clearly, for the RC circuit, $E\left[v(t)^2\right] = \frac{KT}{C}$, and since the resistor noise spectral density is $4KTR$, it follows that the equivalent noise bandwidth is $1/4RC$, a result obtained by integration previously. For the RLC circuit, the integration is quite more formidable, whereas from the Nyquist theorem we have

$$E\left[v(t)^2\right] = KT \lim_{\omega \to \infty} j\omega Z(j\omega) = KT \lim_{\omega \to \infty} j\omega \frac{jL\omega}{1 - LC\omega^2 + \frac{jL\omega}{R}} = \frac{KT}{C},$$

leading to the same equivalent noise bandwidth.

Example: As a more general case, consider the circuit of Figure 5.10. It consists of a noisy resistor R, and a lossless reciprocal (LCM) network. We wish to find the average energy stored in the capacitor produced by the resistor thermal agitation.

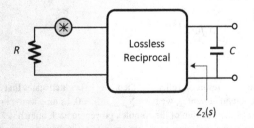

Figure 5.10: Average energy stored in a capacitor

The network output impedance is

$$Z_{OUT}(s) = \frac{\frac{1}{Cs}Z_2(s)}{\frac{1}{Cs}+Z_2(s)} = \frac{Z_2(s)}{1+CsZ_2(s)}.$$

Assume there is no other capacitance in parallel with C, that is, C represents the total capacitance appearing at the output. We have

$$\lim_{s\to\infty} sZ_{OUT}(s) = \lim_{s\to\infty} \frac{sZ_2(s)}{1+CsZ_2(s)} = \frac{1}{C}.$$

From the Nyquist theorem,

$$\overline{v_{OUT}^2} = \frac{KT}{C},$$

where v_{OUT} is the voltage at the network output, appearing across C. Thus the average energy stored in the capacitor is[8]

$$\frac{1}{2}C\overline{v_{OUT}^2} = \frac{1}{2}KT.$$

In a similar fashion and using duality we could show that the average energy stored in an arbitrary inductor is also $\frac{1}{2}KT$. Thus, assigning a degree of freedom associated with a given reactive component, there exits an energy of $\frac{1}{2}KT$. This is consistent with the equipartition law as we described at the very beginning.

5.1.5 MOSFET Thermal Noise

A great deal of work on solid-state devices noise was done by Van der Ziel [6] in the mid-1900s. A detailed discussion of noise in MOS transistors is also presented in [7]. In this section we briefly summarize the results.

From basic semiconductor physics [8], [9], we know that a long channel MOS drain current is

$$I_D = -\mu_n Q(y) \frac{dV(y)}{dy} \approx -\mu_n Q(y) \frac{\Delta V(y)}{\Delta y},$$

where $V(y)$ is the voltage across the channel whose derivative leads to electric field imposed by the drain voltage (Figure 5.11), and $Q(y)$ is the channel inversion charge:

$$Q(y) = -C_{OX}[V_{GS} - V_{TH} - V(y)]W.$$

---

[8] The derivation implies there is no capacitive or inductive loop or cut-set.

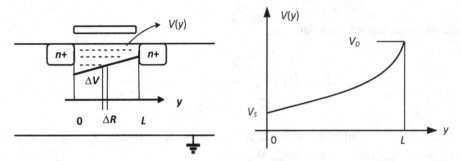

Figure 5.11: An NMOS transistor in strong inversion and its channel voltage

We can therefore assign a small resistance, $\Delta R = \frac{\Delta V}{I_D}$ for a given length of Δy at an arbitrary point in the channel, as follows:

$$\Delta R = \frac{\Delta V}{I_D} = \frac{\Delta y}{-\mu_n Q(y)},$$

This small resistance has some associated thermal noise, which is white, and in turn creates a noise voltage disturbance, ΔV, whose spectral density is

$$S_{\Delta V} = 4KT\Delta R = 4KT \frac{\Delta y}{-\mu_n Q(y)},$$

Note that $Q(y)$ is representing the inversion charge due to electrons in the case of an NMOS, and thus it is a negative quantity. In [9] it is proven that based on the i–v equation of the MOS transistor shown earlier, this small voltage disturbance in turn results in some small current disturbance, Δi, which is

$$\Delta i = \frac{-\mu_n Q(y)}{L} \Delta V,$$

and thus the current noise spectral density is

$$S_{\Delta i} = \left| \frac{\mu_n Q(y)}{L} \right|^2 S_{\Delta V} = 4KT \frac{-\mu_n}{L^2} Q(y) \Delta y.$$

Integrating $S_{\Delta i}$ over the channel leads to the total channel noise power spectral density,

$$S_i = 4KT \frac{-\mu_n}{L^2} \int_0^L Q(y) dy = 4KT \frac{\mu_n}{L^2}(-Q_T),$$

where Q_T is the total inversion charge. If the device is in the triode region, $V_{DS} \approx 0$, and $Q_T = -WLC_{OX}(V_{GS} - V_{TH})$, which leads to

$$S_i = 4KT\mu_n C_{OX} \frac{W}{L}(V_{GS} - V_{TH}) = 4KTg_{DS}.$$

The outcome is not surprising, as in triode region a MOSFET is essentially a resistor, with a channel conductance of $g_{DS} = \mu_n C_{OX} \frac{W}{L}(V_{GS} - V_{TH})$. On the other hand, in saturation, integrating the inversion charge leads to [8], [9]: $Q_T = -\frac{2}{3}WLC_{OX}(V_{GS} - V_{TH})$, and thus

$$S_i = 4KT \frac{2}{3} \mu_n C_{OX} \frac{W}{L}(V_{GS} - V_{TH}) = 4KT \frac{2}{3} g_m,$$

where g_m is the long channel MOS transconductance.[9] For shorter channel devices, the above analysis is not accurate anymore, as there are various short channel impacts such as hot carrier effects that could potentially lead to more noise. For that reason, we choose to express the MOSFET noise in the general form of

$$S_i = 4KT\gamma g_m,$$

where γ is a process- and channel-length-dependent factor, often experimentally obtained. In 40nm CMOS, for example, it appears that γ is close to one for minimum channel devices.

For a long channel MOS,[10] with an arbitrary value of V_{DS} between 0 and $V_{DS,SAT}$, a common expression obtained by integration of the channel total inversion charge is

$$S_i = \frac{8}{3} KT g_{DS} \frac{1+\eta+\eta^2}{1+\eta},$$

where $\eta = 1 - \frac{V_{DS}}{V_{DS,SAT}}$. Thus η is 1 when $V_{DS} = 0$, and approaches 0 when the device enters saturation, and the noise current density is consistent with our previous derivation.

Example: A simulation of a 200nm device in 28nm CMOS technology is shown in Figure 5.12. The value of γ is very close to ⅔ when the device enters saturation and the noise current spectral density is in good agreement with our findings.

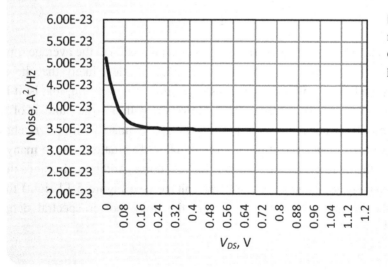

Figure 5.12: Simulated noise of a 200nm long device in 28nm CMOS process

To model the device noise, we may place a parallel current source representing the noise whose spectral density is given above. Equivalently, the device may be modeled with a voltage source at gate whose spectral density is (Figure 5.13)

$$S_v = \frac{4KT\gamma}{g_m}.$$

---

[9] The factor ⅔ is the same ⅔ that shows in the gate-source capacitance, C_{GS}, for a MOS in saturation.
[10] Note that in a long channel MOSFET, $g_m = g_{DS}$.

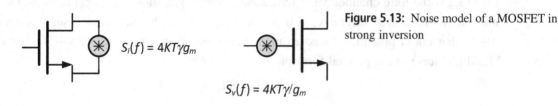

Figure 5.13: Noise model of a MOSFET in strong inversion

5.1.6 Flicker Noise

Besides channel thermal noise, which is Gaussian and white, there is another source of noise in MOSFETs that is frequency dependent. This type of noise is known as flicker or $1/f$ noise (given that the spectral density is approximately inversely proportional to frequency), and is a dominant source of noise at lower frequencies. The noise is still Gaussian, simply as the central limit theorem is valid, but is colored (or frequency dependent).

The origin of the noise may be attributed to the random fluctuations of carriers in the channel due to fluctuations in the surface potential [6]. That in turn is caused by trapping and releasing the carriers near the Si–SiO$_2$ interface with long time constants (Figure 5.14). The traps are a result of imperfection in Si/SiO$_2$ lattice.

The power spectral density of such a waveform is shown to be of the form

$$S_i(f) = \frac{c_t \tau_t}{1 + (2\pi f)^2 \tau_t^2},$$

where c_t is a constant, and τ_t is a characteristic time constant related to the average time between capture and release. The deeper the trap inside the oxide, the less likely that the electron is captured, and thus the longer τ_t will be. The noise spectrum is *Lorentzian*, that is, it is constant at very low frequencies, and decreases as $1/f^2$. However, with a large number of traps with uniform distribution of τ_t values, the spectrum may be approximated with a $1/f$ behavior.

As the noise results from superposition of large number of variations due to many traps, the larger the gate area, the more averaging, and thus the less noise. Additionally, since the trapping and releasing effectively modulates the flat-band voltage (see Figure 5.14 right) through the term $\frac{Q}{C_{OX}}$ (Q is the inversion charge), we expect the noise power spectral density to be proportional to $\frac{1}{C_{OX}^2}$.

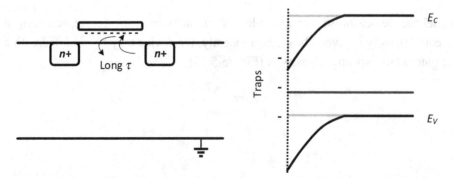

Figure 5.14: MOSFET flicker noise description along with energy bands

In general we can model the noise as a voltage source in series with gate, whose power spectral density is

$$S_v(f) = \frac{K}{WLC_{OX}^2} \frac{1}{f^b},$$

where b is between 0.7 and 1.2, and is often approximated to 1, and K is a process dependent parameter. Consequently, the noise current spectral density at the drain is

$$S_i(f) = g_m^2 S_v(f).$$

The dependence on C_{OX} and $W \times L$ is clear from the qualitative discussion above. Moreover, it is experimentally shown [14] that the noise spectral density as expressed above is not very bias dependent.

A more detailed discussion may be found in [6].

5.1.7 Cyclostationary Noise

The types of noise we have discussed thus far assume that the operating point of the elements is not changing over time. This applies to all linear and time-invariant circuits such as low-noise amplifiers. As we said in Chapter 2, however, the time-varying nature of certain nonlinear circuits such as oscillators results in cyclostationary noise instead [10], [15], whose statistical characteristics vary *periodically*.

Consider a time-varying conductance $G(t)$. As we discussed earlier, the derivation of the resistor thermal noise was based on the assumption that the thermal equilibrium is reached. This will not necessarily be the case if the resistor bias is changing over time. However, if the two conditions below are satisfied [16]:

- the change in the operating points is slow enough so that the dissipative device approximately stays in thermal equilibrium
- the fluctuations caused by noise are not large enough (compared to the deterministic time-varying signal) to alter the operating points

then we could argue that the noise *instantaneous* spectral density is $S_i(f,t) = 4KTG(t)$. In such case the spectral density is time-dependent, and the signal is not stationary anymore. If $G(t)$ is periodic, however, as is the case in most of the circuits we deal with such as mixers or oscillators, the noise current will be cyclostationary.

To generalize, let us define

$$n_{cyclo}(t) = n(t)w(t),$$

where $n(t)$ is stationary (but not necessarily white), $w(t)$ is the periodic modulating function, and $n_{cyclo}(t)$ is the resultant cyclostationary noise.

To gain some insight, let us derive the autocorrelation of the cyclostationary noise. According to definition,

$$R_{n_{cyclo}}(t+\tau,t) = E[n(t+\tau)w(t+\tau)n^*(t)w^*(t)].$$

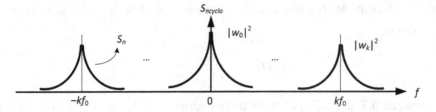

Figure 5.15: Stationary noise sampled by the modulating signal

To simplify, let us represent the periodic function $w(t)$ by its Fourier series:

$$w(t) = \sum_{k=-\infty}^{\infty} w_k e^{j\omega_0 t},$$

where w_k are the Fourier coefficients, and $\omega_0 = \frac{2\pi}{T}$ is the angular frequency. After some straightforward algebra,

$$R_{n_{cyclo}}(t+\tau,t) = R_n(\tau) \sum_{k=-\infty}^{\infty} \sum_{l=-\infty}^{\infty} w_k w_l^* e^{jk\omega_0 \tau} e^{j\omega_0(k-l)t}.$$

Since $n_{cyclo}(t)$ is not stationary, its autocorrelation depends on both t and τ. As we indicated in Chapter 2, it is customary to treat the cyclostationary processes as stationary by taking the time average of the mean and autocorrelation. By doing so, and knowing the t dependent terms average to zero over one cycle, we obtain

$$\overline{R_{n_{cyclo}}}(\tau) = \frac{1}{T} \int_T R_{n_{cyclo}}(t+\tau,t)dt = R_n(\tau) \sum_{k=-\infty}^{\infty} |w_k|^2 e^{jk\omega_0 \tau},$$

and the spectral density is

$$\overline{S_{n_{cyclo}}}(f) = S_n(f) * \sum_{k=-\infty}^{\infty} |w_k|^2 \delta(f-kf_0) = \sum_{k=-\infty}^{\infty} |w_k|^2 S_n(f-kf_0).$$

That is, the modulating signal $w(t)$ samples the stationary noise $n(t)$ at all harmonics of f_0, as shown in Figure 5.15.

If $n(t)$ is stationary and also white, then

$$\overline{S_{n_{cyclo}}}(f) = S_n \sum_{k=-\infty}^{\infty} |w_k|^2,$$

where S_n is the white noise spectral density. Furthermore, we can obtain simple expressions for the instantaneous spectral density if $n(t)$ is white. To do so, let us take the Fourier transform of (with respect to τ) the cyclostationary autocorrelation function ($R_{n_{cyclo}}(t+\tau,t)$)

$$S_{n_{cyclo}}(f,t) = \int_{-\infty}^{\infty} R_{n_{cyclo}}(t+\tau,t) e^{-j2\pi f\tau} d\tau,$$

which can be written as

$$S_{n_{cyclo}}(f,t) = \int_{-\infty}^{\infty} R_n(\tau) \sum_{k=-\infty}^{\infty} \sum_{l=-\infty}^{\infty} w_k w^*{}_l e^{jk\omega_0\tau} e^{j\omega_0(k-l)t} e^{-j2\pi f\tau} d\tau.$$

Rearranging the integral yields

$$S_{n_{cyclo}}(f,t) = \sum_{k=-\infty}^{\infty} \sum_{l=-\infty}^{\infty} w_k e^{j\omega_0 kt} w^*{}_l e^{-j\omega_0 lt} \int_{-\infty}^{\infty} R_n(\tau) e^{jk\omega_0\tau} e^{-j2\pi f\tau} d\tau.$$

Since $n(t)$ is white, the integral $\int_{-\infty}^{\infty} R_n(\tau) e^{jk\omega_0\tau} e^{-j2\pi f\tau} d\tau$ represents its spectral density S_n. Thus,

$$S_{n_{cyclo}}(f,t) = S_n \sum_{k=-\infty}^{\infty} \sum_{l=-\infty}^{\infty} w_k e^{j\omega_0 kt} w^*{}_l e^{-j\omega_0 lt}.$$

This can be further simplified to

$$S_{n_{cyclo}}(f,t) = S_n |w(t)|^2,$$

which is a function of time (but not frequency as $n(t)$ is white).

Time-averaging the spectral density above yields

$$\overline{S_{n_{cyclo}}}(f) = S_n \frac{1}{T}\int_0^T |w(t)|^2 dt.$$

This is identical to the result previously obtained ($\overline{S_{n_{cyclo}}}(f) = S_n \sum_{k=-\infty}^{\infty} |w_k|^2$) given Parseval's energy theorem $\sum_{k=-\infty}^{\infty} |W[k]|^2 = \frac{1}{T}\int_T |w(t)|^2 dt$.

In the case of time-varying conductance (or a MOSFET in strong inversion), $S_n = 4KT$,[11] and $w(t) = \sqrt{G(t)}$. The instantaneous spectral density is $S_{n_{cyclo}}(f,t) = 4KTG(t)$, and the average spectral density as viewed on a *spectrum analyzer* is $\overline{S_{n_{cyclo}}}(f) = \frac{4KT}{T}\int_0^T G(t)dt$.

Example: Let us start with the simple case of Figure 5.16, where a resistor is periodically switched, and the noise current on the other side of the switch is monitored. The output noise waveform is shown on the right side.

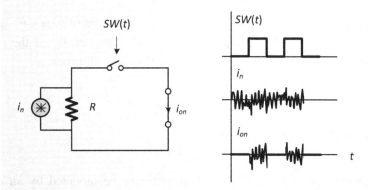

Figure 5.16: An example of a cyclostationary noise

Continued

---

[11] $S_n(f) = 2KT$ if both positive and negative frequencies are considered.

Clearly,

$$S_{i_{on}}(f,t) = \frac{4KT}{R}|SW(t)|^2.$$

If viewed on a spectrum analyzer, the power spectral density will be the time average of $S_{i_{on}}(f)$,

$$\overline{S_{i_{on}}}(f) = \frac{1}{T}\int_0^T S_{i_{on}}(f)dt = \frac{2KT}{R},$$

which intuitively makes sense as half the noise energy is blocked to appear at the output.

Example: Consider the circuit depicted in Figure 5.17, where an input voltage source is sampled by two switches driven by nonoverlapping clock signals, SW_1 and SW_2. As we will explain in Chapter 8, this is essentially a single-ended voltage-mode passive mixer. Let us find the spectral density of the noise at the input of the ideal voltage buffer.

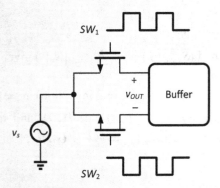

Figure 5.17: Single-ended voltage-mode passive mixer

Using superposition, we can write

$$S_{v_{OUT}}(f,t) = 4KTR_{SW}[|SW_1(t)|^2 + |SW_2(t)|^2] = 4KTR_{SW},$$

where R_{SW} is the switch on resistance. Intuitively, at any given point of time, *one and only one switch* appears in series with the input. Thus, the spectral density of the noise is equivalent to what it would be if we had only one switch on at all times.

5.2 TWO-PORT EQUIVALENT NOISE

From basic circuit theory [1], any linear noisy two-port may be modeled by an equivalent noiseless two-port, and an input-referred pair of voltage and current noise sources as shown in Figure 5.18. The noiseless two-port is identical to the noisy one, except for all the internal noise sources turned off; that is, the current noise sources are opened, and the voltage noise sources are shorted. To obtain the input-referred noise voltage, we need to short the input, thus

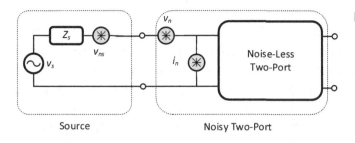

Figure 5.18: A two-port noise equivalent

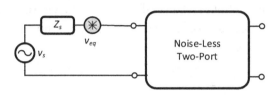

Figure 5.19: Thevenin equivalent circuit

effectively bypassing the input current noise source. The input noise voltage must be such that the output noise spectral density is equal to that of the noisy two-port with a shorted input. The input noise current is obtained similarly, but by performing an open circuit test, and thus bypassing the input noise voltage.

In spite of complicating the analysis, both the input voltage and current noise sources are needed, that is, to support any arbitrary value of source impedance. For instance, if one chooses to model the noisy circuit with only an input-referred current noise, an ideal voltage source whose output resistance is zero results in no output noise, which is clearly not the case.

Assuming the source impedance Z_s is $Z_s = R_s + jX_s$, then we can use a Thevenin equivalent noise source, v_{eq}, whose spectral density is

$$\overline{v_{eq}^2} = \overline{|(v_n + Z_s i_n)|^2} + \overline{v_{ns}^2} = \overline{|(v_n + Z_s i_n)|^2} + 4KTR_s.$$

The equivalent circuit is shown in Figure 5.19.

Since v_n and i_n could arise from the same internal noise source in the noisy two-port but under two different test conditions (that being open and short cases), they may be correlated. Thus, in general, they cannot be separated when calculating the noise variance or spectral density.

Example: As a useful case study, let us calculate the input-referred noise sources of a single MOSFET intended to be used in a common-source configuration at reasonably high frequencies. Since the frequency is assumed to be high, the flicker noise is ignored. We use the high-frequency small signal model of a MOSFET, as shown in Figure 5.20.

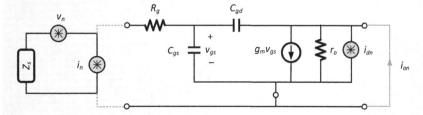

Figure 5.20: A single FET equivalent input noise

Continued

The gate resistance is represented by R_g, arising from the gate poly or metal finite conductivity. It can be minimized to practically negligible values by proper layout using multifinger devices (see Chapter 7 for more details). Hence, we will ignore that. Nevertheless, it appears in series with the gate and the source so it could be absorbed to the source if needed. Also for the moment we will ignore the gate-drain capacitance as it complicates the analysis considerably. We will however discuss its impact on the amplifier noise and input resistance in Chapter 7. For typical design values and process parameters, C_{gd} is roughly one-third to one-fifth of C_{gs} for minimum channel devices. The values of v_n and i_n are found by applying a short and open circuit for Z_s, and find the output noise current. Since v_n appears directly across C_{gs}, we have,

$$\overline{v_n^2} = \frac{\overline{i_{dn}^2}}{g_m^2} = \frac{4KT\gamma}{g_m}.$$

On the other hand, for the open circuit test, v_{gs} is produced by i_n flowing through C_{gs}, and thus

$$\overline{i_n^2} = \frac{\overline{i_{dn}^2}}{\left|\frac{g_m}{j\omega C_{gs}}\right|^2} = \frac{4KT\gamma}{g_m}(\omega C_{gs})^2 = (\omega C_{gs})^2\overline{v_n^2}.$$

We can see that v_n and i_n are completely correlated. Moreover, i_n becomes important only at higher frequencies where the MOSFET input impedance starts to reduce.

Example: For an arbitrary source impedance of Z_s, let us consider the circuit on the left side of Figure 5.21.

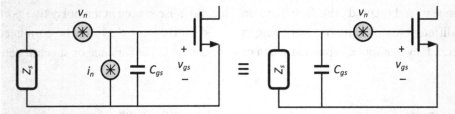

Figure 5.21: MOS equivalent noise model

We can write

$$v_{gs} = \left(\frac{v_n}{Z_s} + i_n\right)\left(Z_s \Big\| \frac{1}{j\omega C_{gs}}\right).$$

Since $\overline{i_n^2} = (\omega C_{gs})^2 \overline{v_n^2}$, we have

$$\overline{v_{gs}^2} = \left(\frac{1}{|Z_s|^2} + (\omega C_{gs})^2\right)\left|Z_s \Big\| \frac{1}{j\omega C_{gs}}\right|^2 \overline{v_n^2} = \overline{v_n^2}.$$

Therefore, the spectral density of the gate-source voltage is equal to $\overline{v_n^2}$, regardless of the source impedance. Consequently, the circuit may be simplified as shown on the right, where only a voltage noise is sufficient to represent it. The simplification arises from the fact that the equivalent input noise voltage and current are completely correlated.

5.3 NOISE FIGURE

Consider the two-port shown in Figure 5.22, where $G_a = \frac{S_o}{S_i}$ is the available power gain, and S_o and S_i are the available output and input signals. From Chapter 3, we showed that $S_i = \frac{\overline{v_s^2}}{8R_s}$. The available noise spectral density was shown to be KT, regardless of the source resistance. Thus, the available noise power due to the source measured at a bandwidth of 1Hz at a certain frequency of interest is KT. Let us assume that the available noise power at the output is N_o, resulted from the source as well the two-port noise contributions. The available noise power referred to the input is then equal to $N_i = \frac{N_o}{G_a}$.

By definition, the noise factor,[12] F, is the signal-to-noise ratio (SNR) at the input, divided by the SNR at the output. Unless the two-port is noiseless, we expect the output SNR to be less than that of the input, as the two-port adds its own noise. Thus, the noise factor is always greater than one (or zero dB).

The noise figure may be expressed in a more familiar format, by the two-port available gain and noise [2]:

$$F = \frac{S_i/KT}{S_o/N_o} = \frac{N_o}{G_a KT} = \frac{N_i}{KT}.$$

Thus the available input-referred noise due to the two-port only is $(F-1)KT$.

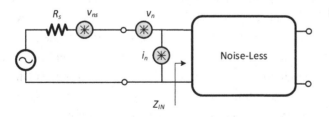

Figure 5.22: Noisy two-port

---

[12] We make the distinction between the noise factor expressed as a number F, and the noise figure in dB as NF.

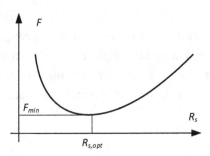

Figure 5.23: Minimum noise figure concept

For a noisy two-port with input-referred voltage and current noise sources as depicted in Figure 5.22, we showed based on the Thevenin equivalent (Figure 5.19) that

$$N_i = \frac{\overline{|(v_n + Z_s i_n)|^2} + 4KTR_s}{4R_s},$$

which allows us to express the noise figure in terms of two-port input-referred noise:

$$F = \frac{\frac{\overline{|(v_n + Z_s i_n)|^2} + 4KTR_s}{4R_s}}{KT} = 1 + \frac{\overline{|(v_n + Z_s i_n)|^2}}{4KTR_s}.$$

Note that we have expressed the results in terms of source complex impedance: $Z_s = R_s + jX_s$.

A few key observations are in order first:

1. The noise figure is a function of source resistance R_s. This does not make the noise figure definition any less valuable, it just simply compares the circuit noise contributors v_n and i_n resulted from internal noise sources, to a universally accepted reference, $4KTR_s$. For almost all practical cases, R_s is agreed to be 50Ω.
2. As expected, the noise figure is a function of the two-port noise voltage and current. If the two-port is noiseless, v_n and i_n are zero, and thus $F = 1$. For any noisy circuit, F is always greater than one, a result we already knew from basic definition.
3. For a source resistance of zero or infinity, noise figure is infinity. This is obvious from the equation, but could be explained intuitively as well: If $R_s = 0$, regardless of how low noise the two-port is, since its noise is always compared to a zero noise reference, the noise figure is infinity. Ironically, the lower the source resistance, despite the fact that a better SNR is achieved, the circuit noise figure is higher! On the other hand, if $R_s = \infty$, no matter how small the input-referred noise current is, it results into infinite noise. This suggests that if for a given circuit i_n were zero, the optimum noise figure would be achieved if source has infinite resistance.

The latter observation is quite critical: For a given two-port, and thus fixed values of v_n and i_n, there must be an optimum value for R_s where noise figure approaches a minimum (Figure 5.23). This is particularly of great importance, as it allows us to systematically optimize the noise figure of a given amplifier. In practice, this is typically done along with optimizing the amplifier as well, and thus varying v_n and i_n. We will discuss this in the next section

Example: We start with a simple example describing a resistive voltage divider as shown in Figure 5.24.

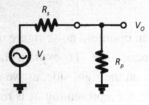
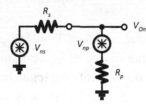

Resistive Voltage Divider   **Voltage Divider Noise Sources**

Figure 5.24: Resistive voltage divider

The noise at the output is

$$v_{on} = v_{ns}\frac{R_p}{R_p + R_s} + v_{np}\frac{R_s}{R_p + R_s}.$$

Since the output noise due to the source only is $v_{ns}\frac{R_p}{R_p + R_s}$, the noise factor becomes

$$F = 1 + \frac{4KTR_p}{4KTR_s}\left(\frac{R_s}{R_p}\right)^2 = 1 + \frac{R_s}{R_p} = \frac{1}{\frac{R_p}{R_p + R_s}}.$$

As expected, removing the shunt resistor leads to a noise factor of 1. Furthermore, the available power gain of the circuit is

$$G_a = \frac{R_s}{R_p \| R_s}\left(\frac{R_p}{R_p + R_s}\right)^2 = \frac{R_p}{R_p + R_s}.$$

So the noise figure is equal to the inverse of the available power gain, or is basically the same as the loss. We will show momentarily that this is true for any passive reciprocal network.

Example: Let us calculate the noise figure of a single FET shown in Figure 5.25. The output may be terminated to some noiseless load or simply shorted to supply.

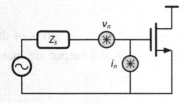
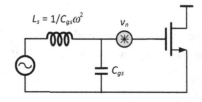

Figure 5.25: Noise figure of a single FET

Continued

From the noise figure equation, and based on the values of v_n and i_n obtained before, we have

$$F = 1 + \frac{\gamma}{g_m R_s}\left[(R_s C_{gs}\omega)^2 + (1 - X_s C_{gs}\omega)^2\right].$$

From the equation above, if $R_s = 0$, and $X_s = 1/C_{gs}\omega$, an optimum noise figure of 0dB is achievable, regardless of the device noise or transconductance. To explain this rather strange outcome, consider the equivalent circuit shown on the right side, as we obtained previously (Figure 5.21). We can see that the series inductor representing X_s is resonating with the gate-source capacitance under optimum noise conditions. Thus, it results in an infinite voltage gain from the source to v_{gs}, assuming the inductor is lossless. Hence, the circuit noise, however large it may be, is entirely suppressed.

Example: Consider the passive lossy circuit of Figure 5.26.

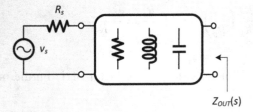

Figure 5.26: Noise figure of a passive lossy network

If the circuit has an output resistance of $R_{OUT} = \text{Re}[Z_{OUT}]$, then the noise spectral density at the output is $4KTR_{OUT}$ according to the Nyquist theorem. The total available noise at the input is the output available noise power, divided by the network available power gain,[13] G_a:

$$N_i = \frac{\frac{4KTR_{OUT}}{4R_{OUT}}}{G_a}.$$

This normalized to the source available power, KT, results in the noise figure:

$$F = \frac{1}{G_a} = \text{Loss}.$$

In Chapter 3 we saw that the available power gain may be expressed in terms of the two-port scattering parameters, and would be equal to $|S_{21}|^2$ if the input and output are matched to Z_0.

---

[13] Since the circuit is passive and lossy, the available power gain, G_a, is less than one.

That is usually the case for many external passive lossy RF components, such as the SAW filters or attenuators.

The previous example's outcome could have been obtained intuitively: The available noise power at the input and output is always equal to KT. As the signal is attenuated due the two-port loss, the SNR degrades, and consequently the noise figure must be equal to the loss.

Thus the noise figure of lossy passive networks is equal to their loss. This is handy when dealing with external filters, such as SAW filters placed at the receiver input. The loss of the filter, typically about 1–2dB, directly translates to its noise figure in dB.

5.4 MINIMUM NF

In the previous section, we showed qualitatively that the noise figure may be optimize by choosing the right source impedance. A systematic approach may be taken by taking the derivative of noise figure versus source impedance, and find the optimum source impedance.

Let us define

$$v_n = v_c + v_u = Z_c i_n + v_u,$$

where $Z_c = R_c + jX_c$ represents the correlation between the input-referred noise voltage and current, and v_u is the uncorrelated portion of the input noise voltage. For convenience, let us assign the following *nonphysical* resistance and conductance to v_u and i_n spectral densities:

$$\overline{v_u^2} = 4KTR_u \quad \text{and} \quad \overline{i_n^2} = 4KTG_n.$$

The noise figure is

$$F = 1 + \frac{\overline{|(v_n + Z_s i_n)|^2}}{4KTR_s} = 1 + \frac{\overline{|(Z_s i_n + Z_c i_n + v_u)|^2}}{4KTR_s} = 1 + \frac{|Z_c + Z_s|^2 \overline{i_n^2} + \overline{v_u^2}}{4KTR_s}.$$

The last step uses the fact that v_u and i_n are uncorrelated. Replacing $\overline{v_u^2}$ and $\overline{i_n^2}$ with their equivalent resistances, and after some simplifications, we arrive at

$$F = 1 + \frac{R_u + G_n\left[(R_s + R_c)^2 + (X_s + X_c)^2\right]}{R_s}.$$

There are two variables, R_s and X_s. To find the optimum noise, we need to take partial derivative of F versus each, and set them to zero. By doing so we obtain

$$R_{s,opt} = \sqrt{\frac{R_u}{G_n} + R_c^2}$$
$$X_{s,opt} = -X_c,$$

and the minimum noise figure when the optimum impedance is used is

$$F_{min} = 1 + 2G_n\left(R_c + \sqrt{\frac{R_u}{G_n} + R_c^2}\right).$$

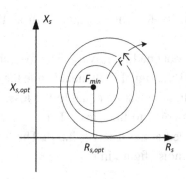

Figure 5.27: Noise circles

Outside the optimum conditions,

$$F = F_{min} + \frac{G_n}{R_s}|Z_s - Z_{s,opt}|^2.$$

Clearly $F \geq F_{min}$. It is insightful to rearrange the equation above as follows:

$$\left(R_s - \left(R_{s,opt} + \frac{F - F_{min}}{2G_n}\right)\right)^2 + (X_s - X_{s,opt})^2 = \frac{F - F_{min}}{G_n}R_{s,opt} + \left(\frac{F - F_{min}}{2G_n}\right)^2.$$

If $Z_s = Z_{s,opt}$, then F must be equal to F_{min}. However, for any other value of F, then the X_s–R_s plane consists of noise circles as shown in Figure 5.27.

Since in a *bilinear transformation* circles are mapped again to circles, the constant noise figure contour will also remain circular in the Smith chart, which depicts the plane of reflection coefficient. As proven in Problem 25, the source reflection coefficient $\Gamma_s = \frac{Z_s - Z_0}{Z_s + Z_0}$, may be described by

$$\left|\Gamma_s - \frac{\Gamma_{s,opt}}{K+1}\right|^2 = \frac{K^2 + K(1 - |\Gamma_{s,opt}|^2)}{(K+1)^2},$$

where $K = \frac{F - F_{min}}{4G_n Z_0}|1 - \Gamma_{s,opt}|^2$ is a positive real number and is a function of noise factor, and $\Gamma_{s,opt} = \frac{Z_{s,opt} - Z_0}{Z_{s,opt} + Z_0}$ is the optimum noise impedance reflection coefficient. The Γ_s circles for a given noise factor, and thus a given K, are depicted on the Smith chart in Figure 5.28.

For $F = F_{min}$, $K = 0$, and clearly $\Gamma_s = \Gamma_{s,opt}$. On the other extreme, if noise figure goes to infinity, Γ_s will reside on $|\Gamma| = 1$ circle. The center of the circles reside on the constant $\angle \Gamma_{s,opt}$ line, since K is real, as shown as a dashed line in Figure 5.28.

For an arbitrary two-port with a given input impedance, the optimum noise condition may be accomplished by placing a lossless matching network to transform the impedance to the desirable value (Figure 5.29). If the loss of the matching network is not significant, it will not affect the input SNR, however it will provide the optimum impedance to minimize the noise figure.

The fundamental drawback of the noise matching approach as presented in Figure 5.29 is that it may conflict with maximum available power requirements. To satisfy the latter condition, the matching network must present an impedance equal to Z_{IN}^* to the circuit, rather than $Z_{s,opt}$, and the two are not necessarily equal. Perhaps one of the biggest challenges of designing a good low-noise amplifier is to satisfy the two conditions simultaneously.

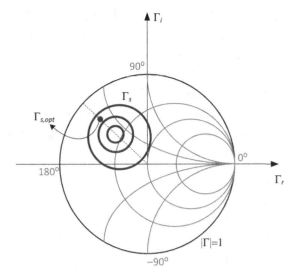

Figure 5.28: The Γ_s circles on the Smith chart

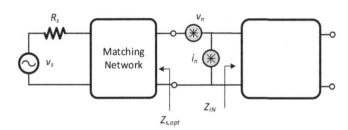

Figure 5.29: Using matching network to achieve minimum noise conditions

Example: Returning to our previous example of the a single FET in Figure 5.25, one can easily show that $v_u = 0$, $Z_c = 1/jC_{gs}\omega$, and $Z_{s,opt} = -1/jC_{gs}\omega$, leading to $F_{min} = 1$, a result obtained previously.

Designing a low-loss matching network that could transform the 50Ω source resistance to a pure reactance of $X_{s,opt} = 1/C_{gs}\omega$ seems to be no easy task at all. As shown in Figure 5.30, this may be accomplished by placing the inductor simply in series with R_s, and hoping that $X_{s,opt} \gg R_s$. Since the 50Ω resistance is relatively large, this may lead to a low-Q inductor, unless the inductance value is very large. This in turn implies that the device C_{gs} (or any parasitic capacitance associated with it, such as routing or pad capacitance) that the inductor is resonating with must be very small.

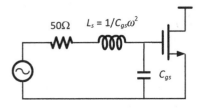

Figure 5.30: Single common-source FET with matching network

This is where the technology scaling could potentially help the RF design, as device transit frequency f_T, which is an indication of $\frac{1}{C_{gs}}$ ($f_T = \frac{1}{2\pi}\frac{g_m}{C_{gs}+C_{gd}} \cong \frac{1}{2\pi}\frac{g_m}{C_{gs}}$), scales favorably. Moreover, in contrast to our initial thought that the minimum noise figure is achieved regardless of the device noise or g_m, one can see that to maximize f_T, the device needs to be biased at a very large gate overdrive voltage. This in turn leads to a poor g_m/I_D, and thus a large bias current. A plot of device f_T versus gate overdrive for 40nm CMOS is shown in Figure 5.31 for the reference. Also shown is the peak f_T for several recent processes indicating the impact of technology scaling.

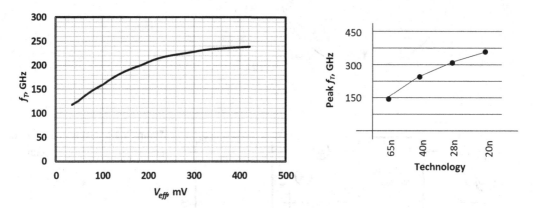

Figure 5.31: NMOS transit frequency

For instance, a C_{gs} of 100fF, a typical value for a minimum channel 40nm device biased at a few mA, leads to a value of 800Ω for $X_{s,opt}$ at 2GHz. The corresponding inductance value is 64nH, which is quite large at especially 2GHz. On the other hand, despite the fact that C_{gs} is small and scales favorably, other parasitics at the amplifier input could dominate. For these reasons, a practical low-noise amplifier may not incorporate a simple common-source FET structure.

Example: Consider a common-gate amplifier shown in Figure 5.32. The fundamental difference between this amplifier and the common-source one is that the input impedance $\frac{1}{g_m} \| \frac{1}{jC_{gs}\omega}$ has a non-zero real part. By performing the same procedure, we find

$$\overline{v_n^2} = \frac{4KT\gamma}{g_m}$$

$$\overline{i_n^2} = \left|\frac{jC_{gs}\omega}{g_m}\right|^2 4KT\gamma g_m,$$

which leads to $Z_c = \frac{-1}{jC_{gs}\omega}$, the same as the common-source example. The minimum noise figure and the optimum source impedance then happen to be identical to the common-source topology.

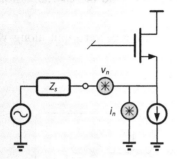

Figure 5.32: A common-gate amplifier

Interestingly, even though a source follower configuration is not considered an amplifier, one can show that (see Problem 1) its minimum noise figure is also zero, much like common-gate and common-source configurations.

As a final remark, if instead of noise voltage, we would have broken the input noise current as

$$i_n = i_c + i_u = Y_c v_n + i_u.$$

In a similar manner we can show (see Problem 7)

$$F = 1 + |Z_s|^2 \frac{G_u + R_n |Y_s + Y_c|^2}{R_s},$$

where $\overline{i_u^2} = 4KTG_u$, $\overline{v_n^2} = 4KTR_n$, $Y_c = G_c + jB_c$, and $Y_s = G_s + jB_s$ is the source admittance.

The optimum noise admittance is

$$G_{s,opt} = \sqrt{\frac{G_u}{R_n} + G_c^2}$$

$$B_{s,opt} = -B_c$$

and the minimum noise figure when the optimum admittance used is

$$F_{min} = 1 + 2R_n \left(G_c + \sqrt{\frac{G_u}{R_n} + G_c^2} \right).$$

Outside the optimum conditions,

$$F = F_{min} + \frac{R_n}{G_s} |Y_s - Y_{s,opt}|^2.$$

This is the dual of the equations derived earlier, obtained by breaking the input noise voltage into correlated and uncorrelated portions. Either set of equations may be used depending on the context and the type of the circuit under analysis.

Example: To lower the noise figure of a matched receiver, a low-noise amplifier with matched input and output is placed before the receiver. Assuming the LNA has $|S_{21}| = 10dB$, and a noise figure of $2dB$, we would like to find the cascaded noise figure if the receiver noise figure is $3dB$.

A simplified block diagram of the source, LNA, and receiver input along with their noise sources is shown in Figure 5.33.

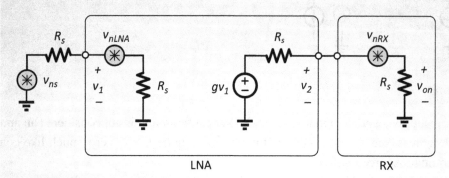

Figure 5.33: An LNA proceeding a receiver to improve the system overall noise figure

The total noise at the receiver input is:

$$v_{on} = \frac{1}{4}gv_{ns} + \frac{1}{4}gv_{nLNA} + \frac{1}{2}v_{nRX}$$

where g is the LNA voltage gain. Thus, the overall noise factor would be:

$$F = F_{LNA} + \frac{F_{RX} - 1}{\left|\frac{g}{2}\right|^2}$$

where $F_{LNA} = 1 + \frac{\overline{v_{nLNA}^2}}{4KTR_s}$ is the LNA noise factor, and $F_{RX} = 1 + \frac{\overline{v_{nRX}^2}}{4KTR_s}$ is the receiver noise factor. On the other hand, for the matched LNA, the available power gain is:

$$G_a = |S_{21}|^2 = \left|\frac{g}{2}\right|^2$$

Consequently, the overall noise factor would be:

$$F = F_{LNA} + \frac{F_{RX} - 1}{|S_{21}|^2} = 1.58 + \frac{3 - 1}{10} = 2.5dB$$

5.5 IMPACT OF FEEDBACK ON NOISE FIGURE

The impact of feedback on an amplifier basic properties such as gain or input impedance is very well understood [1]. In this section we shall study its impact on the amplifier equivalent noise and its minimum noise figure. We consider the case of an ideal feedback, as well as passive lossless feedback. The latter is of great importance and, as we will see in Chapter 7, is employed extensively to favorably lower the noise figure of amplifiers.

5.5.1 Ideal Feedback

Let us start off with Figure 5.34, showing an amplifier in a shunt feedback. We assume the feedback circuitry is ideal, and is particularly noiseless. The latter assumption is not unreasonable, as at radio frequencies we typically incorporate capacitors and inductors to form a low-noise feedback around a noisy amplifier.

We use the Norton equivalent of the source, as is common in the case of shunt-shunt feedback. The closed-loop transimpedance gain is

$$\frac{v_o}{i_s} = \frac{g(Z_s \| Z_{IN})}{1 + fg(Z_s \| Z_{IN})},$$

where $fg(Z_s \| Z_{IN})$ is the loop gain. The closed-loop input impedance is

$$Z_{IN,fb} = \frac{Z_s \| Z_{IN}}{1 + fg(Z_s \| Z_{IN})}.$$

This is the open-loop input impedance reduced by the loop gain, as expected in a shunt feedback network.

Since the circuits are linear we can use superposition and consider each noise source one at the time. By inspection, it is clear that the feedback is not going to change the input-referred noise *current*, as one can slide it out of the feedback amplifier without altering the circuit. The

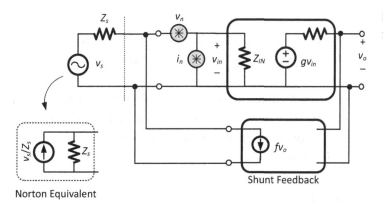

Figure 5.34: Minimum noise figure of a shunt feedback network

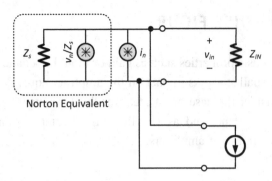

Figure 5.35: Equivalent noise sources of the shunt feedback network

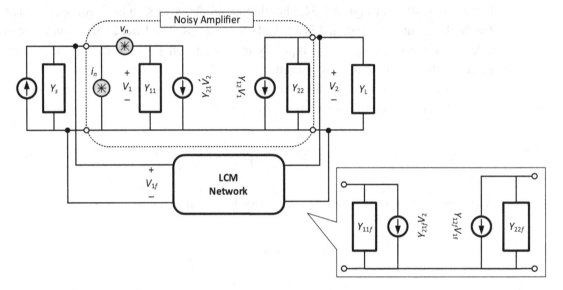

Figure 5.36: Reactive shunt feedback around a noisy amplifier. Both feedback and the amplifier are represented with their Y parameters.

feedback impact on the input-referred noise *voltage* is not as obvious. Turning off the noise current, performing a simple analysis yields

$$v_{in} = \frac{-v_n}{Z_s} \times \frac{Z_s \| Z_{IN}}{1 + fg(Z_s \| Z_{IN})},$$

which is the original noise voltage appearing at the input, but simply reduced by the loop gain. Consequently, the input-referred noise voltage can also be brought outside the feedback, as shown in Figure 5.35.

In conclusion, an ideal noiseless feedback does not affect the circuit input-referred noise, and hence its minimum noise figure. Although our analysis was done for a shunt-shunt feedback network, it can be generalized to other types of feedback too, particularly the *source degeneration*, which is a form of *series–series* feedback.

5.5.2 Passive Lossless Feedback

Next, let us study the impact of reactive feedback (LCM) on an arbitrary amplifier. Shown in Figure 5.36 is the amplifier modeled by its *Y* parameters, along with the feedback network. In

5.5 Impact of Feedback on Noise Figure | 311

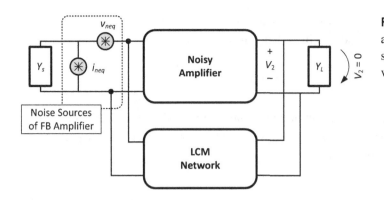

Figure 5.37: Overall feedback amplifier with equivalent input noise sources shown explicitly. The output voltage must be nulled

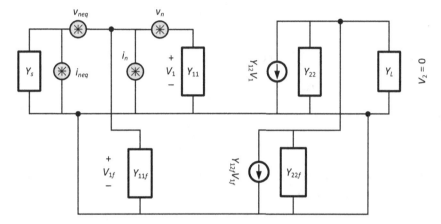

Figure 5.38: Equivalent circuit of the feedback amplifier with all the noise sources when the output voltage is set to zero

this example a shunt feedback configuration is considered, though the analysis is readily extended to any type of feedback (see Problem 18 for the case of series feedback). Also shown is the input-referred noise sources of the open-loop amplifier. The feedback network is clearly noiseless as it is reactive.

We postulate that the equivalent input-referred noise sources of the overall feedback amplifier (v_{neq} and i_{neq}) must be such that (less a negative sign which is unimportant for noise sources) with the amplifier itself noise sources present (v_n and i_n), the output voltage or current must be nulled, as shown in Figure 5.37.

Setting V_2 to zero results in the input controlled current sources ($Y_{12}V_2$ and $Y_{12f}V$) to go away, which brings us to the circuit of Figure 5.38, consisting of the noisy amplifier, the feedback network represented by its Y parameters, and the equivalent noise sources of the overall feedback amplifier. The equivalent noise sources are obtained now, knowing that the output voltage must be zero.

Keeping in mind that V_2 is zero, and writing a KCL at the output, we will have

$$Y_{21}V_1 + Y_{21f}V_{1f} = 0.$$

Since $V_{1f} = V_1 + v_n$, then

$$V_1 = -\frac{Y_{21f}}{Y_{21} + Y_{21f}} v_n.$$

The equivalent noise sources must work for any source impedance, and particularly short and open sources. Setting the source impedance to zero, then

$$v_{neq} = -(v_n + V_1) = -\frac{Y_{21}}{Y_{21} + Y_{21f}}v_n.$$

We can simplify the matters by noticing that in a well-designed feedback system, the open loop amplifier forward gain (Y_{21}) is supposed to be much larger than that of the feedback network (Y_{21f}). Accordingly,

$$V_1 \cong 0,$$

and

$$v_{neq} \cong -v_n.$$

Similarly, by setting the source impedance to infinity,

$$i_{neq} + i_n + Y_{11}V_1 + Y_{11f}V_{1f} = 0,$$

which leads to

$$i_{neq} \cong -(i_n + Y_{11f}v_n).$$

Since $i_n = Y_c v_n + i_u$, the equivalent noise current of the feedback amplifier may be expressed in terms of the correlation admittance:

$$i_{neq} = (Y_C + Y_{11f})v_{neq} - i_u.$$

As we showed in Chapter 3, the Y parameters of any lossless reciprocal network are imaginary. Consequently, an LCM feedback circuitry modifies only the *imaginary* part of the correlation admittance. More importantly, as the minimum noise figure does not depend on the imaginary part of Y_c (recall that $F_{min} = 1 + 2R_n\left(G_c + \sqrt{\frac{G_u}{R_n} + G_c^2}\right)$), an LCM feedback network does not affect the minimum noise figure. Note that for the previous case of an ideal feedback, $Y_{11f} = 0$, and we arrive at the same conclusion.

The above outcome proves to be very helpful in designing low-noise amplifiers, as feedback does change the input impedance of the amplifier. This could be exploited to satisfy the 50Ω matching, and optimum noise figure conditions *simultaneously*. As we will see in Chapter 7, a broad category of low-noise amplifiers rely on this principle.

5.6 NOISE FIGURE OF CASCADE OF STAGES

Since any practical radio consists of several building blocks, it is very helpful to be able to express the overall noise figure as a function of that of the individual blocks. Shown in Figure 5.39, let us consider two stages with noise figure and available power gain of F_1, G_{a1}, and F_2, G_{a2}, respectively. The results can be readily extended to any number of stages. We need not show the input-referred noise sources, as they are already embedded in the noise figures.

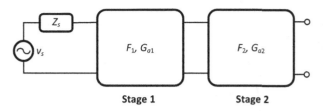

Figure 5.39: Noise figure of cascade of two stages

The input-referred available noise for each stage was shown to be $(F_1 - 1)KT$ and $(F_2 - 1)KT$, respectively. The first stage is directly connected to the source, whereas the input noise of the second stage is attenuated by the available power gain of the first stage when referred to the source. Thus, the total available noise at the input, including the source noise, is

$$N_i = KT + (F_1 - 1)KT + \frac{(F_2 - 1)KT}{G_{a1}},$$

and this normalized to KT is the cascaded noise figure:

$$F = F_1 + \frac{F_2 - 1}{G_{a1}}.$$

The above equation, known as the *Friis equation*, does not require the stages to be matched to one another. However, there is a subtle point that must be clarified: For each stage, the noise figure and the available power gain must be defined with respect to the source driving it. For the first stage, this happens to be the input source, whereas for the second stage, the source is the first stage driving it. Thus F_2 and G_{a2} are defined with respect to R_{OUT1}, that being the output resistance of the first stage.

One important outcome of the Friis equation is that if the gain of the first stage is large, the noise of the second stage is not as important. On the other hand, the noise figure of the first stage directly contributes to the total noise. As a result, it is common to place a *low-noise amplifier* at the very input of the receiver, to minimize F_1 and maximize G_{a1}.

Example: Consider Figure 5.40 representing a practical case in most receivers. The circuit consists of two lossy passive stages with the losses L_1 and L_2, followed by an amplifier with a noise figure of F_3, as shown in the figure. The passive lossy stages may represent filters or switches placed at the receiver input. In most cases such blocks are matched to 50Ω.

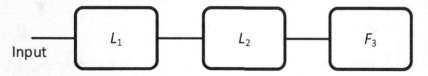

Figure 5.40: Cascade of two lossy stages and an amplifier

Continued

The total noise figure is

$$F = L_1 + \frac{L_2 - 1}{\frac{1}{L_1}} + \frac{F_3 - 1}{\frac{1}{L_1 L_2}} = L_1 L_2 F_3.$$

If expressed in dB the receiver overall noise figure is

$$NF = Loss_1 + Loss_2 + NF_3.$$

So the losses simply add up to the receiver noise figure.

Example: Shown in Figure 5.41 is a two-stage MOS amplifier. For the first stage $g_{m1} = 20\text{mS}$, $R_{L1} = 500\Omega$, and for the second stage $g_{m2} = 5\text{mS}$, $R_{L2} = 2\text{k}\Omega$ (bias details are not shown). Ignore r_o and all the internal capacitances, and assume $\gamma = 1$. To match to 50Ω, we insert the shunt resistor $R_p = R_s = 50\Omega$ at the first stage input. We wish to find the noise figure of the cascaded amplifier.

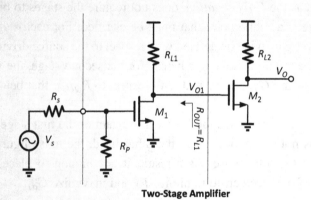

Two-Stage Amplifier

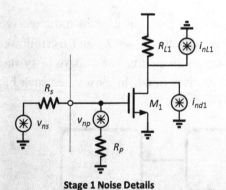

Stage 1 Noise Details

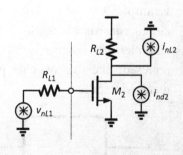

Stage 2 Noise Details

Figure 5.41: Two-stage amplifier noise figure example

The available power gain of each stage is

$$G_{a1} = \frac{R_s}{R_{L1}}\left(\frac{1}{2}g_{m1}R_{L1}\right)^2 = \frac{R_s}{R_{L1}}A_{v1}^2 = 2.5$$

$$G_{a2} = \frac{R_{L1}}{R_{L2}}(g_{m2}R_{L2})^2 = \frac{R_{L1}}{R_{L2}}A_{v2}^2 = 25,$$

where A_{v1} and A_{v2} are the voltage gains of each stage. The factor $\frac{1}{2}$ appearing in G_{a1} signifies the voltage division at the input due to shunt resistance. Note that for the second stage, G_{a2} is calculated with respect to the previous stage output resistance ($R_{OUT1} = R_{L1}$), which is effectively the source driving it. As we pointed out earlier, the same thing must be done for noise figure calculations. The total available power gain is

$$G_a = G_{a1}G_{a2} = \frac{R_s}{R_{L2}}\left(\frac{1}{2}g_{m1}R_{L1}g_{m2}R_{L2}\right)^2 = 62.5.$$

The noise details of each stage are shown in Figure 5.41 as well. Considering the first stage, the total noise referred to the input (source) is

$$v_{ns} + v_{np} + \frac{i_{nd1} + i_{nL1}}{\left(\frac{1}{2}g_{m1}\right)^2}.$$

Note that $R_s = R_p$. Thus, the first stage noise factor is

$$F_1 = 1 + 1 + \frac{4KT\gamma g_{m1} + \dfrac{4KT}{R_{L1}}}{4KTR_s\left(\dfrac{1}{2}g_{m1}\right)^2} = 2 + \frac{4}{g_{m1}R_s}\left(\gamma + \frac{1}{g_{m1}R_{L1}}\right) = 6.4 = 8.1\text{dB}.$$

The noise contribution of the load resistor is suppressed by the gain of the amplifier as expected, and therefore may be ignored.

Similarly, for the second stage

$$F_2 = 1 + \frac{4KT\gamma g_{m2} + \dfrac{4KT}{R_{L2}}}{4KTR_{L1}(g_{m2})^2} = 1 + \frac{1}{g_{m2}R_{L1}}\left(\gamma + \frac{1}{g_{m2}R_{L2}}\right) \approx 1.44 = 1.5\text{dB}.$$

Using the Friis equation,

$$F = F_1 + \frac{F_2 - 1}{G_{a1}} = 2 + \frac{4}{g_{m1}R_s}\left(\gamma + \frac{1}{g_{m1}R_{L1}}\right) + \frac{\dfrac{1}{g_{m2}R_{L1}}\left(\gamma + \dfrac{1}{g_{m2}R_{L2}}\right)}{\dfrac{R_s}{R_{L1}}\left(\dfrac{1}{2}g_{m1}R_{L1}\right)^2} = 6.6 = 8.2\text{dB}.$$

Despite consuming one-fourth of the current in the second stage, the noise figure is mostly dominated by the first stage, given the gain preceding the second stage.

Continued

Note that in the noise figure calculations the system does not need to be matched to the source. In fact, the reader can show that with R_p removed, the noise figure will be

$$F = 1 + \frac{1}{g_{m1}R_s}\left(\gamma + \frac{1}{g_{m1}R_{L1}}\right) + \frac{\frac{1}{g_{m2}R_{L1}}\left(\gamma + \frac{1}{g_{m2}R_{L2}}\right)}{\frac{R_s}{R_{L1}}(g_{m1}R_{L1})^2} = 2.14 = 3.3\text{dB},$$

which is lower than the original value. The price paid here is the absence of a good match at the amplifier input. That is a common trade-off, and in fact the main challenge in designing low-noise amplifiers.

We can rearrange the noise figure equation above as follows:

$$F = F_1 + \frac{\frac{1}{g_{m2}R_s}\left(\gamma + \frac{1}{g_{m2}R_{L2}}\right)}{A_{v1}^2} = F_1 + \frac{F_2|_{R_s} - 1}{A_{v1}^2},$$

where $F_2|_{R_s}$ is the noise factor of the second stage referred to the original source impedance (R_s), and not the first stage output impedance. This noise is then refereed to the input by the first stage voltage gain, and not its power gain. This simplification works only if the amplifier can be modeled by its input equivalent *noise voltage*, only, e.g., a high impedance amplifier like a common-source stage.

5.7 PHASE NOISE

In Chapter 2 we discussed that mixers (or multipliers) are used in modulators to *upconvert* the baseband spectrum to carrier frequency. We also showed that in a similar fashion, we exploit the mixers in receivers to *downconvert* the modulated spectrum around the carrier to a conveniently low frequency, say zero, known as the intermediate frequency or IF (Figure 5.42).

Same as the modulator, the receive mixer is also driven by a local oscillator (or LO) at (or very close to) the carrier frequency. The LO signal could be generally expressed as

$$x_{LO}(t) = A_C \cos(2\pi f_C t + \phi_n(t)).$$

This is a similar notation we used in Chapter 2, except for the additional terms $\phi_n(t)$, representing the noise in the active circuitry in the oscillator, known as *phase noise*. In Chapter 9

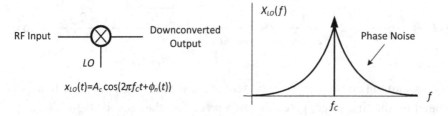

Figure 5.42: A mixer used in a receiver to downconvert the modulated spectrum

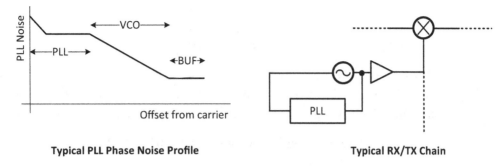

Figure 5.43: Typical PLL phase noise in a typical RX or TX chain

we will discuss the mechanisms responsible for producing this type of noise, but in this chapter as well as Chapter 6 we will study its impact on the system level performance of the transceiver. Since $\phi_n(t)$ represents the noise, we may say $|\phi_n(t)| \ll 1$. Therefore we can use the narrowband FM approximation,

$$x_{LO}(t) \approx A_C[\cos 2\pi f_C t - \phi_n(t) \sin 2\pi f_C t].$$

which indicates that the actual LO signal consists of an impulse representing the ideal cosine, as well as a term representing noise in quadrature, and shaped around the carrier as shown in Figure 5.42. The phase noise magnitude generally decays with a slope of $1/f^2$ moving away from the carrier frequency, as predicted by Leeson [3]. This will be proven in Chapter 9, but we will simply accept it for the moment.

This noisy LO signal is multiplied to the RF signal present at the mixer input. Thus the mixer output, while translated in frequency, contains the phase noise convolved in the frequency domain with the input. From this, we expect that in an otherwise ideal receiver, the LO phase noise set a limit on the signal quality or SNR. Moreover, as we will show in Chapter 6, the phase noise will also limit the receiver sensitivity when exposed to large unwanted signals, known as blockers.

If the oscillator is embedded within a phase-locked loop (PLL), the phase noise profile will change according to the PLL frequency characteristics.

It is well known that [4], [5] the output noise of the VCO, once locked in PLL, will have a phase noise profile as shown in Figure 5.43. It consists of a relatively flat passband set by the PLL components, along with a region of 40dB/Dec (or $1/f^2$) slope set by the VCO noise. At very far-out frequency offsets from the carrier, the noise flattens out as the buffers or other hard-limiting devices following the oscillator dominate. If such VCO is used in the transmit (or receive) chain (Figure 5.43 right), this noise profile is expected to appear at the output. We will show in Chapter 6 that this noise along with the nonlinearity of the TX chain will be limiting factors in the transmitter modulation mask requirements.

5.8 SENSITIVITY

The limit to the minimum detectable signal at the receiver input is readily found from the basic noise figure definition:

$$F = \frac{\text{SNR}_i}{\text{SNR}_o} = \frac{\frac{S_i}{N_i}}{\text{SNR}_o}.$$

We showed that the available noise power at the input is $N_i = KT \approx 4 \times 10^{-21}$ W.s, expressed at a 1 Hz bandwidth. If integrated over a certain bandwidth of interest denoted by B, the noise power will be then KTB. Taking 10 log from each side, we can express noise figure in dB as follows:

$$NF = 10 \log S_i - 10 \log (KTB) - 10 \log \text{SNR}_o.$$

Rearranging, and knowing that $10 \log KT = -174$ dBm/Hz at room temperature,[14] we will have

$$\text{Sensitivity} = -174 + NF + 10 \log B + \text{SNR}.$$

Thus, guaranteeing a minimum SNR (in dB) at the detector output, the minimum detectable signal or sensitivity expressed in dBm would be $-174 + NF + 10 \log B + \text{SNR}$. Clearly, the sensitivity may be improved by lowering the noise figure, or reducing the system bandwidth. The latter directly leads to a lower throughput.

Example: Let us consider a 4G receiver shown in Figure 5.44. When operating in GSM (Global System for Mobile Communications) mode, the standard requires a minimum sensitivity of –102 dBm. Most modern handsets achieve a much better sensitivity, around –110 dBm, corresponding to signal level of 0.7 µV RMS at the antenna. To support multiple bands and the transmit/receive multiplexing, a switch is inserted between the antenna and multiple receivers (as well as the transmitter output though not shown) input. Moreover, each receiver supporting a certain band is preceded by an external SAW filter (or a duplexer in case of 3G or 4G) to attenuate unwanted signals.

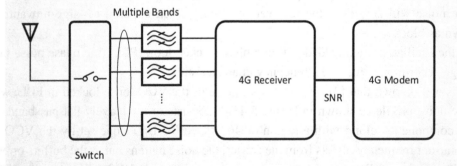

Figure 5.44: A simplified 4G receiver block diagram

The GSM signal is GMSK (Gaussian minimum shift keying) modulated, and takes a bandwidth of about 200 kHz. Typical modems need an SNR of at least about 5 dB to detect this signal properly. To achieve a sensitivity of –110 dBm, the noise figure at the antenna

---

[14] dBm is 10 log of a power quantity when expressed in mW. 1 mW is then 0 dBm. When expressing a power associated with a peak voltage of V over a 50 Ω resistor, then dBm = 10 dB + 20 log V = 10 + dBV.

must be 6dB. Furthermore, assuming a typical insertion loss of 3dB for the switch and SAW filter combined, the receiver must have a minimum noise figure of 3dB. Note that the switch and filter are considered lossy and passive components, so when placed at the very input of the radio, their loss directly adds to the system noise figure.

> **Example:** For the same receiver of previous example, we now assume that the output noise of the receiver after downconverted to IF has a profile as shown in Figure 5.45.
>
>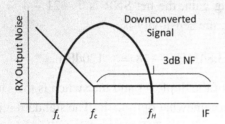
>
> **Figure 5.45:** 4G receiver output noise
>
> It consists of a flat region, corresponding to the same 3dB noise figure, and starts rising at lower frequencies due to the flicker noise. The total output noise may be then expressed as
>
> $$N_o = N_{RX}\left(1 + \frac{f_c}{f}\right),$$
>
> where f_c signifies the *flicker noise corner frequency*, that is, the frequency where the white and flicker noise contributions are equal. N_{RX} when referred to the input must then correspond to the same 3dB noise figure. Assume the downconverted signal occupies a spectrum between f_L and f_H. Then total noise figure including the flicker noise that was previously ignored will be
>
> $$F = \frac{N_{RX}}{G_a KT} \frac{1}{f_H - f_L} \int_{f_L}^{f_H} \left(1 + \frac{f_c}{f}\right) df = \frac{N_{RX}}{G_a KT} \left(1 + f_c \frac{\ln \frac{f_H}{f_L}}{f_H - f_L}\right).$$
>
> The second term is due to the excess noise resulted from flicker noise, while the term $\frac{N_{RX}}{G_a KT}$ corresponds to a 3dB noise figure calculated previously. Assume that the GSM signal is downconverted to a 135kHz IF. If the flicker corner is, say, 100kHz, then the excess noise factor is $1 + 100 \frac{\ln \frac{235}{35}}{200} = 1.95$. Thus the effective sensitivity of the receiver degrades by almost 3dB to −107dBm. Besides the obvious choice of reducing the flicker noise corner, another option is to place the signal at a higher IF. In Chapter 12, where we discuss the receiver architectures, we will show that there are important repercussion of doing so. For narrowband signals like GSM, the only viable option is to ensure the RX 1/f noise corner is low enough. For instance, a flicker corner of 10kHz would result in only $10 \log 1.095 = 0.1$dB degradation in sensitivity.

Example: Suppose the same receiver is operating in 3G mode, where a QPSK modulated signal is received. The 3G signal is about 3.84MHz wide, and since it is spread spectrum, the total SNR required is −18dB. In practice detecting a QPSK signal requires an SNR of about +7dB. However, due to the spread spectrum nature of the 3G signal, there is what is known as the *processing gain*, that is, the ratio of the received signal bandwidth to what it will be after the decoding in the modem, in this case 30kHz. Thus,

$$\text{Processing gain} = \frac{3.84\text{MHz}}{30\text{kHz}} = 128 = 21\text{dB}.$$

Since there is an additional 4dB of coding gain, the net SNR is 7 − 21 − 4 = −18dB. Ignoring the flicker noise for the moment, the sensitivity is

$$-174 + 6\text{dB} + 10\log(3.84\text{MHz}) - 18 = -120\text{dBm}.$$

We assumed the loss of the combination of the duplexer and the switch is the same 3dB we had in the case of GSM, which may be somewhat optimistic. The standard requires −117dBm, that is, a 3dB margin if the receiver has 3dB noise. We will see in the next chapter that there are other factors in the case of a 3G receiver that potentially lead to a worst sensitivity than what we just calculated considering only the receiver thermal noise.

Now let us consider the flicker noise. Suppose we have the same corner frequency of 100kHz, which led to about 3dB degradation in the case of GSM. Also we assume the 3G signal is downconverted to a zero IF. We perform the integration from an arbitrary lower bound of 20kHz, to an upper bound of 1.92MHz where the signal resides. The particular choice of 20kHz lower bound (rather than zero) is due to the fact that in most modulations including 3G, losing the signal energy around DC is acceptable. In fact most receivers incorporate some form of a highpass filter to remove the DC offsets. The filter corner frequency is typically around tens to hundreds of kHz, depending on the actual signal bandwidth. The filter settling would be too slow if the highpass corner were to be pushed too low. With that, the excess noise is $1 + 0.1 \frac{\ln\frac{1.92}{0.02}}{1.9} = 1.23$, and the sensitivity degradation is now only 0.9dB. If we were to integrate from 10kHz instead, the excess noise would have increased to only 1.27, thus somewhat justifying the rather arbitrary choice of the integration lower bound!

The above example points out a very important conclusion that the wideband receivers such as 3G are far more tolerant of the flicker noise even if the signal is downconverted to DC. Thus zero-IF receivers are quite common for most wideband applications.

Example: As our last example, consider Figure 5.46, which shows more circuit-level details of the 4G receiver. It consists of a low-noise transconductance amplifier (LNTA), a current-mode multiplier (or mixer), followed by an active-RC biquad.

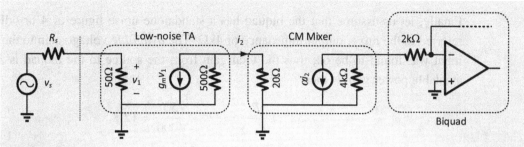

Figure 5.46: Simplified 4G receiver block diagram

The LNTA is matched to 50Ω and has a standalone noise figure of 2dB when properly terminated with a noiseless load representing the mixer input impedance. It produces an output current with a transconductance gain of $g_m = 100$mS, flowing into the current mode mixer. We have simplified the mixer to a linear and time-invariant current mode amplifier with a current gain of $\alpha = 0.5$, and simply assume it translates the signal frequency to a zero-IF. The biquad, whose input stage is only shown, has a passband gain of 10, and presents an input impedance of 2kΩ to the mixer. The receiver total available voltage gain is

$$A_a = g_m \times \frac{500}{500+20} \times \alpha \times (4k\|2k) \times 10 = 640.$$

Since the mixer is current mode, defining a standard noise figure may not be as meaningful. Thus, we instead assume that it can be modeled by an input-referred current noise whose spectral density is $4KT \times 20$mS A²/Hz. The mixer noise when referred to the source will be divided by $g_m/2$, as shown in Figure 5.47. The extra factor of 2 has to do with the voltage division between the source and the LNA input impedances, which are equal.

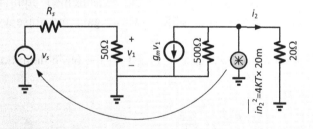

Figure 5.47: Mixer noise referred to the source

If the LNTA noise figure is F_1, then the total noise referred to the source from the LNTA and mixer is

$$(F_1 - 1)4KTR_s + \left(\frac{2}{g_m}\right)^2 \times \overline{i_{n2}^2}.$$

Therefore, the LNTA-mixer combined noise figure is

$$F = F_1 + \left(\frac{2}{g_m}\right)^2 \times \frac{\overline{i_{n2}^2}}{4KTR_s} = 1.58 + 0.16 = 1.74 = 2.4\text{dB}.$$

Continued

Finally, let us assume that the biquad has a standalone noise figure of 4 or 6dB when referred to the mixer output impedance of 4kΩ. The *available* voltage gain to the biquad input was found to be 64, thus the total gain from the source to the biquad is 32. The available power gain is then

$$32^2 \times \frac{R_s}{R_{out2}} = 32^2 \times \frac{50}{4000} = 12.8,$$

and the total noise figure is $1.74 + \frac{4-1}{12.8} = 1.97 \approx 3\text{dB}$.

This is an example of how one can budget different noise and gain values to different stages to meet a certain cascaded noise figure. In our case we started from the sensitivity requirement of −110dBm, which is given to us, and that led to a noise figure of 3dB for the RX. The rest of the numbers were chosen to satisfy the 3dB requirement for the entire receiver.

5.9 NOISE FIGURE MEASUREMENTS

There are two common ways of measuring the noise figure that we will discuss here. Before that, we should emphasize that the noise figure is usually expressed and measured at a given frequency, known as *spot noise figure*. If one is interested, it could be characterized over a range of frequencies of interest of course. Moreover, since noise levels are typically very small, it is not uncommon to perform the noise measurements in isolated shielded chambers immune to outside noise or interference. The noise monitored at the output of the device under test (DUT) must be well above the noise floor of the measurement equipment. As a result, it is often necessary to add pre- or post-amplifiers whose gain and noise are properly de-embedded.

The first method known as *gain method* is directly based on the basic definition of noise figure:

$$F = \frac{N_o}{G_a KT}.$$

If one obtains the DUT available gain and the output noise (when terminated by 50Ω), as KT is known, the noise figure is readily calculated. One potential issue with this type of measurement is that the spectrum analyzer inherently needs to measure the noise *integrated* over a narrow band defined by its resolution bandwidth (RBW). This number may be converted to a spot noise measured per frequency, if one has the *exact* knowledge of the type and shape of the filter. Most modern spectrum analyzers have built-in functions to provide this information with a reasonable accuracy. Also this method relies on two independent exact measurements of gain and noise. Thus, the spectrum analyzer and the signal generator used in the setup must be very well calibrated.

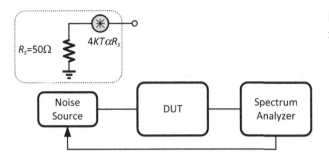

Figure 5.48: Y-factor noise figure measurement

The second method, known as the *Y-factor method*, or hot–cold measurement,[15] only *requires two relative measurements*, and no knowledge of the spectrum analyzer filter is required. Shown in Figure 5.48, the device under test is connected to a very accurate and well-calibrated noise source, capable of producing two different levels of noise at the DUT input.

If the noise source is off, it simply behaves like a 50Ω resistor. Once turned on, it still provides a 50Ω termination but with $\alpha > 1$ times more noise. The output noise of the DUT is measured in each case by a spectrum analyzer or a power meter, and recorded. The input-referred noise of the DUT is always $(F-1)KT$. So the two noise recordings when referred to the input (labeled as N_0 and N_1) will be

$$N_0 = KT + (F-1)KT$$
$$N_1 = \alpha KT + (F-1)KT.$$

The factor α, referred to as the *excess noise ratio* (ENR), is a known quantity, and is given by the noise source manufacturer. By diving the two equations, any dependence on the *absolute accuracy* of the noise measurement or DUT *gain* is eliminated:

$$\frac{N_1}{N_0} = \frac{\alpha + (F-1)}{F}.$$

Solving for F yields

$$F = \frac{\alpha - 1}{\frac{N_1}{N_0} - 1}.$$

So just two *relative* measurements of the DUT noise are needed now, but they do require a well-calibrated noise source. The value of excess noise (α) is typically available over a wide range of frequencies of interest. Most modern network analyzers have built-in functions to use the noise source, and express the noise figure directly.

It is interesting to note that if the device is noiseless, the increase in the output noise when the source is turned on is exactly equal to the excess noise. The noisier the device is, the less increase in the level of the output noise is observed, corresponding to less accuracy. Therefore,

---

[15] The naming convention is from the fact that two levels of noise are injected to the system corresponding to cold and hot conditions given the noise dependence on the absolute temperature.

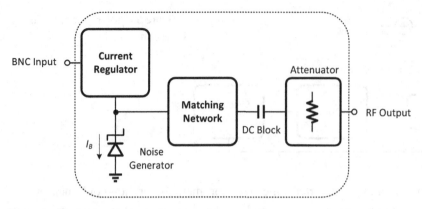

Figure 5.49: Simplified circuit diagram of an RF noise source

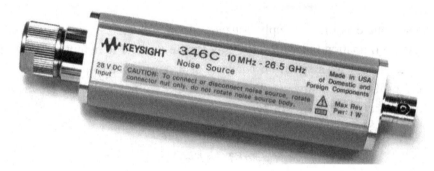

Figure 5.50: Keysight 346C RF noise source

if the device noise figure is expected to be too high compared to the source excess noise, one may consider using a well calibrated *preamplifier*.

The noise source cannot be created using passive only noisy elements (why?), and typically consists of an avalanche diode whose bias current is fixed by a precise current regulator followed by a matching network and an attenuator (Figure 5.49).

The BNC[16] input turns the noise source on and off. When turned on, the avalanche diode whose current I_B is very well regulated, creates a well-defined noise current with the spectral density of $2qI_B$, where $q = 1.6 \times 10^{-19}$C is the electron charge. This excess noise ultimately appears at the output through a buffer and a matching network. The RF output is still matched to 50Ω, but its noise spectral density is greater than that of a 50Ω resistor by an amount defined by the ENR, set by the noise generator circuitry.

An example of a commercially available noise source (Keysight 346C)[17] is shown in Figure 5.50. It has an ENR of 15dB, and works from 10MHz to 26.5GHz.

---

[16] The BNC (Bayonet Neill–Concelman) connector is a miniature radio frequency connector used for coaxial cable.
[17] See https://www.keysight.com/en/pd-1000001300%3Aepsg%3Apro-pn-346C/noise-source-10-mhz-to-265-ghz?pm=PL&nid=-536902744.536880071&cc=US&lc=eng.

5.10 Summary

The important topic of noise was disused in this chapter, and will be extensively referred to in Chapters 7, 8, 9, and 10 in the context of noise properties of RF building blocks such as low-noise amplifiers, mixers, or oscillators.

- Section 5.1 discussed introductory material on the noise, and presented the types of noise in commonly used RF elements such as resistors or transistors. We also discussed the noise of passive reciprocal circuits, as well as cyclostationary noise, which is noise in periodically varying circuits.
- In Section 5.2 we developed the two-port equivalent noise model.
- Noise figure and the minimum noise figure concept were discussed in Sections 5.3 and 5.4. The minimum noise figure is an essential part of designing good low-noise amplifiers.
- The impact of feedback in general, and lossless feedback in particular, on noise were presented in Section 5.5. Most low-noise amplifiers take advantage of lossless feedback in some form to simultaneously satisfy the matching and minimum noise figure.
- Section 5.6 extended the definition of noise figure to more complex structures comprising several sub-blocks.
- In Section 5.7 we briefly discussed the concept of phase noise, which is required to understand reciprocal mixing in amplifiers and receivers. A thorough treatment of phase noise is presented in Chapter 9.
- Sensitivity was defined in Section 5.8, followed by several practical examples.
- The chapter concluded by discussing practical implications of noise measurement.

5.11 Problems

1. Find the minimum noise figure of a FET in a source-follower configuration. Ignore C_{gd}.
   Answer: $F = 1 + \left|\frac{1+j\omega C_{gs1}Z_{TH}}{g_{m1}+j\omega C_{gs1}}\right|^2 \frac{\gamma(g_{m1}+g_{m2})}{\text{Re}[Z_{TH}]}$, and $F_{min} = 0$.

2. For the following circuits, find the input-referred noise voltage. Ignore $1/f$ noise, r_o, and capacitances.

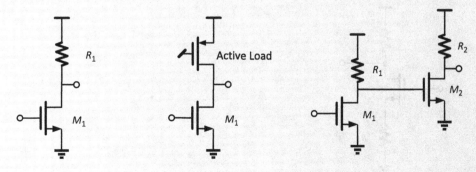

3. Find the equivalent noise bandwidth of an nth-order Butterworth filter.

4. Using Laplace transform differentiation property, prove the initial value theorem, that $z(0^+) = \lim_{s \to \infty} sZ(z)$.

5. For the following current-mode passive mixer, find the noise spectral density at the input of the transimpedance amplifier (TIA). The TIA has a zero input impedance. What is the noise if the input current source is ideal? **Answer:** $S_{i_{OUT}} = \frac{4KT}{R_s + R_{SW}}$.

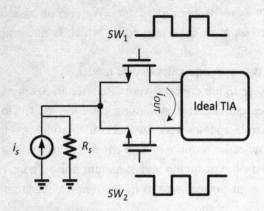

6. Find the impact of an ideal series–series feedback on equivalent noise sources, and the minimum noise figure.

7. Show that instead of the input noise voltage, if we were to break the input noise current into correlated and uncorrelated parts, the noise figure would be

$$F = 1 + |Z_s|^2 \frac{G_u + R_n |Y_s + Y_c|^2}{R_s}.$$

Find the optimum source admittance by taking partial derivatives of the noise equation, and express the optimum noise figure in terms of the noise current parameters and the correlation admittance (Y_c).

8. Find the minimum noise figure of the resistively degenerated common-source amplifier shown below. Ignore all the capacitances, the body effect, and r_o.

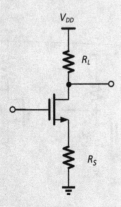

9. For the amplifier shown below, the load is ideal (noiseless). Ignoring C_{gd} and the junction capacitances (but not C_{gs}), find the input impedance of the amplifier. What are the equivalent noise sources and the minimum noise figure? Assume $r_o = \infty$, and ignore the body effect.

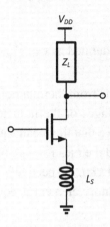

10. Prove the energy conservation for the circuit shown below using Tellegen's theorem, $P = \frac{1}{2} Z_{IN}(j\omega)|I_S|^2 = \frac{1}{2}\sum_i R_i |I_i|^2 + \frac{1}{2}\sum_k j\omega L_k |I_k|^2 - \frac{1}{2}\sum_p \frac{1}{j\omega C_p}|I_p|^2$, where R_i, L_i, and C_i, are the one-port RLC components.

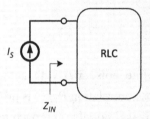

11. Calculate the insertion loss and noise figure (without using the Nyquist theorem) of the following ladder filter, where the inductor has a finite Q, and show that they are equal.
    Answer: $F = 1 + \frac{r}{R_s}$.

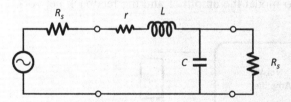

12. Prove the dual of the Nyquist theorem: the current noise spectral density at the output of an RLCM one-port with output admittance $Y(s)$ is $S_{in}(f) = 4KT\,\text{Re}\,[Y(f)]$.

13. Using basic noise definition (and not the Nyquist theorem) show that the noise figure of a passive lossy circuit is equal to its loss. Lump all the loss as a shunt resistor at the output. **Hint:** Use reciprocity.

14. Find the noise figure of the following resistive circuit with and without the Nyquist theorem.

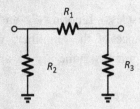

15. Suppose $v(t)$ is white noise, and the random variable s is defined as $s = \int_{-T}^{T} v(t)dt$. Find the variance of s.

16. In the following network, by calculating transfer functions from thermal noise of resistor R modeled as a noise voltage source to the capacitor voltages or the inductor current, and calculating appropriate integrals over $\omega \in (-\infty, +\infty)$, prove that the mean square of thermal noise voltages stored across the capacitors C_1 and C_2, and the mean square of thermal noise current stored in inductor L are equal to kT/C_1, kT/C_2 and kT/L, respectively. Therefore, the mean square of thermal noise energy stored in each memory element is equal to $kT/2$.

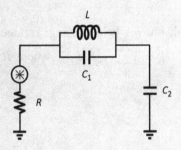

17. Assume that the input for an LTI system is a white noise $x(t)$ with autocorrelation of $R_x(\tau) = \frac{N_0}{2}\delta(\tau)$ and the output is $y(t)$. Prove that cross-correlation $R_{yx}(\tau)$ is proportional to $h(\tau)$, where $h(t)$ is the system impulse response. This is a popular approach to estimate the impulse response of an LTI system by applying a white noise source to the input and correlating the same input with the response of the system.

18. Find the impact of series reactive feedback on the amplifier minimum noise figure depicted below. Use Z parameters to model the amplifier and the feedback network.

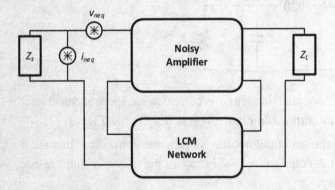

19. A white Gaussian noise $n(t)$ with a two-sided power spectral density of N_0 V^2/Hz is applied to the control voltage of a sinusoidal voltage-controlled oscillator (VCO) with a VCO gain

of K_{VCO}. Prove that the power spectral density of the VCO output voltage is *Lorentzian* given by: $S_V(\omega) = \frac{K_{VCO}^2 \frac{N_0}{2\pi}}{\left(K_{VCO}^2 \frac{N_0}{4\pi}\right)^2 + (\omega - \omega_0)^2}$. **Hint:** Establish the phasor $V(t) = \exp(j(\omega_0 t + \varphi(t)))$, representing the VCO signal, where $\varphi(t) = K_{VCO} \int n(\theta) d\theta$ is the phase perturbation. Show that $R_V(\tau) = e^{j\omega_0\tau} E\left[e^{j(\varphi(t+\tau) - \varphi(t))}\right]$. Define the variable $x(t,\tau) = \varphi(t+\tau) - \varphi(t)$. Knowing that $\varphi(t+\tau)$ and $\varphi(t)$ are jointly Gaussian with a standard deviation σ, show that $E\left[e^{j(\varphi(t+\tau) - \varphi(t))}\right] = e^{-\sigma^2/2} = e^{-(R_\varphi(0) - R_\varphi(\tau))}$. Obtain an expression for $R_\varphi(\tau)$ considering that it is linearly dependent on $n(t)$, and find the VCO voltage spectral density.

20. Drive the noise figure of cascade of three stages, each stage represented only by its input-referred noise voltage. The stages are assumed to have an infinite input impedance, and thus zero input-referred noise current.

21. Not ignoring the base resistance, find the low-frequency input-referred noise voltage and current of a BJT. How does that compare to a FET biased at the same current?

22. In the following g_m-C resonator, the noise of the transconductors is modeled as current sources i_{n1} and i_{n2} at their outputs. Derive transfer functions from these current sources to the voltage across C_1 and prove that one of the transfer functions is a lowpass and the other one is a bandpass. Assuming one-sided power spectral density of these noise currents are $4kTFg_m$, prove that the mean square of the noise voltage across C_1 is equal to $\frac{kTF}{C_1}g_m R$ and $\frac{kTF}{C_2}g_m R$ contributed by noises i_{n1} and i_{n2}, respectively.

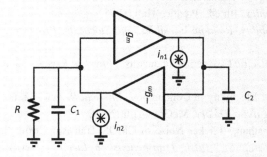

23. Find the input impedance and the input-referred noise of the following active gyrator. The transconductors consist of a single FET with ideal active load. Ignore the device capacitances.

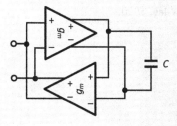

24. Repeat Problem 23, where the transconductors are realized by inverters with identical N and P devices.

25. Using the bilinear relation between impedance and reflection coefficient ($Z_{s/s,opt} = Z_0 \frac{1+\Gamma_{s/s,opt}}{1-\Gamma_{s/s,opt}}$, where Z_0 is the characteristics impedance), map the noise factor equation into the Smith chart. **Hint:** Describe the noise factor equation in terms of the reflection coefficient as follows: $F = F_{min} + 4G_n Z_0 \frac{|\Gamma_s - \Gamma_{s,opt}|^2}{(1-|\Gamma_s|^2)|1-\Gamma_{s,opt}|^2}$. For a given amplifier, G_n, F_{min}, and $\Gamma_{s,opt}$ are known. Thus, we can rearrange $\left|\Gamma_s - \frac{\Gamma_{s,opt}}{K+1}\right|^2 = \frac{K^2 + K(1-|\Gamma_{s,opt}|^2)}{(K+1)^2}$. For a given F, and hence K, Γ_s is described by a circle on the Smith chart.

5.12 References

[1] P. R. Gray and R. G. Meyer, *Analysis and Design of Analog Integrated Circuits*, John Wiley, 1990.
[2] H. Friis, "Noise Figures of Radio Receivers," *Proceedings of the IRE*, 32, no. 7, 419–422, 1944.
[3] B. Leeson, "A Simple Model of Feedback Oscillator Noise Spectrum," *Proceedings of the IEEE*, 54, 329–330, 1966.
[4] F. M. Gardner, *Phaselock Techniques*, John Wiley, 2005.
[5] D. H. Wolaver, *Phase-Locked Loop Circuit Design*, Prentice Hall, 1991.
[6] A. Van der Ziel, *Noise in Solid State Devices and Circuits*, John Wiley, 1986.
[7] Y. Tsividis and C. McAndrew, *Operation and Modeling of the MOS Transistor*, Oxford University Press, 2010.
[8] S. M. Sze and K. K. Ng, *Physics of Semiconductor Devices*, John Wiley, 2006.
[9] B. G. Streetman, *Solid State Electronics*, 4th ed., Prentice Hall, 1995.
[10] A. Papoulis and S. U. Pillai, *Probability, Random Variables, and Stochastic Processes*, McGraw-Hill, 2002.
[11] H. Nyquist, "Thermal Agitation of Electric Charge in Conductors," *Physical Review*, 32, no. 1, 110–113, 1928.
[12] J. B. Johnson, "Thermal Agitation of Electricity in Conductors," *Physical Review*, 32, no. 1, 97, 1928.
[13] C. A. Desoer and E. S. Kuh, *Basic Circuit Theory*, McGraw-Hill, 2009.
[14] J. Chang, A. Abidi, and C. Viswanathan, "Flicker Noise in CMOS Transistors from Subthreshold to Strong Inversion at Various Temperatures," *IEEE Transactions on Electron Devices*, 41, no. 11, 1965–1971, 1994.
[15] W. Gardner and L. Franks, "Characterization of Cyclostationary Random Signal Processes," *IEEE Transactions on Information Theory*, 21, no. 1, 4–14, 1975.
[16] A. Demir and A. Sangiovanni-Vincentelli, *Analysis and Simulation of Noise in Nonlinear Electronic Circuits and Systems*, Kluwer, 1998.
[17] D. Halliday, R. Resnick, and J. Walker, *Fundamentals of Physics*, John Wiley, 2013.
[18] R. Eisberg, *Fundamentals of Modern Physics*, John Wiley, 1990.

6 Distortion

In Chapter 5 we showed that the thermal noise of a receiver sets a *lower limit* on the signal detectable at the receiver input. The *upper limit* to the maximum signal a receiver can handle is set by the distortion, arisen from nonlinearity present in the active circuits making up the receiver. However, handling a desired large input is generally not an issue as it is typically managed by the proper gain control in the receiver. On the other hand, we will show in this chapter that the distortion has a far more detrimental impact on the receiver, when subject to large unwanted signals, known as blockers. Similar to noise, the blockers will also set a lower limit on the detectable signal.

We start this chapter with general description of blockers and their profile in wireless systems. We then present four distinct mechanisms that can impact the receiver performance in the presence of large blockers: small signal nonlinearity, large signal nonlinearity, reciprocal mixing, and harmonic mixing. We discuss the appropriate figures of merit for each case, and describe their practical impacts on modern receivers.

The specific topics covered in this chapter are:

- General description of blockers in radios
- Full duplex systems
- 2nd-, 3rd-, and 5th-order intercept point
- Compression and desensitization
- Reciprocal mixing
- Harmonic distortion
- Noise and linearity in transmitters
- AM–AM and AM–PM distortion
- Pulling in transmitters and its impact on performance

While this chapter may be very appealing for RF circuit and system engineers, for class use we recommend focusing only on a summary of Sections 6.1 and 6.2 and covering Sections 6.3.1, 6.3.3, 6.3.7, and 6.4. The rest of the material may be assigned as reading. The chapter includes many practical examples of deriving blocker requirements for both receivers and transmitters for various applications, such as cellular or wireless LAN.

Distortion

6.1 BLOCKERS IN WIRELESS SYSTEMS

Let us consider a cellular network where the area under coverage is divided into hundreds of *cells*. In a typical metropolitan environment, each cell is a few miles wide, where a simple conceptual graph for the purpose of demonstration is shown in Figure 6.1.

To every cell is assigned a base station, and the base stations are connected to each other by wire. Mobile handsets do not communicate directly, rather each handset residing in a given cell communicates only with the corresponding base station, which is a similar radio in nature as the handset, but with somewhat more stringent requirements.

It is well known that the energy of an electromagnetic wave decays in free space by $\frac{1}{d^2}$, d being the distance between the transmitter and the receiver (Figure 6.2). In metropolitan areas, on the other hand, the wave decays with much faster slope of $\frac{1}{d^n}$, where n > 2, and could be as large as 4 [1], [2].

There are several reasons behind this which could be attributed to phenomena such as multipath fading, or blocking as shown in Figure 6.2. In the case of fading, for instance, the transmitted signal could be subtracted entirely from its reflected replica, if the delays between the two paths are such that a 180° phase shift is created. Since the mobile users are moving, the signal strength can be changed dynamically in either direction.

Now consider the cellular network of Figure 6.1, where a given handset is subject to a very weak signal from its own intended base station for the reasons mentioned, but happens to be receiving strong signals from the nearby base stations (Figure 6.3). The receiver must be still able to detect the weak desired signal properly, despite being subject to such large undesirable signals from other base stations.

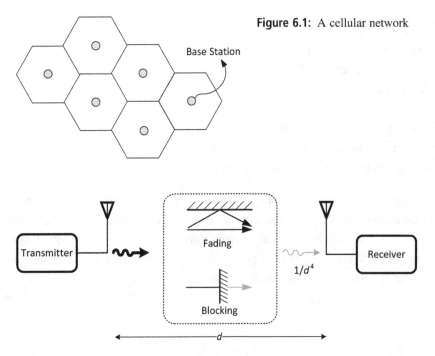

Figure 6.1: A cellular network

Figure 6.2: Electromagnetic waves decaying in metropolitan areas

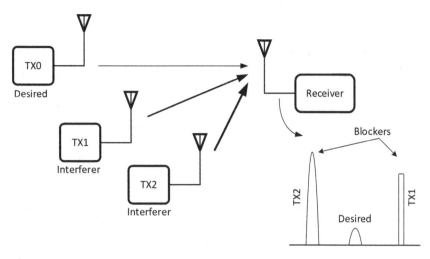

Figure 6.3: A weak desired signal accompanied by a strong interferer

If the thermal noise were the only source of nonideality, this would not have been an issue. As we will show shortly, however, other imperfections in the receiver such as nonlinearity or phase noise would impact the weak desired signal drastically. These unwanted signals created by the other base stations are commonly known as *blockers or interferers*. Typically in a given standard, their strength and frequency profile are provided to the RF designers. We should emphasize that statistically speaking, there is always a probability, however small, for the receiver to fail under very extreme conditions, but the standards are generally evolved to minimize that as much as practically possible.

The blockers are not only limited to the other base stations in the case of a cellular network. Any other wireless device that happens to be in the vicinity of the handset of interest may be a potential interferer. These blockers are known as *out-of-band blockers*, which fall outside the band of interest for a given standard, as opposed to *in-band blockers* discussed earlier. The standard has provisions for these out-of-band blockers as well, although since they are generally not as much under the control of the given network of interest, they happen to be more challenging. On the other hand, since they are outside the band of operation, an RF filter placed right at the input of the receiver could help attenuate them. Such a filter may not be as much practical for the in-band blockers, as it would be too narrow. Moreover, it must be tunable to ensure that as the receiver tunes to different channels (or effectively connects to different base stations operating at different frequencies), the desired signal falls in the filter passband. An example of GSM blocker profile for 1900MHz band is shown in Figure 6.4.

Each channel is 200kHz wide, and the channel spacing is 200kHz. The total band is 60MHz wide, from 1930 to 1990MHz, and thus it contains a total of 300 channels. The desired signal is specified to be 3dB above the reference sensitivity, that is at −99dBm. The in-band blockers strength vary from −43dBm to −26dBm, ranging from 600kHz offset from the desired signal to 3MHz and beyond. The out-of-band blockers, however, may be as large as 0dBm. The reason that the blocker level progressively increases as the offset frequency from the desired signal goes up has to do with the way the frequency is assigned to each cell. Typically adjacent cells have frequencies that are relatively far from that of the desired. Their close vicinity, however,

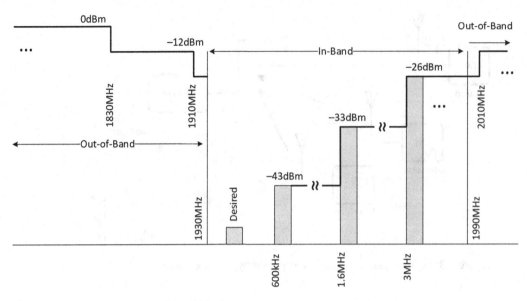

Figure 6.4: GSM in- and out-of-band blockers

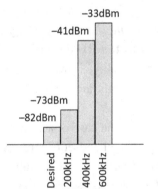

Figure 6.5: GSM adjacent blockers

results in strong blockers. The second adjacent cells that potentially result in somewhat weaker blockers have frequencies closer and so on. This helps the receiver filter design as well as phase noise requirements as we will discuss shortly.

To point out the challenge of applying a filter at the front end to attenuate the in-band blockers, consider that a bandwidth of 200kHz, centered at around 1960MHz, needs to be covered. This leads to a center frequency to bandwidth ratio of 9800. Whereas to only attenuate the out-of-band blockers, the ratio is roughly 33, much more practical. We also note that there is a 20MHz guard band on each end that blockers are still specified to be −26dBm. They do not increase to 0dBm until 80MHz away from the edges of the band.

In addition to in-band blockers that are specified from 600kHz and beyond, there are also adjacent blockers shown in Figure 6.5 that are closer in frequency but much weaker. The desired signal is specified at −82dBm, well above the sensitivity. Thus it is possible to reduce the receiver gain to some extent to enhance the blocker tolerance. While only shown on the right side of the desired signal in the figure, the blockers could be located on either side with the same frequency offset and strength.

6.2 FULL-DUPLEX SYSTEMS AND COEXISTENCE

Apart from the in- and out-of-band blockers, there is yet another blocker mechanism present in *full-duplex division* (FDD) transceivers, such as 3G or LTE (long-term evolution) radios. In contrast to *time-duplex division* (TDD) systems, in FDD radios, the receiver and the transmitter operate concurrently. Shown in Figure 6.6, the receive and transmit chains are typically isolated from one another by a *duplexer*, which could be thought of two SAW filters, each tuned to the corresponding receiver and transmit bands (see Chapter 4 for more details).

In practice, the duplexer has a finite isolation, somewhere around 45–55dB, depending on its size and cost. Thus, the output of the transmitter leaks to the receiver input and acts as a blocker. In 3G radios, for instance, the transmitter output is about +27dBm (24dBm at the antenna according to the standard, after experiencing 3dB loss for the duplexer and switch). A 50dB isolation leads to an always present −23dBm blocker. This blocker may be accompanied by other in- and out-of-band blockers discussed before, further exacerbating the issue (Figure 6.6).

In TDD systems on the other hand, only one of the TX or RX is on at a given point of the time. Thus, all that is needed is a switch to connect each to the antenna (Figure 6.7), and there is no such blocker concern. Examples of TDD systems are Bluetooth, Wi-Fi, and GSM radios.

Another similar concern arises from platforms where there are multiple radios present supporting various applications. These radios may be integrated all on the same die, or be on separate chips, but still at close vicinity to each other. An example is a smart handset consisting of several radios for Bluetooth, Wi-Fi, cellular, and other applications. Each radio comprises its

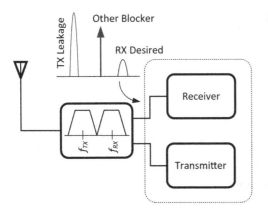

Figure 6.6: Full-duplex transceivers

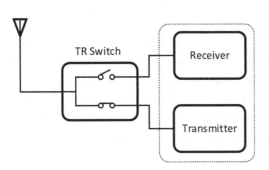

Figure 6.7: TDD radios

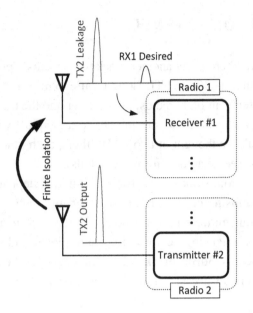

Figure 6.8: Coexistence of several radios in one platform

own antenna in general, but the antennas enjoy only a finite isolation of less than 20dB. In some applications, antennas are shared and radios are separated by a duplexer, as in Figure 6.6. Since in general all these applications may run simultaneously, we could have scenarios where one transmitter acts as a blocker to the other receiver (Figure 6.8), leading to a very similar situation as we observed in FDD transceivers.

With this background, let us now discuss the different mechanisms by which these blockers can harm the receiver.

6.3 SMALL SIGNAL NONLINEARITY

Consider the circuit shown in Figure 6.9, where we assume the output–input characteristics is not linear and is described below:

$$y = a_1 x + a_2 x^2 + a_3 x^3.$$

For the moment, let us consider only up to 3rd-order nonlinearity. Also we have ignored any DC offset associated (the term a_0), as it will not affect our discussion.

When a sinusoid input applied, that is, when $x = A \cos \omega t$, the output will be

$$y = \frac{a_2 A^2}{2} + \left(a_1 A + \frac{3 a_3 A^3}{4} \right) \cos \omega t + \frac{a_2 A^2}{2} \cos 2\omega t + \frac{a_3 A^3}{4} \cos 3\omega t.$$

As a result of the 2nd-order nonlinearity, a DC term is created, despite the fact that we assumed the DC term a_0 is equal to zero. Additionally, the output consists of all the harmonics of the fundamental frequency. This is known as *harmonic distortion*, and in certain applications, such as audio amplifiers, it may be problematic, as the low-frequency audio signal is distorted due to the presence of the unwanted harmonics. In most narrowband RF applications, however, this is not a major concern as these harmonics are far away, and subject to filtering (Figure 6.10).

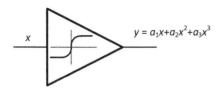

Figure 6.9: A generic nonlinear circuit

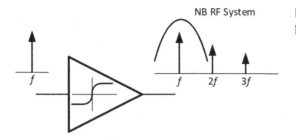

Figure 6.10: NB RF system filtering the undesired harmonics

6.3.1 Input Intercept Point

Now suppose the input consists of two sinusoids, $x = A_1 \cos(\omega_1 t) + A_2 \cos(\omega_2 t)$, where ω_1 and ω_2 are relatively close. Once experiencing the nonlinearity, the output will consist of many components; of those we consider only the ones that will fall around the desired components at ω_1 and ω_2, knowing that the rest are subject to filtering:

$$y = a_1(A_1 \cos(\omega_1 t) + A_2 \cos(\omega_2 t)) + \frac{3a_3 A_1^2 A_2}{4} \cos(2\omega_1 - \omega_2)t$$

$$+ \frac{3a_3 A_2^2 A_1}{4} \cos(2\omega_2 - \omega_1)t + \cdots.$$

If ω_1 and ω_2 are close to each other, then the two products located at $2\omega_1 - \omega_2$, and $2\omega_2 - \omega_1$, known as *intermodulation products* (IM), will also be close. Consequently, as shown in Figure 6.11, they will not experience any significant front-end filtering, and may degrade the receiver performance.

Particularly, this could be quite problematic if the either one of the intermodulation (IM) products happen to be close to the desired signal. Consider the example of a GSM system, where two in-band blockers are located at $f_1 = f_0 + N\Delta f$ and $f_2 = f_0 + 2N\Delta f$; N is the channel number; and $\Delta f = 200$kHz is the channel spacing. For the case of the 1900MHz band example we showed earlier, N could be anything from 0 to 149 ($2N$ is between 0 and 299). If the desired signal happens to be at f_0, the IM$_3$ product[1] will fall on the desired signal (Figure 6.12).

Same as noise and noise figure, we would like to attach a universal figure of merit describing the 3rd-order intermodulation performance of a radio. For that, we define the 3rd-order *input intercept point* (IIP$_3$) as follows: Suppose we apply two tones with equal amplitude of A. At the output, we have fundamentals with amplitude of roughly $a_1 A$, and IM$_3$ components with an amplitude proportional to $|a_3 A^3|$. The value of the input amplitude A for which the two curves intercept is known as IIP$_3$ (Figure 6.13).

---

[1] The subscript 3 emphasizes the fact that these products are created due to *3rd-order nonlinearity*.

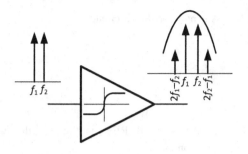

Figure 6.11: Third-order nonlinearity when two tones are applied

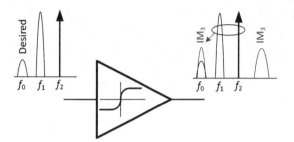

Figure 6.12: Intermodulation in a GSM receiver

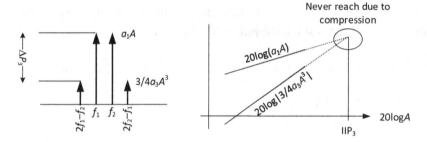

Figure 6.13: Illustration of input intercept point

The fundamentals increase proportional to A (or a slope of 20dB/Dec), whereas the IM products increase proportional to A^3 (or a slope of 60dB/Dec). Therefore, they are expected to intercept eventually, although in practice this never happens. This has to do with the fact that almost all practical nonlinear systems are *compressive*, that is, the gain (or the slope) eventually reaches zero for very large inputs. This in turn is due to the fact that large inputs could turn off some transistors or that eventually the output is limited by the supply. In other words, the characteristics polynomial has other terms beyond a_3 that we have ignored. Nevertheless, the two curves can be extrapolated, and at the interception point we can write

$$|a_1 A| = \left|\frac{3 a_3 A^3}{4}\right|,$$

which leads to

$$A_{\text{IIP3}} = \sqrt{\frac{4}{3} \frac{|a_1|}{|a_3|}}.$$

A perfectly linear system thus has an IIP$_3$ of ∞.

6.3 Small Signal Nonlinearity

To measure the IIP_3 properly, we must apply large enough inputs that the IM products are well above the radio or the measurement equipment noise floor. However, they must be small enough to avoid the compression. A practical sanity check is to ensure that a well-behaved slope of $1\times$ and $3\times$ is observed for the desired and IM components.

Let us denote the *difference* between the fundamental and IM components in dBm (or dBV) by ΔP_3 (Figure 6.13). We have

$$\Delta P_3 = 20 \log \left| \frac{a_1 A}{\frac{3 a_3 A^3}{4}} \right| = 2 IIP_3 (\text{in dBm}) - 40 \log A = 2 IIP_3 - 2P,$$

where $P = 20 \log(A)$ is the input in dBV.[2] Rearranging the equation above yields

$$IIP_3 = P + \frac{\Delta P_3}{2}.$$

This leads to a more convenient way of measuring IIP_3, as long as we choose the input such that a slope of 3:1 is guaranteed.

Now suppose an otherwise ideal 3rd-order nonlinear receiver is subject to a 2-tone blocker with an amplitude of P_B in dBm. This results in an undesirable IM component whose level in dBm is

$$IM_3 (\text{in dB}) = P_B - \Delta P_3 = 3 P_B - 2 IIP_3.$$

If the intermod product falls on the desired signal, it must be at least an SNR below to detect the signal properly. Note that the SNR was earlier defined as the detector response to the noise. However, since the IM product is resulting from a randomly modulated blocker, it is safe to assume that the modem requires the same SNR. Under such assumption, the minimum input that satisfies this equation defines the receiver sensitivity. Therefore,

$$\text{Sensitivity} = 3 P_B - 2 IIP_3 + SNR.$$

The receiver is free of any noise, and the limit on the sensitivity is solely due to the response of the nonlinear receiver to large blockers.

Example: Consider the GSM receiver. The standard specifies a 2-tone blocker at –49dBm located at 800kHz and 1.6MHz away from the desired input at –99dBm. Suppose we would like to achieve an SNR of 10dB. Even though 5dB is typically sufficient for most modems, we would like to leave room for other nonidealities and particularly the noise. Then the IIP_3 may be calculated as

$$-99 = 3 \times -49 - 2 IIP_3 + 10,$$

Continued

---

[2] The equation is equally valid if everything is expressed in dBm also, where dBm = 10 + dBV, with A representing the *peak* voltage.

leading to an IIP$_3$ of −19dBm for the receiver. Every dB increase in the blocker level results in 3dB degradation in the sensitivity. For the same IIP$_3$, if the blockers are now −47dBm, the SNR degrades to only 4dB, and the receiver fails.

Example: We shall calculate the IIP$_3$ of a single FET. If we use the long channel I-V characteristic that is square-law, the IIP$_3$ will be infinite as there is no 3rd-order nonlinearity. To obtain a more meaningful result, let us assume a more realistic model that incorporates both velocity saturation and mobility degradation due to the gate vertical field [3]

$$I_D = \frac{1}{2}\mu_0 C_{OX} \frac{W}{L} \frac{(V_{GS} - V_{TH})^2}{1 + \left(\frac{\mu_0}{2v_{sat}L} + \theta\right)(V_{GS} - V_{TH})},$$

where v_{sat} is the saturated velocity, and μ_0, C_{OX}, W, and L are the device parameters. The parameter θ captures the impact of the vertical field imposed by the gate. Let us define

$$\theta' = \frac{\mu_0}{2v_{sat}L} + \theta.$$

If $\theta' = 0$, the device will be square-law. In practice however, the velocity saturation (the first term), as well as the vertical field mobility degradation factor (the second term) result in third and higher order nonlinearity terms.

For a given nonlinear function $y = f(x)$, Taylor series [4] may be employed to find the coefficients of the nonlinear function expanded around $x = 0$:

$$a_1 = \left.\frac{\partial y}{\partial x}\right|_{x=0}, \quad a_2 = \left.\frac{1}{2}\frac{\partial^2 y}{\partial x^2}\right|_{x=0}, \quad a_3 = \left.\frac{1}{6}\frac{\partial^3 y}{\partial x^3}\right|_{x=0}, \ldots$$

Performing these derivatives on the I_D–V_{GS} function, and considering that V_{GS} consists of a fixed DC bias as well as an AC small signal component, we can obtain a_1 and a_3. Accordingly, the IIP$_3$ (in volts) is

$$A_{\text{IIP3}} = (1 + \theta' V_{\text{eff}})\sqrt{\frac{4}{3}\left(V_{\text{eff}}^2 + \frac{2V_{\text{eff}}}{\theta'}\right)},$$

where $V_{\text{eff}} = V_{GS} - V_{TH}$ is the gate overdrive voltage. This derivation only considers I_D–V_{GS} nonlinearity and the impact of channel length modulation and other 2nd-order effects are ignored. Clearly the IIP$_3$ monotonically improves as the overdrive voltage increases. On the other hand the impact of θ' is not as obvious. If it is very small, then we have

$$A_{\text{IIP3}} \approx \sqrt{\frac{8}{3}\frac{V_{\text{eff}}}{\theta'}}.$$

However, for most modern CMOS processes θ' is not negligible even for longer channel devices and the above approximation may not hold.

6.3 Small Signal Nonlinearity

Example: For a BJT, the I_C–V_{BE} characteristics are exponential, and upon expanding, the IIP₃ is readily found to be constant and equal to

$$A_{\text{IIP3}} = 2\sqrt{2}V_T = \frac{2\sqrt{2}KT}{q},$$

which is about 70mV at room temperature.

It is well known that MOSFETs have exponential I–V characteristics in weak inversion, similar to BJTs as shown below,

$$I_D \approx \mu_n C_D \frac{W}{L}\left(\frac{KT}{q}\right)^2 \left(\frac{n_i}{N_A}\right)^2 e^{\frac{V_{GS}-V_{GS}^*}{nKT/q}},$$

where C_D is the depletion capacitance and $n = 1 + \frac{C_D}{C_{OX}}$ is roughly equal to 2 for typical processes, n_i is the Si intrinsic electron/hole density, and N_A is the substrate doping. Thus the IIP₃ is constant and equal to $2\sqrt{2}nV_T$.

Shown in Figure 6.14 is the simulated IIP₃ of a single FET for two different channel lengths, corresponding to two different values of θ'. The drain voltage is kept constant at V_{DD}, and the output current is monitored, thus eliminating the potential impact of r_o nonlinearity. Also shown is the calculated IIP₃ in dashed line for each case. For $V_{GS} < V_{TH}$, the device is in weak inversion and the IIP₃ is constant at about 104mV. At the onset of device turning on, the IIP₃ peaks. This is attributed to a discontinuity for the second derivative of the drain current. Although the phenomenon is real, it may not be much use in practice as it happens in a very narrow region, and is likely to vary over process or temperature variation. Ignoring this region (which is clearly not captured by the simple model), we observe a good agreement between our hand calculations and the simulated IIP₃. It is interesting that for modest to large values of overdrive voltage, a shorter channel length, corresponding to a larger value of θ', is more linear.

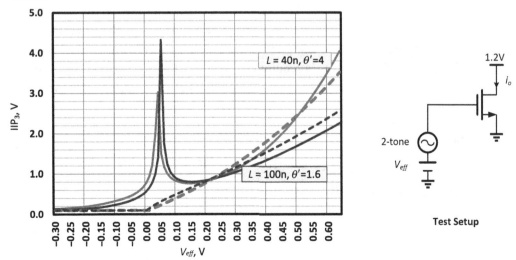

Figure 6.14: Simulated IIP₃ of a FET

Distortion

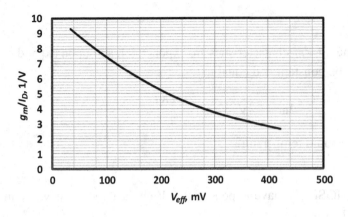

Figure 6.15: g_m/I_D characteristics of a 40nm NMOS

To achieve a reasonable linearity, an overdrive of 100–200mV is typical for most designs. A higher overdrive leads to a better IIP_3 of course, but the device g_m and noise suffer for a given bias current. To understand the trade-off, we show in Figure 6.15 g_m/I_D characteristics of a 40nm NMOS transistor versus the overdrive voltage. For an overdrive voltage of 100mV, g_m/I_D is about $7V^{-1}$. Note that g_m/I_D is expected to flatten at roughly $1/nV_T = 20V^{-1}$ (that is, $n = 2$) in weak inversion.

Example: Let us obtain the IIP_3 of a 3G receiver. We consider the in-band blockers first. According to the standard QPSK modulated blockers at 10 and 20MHz away, each with an amplitude of −46dBm accompany the desired signal 3dB above the reference sensitivity, as shown in Figure 6.16.

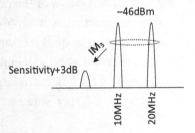

Figure 6.16: 3G in-band IIP_3

Assume we wish the resultant IM_3 to be 10dB suppressed with respect to the signal less SNR to provide sufficient margin. The signal is at −117dBm + 3dB for band I for instance,[3] and thus the IM_3 is required to be at $-117 + 3 - (-18) - 10 = -106$dBm.

If the two blockers have an equal magnitude of P_B, we calculated before:

$$IM_3 = 3P_B - 2IIP_3.$$

Thus

$$-106 = 3 \times -46 - 2IIP_3,$$

leading to an in-band IIP_3 of −16dBm.

---

[3] The reference sensitivity and some other requirements may vary by a few dB for different bands in both 3G and LTE cases.

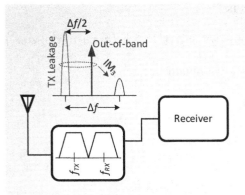

Figure 6.17: 3G out-of-band IIP$_3$

The second scenario is the case where the TX leakage mixes with an out-of-band blocker at exactly half the frequency between the RX and TX signals as shown in Figure 6.17. This results in an IM$_3$ component falling on the desired RX signal.

Same as the in-band scenario, the signal is 3dB above the reference sensitivity, and we assume the same 10dB margin. Hence, we wish to have the IM$_3$ to be no more than −106dBm.

For this case, the two blockers are not necessarily equal in amplitude, and the IM$_3$ equation is modified to

$$\text{IM}_3 = P_{B1} + 2P_{B2} - 2\text{IIP}_3,$$

where P_{B2} corresponds to the blocker that is closest to the desired signal, in this case that would be the out-of-band blocker. Thus, P_{B1} will represent the TX leakage. The out-of-band blockers are specified to be at −15dBm for 3G. Assuming a duplexer isolation of 50dB, and a duplexer filtering of 30dB on the blocker, we have

$$-106\text{dBm} = (27\text{dBm} - 50) + 2 \times (-15\text{dBm} - 30) - 2\text{IIP}_3.$$

The required out-of-band IIP$_3$ will be −3.5dBm.

6.3.2 IIP$_3$ of Cascade of Stages

Same as noise figure, we would like to express the receiver IIP$_3$ in terms of that of its sub-blocks. Consider the cascade of two nonlinear stages, each with input–output characteristics shown in Figure 6.18.

The small signal gain of each stage is a_1 and b_1, and the input IIP$_3$ is $A_{\text{IIP3},1} = \sqrt{\frac{4}{3}\left|\frac{a_1}{a_3}\right|}$, and $A_{\text{IIP3},2} = \sqrt{\frac{4}{3}\left|\frac{b_1}{b_3}\right|}$.

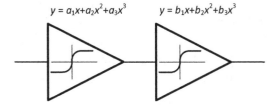

Figure 6.18: Cascade of two nonlinear stages

The cascade input–output characteristics is

$$y = b_1(a_1x + a_2x^2 + a_3x^3) + b_2(a_1x + a_2x^2 + a_3x^3)^2 + b_3(a_1x + a_2x^2 + a_3x^3)^3.$$

After performing a set of tedious derivatives, we obtain

$$y'(0) = a_1 b_1$$

$$\frac{y''(0)}{2} = a_2 b_1 + a_1^2 b_2$$

$$\frac{y'''(0)}{6} = a_3 b_1 + b_3 a_1^3 + 2 a_1 a_2 b_2.$$

If we ignore the last term involved in $y'''(0)$ expression, assuming that the system is *dominantly 3rd-order*, we obtain the cascade IIP$_3$ as follows:

$$A_{\text{IIP3},T} = \sqrt{\frac{4}{3} \left| \frac{a_1 b_1}{a_3 b_1 + b_3 a_1^3} \right|}.$$

Once rearranged, we obtain the well-known expression

$$\frac{1}{A_{\text{IIP3},T}^2} = \frac{1}{A_{\text{IIP3},1}^2} + \frac{a_1^2}{A_{\text{IIP3},2}^2},$$

indicating that the IIP$_3$ of the second stage is dominant if the first stage has a large gain, opposite of the noise figure equation. This points out that the gain of the first stage, often the LNA, has to be carefully chosen to compromise the optimum noise figure and optimum IIP$_3$ of the overall system.

In addition to ignoring the third term in $y'''(0)$ equation, the above formula for cascaded IIP$_3$ is correct only if both stages are compressive, or both are expansive. This is due to the fact that to derive the equation, we have assumed same signs for a_1/a_3 and b_1/b_3 terms. If they had opposite signs, it could have led to cancellations of nonlinearity. This outcome is generally exploited in the context of pre-distortion linearization, as we will discuss in Chapter 11.

6.3.3 Second-Order Distortion

Similar to the 3rd-order nonlinearity, other order nonlinearities are expected to harm the receiver. Of special interest is the 2nd-order nonlinearity, resulted from the term $a_2 x^2$ is the general input–output function described before. If the input consists of a 2-tone sinusoid signal, that is, if $x = A_1 \cos \omega_1 t + A_2 \cos \omega_2 t$, considering only the 2nd-order nonlinearity, we have

$$y = a_1(A_1 \cos \omega_1 t + A_2 \cos \omega_2 t) + a_2 A_1 A_2 \cos(\omega_1 - \omega_2)t + \frac{a_2 A_1^2}{2} + \frac{a_2 A_2^2}{2} + \cdots,$$

where the terms subject to filtering (e.g., at $\omega_1 + \omega_2$ or $2\omega_1$) are not shown. The term $a_2 A_1 A_2 \cos(\omega_1 - \omega_2)t$ may be problematic if the desired signal is eventually downconverted

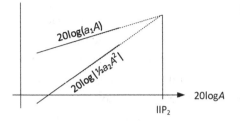

Figure 6.19: IIP$_2$ definition

to or close to DC. Similar to IIP$_3$, we can plot the two components (Figure 6.19), the desired with a slope of ×1, and the 2nd-order term with a slope of ×2, where $A_1 = A_2 = A$.

The intercept point will be the IIP$_2$ (in volt):

$$\text{IIP}_2 = \left|\frac{a_1}{a_2}\right|.$$

Furthermore, similar to IIP$_3$, we can find out the IIP$_2$ of the cascade of several stages. Since we had obtained the second derivative already, we arrive at a similar expression for the cascade of two stages IIP$_2$:

$$\frac{1}{A_{\text{IIP2},T}} = \frac{1}{A_{\text{IIP2},1}} + \frac{a_1}{A_{\text{IIP2},2}}.$$

As expected, if the gain of the first stage is large, the second stage 2nd-order nonlinearity dominates.

Example: We shall find the IIP$_2$ of an LTE (or 4G) receiver. Since in most 4G receivers the desired signal is usually downconverted to DC, of several different mechanisms, the most problematic one will be due to the amplitude demodulation of the transmitter leakage as shown in Figure 6.20.

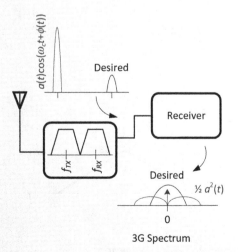

Figure 6.20: IIP$_2$ of a 4G receiver

Continued

Transmitter leakage is amplitude modulated and thus when *squared* results in a spectrum that is roughly twice as wide as the receiver signal. Once passed through the receiver filter, the integrated energy will be somewhat reduced, depending on the actual modulation properties of the 4G signal. We denote this attenuation factor generally as α.

It is easy to show that for the second-order nonlinearity we have

$$IM_2 = 2P_B - IIP_2.$$

Since the desired signal is at sensitivity, we would like for the IM_2 component to be well below the desired signal less SNR, say 10dB or more. Thus

$$IM_2 = -117 - (-18) - 10 + \alpha = -109\text{dBm} + \alpha.$$

The blocker level (P_B) is the TX leakage, which came out to be -23dBm for 50dB duplexer isolation. Thus

$$IIP_2 = 2 \times -23 + 109 - \alpha = 66 - \alpha.$$

It turns out that for 3G where is the signal is QPSK modulated, the factor α is about 13dB. In the case of LTE that the signal is OFDM (orthogonal frequency division multiplexing), it is about 7dB. The corresponding required IIP_2 at the TX frequency then will be 53/59dBm for 3G/LTE cases.

One may wonder what the exact nature of the correction factor α is. As we pointed out, it largely depends on the statistical characteristics of the modulated spectrum appearing at the receiver input.

As a thought experiment, let us consider Figure 6.21, where a band-limited white noise is applied to a system with 2nd-order nonlinearity. We shall find the spectral density of the output $y(t) = x(t)^2$.

Shown in Figure 6.21, the spectrum of the band limited white noise may be broken into infinitesimal bins of width Δf. In the limit case, each bin may be approximated by an impulse in the frequency domain. Assuming a bandwidth of B for input noise, there are a total of N bins, where $N \approx \frac{B}{\Delta f}$. The spectral density of the input then may be expressed as

$$S_x(f) = \frac{\eta}{2} \sum_{n=0}^{N} n\Delta f.$$

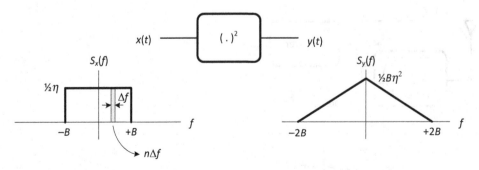

Figure 6.21: Band-limited white noise passing through 2nd-order nonlinearity

In Chapter 2 we showed that a randomly phase sinusoid,

$$v(t) = A\cos(\omega_0 t + \phi),$$

has the spectral density of

$$S_v(f) = \frac{A^2}{4}\delta(f - f_0) + \frac{A^2}{4}\delta(f + f_0),$$

where ϕ is a random variable uniformly distributed between 0 and 2π. Consequently, the band-limited noise in time domain may be expressed as

$$x(t) = A\sum_{n=0}^{N}\cos(n2\pi\Delta f t + \phi_n),$$

where all ϕ_n are independent. If N is large (or Δf is small),

$$\frac{\eta}{2}\Delta f = \frac{A^2}{4},$$

or $A = \sqrt{2\eta\Delta f}$. The output signal, $y(t) = x(t)^2$, may be constructed accordingly:

$$y(t) = \frac{A^2}{2}\sum_{i=0}^{N}\sum_{j=0}^{N}\left[\cos((i+j)2\pi\Delta f t + \phi_i + \phi_j) + \cos((i-j)2\pi\Delta f t + \phi_i - \phi_j)\right].$$

Since ϕ_i and ϕ_j ($i \neq j$) are independent, we can show that $\phi_i + \phi_j$ and $\phi_i - \phi_j$ are also independent. For the limit case $\Delta f \to 0$, the output spectral density may be found to be (see Problem 16)

$$S_y(f) = \left(\frac{\eta}{2}\right)^2 2B\left(1 - \frac{|f|}{2B}\right),$$

for $|f| \leq 2B$, and zero otherwise (Figure 6.21).

The noise energy of the input within the bandwidth of $\pm B$ is clearly ηB. For the same band, the energy of the squared signal is $\frac{3}{4}(\eta B)^2$. Assuming $\eta = \frac{1}{B}$ such that the input energy is normalized to one, the output energy will be three-fourths, or 1.3dB less.

To arrive at this conclusion, we assumed the phase components of each bin, ϕ_is, are completely uncorrelated. For a general modulated signal, however, this is not necessarily the case. Consequently, there will be additional components resulted from cross-correlation of the dependent terms that alter the final outcome. It turns out that for a 3G signal, for example, the squared output once integrated has about 13dB less energy. If the signal would have been spread uniformly across twice the bandwidth, we would have expected a factor of 2 loss. For white noise the output is actually a triangle, and hence α is 1.3dB. For 3G, on the other hand, the squared signal has little energy around DC (Figure 6.22), which results in a bigger reduction of the output energy.

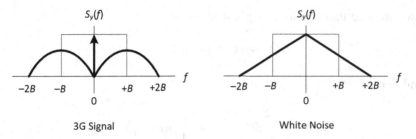

Figure 6.22: A comparison between a 3G signal and white noise when squared

Example: Consider a Wi-Fi signal at sensitivity of 2472MHz accompanied by a band 40 LTE blocker located at 2510MHz. The LTE signal is 20MHz wide, and it is assumed that, once squared, it will have a uniformly distributed power spectral density as shown in Figure 6.23. This can be verified by system simulations. We assume the signals are ultimately downconverted to zero.

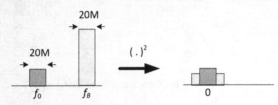

Figure 6.23: A WLAN signal accompanied by a 20MHz LTE blocker

Since both the LTE and WLAN signals are 20MHz wide, after the second-order nonlinearity, the strength of the IM_2 component once filtered over the 20M bandwidth is reduced by 3dB.

Assuming a NF of 3dB for the system with no blocker, and an SNR of 22dB for 64QAM, the Wi-Fi sensitivity is found to be

$$-174 + NF + 10\log(20M) + SNR = -76 \text{dBm}.$$

Assuming the LTE blocker level is −20dBm, to induce 3dB desensitization due to the 2nd-order nonlinearity, the IM_2 level must be equal to the thermal noise floor, and is found to be

$$IM_2 = -174 + NF + 10\log(20M) = -98 \text{dBm}.$$

The corresponding system IIP_2 is

$$IIP_2 = 2P_B - IM_2 - 3 = 55 \text{dBm}.$$

The 3dB is subtracted to take into account the spreading of the IM_2 component, as shown in Figure 6.23.

For every dB increase to the blocker level, the system IIP_2 must improve by 2dB to allow the same level of desensitization. A −15dBm blocker demands an IIP_2 of 65dBm, for instance.

As an interesting case study, consider the cascade of two amplifiers with second-order nonlinearity as shown in Figure 6.24. We will show that even though the amplifiers do not

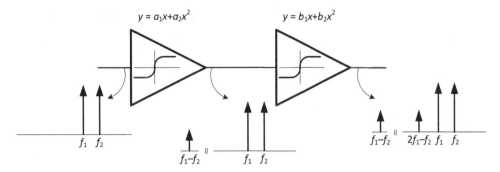

Figure 6.24: Cascade two amplifiers with second-order nonlinearity

have any 3rd-order nonlinearity of their own, the IIP$_3$ of the cascaded block is finite. Intuitively this can be explained examining what happen to the system when two interferers at the frequencies of f_1 and f_2 are applied to the input as graphically illustrated in the figure. Due to the 2nd-order nonlinearity of the first stage, an undesired tone at $f_1 - f_2$ appears at its output. This tone along with the original two-tone blocker at f_1 and f_2 once passed through the second stage create a tone at $2f_1 - f_2$, which resembles the one created by the 3rd-order nonlinearity (Figure 6.11).

Mathematically speaking, we can express the second stage output as

$$y = b_1(a_1x + a_2x^2) + b_2(a_1x + a_2x^2)^2,$$

where x is the first stage input. After expanding and regrouping:

$$y = a_1b_1x + (a_2b_1 + b_2a_1^2)x^2 + 2a_1a_2b_2x^3 + b_2a_2^2x^4.$$

The first term is of course the desired one, showing a linear gain of a_1b_1. The second term leads to the cascaded IIP$_2$ as was discussed at the beginning of the section. However, the third term shows that the overall system exhibits 3rd-order nonlinearity, with the effective IIP$_3$ of

$$A_{\text{IIP3},T} = \sqrt{\frac{4}{3}\left|\frac{a_1b_1}{2a_1a_2b_2}\right|} = \sqrt{\frac{2}{3}\left|\frac{b_1}{b_2}\right|\left|\frac{1}{a_2}\right|},$$

or in dB,

$$\text{IIP}_{3,T} = -3.5 - 10\log|a_2| + \frac{1}{2}\text{IIP}_{2,2},$$

where IIP$_{2,2}$ is the IIP$_2$ of the second stage in dB.

There are two ways that this could be problematic in a system: First is if the amplifiers are wideband enough such that the undesired signal at $f_1 - f_2$ experiences little attenuation. That is typically not the case. The second and more dominant cause would be if the low-frequency IM$_2$ signal leaks to the second amplifier through supply or ground or other paths alike. This can happen even within an amplifier. For instance, consider the cascode amplifier shown in Figure 6.25.

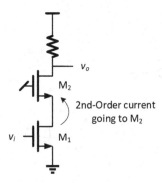

Figure 6.25: A cascode amplifier showing 3rd-order nonlinearity

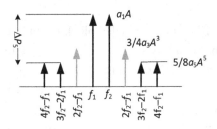

Figure 6.26: A two-tone blocker leading to undesired sidebands due to the 5th-order nonlinearity

The input transistor M_1 experiences a relatively strong 2nd-order nonlinearity as is the case for MOS devices. The IM_2 current produced sees little filtering once going to the cascode transistor M_2. If this transistor creates a strong 2nd-order nonlinearity, overall could lead to worse than expected 3rd-order nonlinearity for the cascode amplifier. This however in practice may be as problematic, as typically the dominant source of the 3rd-order nonlinearity is the input transistor itself, which could become an issue in certain cases.

6.3.4 Fifth-Order Intercept Point

Although not as commonly discussed as 2nd- or 3rd-order distortion, 5th-order distortion could be problematic in certain cases. Shown in Figure 6.26, if there is a two-tone blocker at the frequencies of f_1, and f_2, there will be additional sidebands due to the 5th-order terms. Also shown in gray are the close-by sidebands due to the 3rd-order nonlinearity.

The system is generally characterized as

$$y = a_1 x + a_2 x^2 + a_3 x^3 + a_4 x^4 + a_5 x^5,$$

where higher order nonlinear terms up to fifth have been included. With

$$x = A_1 \cos \omega_1 t + A_2 \cos \omega_2 t,$$

among many terms, the most notable ones due to the 5th-order nonlinearity are

$$y = \cdots + \frac{5 a_5 A_1^3 A_2^2}{8} \cos(3\omega_1 - 2\omega_2)t + \frac{5 a_5 A_1^4 A_2}{8} \cos(4\omega_1 - \omega_2)t$$
$$+ \frac{5 a_5 A_1^2 A_2^3}{8} \cos(3\omega_2 - 2\omega_1)t + \frac{5 a_5 A_1 A_2^4}{8} \cos(4\omega_2 - \omega_1)t + \cdots,$$

6.3 Small Signal Nonlinearity

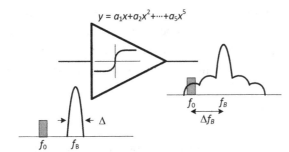

Figure 6.27: A close-by blocker creating interference to the desired signal due to the 5th-order nonlinearity

while the desired tones are $a_1(A_1\cos(\omega_1 t) + A_2\cos(\omega_2 t))$. For $A_1 = A_2 = A$, the 5th-order input intercept point is

$$A_{\text{IIP5}} = \sqrt[4]{\frac{8}{5}\left|\frac{a_1}{a_5}\right|}.$$

Similar to IIP_3, a short-cut method may be applied, in which case

$$\text{IIP}_5 = P + \frac{\Delta P_5}{4},$$

where $P = 20\log(A)$ is the input in dBV, and IIP_5 is now in dBV. ΔP_5 is the distance between the desired and 5th-order terms in dB.

Since often $|a_5| \ll |a_3|$, $A_{\text{IIP3}} \gg A_{\text{IIP5}}$, and thus IIP_5 not as important. There are exceptions to this, however. Consider a modulated blocker sitting close to the desired signal as depicted in Figure 6.27.

Once passed through the nonlinear amplifier, the blocker will experience spectral regrowth, as shown in Figure 6.27. The root cause of spectral regrowth will be examined more closely during our TX discussion later in this chapter (Figure 6.48), but simply put, if a modulated signal has a bandwidth of say Δ, the 3rd-order nonlinearity leads to a bandwidth expansion of 3Δ, or the 5th-order nonlinearity will create 5Δ expansion. From this point of view, the 5th-order nonlinearity may be more troublesome. While it could be less intense, the wider expansion (up to $5\times$) will create problems that would have not existed otherwise due to the 3rd-order nonlinearity alone depending on the blocker spacing.

Example: Suppose a Bluetooth signal located at the edge of the ISM band (2.484GHz) is accompanied by a Band 41 10MHz LTE blocker. Band 41 ranges from 2496MHz to 2690MHz, so the closest the blocker may be is at a center frequency of 2501MHz. After spectral expansion due to the 3rd-order nonlinearity alone, the blocker edge will be at 2486MHz ($2501M - 1.5 \times 10M$), not affecting the desired BT signal. On the other hand, the 5th-order nonlinearity will cause the blocker to leak down to 2476MHz, overlapping the entire desired signal. For either case of 3rd- or 5th-order nonlinearity however, a 20MHz LTE blocker (sitting at 2506MHz) will be problematic. In this case, one can argue the 3rd-order nonlinearity is perhaps more of an issue given that $|a_3| \gg |a_5|$.

In addition to wider spectral regrowths, the 5th-order nonlinearity could be still problematic, considering that it grows much more rapidly (with a slope of 5×) as the blocker grows. We showed earlier that for a blocker power of P_B (dBm), the undesired 3rd-order intermod signal is $\text{IM}_3 = 3P_B - 2\text{IIP}_3$. One can show that due to the 5th-order nonlinearity, the 5th-order intermod strength is

$$\text{IM}_5 = 5P_B - 4\text{IIP}_5.$$

Even if IIP_5 is larger than IIP_3 given a smaller 5th-order coefficient, the undesired 5th-order intermod may be still significant, which ultimately is what matters. For a given blocker level, for the 3rd- and 5th-order IM components to be comparable, that is, to have comparable contributions from the 3rd- and 5th-order distortion, then

$$\text{IIP}_5 = \frac{P_B + \text{IIP}_3}{2}.$$

Accordingly, the stronger the blocker is, the higher IIP_5 is demanded.

6.3.5 Cross-Modulation

In the previous section it was shown that a finite IIP_3 may affect a nonlinear system if there happens to be a two-tone blocker with the frequency spacing such that the intermodulation term falls on the desired signal. One may argue, however, that the likelihood of having two strong blockers with a certain specific frequency separation may be low. There is another mechanism that could affect a 3rd-order nonlinear system even if there is only a single blocker present.

Consider Figure 6.28, where a weak desired signal is accompanied by a *single modulated blocker* at an offset frequency of Δf_B away from the signal. The amplifier input may be expressed as

$$x = A_0 \cos(\omega_0 t) + a_B(t) \cos(\omega_B t + \phi_B(t)),$$

where $a_B(t)$ and $\phi_B(t)$ denote the amplitude and frequency modulation of the blocker. For simplicity, the desired signal is represented as a tone, though it will not affect the general conclusions drawn later.

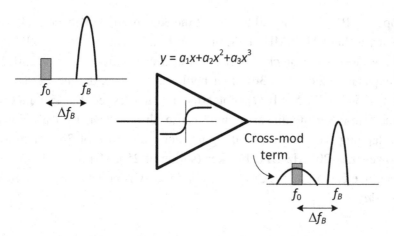

Figure 6.28: A weak desired signal accompanied by a large modulated blocker may be corrupted due to cross-modulation

6.3 Small Signal Nonlinearity

The term due to the 3rd-order nonlinearity, $a_3(A\cos(\omega_0 t) + a_B(t)\cos(\omega_B t + \phi_B(t)))^3$, after expansion, leads to the following unwanted signal that falls right on the desired signal:

$$\frac{3}{2} a_3 A_0 a_B(t)^2 \cos(\omega_0 t).$$

This undesired signal, known as the *cross-modulation* term, is independent of the blocker frequency, and has a bandwidth set by $a_B(t)^2$, likely about two times the blocker bandwidth. The term due to the desired signal is

$$a_1 A_0 \cos(\omega_0 t).$$

Thus, treating the cross-modulation like noise, it sets the signal-to-noise ratio at the output, which is independent of the desired signal strength,

$$\text{SNR} = \frac{\frac{2}{3}\left|\frac{a_1}{a_3}\right|}{\overline{a_B(t)^2}} = \frac{A_{\text{IIP3}}^2}{2\overline{a_B(t)^2}},$$

where $\overline{a_B(t)^2}$ denotes the blocker amplitude squared mean value. If the blocker is constant envelop, then cross-modulation leads to a DC term that may not be as important in many applications. Interestingly, the cross-modulation term raises with the blocker level squared (not cubed), despite the fact that it is created due to the 3rd-order nonlinearity, again similar to the 2nd-order nonlinearity.

Given that the cross-modulation itself is dependent on the blocker modulation, it is common to quantify its impact through applying a nearby two-tone blocker as depicted in Figure 6.29. The exact frequency of the blockers (f_{B1} and f_{B2}) is not important, and matters only if there is any prior filtering. Their spacing (Δf in the figure) must be such that the cross-modulation components that are $\pm \Delta f$ away from the desired signal lie in band. That is usually the case for any arbitrary modulated blocker.

The input to the nonlinear amplifier may be expressed as

$$x = A_0 \cos(\omega_0 t) + A_{B1} \cos(\omega_{B1} t) + A_{B2} \cos(\omega_{B2} t),$$

leading to the following nearby components at the output:

$$\frac{3}{2} a_3 A_0 A_{B1} A_{B2} \cos((\omega_0 \pm \Delta\omega)t).$$

With $A_{B1} = A_{B2} = A_B$, the amount of blocker it takes to make the cross-modulation component (XIM$_3$) equal amplitude to the desired signal ($a_1 A_0 \cos(\omega_0 t)$) is

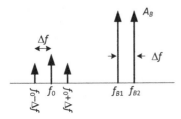

Figure 6.29: Cross-modulation due to a nearby two-tone blocker

Distortion

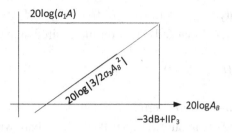

Figure 6.30: Cross-modulation component as a function of the blocker power

$$A_B = \sqrt{\frac{2}{3}\left|\frac{a_1}{a_3}\right|} = \frac{1}{\sqrt{2}}A_{\text{IIP3}}.$$

So the cross-modulation induced IIP$_3$ is 3dB worse than the standard two-tone IIP$_3$. This is illustrated in Figure 6.30.

Example: A WLAN signal at 2.412GHz is accompanied by a −20dBm 10MHz LTE interferer close by. Due to cross-modulation, the blocker creates an undesired cross-modulation signal with about 20MHz bandwidth falling on the desired WLAN signal, which is also 20MHz wide. Now the signal-to-noise ratio in dB is

$$\text{SNR(dB)} = -6 + 2\text{IIP}_3 - 2P_B,$$

where P_B is the power of the blocker squared signal. If for simplicity we assume this power remains to be the same as the original blocker power of −20dBm, then the SNR is calculated to be 34dB for a system IIP$_3$ of 0dBm. This level is signal independent, but is well above what is required to successfully demodulate an 802.11g signal (about 22dB).

6.3.6 Impact of Feedback on Linearity

We showed that if a noiseless feedback is incorporated, the amplifier minimum noise figure is not going to be affected, although the input impedance changes according to the type of feedback used. We wish to find the impact of the feedback on IIP$_2$ and IIP$_3$ here. From basic analog design, we expect the *feedback to improve linearity*, and we shall prove that quantitatively.

Consider the general nonlinear amplifier with an ideal feedback around it as shown in Figure 6.31. We have

$$y = a_1(x - \beta y) + a_2(x - \beta y)^2 + a_3(x - \beta y)^3.$$

We will not attempt to find the closed form solution for y as a function of x; however, we can exploit the Taylor series expansion to find the coefficients.

Upon taking derivatives up to the 3rd order, we will have

$$y = \frac{a_1}{1 + a_1\beta}x + \frac{a_2}{(1 + a_1\beta)^2}x^2 + \frac{a_3(1 + a_1\beta) - 2\beta a_2^2}{(1 + a_1\beta)^5}x^3.$$

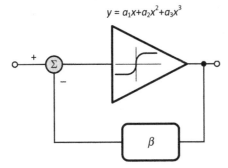

Figure 6.31: Nonlinear amplifier in a feedback loop

The linear gain is reduced by the loop gain: $1+a_1\beta$ as expected. More importantly the 2nd- and 3rd-order coefficients are decreased more steeply, leading to a net improvement for both IIP_2 and IIP_3. We can thus write

$$\text{IIP}_{2,fb} = (1+a_1\beta) \times \text{IIP}_2,$$

and if the system is dominantly 3rd-order (that is, ignoring $2\beta a_2^2$ term),

$$\text{IIP}_{3,fb} \approx (1+a_1\beta)^{\frac{3}{2}} \times \text{IIP}_3.$$

It is interesting to point out that if the system is only 2nd-order, the presence of 2nd-order nonlinearity in the feedforward path still leads to a finite IIP_3 (see also Problem 4).

Example: We showed before that the IIP_3 of a BJT is constant and equal to $2\sqrt{2}V_T$. If degenerated, however (Figure 6.32), we can treat the degeneration as a local series feedback, where the loop gain is $1+g_m R = 1 + \frac{RI_C}{V_T}$.

Thus the IIP_3 of a degenerated BJT is

$$\text{IIP}_3 = 2\sqrt{2}V_T \left(1 + \frac{RI_C}{V_T}\right)^{\frac{3}{2}},$$

which is not constant anymore, and can be improved by raising the bias current, or in general the loop gain.

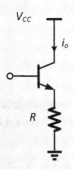

Figure 6.32: A degenerated BJT

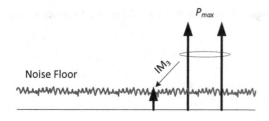

Figure 6.33: Spurious free dynamic range illustration

6.3.7 Dynamic Range

With the lower end of a receiver acceptable input set by sensitivity, and the upper end determined by linearity, we can specify the dynamic range. A widely accepted figure of merit is *spurious free dynamic range*, defined as follows:

The maximum 2-tone input that a receiver may take is such that the corresponding IM_3 falls at the receiver noise floor as shown in Figure 6.33.

The receiver noise floor is $-174 + NF + 10\log BW$, and the IM_3 as a result of a 2-tone blocker with a power of P_{max} is $IM_3 = 3P_{max} - 2IIP_3$. This results in

$$P_{max} = \frac{2IIP_3 + \textit{Noise floor}}{3} = \frac{2IIP_3 + -174 + NF + 10\log BW}{3}.$$

With the lower limit being the sensitivity itself, the spurious free dynamic range (SFDR) is then[4]

$$\text{SFDR} = P_{max} - \textit{Sensitivity} = \frac{2(IIP_3 - \textit{Noise floor})}{3} - \text{SNR}.$$

For our example of the GSM receiver, a noise figure of 3dB leads to a noise floor of −118dBm at the receiver input. We found the required IIP_3 to be −19dBm, leading to a SFDR of 61dB for 5dB SNR.

As we will show, this definition however does not capture all the various scenarios that the blockers may affect the receiver. In many applications, the upper end of the acceptable blocker may very well be worse than what is determined due to the 3rd-order nonlinearity alone.

6.4 LARGE SIGNAL NONLINEARITY

We earlier showed that the response of a nonlinear system to a given input with an amplitude of A is

$$y = \left(a_1 A + \frac{3a_3 A^3}{4}\right)\cos \omega t + \cdots,$$

where only the fundamental component is shown as the other harmonics are subject to filtering. The fact that most practical amplifiers are *compressive* implies that a_3 must be negative.

---

[4] In some books the SNR does not appear in the equation, and sensitivity is simply defined as the noise floor. This potentially leads to a more generic definition, as SNR is a standard dependent parameter.

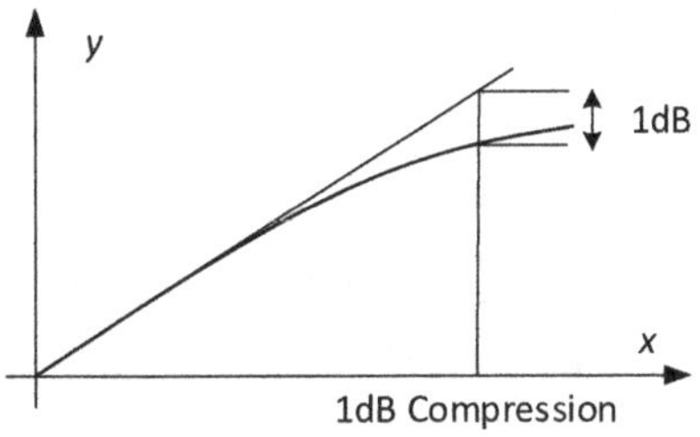

6.4 Large Signal Nonlinearity

Figure 6.34: Receiver gain compression

Consequently, as the input increases, we expect the gain to reduce and eventually compress [5] (Figure 6.34).

To describe the compression quantitatively, we define the input 1dB compression point corresponding to the input level that causes the linear gain to drop by 1dB. Since 1dB reduction means 0.89 times less, we have

$$0.89 a_1 A = a_1 A + \frac{3 a_3 A^3}{4},$$

which yields

$$A_{1\text{dB}} = \sqrt{1 - 0.89} \sqrt{\frac{4}{3} \left|\frac{a_1}{a_3}\right|} = 0.33 \sqrt{\frac{4}{3} \left|\frac{a_1}{a_3}\right|}.$$

If the IIP_3 and 1dB compression are caused by the same type of nonlinearity, then the equation above indicates that 1dB compression point in dB is $20\log 0.33 = -9.6$dB smaller than IIP_3 in dB. This, however, is not always true as the compression and IIP_3 could be caused by different mechanisms.

Example: Consider a long channel MOS common-source amplifier as shown in Figure 6.35.

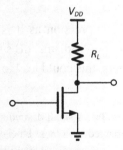

Figure 6.35: A long channel MOS CS amplifier

Assuming the device is square-law and is biased at a DC gate-source voltage of V_{GS0}, the output voltage will be

$$V_o = \left(V_{DD} - \frac{1}{2} R_L \beta (V_{GS0} - V_{TH})^2 \right) - R_L \beta (V_{GS0} - V_{TH}) v_{IN} - \frac{1}{2} R_L \beta v_{IN}^2,$$

Continued

where $\beta = \mu C_{OX} \frac{W}{L}$. The IIP_2 is

$$\text{IIP}_2 = 4(V_{GS0} - V_{TH}) = 4V_{\text{eff}},$$

and clearly the IIP_3 is infinite. On the other hand, the device will turn off if the input swing exceeds $V_{GS0} - V_{TH} = V_{\text{eff}}$, which will yield a finite 1dB compression.

The compression is typically not a concern in response to the desired signal only, as proper gain control may be incorporated to adjust the receiver gain as the signal increases. The issue will be when small desired signal is accompanied by a large blocker. As shown in Figure 6.36, the large blocker drives the amplifier into compression, and thus effectively reduces the gain (or slope of the curve), despite the fact that the small signal is residing in the linear region. If the amplifier is expected to provide gain to suppress the noise of the following stages, this effective reduction of gain results in noise figure degradation. As a result, the sensitivity suffers, and the receiver is said to be *desensitized*.

To quantify this, we can expand the y–x nonlinear function in response to the signal and blocker, where $x = A_D \cos \omega_D t + A_B \cos \omega_B t$. Assuming the signal level is much smaller than blocker, $A_D \ll A_B$, and taking only the components at the desired signal frequency, as all the other terms are subject the filtering, we have

$$y = a_1 A_D \cos \omega_D t + 3a_3 A_D \cos \omega_D t (A_B \cos \omega_B t)^2 + \cdots \approx \left(a_1 + \frac{3a_3}{2} A_B^2\right) A_D \cos \omega_D t.$$

Thus we have

$$\frac{\text{Desensitized gain}}{\text{Linear gain}} = 1 + \frac{3a_3}{2a_1} A_B^2.$$

Since a_3 is negative, as the blocker increases the gain is expected to reduce.

All the blocker scenarios we have discussed could potentially lead to receiver compression, but we will discuss a few more common cases here:

– The 3MHz blocker in GSM (Figure 6.4) could cause gain compression as it is subject to little front-end filtering. For low bands the blocker is specified to be −23dBm, 3dB stronger than high band cases. In traditional voltage-mode low-noise amplifiers, this could lead to substantial

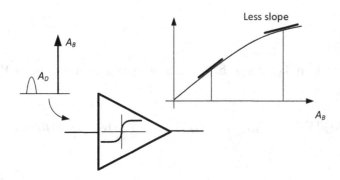

Figure 6.36: A small desired signal accompanied by a large blocker passing through a compressive amplifier

compression at the LNA output. We will show in Chapter 8 how reciprocity of a passive mixer could be exploited to ameliorate this issue. Since the signal is specified to be 3dB above the reference sensitivity, one may suggest to take advantage of some front-end gain reduction to improve the compression point. This, however, is not a viable solution for the following reason: To boost the receiver throughput as the desired signal increases, more complex modulation schemes are exploited that require higher SNR. For instance, for the EDGE (enhanced data rates for GSM evolution) MCS9 (modulation and coding schemes) case, most advanced receivers are required to achieve a sensitivity of about −95dBm, which coincides with a −99dBm blocker case. The receiver has no knowledge of which type of signal is being received a priori, and thus any noise degradation as a result of gain control is not desired.

- The 0dBm out-of-band blocker, which is as close as 20/80MHz for low/high band cases, could cause significant compression. Since the blocker is out-of-band, however, a front-end filter may be used. If the filter provides an attenuation of 23dB or more, these blockers will not cause any bigger threat than the in-band 3MHz ones. Accordingly, the receiver is expected to function properly, as it is already designed to handle the −23dBm blockers. Most practical SAW filters provide more attenuation. Recently, to reduce cost, several topologies have been proposed that allow the removal or relaxation of the front-end filtering, leading to a more challenging linearity requirement. We will discuss those architectures in Chapter 12.
- The transmitter leakage in 3G or LTE receivers is about −23dBm for a 50dB duplexer isolation similar to GSM 3MHz blocker. Although the TX signal is further away (anywhere from 35MHz to 400MHz depending on what band of 3G or LTE is used), the received signal is at sensitivity and any level of noise degradation is not acceptable.
- The GSM 400kHz adjacent blocker (Figure 6.5) specified at −41dBm is not strong enough to compress the receiver front-end. However, since it is very close to the desired signal, it may potentially create issues for later stages of the receivers (for instance, channel-select filter or the ADC) as it is subject to very little filtering. For the reasons mentioned before, even though the desired signal is at −82dBm, receiver noise figure degradation is generally not desirable, limiting the gain control only to the later stages of the receiver.

We will revisit some of these cases more closely when we discuss the receiver architectures in Chapter 12.

6.5 RECIPROCAL MIXING

In Chapter 5 we discussed that the LO signal used to shift the frequency of the incoming RF signal is noisy, and may be generally expressed as

$$x_{LO}(t) = A_C \cos(2\pi f_C t + \phi_n(t)),$$

where the term $\phi_n(t)$ represents the noise caused by the active circuitry in the oscillator, known as *phase noise*. We also showed that using the narrowband FM approximation [6], the actual LO signal consists of an impulse representing the ideal cosine, as well as a term representing noise in quadrature, shaped around the carrier as shown in Figure 6.37.

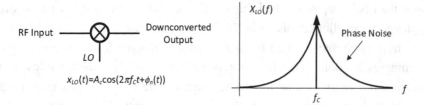

Figure 6.37: Downconversion mixer driven by a noisy LO

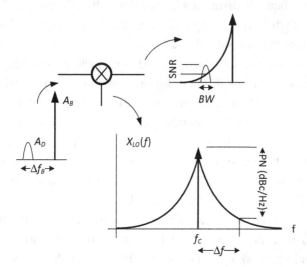

Figure 6.38: Reciprocal mixing in a receiver

Thus the mixer output, while translated in frequency, contains the phase noise convolved in the frequency domain with the input. We shall show that in an otherwise ideal receiver, the LO phase noise limits the receiver sensitivity when exposed to blockers.

Consider Figure 6.38, where a small desired signal accompanied by a large blocker appear at the input of an ideal mixer, driven by a noisy LO. At any given frequency offset Δf from the carrier, the phase noise is typically expressed in dBc/Hz, with the noise integrated over a 1Hz bandwidth at that offset, normalized to the carrier amplitude (Figure 6.38). For instance, if the oscillator has an RMS amplitude of 1V, and the measured spot *phase* noise at a certain offset is $100\text{nV}\sqrt{\text{Hz}}$, then the phase noise is -140dBc/Hz.

As shown in Figure 6.38, when the noisy LO is convolved with the blocker, it results in the skirt of the noise on the left sideband appearing on top of the desired signal. Note that although not shown, the desired signal also convolves with the LO, but since its magnitude is small, the resultant noise is negligible. If the phase noise of the LO at the offset frequency of Δf_B, that is, the frequency where the blocker and desired signal are separated, is PN, then the total noise (in dBm) integrated over the signal bandwidth is

$$P_B + PN + 10\log BW,$$

where P_B is the blocker power in dBm. Note that the phase noise is measured relative to the carrier, and thus the noise itself is normalized by the blocker power. Furthermore, the multiplier (or mixer) amplitude (A_C) is inconsequential, as phase noise is normalized to carrier power, and both signal and noise are scaled by the same carrier amplitude at the multiplier output. Also we made the approximation that the noise is relatively flat over the signal bandwidth. The desired signal can be successfully detected if this noise is an SNR below. Thus

$$\text{Sensitivity} = P_B + PN + 10\log BW + \text{SNR},$$

indicating that an otherwise ideal receiver sensitivity is set by the phase noise of the LO signal in the presence of large blockers. This phenomenon is often known as *reciprocal mixing*, and sets the requirements of the LO phase noise depending on the blocker profile for a certain standard.

Example: Consider the blocker profile shown in Figure 6.4 corresponding to GSM standard. Suppose an SNR of 10dB is required to achieve sufficient margin. Since the desired signal is at −99dBm, that is the sensitivity, and that the GSM bandwidth is 200kHz, a 600kHz blocker at −43dBm requires a phase noise of no higher than −99 − (−43) − 10log(200kHz) − 10 = −119dBc/Hz at 600kHz offset. For the 3MHz blocker on the other hand, as it is 17dB stronger, we expect a phase noise of −136dBm for the LO to achieve the same SNR of 10dB. Assuming the phase noise drops with a slope of $1/f^2$ away from the carrier, achieving −136dBm/Hz phase noise at 3MHz, will automatically result in a phase noise of −136 + 20log(3MHz/600kHz) = −122dBm/Hz at 600kHz. Thus meeting the 3MHz phase noise is sufficient to pass the 600kHz blocker case. Considering all other blockers, it turns out that the 3MHz is the most stringent one.

The blocker profile shown in Figure 6.4 typically leads to a phase noise profile shown in Figure 6.39. Only one side is shown, but since the blockers could be present at either side, the required phase noise profile is symmetric around the carrier.

For a given phase noise specified at a certain offset, and for a given blocker, we can calculate the receiver noise figure as a result of reciprocal mixing: the spot noise for a blocker power of P_B and a phase noise of PN is $P_B + PN$ in dBm/Hz. Since the available noise power is KT or −174dBm/Hz, by definition, the noise figure is the ratio of the two, resulting in subtraction of the two terms in dB,

$$NF_B = 174 + P_B + PN,$$

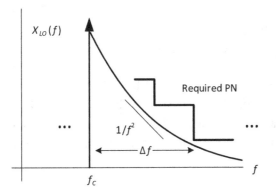

Figure 6.39: Required LO phase noise according to the blocker profile

where NF_B expressed in dB signifies the *blocker induced noise figure*, or in short, the blocker noise figure.

Example: Consider the LTE receiver designed in Chapter 5 to achieve a GSM sensitivity of −110dBm as shown in Figure 6.40. We calculated a noise figure of 3dB at the receiver input, and 6dB at the antenna assuming 3dB loss due to the filter and switch. Now suppose this receiver is subject to a −26dBm 3MHz blocker at the antenna. At the receiver input, the blocker is 3dB weaker, and given a phase noise of −136dBm/Hz for the LO, the receiver blocker noise figure is $174 + -29 + -136 = 9$dB. Once the 3dB thermal noise of the receiver is added, the total noise figure becomes 10dB at the receiver input, or 13 at the antenna.

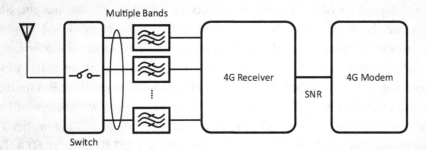

Figure 6.40: A 4G receiver

The minimum detectable signal, given a minimum SNR of 5dB needed, will be then about −103dBm, leaving us 4dB of margin. This calculation is, however, somewhat optimistic, as the receiver gain compression may cause worse than 3dB noise figure, and a composite noise figure of higher than 13dB.

Note that the front-end passive loss to the 1st-order does not affect the reciprocal mixing induced noise figure.

Example: Consider a Wi-Fi receiver with 3dB base noise figure and a far-out phase noise of −155dBc/Hz. Let us first find the receiver total noise figure in the presence of a −20dBm blocker. The receiver thermal noise floor is −171dBm, whereas the noise floor due to the blocker reciprocal mixing is

$$-155\text{dB/Hz} - 20\text{dBm} = -175\text{dBm/Hz}.$$

When the two added up, the total noise floor is −169.5dBm/Hz, leading to a noise figure of 4.5dB. Now suppose the receiver is preceded by an amplifier with a gain of 10dB and noise figure of 2dB. When referred to the amplifier input, the new noise floor due to the blocker is −165dBm/Hz as the blocker is 10dB stronger, and the total receiver noise figure becomes 9.8dB. Once referred to the amplifier input, the noise factor according to Friis formula is

$$1.58 + \frac{9.55-1}{10} = 2.44,$$

or 3.86dB. Without the blocker, the cascaded noise figure would have been

$$10\log\left(1.58 + \frac{2-1}{10}\right) = 2.25\text{dB}.$$

So the amount of desensitization due to the blocker in either case is about the same (~1.5dB).

6.6 HARMONIC MIXING

Consider the receiver shown in Figure 6.41, where the low-noise amplifier is assumed to have kth-order nonlinearity as denoted by IIP_k [7].

The mixer is driven by an LO signal at a frequency of f_{LO}. As we will show in Chapter 8, most practical mixers are designed based on the hard switching concept. Hence, effectively, the LO signal driving them is assumed to be a square-wave, consisting of all odd harmonics as shown in Figure 6.41. As a result, we expect for instance a blocker located at exactly $3f_{LO}$ to be also downconverted, falling on top of the desired signal, but only $20\log(1/3) \approx -10$dB weaker.

Suppose the LO is on the low-side of the desired signal, that is, the desired signal is located at $f_{LO}+f_{IF}$, where f_{IF} signifies the intermediate frequency. This is known as *low-side injection*. When the LO is at the high side of the desired signal, it is called *high-side injection*.

Any blocker at the frequency of $(nf_{LO} \pm f_{IF})/k$ results in an unwanted signal at the LNA output at $nf_{LO} \pm f_{IF}$ due to its kth-order nonlinearity. This signal is subsequently mixed with any of the LO harmonics at nf_{LO} (n is a positive integer), appearing on the top of the desired signal after downconversion. If the mixer is differential, n is expected to be odd, like the example shown. In practice, however, due to the mismatches the even harmonics of the LO may also be

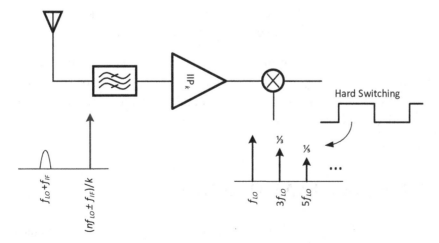

Figure 6.41: Impact of harmonic mixing in receivers

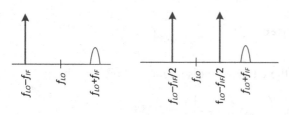

Figure 6.42: Image and half-IF blockers

responsible for the undesirable harmonic downconversion. The situation may be alleviated in one of the following ways:

- The blocker may be attenuated by an RF filter placed before the LNA. The filter attenuation depends on the IF, as well as different combination of values of n and k.
- The LNA linearity may be improved.
- The LO harmonics may be suppressed by manipulating the mixer. This will be discussed in Chapter 8.

Let us consider a few common special cases.

1. $n = k = 1$. For a low-side injection, this blocker is located an IF below the LO (or two IF below signal), commonly known as the *image blocker*. Since the LNA is perfectly linear in this case, the image blocker can be attenuated only by filtering. However, if IF is small enough, as we will show in Chapter 12, a quadrature receiver may be exploited to remove the image.
2. $n = k = 2$. Known as the *half-IF blocker* as is located at: $f_{LO} \pm \frac{1}{2}f_{IF}$, the blocker is problematic due to the LNA 2nd-order nonlinearity as well as the LO second harmonic. A fully differential design is then expected to help. Compared to image case, on one hand the blocker is caused due to the 2nd-order effects, and thus may not be as troubling. On the other hand, it is only half-IF away from the desired signal, as opposed to twice the IF in the case of image. Thus the half-IF blocker is subject to less filtering. The image and half-IF blockers scenarios are illustrated in Figure 6.42.
3. Harmonic blockers located at nf_{LO} are downconverted despite the fact that a linear LNA is used ($k = 1$), unless a *harmonic rejection mixer* [8] is used (Chapter 8).

Aside from harmonic blockers that may be attenuated only by filtering, or through using harmonic rejection mixers, the majority of other blockers largely depend on the choice of IF. As we will discuss in Chapter 12, the choice of receiver architecture to a large extent relies on the proper choice of IF, and the issues associated therewith.

6.7 TRANSMITTER NONLINEARITY CONCERNS

The majority of Chapter 5 and most of this chapter so far have been dedicated to the receivers. The transmitters suffer from very similar issues whose principle was introduced. In fact, we could loosely consider transmitters as the *dual* of the receivers, both from the architecture point of view as well as issues related to noise and distortion.

We categorize the transmitter requirements into three groups, output power, modulation mask, and signal quality, and discuss each in this section.

6.7.1 Output Power

The transmitter output power along with the receiver sensitivity define the *range of coverage*. In mobile applications, as a longer range is required, the transmitter output power needs to be higher. In GSM for example, the low bands must put out 33dBm at the antenna (that is, 2W), while for high bands the required output power is 30dBm. LTE, on the other hand, requires less output power of about 23dBm for most of the bands as it supports a wider bandwidth and requires a more stringent linearity.

For LAN (local area network) applications where a higher throughput is supported, the output power is lower, generally in the range of the high teens. Furthermore, the receiver sensitivity is worse, given the wider bandwidth, and higher SNR. In general, we can say that the range of coverage is always traded for throughput.

The physical limit to raising the transmitter output power arises from a number of issues:

- High power transmitter imposes itself as a strong interferer to the other receivers.
- Higher throughput requires a higher fidelity in the TX, or better SNR. Achieving a higher SNR typically requires a *more linear power amplifier*, as we will discuss next. In Chapter 11 we will show that linear power amplifiers tend to be less efficient, and ultimately more power consuming. Given the battery size and cost concerns, high-SNR applications such as WLAN tend to support less output power. A GSM power amplifier, on the other hand, is considerably more efficient as GSM uses constant envelope modulation.

6.7.2 Transmitter Mask

The transmitter mask is generally defined to limit the amount of interference that a TX can potentially cause for the nearby receivers. So transmitter mask requirements can be viewed as dual of the receiver blocker requirements. The TX mask is limited by either the LO phase noise or the transmitter chain nonlinearity. We discuss each case here.

In Chapter 5 we showed that if the oscillator is embedded within a phase-locked loop (PLL), the phase noise profile will change according to the PLL characteristics, and often looks like what is shown in Figure 6.43.[5] It consists of a relatively flat passband set by the PLL, along with a region of 20dB/Dec (or $1/f^2$) slope set by the VCO characteristics. At very far-out frequency offsets from the carrier, the noise flattens as the buffers or other hard-limiting devices following the oscillator dominate.

When multiplied by the TX baseband signal, this noise profile appears directly at the transmitter output as shown in Figure 6.44. In practice, any other part of the transmitter chain

---

[5] We shall analytically prove this in Chapters 9 and 10, but will accept it for the moment.

366 Distortion

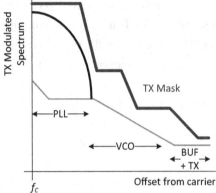

Typical PLL Phase Noise Profile

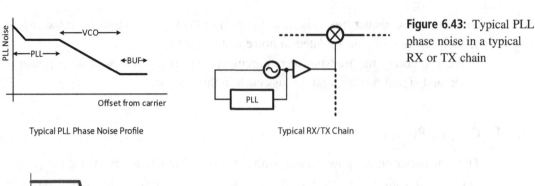

Figure 6.43: Typical PLL phase noise in a typical RX or TX chain

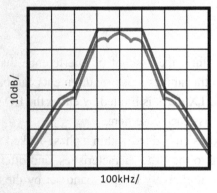

Figure 6.44: Transmitter mask limited by the LO phase noise

may contribute to this noise as well, although unless for far-out frequencies, the LO phase noise is dominant.

Example: Let us consider GSM transmitter modulation mask requirements as shown in Figure 6.45.

Figure 6.45: GSM modulation mask requirements

Since GSM uses GMSK modulation, which is constant envelope, the transmitter chain linearity is not a major concern, and mask is typically limited by the phase noise. We will show in Chapter 12 that this is not always true, especially if a Cartesian transmitter is chosen. Ignoring the linearity impact for the moment, a key requirement is a 400kHz mask that is specified to be 60dBc below the carrier measured in a 30kHz *resolution bandwidth*. Suppose we would like to have a 3dB production margin, and leave another

2dB for process and temperature variations, giving us a typical requirement of −65dBc. If the phase noise at 400kHz offset is PN, then the relative noise in dBc is

$$PN + 10\log 30\text{kHz} + 10\log \frac{200\text{kHz}}{30\text{kHz}}.$$

The last term arises due to the fact that the GSM signal is roughly 200kHz wide, and as the measurement is performed in a 30kHz bandwidth, the signal energy is subject to some filtering that must be adjusted. Accordingly, the required 400kHz phase noise is

$$PN = -65 - 10\log 30\text{kHz} - 10\log \frac{200\text{kHz}}{30\text{kHz}} = -118\text{dBc/Hz}.$$

Recall that for the GSM receiver, the 3MHz offset was the most critical requirement (−136dBc/Hz). Assuming a 20dB/Dec roll-off, the equivalent 400kHz receiver phase noise would have been −118.5dBc/Hz, coincidentally almost identical to the TX requirement.

The 200kHz and 250kHz mask requirements of −30dBc and −33dBc, respectively, are generally limited only by the modulator and the filtering incorporated, and the phase noise does not play a major role. Also similar to the case of RX blockers, with a 20dB/Dec roll-off assumed for the phase noise, meeting the 400kHz mask typically automatically guarantees the other far-out frequencies.

There is yet another very stringent noise requirement for the GSM transmitter explained as follows: the EGSM transmitter for instance occupies a band of 880–915MHz, whereas the receiver takes 935–960MHz. Suppose a given handset is transmitting +33dBm at the upper edge of the band, and a nearby handset is receiving at the lower edge as shown in Figure 6.46.

For this transmitter not to impose a problem for the nearby receiver, the noise at 20MHz away (that is the worst case spacing between the two) is specified to be −79dBm or better when measured at a 100kHz bandwidth. The rationale behind −79dBm number has to do with the fact that if the isolation between the two handsets is say 30dB or better, the noise created by the TX will be low enough not to affect the −102dBm sensitivity of the receiver. An isolation of worse than 30dB requires the two handsets to be very close, within perhaps a few centimeters. The noise is then equal to

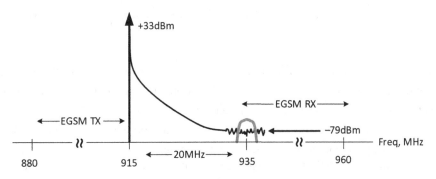

Figure 6.46: GSM 20MHz noise requirement

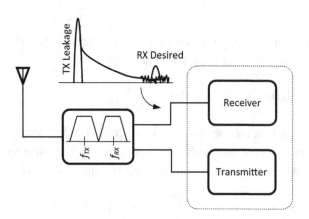

Figure 6.47: A 3G or LTE transmitter noise requirement

$$-79 - 10\log 100\text{kHz} - (+33\text{dBm}) = -162\text{dBc/Hz}$$

Typical power amplifiers have a noise floor of about −83 to −84dBm, and to have enough margin, the radio must achieve an out-of-band (OOB) noise level of −165dBc/Hz or better. Since this noise level is very low, not only the VCO and LO contribute, but also any other component of the transmitter chain could potentially be a problem. This greatly affects the choice of the GSM transmitter architecture, as we will discuss in Chapter 12.

A very similar noise requirement exists in 3G and LTE transmitters due to their full duplex nature, where the 3G transmitter itself is the source of the noise, as opposed to the nearby handset in the case of GSM (Figure 6.47). This could also exist in platforms (such as mobile handsets) where several radios of different standards exist.

The noise requirement may be calculated similarly, and is illustrated as an example below:

> **Example:** Assuming we would like to have less than say 0.5dB degradation is the LTE receiver noise, we would want the TX induced noise to be −182dBm/Hz or less. Note that a 3dB noise figure of receiver leads to an input-referred noise floor of −171dBm/Hz. Thus the TX noise in dBc/Hz at the receiver frequency is
>
> $$-182\text{dBm/Hz} - (+27\text{dBm} - 50) = -159\text{dBc/Hz},$$
>
> where a duplexer isolation of 50dB is assumed. The PA output power is 27dBm.

Note that in the case of 3G or LTE, the TX noise is not caused by an *infrequent event* of having two handsets too close, and thus is somewhat more critical. On the other hand, the lower output power compared to GSM is a helpful factor here.

The transmitter nonlinearity and *spectral regrowth* could be another source of mask violation. This is intuitively explained as shown in Figure 6.48.

For a given modulated signal, let us break it into several tones occupying the band of interest, say ±1.9MHz for the case of 3G. Every pair of two arbitrary tones then create IM_3, IM_5, ... components caused by the transmitter 3rd-, 5th-, ... order nonlinearity. Since the frequency of

6.7 Transmitter Nonlinearity Concerns

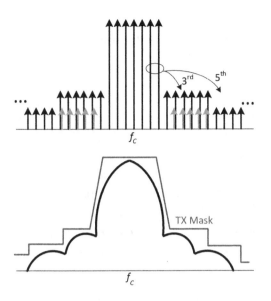

Figure 6.48: Transmitter nonlinearity leading to spectral regrowth, and potentially mask violation

the IM_3 components is two times frequency of one component subtracted from the other, the 3rd-order nonlinearity-induced spectrum is as wide as three times the signal bandwidth, while the 5th-order nonlinearity extends the spectrum to a five times larger bandwidth. Overall the TX signal will look something like the one shown on the bottom of Figure 6.48.

For 3G (or LTE) it is common to specify *adjacent channel leakage ratio* (ACLR) measured at ± 1.9MHz bandwidth at 5 or 10MHz away. Translating the required ACLR to a given IP_3 or IM_3 requirement is a function of the modulation and its statistical properties, and there is no closed equation.

Example: The ACLR of a 3G transmitter at 5MHz away versus IM_3 is shown in Figure 6.49, obtained from system-level simulations.

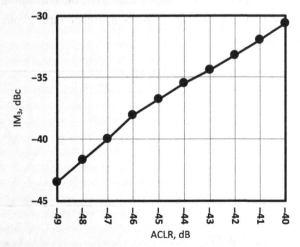

Figure 6.49: An example of 3G transmitter linearity requirement

The standard requires an ACLR of better than −33dBc at 5MHz away. Assuming the power amplifier ACLR is −37dBc, with some margin included, a typical ACLR of −40dBc may be considered for the radio.

Continued

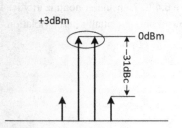

Figure 6.50: 3G transmitter OIP$_3$ calculations

This then translates to an IM$_3$ of better than −31dBc for the TX deduced from Figure 6.49 plot. For a 2-tone measurement at the transmitter output with each tone at 0dBm (corresponding to a total power of +3dBm), the output IP$_3$ (OIP$_3$) is then 15.5dBm (Figure 6.50).

This leads to an output compression point of about 15.5 − 9.6 = 5.9dBm. Thus to ensure the ACLR is met, the transmitter output peak power is roughly 3dB below the compression.

In the case of 3G signal, which is QPSK modulated, the transmitter average power is known to be about 3dB less than the peak power. This is often known as peak-to-average ratio (PAPR), and is a function of the type of modulation used. Typically, the more complex modulation schemes correspond to a larger PAPR. To ensure that the modulation mask is met, as a rule of thumb, the transmitter output power must be *backed off* from the compression point by a value that is comparable with PAPR. Therefore, a more complex modulation scheme that provides higher throughput for the same bandwidth comes at the expense of a larger *back-off*. Consequently, a worse efficiency is expected, as was pointed out. Note for a pure sinusoidal signal, the peak is exactly 3dB higher than average (or the RMS value), coincidentally very similar to that of 3G.

Example: Shown in Figure 6.51 are the baseband IQ signals corresponding to LTE 20MHz, as well as 3G voice in time domain. The sample rate is 61.44MHz, and the plots shown contain 100 samples. The LTE signal that supports a higher throughput clearly contains more fluctuations more frequently.

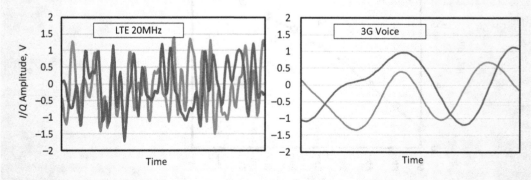

Figure 6.51: Examples of 3G voice and LTE PAPR

Statistically speaking, the portability of peak versus the average power is shown in Table 6.1. For instance, there is a 1% probability that the 3G signal peaks as high as 2.6dB above the average, while for the same probability the LTE signal may peak as much as 5dB.

Table 6.1: **3G and LTE peak probability**

Probability	3G Voice	LTE 20MHz
10%	1.7dB	2.8dB
1%	2.6dB	5dB
0.1%	3.2dB	6.4dB
0.01%	3.4dB	7.4dB

Leaving a margin of 2dB for power control, the total back-off for the LTE is about 8.4dB (for less than 0.1%), which is considerably higher for what is needed for a 3G voice signal. This demand for better linearity directly translates to a worse efficiency.

Finally, same as the receivers, we can define the OIP$_3$ of the chain as sum of each individual block as shown below:

$$\frac{1}{A_{OIP3,T}^2} = \frac{1}{A_{OIP3,1}^2} + \frac{a_1^2}{A_{OIP3,2}^2} + \cdots.$$

To maintain the relative noise floor and other potential unwanted signals low, typically transmitter signal is large from the very beginning of the chain. For that reason, almost all blocks have a gain of close to unity at maximum output power, and they may all contribute to the nonlinearity.

Example: Suppose we have a WLAN transmitter at 20dBm average output power. This signal leaks to a nearby LTE receiver with 3dB noise figure (Figure 6.8). Assuming 40dB of isolation between the WLAN TX and LTE RX, we shall find the transmitter out-of-band noise such that the LTE receiver is desensitized by no more than 1dB.

Since the receiver has a noise figure of 3dB, its input referred noise is −171dBm/Hz. To desensitize by no more than 1dB, the TX noise leaking to the receiver should be about 6dBless, that is, −177dBm/Hz. So at the TX output it translates to −137dBm/Hz considering 40dB of isolation. Thus, the noise required will be −137 − 20 = −157dBc/Hz. Note that the PAPR of the TX signal signifies the amount of back-off from the peak transmitter output, leading to a more stringent dynamic range for TX.

6.7.3 Transmitter Signal Quality

The transmitter signal must maintain a high enough signal-to-noise ratio to ensure that once received at the other end, the signal quality is acceptable. This is typically characterized by EVM (error vector magnitude) or phase error, and is directly related to the SNR. At the receiver end, as the signal is typically weak, the SNR is usually limited by the receiver thermal noise, or

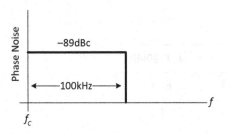

Figure 6.52: GSM transmitter in-band phase noise

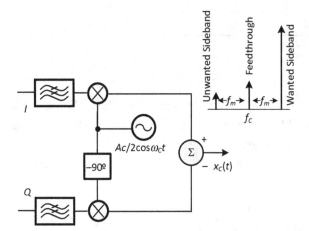

Figure 6.53: A single-sideband modulator

the aforementioned blocker induced interference. However, as the receiver signal increases, and in the absence of the blocker, the SNR of the receiver is also expected to improve to guarantee a high throughput.

The EVM, measured in percent or dBc, is limited by any one of the three following mechanisms:

The LO in-band phase noise — As shown in Figure 6.44, the transmitter signal once upconverted sits directly on top of the in-band noise of the PLL. This noise when integrated over the band of interest leads to a certain phase error that sets the lower limit of SNR or EVM. A good example is GSM, which uses constant envelope, so the signal is usually not affected by other mechanisms. The GSM specifies a phase error of better than 5°, whereas most typical radios target for 1–2° to leave room for process or temperature variations and production related variations. Consider Figure 6.52, which shows the in-band phase noise of a GSM transmitter as an example. Let us assume that the noise is flat over the ±100kHz band of interest.

The integrated noise over GSM band is −89dBc + 10log200kHz = −36dBc. Since −36dBc is 0.0158 radians or 0.9°, the phase error is 0.9°. For applications where a high throughput is supported such as LTE or WLAN family, the in-band noise requirement is more stringent, and usually an integrated phase noise of a few tenths of a degree is targeted.

TX chain unwanted interferers — Consider the simplified single-sideband modulator shown in Figure 6.53.

Suppose for the moment only a single tone at the upper sideband ($+f_m$) is being transmitted. If the 90° phase shift has some error, or the quadrature signals produced at the baseband are not perfectly 90° out of phase, a residual unwanted sideband (or image, same as the receiver) will appear at $-f_m$. Moreover, the carrier may directly feed through and appear as an unwanted tone

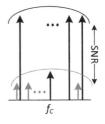

Figure 6.54: EVM limited by quadrature mismatches and LO feed-through

at f_C. Once a modulated signal is applied, the image components and the feed-through set a floor that limits the SNR (Figure 6.54).

Once the impact of the in-band phase noise is also included, the EVM will be approximately

$$\text{EVM} \approx \sqrt{10^{\text{IQ}} + 10^{\text{LOFT}} + 10^{\text{PN}}},$$

where IQ is the TX image suppression, LOFT is the LO feed-through, and PN is the in-band phase, all expressed in dB.

Note the above equation is not generally true in all modulation schemes; it only gives us an idea of how to deal with various contributors. The first term shows the contribution of the quadrature imbalance, the second term is the feed-through, and the last term is the phase noise. For 1% quadrature inaccuracy, −40dBc feed-through (also 1%), and an in-band noise of −40dBc, the EVM will be $\sqrt{1+1+1} = 1.7\%$. Most practical transmitters achieve a typical EVM of better than 3–5%.

Nonlinearities — This is very similar to the case of spectral regrowth as was demonstrated in Figure 6.48. As the IM_3 components will also appear within the band of interest, they set a noise floor that ultimately limits the SNR. However, for many applications, this not a dominant factor. For instance, in our example of 3G transmitter, an IM_3 of better than −31dBc was required to guarantee a −40dB ACLR. This results in a negligible degradation of the EVM. We shall have a more detailed discussion on the impact of nonlinearity on EVM and constellation shortly in Section 6.7.5. On the other hand, in an 802.11ac WLAN application for a 1024QAM signal, given the very stringent signal fidelity needed, the transmitter nonlinearity could be a factor.

Similar mechanisms limit the receiver SNR when the signal is strong enough. The receiver EVM requirement is usually very similar to that of the TX, except for it is meaningful only when the desired signal is strong enough so as not limited by the blockers or receiver thermal noise. In the case of the TX, the signal is always strong, at least for maximum output power.

6.7.4 Switching Spectrum and Time-Domain Mask

In addition to the modulation mask and frequency domain specifications described before, there is yet another requirement for the TDMA (time division multiple access) systems, where different users are assigned different time slots. In contrast to TDMA, in FDMA systems users are assigned different frequencies. The examples of the latter are FM radio and satellite TV, where each station transmits at a fixed known frequency. GSM and EDGE radios are on the other hand examples of TDMA-based systems. Consequently, a TDMA signal often consists of frames, with each frame composed of several time slots. Shown in Figure 6.55, is a

Distortion

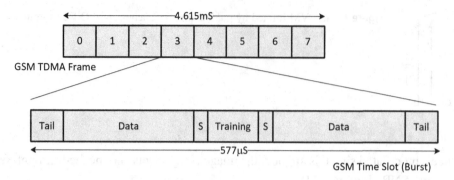

Figure 6.55: GSM frame and time slots

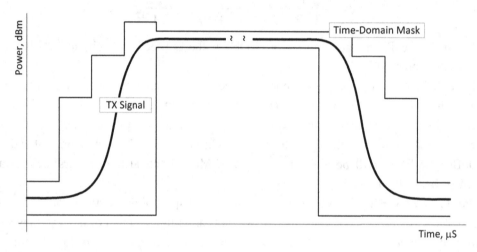

Figure 6.56: Example of GSM time-domain mask

GSM/EDGE frame for instance, where each time slot is about 577µS long, and one frame consists of eight time slots.

In consequence, it is expected for the transmitter to produce an output signal that smoothly *ramps* up and down to avoid sharp time-domain transitions, and hence undesired high-frequency contents in the spectrum. Accordingly, a *time-domain mask* is defined within which the transmitter output must fit properly, as shown in Figure 6.56.

Along with the time-domain mask, 2G transmitters must also satisfy a frequency-domain mask measurement known as *switching spectrum*. Whereas a modulation mask of better than −60dBc measured in a 30kHz resolution bandwidth is required in GSM for instance, a switching spectrum of lower than −23dBm measured at the same bandwidth is specified as well. There are two distinct differences between the modulation mask and the switching spectrum measurements:

- The switching spectrum is an absolute measurement specified in dBm (not dBc). For the full output power of +33dBm, it translates a requirement of −56dBc. If the transmitter power is backed off, however, the requirement relaxes proportionally.
- While modulation mask is an average measurements, with often 50 bursts taken, the switching spectrum is a *MAX hold* measurement. That is, for the same 50 bursts measured for instance, the spectrum analyzer is set to take the maximum instance of the 50 bursts at each point of the time

Consequently, in a well-designed transmitter, often meeting the modulation mask is sufficient to pass the switching spectrum test. In the case of the worst case maximum power, the switching spectrum is 4dB relaxed, although the MAX hold measurement could offset that. On the other hand, if the transmitter time-domain output has sharp transitions, for instance due to improper design of the ramping phase, the switching spectrum is expected to be affected, even though the modulation mask may be met.

6.7.5 AM–AM and AM–PM in Transmitters

In addition to noise and static nonlinearities, there is yet another mechanism in transmitters, and especially power amplifiers causing EVM and mask degradation.

Let us start with a more detailed analysis of static nonlinearity presented earlier in the context of IM_3, also known as *AM–AM nonlinearity*.

Consider the following narrowband RF signal,

$$x(t) = i(t)\cos \omega_c t - q(t)\sin \omega_c t,$$

where $i(t)$ and $q(t)$ are baseband signals with *zero* mean and ω_c is the carrier frequency. The autocorrelation $R_x(\tau)$ is found to be

$$R_x(\tau) = \frac{1}{2}R_i(\tau)\cos \omega_c \tau + \frac{1}{2}R_q(\tau)\cos \omega_c \tau + \frac{1}{2}\left(R_{i,q}(\tau) - R_{i,q}(-\tau)\right)\sin \omega_c \tau,$$

in which $R_i(\tau)$ and $R_q(\tau)$ are autocorrelations of $i(t)$ and $q(t)$, respectively, and $R_{i,q}(\tau)$ is their cross-correlation. Therefore, for frequencies around $+f_c$ the power spectral density of $x(t)$ is given by the following expression:

$$S_x(f) = \frac{1}{4}S_i(f - f_c) + \frac{1}{4}S_q(f - f_c) + \frac{1}{2}\text{Im}\{S_{i,q}(f - f_c)\}.$$

For two frequencies, $f_c \pm f_m$, where f_m is a small offset, the first two terms in $S_x(f)$ would lead to identical positive numbers, i.e., $S_i(f_m) = S_i(-f_m) > 0$, $S_q(f_m) = S_q(-f_m) > 0$. However, the third term would return negative values, i.e., $\text{Im}[S_{i,q}(f_m)] = -\text{Im}[S_{i,q}(-f_m)]$. Consequently, any correlation between the quadrature components of $x(t)$ can potentially cause an *asymmetric* power spectral density for $x(t)$. This is because on one side the third terms adds up constructively, whereas they add destructively on the other side. In contrast, if the two quadrature components are statistically independent, the power spectral density of $x(t)$ is symmetric. Later we will make use of this conclusion to study power spectral densities caused by AM–AM and AM–PM nonlinearities.

If $i(t)$ and $q(t)$ are partially correlated, $q(t)$ can be written as $q(t) = c(t) + u(t)$, where $c(t)$ is the correlated part, and $u(t)$ is the uncorrelated part. $c(t)$ can be written as $f(i(t))$, in which the function f can be linear or nonlinear or a combination. It is readily proven that $R_{i,q}(\tau)$ is equal to $R_{i,c}(\tau)$. For example, if $c(t)$ is equal to $i(t) * h(t)$, where $h(t)$ is some impulse response, $R_{i,c}(\tau)$ becomes equal to $R_i(\tau) * h(-\tau)$ (see Chapter 2). As a result, $S_{i,q}(f) = S_i(f)H^*(f)$. Hence, $\text{Im}[S_{i,q}(f)] = S_i(f)\text{Im}[H^*(f)]$ given that $S_i(f)$ is real and positive.

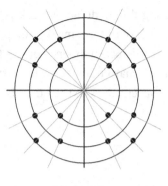
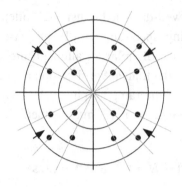

Figure 6.57: Impact of AM–AM nonlinearity on constellation

Ideal 16QAM   **16QAM Experiencing AM-AM**

Now, consider a modulated signal $x(t) = A(t)\cos(\omega_c t + \phi(t))$, where $A(t)$ and $\phi(t)$ are amplitude and phase information. For this signal the quadrature components are given by the following expressions: $i(t) = A(t)\cos\phi(t)$, and $q(t) = A(t)\sin\phi(t)$. Assuming the modulation is symmetric around the center of the constellation, and that all of the constellation points are covered with equal probability, $i(t)$ and $q(t)$ are *statistically independent*. Therefore, the power spectral density of $x(t)$ is symmetric around the carrier.

Now, let us study the impact of AM–AM on the constellation diagram, and symmetry of the power spectral density. Consider a system whose input is $x(t) = A(t)\cos(\omega_c t + \phi(t))$ and with an output equal to $y(t) = F(A(t))\cos(\omega_c t + \phi(t))$. The function $F(A)$ is the AM–AM characteristic of the system, represented by a nonlinear transfer function as we have seen before. For a distortionless system, $F(A)$ must be proportional to A. In reality, power amplifiers and in general transmitters are nonlinear and mostly compressive. As shown in Figure 6.57, each point in the constellation with an amplitude of A and a phase of θ is mapped to another point with the amplitude $F(A)$ but with the same phase θ. Therefore, the constellation points are shifted radially inward[6] to origin. The quadrature components of the resulting signal are $i(t) = F(A)\cos\phi$ and $q(t) = F(A)\sin\phi$, which can be proven to be statistically independent (if we loosely assume that A and ϕ are statistically independent). Note that the two quadrature components occupy more bandwidth as a result of the spectral regrowth due to the AM–AM nonlinearity, but they still remain statistically independent. This is the main reason for mask degradation.

In the *memoryless nonlinear* systems we have discussed thus far, input amplitude variation only causes output *amplitude* distortion. In practice, the nonlinear parasitic capacitors often lead to phase distortion as well. Known as *AM–PM nonlinearity*, we shall study its impact on the constellation diagram and modulation spectrum.

The system is assumed to take the input of $x(t) = A(t)\cos(\omega_c t + \phi(t))$ and produce an output of $y(t) = A(t)\cos(\omega_c t + \phi(t) + F(A(t)))$, where $F(A)$ is the AM–PM characteristic of the system. Ideally $F(A)$ must be zero for a distortionless system. As shown in Figure 6.58, each constellation point with an amplitude of A and a phase of ϕ is mapped to another point with the same amplitude A, but phase of $\phi + F(A)$. Therefore, the constellation points are *rotated* by the amount equal to $F(A)$. The quadrature components of the resulting signal are $i(t) = A\cos(\phi + F(A))$ and

---

[6] For a compressive system of course.

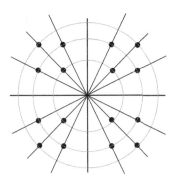 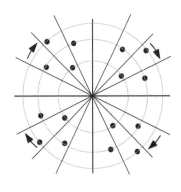

Figure 6.58: Impact of AM–PM nonlinearity on constellation

Ideal 16QAM   16QAM Experiencing AM-PM

$q(t) = A \sin(\phi + F(A))$, which are not necessarily independent. Consequently, in general the spectrum is not expected to remain symmetric around the carrier. Furthermore, the signal bandwidth widens as a result of the AM–PM nonlinearity.

Example: Consider the following format of AM–PM distortion in a communication system: $y(t) = A \cos(\omega_c t + \phi + \beta A)$, where $F(A) = \beta A$. Assume that the input is an AM-modulated signal given by $x(t) = \sin(\omega_m t) \cos(\omega_c t)$. The resulting output will be equal to $y(t) = \sin(\omega_m t) \cos(\omega_c t + \beta \sin \omega_m t)$, which contains both AM and PM modulations. Let us use the following two identities as we introduced in Chapter 2:

$$\cos(\omega_c t + \beta \sin \omega_m t) = \sum_{n=-\infty}^{+\infty} J_n(\beta) \cos(\omega_c + n\omega_m)t$$

$$\sin(\omega_c t + \beta \sin \omega_m t) = \sum_{n=-\infty}^{+\infty} J_n(\beta) \sin(\omega_c + n\omega_m)t$$

The function $J_n(\beta)$ is the Bessel function, and is described below (for positive n):

$$J_n(\beta) = \frac{\beta^n}{n! 2^n} \left(1 - \frac{\beta^2}{2(2+2n)} + \frac{\beta^4}{(2)(4)(2+2n)(4+2)} - \cdots \right).$$

One can prove that $y(t)$ can be written as

$$y(t) = \sum_{n=-\infty}^{+\infty} \frac{1}{2} \{ J_{n-1}(\beta) - J_{n+1}(\beta) \} \sin(\omega_c + n\omega_m)t.$$

Knowing that $J_{-n}(\beta) = (-1)^n J_n(\beta)$, the resulting spectrum in this case is symmetric around the carrier ω_c.

To illustrate how AM–PM could lead to spectral asymmetry, let us consider a more realistic case shown in Figure 6.59. Depicted in the figure is the envelope signal varying at a frequency of f_m, and the corresponding phase variations caused due to AM–PM nonlinearity. As the phase variations occur at every amplitude maxima or minima, they have a rate of twice the fundamental frequency [9]. Thus, $F(A) \approx \beta A^2$.

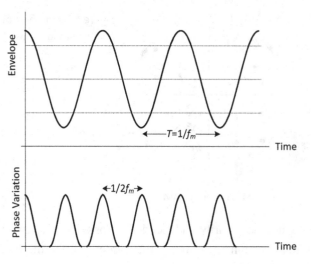

Figure 6.59: Envelope and corresponding AM–PM phase variations

We shall also assume there is a 3rd-order nonlinearity on the envelope contents of the RF signal, representing the AM–AM. For the case of a tone modulating the envelope, $A(t) = \cos \omega_m t$, and the RF output may be generally expressed as

$$y(t) = (a_1 \cos \omega_m t + a_3 \cos 3\omega_m t) \cos\left(\omega_C t + \frac{\beta}{2} + \frac{\beta}{2} \cos 2\omega_m t\right).$$

For simplicity, we ignore the constant phase shift $\frac{\beta}{2}$ as it has no impact on the outcome. It is clear that the RF signal has additional unwanted components at $\pm 3\omega_m, \pm 5\omega_m, \ldots$ offset, leading to spectral regrowth. For the moment, let us assume $\beta \ll 1$. The general case may be analyzed similarly using Bessel series as shown in the previous example (see Problem 26). Using narrowband FM approximation, and focusing only on IM$_3$ sidebands, upon expanding the RF signal we obtain

$$y(t) = \frac{a_3}{2}[\cos(\omega_C + 3\omega_m)t + \cos(\omega_C - \omega 3_m)t] - \frac{a_1 \beta}{8}[\sin(\omega_C + 3\omega_m)t + \sin(\omega_C - 3\omega_m)t].$$

Even assuming no 3rd-order nonlinearity on the envelope, that is $a_3 = 0$, there is still 3rd-order distortion caused by AM–PM nonlinearity, leading to spectral regrowth. The spectrum, however, is still symmetric at $\pm 3\omega_m$ away from the carrier. The spectral asymmetry happens if we introduce a finite delay between the envelope and phase components. Assuming a phase shift of $\Delta = \omega_m \tau$ for a delay of τ between the envelope and phase, the RF signal may be expressed as

$$y(t) = (a_1 \cos(\omega_m t - \Delta) + a_3 \cos(3\omega_m t - 3\Delta)) \cos\left(\omega_C t + \frac{\beta}{2} \cos 2\omega_m t\right).$$

Upon expanding, we arrive at various components, of which we focus on only the terms appearing at IM$_3$ positive and negative sidebands, that is, the ones at $\omega_C \pm 3\omega_m$, representing the output distortion.

For positive sideband we can write

$$\left[\frac{a_3}{2} \cos 3\Delta + \frac{a_1 \beta}{8} \sin \Delta\right] \cos(\omega_C + 3\omega_m)t + \left[\frac{a_3}{2} \sin 3\Delta - \frac{a_1 \beta}{8} \cos \Delta\right] \sin(\omega_C + 3\omega_m)t,$$

while for the negative sideband we have

$$\left[\frac{a_3}{2}\cos 3\Delta - \frac{a_1\beta}{8}\sin\Delta\right]\cos(\omega_C - \omega 3_m)t + \left[-\frac{a_3}{2}\sin 3\Delta - \frac{a_1\beta}{8}\cos\Delta\right]\sin(\omega_C - 3\omega_m)t.$$

Therefore the magnitudes of the IM$_3$ signal residing at positive and negative sidebands are

$$|IM3_{P/N}| = \sqrt{\left(\frac{a_3}{2}\right)^2 + \left(\frac{a_1\beta}{8}\right)^2 \mp \frac{a_1 a_3 \beta}{8}\sin 2\Delta}.$$

If the phase shift $\Delta = \omega_m \tau$ is significant, clearly the two sidebands are not equal, leading to spectrum asymmetry. There are several reasons that such delay may exist. For instance, supply rail voltage variations induced by varying current in a class AB power amplifier could result in time constants comparable to the envelope variations.

6.7.6 Pulling in Transmitters

In many applications the strong signal at the power amplifier output may disturb the transmitter VCO used to generate the carrier. Known as *pulling*, this is particularly problematic if the carrier and PA output frequencies are the same, which is the case in a *direct-conversion transmitter*.[7] Shown in Figure 6.60 are two examples of a direct-conversion transmitter, one with the VCO at the output frequency, and the other at twice that. The latter approach puts the VCO frequency far away from the output, but as we will discuss shortly, it may be still pulled by the PA second harmonic. In both cases, the problem arises from the fact that the VCO output is a pure tone, at least ideally, whereas the PA output is a modulated spectrum.

The exact nature of pulling and its impact on the transmitters is beyond the scope of this book. While details can be found in [10], [11], we shall recast a summary here to gain some perspective for our discussion of transmitters in Chapter 12.

Generally, the pulling impact on an LC oscillator can be analyzed by solving what is known as Adler's differential equation [12],

$$\frac{d\theta}{dt} = \omega_0 + \frac{\omega_0}{2Q}\frac{I_{inj}}{I_S}\sin(\theta_{inj} - \theta),$$

where θ describes the oscillator phase, ω_0 is the un-pulled frequency of oscillation, Q is the LC tank quality factor, I_S is the oscillator current, and $I_{inj}e^{j\theta_{inj}}$ is the external source injected to the oscillator in complex format. If the injection is a tone, the nonlinear differential equation has a closed solution, but for most general cases, one may need to resort to numerical methods.

The pulling strength, η, is defined as follows:

$$\eta = \frac{\omega_0}{2Q}\frac{I_{inj}}{I_S}\frac{1}{|\omega_0 - \omega_{inj}|}.$$

---

[7] Various transmitter topologies will be discussed in Chapter 12.

Distortion

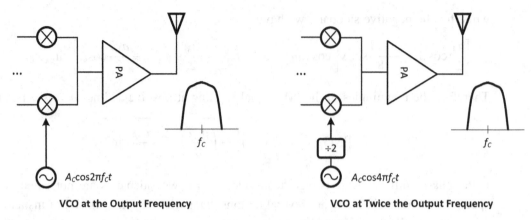

Figure 6.60: Illustration of pulling in direct-conversion transmitters

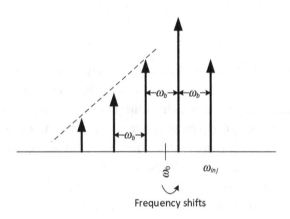

Figure 6.61: Oscillator pulled by an external source

What the pulling strength signifies is the fact that the further the frequency of the injected source is from the oscillator frequency, the weaker the pulling. In contrast, even if the injected current is small compared to the oscillator quiescent current, close-by injections may be quite troublesome. Thus as shown in Figure 6.60, designing the VCO at twice the frequency may be a more attractive choice.[8] We shall thus focus on the more common case of VCO followed by the divider. Although the pulling strength will be very small, presence of the 2nd-order nonlinearity still creates unwanted components at the VCO frequency.

If $\eta > 1$, the oscillator is locked to the injected source, that is, its frequency of oscillation changes to that of the injected source. In most cases $\eta < 1$, and solving the Adler's equation [12] reveals that the oscillation frequency shifts toward ω_{inj}, but its spectrum is corrupted by unwanted sidebands resulted from the injection source, as shown in Figure 6.61.

In the case of a direct-conversion transmitter, first the injection source is not arbitrary, as what is used to upconvert the baseband components is the VCO itself. Second, the VCO is not free running, rather is locked inside a PLL to a reference (Figure 6.62).

---

[8] There are other advantages, such as convenient quadrature generation. See Chapter 4.

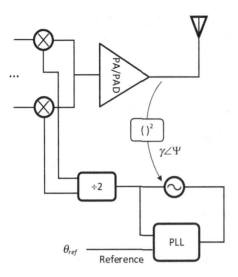

Figure 6.62: Locked VCO pulled by its transmitter

Thus, the VCO phase is affected by its own transmitter (unwanted), as well as the PLL interaction (wanted). Assuming the injection source is not too strong, which is a reasonable assumption, the Adler differential equation may be modified as follows to capture these effects,

$$\frac{d\theta}{dt} = \omega_0 + K_{VCO}\frac{I_{CP}}{2\pi}\left(\theta_{ref} - \frac{\theta}{N}\right) * h_{LF}(t) + \frac{\omega_0}{2Q}\frac{\gamma A_{BB}^2}{I_S}\sin(2\theta_{BB} - \Psi),$$

where K_{VCO} is the VCO gain, I_{CP} is the charge pump current, h_{LF} is the loop filter impulse response, θ_{ref} is the reference phase, N is the PLL divide ratio, and $A_{BB} \angle \theta_{BB}$ is the polar representation of the baseband signal fed to the upconversion mixer. The parameter $\gamma \angle \Psi$ is a complex number that captures the unwanted 2nd-order nonlinearity, and the finite isolation between the transmitter output and the VCO. Clearly, if $\gamma \angle \Psi = 0$, then the unwanted pulling term (the last term) goes away, and the VCO phase is locked to that of the reference as expected.

The above equation has a closed form solution. Defining $\theta' = \theta - \omega_0 t$ as the *phase perturbation*, then the frequency domain solution is

$$\Theta'(j\omega) = \frac{\omega_0}{2Q}\frac{1}{I_S}\frac{\mathcal{F}\{\gamma A_{BB}^2 \sin(2\theta_{BB} - \Psi)\}}{j\omega + \frac{K_{VCO}}{N}\frac{I_{CP}}{2\pi}H_{LF}(j\omega)}.$$

Apart from improving isolation, the pulling may be reduced by employing a higher Q resonator or raising the VCO current. Moreover, as $H_{LF}(j\omega)$ is typically lowpass, the transfer function of the denominator of the equation above is bandpass, typically peaking around the PLL 3dB bandwidth. This indicates that the baseband frequency contents falling around the PLL cutoff cause the most amount of pulling.

The equation above may be used to find the impact of pulling on EVM and modulation mask [11]. It is clear that as pulling only perturbs the VCO phase, it will not alter the magnitude of the constellation points; rather it would spread the constellation rotationally, a very similar impact as phase noise. Moreover, as the pulling is a direct function of the baseband signal, $A_{BB} \angle \theta_{BB}$,

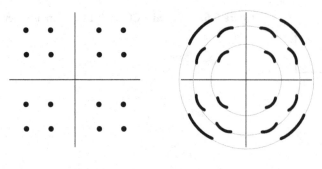

Figure 6.63: Impact of pulling on constellation

Ideal 16QAM   16QAM w/ Pulled VCO

the amount of rotational movements for each symbol of the constellation is proportional to the distance of that symbol from the center, as shown in Figure 6.63.

Similarly, we can show that the pulling causes spectral regrowth, an impact very similar to the 3rd-order nonlinearity. This is due to the fact that the pulling phase perturbation results in an error term that is twice as wide as the spectrum (given the square function cause by the 2nd-order nonlinearity), subtracted from the signal. This leads to a spectrum as wide as three times the wanted signal spectrum.

6.8 Summary

This chapter discussed various distortion mechanisms present in receivers and transmitters. This, along with noise, defines the system dynamic range.

- Section 6.1 presented a summary of blockers and their implications in wireless systems.
- Full duplex operation and its impact on the transceiver were discussed in Section 6.2.
- Section 6.3 mainly dealt with the receiver small signal nonlinearity, including 2nd-, 3rd-, and 5th-order nonlinearity, and cross-modulation. The impact of feedback on nonlinearity and dynamic range definition was also presented in this section.
- Receiver large signal nonlinearity and compression issues were discussed in Section 6.4.
- Section 6.5 dealt with the reciprocal mixing issue, which is primarily a receiver concern.
- Harmonic mixing in receiver, including image and half-IF concerns, was discussed in Section 6.6.
- Section 6.7 dealt with a variety of issues in transmitters such as spectral mask, out-of-band noise, phase error and EVM, AM–AM and AM–PM, and pulling.

6.9 Problems

1. Find the impact of the lowpass filter shown below on the IIP_3 of a nonlinear system following the filter:

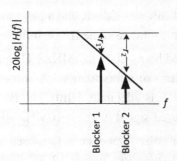

The blockers are subject to an attenuation of r_1 and r_2 (in dB) respectively.

2. Calculate the IIP_2, IIP_3, and 1dB compression points of a single BJT. **Answer:** At room temperature, the 1dB compression point is 24mV.

3. Prove the equation $A_{IIP3} = (1 + \theta'V_{eff})\sqrt{\frac{4}{3}\left(V_{eff}^2 + \frac{2V_{eff}}{\theta}\right)}$ for the IIP_3 of a single FET. Also find the IIP_2 of a single FET using the short channel I_D-V_{GS} equation.

4. Find the IIP_3 of a 2nd-order system $y = a_1x + a_2x^2$, with and without feedback. Show all the steps.

5. Find the IIP_3 of an inductively degenerated FET.

6. Find the cascaded IIP_2 of two nonlinear stages represented by up to 3rd-order nonlinearity.

7. A desired WLAN signal with 20MHz bandwidth is located at 2412MHz. A 20MHz LTE blocker located in band 40 (2300–2400MHz) is sitting at 2380MHz. Considering the spectral regrowth due to 3rd- or 5th-order distortion, show how this blocker may be problematic. Label all the frequencies.

8. For the previous problem, what is the minimum blocker spacing such that the 3rd-order distortion will not be an issue? What is the minimum for the 5th-order distortion not to be problematic?

9. A desired Bluetooth signal with 1MHz bandwidth is located at 2482MHz, and is accompanied by a 20MHz LTE blocker at 2510MHz. The system has an IIP_3 of 5dBm, and the blocker power is −15dBm. Find the minimum IIP_5 such that the IM_5 component created due to the spectral regrowth is at least 10dB below the IM_3 component.

10. Similar to the steps taken in Section 6.3.5 (Figure 6.29), work out the cross-modulation formulas in the presence of the 5th-order nonlinearity. Find the equivalent signal-to-noise ratio in terms of the system IIP_5. **Answer:** $XIM_5 = \frac{15}{4}a_5A_0A_{B1}^2A_{B2}^2\cos((\omega_0 \pm 2\Delta\omega)t)$.

11. A two-tone blocker at 2510 ± 1MHz accompanies a WLAN desired signal at 2472MHz. Assuming the system has nonlinearity up to 5th order, identify all the undesired sidebands around the signal at the output.

12. Consider a Bluetooth receiver with a noise figure of 3dB. Assuming an SNR of 14dB, and a bandwidth of 1MHz, find:
    a. The sensitivity.
    b. The equivalent noise floor at the input in dBm.
    c. Required IIP_2 for 3dB desensitization, assuming a −20dBm 20M LTE blocker is present (follow Figure 6.23 for blocker 2nd-order induced calculations).

d. Receiver SNR (including the thermal noise) if $IIP_3 = -2$dBm, and a two-tone blocker with 100kHz spacing and -15dBm level is present. Assume IIP_2 is infinite for this part.

13. A Bluetooth signal at 2.402GHz is accompanied by a -20dBm 5MHz LTE interferer close by. Find the receiver SNR set by the blocker cross-modulation. Assume the receiver $IIP_3 = -2$dBm. Argue whether the receiver IIP_3 is adequate. **Hint:** The BT signal has a bandwidth of 1MHz, whereas the blocker squared signal is about 10MHz wide.

14. The cross-modulation could elevate the receiver noise floor even in the absence of a desired signal due to the LO feedthrough. As we will show in Chapter 12, in direct-conversion receivers, an undesired tone at the frequency of the signal can leak to the receiver input. Explain how this could cause SNR degradation.

15. A two-tone nearby blocker at -20dBm is accompanied by a desired signal at -70dBm as shown below. Assume the receiver has an IIP_3 of 0dBm, and an IIP_2 of 60dBm. Find which is more problematic, the 2nd-order nonlinearity or cross-modulation?

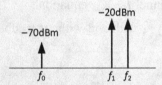

16. For the band-limited white noise example, show the output is $y(t) = \frac{A^2}{2}\sum_{i=0}^{N}\sum_{j=0}^{N}$ $\left[\cos\left((i+j)2\pi\Delta ft + \phi_i + \phi_j\right) + \cos\left((i-j)2\pi\Delta ft + \phi_i - \phi_j\right)\right]$. Show that for an arbitrary bin located a point $p\Delta f$ (p spans from 0 to $2N$), the number of i and j that satisfy $i+j=p$ is $\begin{cases} p+1 & p \leq N \\ 2N-p+1 & N < p \leq 2N \end{cases}$. Similarly, the number of i and j that satisfy $|i-j|=p$ is $\begin{cases} 2(N-p+1) & p \leq N \\ 0 & N < p \leq 2N \end{cases}$. Hence, the total is roughly $2N-p$, if N is large. Consequently find the area under the bin, and find the output spectral density. **Answer:** The area under the bin $p\Delta f$ is $\frac{1}{4}\left(\frac{A^2}{2}\right)^2(2N-p) = \left(\frac{\eta}{2}\right)^2 2B\left(1-\frac{p}{2N}\right)\Delta f$.

17. Calculate the required phase noise for a GSM receiver for a 0dBm blocker at 20MHz offset from the desired signal at -99dBm. What about a 1.6MHz blocker at -33dBm? Which one is more important if the phase noise has a roll-off of 40dB/Dec? **Answer:** -125dBc/Hz at 1.6MHz for 1.6MHz blocker.

18. A Bluetooth receiver has a base noise figure of 3dB, and a far-out phase noise of -150dBc/Hz. Assume the receiver is accompanied by an close-by -25dBm blocker.
    a. Find the overall noise figure of the receiver.
    b. If the receiver is preceded with an amplifier with 6dB of gain and 3dB noise figure, find the overall noise figure with and without the blocker.

19. A matched LNA has a noise figure of 2dB, and a nonlinear available voltage of $y = 10x - 10x^3$. Find IIP_3 and 1dB compression. What is the input-referred noise voltage of the mixer to achieve a total noise figure of 3dB? What is the SFDR? **Answer:** SFDR $= 86$dB.

20. Find the receiver noise figure, when a 0dBm blocker is present at the input of the receiver of Problem 19. The LO has a phase noise is of -160dBc/Hz at the blocker offset.

21. Find the OIP$_3$ and output 1dB compression of a 3G transmitter if an ACLR of –46dBc at 5MHz is desired. **Answer:** The output 1dB compression is 9dBm.

22. A WLAN transmitter has 26dBm of average output power, with 6dB PAPR. Assuming 40dB of isolation between the WLAN TX and a nearby LTE RX with 3dB noise figure, find the transmitter out-of-band noise such that the LTE receiver is desensitized by no more than 3dB.

23. Find the PAPR of the following signal: $x(t) = A_1 \cos \omega_1 t + A_2 \cos \omega_2 t$.

24. A 3G radio has a thermal noise figure of 3dB, IIP$_2$ of 45dBm, and the transmitter phase noise of –158dBc/Hz at the receiver frequency. There is a total front-end loss of 3dB from the duplexer to the radio. Find the minimum duplexer isolation to meet the sensitivity.

25. Prove the autocorrelation equation of the following signal: $x(t) = i(t) \cos \omega_c t - q(t) \sin \omega_c t$ discussed in the AM–PM section.

26. Using Bessel series, find the IM$_3$ and IM$_5$ components for the following RF signal: $y(t) = (a_1 \cos(\omega_m t - \Delta) + a_3 \cos(3\omega_m t - 3\Delta)) \cos(\omega_C t + \frac{\beta}{2} + \frac{\beta}{2}\cos 2\omega_m t)$.
Answer (IM$_3$ case): For $\cos((\omega_C \pm 3\omega_m)t + \frac{\beta}{2})$: $\left[\left(\frac{-a_1 J_2(\frac{\beta}{2})}{2} \cos\Delta + \frac{a_3 J_0(\frac{\beta}{2})}{2} \cos 3\Delta \right) \pm \left(\frac{a_1 J_1(\frac{\beta}{2})}{2} \sin\Delta + \frac{a_3 J_3(\frac{\beta}{2})}{2} \sin 3\Delta \right) \right]$. For $\sin((\omega_C \pm 3\omega_m)t + \frac{\beta}{2})$: $\left[\left(\frac{-a_1 J_1(\frac{\beta}{2})}{2} \cos\Delta + \frac{a_3 J_3(\frac{\beta}{2})}{2} \cos 3\Delta \right) \pm \left(\frac{a_1 J_2(\frac{\beta}{2})}{2} \sin\Delta + \frac{a_3 J_0(\frac{\beta}{2})}{2} \sin 3\Delta \right) \right]$. If: $\frac{\beta}{2} \ll 1$, $J_0(\frac{\beta}{2}) \approx 1$, $J_1(\frac{\beta}{2}) \approx \frac{\beta}{8}$, $J_2(\frac{\beta}{2}) \approx J_3(\frac{\beta}{2}) \approx 0$, and we arrive at the previous results.

6.10 References

[1] J. G. Proakis, *Digital Communications*, McGraw-Hill, 1995.

[2] K. S. Shanmugam, *Digital and Analog Communication Systems*, John Wiley, 1979.

[3] P. R. Gray and R. G. Meyer, *Analysis and Design of Analog Integrated Circuits*, John Wiley, 1990.

[4] R. Ellis and D. Gulick, *Calculus, with Analytic Geometry*, Saunders, 1994.

[5] R. G. Meyer and A. K. Wong, "Blocking and Desensitization in RF Amplifiers," *IEEE Journal of Solid-State Circuits*, 30, 944–946, 1995.

[6] H. Taub and D. L. Schilling, *Principles of Communication Systems*, McGraw-Hill, 1986.

[7] E. Cijvat, S. Tadjpour, and A. Abidi, "Spurious Mixing of Off-Channel Signals in a Wireless Receiver and the Choice of IF," *IEEE Transactions on Circuits and Systems II: Analog and Digital Signal Processing*, 49, no. 8, 539–544, 2002.

[8] J. Weldon, R. Narayanaswami, J. Rudell, L. Lin, M. Otsuka, S. Dedieu, L. Tee, K.-C. Tsai, C.-W. Lee, and P. Gray, "A 1.75-GHz Highly Integrated Narrow-Band CMOS Transmitter with Harmonic-Rejection Mixers," *IEEE Journal of Solid-State Circuits*, 36, no. 12, 2003–2015, 2001.

[9] S. Cripps, *RF Power Amplifiers for Wireless Communications*, Artech House, 2006.

[10] A. Mirzaei and H. Darabi, "Mutual Pulling between Two Oscillators," *IEEE Journal of Solid-State Circuits*, 49, no. 2, 360–372, 2014.

[11] A. Mirzaei and H. Darabi, "Pulling Mitigation in Wireless Transmitters," *IEEE Journal of Solid-State Circuits*, 49, no. 9, 1958–1970, 2014.

[12] R. Adler, "A Study of Locking Phenomena in Oscillators," *Proceedings of the IEEE*, 61, no. 10, 1380–1385, 1973.

7 Low-Noise Amplifiers

In Chapter 5 we showed that according to the Friis equation the lower limit to a receiver noise figure is set by the first block. Moreover, the gain of the first stage helps reduce the noise contribution of the subsequent stages. Hence, it is natural to consider a *low-noise amplifier* at the very input of a receiver. A first look then will tell us that for a given application, the higher the gain of the amplifier the better. However, the limit to that is typically imposed by the distortion caused by the blocks following the low-noise amplifier. As we stated earlier, there is a compromise between the required gain, noise, and linearity for a given application, and a certain cost budget. These requirements may be different for a different application or standard. In this chapter we assume that these requirements are given to us, and our goal is to understand the design trade-offs. In Chapter 12 we will study receiver architectures and various trade-offs associated with each choice of architecture.

What makes a low-noise amplifier (LNA) different from other typical amplifiers designed for analog applications is the need for the LNA to present a well-defined typically 50Ω input impedance to the outside world. That, along with a relatively higher frequency of operation limit the choice of topologies to only a handful. Most LNAs in practice use a variation thereof, one way or another.

We will start this chapter with a review of some of the basic concepts we presented in Chapter 5 and expand those a little further. We then describe three general categories of the LNAs, namely shunt feedback LNAs, series feedback LNAs, as well as feedforward LNAs. We close this chapter by considering some practical aspects of LNA design regardless of its topology. A simple case study in 16nm CMOS is presented as well. Also included in this chapter is an introductory discussion on signal and power integrity.

The specific topics covered in this chapter are:

- LNA matching concerns
- Amplifier design concerns for radio frequencies
- Series and shunt feedback LNAs
- Feedforward LNAs
- Impact of substrate and gate resistance on LNAs
- LNA layout and other practical considerations
- Signal and power integrity

For class teaching, we recommend covering most of this chapter, possibly with the exception of Sections 7.7 and 7.9, which may be assigned as reading. Sections 7.6 and 7.8 may be skipped as well if time does not permit, but we recommend presenting a brief summary at least.

7.1 MATCHING REQUIREMENTS

Consider a 50Ω source with an RMS amplitude of V_s, connected to the input of an amplifier through a lossless matching network comprising inductors, capacitors, or transformers as shown in Figure 7.1. The source could simply represent a simple model of the antenna connected to the receiver input. Shown in Figure 7.1 is also the Thevenin representation of the source and matching network through an equivalent voltage source (V_{TH}), an equivalent impedance (Z_{TH}), and a noise source modeling the equivalent noise of R_s seen by the amplifier.

The Thevenin equivalent may be obtained as depicted in Figure 7.2, and is described as follows:

First, let us represent the source with a Norton equivalent consisting of a shunt resistor and a current source I_s. Suppose the voltage appearing at the output of the matching network is $V(f)$, as the response to the source current I_s. We assume the transfer function from I_s to $V(f)$ is $H(f)$. Since the circuit consists of passive elements, it is *reciprocal*, and consequently, if we inject a

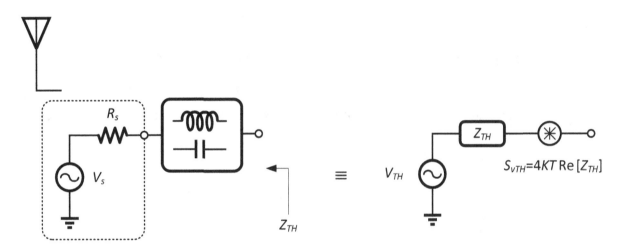

Figure 7.1: Source driving an amplifier through a lossless matching network

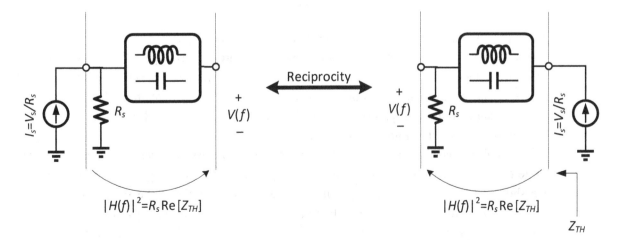

Figure 7.2: Finding the source Thevenin equivalent using reciprocity

current I_s to the *output* of the matching network, we expect the same voltage $V(f)$ appearing across the resistor as shown on the right side of the figure. Let us assume that the total impedance looking into the output of the matching network is Z_{TH}. The total power delivered to the matching network by the current I_s is then: $|I_s|^2 \operatorname{Re}[Z_{TH}]$, and the total power dissipated in the resistor is $\frac{|V|^2}{R_s}$.

Since the matching network is *lossless*, the two powers must be equal, which yields

$$|H(f)|^2 = \frac{|V|^2}{|I_s|^2} = R_s \times \operatorname{Re}[Z_{TH}].$$

From this, it follows that the total noise spectral density appearing at the matching network output is $4KT\operatorname{Re}[Z_{TH}]$, which was already proven in Chapter 5 (Nyquist theorem). Thus, the entire circuit may be represented by a Thevenin equivalent consisting of a voltage source V_{TH}, where

$$|V_{TH}|^2 = |V_s|^2 \frac{\operatorname{Re}[Z_{TH}]}{R_s},$$

an equivalent impedance Z_{TH}, and a noise source whose spectral density is

$$S_{v_{TH}} = 4KT\operatorname{Re}[Z_{TH}].$$

From the equivalent circuit model it readily follows that the total available power delivered to the matching network output is

$$P_a = \frac{|V_{TH}|^2}{4\operatorname{Re}[Z_{TH}]} = \frac{|V_s|^2}{4R_s},$$

and the available input noise is

$$N_a = \frac{4KT\operatorname{Re}[Z_{TH}]}{4\operatorname{Re}[Z_{TH}]} = KT.$$

Thus, the *lossless* matching network does not affect the available power and noise as expected. We shall use this Thevenin equivalent circuit in Figure 7.1 as a general representation of the source throughout this chapter. In addition, all the reactive components associated with package, pad, ESD, routing, and other parasitic capacitances or inductances that are generally low-loss can be lumped as part of the matching network. As shown in Figure 7.3, they may be represented by the Thevenin equivalent circuit.

Although we have assumed the matching network is lossless, in practice its elements have a finite Q, especially if implemented on-chip. In this case, still the Nyquist theorem is valid, as we showed in Chapter 5, and the total noise seen at the output of the matching network is $4KT\operatorname{Re}[Z_{TH}]$. Moreover, as a narrowband approximation, we can always bring out a parallel noisy resistor at the amplifier side to account for this loss. This will clearly result in noise figure degradation. We shall detail this further at the end of this chapter.

7.1 Matching Requirements

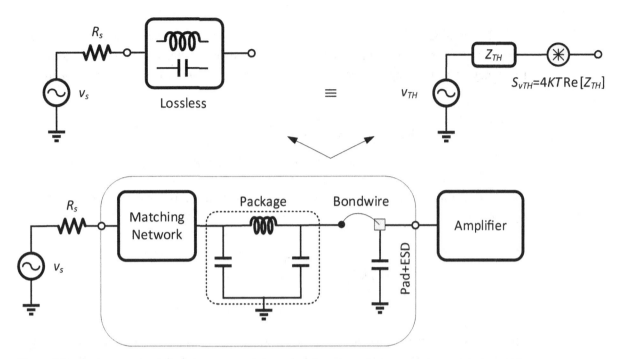

Figure 7.3: Equivalent circuit lumping the matching network, pad, package, and other reactive parasitic

Example: Let us consider the matching circuit of Figure 7.4, where we showed in Chapter 3 that it can be used to match the amplifier input impedance of R_{IN} to the source. Using parallel–series conversion we showed

$$L = \frac{R_{IN}}{\omega_0}\sqrt{\frac{R_s}{R_{IN} - R_s}}$$

$$C = \frac{1}{L_s \omega_0^2} = \frac{1}{\omega_0 \sqrt{R_s(R_{IN} - R_s)}}.$$

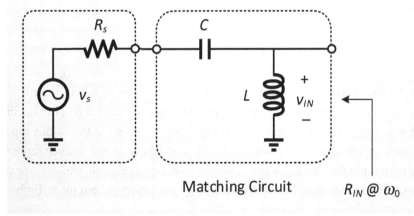

Figure 7.4: LC circuit to match amplifier input impedance to source

Continued

With the values of L and C given above, then we expect to see an impedance of R_{IN} looking into the right side of the matching circuit at exact frequency of ω_0. At an arbitrary frequency of ω, the output impedance of the matching network has a real component of

$$\text{Re}[Z_{OUT}] = R_s \frac{\left[\frac{R_{IN}}{R_{IN}-R_s}\left(\frac{\omega}{\omega_0}\right)^2\right]^2}{\left[1 - \frac{R_{IN}}{R_{IN}-R_s}\left(\frac{\omega}{\omega_0}\right)^2\right]^2 + \frac{R_s}{R_{IN}-R_s}\left(\frac{\omega}{\omega_0}\right)^2}.$$

Thus, according to the Nyquist theorem we expect the noise spectral density at the output to be

$$4KTR_s \frac{\left[\frac{R_{IN}}{R_{IN}-R_s}\left(\frac{\omega}{\omega_0}\right)^2\right]^2}{\left[1 - \frac{R_{IN}}{R_{IN}-R_s}\left(\frac{\omega}{\omega_0}\right)^2\right]^2 + \frac{R_s}{R_{IN}-R_s}\left(\frac{\omega}{\omega_0}\right)^2}.$$

Specifically, at $\omega = \omega_0$, the noise spectral density becomes $4KTR_{IN}$ as expected. Alternatively, let us find the transfer function from the source to the input of the amplifier:

$$\frac{V_{IN}}{V_s} = H(\omega) = -\frac{\frac{R_{IN}}{R_{IN}-R_s}\left(\frac{\omega}{\omega_0}\right)^2}{1 - \frac{R_{IN}}{R_{IN}-R_s}\left(\frac{\omega}{\omega_0}\right)^2 + j\sqrt{\frac{R_s}{R_{IN}-R_s}}\frac{\omega}{\omega_0}}.$$

Thus the output noise spectral density is

$$4KTR_s|H(\omega)|^2 = 4KTR_s \frac{\left[\frac{R_{IN}}{R_{IN}-R_s}\left(\frac{\omega}{\omega_0}\right)^2\right]^2}{\left[1 - \frac{R_{IN}}{R_{IN}-R_s}\left(\frac{\omega}{\omega_0}\right)^2\right]^2 + \frac{R_s}{R_{IN}-R_s}\left(\frac{\omega}{\omega_0}\right)^2},$$

which is the same as the result obtained by applying the Nyquist theorem. Moreover, at $\omega = \omega_0$, $|H(\omega)|^2 = \frac{R_{IN}}{R_s}$, which is again consistent with $\frac{\text{Re}[Z_{OUT}]}{R_s}$ predicted by the Nyquist theorem.

Example: An alternative to the previous matching network is shown in Figure 7.5. Both matching circuits downconvert the amplifier input impedance ($R_{IN} > R_s$), but the circuit of Figure 7.5 is lowpass, whereas Figure 7.4 is highpass. Another feature of this topology is that there is a series inductor to the source, which may be combined or even replaced with the bondwire or input routing inductance, whereas the input parasitic capacitance could be lumped in C. The reader may find more details by analyzing the circuit in Problem 2.

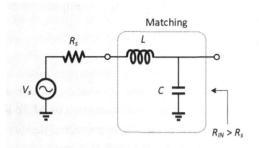

Figure 7.5: Lowpass LC matching circuit to downconvert the amplifier input impedance

As stated before, if the amplifier has an input capacitance, it can be simply lumped within the matching inductor. This is shown in Figure 7.6, where the matching inductor consists of two parallel legs, one that resonates with amplifier input capacitance C_{IN}, and the other whose value was obtained before and converts R_{IN} to R_s.

Furthermore, the loss of the inductor is shown as a parallel resistor, R_p, which effectively modifies the amplifier input resistance, and must be taken into account in the design of the matching network.

In Chapter 5 we also showed that for a given amplifier represented by its input-referred noise currents and voltages, there is an optimum value of the source impedance, $Z_{TH} = Z_{s,opt}$, to minimize the noise. On the other hand, source matching requires $Z_{TH} = Z_{IN}^*$, where Z_{IN} is the amplifier input impedance (Figure 7.7). As we discussed in Chapter 5, $Z_{s,opt}$ is not a physical resistor but rather a function of the device noise parameters, whereas Z_{IN} is a physical

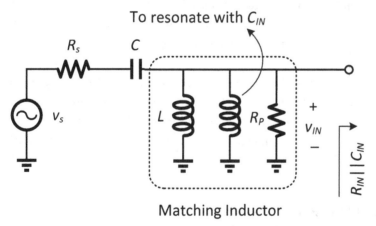

Figure 7.6: Matching an amplifier input impedance with reactive component

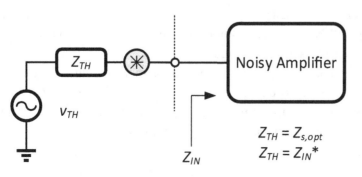

Figure 7.7: Noisy amplifier connected to the source

impedance, and thus in general there is no reason for the two condition to be satisfied simultaneously.

> **Example:** To understand the above limit better, let us assume that the input impedance of the amplifier in Figure 7.7 has a noise voltage with a spectral density of $4KT\alpha\,\text{Re}[Z_{IN}]$, where α is the noise factor. The total input-referred noise voltage spectral density is then $4KT\,\text{Re}[Z_{TH}] + 4KT\alpha\,\text{Re}[Z_{IN}]$, and by definition the noise factor is
>
> $$F = 1 + \alpha \frac{\text{Re}[Z_{IN}]}{\text{Re}[Z_{TH}]}.$$
>
> The matching requires $Z_{IN} = Z_{TH}{}^*$, and the overall noise factor simply becomes
>
> $$F = 1 + \alpha.$$
>
> Thus, for the simple case of realizing the real part of the input impedance by a physical resistor, $\alpha = 1$, $NF = 3\text{dB}$, which is often considered relatively poor. An example of such realization is a common-gate topology (which will be discussed in the next section) that has a subpar noise figure.

To overcome this fundamental drawback, two topologies may be sought. The first topology relies on the feedback principle. As we showed in Chapter 5, while a noiseless feedback does not affect the circuit minimum noise figure, it may be exploited to alter the amplifier input impedance favorably. We will discuss both local shunt and series feedback topologies. In either case, the feedback attempts to provide a noiseless (ideally) real part of the input impedance.

Another scheme relies on noise cancellation, where intentionally the optimum noise conditions are not met in favor of a 50Ω matching. The excess noise is only subsequently canceled at the output through a feedforward path. Before presenting the LNA topologies, we shall have a general discussions on the RF amplifier design, as it differs from the conventional analog amplifiers in certain aspects.

7.2 RF TUNED AMPLIFIERS

To emphasize the challenge of realizing a low-noise 50Ω-matched amplifier, let us first take a closer look at the two well-known analog amplifiers, namely the common-source (CS) and common-gate (CG) topologies as shown in Figure 7.8.

Both topologies are simple enough that given the high f_T of current modern CMOS technologies, they can operate at several GHz frequencies. A choice of active load for either structure is not common for two reasons: First, it limits the amplifier bandwidth, and second, it adds noise. To clarify the latter, consider a CS stage for instance, using active load as shown in Figure 7.9.

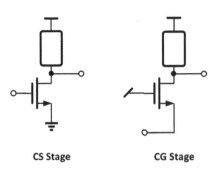

Figure 7.8: Common-source and common-gate stages. Bias details not shown.

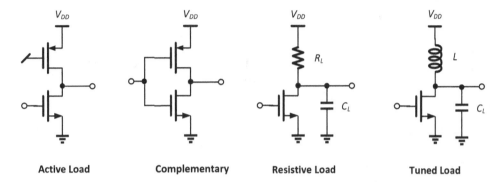

Figure 7.9: Variations of CS amplifier load

The input-referred noise voltage can be shown to be

$$\overline{v_n^2} = \frac{4KT\gamma}{g_{mn}}\left(1 + \frac{g_{mp}}{g_{mn}}\right) = \frac{4KT\gamma}{g_{mn}}\left(1 + \frac{V_{effn}}{V_{effp}}\right),$$

where V_{eff} is the gate overdrive voltage. To minimize the noise, one may suggest increasing the PMOS overdrive voltage (V_{effp}) to minimize $\frac{V_{effn}}{V_{effp}}$. On the other hand, for both transistors to stay in saturation, $V_{DD} > V_{effn} + V_{effp}$. Consequently, given the low supply voltages in most modern CMOS processes, the PMOS overdrive must be kept small and perhaps comparable to that of the NMOS device, effectively almost doubling the input-referred noise.

This problem is resolved if a complementary stage [1] is adopted as shown in Figure 7.9, where the P device not only adds noise, but also contributes to the overall transconductance if sized properly. In fact this structure leads to the same noise and transconductance at half the bias current. As traditionally PMOS devices have about two times worse mobility, sizing them up to achieve a similar g_m as the NMOS device, leads to extra capacitance and worse bandwidth which may not be acceptable. Moreover, at low supply voltage the headroom becomes an issue that potentially leads to further degradation of the bandwidth, and of course the linearity. In more recent technologies, the PMOS devices appear to be much closer to the NMOS transistors, and thus using a complementary structure may be more advantageous. As a comparison, Figure 7.10 shows the DC gain of minimum channel N and P devices versus transistor width for a constant bias current of 1mA for 40nm processes. Evidently, in 40nm P devices have about 50% less DC gain biased at the same overdrive voltage.

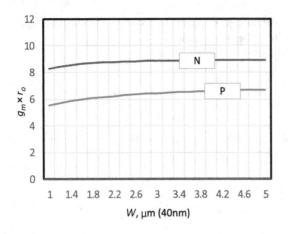

Figure 7.10: DC gain of N and P devices in 40nm CMOS

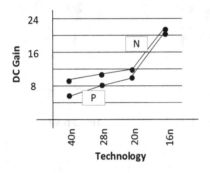

Figure 7.11: Peak DC gain of N and P MOS devices versus technology

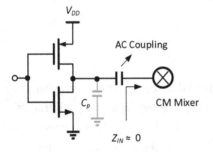

Figure 7.12: Complementary amplifier driving a current-mode mixer

The technology trend for the recent CMOS processes is shown in Figure 7.11. Not only the DC gain has been improving, but also the gap between N and P devices is becoming smaller. The sudden increase in the DC gain of 16nm devices may be attributed to the FinFET structure adopted.

In addition to the trend pointed out above, the complementary structure may become much more attractive in cases where the amplifier is connected to a device with low input impedance, such as a current-mode mixer (see the next chapter for more details), as shown in Figure 7.12. Both headroom and bandwidth concerns will be less important if the mixer input impedance is low enough to keep the swing at the amplifier output low, and prevent the amplifier current from flowing into parasitic capacitances (for example, into C_p in Figure 7.12). In this case the amplifier behaves more like a *transconductance* stage where the net g_m is the sum of the N and P devices' transconductances.

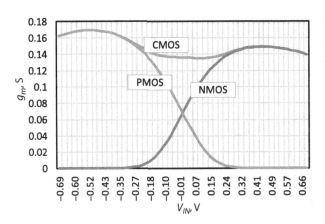

Figure 7.13: Transconductance of the complementary amplifier versus input voltage

Besides the features mentioned for the complementary amplifier of Figure 7.12, there are also some interesting linearity advantages. First, if the devices are sized properly, even in a single-ended design, the 2nd-order nonlinearity of the N and P devices cancel to the first order, leading to high IIP_2. Second, when the amplifier is driven by a large input, the push–pull operation of the N and P devices helps maintain a constant overall transconductance, leading to superior 1dB compression and IIP_3. Shown in Figure 7.13 is the large signal g_m for N and P devices as well as the net total transconductance for a 40nm design versus the input signal level. Even though the N and P devices alone experience substantial distortion, the net transconductance of the complementary structure stays relatively flat for a wide input range.

Turning back to various load structures in Figure 7.9, another possibility for the load may be a linear resistor, which could be superior in terms of bandwidth for applications where the amplifier is not connected to a low-input impedance stage, such as a voltage-mode amplifier. The input-referred noise voltage is

$$\overline{v_n^2} = \frac{4KT}{g_m}\left(\gamma + \frac{1}{g_m R_L}\right),$$

which is clearly less than the case of active load if the amplifier gain, that is $g_m R_L$, is large.

Assuming the DC drop across the resistor is half the supply voltage, the amplifier low-frequency gain is

$$g_m R_L \approx \frac{I}{V_{eff}} \times \frac{\frac{V_{DD}}{2}}{I} = \frac{V_{DD}}{2V_{eff}},$$

where I is the device bias current, and we assume the transconductance of short channel devices is $g_m \approx \frac{I}{V_{eff}}$. To achieve a reasonable linearity, from our discussion in the previous chapter, we would like to maintain an overdrive voltage of say no less than 100mV, which leads to a total gain of only 6 for a 1.2V supply device. Furthermore, if the device transconductance is chosen to be say 50mS,[1] the bias current will be 5mA, and the load

---

[1] To justify this particular value of g_m, we note that a transconductance of 20mS corresponds to an input-referred noise voltage roughly equivalent to that of a 50Ω resistor. Thus a g_m of 50mS may be sufficient to guarantee a low noise figure. We will present an accurate discussion shortly.

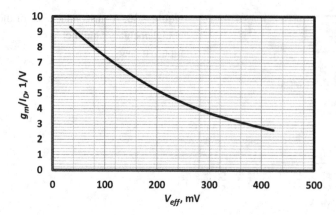

Figure 7.14: NMOS g_m/I_D versus overdrive voltage

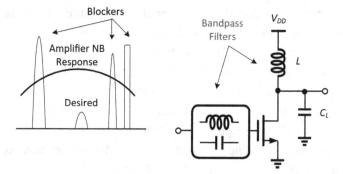

Figure 7.15: Filtering resulted from the amplifier tuned load

resistance is 120Ω. To maintain a flat passband for up to 2GHz for instance, the bandwidth may be set to 5GHz, in which case the total parasitic capacitance at the amplifier output cannot exceed 265fF. To find out if our hand calculations check out, shown in Figure 7.14 is the device g_m/I_D for 40nm process versus the gate overdrive voltage. For an overdrive voltage of 100mV, g_m/I_D is about 7.5V^{-1}, and hence a transconductance of 50mS requires a bias current of 6.7mA somewhat higher than what we predicted, but not too far. Also note that a gain of 6 is somewhat optimistic, if the device DC gain is around 10. This, however, may be improved if a cascode structure is used.

The amplifier gain may dramatically improve if it is loaded by an inductor, resonating with the total parasitic capacitance at the output. Considering our previous example, if the total parasitic capacitance is for instance 500fF, which is well above the 265fF limit for the resistive load, an inductance of 12.7nH is needed to resonate at 2GHz. A Q of only 5 leads to an effective parallel resistance of $L\omega_0 Q = 796\Omega$, leading to gain of roughly 40.[2] Furthermore, the device drain is now biased at roughly the supply voltage,[3] leading to better linearity and less junction capacitance at the drain. In addition, the relatively narrowband (NB) bandpass nature of the load (as well as the input matching) provides some filtering which attenuates the large far out-of-band blockers (Figure 7.15).

---

[2] We can show that there is an overall improvement of Q times in gain, Q being the inductor quality factor. See Problem 19 for more details.

[3] For a modest Q of 5, the inductor series resistance is 32Ω, leading to a DC drop of 160mV for a 5mA bias current, as opposed to 600mV for resistive load.

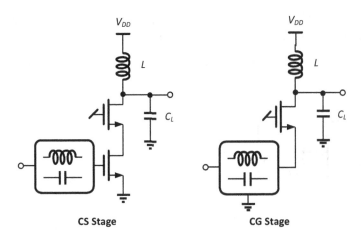

Figure 7.16: RF common-source and common-gate amplifiers

The main disadvantage is the extra silicon area as the inductors are relatively bulky compared to transistors or resistors. Nonetheless, in many applications the aforementioned advantages are overwhelming enough that make the tuned amplifiers a popular choice, despite the extra cost.

Finally, we may consider using *cascode* structure to improve the DC gain as well as the reverse isolation,[4] which is important especially at radio frequencies.

7.3 COMMON-SOURCE AND COMMON-GATE LNAs

Inserting some form of LC matching network at the input, Figure 7.16 presents the common-source and common-gate *RF equivalent* of the analog amplifiers originally shown in Figure 7.8. The two structures shown in Figure 7.16, or some variations thereof, are popular choices for RF LNAs. Note that in the case of CG circuit, the cascode may not be needed for isolation reasons, but still may be used to improve the amplifier gain. Also the matching network is expected to provide a convenient DC current path to ground to avoid biasing by a current source for better noise performance.

So far we have considered only the gain and linearity concerns, and we are yet to discuss the noise properties of the two amplifiers shown in Figure 7.16. Let us start with the CS stage. We include C_{gs} as a part of the matching network as well as any other reactive component associated with the design, and we will ignore C_{gd} for now.[5] If C_{gd} is negligible, then the amplifier input is purely capacitive, and it can be matched to 50Ω only if a parallel resistor is inserted at the input as shown in Figure 7.17. This resistor need not be a separate component and may be combined with the matching network, if its loss is high enough.

Using the equivalent Thevenin model for the source, as also shown in the figure, the noise figure can be readily calculated to be

---

[4] That is, to reduce the gain from the output to the input.
[5] The gate-drain capacitance creates a local shunt feedback that we will study in the next section.

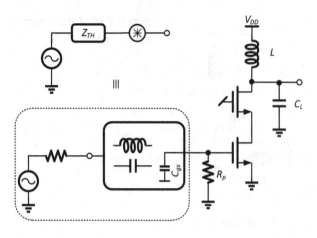

Figure 7.17: CS LNA matched by a shunt resistor

$$F = 1 + \frac{|Z_{TH}|^2}{R_p \, \text{Re}[Z_{TH}]} + \left|1 + \frac{Z_{TH}}{R_p}\right|^2 \frac{\gamma + \frac{1}{g_m R_L}}{g_m \, \text{Re}[Z_{TH}]},$$

where we assume R_L represents the equivalent parallel resistance due to the finite Q of the load inductor when resonating with C_L at the frequency of interest. The noise equation may be greatly simplified when realizing that to satisfy the matching requirement, $Z_{TH} = Z_{IN}{}^* = R_p$. This leads to

$$F = 2 + \frac{4}{g_m R_p}\left(\gamma + \frac{1}{g_m R_L}\right) > 2.$$

Even if g_m and R_L are made arbitrarily large, the amplifier noise figure never goes below 3dB.

This brings us to a very important conclusion that will be a recurring theme of this chapter: As long as the *real part* of the amplifier input impedance is made up of a *noisy resistor*, the noise figure 3dB barrier cannot be broken. This can be generally proven (see also the example discussed in Section 7.1), although we showed here only a more special case of a CS stage only.

In our analysis above, we assumed that the LNA needs to be perfectly matched to 50Ω. Not only is this not possible in practice, in most cases it is not even necessary. For most applications, an input return loss of −10dB or better is acceptable.[6] The input return loss may be traded for a better noise figure.

> **Example:** For the CS LNA, let us assume $Z_{TH} = \frac{R_p}{2}$. This implies that the parallel resistor is larger (to be precise twice as large) than what is needed for a perfect 50Ω match. The resultant input return loss (or reflection coefficient) is
>
> $$|S_{11}| = \left|\frac{R_p - Z_{TH}}{R_p + Z_{TH}}\right| = \frac{1}{3} \approx -10\text{dB}.$$

---

[6] There is generally no hard requirement for S_{11}. For most external filters or duplexers the stop band rejection or passband loss is affected minimally if $|S_{11}|$ is around 10dB or better.

If the load noise contribution is small, and assuming $g_m R_p \gg 1$, the noise figure will be

$$F = 1.5 + \frac{4.5}{g_m R_p}\left(\gamma + \frac{1}{g_m R_L}\right) \approx 1.5 = 1.76\text{dB},$$

which is a far more acceptable value. While there are more elegant ways of realizing LNAs with superior noise figure, this CS design may be adequate for many applications. In fact, if the matching network is on-chip, the shunt resistor R_p may be simply incorporated as a part of the matching circuitry loss.

To elaborate further on the example, let us obtain the noise figure as a general function of S_{11}. To do so, we need to obtain Z_{TH} as a function of R_p and S_{11} ($Z_{TH} = R_p$ only if $S_{11} = 0$), and use the general noise equation obtained earlier. For simplicity, we assume S_{11} is real. In that case we can show:

$$F = \frac{2}{1+S_{11}} + \frac{4}{(1+S_{11})^2} \frac{\gamma + \frac{1}{g_m R_L}}{g_m \operatorname{Re}[Z_{TH}]}.$$

If perfectly matched, $|S_{11}| = 0$, and we arrive at the same equation as before where $F > 2$. On the other hand, if we let R_p approach infinity, then $|S_{11}|$ approaches 1, and we have

$$F = 1 + \frac{\gamma + \frac{1}{g_m R_L}}{g_m \operatorname{Re}[Z_{TH}]},$$

which is expected as only the device and load contribute noise. Clearly, to improve noise figure we would like S_{11} to be positive, which is to say, we would like $R_p > Z_{TH}$.

A plot of noise figure versus the magnitude of S_{11} for two values of $g_m \operatorname{Re}[Z_{TH}]$ of 10 and 20 is shown in Figure 7.18. The load noise is ignored, and $\gamma = 1$.

The common-gate amplifier may seem more favorable as it does provide a real input impedance inherently as a part of the design. However, we shall prove shortly that it suffers from a similar fundamental drawback. Consider the CG LNA shown in Figure 7.19, where, the same as before, C_{gs} along with other reactive elements are considered a part of the matching network. Moreover, we assume the matching network creates a DC path to ground, say through a shunt inductor. Otherwise, a current source is needed to bias the amplifier, which results in

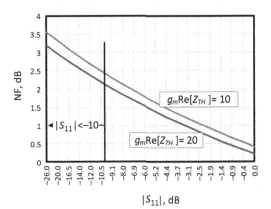

Figure 7.18: Noise figure versus return loss

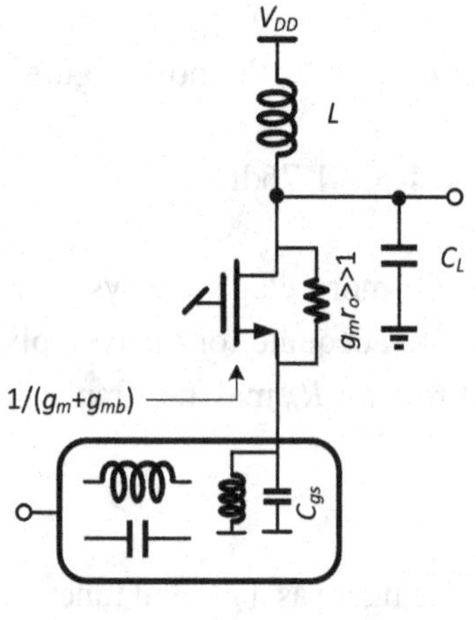

Figure 7.19: Common-gate LNA

unnecessary excess noise. C_{gd} is included as a part of C_L, and we assume that $g_m r_o$ is large, so the impact of r_o is ignored.

Under those assumptions, the LNA input impedance is purely real, and equal to

$$Z_{IN} = \frac{1}{g_m + g_{mb}},$$

where $g_{mb} = \frac{\frac{\sqrt{2\epsilon q N_A}}{C_{OX}}}{2\sqrt{2\phi_B + V_{BS}}} g_m = \chi g_m$ represents the body effect as a function of the process parameters (N_A, ϕ_B, and C_{OX}). For reasons that we will outline shortly, we choose not to ignore the body effect. The noise figure general expression is

$$F = 1 + \frac{1}{(g_m + g_{mb})\operatorname{Re}[Z_{TH}]}\left(\gamma\frac{g_m}{g_m + g_{mb}} + \frac{|1 + (g_m + g_{mb})Z_{TH}|^2}{(g_m + g_{mb})R_L}\right),$$

and since $Z_{TH} = Z_{IN}^* = \frac{1}{g_m + g_{mb}}$, it becomes

$$F = 1 + \gamma\frac{g_m}{g_m + g_{mb}} + \frac{4}{(g_m + g_{mb})R_L} \approx 1 + \gamma\frac{g_m}{g_m + g_{mb}}.$$

Ignoring the body effect, and assuming $\gamma \approx 1$, we arrive at the same 3dB noise figure, simply because the input resistance is created by a *noisy component*, in this case the transistor itself. Despite the same 3dB noise figure, the CG LNA is still superior to the CS stage shown in Figure 7.17 for two reasons: First, unlike the CS LNA where g_m needs to be large to ensure the transistor noise contribution is small, for the CG LNA g_m is given, roughly set to 20mS, for a 50Ω match. For an overdrive voltage of 100mV, this leads to a considerably lower bias current of 2mA. Second, the body effect helps improve the noise figure to some extent.[7] In a typical 40nm design, at 0V body-source bias, g_{mb} comes out to be roughly one-fifth of g_m.

Furthermore, considering the fact that γ is also close to 0.9 to be exact and not one for minimum channel devices, the noise figure becomes 2.4dB. This again has to do with the fact that g_{mb}, which is contributing to the input impedance, is noiseless.

---

[7] That is no longer the case in recent *FinFET* topologies, however, where the gate tight control of the channel effectively diminishes the body effect.

Example: We shall work out the common-gate LNA noise figure as a function of its return loss.

Since $S_{11} = \frac{\frac{1}{g_m+g_{mb}} - Z_{TH}}{\frac{1}{g_m+g_{mb}} + Z_{TH}}$, we have

$$Z_{TH} = \frac{1}{g_m + g_{mb}} \frac{1 - S_{11}}{1 + S_{11}}.$$

Ignoring the load noise, we can use the original noise equation derived above, and simply replace Z_{TH} with its S_{11} equivalent:

$$F \approx 1 + \frac{1}{(g_m + g_{mb}) \operatorname{Re}[Z_{TH}]} \frac{\gamma g_m}{g_m + g_{mb}} = 1 + \frac{\gamma g_m}{g_m + g_{mb}} \frac{1}{\operatorname{Re}\left[\frac{1 - S_{11}}{1 + S_{11}}\right]}.$$

If a return loss of −10dB is acceptable (i.e., $S_{11} \approx \frac{-1}{3}$), for instance, ignoring the load noise, and with $\gamma = 0.9$, and $g_{mb} = \frac{g_m}{5}$, the noise figure becomes 1.4dB, all achieved at a relatively small bias current.

The 3dB noise barrier (for a perfectly matched design at least) in the CS and CG LNAs may be broken if the real part of the LNA input impedance is made up of a noiseless resistor. This can be done by incorporating noiseless feedback, where the LNA input impedance is reduced/increased by applying shunt/series feedback favorably. We shall study both cases next.

7.4 SHUNT FEEDBACK LNAs

Consider the amplifier shown in Figure 7.20, where Z_F represents the local shunt feedback. We assume both the feedback and load impedances are complex and noisy, and represent them with their associated noise currents.

We include the C_{gs} as a part of the matching network, and C_{gd}, as a part of Z_F. The feedback circuitry, that is the impedance Z_F, can be represented by a Y network as is common in shunt

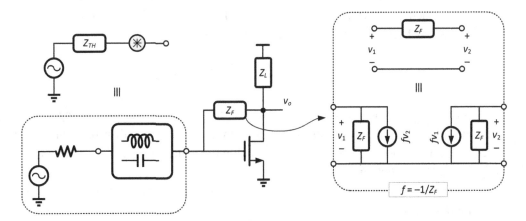

Figure 7.20: Amplifier with local shunt feedback

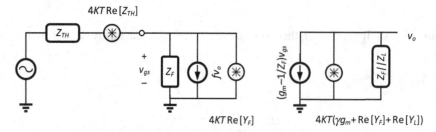

Figure 7.21: Equivalent circuit of the shunt feedback amplifier

feedback networks and is shown on the right side of Figure 7.20. Doing so, we arrive at the equivalent circuit of Figure 7.21 representing the feedback amplifier. Although commonly the *feedforward gain* of the feedback circuit is ignored, we have not done so here, and simply modified the amplifier transconductance gain to $g_m - \frac{1}{Z_F}$. At the particular frequency when $g_m = \frac{1}{Z_F(s)}$, the feedforward gain leads to a zero in the transfer function, as is evident from the frequency response analysis of a CS amplifier [2].

The loop gain is

$$\left(g_m - \frac{1}{Z_F}\right)(Z_F \| Z_L).$$

Thus the amplifier input impedance will be

$$Z_{IN} = \frac{Z_F}{1 + \left(g_m - \frac{1}{Z_F}\right)(Z_F \| Z_L)} = \frac{Z_F + Z_L}{1 + g_m Z_L},$$

and the amplifier noise figure is

$$F = 1 + \frac{\gamma g_m + \text{Re}[Y_L]}{\text{Re}[Z_{TH}]} \left|\frac{Z_F + Z_{TH}}{g_m Z_F - 1}\right|^2 + \frac{\text{Re}[Y_F]}{\text{Re}[Z_{TH}]} \left|\frac{g_m Z_{TH} + 1}{g_m Z_F - 1}\right|^2$$

To simplify and gain some insight, let us assume the noise of the feedback circuit can be ignored (the last term). This is usually a reasonable assumption in a well-designed feedback amplifier. Then, satisfying the matching condition, $Z_{TH} = Z_{IN}^*$, and given that the input impedance has been already calculated, we arrive at

$$F = 1 + \left|\frac{Z_F + Z_{TH}}{g_m Z_F - 1}\right|^2 \frac{\gamma g_m + \text{Re}[Y_L]}{\text{Re}[Z_{IN}]}.$$

The term $\left|\frac{Z_F + Z_{TH}}{g_m Z_F - 1}\right|^2$ arises from the impact of feedback impedance loading at the input. If the feedback impedance consists of a capacitance in parallel with a resistance ($R_F \| C_F$), the feedback capacitance (C_F) may be lumped within the matching circuit at the input, and the load capacitance at the output.

Now let us consider a few commonly used LNAs that can be treated as special cases of our general derivation.

7.4.1 Resistive Feedback with Large Loop Gain

This is the common shunt feedback amplifier case, and it assumes $|g_m Z_L| \gg 1$. If the feedback resistor is large, then

$$\left|\frac{Z_F + Z_{TH}}{g_m Z_F + 1}\right|^2 \approx \frac{1}{g_m^2},$$

and since

$$Z_{TH}^* = Z_{IN} = \frac{Z_F + Z_L}{1 + g_m Z_L} \approx \frac{1}{g_m}\left(1 + \frac{Z_F}{Z_L}\right),$$

the noise equation then simplifies to

$$F = 1 + \frac{\gamma + \frac{\text{Re}[Y_L]}{g_m}}{\text{Re}\left[1 + \frac{Z_F}{Z_L}\right]} \approx 1 + \frac{\gamma}{\text{Re}\left[1 + \frac{Z_F}{Z_L}\right]},$$

which shows that it is indeed possible to break the 3dB noise limit by maximizing $\text{Re}\left[1 + \frac{Z_F}{Z_L}\right]$. For a purely resistive feedback, and thus ignoring C_{gd}, we have

$$\text{Re}\left[1 + \frac{Z_F}{Z_L}\right] = \text{Re}\left[1 + R_F\left(\frac{1}{R_L} + j\omega C_L\right)\right] = 1 + \frac{R_F}{R_L},$$

where we assumed the load impedance Z_L, consists of R_L and C_L. Thus

$$F = 1 + \frac{\gamma}{1 + \frac{R_F}{R_L}},$$

and achieving a sub-3dB noise figure is fairly straightforward if the load and feedback resistances are chosen properly.

Note that our initial assumption has been $|g_m Z_L| \gg 1$, as it should be, otherwise the amplifier gain and bandwidth will suffer.

For many applications where $R_F C_{gd} \omega$ can be kept small (this assumes the transistor C_{gd} is small), a good design may incorporate a complementary structure with resistive feedback as shown in Figure 7.22.

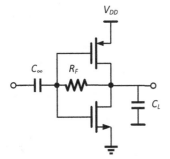

Figure 7.22: Complementary shunt feedback LNA

Clearly, C_L needs to be minimized as much as possible. Moreover, there may be no need for an explicit R_L, as it can be realized by the inherent r_o of the devices (which are noiseless), as long as the condition $R_F > r_{on}\|r_{op}$ is met.

One important drawback of this scheme is the direct dependence of noise figure and input impedance on the LNA load impedance. Thus, either the LNA load must be very well characterized, and behave favorably over process or temperature variations, or a voltage buffer may be needed to isolate the LNA output from the possible variations of the mixer input impedance.

7.4.2 CS Cascode LNA

A second special case is a CS cascode LNA that was discussed before, but this time we include C_{gd}, and denote that more generally as C_F in case one choses to include additional capacitance. The rationale behind this will be clear shortly. Without any feedback capacitance, we saw that the LNA can be matched only through a lossy matching network, leading to $> 3\text{dB}$ noise figure, unless the return loss is traded for a better noise. Since the feedback is purely capacitive, it is noiseless, and moreover, the feedback loading consists of a capacitance C_F at the input and the output of the amplifier. The former will be lumped into the matching network, where the latter is added to the load capacitance.

Let us assume that the total capacitance at the cascode node is C_p, and that the main and cascode devices have the same transconductance. Moreover, we assume that $\frac{g_m}{\omega C_p} \gg 1$. Thus the signal and noise pass through the cascode device and appear at the output without any loss.

At the cascade node, the load impedance including C_F is

$$Z_L = \frac{1}{j(C_F + C_p)\omega} \Big\| \frac{1}{g_m} = \frac{1}{g_m + j\omega(C_F + C_p)}$$

where $f = -j\omega C_F$ is the feedback factor. The input impedance is

$$Z_{IN} = \frac{-1}{f(g_m + f)Z_L} = \frac{1 + \frac{C_p}{C_F} + \frac{g_m}{j\omega C_F}}{g_m - j\omega C_F}.$$

Note that C_F appearing at the input (Figure 7.21) has been already lumped in the matching network, and thus our expression for the input impedance does not include it. The input impedance has a real part, which is

$$\text{Re}[Z_{IN}] = \frac{g_m\left(2 + \frac{C_p}{C_F}\right)}{|g_m + j\omega C_F|^2},$$

and the noise factor is

$$F = 1 + \frac{2\gamma gm}{|gm - j\omega C_F|^2 \text{Re}[Z_{TH}]} = 1 + \frac{2\gamma gm}{|gm - j\omega C_F|^2 \text{Re}[Z_{IN}]} = 1 + \frac{\gamma}{1 + \frac{C_p}{2C_F}}$$

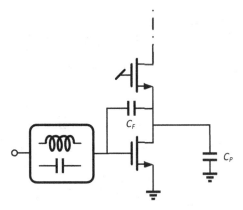

Figure 7.23: Cascode LNA incorporating shunt capacitive feedback

Note the cascode device noise is assumed to be $4KT\gamma g_m$, and has not been ignored. The equation above shows that the noise figure 3dB limit may be now broken if $\frac{C_p}{2C_F} \gg 1$. Even if C_F includes only the gate-drain capacitance, this condition is fairly comfortably met, as C_p consists of the source and drain junction capacitances of the cascode and main transistors, as well as the cascode device C_{gs}.

It is constructive to take a look at the imaginary part of the input impedance as well:

$$\text{Im}[Z_{IN}] = \frac{\frac{-g_m^2}{\omega C_F} + \omega(C_F + C_p)}{|g_m + j\omega C_F|^2} \approx \frac{-1}{\omega C_F}$$

If the feedback consists of only C_{gd}, then the amplifier input impedance consists of a real component in series with a relatively small capacitance, and thus it may be very hard to match. This also leads to a relatively large gain from the source to the amplifier input, which degrades the linearity and makes the matching too sensitive. Moreover, the loss of the large matching inductance could potentially lead to a lower bound on the noise figure that can be practically achieved in this topology. Nonetheless, the fact that no additional inductance is required to produce a real part for Z_{IN} may still make this topology quite attractive.

Ironically, if there is no parasitic capacitance at the cascode node, $\text{Re}[Z_{IN}] = \frac{1}{2g_m}$, and the noise factor will be $F = 1 + \gamma$!

7.5 SERIES FEEDBACK LNAs

A *local series feedback* may be incorporated by applying degeneration to a CS amplifier as shown in Figure 7.24 [2]. Let us assume that the degeneration is generally represented by a noisy impedance of Z_s, not to be confused with the input source impedance, and for simplicity, we shall ignore the body effect. Moreover, we ignore C_{gd}, otherwise the calculations will become too complicated. The impact of C_{gd} has been already discussed in details in the previous section.

Knowing that the total output current, i_o, must inevitably flow through the degeneration impedance, we use the Thevenin equivalent shown in Figure 7.24 to analyze the circuit. The

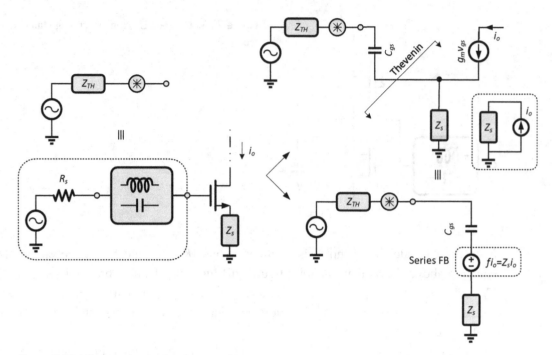

Figure 7.24: Local series feedback in a CS amplifier

feedback factor then is simply equal to Z_s. The open-loop transconductance gain is $\frac{\frac{1}{j\omega C_{gs}}}{\frac{1}{j\omega C_{gs}}+Z_s}g_m = \frac{g_m}{1+j\omega C_{gs}Z_s}$, and the open-loop input impedance is $\frac{1}{j\omega C_{gs}}+Z_s$. Thus, the closed-loop input impedance is

$$Z_{IN} = \left(\frac{1}{j\omega C_{gs}}+Z_s\right)\left(1+\frac{g_m Z_s}{1+j\omega C_{gs}Z_s}\right) = \frac{1}{j\omega C_{gs}}+Z_s+\frac{g_m Z_s}{j\omega C_{gs}}.$$

To find the noise figure, we ignore r_o, which is a fair assumption especially if a cascode structure is used. We can show

$$F = 1 + \frac{4KT\,\mathrm{Re}[Z_s] + 4KT\gamma g_m \left|\frac{1+j\omega C_{gs}Z_s+j\omega C_{gs}Z_{TH}}{g_m}\right|^2}{4KT\mathrm{Re}[Z_{TH}]}.$$

Let us assume $Z_s = r+jX$. To satisfy matching,

$$\mathrm{Re}[Z_{TH}] = \mathrm{Re}[Z_{IN}] = \mathrm{Re}\left[r+\frac{g_m X}{\omega C_{gs}}+\frac{1+g_m r}{j\omega C_{gs}}+jX\right] = r+\frac{g_m X}{\omega C_{gs}}.$$

We can see that both the real and imaginary parts of the degeneration impedance contribute to the real part of the LNA input impedance. Replacing Z_{TH} from the equation above into the noise equation yields

$$F = 1 + \frac{r+\frac{\gamma}{g_m}\left[(g_m r)^2 + (g_m X + 2\omega C_{gs} r)^2\right]}{r+\frac{g_m X}{\omega C_{gs}}}.$$

It is easy to show that the noise figure monotonically increases with $r = \mathrm{Re}[Z_s]$, as shown in Figure 7.25.

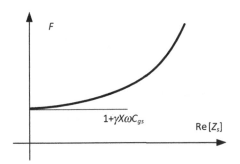

Figure 7.25: Noise figure versus r for degenerated CS LNA

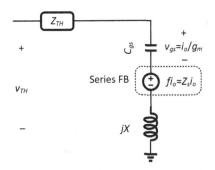

Figure 7.26: CS LNA Thevenin model to calculate the input gain

Thus, despite the fact that r does contribute to the real part of LNA input impedance, since it is a noisy resistor, we are better off not having it. In fact in the extreme case that $X = 0$, we have

$$F = 1 + \frac{r + \frac{\gamma}{g_m}\left[(g_m r)^2 + (2\omega C_{gs} r)^2\right]}{r} > 2,$$

which leads to a noise figure of well above 3dB, as expected.

On the other hand, we notice that the imaginary part of the degeneration impedance, X, also contributes to the real part of LNA input impedance through the term $\frac{g_m X}{\omega C_{gs}}$. We will not consider a capacitive degeneration as it leads to a positive feedback and negative real input impedance,[8] so it leaves us with the choice $Z_s = jX = j\omega L_s$. The LNA input impedance then becomes

$$Z_{IN} = \frac{g_m L_s}{C_{gs}} + \frac{1}{j\omega C_{gs}} + j\omega L_s,$$

and the noise figure is

$$F = 1 + \gamma L_s C_{gs} \omega^2.$$

From this we see that fundamentally there is no limit to reduce the noise figure, although there are practical limitations as how far low the noise figure could be. For a given frequency, to improve the noise figure we can either reduce L_s or C_{gs}. To understand the lower limit for the L_s, consider Figure 7.26, which illustrates the LNA Thevenin model we discussed earlier.

---

[8] We will consider a capacitive degeneration to realize an oscillator in Chapter 8.

Writing KVL yields

$$v_{TH} = ri_o + \frac{i_o}{g_m} + (Z_{TH} + jX) j\omega C_{gs} \frac{i_o}{g_m},$$

and since $Z_{TH} = Z_{IN}^*$, we find the *available* voltage gain from the source to the voltage across C_{gs} to be

$$\frac{v_{GS}}{v_{TH}} = \frac{1}{jXg_m},$$

and the effective available transconductance gain of the LNA is

$$\frac{i_o}{v_{TH}} = \frac{1}{jX}.$$

We can see that reducing degeneration inductance or X leads to a larger gain to the amplifier input. This is good as it suppresses the noise and thus leads to a lower noise figure, but results in a poor linearity.

Ironically, the optimum noise figure is realized when both r and X are equal to zero. This in fact is the case of a simple FET with no degeneration, which we discussed in Chapter 5, and concluded that the minimum noise figure is indeed equal to zero. However, that can be accomplished only through a large source matching inductance that provides an infinite gain to the FET input.

Similarly, reducing C_{gs} leads to unreasonably large values for the matching inductance, at which point its loss will start limiting the noise figure. Nevertheless, the CS LNA with inductive degeneration provides perhaps the best noise figure at GHz frequency applications, but does come at the cost of an extra inductor for source degeneration. It possible to incorporate the degeneration inductance using bondwires (in cases where bondwires are used for packaging), which leads to a more complex, but lower area design.

Finally, we can find the impact of the finite Q of the degeneration inductance on the LNA noise figure using the general noise equation we derived earlier, where $r = \frac{X}{Q}$. We can easily show that if Q is reasonably large, say more than 5, the inductance loss has little impact on the noise figure.

A complete schematic of the LNA is shown in Figure 7.27. A cascode scheme is very desirable to improve the isolation.

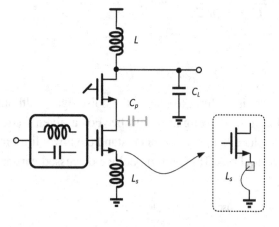

Figure 7.27: CS LNA complete schematic

We have ignored the noise contribution of the cascode transistor, but in most cases it has a negligible impact. Assuming the total parasitic capacitance at the cascode node is C_p (Figure 7.27), from basic analog design we know that the signal loss, as well as the noise contribution of the cascode transistor is negligible if $\frac{g_m}{\omega C_p} \gg 1$. This can be accomplished to a large extent by a proper layout as we will discuss later on.

Example: As a related case study, let us find the optimum noise figure of the inductively degenerated CS LNA. First with the aid of Figure 7.28 we will find the equivalent input noise sources of the LNA.

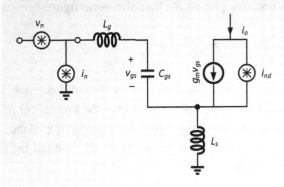

Figure 7.28: Simplified CS LNA noise model for equivalent input noise calculations

As we did earlier, with both the noise sources and the amplifier noise itself present, nulling the output current for the short and open conditions leads to the negative of the noise values, but the actual sign is unimportant. Doing so (see Problem 10 for more details) yields

$$\overline{v_n^2} = 4KT\gamma \frac{|1 - \omega^2 C_{gs}(L_g + L_s)|^2}{g_m}$$

$$\overline{i_n^2} = 4KT\gamma \frac{|\omega C_{gs}|^2}{g_m}.$$

Note that under source matching conditions $\omega^2 C_{gs}(L_g + L_s) = 1$, and $\overline{v_n^2} = 0$, but as we will see shortly this is inconsequential. The input equivalent voltage and current noise sources are fully correlated, and we therefore have

$$R_u = 0$$

$$Z_c = \frac{1 - \omega^2 C_{gs}(L_g + L_s)}{j\omega C_{gs}}$$

$$G_n = \gamma \frac{|\omega C_{gs}|^2}{g_m}.$$

The optimum noise impedance will be reactive, and is equal to

$$Z_{s,opt} = -\frac{1 - \omega^2 C_{gs}(L_g + L_s)}{j\omega C_{gs}},$$

Continued

and the minimum noise figure is

$$F_{min} = 1 + 2G_n\left(R_c + \sqrt{\frac{R_u}{G_n} + R_c^2}\right) = 1.$$

Under source matching conditions $Z_{s,opt} = 0$, but this does not affect the minimum noise figure as the optimum noise impedance is purely imaginary. This implies that to achieve a minimum noise figure of 0dB, the LNA must be driven with an ideal voltage source that is obviously not practical. The reader can verify that if the LNA is indeed driven by an ideal voltage source, its output noise is zero!

If the source impedance has a resistive part of R_s, then the noise figure becomes

$$F = F_{min} + \frac{G_n}{R_s}|Z_s - Z_{s,opt}|^2 = 1 + \gamma\frac{|\omega C_{gs}|^2}{g_m}R_s.$$

Assuming the LNA is matched, $\frac{g_m L_s}{C_{gs}} = R_s$, and the noise figure becomes $F = 1 + \gamma L_s C_{gs}\omega^2$, as previously reported. This is obviously higher than the theoretical minimum noise figure; nevertheless, it is quite low while providing the required impedance matching at the same time. In practice, the LNA noise figure is usually limited by parasitic elements and the matching network loss.

7.6 FEEDFORWARD LNAs

In our discussion earlier we saw that creating the 50Ω matching through a noisy resistance leads to a noise figure always higher than 3dB, which may not be acceptable for many applications. Also we saw that the majority of topologies that rely on feedback to break this trade-off result in a relatively narrowband design. To realize a broadband matching, one alternative is to create a noisy but wideband 50Ω match, for instance by using a CG topology, but cancel or reduce this noise subsequently through a feedforward (FF) path [3]. The basic concept underlying this principle is shown in Figure 7.29.

To create the matching, we insert a noisy resistor R_{IN}, whose value if chosen to be equal to R_s results in wideband but noisy match. Unlike the conventional amplifiers, let us monitor *both* the

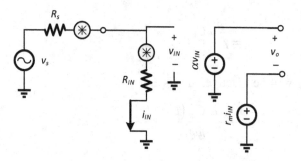

Figure 7.29: Noise cancellation concept

voltage across, and the current flowing through R_{IN} as shown in Figure 7.29. The corresponding signals, that is, v_{IN} and i_{IN}, are fed to the output through dependent voltage sources, creating a differential voltage. Let us first find the gain from the source to the output. We have

$$v_{IN} = \frac{R_{IN}}{R_{IN}+R_s} v_s$$

and

$$i_{IN} = \frac{v_s}{R_{IN}+R_s}.$$

Thus, the total gain is

$$\frac{v_o}{v_s} = \frac{\alpha R_{IN} - r_m}{R_{IN}+R_s}.$$

Next, we will find the noise figure. The source noise appears directly at the output, as would the source voltage v_s, whose transfer function we found previously. However, the noise of R_{IN}, v_{nIN}, produces signals at the output with opposite polarities as the ones calculated for the source. Turning off the source and its noise, the output noise voltage due to the R_{IN} noise is

$$v_{on} = \frac{\alpha R_s + r_m}{R_{IN}+R_s} v_{nIN},$$

and from this the noise figure is readily obtained:

$$F = 1 + \frac{R_{IN}}{R_s} \left| \frac{\alpha R_s + r_m}{\alpha R_{IN} - r_m} \right|^2.$$

To satisfy matching $R_{IN}=R_s$, and the noise figure simplifies to

$$F = 1 + \left| \frac{\alpha R_s + r_m}{\alpha R_s - r_m} \right|^2.$$

A plot of noise figure and normalized gain versus $\frac{r_m}{\alpha R_s}$ is shown in Figure 7.30.

Evidently, if $r_m = -\alpha R_s$, a 0dB noise figure is obtained, while the differential gain will be α, twice as much as a voltage-mode amplifier.

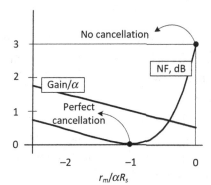

Figure 7.30: Feedforward noise canceling LNA noise figure and gain

Low-Noise Amplifiers

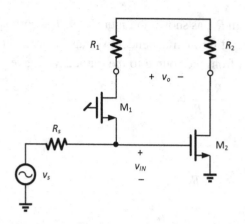

Figure 7.31: An example of a noise-canceling LNA

The transistor level implementation could be done by incorporating a CS-CG topology as shown in Figure 7.31.

We expect the common-gate device, M$_1$, to provide the matching resistor, $R_{IN} = 1/g_{m1}$ (ignoring body effect and channel length modulation). Accordingly, the noise of M$_1$ is subject to cancellation if the parameters are chosen properly. The common-source device, M$_2$, provides the voltage monitoring, and thus we expect $\alpha = g_{m2}R_2$, while M$_1$ performs current monitoring, and hence $r_m = -R_1$. Thus, if we choose $\frac{R_1}{R_s} = g_{m2}R_2$, the noise of M$_1$ is subject to perfect cancellation. The CS transistor still contributes noise; however, this noise can be made arbitrarily low by raising g_{m2} without worrying about matching.

The total gain from the source to the output is

$$\frac{v_o}{v_s} = \frac{g_{m1}R_1 + g_{m2}R_2}{1 + g_{m1}R_s}.$$

The general expression for the noise figure is

$$F = 1 + \frac{\gamma}{R_s}\frac{g_{m1}|R_1 - g_{m2}R_2R_s|^2 + g_{m2}|R_2(1+g_{m1}R_s)|^2}{|g_{m1}R_1 + g_{m2}R_2|^2} + \frac{(R_1+R_2)|1+g_{m1}R_s|^2}{R_s|g_{m1}R_1 + g_{m2}R_2|^2}.$$

The last term shows the load resistors' noise contribution. With matching and noise cancellation conditions both met, the noise figure simplifies to

$$F = 1 + \frac{\gamma}{g_{m2}R_s} + \frac{R_1+R_2}{R_s|g_{m2}R_2|^2} \approx 1 + \frac{\gamma}{g_{m2}R_s},$$

and the gain is $g_{m2}R_2$. The noise figure can be made arbitrarily low if g_{m2} is increased without sacrificing the matching. Another advantage of this LNA is that it provides a differential output from a single-ended input. Keeping the capacitances low, a very wideband and low-noise amplifier can be realized at a reasonable power consumption.

A summary of various LNA topologies is shown in Table 7.1.

Obviously we have made a number of arbitrary assumptions, but one may be surprised that in a well-designed LNA, in most available modern CMOS technologies used today, performance metrics will not be drastically different from the ones shown above.

Table 7.1: **LNA topologies summary**

LNA Type	CS	CG	CS w/Cas.	Complementary	CS w/Deg.	FF
Topology	Lossy matching	Common-gate	Shunt FB	Shunt FB	Series FB	Noise cancellation
Typical NF	2–3dB[a]	2–3dB[a]	2dB	2dB	< 2dB	2dB
Linearity	Moderate	Good	Moderate	Good	Poor	Good
Current	5mA	2mA	5mA	2mA	3mA	4mA
Bandwidth	Narrow	Wide	Narrow	Wide	Narrow	Wide

[a] We assume an S_{11} of −10dB is acceptable.

7.7 LNA PRACTICAL CONCERNS

In this section we discuss some practical design aspects of the LNA, which to a large extent are applicable to any of the topologies described before.

7.7.1 Gate Resistance

The gate material has typically a large sheet resistance (10–20Ω/□ or even higher in recent technologies). As the gate resistance usually appears directly at the input, it could potentially lead to a large noise degradation. Fortunately it can be minimized by proper layout. Shown in Figure 7.32, as far as noise is concerned gate resistance may be modeled by a resistor R_g whose value is

$$R_g = \frac{1}{3}\frac{W}{L}R_\square,$$

where $R_\square$ is the gate sheet resistance, W, and L are the devices dimensions, and the factor 1/3 arises from the distributed nature of the gate resistance [4] (see Problem 15 for the proof). As the LNA devices tend to use short channel length and wide channel width, this resistance could be very large. For instance, a sheet resistance of 10Ω/□, and a devices size of 20/0.04 leads to a gate resistance of 1.6kΩ, appearing directly in the signal path.

The resistance may be reduced substantially if multifinger layout is used as shown in Figure 7.32. If the transistor is broken into 40 smaller devices of 0.5/0.04, the total size remains the same but the net gate resistance is now about 1Ω only.

7.7.2 Cascode Noise Degradation and Gain Loss

We briefly discussed the cascode noise earlier and concluded that its noise contribution is typically small. Let us have a closer look here. Consider Figure 7.33, showing the cascode

Low-Noise Amplifiers

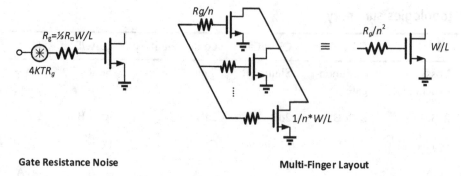

Gate Resistance Noise **Multi-Finger Layout**

Figure 7.32: Gate resistance noise and its minimization

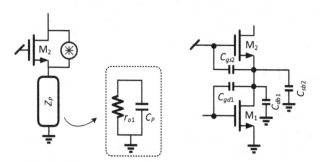

Figure 7.33: Impact of cascode device noise

transistor noise current source, while the main device has been replaced with a resistor modeling the channel length modulation, and a net parasitic capacitance of C_p.

If the cascode device r_o is ignored, then the noise current density appearing at the output is

$$\overline{i_{on}^2} = 4KT\gamma g_{m2} \frac{1}{|1+g_{m2}Z_P|^2}.$$

Since $g_m r_o \gg 1$, at frequencies of interest the impact of C_p is more dominant. The breakdown of the device capacitances is also shown in Figure 7.33, leading to

$$C_p \approx C_{gs2} + C_{gd1} + C_{sb2} + C_{db1} \approx 2C_{gs}.$$

In most modern technologies the gate-drain and junction capacitances are about one-third of the gate-source capacitance, and thus the total parasitic at the cascode node is on the order of two times gate-source capacitance. Knowing that the device transit frequency is roughly $\omega_T \approx \frac{g_{m2}}{C_{gs2}}$, we conclude that the cascode noise is suppressed by roughly $\left|\frac{\omega_T}{2\omega}\right|^2$. Unless the frequency of operation is too close to transit frequency, cascode noise contribution is indeed negligible.

If needed, the cascode node parasitic may be reduced if a *dual-gate layout* as shown in Figure 7.34 is incorporated.

If the size of the main and cascode devices are the same, assuming no external access to the cascode node is required,[9] then there is no need to insert contacts at the source of the cascode

---

[9] An example of a situation where access to cascode node is needed is when a current steering gain control is implemented at the cascode node.

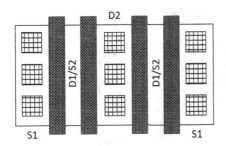

Figure 7.34: Dual-gate layout to minimize cascode node parasitic capacitance

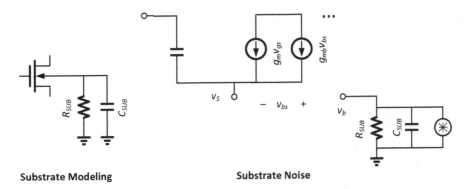

Substrate Modeling **Substrate Noise**

Figure 7.35: Substrate impact

device, minimizing the area and thus reducing the junction capacitances. This also eliminates the metal capacitances due to contacts and routing. Moreover, by sharing the drain of two devices, the output node junction capacitance is halved.

7.7.3 Substrate Impact

At radio frequencies, the substrate noise or any other harmful impact it may produce is typically negligible. A simple model of the device including the substrate resistance is shown in Figure 7.35.

The exact value of R_{SUB} is very layout dependent, and may be found experimentally to be included in the device small signal model. Shown on the right, even if the device source is connected to ground, the substrate resistance noise appears at the output through body transconductance. If R_{SUB} is known, so is its thermal noise, and the small signal model shown in Figure 7.35 may be used to find the noise impact.

A common layout technique to minimize the impact of substrate is shown in Figure 7.36. The device is broken into smaller fingers, with each pair forming a dual-gate layout, and put into an island surrounded with sufficient substrate contacts connected to a clean ground. This not only reduces the substrate noise, but also helps picking up any noise injected to the substrate from close by noisy circuits. The drawback is a somewhat bigger structure and more parasitic capacitance.

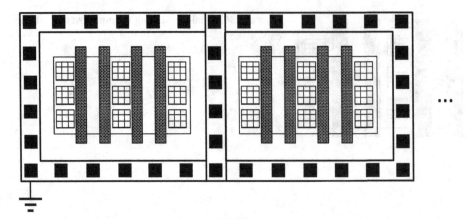

Figure 7.36: Proper layout to minimize substrate impact

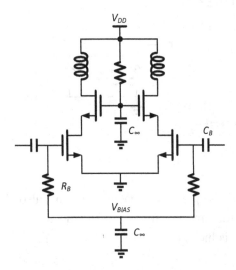

Figure 7.37: Fully differential LNA with biasing details

7.7.4 LNA Biasing

The LNA biasing circuitry is potentially a noise contributor, but fortuitously can be greatly minimized by simple circuit techniques. Shown in Figure 7.37 is an example of a fully differential structure with bias details.

The cascode device is conveniently biased at V_{DD} by a resistor. This resistor along with a bypass capacitor at the cascode gate (labeled as C_∞, a sufficiently large capacitance at the frequency of interest) helps filter any potential noise at supply. If the *cascode gate* is not properly bypassed, looking into it, the input impedance may have a negative real component, leading to potential *instability*.

The main device bias may be applied through a large resistor (several tens or hundreds of kΩ), and the same as cascode, needs to be bypassed properly. The input itself is typically AC coupled. The size of the AC coupling capacitor depends on the frequency of operation, but usually a few pF is sufficient for most RF applications.

Most of these concerns are greatly alleviated if a differential design is chosen. Still, if a *large blocker* is present at the LNA input, the *bias devices* (attached to the node V_{BIAS} in the figure) can substantially contribute *noise* due to nonlinearities. For this reason, proper bypassing is always very critical. Since the LNA frequency of operation is high, the bypass capacitances are typically small, ranging from a few to a few tens of pF.

7.7.5 Linearity

The LNA linearity is set by two factors:

- The device inherent nonlinearity, which was characterized in Chapter 6. Generally the IIP_3 tends to improve if the gate overdrive is raised, but that comes at the expense of a higher current if the device g_m is desired to be kept constant.
- The matching network gain, although it improves the noise, results in a stronger signal at the device input, and thus worse linearity. This also clearly leads to a trade-off between noise and IIP_3 if the bias current is kept constant.

The LNA 2nd-order distortion is generally not a major concern as the IM_2 components created due to 2nd-order nonlinearity appear at a very low frequency and are subject to filtering. They, however, may leak to the output in the presence of mismatches in the mixer stage following the LNA. This will be analyzed in Chapter 8.

Example: A common scenario where 2nd-order distortion is important is when a blocker around half the desired frequency appears at the LNA input (Figure 7.38). Although subject to substantial filtering at the LNA input if narrowband matching is employed, the 2nd-order nonlinearity in the LNA leads to a blocker close to the desired frequency at the output.

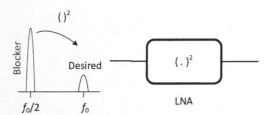

Figure 7.38: Second-order distortion in LNAs

For that reason, it is common to specify an IIP_2 requirement for the LNA at half the desired frequency. This is clearly a bigger concern in wideband and/or single-ended LNAs.

Example: Shown in Figure 7.39 is a common-gate LNA with an inductive load. Let us assume that at the frequency of interest the load inductance resonates with the output capacitance, effectively giving a net resistance of R_D. The gate is biased at V_B, and the input carries a DC voltage of V_{DC}. We would like to explore the LNA 1dB compression.

Continued

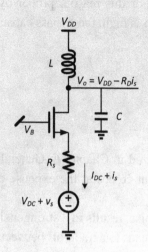

Figure 7.39: A common-gate LNA and its compression mechanisms

Assuming the LNA carries a DC current of I_{DC}, a KVL at the input yields

$$V_B = \sqrt{\frac{2(I_{DC}+i_s)}{\beta}} + V_{TH} + V_{DC} + v_s + R_s(I_{DC}+i_s),$$

where V_{TH} is the transistor threshold voltage, and $\sqrt{\frac{2I_{DC}}{\beta}}$ is the effective voltage at quiescent point (assuming square-law characteristics for the FET). For no AC input ($v_s = 0$), there will be no AC current ($i_s = 0$), and thus

$$V_B = \sqrt{\frac{2I_{DC}}{\beta}} + V_{TH} + V_{DC} + R_s I_{DC}.$$

Combining the two equations yields

$$v_s = -\left(\sqrt{\frac{2I_{DC}}{\beta}}\left(\sqrt{1+\frac{i_s}{I_{DC}}}-1\right) + R_s i_s\right),$$

which after expanding $\sqrt{1+\frac{i_s}{I_{DC}}}$ becomes

$$v_s = -\left[\sqrt{\frac{2I_{DC}}{\beta}}\left(\frac{i_s}{2I_{DC}} - \frac{1}{4}\left(\frac{i_s}{I_{DC}}\right)^2 + \cdots\right) + R_s i_s\right].$$

This allows us to express the input AC voltage in terms of the output AC current. We make the conscience judgment that given the high gain of the amplifier, the LNA compression is likely to be dominated by the output. If so, the term $\left(\frac{i_s}{I_{DC}}\right)^2$ may be ignored, leading to

$$v_s \approx -\left(\sqrt{\frac{1}{2\beta I_{DC}}} + R_s\right) i_s.$$

Since $\sqrt{2\beta I_{DC}}$ is the input device g_m, and due to the matching requirements, it is the same as $\frac{1}{R_s}$. Thus

$$v_s \approx -2R_s i_s,$$

a result that could have been obtained by the small signal analysis as well. In Problem 17, the reader will explore the situation where the LNA is compressed at the input.

For the output not to be compressed, the transistor must stay in saturation, hence

$$V_o - (V_B - V_{gs}) > V_{DS,\,SAT},$$

or

$$V_{DD} - R_D i_s > V_B - V_{gs} + V_{DS,\,SAT}.$$

Since, $V_{gs} - V_{DS,\,SAT} \approx V_{TH}$, we have

$$i_s < \frac{V_{DD} - V_B + V_{TH}}{R_D}.$$

Translated to the input,

$$|v_s| < \frac{2R_s}{R_D}(V_{DD} - V_B + V_{TH}).$$

Note that the compression occurs only at the input downswing, given that a common-gate amplifier is noninverting. Since the amplifier gain is $\frac{g_m R_D}{2} = \frac{R_D}{2R_s}$, then the input compression in volts is roughly

$$P_{1dB} \approx \frac{V_{DD} - V_B + V_{TH}}{\text{Gain}},$$

or equivalently the output compression is roughly $V_{DD} - V_B + V_{TH}$. On the other hand, for the transistor to turn on,

$$V_B > V_{TH} + R_s I_{DC},$$

So the upper bound for the output compression is $V_{DD} - R_s I_{DC}$. Putting an inductor at the source to carry the DC current (similar to the LNA of Figure 7.19) will eliminate the term $R_s I_{DC}$, and the upper bound for the LNA output compression would increase up to V_{DD} roughly, which is expected.

7.7.6 Magnetic Coupling between the Inductors

Most LNAs feature two or more inductors that if integrated on-chip will interact with each other through magnetic coupling given their close vicinity. In this section, we study this in the context of an example.

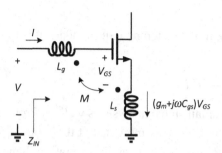

Figure 7.40: Common-source LNA with coupling between the source and gate inductors

Shown in Figure 7.40 is an inductively degenerated CS LNA, where it is assumed the gate and source inductors magnetically couple to each other. The case of source and output inductor coupling is studied in Problem 14.

We shall find the input impedance first. The input voltage consists of the two inductors' voltages, as well as the transistor V_{GS}. Thus,

$$V = [j\omega L_g I + j\omega M(g_m + j\omega C_{gs})V_{GS}] + V_{GS} + [j\omega MI + j\omega L_s(g_m + j\omega C_{gs})V_{GS}]$$

Since the input current I is equal to $j\omega C_{gs} V_{GS}$, then the input impedance is

$$Z_{IN} = j\omega(L_s + L_g + 2M) + \frac{1}{j\omega C_{gs}} + \frac{g_m}{C_{gs}}(L_s + M).$$

Compared to the earlier derivation, the inductive part has an extra term $2M$, whereas the resistive part has the extra term $\frac{g_m}{C_{gs}}M$. This makes sense as both inductors carry a common current of $j\omega C_{gs}$, and hence the addition of $2j\omega M$ to the inductive part, though the source inductor has an additional current component of $g_m V_{GS}$, which leads to the additional resistive term $\frac{g_m}{C_{gs}}M$.

To find the LNA gain, we can similarly write

$$\frac{V_{GS}}{V} = \frac{1}{1 - \omega^2 C_{gs}(L_s + L_g + 2M) + g_m j\omega(L_s + M)}.$$

If C_{gs} resonates with the inductive part, then

$$\frac{V_{GS}}{V} = \frac{1}{g_m j\omega(L_s + M)},$$

which is a similar expression to what we had before, except for L_s changes to $L_s + M$.

Note that M could be positive or negative, and its impact, if understood and modeled properly, is generally not an issue. In fact, one can take advantage of a positive M for instance, to effectively use smaller L_s and L_g.

7.7.7 Gain Control

As we discussed in Chapter 6, the LNA gain is desired to vary if for instance the wanted signal is too large, or in some cases to improve linearity in the presence of very large blockers. Of various structures used, two of the more common techniques are shown in Figure 7.41.

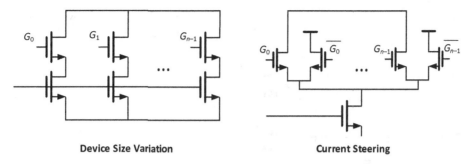

Figure 7.41: LNA gain control

In the first method shown on the left, the device is broken into smaller pieces that can be conveniently done given that multifinger layout with many fingers is desired for the reasons we mentioned earlier. All the nodes are shorted, except for the gate of cascode devices that are connected to on (V_{DD}) or off (V_{SS}) voltages. Thus the device size is effectively chosen arbitrarily, and hence so is the gain. If laid out properly, the sections are well matched and thus the gain steps will be very accurate. The main drawback of this scheme is that the input matching is inevitably affected as the gate-source capacitance varies. For small gain changes this may be acceptable, but larger gain steps typically result in S_{11} degradation.

The second scheme relies on current steering where part of the signal is steered to V_{DD} by selecting the appropriate gate voltage of cascode branches. There are two main drawback with this scheme: First, the parasitic at the cascode node is more than doubled, and moreover dual-gate layout is not possible anymore. Second, as the gain reduces, the LNA bias current while scales in the first approach, in current steering method stays constant. However, the matching is not affected much by the gain control.

Perhaps it may be prudent to use a combination of the two approaches, and apply current steering at only very lower gain settings.

7.8 LNA POWER–NOISE OPTIMIZATION

In this section we discuss some of the trade-offs between noise, linearity, and the LNA power consumption in a somewhat more methodical way. Let us start with a simple CS LNA as shown in Figure 7.42, and let us assume that the matching is performed by an explicit resistor of R_p at the input.

This resistor, as indicated before, may be treated as the loss of the matching network. We make the sensible assumption that R_p is larger than the source resistance R_s, and thus we use the shunt-L, series-C matching network discussed earlier to downconvert the impedance. Moreover, we lump all the parasitic capacitances into one capacitor C_{IN}, which is also treated as a part of the matching circuit. We showed that this amplifier has a noise figure of greater than 3dB, unless input returns loss is allowed to be traded for noise. So we shall assume an input S_{11} of about −10dB across the band is acceptable.

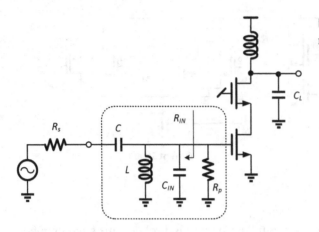

Figure 7.42: CS LNA with lossy matching network

For an input return loss of $|S_{11}| = \frac{1}{3} \approx -9\text{dB}$, we showed that the noise figure is

$$F = 1.5 + \frac{4.5}{g_m R_p}\left(\gamma + \frac{1}{g_m R_L}\right) \approx 1.5 + \frac{4.5}{g_m R_p}.$$

To satisfy the noise figure shown above, and the corresponding $|S_{11}| = \frac{1}{3}$, then looking into the right side of the matching network (Figure 7.42), the resistance seen must be

$$R_{IN} = \frac{R_p}{2}.$$

Or alternatively, on the source side looking into the left side of the matching network the equivalent resistance must be $2R_s$. This implies that the value of the parallel resistance is higher than what would have been needed for perfect matching conditions. This in turn is the very reason for a noise figure of lower than 3dB.

From the noise equation, all that can be optimized is the product of $g_m R_p$. The g_m may be raised to lower noise, but at the expense of more power consumption. Alternatively, for a given g_m, R_p may be increased, and converted to the desired resistance (in our example $2R_s$) by adjusting the matching network. To find the upper limit for R_p, we need to understand what are the consequences of increasing it:

A higher R_p results in more gain from the source to the amplifier input, and thus worse linearity.

Increasing R_p raises the matching network effective Q, which results in a smaller bandwidth and more sensitivity to variations. Ultimately R_p is limited by the inherent Q of the inductance used.

We showed before that to obtain an input resistance of R_{IN} looking into the matching network (Figure 7.42), the matching inductance is

$$L = \frac{R_{IN}}{\omega_0}\sqrt{\frac{R_s}{R_{IN} - R_s}}.$$

If we assign a quality factor of $Q \gg 1$ to the inductor, the total parallel resistance as a result is $L\omega_0 Q$. To avoid using an explicit resistor, we let the inductor loss be equal to R_p. Thus

$$R_p = L\omega_0 Q = QR_{IN}\sqrt{\frac{R_s}{R_{IN} - R_s}}.$$

To satisfy noise and matching we showed $R_{IN} = \frac{R_p}{2}$. This yields

$$R_p = \frac{Q^2 + 4}{2}R_s.$$

We also showed the total voltage gain from the source to the input is $\frac{Q-1}{2}$. So the upper limit to R_p is simply determined by what kind of Q is acceptable or perhaps achievable for the matching network.

If we choose a Q of 10, to allow the bandwidth to be roughly 10% of the center frequency, then R_p is found to be 2.6kΩ. Moreover, there is a gain of 4.5 or 13dB from the source to the input. Letting $g_m R_p = 40$, and assuming a short-channel device with an overdrive voltage of 200mV to allow sufficient linearity (especially considering the large gain up front), the LNA bias current is 3mA, and the noise figure is 2dB. In Chapter 6 we showed that with 200mV overdrive the IIP$_3$ of a single device is about 1V or 10dBm. Given the *available voltage gain* of $13 + 6 = 19$dB in front, then we expect an IIP$_3$ of about -9dBm for the LNA assuming it is dominated by the input stage nonlinearity.

Note that the Q of 10 is not necessarily what is fundamentally achievable but what is acceptable given the linearity and bandwidth requirements. The actual matching inductor in fact consists of two components in parallel, the one that converts the impedance, and was already calculated, and another part that needs to resonate with C_{IN}, which we call L_{IN}. Assuming $C_{IN} = 1$pF, then at 2GHz, $L_{IN} = 6$nH. The total inductance is

$$\frac{1}{L_{tot}} = \frac{1}{L_{IN}} + \frac{1}{L},$$

which results in $L_{tot} = 4.65$nH. A parallel resistance of 2.6kΩ then implies an inherent Q of 45 for the inductor (substantially higher than 10), which is on the order of what external inductors of this size can provide at 2GHz.

In this example, the matching network loss dominates the noise figure, and effectively it matches its own loss. The 2dB noise figure may be improved if one of the feedback schemes introduced earlier is incorporated to make up part of R_p as a noiseless resistor.

Let us assume there is a shunt capacitive feedback of C_F between the input and cascode node. The input impedance was calculated before, and is shown in Figure 7.43 (assuming $\frac{g_m}{\omega C_F} \gg 1$).

Using series–parallel conversion we have

$$Y_{IN} = j\omega C_{IN} + \frac{1}{R_P} + j\omega C_F + \frac{2 + \frac{C_p}{C_F}}{g_m}(\omega C_F)^2,$$

and we arrive at the equivalent model of the amplifier input also shown at the bottom of Figure 7.43.

The LNA noise figure is

$$F = 1 + \frac{|Z_{TH}|^2}{R_P \operatorname{Re}[Z_{TH}]} + \left|\frac{Z_{TH}}{(g_m + f)(R_P \| Z_{TH} \| Z_F)}\right|^2 \frac{2\gamma g_m}{\operatorname{Re}[Z_{TH}]},$$

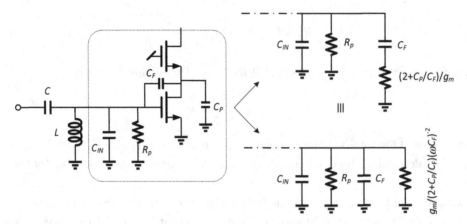

Figure 7.43: LNA with lossy matching and capacitive shunt feedback

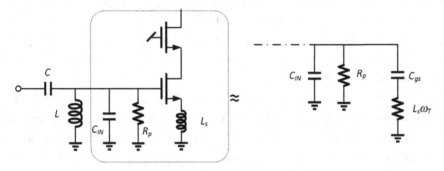

Figure 7.44: LNA with lossy matching and inductive degeneration

where $f = -j\omega C_F$ is the feedback factor. As C_F may be combined with the rest of C_{IN}, and assuming $g_m \gg \omega C_F$, we have

$$F = 1 + \frac{|Z_{TH}|^2}{R_p \, \text{Re}[Z_{TH}]} + \left|1 + \frac{Z_{TH}}{R_P}\right|^2 \frac{2\gamma}{g_m \, \text{Re}[Z_{TH}]}.$$

Z_{TH} is the impedance looking into the matching network as a result of R_s upconversion (R_{IN} in Figure 7.42). This value is fixed by the bandwidth and linearity concerns as we pointed out earlier. The first term in the noise equation is the impact of the matching network loss, and may be reduced if R_p can be raised. To satisfy the S_{11} condition, the increase in R_p must be then compensated by the term $\frac{2 + \frac{C_p}{C_F}}{g_m}(\omega C_F)^2$ in the input impedance equation, which is a result of the feedback. This assumes that a higher Q for the matching inductor is feasible. Now let us assume the same current budget of 3mA, and hence a g_m of 15mS. Furthermore, assuming a value of 25fF for C_F, which is mostly made up of gate-drain capacitance, and that $\frac{C_p}{C_F} = 6$ (see Figure 7.33), the parallel resistance as a result of the feedback is about 19kΩ. When compared to the original values of 2.6kΩ for R_p, this indicates that the noise figure improvement is only marginal unless C_F can be raised substantially, or a higher gain for the matching network is acceptable.

For a CS amplifier using inductive degeneration, the equivalent input is shown in Figure 7.44.

Both C_{gs} and $L_s\omega_T$ are expected to be larger than the corresponding values for the shunt feedback, thus we expect a more favorable situation. For instance, if $C_{gs} = 50\text{fF}$, $L_s = 0.5\text{nH}$, and $f_T = 100\text{GHz}$, the net parallel resistor is now 8.5kΩ, which is roughly half of what we had in the shunt feedback LNA (Figure 7.43), and thus a better noise figure is expected. Thus, the total 2.6kΩ resistor is now made of a *noiseless* 8.5kΩ part in parallel with a *noisy* 3.75kΩ (as opposed to 2.6kΩ before). Consequently, we expect a reduction in the LNA noise without increasing the bias current.

7.9 SIGNAL AND POWER INTEGRITY

We have dedicated much of this chapter to dealing with the thermal noise of the resistors and transistors, and ways to minimize this. Unfortunately, in a real environment this kind of noise (call it *intrinsic noise*) is not the only source contaminating the LNA. In addition to the intrinsic noise, there is usually *extrinsic noise* as well originated from blocks in the vicinity of the LNA such as switching power supplies, crystal oscillators, or any digital circuitry with large activity. This kind of noise can contaminate the sensitive block of interest through parasitic capacitive or magnetic coupling, or through the power lines transient, effectively degrading the system signal-to-noise ratio.

To remedy these issues, it is common among RF designers to use decoupling capacitors on supplies, or use shielded lines to protect the sensitive blocks such as low-noise amplifiers or oscillators. What is very important though is that if these techniques are not done properly, very quickly they become ineffective or even harmful. For this reason we dedicate this section to a deeper understanding of what is known as the *signal* or *power integrity*, essentially a measure of the quality of the signal or power rails in the presence of the aforementioned interference. For further reading see [5], [6], [7], [8], [9], [10], [11], [12]. Especially [5] has very detailed discussions on coupling and shielding. In addition to noise pickup and shielding concerns, an overview of electromagnetic waves and their impact on radios can also be found in [6].

7.9.1 Power Lines

Shown in Figure 7.45, one may consider the following arrangement to connect the signal and power lines to the amplifier. The source is not necessarily close to the LNA. Often, it may be external, say the antenna input that will be brought on-chip. This routing will have some inductance and resistance associated with it. Similarly, the supply will be connected to the LNA through some impedance, which is partially on-chip, and partially off-chip. These impedances are represented with the gray RL circuits in the figure.

Often, RF designers forget that the notion of *ground* is nothing but a common reference point for the entire circuit as taught to us in basic circuit theory.[10] An alternative and more proper

---

[10] Not to be confused with *earth* connection, which is primarily for safety reasons.

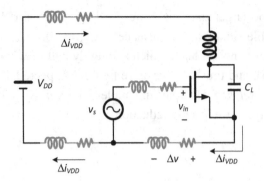

Figure 7.45: Improper connection of the supply to an amplifier input

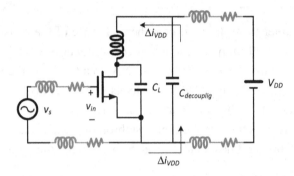

Figure 7.46: Properly connected supply along with decoupling capacitor

definition for the ground would be a *low impedance path for the current to return back to the source*. As such, as far as this discussion is concerned, the ground (or V_{DD}) is simply treated like any other signal connecting to the LNA.

The amplifier naturally drains some transient high-frequency current from the supply, labeled as $\Delta i_{V_{DD}}$. This current must somehow go back to the supply, and KCL tells us that it must follow the path shown in the figure (for the moment we ignore the current drained at the amplifier input from the source). As a result of this, there will be some voltage drop on the RL circuit associated with the source (Δv in the figure), which will make the amplifier input (v_{in}) be different from the actual source voltage (v_s). In other words, the supply transient (noise) will also appear at the input, and amplified along with the desired signal.

To avoid this, a better arrangement shown in Figure 7.46 may be considered. This way the supply transient will not travel through the source parasitic impedance. Still, however, the transient current will create voltage drop on the supply parasitic impedances, and depending on the LNA supply rejection, it will translate to noise. To remedy this, one could consider adding a *decoupling capacitor* placed physically close to the LNA (while the actual supply is often not). Now the transient current will be supplied by the capacitor as long as its size is chosen properly, rather the actual supply itself, and hence avoids the undesired voltage drop.

The decoupling capacitor itself must be carefully designed and modeled. It is obviously naïve to expect having an ideal capacitor at GHz range frequencies. The parasitic inductance (especially if the distance between the supply and ground ends of the LNA is long), and the resistance associated with the routing could be problematic, and usually the notion of the larger the decoupling capacitor, the better is not necessarily correct. If appropriate, one may consider choosing the size of the capacitor to resonate with the parasitic inductance in series, creating a near ideal short at the frequency of interest.

We are yet to deal with the source parasitic impedance connecting to the LNA input. Unfortunately, there is no easy way of eliminating that aside from common practices to minimize the inductance and resistance. This often implies short and wide routing as much as possible. Nonetheless, the impedances are present, and must be carefully modeled and ultimately treated as a part of the amplifier, for instance, lumped inside the matching network if applicable. Some performance degradation in general is common, though all the parasitic must be carefully modeled and minimized. A co-design with package/PCB may be often needed to properly model the entire path.

7.9.2 Coupling and Shielding

There are two types of shielding common in RF:

– Shielding a noisy line to avoid it coupling to sensitive blocks
– Shielding a sensitive line to avoid picking up noise from the interfering blocks surrounding it

The concepts are similar, and will be briefly explained here. We will consider both the capacitive and magnetic coupling.

7.9.2.1 Capacitive Coupling

Shown in Figure 7.47 is simple demonstration of capacitive coupling from a noise source to the desired block (represented by Z_{IN}). This could arise from a number of reasons, most commonly due to the signal and noise lines running side by side, with the coupling occurring due to the stray capacitances. A simplified equivalent circuit is also shown in the figure.

The obvious way of reducing the coupled noise is to minimize the coupling capacitance, which often requires increasing the physical separation between the noise source and the victim. This, however, is not always possible due to floor planning constraints. Alternatively, the line may be protected by placing it inside a metallic cage completely surrounding it, often known as a *Faraday cage*, as shown in Figure 7.48. Similarly, the source of the noise may be shielded if placed inside the Faraday cage. If the cage is completely closed, it need not be grounded, as the electric field inside the perfect conductor is always zero.

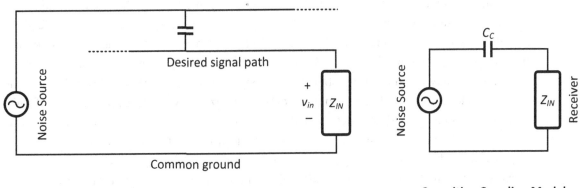

Figure 7.47: Electrical coupling demonstration and a simple model

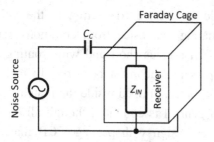

Figure 7.48: Shielding the desired signal using a Faraday cage

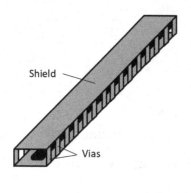

Figure 7.49: Coaxial shield implemented in an integrated circuit

If the cage has openings, as would be the case in an on-chip implementation, then the amount of noise pickup would depend on its size, or its effective capacitance to ground [6]. If the cage is physically large, it effectively acts like a ground, as it has a very large supply of charge that can easily respond to an external disturbance. Otherwise, the cage may be grounded, in which case it allows the charges to flow freely on and off of the shield to maintain zero voltage on it. Consequently, with the potential of the shield constant, the voltage pickup on the sensitive line remains zero.

The Faraday cage is obviously quite impractical to be implemented on-chip, though it can be placed externally to protect the radio chip or other sensitive blocks. A common on-chip implementation is shown in Figure 7.49, where the line to be shielded is sandwiched between two metal layers connected to each other through a wall of vias. Since there is a minimum spacing required between adjacent vias, the shield will not be perfect, and will have holes.

7.9.2.2 Magnetic Coupling

The side-by-side lines could also experience magnetic coupling, when the current of the noisy line induces an electromotive force in the desired signal line through their mutual inductance. Placing a conductive *nonmagnetic* shield around the desired line is absolutely ineffective at shielding the magnetic field as no current flows in the shield. However, if the shield is grounded in two points as shown in Figure 7.50, then given Faraday's law, the noise current (i_n) effectively creates an equal current but in the opposite direction in the shield (i_2). The direction of the shield current is determined according to Lenz's law to be opposite, and if shield resistance is small, it will be exactly equal to noise current. Note that it is this induced current that effectively shields the signal, and not the shield itself. If the shield resistance is nonzero (R_{shield}), the induced noise voltage is shown [6] to be around $R_{shield} i_n$ at RF.

Note that in the Figure 7.50 setup, not only is the amplifier return path to the source coming through the shield, but there is also a common ground connection that allows the shield current

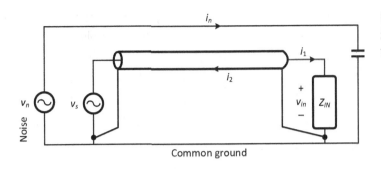

Figure 7.50: Shielded amplifier input signal with connection at two ends

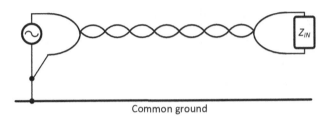

Figure 7.51: Twisted pair with balanced currents

i_2 to flow. This common ground is often partially on-chip, and partially on PCB depending on how far the source is. One potential concern is that this additional return path creates a large loop (between the shield and the common ground), and is prone to pickup-interfering magnetic fields. Consequently, a noise current is created in the shield that can potentially couple to the signal line. This must be studied carefully depending on where the nearby interfering signals are, how the coupling is created, and the situation of the common ground path. Another important reason for not grounding at the other end is that we would like to force the return current to be through the shielded line, and not through another lower impedance path.

Besides shielding, another common isolation technique is twisting the signal and its return path, as shown in Figure 7.51. This is very effective as it reduces the area of the loop to which the interfering magnetic field can couple. Grounding both ends of the twisted pair may not be wise, as it increases the area of the loop formed by the common ground line and the return path, and negates the main benefit.

The main limitation of the twisted pair wiring arises from the practical implementation, and frequent cutoff throughout the line. It must be noted that the twisted pair can be further put inside a shield, and thus enjoying extra benefit of the shield as well (grounded at both ends like shown in Figure 7.50).

Before we conclude, it must be emphasized that there is generally no unique recipe as to how address the coupling and the shielding issues. Often the designer must painstakingly trace the signal and return currents (basically the current flowing through the ground line), and identify which path is more desirable for the current to flow, and force it accordingly by proper placement of shield, and the relevant supply and ground connections.

7.9.3 Inductor Shield Case Study

Consider Figure 7.52, showing a 2-turn differential inductor using the top AP metal layer. The standalone simulated differential inductance and quality factor are shown in Figure 7.53,

Low-Noise Amplifiers

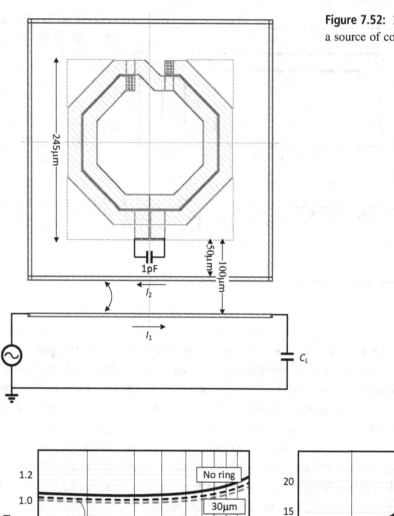

Figure 7.52: 1nH differential inductor with a source of coupling

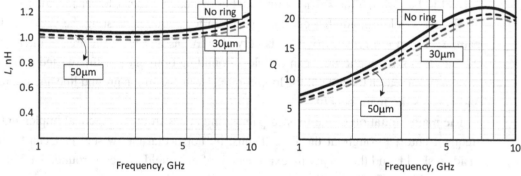

Figure 7.53: Inductance and Q for the cases of inductor with and without shield

denoted by the black solid curves. The inductance at 5GHz is 1.06nH, and the EMX simulated quality factor is 20.4.

Also shown in Figure 7.52 is a 5μm wide AP line running in parallel to the inductor 100μm away, carrying an AC interfering signal. The 1nH inductor is resonating with a 1pF capacitance at 5GHz as shown in the figure. The parallel line is also terminated by $C_L = 100f$, carrying an AC current of $I_1 = 2\pi \times 5 \times 10^9 \times 100f$, assuming a 1V AC source. Ignoring the metal ring around the inductor for the moment, the interfering current in turn induces a magnetic field, and consequently another current in the inductor that is measured differentially at the inductor

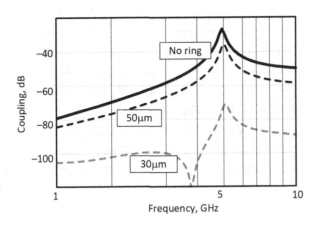

Figure 7.54: Simulated coupling for the cases of inductor with and without shield

terminals. Obviously the larger the C_L, the larger the induced current, and hence the differential voltage at the inductor output.

Shown in Figure 7.54 is the simulated differential voltage at the inductor output, which is about −27dBc relative to the 1V AC voltage source, effectively representing the coupling form the aggressor to the differential inductor, or the victim.

Next, we include a 5μm-wide AP layer shield around the inductor. Two cases are simulated, the ring 50μm away from the inductor edge, and another case of only 30μm away (only the 50μm ring is shown in the figure). The EMX simulations of the inductance and the Q with the ring are also depicted in Figure 7.53. The inductance is reduced by as much as about 8%, whereas the Q drops to 18.1 for the case of the ring 30μm away, which is still a respectable Q at 5GHz. What is more interesting is the impact of the ring on the coupling. Due to the presence of the ring, the interfering current I_1 induces a magnetic field, which according to Faraday's law creates an emf in the ring. The electromotive force when divided by the line resistance creates another current in the ring of I_2 in the opposite direction according to Lenz's law. This current creates a magnetic field in the opposite direction, which partially cancels the interfering field and shields the inductor. The coupling for the two cases with the ring is also shown in Figure 7.54. If the ring is placed close enough to the inductor, the two fields almost perfectly cancel each other, reducing the coupling significantly. The compromise is in the Q reduction and in many cases may be well worth it. Essentially, the main trade-off the designer facing is how close the ring needs to be, which primarily defines the Q degradation versus the coupling improvement. There may be of course physical area constraints as well depending on the overall chip floorplan.

The shield may or may not need to be connected to the local ground. Of course depending on where it is grounded, and how many taps are there, the results could be affected, but it won't change the overall positive impact the shield has in protecting the inductor from picking up nearby interfering signals. In most cases, connecting the shield to the ground is unnecessary.

7.10 LNA DESIGN CASE STUDY

To conclude this chapter, an inductively degenerated common-source 5.5GHz LNA in 16nm CMOS is presented as a case study.

Shown in Figure 7.55 is the LNA schematic with the components values. The transistors use a channel length of 20nm (as opposed to minimum 16nm), and are biased at 3mA, corresponding to $g_m = 40$mS, and $C_{gs} = 35$fF. The transistors are biased close to weak inversion (an overdrive voltage of about 65mV); nonetheless, they enjoy a transit frequency of about $f_T = 180$GHz. The cascode transistor is biased at $V_{DD} = 800$mV, and given $V_{TH} = 320$mV, leaves about 400mV headroom for each NMOS. The combination of 10k resistor and 10p capacitor provides supply filtering. Obviously the gate of the cascode NMOS must be well bypassed to avoid any instability, which is accomplished by the 10p capacitor to ground.

The core transistor is biased through a replica current mirror (not shown in the figure), along with the 100k resistor and 4p AC coupling capacitor. The value of resistance is chosen large to minimize its noise contribution.

The LNA simulated performance is shown in Figure 7.56.

With $L_s = 400$pH, the effective real part of the input impedance is about $\frac{g_m L_s}{C_{gs}} = 457\Omega$, which comes in series with C_{gs}. The simulated input impedance of the LNA is $160\Omega - j470\Omega$, however. The difference is attributed to the main transistor C_{gd}, which is about 15fF, and

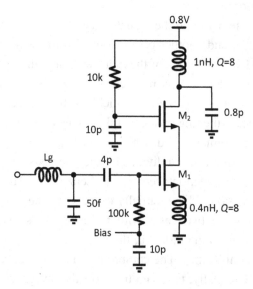

Figure 7.55: Inductively degenerated common-source LNA in 16nm CMOS

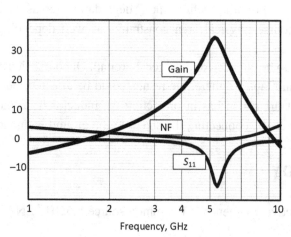

Figure 7.56: Simulated performance of the LNA of Figure 7.55

creates a shunt feedback around the transistor, effectively lowering the input impedance. This impedance is matched to 50Ω through a shunt C-series L matching network. The matching network is left ideal to show the LNA inherent performance. The simulated noise figure is 0.26dB, but it will be far worse once all the parasitic capacitances and a more realistic matching network are utilized. Nevertheless, it demonstrates the superior noise figure of this topology. The theoretical noise factor $F = 1 + \gamma L_s C_{gs} \omega^2$ leads to a noise figure of about 0.1dB, but there are several factors raising this to the actual simulated value: noise contribution of the cascode device, the finite Q of the load and source inductances, as well as other parasitic elements within the transistor.

The simulated gain of the LNA is high, about 33dB, which is set by the load inductance, and the gain of the matching network. The latter is relatively high, about 15dB, which justifies the relatively narrow S_{11}. This can be improved by raising the source inductance to increase the effective real part of the impedance, but comes at the expense of worse noise figure. It must be noted that the design is not optimized, and is just an example of how the theory and practice match in reality.

7.11 Summary

This chapter has discussed the theory and design procedure behind low-noise amplifiers.

- Section 7.1 discussed the matching networks and their properties. Matching is an essential part of any RF amplifier, and especially the LNAs.
- In Section 7.2 we gave a brief overview of RF amplifiers topologies and their pros and cons.
- Sections 7.4 and 7.5 discussed the general categories of the LNAs, namely shunt and series feedback low-noise amplifiers. In both topologies, lossless feedback is exploited to satisfy the conflicting impedance and noise matching requirements.
- Feedforward LNAs were presented in Section 7.6.
- Practical design and layout concerns in LNAs were discussed in Section 7.7.
- A general discussion on the LNA power consumption and the trade-offs involved was offered in Section 7.8.
- Practical concerns originating from signal and power integrity of the amplifier were discussed in Section 7.9. While this discussion was presented in the context of low-noise amplifiers, the conclusions are general and apply to any building block of the radio.
- The chapter was concluded in Section 7.10 by presenting a simple case study highlighting the design details of an inductively degenerated CS LNA.

7.12 Problems

1. Find the transfer function, noise spectral density of the following matching network with and without using the Nyquist theorem. **Answer:** $S_{v_{IN}}(\omega) = 4KTR_s \frac{(L\omega)^2}{(L\omega)^2 + R_s^2}$.

434 Low-Noise Amplifiers

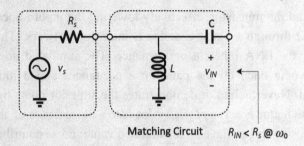

Matching Circuit $\quad R_{IN} < R_s \,@\, \omega_0$

2. For the matching network of Figure 7.5:
   a. Calculate the values of L and C to match $R_{IN} > R_s$ to the source.
   b. Find the transfer function and output noise spectral density of the matching network.
   c. Argue how that differs from the matching circuit of Figure 7.4.

3. Find the noise figure of the following complementary shunt-FB LNA. Ignore the capacitances, but not the inverter r_o. R_F is noisy. Also for simplicity, assume $g_m r_o \gg 1$, and $g_m R_F \gg 1$, where g_m is the inverter transconductance, and r_o is its output resistance.
   Answer: $F \approx 1 + \frac{\gamma}{1 + \frac{R_F}{r_o}}$.

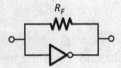

4. Find the noise figure of a common-gate LNA with an ideal transformer of $1:n$ at its input. Use the g_m–I_D curve provided in Chapter 6 to determine the device DC current. Propose a proper LC matching circuit to approximate the transformer. $n = 4$, and the frequency of operation is 2GHz.

5. Repeat Problem 4, with a $4:1$ transformer. Discuss the pros and cons of each case.

6. Derive an expression for the noise figure of a common-gate LNA versus the input reflection coefficient.

7. Find the minimum noise figure and the optimum noise source impedance for a common-gate LNA.

8. Calculate the noise figure of a resistive shunt feedback LNA without using feedback method.

9. Find the noise figure and gain of a cascode common-gate LNA, without ignoring the cascode noise and gain loss. Assume no body effect and r_o for the cascade, but nonzero capacitances.

10. Find the equivalent input noise sources of an inductively degenerated CS LNA. Assume the LNA is matched to 50Ω. **Hint:** Use the circuit of Figure 7.28, and find the input noise sources such that the output current is nulled.

11. Show that if the inductively degenerated CS LNA (Figure 7.28) is driven by an ideal voltage source, its output noise is zero.

12. Ignoring the cascode transistor and load noise, show that if an inductively degenerated CS LNA is driven by an ideal voltage source, the output noise is zero.

13. Using the g_m–I_D and IIP$_3$ curves provided in Chapter 6, design a cascode LNA with proper matching at 2GHz. The matching network available gain is 4. Assume $C_{db} = C_{sb} = C_{gd} = 0.5C_{gs}$, and the devices have an overdrive voltage of 100mV, and f_T of 20GHz. Assume matching network components have infinite Q. What is the LNA IIP$_3$?

14. As discussed in Section 7.7, in most practical inductively degenerated LNAs, due to the close vicinity, the source (L_1) and drain (L_2) inductors are often mutually coupled to each other. In this problem, we explore the impact of coupling between the source and drain inductors on the LNA gain, input impedance, and tuning. Consider both cases of negative and positive mutual inductance (M). Intuitively, explain the variation of gain and input impedance for positive and negative coupling based on the feedback theory. Neglect r_o and all the capacitances expect for the main device gate-source and the load capacitance. The degeneration inductor is otherwise ideal, whereas the drain inductor has a finite Q. **Hint:** To simplify the analysis, assume the coupling is small, and thus its impact on the inductors' original current (with no coupling), I_{L_1}, and I_{L_2} is negligible.

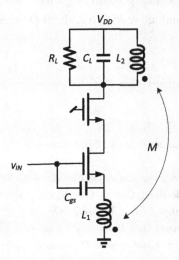

15. To find the power spectral density of distributed gate resistance, we break the transistor into N identical devices, each with N times lower g_m and gate resistance, as shown below [4].

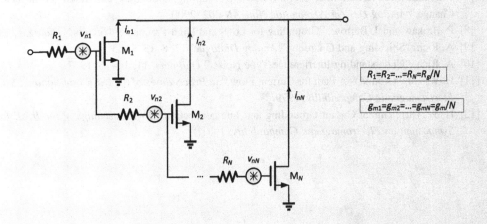

Show that for a given segment, $i_{ni} = g_{mi}\sum_{k=1}^{i} v_{nk}$ ($i = 1, \ldots, N$). Assuming that the input-referred noise sources are uncorrelated, find the total output current noise spectral density. What is the equivalent input-referred noise voltage if $N \to \infty$?

16. Using current steering, design a variable-gain LNA with gain steps of −3, −6, and −12dB from the maximum. Find the ratio between the device sizes.

17. For the common-gate LNA, derive simple expressions for the case that the LNA compression is set by the input. Compare the results with the case where the compression is set by the output as derived in Section 7.5. Which one is likely to dominate? **Hint:** For the LNA to be compressed at the input, $I_{DC} + i_s = 0$.

18. In the shielded line of Figure 7.50, derive expressions for the shield current versus the noise current based on the mutual inductances between the noisy line, the shield, and the signal line. Find the induced noise voltage for both a zero resistance shield, and the case of a nonzero resistance.

19. Consider a CS LNA with two cases of RC and LC loads. Assuming a fixed total capacitance at the LNA output for each case, show that the inductive load has about Q times higher gain than the RC load.

7.13 References

[1] B. Nauta, "A CMOS Transconductance-C Filter Technique for Very High Frequencies," *IEEE Journal of Solid-State Circuits*, 27, no. 2, 142–153, 1992.

[2] P. R. Gray and R. G. Meyer, *Analysis and Design of Analog Integrated Circuits*, John Wiley, 1990.

[3] F. Bruccoleri, E. Klumperink, and B. Nauta, "Wide-Band CMOS Low-Noise Amplifier Exploiting Thermal Noise Canceling," *IEEE Journal of Solid-State Circuits*, 39, no. 2, 275–282, 2004.

[4] B. Razavi, R.-H. Yan, and K. Lee, "Impact of Distributed Gate Resistance on the Performance of MOS Devices," *IEEE Transactions on Circuits and Systems I: Fundamental Theory and Applications*, 41, no. 11, 750–754, 1994.

[5] H. W. Ott, *Noise Reduction Techniques in Electronic Systems*, vol. 442, John Wiley, 1988.

[6] A. M. Niknejad, *Electromagnetics for High-Speed Analog and Digital Communication Circuits*, Cambridge University Press, 2007.

[7] P. Brokaw, "An IC Amplifier User's Guide to Decoupling, Grounding, and Making Things Go Right for a Change," *Analog Devices Application Note AN-202*, 2000.

[8] P. Brokaw and J. Barrow, "Grounding for Low- and High-Frequency Circuits," *Dialogue*, 18, 1, 1984.

[9] A. Rich, "Shielding and Guarding," *Analog Dialogue*, 17, 8–13, 1983.

[10] A. Rich, "Understanding Interference-Type Noise," *Dialogue*, 11, 10–16, 1977.

[11] H. W. Ott, "Ground – A Path for Current Flow," in *Proceedings of the IEEE International Symposium on Electromagnetic Compatibility*, 1979.

[12] H. W. Ott, "Digital Circuit Grounding and Interconnection," in *Proceedings of the IEEE International Symposium on Electromagnetic Compatibility*, 1981.

8 Mixers

By this point we have established the need for mixers in a radio, which is to provide frequency translation, and consequently to ease analog and digital signal processing by means of performing them at a conveniently lower frequency. Since the frequency translation is created due to either time variance or nonlinearity, or often both, the small signal analysis performed typically on linear amplifiers does not hold. This makes understanding and analysis of the mixers somewhat more difficult. Although exact methods have been presented, in this chapter we resort to more intuitive yet less complex means of analyzing the mixers. In most cases this leads to sufficient accuracy but more physical understanding of the circuit.

We present both active and passive mixers, and in each case discuss the noise and linearity trade-offs based on physical models, mostly from the receiver point of view. We dedicate a section at the end to upconversion mixers, although very similar in principle, as well.

The specific topics covered in this chapter are:

- Mixers' basic requirements
- Active mixer operation
- Noise, linearity, and second-order distortion in active mixers
- Passive mixer operation
- Noise, linearity, and second-order distortion in passive mixers
- LO duty cycle optimization
- *M*-phase and harmonic-rejection mixers
- Transmitter mixers
- LNA/mixer practical design example

For class teaching, we recommend covering Section 8.1, as well as active and passive mixer basic operation (initial parts of Sections 8.3 and 8.4). A qualitative discussion may also be offered from the material presented in Section 8.8.

8.1 MIXERS FUNDAMENTALS

Before getting into the details of mixer circuit realization, let us first have a short introductory discussion of the basis of mixer operation from a black box system perspective. We shall also discuss the general realization and the basic definitions in Section 8.1.2.

8.1.1 Mixer Operation from System Point of View

The exact role of a mixer in radios is best understood from the earlier discussion in Chapter 2. We shall recast a summary first in here, and present a somewhat broader perspective.

As almost all wireless transmitters produce signals around a high-frequency carrier (for the reasons discussed in Chapter 2), a *shift in frequency* is often needed to delegate as much analog and digital signal processing as possible to a conveniently lower frequency. As we showed, a shift in frequency is equivalent to multiplying by $e^{\pm j2\pi f_0 t}$ in the time domain, that is,

$$x(t)e^{\pm j2\pi f_0 t} \leftrightarrow X(f \mp f_0),$$

where $x(t)$ (either a voltage or a current) is the baseband signal. Shown in Figure 8.1 is a typical RF signal spectrum, located around a carrier frequency of f_C. Since the signal is real, its spectrum is symmetric with respect to the y-axis,[1] or positive and negative frequencies. However, most practical RF signals are not symmetric with respect to the carrier frequency,[2] and so it has been shown as such. If this signal is multiplied by $e^{-j2\pi f_0 t}$, it experiences a shift to the left, as shown in the figure. That is due to the fact that the spectrum is convolved in the frequency domain with an impulse at $-f_0$, representing the Fourier transform of $e^{-j2\pi f_0 t}$.

A lowpass filter wide enough to pass the entire shifted or *downconverted* spectrum now located at $f_C - f_0$, known as *intermediate frequency* or IF, is often applied to remove the component at $-f_C - f_0$. This leaves us with a now lower frequency version of the original RF signal, located at the intermediate frequency, as desired.

The main challenge of actually realizing the simple concept described above is the fact that the signal $e^{-j2\pi f_0 t}$ needed to shift the RF spectrum is not real. Using the Euler identity, one may write

$$e^{-j2\pi f_0 t} = \cos 2\pi f_0 t - j \sin 2\pi f_0 t.$$

Accordingly, the shifting signal, $e^{-j2\pi f_0 t}$, is composed of a sine and a cosine, both of which are real, and readily synthesizable. Depicted in Figure 8.2 is the RF signal multiplied (in the time domain) by individual sine and cosine functions.

Either signal consists of two impulses at $\pm f_0$, with half the magnitude now. The half magnitude results in signal loss, and thus performance degradation in general, but may or may not be an issue, as it is often retrievable by employing either RF and/or IF amplification. The presence of two impulses, however, complicates the matters, as around DC a folded replica of the shifted RF spectrum, basically its exact mirror *image*, presides as well. If $f_C - f_0$ is small with respect to the signal bandwidth,[3] the two replicas overlap, and are no longer distinguishable. Consequently, unless the IF is high enough to avoid the signal mixing with its image, shifting or downconverting by either a sine or cosine alone is not possible. To avoid this issue, the arrangement depicted in Figure 8.3 is often employed, leading to two versions of the downconverted spectrum, labeled as *I* (in-phase), and *Q* (quadrature). Equivalently, this scheme

---

[1] If $x(t)$ is real, then $X(-f) = X^*(f)$.
[2] Even though it may appear to be so when viewed on a spectrum analyzer. As discussed in Chapter 2, a symmetric spectrum around the carrier leads to spectrum and transmission power inefficiency.
[3] Many modern receivers employ $f_C - f_0 = 0$, known as zero-IF or direct-conversion receivers.

8.1 Mixers Fundamentals | 439

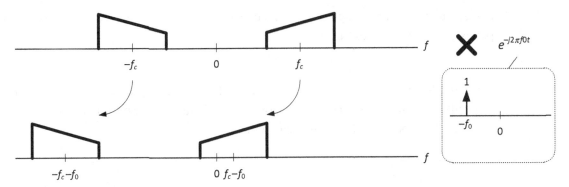

Figure 8.1: RF signal centered around carrier frequency of f_C, multiplied by $e^{-j2\pi f_0 t}$

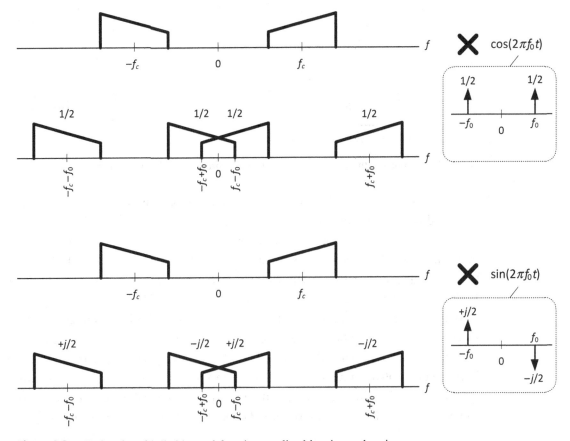

Figure 8.2: RF signal multiplied by real functions realized by sine and cosine

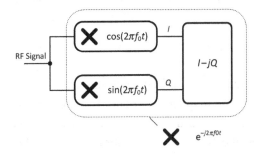

Figure 8.3: Quadrature downconversion employed in most modern radios to avoid signal mixing with its image

realizes an ideal shift through $e^{-j2\pi f_0 t}$, if a collective combination of the I and Q signals (or to be precise, $I \pm jQ$) is processed.

> **Example:** As a useful exercise, let us consider Figure 8.4, where the I and Q channels resulting from the quadrature downconversion (Figure 8.3) are further processed through a pair of Hilbert transform units and adders.
>
>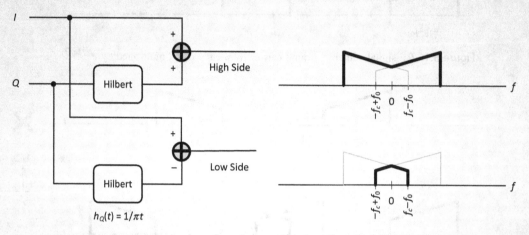
>
> **Figure 8.4:** Extracting the low- and high-sidebands through a Hilbert transform
>
> We assume the original RF signal is of the general form shown already in Figure 8.1. As discussed earlier, the Hilbert transform multiplies the positive frequency contents of the spectrum by $-j$, and the negative components by $+j$. Hence, for instance, the Hilbert transform of a cosine becomes a sine, as is evident from Figure 8.2 as well. Consequently, the outputs after addition/subtraction look like what is shown on the right side of Figure 8.4. Both spectrums are real, and thus symmetric, as they should be. Addition however results in only the high side (with respect to carrier or f_C) of the ideal shifted spectrum (see Figure 8.1), whereas subtraction extracts the low side of the spectrum. Compared to the spectrum of the ideal shifted signal (Figure 8.1), both spectrums in Figure 8.4 are real. Furthermore, unlike shifting by sine or cosine only, the signal and its mirror image no longer overlap. Even though the Hilbert transform is noncausal in general, as pointed out in Chapter 2, it could be well approximated over a limited range of frequencies.

The above example is the basis of what is known as *image-reject* receiver, which we shall analyze in more detail in Chapter 12.

8.1.2 Mixer Basic Circuit Operation

Ideally, a mixer needs to preserve the signal in its entirety, without adding any noise or distortion of its own, and translate it only in the frequency domain. In practice, since almost all mixers are *transistor-based*, they add both noise and distortion. As in any LTI system, the input and output frequencies are the same; the frequency translation may only be achieved through time variance or nonlinearity.

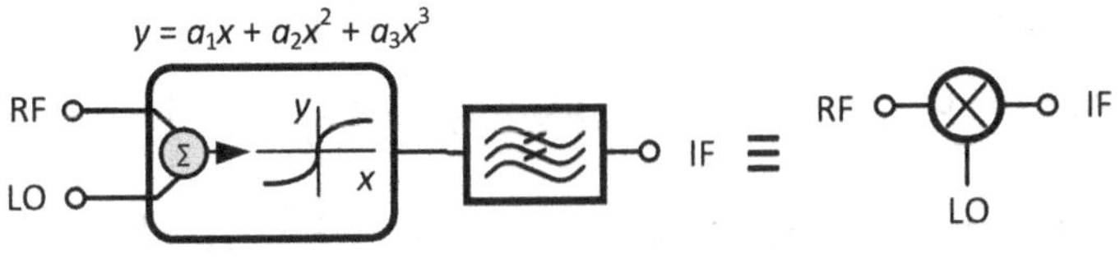

8.1 Mixers Fundamentals

Figure 8.5: A nonlinear system acting as a mixer

Given the inherent nonlinearity of any transistor-based circuitry, it is not difficult to conceive how one can create the frequency translation needed. Consider Figure 8.5, which represents a general nonlinear system where two arbitrary inputs are applied.[4]

Suppose the signal applied to the port labeled RF is $A_1 \cos(\omega_1 t)$, and the other signal to the LO port is $A_2 \cos(\omega_2 t)$. The signals are added and passed through a nonlinear system whose input–output characteristic is $y = a_1 x + a_2 x^2 + a_3 x^3$. Then $x = A_1 \cos(\omega_1 t) + A_2 \cos(\omega_2 t)$. As we established in Chapter 5, the output consists of several terms, of which most will diminish after lowpass filtering, except for the difference term created due to the 2nd-order nonlinearity assuming that $\omega_1 - \omega_2 \ll \omega_1$ and ω_2:

$$y = \cdots + a_2(A_1 \cos(\omega_1 t) + A_2 \cos(\omega_2 t))^2 + \cdots = \cdots + a_2 A_1 A_2 \cos(\omega_1 \mp \omega_2)t + \cdots.$$

Thus, only a component at $\omega_1 - \omega_2$, that is, the difference between the RF and LO frequencies, passes through the narrow LPF, and appears at the output. This is exactly what is needed to realize a *downconversion* mixer. The mixer gain would be the ratio of the output amplitude to that of the RF port, that is, $a_2 A_2$. Thus, it depends not only on how strong the 2nd-order nonlinearity is (i.e., a_2), but also on the signal applied to the LO port.

Example: Any circuit with strong 2nd-order nonlinearity, such as a long channel FET (Figure 8.6), could potentially be a mixer.

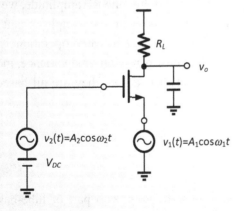

Figure 8.6: A long-channel FET operating as a mixer

If the device is square-law, ignoring the bias details, the AC component of the output voltage is

Continued

---

[4] A more generalized case would be in the form of $y = f(\text{RF}, \text{LO})$, as opposed to $y = f(\text{RF} + \text{LO})$. In that case the multivariable Taylor series must be used instead.

$$v_o = \frac{1}{2}\mu C_{OX}\frac{W}{L}R_L A_1 A_2 \cos(\omega_1 - \omega_2) = \left(\frac{1}{2}\mu C_{OX}\frac{W}{L}R_L A_2\right)A_1 \cos(\omega_1 - \omega_2).$$

Thus, the circuit is indeed behaving as a mixer whose gain is $\frac{1}{2}\mu C_{OX}\frac{W}{L}R_L A_2$. Since an *exact* squarer function is not needed to realize the mixer, the FET could have been replace by a BJT, in which case the gain would have been $\frac{1}{2}\frac{I_S}{V_T^2}R_L A_2$, where I_S is the BJT saturation current and $V_T = \frac{KT}{q}$. The gain dependence on the LO amplitude as would be the case for the simple mixer of Figure 8.6 is not desirable, and as we will show momentarily, in most cases is eliminated by proper design of the LO port.

If instead of a LPF, a BPF tuned to $\omega_1 + \omega_2$ were to be used to select the component at the sum frequency, the circuit would have acted as an *upconversion* mixer.

We expect the mixer then to be a three-port network, two inputs that are mixed, and one output that is the result of frequency conversion. We label the main input as RF (radio frequency), and the output as IF (intermediate frequency), whereas the second input terminal would be the LO (local frequency). However, this convention applies only to downconversion mixers.

As we saw in Chapter 6, the nonlinearity is often what we try to avoid, and thus it is often the case that the mixer would rely on time variance, rather than nonlinearity. A *good* mixer is the one that performs the frequency translation without imposing much noise or distortion on the signal. How good the mixer needs to be is entirely a function of the radio specifications and the application, as long as we can represent the mixers with typical known circuit properties such as noise figure or IIP_3.

We often treat a *good* mixer, then, as a *linear but time-variant* circuit with respect to its input (RF) and output (IF) ports. That is, for example when doubling the RF amplitude, we expect the amplitude of the IF signal to double as well. Obviously for the simple circuit shown in Figure 8.5, this would be the case in spite of the fact that its very operation is based on nonlinearity. As pointed out, however, most modern mixers rely on time variance, more so than nonlinearity, to create the frequency translation. For that reason, they are all based on some form of *voltage or current switching*.

8.2 EVOLUTION OF MIXERS

Mixers have been around before the invention of transistors, as a part of tube-based radios. Shown in Figure 8.7, a structure commonly known as a *diode-ring mixer* relies on Schottky diodes to act as fast switches, which result in frequency translation given the inherent *time variance* of any switching circuitry.

The input and output signals are attached to transformers connected to the ring, whereas a differential LO signal is applied to the center tap of the transformers. For the one half cycle where LO is high, D_2 and D_4 are on, but D_1 and D_3 are off, and the output is equal to the input, whereas for the next half cycle, where D_1 and D_3 are on, the positive–negative terminals of the input and output are reversed. Thus, the output will be equal to the input signal, multiplied by

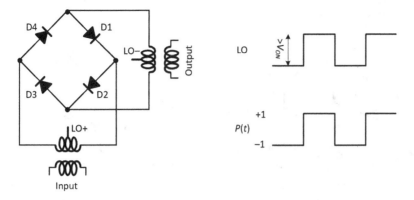

Figure 8.7: A diode-ring mixer

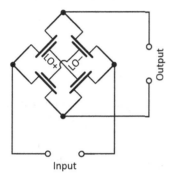

Figure 8.8: A FET-based mixer

$P(t)$ as shown in Figure 8.7. The minimum required amplitude of the LO is then equal to diode V_{ON}, to ensure that they turn on properly. Expanding $P(t)$ using Fourier series we have

$$P(t) = \frac{4}{\pi}\left(\sin \omega_{LO}t - \frac{1}{3}\sin 3\omega_{LO}t + \frac{1}{5}\sin 5\omega_{LO}t - \cdots\right),$$

which consists of a fundamental at the LO frequency, as well as its odd harmonics. Assuming the input signal is $A \sin \omega_{RF}t$, and that the output is tuned to $|\omega_{RF} - \omega_{LO}|$ to filter all undesired terms, the mixer output is

$$\frac{2}{\pi}A\cos(\omega_{RF} - \omega_{LO})t.$$

Thus, the mixer has a conversion gain of $\frac{2}{\pi}$, which is less than one, but is *independent of the LO* amplitude, as long as it is large enough to turn the diodes on and off comfortably. In fact the LO signal does not need to be a square-wave, if its slope is sharp enough during the transition from one half cycle to the other. Moreover, the circuit is linear with respect to the input–output ports.

It wasn't until 1968, well after the invention of the transistors and integrated circuits, that a group of researchers at MIT proposed the idea of using a FET as a switch rather than the diode [1] as shown in Figure 8.8.

The mixer operates exactly the same way as the diode ring circuit, except for the LO signals are conveniently applied directly to the gate of the FETs. Moreover, the FET is a better switch and thus results in a more linear mixer. Particularly, unlike diodes, one can completely shut

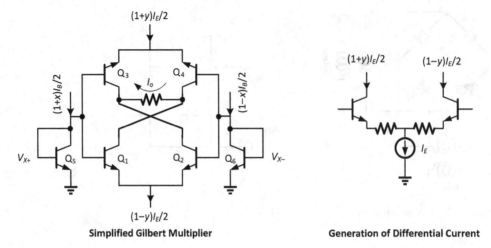

Figure 8.9: Gilbert four-quadrant multiplier

down the FET if the proper voltage is applied. All passive mixers are fundamentally a variation of this topology.

Also in 1968, Barrie Gilbert published his landmark paper [2], describing a precise *four-quadrant analog multiplier* whose simplified diagram is shown in Figure 8.9.

Two sets of differential currents are applied to the multiplier on the left, where each may be produced by a linear differential pair as shown on the right. We assume that the devices are matched, that they have a perfect exponential characteristics, and that β is high. By writing KVL we have

$$V_{BE1} - V_{BE2} = V_{BE3} - V_{BE4} = V_{BE5} - V_{BE6}.$$

Since for each transistor $V_{BE} = \frac{KT}{q} \ln \frac{I_E}{I_S}$, we arrive at

$$\frac{I_{E1}}{I_{E2}} = \frac{I_{E3}}{I_{E4}} = \frac{I_{E5}}{I_{E6}} = \frac{1+x}{1-x}.$$

Writing KCL yields

$$I_{E1} + I_{E2} = (1-y)\frac{I_E}{2}$$

$$I_{E3} + I_{E4} = (1+y)\frac{I_E}{2}.$$

Solving for the three equations above results in Q_{1-4} as a function of input currents fed to the multiplier:

$$I_{E1/2} = (1 \pm x)(1-y)\frac{I_E}{4}$$

$$I_{E3/4} = (1 \pm x)(1+y)\frac{I_E}{4}.$$

Since the differential output current is

$$I_o = (I_{E2} + I_{E3}) - (I_{E1} + I_{E4}),$$

defining $z = \frac{I_o}{I_E}$ as the normalized output current, we arrive at

$$z = x \times y.$$

What is remarkable about this design, and perhaps the reason for its longevity, is that a very nonlinear *I–V* characteristic of a BJT is exploited to create a perfectly linear multiplier. If x and y are the RF and the LO signals, the output z after lowpass filtering will be the IF signal, creating a mixer. The dependence on the LO voltage is eliminated by applying the proper signal to the LO port as we will show in the next section. Almost all active mixers used today are based on this topology or some variation thereof.

The main distinction between the passive and active mixers as described in Figure 8.8 and Figure 8.9 has to do with the fact that in one case an AC voltage (or current) is being commutated by the switches, whereas in the other case a combination of small AC and a large DC current is applied to the switching circuitry. A good analogy to this is static CMOS logic to the current-mode logic (CML), where the latter is found to be much more superior in terms of speed. This more or less applies to the mixers, as the active mixers tend to operate at higher frequencies, since the passive-based topologies require sharp rail-to-rail LO signals for proper operation. However, as we will show, passive mixers appear to be more superior in terms of noise and linearity, as long as the switches operate properly.

8.3 ACTIVE MIXERS

Active mixers rely on the concept of *current switching*. Let us start with the rearrangement of the Gilbert multiplier of Figure 8.9 redrawn in Figure 8.10.

Instead of feeding a voltage created by the diode-connected pair Q_{5-6} to the base of core transistors (nodes V_{X+} and V_{X-}), we apply a square-wave-like signal as shown on the right, large enough to steer the emitter current abruptly from one device to the other at each zero crossing. To understand this better, consider Figure 8.11, showing an emitter-coupled pair whose *I–V* characteristics are shown on the right.

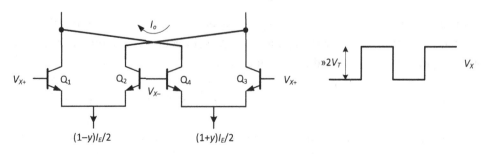

Figure 8.10: A simplified Gilbert mixer

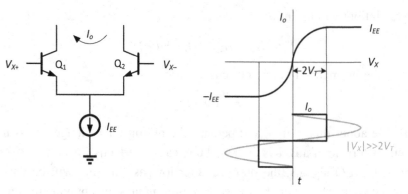

Figure 8.11: An emitter-coupled pair and its I–V curve

From basic analog design [3], we know

$$I_o = I_{EE} \tanh \frac{-V_X}{2V_T},$$

where $V_X = V_{X+} - V_{X-}$ is the differential input, and $V_T = \frac{KT}{q}$. For values of $|V_X| > 2V_T$, most of the current is steered from one device to the other due to the tanh sharp transition, and the output current approaches I_{EE}. Thus, in the time domain, once a voltage applied to the differential pair with an amplitude large enough compared to $2V_T \approx 50m$, an almost square-wave current is created at the output. The current toggles between $\pm I_{EE}$, regardless of the shape and amplitude of V_X. Thus under such conditions

$$I_o \approx \frac{4}{\pi} I_{EE} \left(\sin \omega_{LO} t - \frac{1}{3} \sin 3\omega_{LO} t + \frac{1}{5} \sin 5\omega_{LO} t - \cdots \right),$$

where V_X is assumed to be at the LO frequency. In practice, the emitter current consists of a large DC component, along with a small AC signal as a result of the input RF voltage,

$$I_{EE} = (1-y)\frac{I_E}{2} = (1 - A \cos \omega_{RF} t)\frac{I_E}{2},$$

where we assume $y = A \cos \omega_{RF} t$ is the RF input (note that A is unit-less). If the DC quiescent current is large, or equivalently if the amplitude of the RF voltage is small, as is the case in most practical situations, the switching is mostly unaffected. Thus, ignoring other harmonics except for the fundamental, the output current is

$$I_o = \frac{4}{\pi}(1-y)\frac{I_E}{2} \sin \omega_{LO} t,$$

and once the second pair consisting of Q_{3-4} is added (Figure 8.10) to realize the fully differential circuit, the terms at f_{LO} cancel as a result of subtraction at the output, whereas the terms consisting of the RF inputs add up, leading to

$$z = \frac{I_o}{I_E} = \frac{4}{\pi} y \times \sin \omega_{LO} t = \frac{2}{\pi} y [\cos(\omega_{RF} - \omega_{LO})].$$

Consequently, the linear mixer function whose gain is independent of the LO signal is created. The mixer is said to be *double-balanced*, as neither the LO nor the RF inputs feed through the

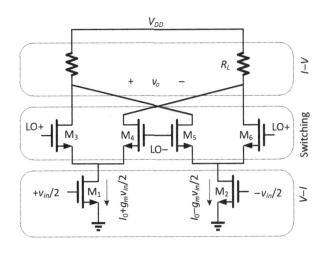

Figure 8.12: A double-balanced MOS active mixer

output. If the pairs are not matched, or the input is not perfectly differential, there is a small feed through, which is usually unimportant as it is subject to output filtering since in most cases: $\omega_{RF} - \omega_{LO} \ll \omega_{RF}$ and ω_{LO}.

An MOS version of the Gilbert mixer is created upon the same principle, and is shown in Figure 8.12.

Keeping the same notation as the BJT pair, for a MOS differential pair [3]

$$I_o = \frac{1}{2}\mu C_{OX}\frac{W}{L}V_X\sqrt{\frac{4I_S}{\mu C_{OX}\frac{W}{L}} - V_X^2},$$

where I_S is the source-coupled pair current. The current is completely steered from one device to the other if

$$|V_X| = \sqrt{\frac{2I_S}{\mu C_{OX}\frac{W}{L}}} = \sqrt{2}V_{eff}.$$

Thus, as long as the LO voltage is large enough compared with $\sqrt{2}V_{eff}$, the mixer functions as described before, and the conversion gain is

$$\frac{2}{\pi}g_m R_L,$$

where g_m is the input pair (M_{1-2}) transconductance.

The mixer operation may be understood better if it is broken into three distinct functions:

1. The input V–I converter, which consists of a pseudo-differential pair. Any variation of that, such as one with degeneration or a true source-coupled pair with a current source, may be chosen (Figure 8.13), although the one shown in Figure 8.12 is the most common choice for its simplicity, and gives the best compromise between headroom, noise, and linearity.
2. The switching quad (devices M_{3-6}) commutates the RF current.
3. The load performs the I–V conversion. The load also may be active or passive.

Each of these stages potentially contributes to the mixer noise and nonlinearity, and will be discussed next.

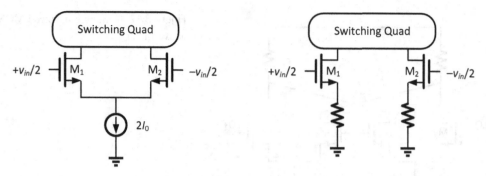

Figure 8.13: Different variations of the input V–I converter

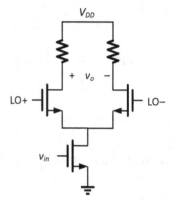

Figure 8.14: A single-balanced active mixer

One final note: as most antennas are single-ended, producing a differential RF voltage, whether at the LNA input or output, requires an extra transformer that comes at a cost. If the LNA output happens to be single-ended, then a single-ended version of the Gilbert mixer known as a single-balanced mixer may be used (Figure 8.14). The output is still differential, and in perfect matching conditions, the RF feedthrough is eliminated.

However, there is an LO feedthrough appearing at the output that is to the first order independent of the LO, as we already calculated before,

$$\frac{4}{\pi} R_L I_0 \sin \omega_{LO} t,$$

where I_0 is the input device bias current. The feedthrough may be removed only through IF filtering. Another drawback of the single-balanced scheme is that any noise or interferer at the LO port appears at the output.

8.3.1 Active Mixers Linearity

We will present here only a qualitative discussion on active mixers linearity and IIP_3. A more rigorous analysis may be found in [4].

Consider the simplified mixer schematic shown in Figure 8.15. The load is typically not a major nonlinearity contributor if designed properly, and particularly if it is resistive. The switches also do not contribute much to the overall nonlinearity provided that the output and

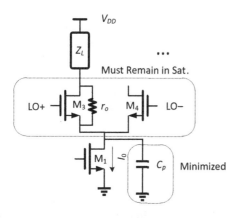

Figure 8.15: Active mixer nonlinearity contributors

the LO DC biases are chosen such that for the half cycle that a given switch is on, it remains in saturation to ensure that the lower transconductance devices are well isolated from the output.

The parasitic capacitance at the common source of each switching pair (C_p in Figure 8.15) must be minimized. Specifically, one must ensure that $\frac{g_{m3-4}}{C_p} \gg \omega_{LO}$. In modern nanometer CMOS processes this condition is comfortably met for most GHz RF applications.

The transconductance stage appears to set the bottleneck. A detailed analysis of IIP_3 of a FET was presented in Chapter 6. To a 1st-order approximation, if designed properly, and assuming that the devices stay in the intended region of operation, that will set the mixer IIP_3. The latter condition may be difficult to be met, as the active mixer consists of at least three stages stacked, and the headroom may be a challenge. Nevertheless, achieving an IIP_3 of better than 1V (or 10dBm) is possible, and has been reported for 1.2V supply voltages in the literature.

8.3.2 Active Mixers 1/f Noise Analysis

We deal with the noise contribution of each section of the mixer separately, and offer a physical model to capture the noise with sufficient accuracy [5].

Load – Noise in the output load does not suffer frequency translation, and competes directly with the signal. Fortunately, this noise may be lowered in one of many ways. If a PMOS active load is used, the device channel length must be chosen sufficiently large to reduce the $1/f$ noise corner below the frequency of interest. Alternatively, at the expense of some voltage headroom the mixer may be loaded with polysilicon resistors, which are free of flicker noise.[5]

Input FETs – Noise in the lower transconductance FETs accompanies the RF input signal, and is translated in frequency just like the signal is. Therefore, flicker noise in these FETs is upconverted to ω_{LO} and other odd harmonics, while white noise at ω_{LO} (and other odd harmonics) is translated to DC. If the output of interest lies at or around zero IF, then the transconductance FETs contribute only white noise after frequency translation, since the flicker corner of these devices is usually much lower than the LO frequency. Due to mismatches in the

---

[5] This is not generally true, as the polysilicon resistors have been shown to have $1/f$ noise [28], [29]. However, in most cases it may be ignored, as we choose to do throughout this chapter and the entire book.

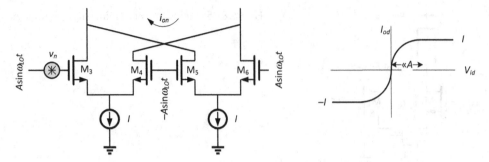

Figure 8.16: 1/f noise model for mixer switches

switching transistor quad, a small amount of the flicker noise in the transconductance FETs can appear at the loads. This is dealt with at the end of this section.

Switches – Consider the simplified double-balance balanced mixer in Figure 8.16. The switch FETs M_{3-4} and M_{5-6} carry a bias current varying at a frequency ω_{LO}. Flicker noise arises from traps with much longer time constants than the typical period of oscillation at RF, and it may be assumed that the *average* inversion layer *charge* in the channel determines the RMS flicker fluctuations. Thus, these charge fluctuations are referred to as a voltage to the gate of one FET (say M_3), with a non-time-varying RMS value and a spectral density varying as $1/f$ (v_n in Figure 8.16). Roughly speaking, this equivalent voltage may be thought of as a slowly varying offset voltage associated with differential pair. It should be noted that based on the carrier-density fluctuation model, the input-referred flicker noise of MOSFETs is independent of the gate voltage. This has been experimentally verified for NMOS transistors [6], [7].[6]

To simplify analysis, it is also assumed that a sine-wave drives the switches, and that the circuit switches sharply, that is, a small differential voltage excursion (V_{id}) causes the current (i_{od}) to completely switch from one side of the differential pair to the other side (Figure 8.16). Furthermore, as the noise transfer function to the output is *linear* (but time-variant), the *superposition* holds. Thus, it is sufficient to analyze the noise of one switch, and simply multiply the result by four as all the noise sources are *uncorrelated*.

First consider the direct effect of the switch noise at the mixer output. The transconductance section of the mixer may be replaced by a current source, I, at the tail. In the absence of noise, for positive values of LO signal, M_{3-6} switch on and M_{4-5} switch off, and a current equal to I appears at each branch, although no net current appears at the output due to the subtraction of M_3 and M_6 currents. In the next half period, M_{4-5} are switched on and carry a current I. Once noise is included though, the slowly varying v_n modulates the time at which the pair M_{3-4} switches (Figure 8.17). Now at every switching event the skew in switching instant modulates the differential current waveform at the mixer output. The *height* of the square-wave signal at the output remains constant, but noise modulates its *zero-crossing*, by $\Delta t = \frac{v_n}{S}$, where S is the

---

[6] In more recent CMOS processes this may not hold anymore. See [29] for more details. Nonetheless, to gain some insight we make the assumption that the 1/f noise is not a strong function of the bias, which seems to be reasonably accurate.

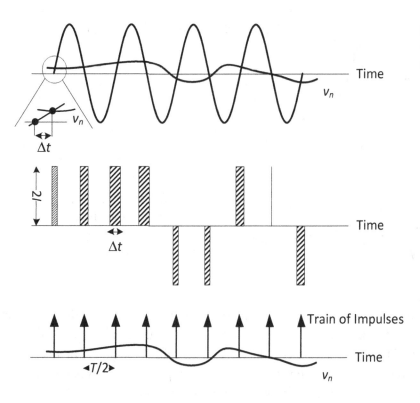

Figure 8.17: Switch input voltage and mixer output current

slope of LO signal at zero-crossing. The waveform at the mixer output then consists of a train of pulses with width of Δt and amplitude of $2I$ and a frequency of $2\omega_{LO}$ (Figure 8.17).

Over one period the output current has an average value equal to

$$i_{on} = \frac{2}{T} \times 2I \times \Delta t = 4I \frac{v_n}{S \times T},$$

where T is the period of LO equal to $2\pi/\omega_{LO}$. This means that the low-frequency noise at the gate of switch, v_n, is directly transferred to the output and will compete with a signal down-converted to zero IF. The zero-crossing offset in the time domain, Δt, depends on the low-frequency noise, v_n, and the LO sine-wave slope, S. For a sine-wave LO signal, the slope is equal to $2A\omega_{LO}$, where A is the amplitude of the LO. Thus, the base-band $1/f$ component appearing at the output can be easily calculated to be

$$i_{on} = \frac{I}{\pi A} v_n.$$

To find the complete spectrum of the mixer output noise, we notice that since $\frac{\Delta t}{T} \ll 1$, the pulses may be approximated with ideal impulses of amplitude of $\frac{2I\Delta t}{S}$, and twice the LO frequency, sampling the mixer noise (Figure 8.17). The resulting spectrum at the mixer output is then easily obtained based on sampling theory, and is shown in Figure 8.18. It should be noted that since noise pulses have a frequency of twice the LO frequency, noise spectrum is repeated at $2\omega_{LO}$ and its other even harmonics.

If such a mixer is used for upconversion, then there will be no flicker noise contribution from switches, as it can be seen from Figure 8.18, but the flicker noise of transconductance stage will be directly upconverted and appear at ω_{LO}.

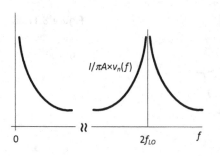

Figure 8.18: Mixer output noise spectrum

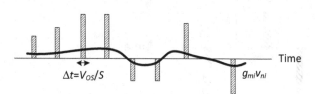

Figure 8.19: Flicker noise of input devices due to switching pair offset

This analysis is also used to answer the earlier question about flicker noise originating in the transconductance devices, which leaks through switch FETs unbalanced by an offset. The noise voltage v_n is replaced with an offset voltage, V_{OS}, and the fluctuations of interest are in the current I. The mixer output current is now a train of pulses of *constant offset in its zero-crossings*, V_{OS}/S, whose *height* is modulated by the noisy current, $g_m v_{ni}(t)$, where v_{ni} is the input-referred flicker noise of the transconductance FET, and $g_m \approx \frac{I}{V_{eff}}$ is the short-channel transconductance of input FETs (Figure 8.19).

The low-frequency noise component is equal to

$$i_{on} = 4I \frac{V_{OS}}{S \times T} = 4 g_m v_{ni} \frac{V_{OS}}{S \times T} = \frac{I}{\pi A} v_{ni} \frac{V_{OS}}{V_{eff}}.$$

If it is assumed in comparing the two noise equations that the noise voltages are of a similar order of magnitude, then switch noise is clearly more important as usually $V_{OS} \ll V_{eff}$. Note that in the equation v_{ni} can also represent any kind of low-frequency interferer coupled to the mixer input, such as the low-frequency noise on the ground line.

If the slope of the LO is increased to infinity, that is, if a perfect square-wave is applied, the mixer flicker noise does not diminish entirely, rather is limited by the parasitic capacitance at the common source of each switching. This *indirect mechanism* is analyzed in [5], and is typically not a dominant factor as the slope of the LO in reality is never that large unless at very low frequencies where the parasitic capacitances are not important.

8.3.3 Active Mixers White Noise Analysis

A similar analysis as the one presented for $1/f$ noise of switches may be extended to the white noise.

Intuitively, we know that switches contribute noise at the mixer output in the time interval when they are both on. That is because when one switch is on and the other one is off, clearly the off switch contributes no noise. The on switch also cannot produce any noise current at the

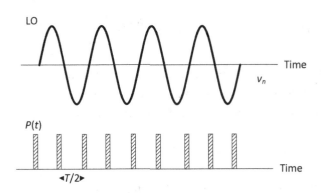

Figure 8.20: Mixer noise due to switches' white noise

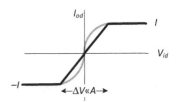

Figure 8.21: Simplified switching pair I–V response

mixer output as its drain current is fixed by the bias current I produced by the input devices (Figure 8.16). Thus, the output noise is effectively a result of the input-referred noise of each switch, multiplied by a sampling function $P(t)$, as shown in Figure 8.20.

The width of $P(t)$ is approximately the time window that both switches are on, and its frequency is twice the LO frequency. The exact shape of $P(t)$ depends on the LO signal characteristics, but can be well approximated if the LO amplitude is large, and the switching is sharp, as is the case in most switching mixers. We can approximately find the width and height of the pulses if we assume a linear response for each switching pair during which they are open as shown in Figure 8.21.

If the excursion that the switches are both open is ΔV, then in time domain the pulse width must be $\Delta V/S$, S being the slope of the LO signal. To keep the area of each pulse constant, the height is then $2I/\Delta V$. This approximates the I–V response of the pair (Figure 8.16) fairly well, and may be employed to gain some insight into mixer noise (and the 2nd-order nonlinearity as we will discuss in the next section) mechanisms. However, as we will show momentarily, the exact shape of the LO, and the switching pair I–V response is not as important as long as the switching is abrupt.

To generalize, we can say

$$i_{on}(t) = P(t) \times v_n(t).$$

Since the switches' input-referred noise is white, and $P(t)$ is a deterministic periodic function, the resultant mixer output noise is white and *cyclostationary* (see Chapter 5). Consequently, consistent with our derivations in Chapter 5, the spectral density of each switch noise will be $4KT\gamma$, sampled by $|P(t)|^2 = g_m(t)$, where $g_m(t)$ is the effective transconductance of each switching pair (see Problem 9 for more details.). The average output noise spectral density is then

$$S_{i_{on}} = 4KT\gamma \frac{1}{T}\int_T G_m(t)dt = \frac{8KT\gamma}{T}\int_{T/2}\frac{\partial i_o}{\partial v_{LO}}dt = \frac{8KT\gamma}{T}\int_{T/2}\frac{\partial i_o/\partial t}{\partial v_{LO}/\partial t}dt.$$

Without loss of generality, let us choose our reference time to be at $t = 0$. In the case of abrupt switching, the mixer output current is a square-wave toggling between $\pm I$ at every zero crossing. Thus, for the particular zero crossing at $t = 0$, $\frac{\partial i_o}{\partial t}$ may be approximated as

$$\frac{\partial i_o}{\partial t} \cong 2I\delta(t),$$

where $\delta(t)$ is the *Dirac impulse* function. Consequently, the integral simplifies to

$$S_{i_{on}} = \frac{8KT\gamma}{T}\int_{T/2}\frac{2I\delta(t)}{\partial v_{LO}/\partial t}dt = 4KT\gamma \frac{4I}{S \times T},$$

where $S = \frac{\partial v_{LO}}{\partial t}\big|_{t=0}$ is the slope of the LO at zero crossing as defined earlier. The switch white noise spectral density comes out to be independent of LO waveform shape or switch size, assuming abrupt switching. This is very similar to the 1/f noise results calculated earlier based on the approach taken from [5]. For a sinusoidal LO, the spectral density of the output noise is

$$S_{i_{on}} = 4KT\gamma \frac{I}{\pi A}.$$

A more rigorous approach presented in [8] leads to a similar result assuming the LO amplitude is large.

Next, we shall find the noise contribution of the input FETs. The white noise of the FETs downconverts to the IF along with the signal, as shown in Figure 8.22.

For the case of high-side injection, for instance, the signal is present only at $f_{LO} - f_{IF}$, whereas noise appears at both sidebands of the fundamental, $f_{LO} \pm f_{IF}$, as well as all its odd harmonics as shown in Figure 8.22. If we ignore the higher order harmonics for the moment, the presence of signal at only one sideband leads to a factor of 2 or 3dB worse SNR after the downconversion, as the *white* noise is inevitably present at both sides. This is known as *single side-band* (SSB) noise figure. On the other hand, if a receiver is capable of rejecting the image at the unwanted side-band after downconversion, such as a quadrature receiver, the noise is removed as well, and the SNR is preserved. This is referred to as *double side-band* noise figure.

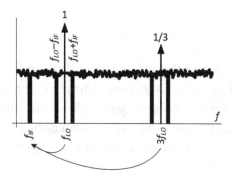

Figure 8.22: Mixer noise due to input FETs' white noise

For the general case, the total output noise spectral density due to one input FET is

$$S_{i_{on}} = 4KT\gamma g_m \left(\frac{2}{\pi}\right)^2 \left[2 \times \left(1 + \frac{1}{3^2} + \frac{1}{5^2} + \cdots\right)\right] = 4KT\gamma g_m,$$

where $4KT\gamma g_m$ is the noise current spectral density of the input FET, $\frac{2}{\pi}$ is the current conversion gain, and the last term signifies the contribution of all the harmonics: the white noise at $\omega_{LO} \pm \omega_{IF}$ downconverted by the main harmonic of the LO, the white noise at $3\omega_{LO} \pm \omega_{IF}$ downconverted by the third harmonic of the LO whose amplitude is one-third of the main harmonic, and so forth. Notice that LO is not necessarily a square-wave signal; however, due to the hard switching action in the mixer, it will effectively perform as a square-wave. Particularly, the factor 2 arises from SSB noise. It is interesting to note that the noise current of the input FET appears exactly intact at the mixer output, whereas the signal is subject to a loss of $\frac{2}{\pi}$. This is because the *square-wave-like LO* downconverts the white noise at all harmonics, whereas the signal of interest lies only at a specific sideband of the fundamental. This, along with the extra noise contribution of the switches, is why mixers tend to be noisier than linear amplifiers.

The output noise spectral density could have been obtained intuitively without any of the calculations above. Due to the hard switching, the output noise current is always

$$i_{on} = \pm 1 \times i_{n,\,gm}.$$

Thus

$$S_{i_{on}} = |\pm 1|^2 S_{i_{n,\,gm}} = S_{i_{n,\,gm}} = 4KT\gamma g_m.$$

When all the noise contributors added up, the mixer output noise spectral density will be

$$S_{i_{on}} = 2 \times 4KT\gamma g_m + 4 \times 4KT\gamma \frac{I}{\pi A} + 2 \times \frac{4KT}{R_L},$$

where we assume the mixer is loaded with a resistance of R_L. Since the current conversion gain is $\frac{2}{\pi} g_m$, the noise figure with respect to a source resistance of R_S is

$$F = 1 + \frac{\frac{\pi^2}{4}}{g_m R_S}\left(\gamma + \frac{2\gamma}{g_m}\frac{I}{\pi A} + \frac{1}{g_m R_L}\right).$$

Assuming a transconductance of $g_m \approx \frac{I}{V_{eff}}$ for the short-channel input FET, and ignoring the passive load noise contribution,

$$F \approx \frac{\pi^2}{4}\frac{V_{eff}}{R_S I}\gamma\left(1 + \frac{2V_{eff}}{\pi A}\right).$$

The relative noise contribution of the switches to that of the transconductance FET is equal to $\frac{2V_{eff}}{\pi A}$. This means that as the overdrive of transconductance FET becomes close to LO amplitude, switches contribute a comparable noise to what is contributed by the transconductance stage. This clearly displays the trade-off between noise and linearity in active mixers: In linear mixers a large overdrive (to enhance the linearity of the transconductance FET) with approximately small LO swing (to keep the switch transistors in saturation) is desirable, which boosts the relative noise contribution of the switches.

Example: To improve linearity and enhance gain, and to avoid headroom issues, one may choose an active load for the mixer as shown in Figure 8.23.

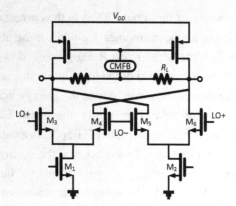

Figure 8.23: Mixer with active load

The device channel length must be large enough to lower the $1/f$ noise corner of the PMOS load sufficiently. Nevertheless, the load may contribute significant white noise. The modified noise figure is (noise of R_L ignored)

$$F \approx \frac{\pi^2}{4} \frac{V_{eff}}{R_S I} \gamma \left(1 + \frac{V_{eff}}{V_{effp}} + \frac{2 V_{eff}}{\pi A}\right),$$

where V_{effp} is the PMOS load overdrive voltage. At low supply voltages, the PMOS overdrive must be kept low to ensure it stays in saturation, and as a result the active load will add as much or possibly more noise than the input FETs.

While the mixer thermal noise figure enjoys similar trade-offs as those of linear amplifiers, the switches' flicker noise can be improved only by either increasing the LO amplitude or reducing the switches input-referred noise. There are fundamental limitations to either option: the LO amplitude is limited by the supply, while the switches' input noise may be improved only by using bigger devices, which has frequency response consequences.

Example: Considering that the output $1/f$ noise is

$$i_{on} = \frac{I}{\pi A} v_n.$$

One may suggest lowering noise through reducing the mixer bias current. However, this is not a viable choice, as lowering the current degrades the signal as well. On the other hand, if the switches' bias current is lowered without considerably affecting the signal, we can break the fundamental barrier of lowering the $1/f$ noise.

This goal may be accomplished by either one of the two active mixer topologies shown in Figure 8.24, which we will discuss as case studies.

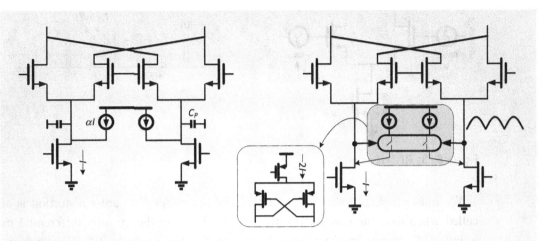

Figure 8.24: Active mixer topologies with enhanced $1/f$ noise performance

In the left scheme, two constant currents of αI ($\alpha < 1$) are fed to the input FETs, which are biased at a constant current of I. This is not going to impact the input FET transconductance, so we expect the signal gain to remain the same. However, the switches' DC current is lowered to $(1 - \alpha)I$, and so is their $1/f$ noise contribution. There are two drawbacks to this approach: First, the current sources add their own *white noise*, although they reduce $1/f$ noise. If they have a similar overdrive as the input FETs, as a result the mixer white noise is raised by a factor of $(\alpha + 1)$. Second, reducing the switches' bias effectively lowers their transconductance, and thus part of the input current produced by the input FET is going to be wasted in the parasitic capacitance at the common source of the switches (C_P in Figure 8.24).

A better approach is to reduce the bias current of the switches only when needed through a dynamic switching scheme as shown on the right side of Figure 8.24 [9]. We notice that the switches contribute noise only when they are both on, or right around the zero-crossings. Thus, we inject a *dynamic current* equal to I only at every zero-crossing. To do so, we take advantage of the full-wave rectified input voltage present at the common source of the switches as shown in Figure 8.24. This voltage is low right around zero-crossing, and once fed to a cross-coupled PMOS pair, it turns on the top current source, and a current of I is injected to each switching pair. Away from zero-crossings, the voltage is high, and consequently the PMOS devices are off. Fabricated in 130nm CMOS, and biased at 2mA from a 1.2V supply, this mixer achieves 11dB white noise figure when referred to 50Ω, and a $1/f$ noise corner of as low as 10kHz [9]. The input IP_3 is 10dBm, showing the fundamental limits of active mixers given the headroom and bias current trade-offs.

8.3.4 Active Mixers 2nd-Order Distortion

The mixer 2nd-order distortion can be analyzed very much through the same approach as we took for $1/f$ noise. Besides the input stage and switches as potential contributors, there is yet another mechanism caused by the feedthrough of the input voltage to LO ports. We shall study each of these cases below, and defer to [10] for a more detailed analysis.

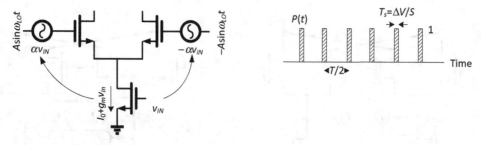

Figure 8.25: RF self-mixing

We must emphasize that, in a double-balanced mixer, 2nd-order distortion is always cancelled when the output is taken differentially. Thus, in theory fully differential mixers have infinite IIP_2. In practice, however, the mismatches cause a finite IIP_2. Nonetheless, we expect the IIP_2 to vary from one sample to the other, and thus it must characterized statistically.

RF Self-Mixing — Due to parasitic capacitive or magnetic coupling, the RF input voltage to the mixer leaks to the LO port and appears as a differential voltage in series with the LO as shown in Figure 8.25. Thus, the mixer downconverts the input RF signal with a composite LO consisting of the original desired LO, as well as the superimposed RF leakage. Note that the coupling from RF to LO is not generally differential, but given the common-mode rejection, we would consider only a differential coupling here.

Let us assume that the coupling coefficient is $\alpha \ll 1$. With a similar rationale as the one we used for the mixer noise, it is clear that self-mixing happens only during the window that both switches are on, in which case the switching pair acts as a linear amplifier, and whose current is being modulated by the input RF voltage.

Thus, we define the sampling function, $P(t)$, as shown in Figure 8.25 exactly as we had before (Figure 8.20). During the *on window*, the effective transconductance gain of the pair is

$$\frac{2I}{\Delta V} = \frac{2}{\Delta V}(I_0 + g_m v_{in}),$$

where ΔV is the switching pair input excursion (Figure 8.21). Thus, the output current caused by self-mixing will be

$$i_o = \left[\frac{2}{\Delta V}(I_0 + g_m v_{in})\right](2\alpha v_{in}).$$

However, this current is scaled by the window that the switches are on, that is, $\frac{T_s}{T/2}$. Thus the 2nd-order coefficient of the output current is

$$a_2 = \frac{4\alpha}{\Delta V} g_m \frac{2T_s}{T} = \frac{8\alpha}{S \times T} g_m.$$

The fundamental coefficient is $a_1 = \frac{2}{\pi} g_m$, as before. Thus, the IIP_2 is

$$IIP_2 = \frac{a_1}{a_2} = \frac{S \times T}{4\pi\alpha},$$

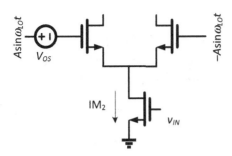

Figure 8.26: Transconductance stage 2nd-order distortion as a result of offsets in switches

and for a sinusoidal LO of amplitude A,

$$\text{IIP}_2 = \frac{A}{\alpha}.$$

Besides increasing the LO amplitude or its slope, to a 1st-order all that can be done is to improve isolation to reduce α.

Input FET 2nd-Order Nonlinearity — As it is very common to incorporate a pseudo-differential pair for the input transconductance stage, a strong 2nd-order component is expected to be present. The IM_2 component, however, is at DC or around $2f_{LO}$, and thus ideally appears at f_{LO} at the output when mixed with the LO. However, very similar to the $1/f$ noise argument, any mismatches in the switching pair results into low-frequency IM_2 current to leak to the output, and to ultimately set an upper limit for IIP_2 (Figure 8.26).

In Chapter 6 we showed that a square-law FET creates a 2nd-order component,

$$\frac{1}{2}\beta v_{in}^2,$$

where $\beta = \mu C_{OX} \frac{W}{L}$. The leakage current gain as a result of offset was found to be $4\frac{V_{OS}}{S \times T}$ in the previous section. Thus the 2nd-order coefficient is

$$a_2 = 2\beta \frac{V_{OS}}{S \times T},$$

and the IIP_2 is

$$\text{IIP}_2 = \frac{V_{eff}}{V_{OS}} \frac{S \times T}{\pi} = 4A \frac{V_{eff}}{V_{OS}},$$

where V_{eff} is the input FET overdrive voltage. Evidently, we face very similar trade-offs as with the $1/f$ noise.

Switching Quad — The rigorous analysis of the switches is very lengthy in math, and is presented in [10]. To summarize, the switches contribute mainly as a result of the parasitic capacitances at their common source, and in the presence of mismatches. So their contribution is dominant only at high frequencies, and can be improved by reducing the parasitic capacitances at the common source of each pair.

Example: As another case study, we will discuss two schemes that are variations of the original Gilbert cell, and are intended to boost IIP$_2$ and $1/f$ noise performance of the mixer [11], [12] (Figure 8.27). Both circuits are fully differential, while only half are shown.

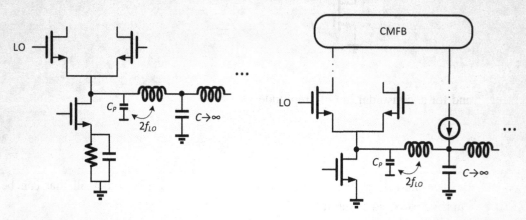

Figure 8.27: Active mixer topologies with enhanced IIP$_2$

In the left scheme, the RF self-mixing is improved by proper layout, and to avoid the input FET 2nd-order nonlinearity, the device has been heavily degenerated. However, in order not to affect the mixer gain and noise figure, the source degeneration is bypassed at the frequency of interest by a shunt capacitor. With those precautions, the main contributor remains to be the switches' high-frequency 2nd-order distortion. That has been handled by resonating the switching pair common-source parasitic capacitance (C_P) with an inductor at $2f_{LO}$. The reason for choosing $2f_{LO}$ is that there are two zero-crossings for every cycle, and thus, similar to $1/f$ noise, we expect the IIP$_2$ events to be at a rate of $2f_{LO}$. The main drawback of this approach is the need for additional inductors, and that the resistive degeneration takes a large headroom. Nevertheless, the mixer achieves an IIP$_2$ of +78dBm [11]. The latter issue is resolved by incorporating a common-mode feedback that monitors the output 2nd-order distortion and injects a low-frequency current to the switches to negate that. Since the common-mode current is heavily bypassed, it does not affect the mixer regular performance.

8.4 PASSIVE CURRENT-MODE MIXERS

The passive mixers rely on on/off switches that dissipate no power. The *notion of passive*, however, is somewhat misleading, as there are typically several active circuits associated with the passive mixer that lead to a nonzero power dissipation.[7] Moreover, to enhance the mixer

---

[7] In most RF applications, given the stringent performance requirements, this is true. However, using the passive gain and employing resonance, it is possible to realize a true passive mixer [31]. This has been exploited in certain applications such as wireless sensors, where extremely low power transceivers are needed.

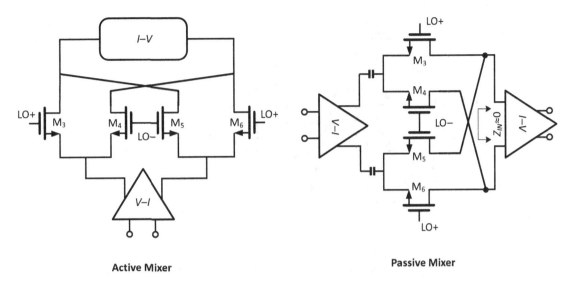

Figure 8.28: Active mixer to passive current-mode mixer evolution

performance, the switches require sharp, rail-to-rail waveforms that generally lead to more power consumption in the LO delivery circuits compared to active mixers.

As we discussed earlier, there are two fundamental drawbacks with the active mixers: modest linearity due to several devices stacked and poor $1/f$ noise. Both of these issues can be avoided if the active mixer is redesigned as shown in Figure 8.28.

As we said before, the active mixer consists of a transconductor stage that coverts the voltage to current, the switching quad, and a load (passive or active) that converts the current back to voltage. All three stages are stacked, which not only results in poor linearity, but also forces a DC current passing through the switches, leading to poor $1/f$ noise. If the three stages are cascaded instead as shown in Figure 8.28, both of these issues are resolved. Particularly, placing a blocking capacitor in the signal path ensures that the switches are biased at zero DC current, and thus for most practical purposes, this passive mixer arrangement is free of flicker noise. Thus, a fundamental difference between the active and passive mixers is that the latter commutate the AC signal (whether it is a current or voltage) only.

If desired, the blocking capacitor may be removed and a common-mode feedback (CMFB) circuitry may be incorporated to ensure that the DC level of the two sides of the switches are equal, and thus they carry no DC current (Figure 8.29). Moreover, the V–I stage may be combined with the low-noise amplifier (known as low-noise transconductance amplifier or LNTA for short) as long as it can be ensured that the amplifier operates in the current mode. This largely has to do with the specific design of the stage following the mixer switches that convert the current to voltage.

Ideally, a transimpedance amplifier (TIA) is needed to take the input current and convert to voltage, and, more importantly, to present a very low input impedance to the mixer switches (Figure 8.29). This is required to guarantee that the mixer operates in current domain. As we will discuss in Section 8.5, in the case of a *voltage-mode* passive mixer, the situation is the opposite: the buffer following the mixer needs to present a high impedance circuit to ensure voltage-mode operation.

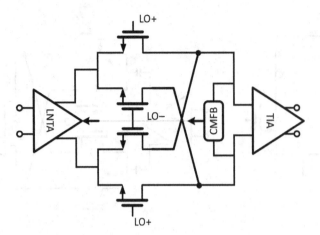

Figure 8.29: Complete schematic of a current-mode passive mixer using CMFB

In the next few sections, we will study the mixer performance more closely. Before that, however, we would like to have a general discussion on the type and specifically the duty cycle of the LO signal used to drive the mixer. There are several options, some of which are not as optimum or as powerful. Consequently, once the less common ones are ruled out, we will focus only on the more practical structures throughout the rest of this chapter.

8.4.1 LO Duty Cycle Concerns

Consider a single-ended ideal current source driving a current mode mixer connected to an ideal TIA as shown in Figure 8.30.

Let us assume that the general waveform shown in Figure 8.31 is applied to the gate of each switch. We assume the waveforms are periodic with a period of T, and are differential, meaning that LO_2 is half period shifted with respect to LO_1. We assume that the LO waveforms may have an overlap time of τ. If τ is negative, it will represent a case of nonoverlapping or *negative overlapping* LO signals as shown in the figure.

In the case of overlapping clocks, for the period of time $T/2 < t < T/2 + \tau$, both switches turn on, resulting in the RF current splitting equally in half and appearing at both branches of the output as shown in the right side of Figure 8.30. When taken differentially, this results in no output current. On the other hand, for the nonoverlapping LO case, there is the same time duration τ, where both switches are off and no current appears at the output.[8] Thus, in either case the output current i_o is equal to the source RF current, i_s, multiplied by the effective LO waveform shown in Figure 8.31. A simple Fourier analysis shows that the fundamental of the effective LO signal is

$$\frac{4}{\pi} \cos \frac{\omega_{LO}\tau}{2} \sin\left(\omega_{LO}t - \frac{\omega_{LO}\tau}{2}\right),$$

---

[8] Although it appears that this is in violation of KCL, in practice the current flows through the parasitic paths at the mixer input. We shall discuss this shortly.

8.4 Passive Current-Mode Mixers

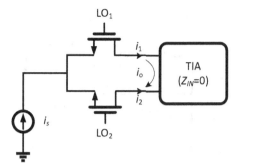
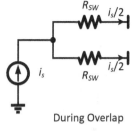

Figure 8.30: Single-balanced current-mode mixer

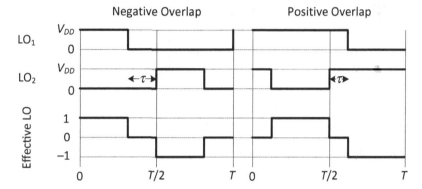

Figure 8.31: Mixer LO waveforms with arbitrary overlap

where $\omega_{LO} = \frac{2\pi}{T}$ is the LO angular frequency. Thus, when multiplied by a sinusoid RF input and lowpass filtered at the output to take only the difference frequency component, the mixer conversion current gain, A_I, will be

$$A_I = \frac{2}{\pi} \cos \frac{\omega_{LO} \tau}{2},$$

which is maximum for zero overlap, or equivalently for an LO with exactly 50% duty cycle. A lower or higher duty cycle will result in the RF current to either not appear at all, or appear as common-mode at the output.

The situation will be different if the mixer is used in a quadrature receiver. In Chapter 2 we saw that single-sideband generation in transmitters require quadrature mixers to upconvert the baseband quadrature components around carrier. Similarly, as we have discussed already and will show in more detail in Chapter 12, detection of such signals requires a *quadrature downconversion* path, as is shown in Figure 8.32.

The corresponding LO waveforms are shown in Figure 8.33. Each LO signal is shifted by $T/4$, or $\pi/2$, and thus the pair LO_{1-3} represent the differential I LO signals, whereas the pair LO_{2-4} represent the differential Q LO signals. As before, we have assumed an arbitrary overlap time of τ.

For negative τ or effectively less than 25% duty cycle for each signal, the analysis has been already carried out. Given there is a period of time that all four switches are off, the mixer conversion gain is expected to go down.

Let us then consider the case that $\tau > 0$, that is, there is a nonzero overlap as shown in Figure 8.33. Consider the differential current flowing through the *I*+ branch TIA: For $\tau < t < T/4$,

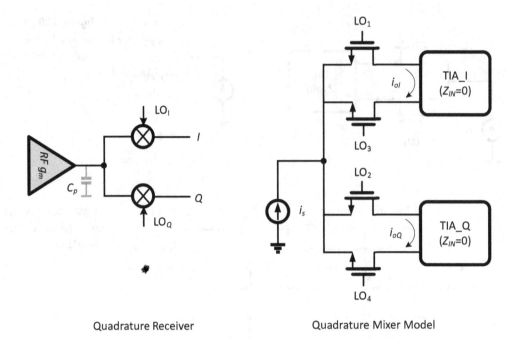

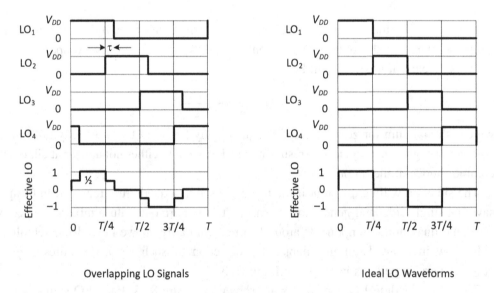

Figure 8.33: LO waveforms in a quadrature mixer

only LO_1 is on, resulting in the entire RF current to appear at the TIA. At $t = T/4$, LO_2 turns on, and from this point on until $t = T/4 + \tau$, both $I+$ and $Q+$ switches are on, resulting in the RF current split in half into each branch. At $t = T/4 + \tau$, the I switch turns off and no current appears at the output till the rest of the period. For the beginning portion, $0 < t < \tau$, the RF current also splits in half, but this time between the $I+$ and $Q-$ branches. A similar situation happens for the $I-$ branch for the time period $T/2 < t < 3T/4 + \tau$, resulting in an effective LO waveform shown in Figure 8.33 for the *I* branch. Fourier series analysis of this waveform results in a fundamental LO signal of

$$\frac{2\sqrt{2}}{\pi} \cos \frac{\omega_{LO}\tau}{2} \cos\left(\omega_{LO}t - \frac{\omega_{LO}\tau}{2} - \frac{\pi}{4}\right),$$

and thus the current conversion gain is

$$A_I = \frac{\sqrt{2}}{\pi} \cos \frac{\omega_{LO}\tau}{2}.$$

A 50% duty cycle LO in this case corresponds to an overlap of $\tau = T/4$, which leads to a conversion gain of $1/\pi$. It is clear that the conversion gain is maximized for $\tau = 0$, or a 25% duty cycle LO signal, resulting in waveforms with a quarter of a period duration, progressively shifted by quarter of a period, as shown on the right side of Figure 8.33. The mixer gain and consequently noise figure both improve by 3dB as a result [13]. Note that the total area under each waveform is the same, but the *staircase-like* shape of the overlapping LOs results in a smaller portion of the sine being integrated, thus less conversion gain. Alternatively we could say that the overlap cancels the signal, but not the noise.

A simulation of an actual 40nm passive mixer for each of the cases discussed is shown in Figure 8.34. There is close agreement between our findings and the simulations. However, in the case of less than 25% duty cycle for the IQ case (or less than 50% for the *I* only case), the mixer gain does not drop as steep as what would have been predicated by the equation.

The explanation for this is as follows: During no overlap time, when all four switches are disconnected, in the ideal setup, the RF current has nowhere to go, and the signal is entirely lost. In a practical case, however, there is a parasitic capacitance (as well as a resistance) at the output of the current source due to the switches' junction or gate-source capacitance, as well as the g_m cell output parasitic (Figure 8.35). During this time, the RF current source charges this capacitance almost linearly, and once any of the switches are turned back on, this charge is delivered to the output, retrieving some of the signal loss.

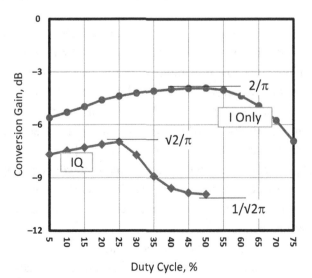

Figure 8.34: Simulated mixer conversion for various LO duty cycles

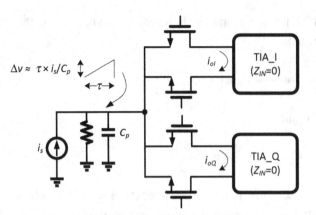

Figure 8.35: Parasitic capacitance at the g_m output leading to less conversion gain loss

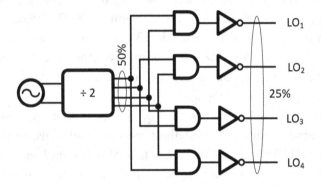

Figure 8.36: Using AND gates to create 25% LO signals

There are several other disadvantages associated with overlapping LO signals, and 50% duty cycle waveforms. Generally speaking, the overlap results in a cross-talk between the I and Q branches, or effectively appearance of an *image current* circulating between the two paths. This cross-talk will cause further degradation in the mixer noise and linearity beyond the 3dB loss of gain showed earlier. For these reasons, throughout the rest of this chapter we will consider only passive mixers with exact no overlapping clocks, which are practically used in almost all transceivers. A detailed discussion of 50% mixers and the issues associated therewith can be found in [14].

One drawback of 25% mixers is the need for 25% quadrature LO signals whose waveforms were shown in Figure 8.33. This in practice may be done by applying the AND function to any of the two consecutive signals available in a 50% quadrature LO signal as shown in Figure 8.36. A shift register can also be used with less delay between clock edge and LO edge, which is fundamentally better for phase error and jitter [15], [16]. While the simplicity of the structure shown in Figure 8.36 may justify its popularity for a 4-phase mixer, a shift register may be more prominent if more than four phases are desired, as we will discuss next [17].

Although a more complex LO generation potentially leads to more power consumption, the advantages of the 25% mixers are overwhelming enough to justify such extra power dissipation for most cases. Moreover, the additional LO generation circuitry comprising AND gates or shift registers is completely scalable by the technology.

Finally, the size of the inverter buffers driving the switches' gate must be chosen such that the overlap is minimized including the process and temperature variations. A small amount of

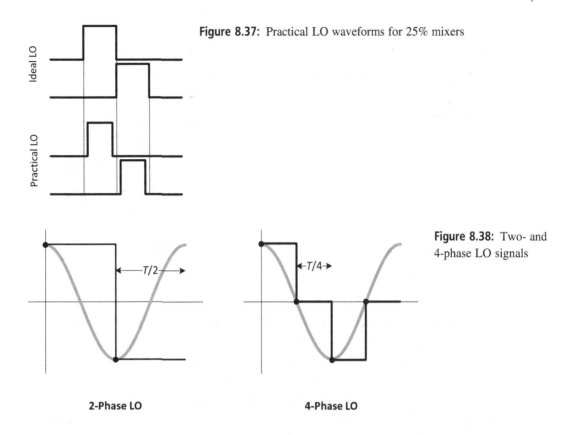

Figure 8.37: Practical LO waveforms for 25% mixers

Figure 8.38: Two- and 4-phase LO signals

no overlap where all switches are disconnected is generally acceptable, as the signal loss will be very small. On the other hand, overlapping must be avoided to prevent the aforementioned cross-talk and image issues (Figure 8.37).

8.4.2 *M*-Phase Mixers

In Chapter 6, we saw that the square-wave-like nature of the LO waveforms results in a harmonically rich signal, leading to down-conversion of certain unwanted blockers located at or around the LO harmonics.

To understand the root cause of this issue better, let us start with a single mixer clocked by a 50% differential LO shown in the left side of Figure 8.38. Such an LO signal effectively samples an ideal sine-wave with only two points, 1 and −1, as shown, and thus resembles a differential square-wave toggling rail-to-rail. For that, we call this mixer arrangement a 2-phase mixer.

As we are interested in only the fundamental, once filtered, the mixer output is $\frac{2}{\pi}$ times less than the input. The nature of the square-wave LO results in all odd harmonics (odd because still the LO is differential, or 2-phase) to downconvert as well, unless filtered beforehand.

As we have shown before, almost all transceivers rely on quadrature mixing to select only one sideband. In the case of receiver, this corresponds to rejecting the unwanted sideband on the image side. To do so, we showed that we need a 4-phase LO signal whose effective LO was derived as in Figure 8.33. This is equivalent to sampling the ideal sinusoid with four points as

468 Mixers

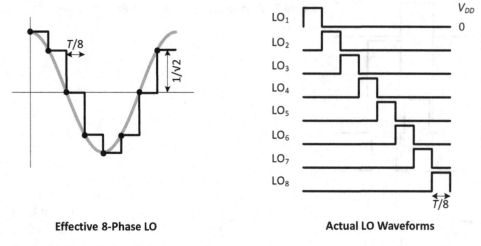

Figure 8.39: An 8-phase LO

shown on the right side of Figure 8.38. The combined output consisting of both *I* and *Q* channels will have a net gain of $\frac{2\sqrt{2}}{\pi}$. The increase in the mixer conversion gain is a direct result of 4-point sampling, that is, the unwanted sideband at $-f_{LO}$ is subject to cancellation, thus more energy is focused in the fundamental component.

This concept may be extended to as many phases as desired. An example of 8-phase LO is shown Figure 8.39, where the ideal sinusoidal LO is sampled by eight points at 1, $1/\sqrt{2}$, 0, $-1/\sqrt{2}$, -1, $-1/\sqrt{2}$, 0, and $1/\sqrt{2}$. The actual LO waveforms are shown on the right side and are pulses from 0 to V_{DD}, each with one-eighth of a period duration, and shifted by one-eighth with respect to one another. As the LO signals are square-wave, the effective scaling factor of $1/\sqrt{2}$ is provided by changing the relative gain of the corresponding TIAs. Negative scaling is easily accomplished if the design is differential. By employing an 8-phase LO, all the harmonics up to the 7th are also rejected, leading to a more fundamental gain, but also less noise aliasing and more immunity to harmonic blockers. The drawback is the need for generating LO signals that are two times narrower compared to the 4-phase mixer. This is effectively done by placing the fundamental at four times the LO, and using a divide by 4 along with AND gates, a scheme similar to that shown in Figure 8.36 for 4-phase. As we mentioned earlier, employing a shift register may also be another attractive option.

The concept of 8-phase mixer to suppress higher harmonics was first introduced in 2001 in transmitters using active mixers [18], but can be generalized to either active or passive mixers, receivers, or transmitters.

To present this concept mathematically, we shall consider the general case of an *M*-phase mixer shown in Figure 8.40 [19]. Given the practical implementation of the LO signals using dividers, we assume *M* is two to the power of an integer, although the concept is readily extended to any integer values of *M*.

Since the cosine is sampled by *M* points at time intervals of $t = n \times \frac{T}{M}$, $n = 0, 1, \ldots$, *M*−1, dictated by the effective LO signal consisting of *T/M* pulses, the corresponding scaling factor for the output v_n is $\cos\left(n\frac{2\pi}{M}\right)$, as shown in Figure 8.40. Thus there are *M* outputs that are each phase shifted by $\frac{2\pi}{M}$, with respect to each other; of those, only the

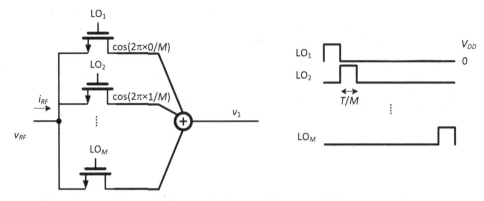

Figure 8.40: *M*-phase mixer and the corresponding LO signals

first output is shown in Figure 8.40. This is performed by employing M banks of M-input, 1-output TIAs with the proper gain.

To obtain the effective LO signal, let us first find the Fourier series representation of each of the LO signals driving the branches,

$$S_k(t) = \sum_{n=-\infty}^{\infty} a_n e^{-jk\frac{n2\pi}{M}} e^{jn\omega_{LO}t},$$

where S_k is the effective LO switching the kth branch, $k = 1, \ldots, M$, and

$$a_n = \frac{e^{-jn\frac{\pi}{M}}}{M} \operatorname{sinc}\left(\frac{n}{M}\right).$$

Note that consistent with our definition in Chapter 2, $\operatorname{sinc}\left(\frac{n}{M}\right) = \frac{\sin\frac{n\pi}{M}}{\frac{n\pi}{M}}$. It is obvious that the effective LO signals are identical, but shifted only by T/M with respect to each other. Now the net effective LO, $S(t)$, consists of the effective LO signals of each branch, scaled by $\cos\left((k-1)\frac{2\pi}{M}\right)$, and added together:

$$S(t) = \sum_{k=1}^{M} \cos\left((k-1)\frac{2\pi}{M}\right) S_k(t) = \sum_{n=-\infty}^{\infty} a_n e^{jn\frac{2\pi}{M}} e^{jn\omega_{LO}t} \sum_{k=0}^{M-1} \cos\left(k\frac{2\pi}{M}\right) e^{-j\frac{nk2\pi}{M}}.$$

Using simple algebraic unity, we can show that

$$\sum_{k=0}^{M-1} \cos\left(k\frac{2\pi}{M}\right) e^{-j\frac{nk2\pi}{M}} = \frac{M}{2}$$

for $n = pM \pm 1$, $p \in Z$, and zero otherwise.

For the main harmonic then, $n = \pm 1$, and the effective LO for say the output v_1 becomes

$$\operatorname{sinc}\left(\frac{1}{M}\right) \cos\left(\omega_{LO}t - \frac{\pi}{M}\right).$$

The effective LO amplitude for different harmonics is graphically shown in Figure 8.41, which is consistent with the sampling theory.

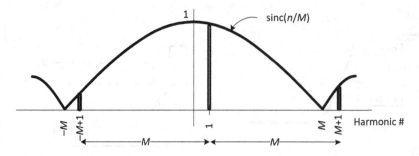

Figure 8.41: *M*-phase mixer effective LO harmonic content assuming a complex (*I*+*jQ*) output

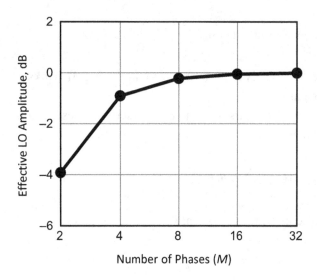

Figure 8.42: Mixer conversion gain versus the number of phases

Let us point out a few key observations:

- With the number of phases (*M*) approaching infinity, the effective LO approaches an ideal cosine, despite the fact that the switches are still clocked by square-wave signals. The effective LO amplitude approaches unity, and there is no harmonic downconversion.
- Any signal at the main harmonic, as well as any harmonic separated by $p \times M$ (p is an integer) folds on top of the desired signal. Thus, we expect the (*M*+1)th and (−*M*+1)th as the closest harmonics to be downconverted along with the desired signal.

As *M* increases, the effective LO amplitude approaches unity very quickly, as shown in Figure 8.42. Given the overhead of clock generation for large number of phases, perhaps $M = 8$ is an optimum choice, both from gain/noise point of view, and the fact that only the 7th harmonic and beyond are problematic. The 7th harmonic is far enough to receive reasonable amount of filtering. In the case of 8-phase, the effective LO amplitude is

$$\frac{8}{\pi} \sin\left(\frac{\pi}{8}\right) \approx 0.97,$$

which is very close to unity.

Setting $M = 4$, we arrive at the same derivations as we had presented before for the 25% mixer, where the conversion gain was calculated to be $\frac{\sqrt{2}}{\pi}$. Note that in the case of a 4-phase mixer, the first and third branches, each with a gain of $\pm\frac{\sqrt{2}}{\pi}$, are subtracted, resulting in a net gain for the effective signal (see Figure 8.40).

In the remainder of this section, we will mainly focus on 4-phase mixers as is the most popular choice for narrowband receivers, given the complexity and LO generation overhead for higher values of M. In Chapter 12, we will discuss wideband receivers consisting of 8-phase mixers, primarily used for harmonic rejection purposes. The mixer operation is readily extendable to any number of phases desired however.

8.4.3 Passive Mixer Exact Operation

To analyze the mixer operation, let us consider Figure 8.43, where we assume each TIA has a single-ended input impedance of Z_{BB}, and the mixer LO voltages and the other corresponding signals are labeled as shown in Figure 8.43.

The effective LO signals are denoted as $S_1(t)$, $S_2(t)$, ... which are identical to LO_1, LO_2, ... except for they toggle from 0 to 1. Just as before, we take a single-ended input mixer for simplicity, though the results derived are readily extended to a fully differential structure.

As done previously, we represent the effective LO signals using Fourier series

$$S_1(t) = \sum_{n=-\infty}^{\infty} a_n e^{jn\omega_{LO}t}$$

$$S_2(t) = \sum_{n=-\infty}^{\infty} a_n e^{-jn\frac{\pi}{2}} e^{jn\omega_{LO}t}$$

$$S_3(t) = \sum_{n=-\infty}^{\infty} a_n e^{-jn\pi} e^{jn\omega_{LO}t}$$

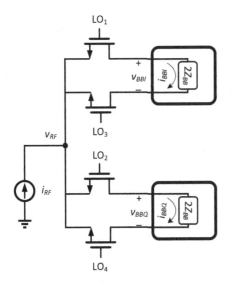
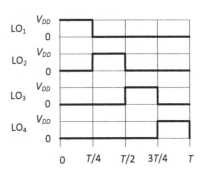

Figure 8.43: Current-mode mixer operation

$$S_4(t) = \sum_{n=-\infty}^{\infty} a_n e^{+jn\frac{\pi}{2}} e^{jn\omega_{LO}t},$$

where $a_n = \frac{e^{-jn\frac{\pi}{4}}}{4} \text{sinc}\left(\frac{n}{4}\right)$.

Since a given switch is on only at one quarter of a cycle, for the corresponding branch, say the I switch toggled by LO_1, we have

$$i_{BBI+} = S_1(t) i_{RF}(t),$$

and so on. This baseband current in turn creates a voltage across the input of the TIA,

$$v_{BBI+} = [S_1(t) i_{RF}(t)] * z_{BB}(t),$$

where * denotes convolution. This baseband voltage is only transparent to the RF input when S_1 is 1, and thus is effectively multiplied by S_1 when monitoring the RF voltage.

Once all four branches are considered, including the switch resistance, R_{SW}, which always appears in series with the RF current, as one and only one switch is on at the time, we have

$$v_{RF}(t) = R_{SW} i_{RF}(t) + \sum_{k=1}^{4} S_k(t) \times \{[S_k(t) i_{RF}(t)] * z_{BB}(t)\}.$$

The analysis may be carried on more easily in the frequency domain, if the Fourier transform is utilized. By applying straightforward algebra, we can show the Fourier transform of $S_k(t) \times \{[S_k(t) i_{RF}(t)] * z_{BB}(t)\}$ is

$$\sum_{m=-\infty}^{\infty} \sum_{n=-\infty}^{\infty} e^{-j(n+m)(k-1)\frac{\pi}{2}} a_n a_m I_{RF}(\omega - (n+m)\omega_{LO}) Z_{BB}(\omega - n\omega_{LO}),$$

where $k = 1, 2, 3, 4$ corresponds to any of the branches switched by the LO_k signal, and a_n is the Fourier series coefficient obtained previously. Thus we arrive at

$$V_{RF}(\omega) = R_{SW} I_{RF}(\omega) + 4 \sum_{m=-\infty}^{\infty} \sum_{n=-\infty}^{\infty} a_n a_m I_{RF}(\omega - (n+m)\omega_{LO}) Z_{BB}(\omega - n\omega_{LO}),$$

where $n+m$ is restricted to an integer multiple of 4. From this, we conclude that if for instance the RF current is a sinusoid at a frequency of $\omega_{LO} + \omega_m$, then the RF voltage has its main component at $\omega_{LO} + \omega_m$, and the rest reside at $3\omega_{LO} - \omega_m$, $5\omega_{LO} + \omega_m$, and so forth. Ignoring the frequency components at third and higher harmonics,[9] knowing that the receiver provides some modest filtering, then, $n+m=0$, and we have

$$V_{RF}(\omega) = R_{SW} I_{RF}(\omega) + 4 \sum_{n=-\infty}^{\infty} |a_n|^2 I_{RF}(\omega) Z_{BB}(\omega - n\omega_{LO}).$$

---

[9] With typical design values, it can be shown that [23] the harmonics have a negligible impact on the mixer performance and its transfer function. As the goal of this chapter is to provide a sufficiently accurate yet intuitive description of the mixers, we shall ignore their impact throughout the rest of this chapter.

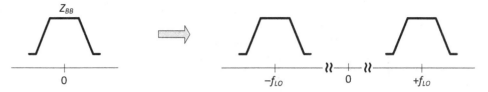

Figure 8.44: Passive mixer impedance transformation

As $V_{RF}(\omega)$ now becomes only a function of $I_{RF}(\omega)$, we can effectively define the impedance seen at the RF input node as

$$Z_{IN}(\omega) = R_{SW} + 4 \sum_{n=-\infty}^{\infty} |a_n|^2 Z_{BB}(\omega - n\omega_{LO}),$$

which indicates that the TIA input impedance appears at RF, at around the LO frequency and its harmonics. In the vicinity of ω_{LO}, we have

$$Z_{IN}(\omega) \approx R_{SW} + \frac{2}{\pi^2}[Z_{BB}(\omega - \omega_{LO}) + Z_{BB}(\omega + \omega_{LO})].$$

Thus the baseband impedance is *transformed* to the input around $\pm\omega_{LO}$. Accordingly if the baseband impedance is lowpass, the corresponding RF input sees a bandpass impedance, whose center frequency is precisely set by the LO (Figure 8.44). This property may be utilized to create low-noise narrowband filters, often known as *N*-path filters, something that active filters fail to provide as we discussed in Chapter 4.

Intuitively, we showed in the previous section that a passive mixer with 25% LO has a conversion gain of $\frac{\sqrt{2}}{\pi}$. Since the passive mixer is reciprocal (ignoring the capacitances), then we expect the baseband impedance to be upconverted to RF with a conversion factor of $\left(\frac{\sqrt{2}}{\pi}\right)^2 = \frac{2}{\pi^2}$, in agreement with our findings. This statement, however, is not accurate as we had to ignore the impact of all the harmonics.

We showed already that

$$i_{BBI} = [S_1(t) - S_3(t)]i_{RF}(t) = 2i_{RF}(t)\sum_{n=-\infty,\,odd}^{\infty} a_n e^{jn\omega_{LO}t}$$

$$i_{BBQ} = [S_2(t) - S_4(t)]i_{RF}(t) = 2i_{RF}(t)\sum_{n=-\infty,\,odd}^{\infty} a_n e^{-jn\frac{\pi}{2}}e^{jn\omega_{LO}t}.$$

Around the fundamental, $n = 1$, and $a_1 = \frac{e^{-j\frac{\pi}{4}}}{4}\text{sinc}\left(\frac{\pi}{4}\right) = \frac{e^{-j\frac{\pi}{4}}}{\sqrt{2}\pi}$. Thus the mixer has a conversion current gain of $\frac{\sqrt{2}}{\pi}$ to each of the I and Q channels, as derived before.

With the switches' input impedance and current gain already calculated, it is now straightforward to find the overall conversion gain of the mixer including the RF g_m cell. A simple model of the mixer is shown in Figure 8.45, and can be easily extended to a fully differential topology.

The g_m cell is modeled by an ideal current source loaded with an impedance of Z_L. A capacitor C isolates the switches from the output of the g_m stage. This is done to prevent undesired low-frequency components at the g_m output, such as IM$_2$ or $1/f$ noise, to leak into the

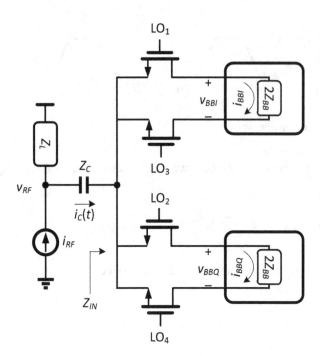

Figure 8.45: Simple model of a current-mode passive mixer

switches' output in the presence of mismatches. We shall assume the capacitor has an impedance of Z_C, which is not necessarily a short at the frequency of interest. In fact, we will show that to maximize the gain, there is an optimum value for C.

Suppose the RF current has a magnitude of I_{RF}, and a frequency of $\omega_{LO} + \omega_m$, thus residing at the high side of LO. We also assume that $\omega_m \ll \omega_{LO}$, which is a reasonable assumption in most narrowband radios.

Given the mixer input impedance of $Z_{IN}(\omega) \approx R_{SW} + \frac{2}{\pi^2}[Z_{BB}(\omega - \omega_{LO}) + Z_{BB}(\omega + \omega_{LO})]$ looking into switches, the RF current is subject to a current division, and thus

$$I_C = \frac{Z_L(\omega_{LO} + \omega_m)}{Z_L(\omega_{LO} + \omega_m) + Z_C(\omega_{LO} + \omega_m) + Z_{IN}(\omega_{LO} + \omega_m)} I_{RF}.$$

Since Z_L and Z_C are almost constant around ω_{LO}, and Z_{BB} is lowpass, we have

$$I_C \approx \frac{Z_L(\omega_{LO})}{Z_L(\omega_{LO}) + Z_C(\omega_{LO}) + R_{SW} + \frac{2}{\pi^2} Z_{BB}(\omega_m)} I_{RF}.$$

We showed the current gain from the switch input to say I channel TIA is $\frac{\sqrt{2}}{\pi} e^{-j\frac{\pi}{4}}$. Thus we have

$$I_{BBI} = I_{BBQ} e^{j\frac{\pi}{2}} = \frac{\sqrt{2}}{\pi} e^{-j\frac{\pi}{4}} \frac{Z_L(\omega_{LO})}{Z_L(\omega_{LO}) + Z_C(\omega_{LO}) + R_{SW} + \frac{2}{\pi^2} Z_{BB}(\omega_m)} I_{RF}.$$

If the load impedance Z_L is large, that is, the g_m cell is an ideal current source, we arrive at our previous results. In practice, the g_m cell is likely to be combined with the LNA, and its output is either a parallel RLC circuit, or is RC.

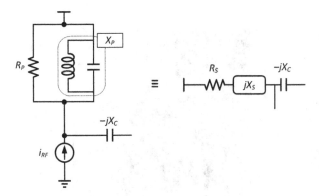

Figure 8.46: Series equivalent of the g_m cell load

As we saw in Chapter 7, RF amplifiers commonly use an inductive load. To gain some insight, let us assume Z_L is a parallel RLC network. We also assume Z_{BB} is mostly resistive and equal to R_{BB} at the frequency of interest, which is a reasonable assumption. Ideally, Z_{BB} must be zero. Since Z_C is representing a capacitor, $Z_C = -jX_C$, and as R_{SW} and Z_{BB} are real, to maximize the conversion gain, that is, to minimize the denominator of the equation above, Z_L must resonate with C at the frequency of interest. Now consider Figure 8.46, where we use the parallel–series conversion to arrive at the equivalent circuit shown on right.

In most typical designs, the quantity $\left|\frac{R_p}{X_p}\right|$, which is an indication of the inductor Q, is large,[10] and thus let us assume

$$\left|\frac{R_p}{X_p}\right| \gg 1.$$

Thus the equivalent series resistance is

$$R_p \frac{X_p^2}{R_p^2 + X_p^2} \approx \frac{X_p^2}{R_p},$$

and the series reactance is

$$X_p \frac{R_p^2}{R_p^2 + X_p^2} \approx X_p,$$

which must resonate with the capacitor C, and thus $X_p \approx X_C$.

From this the conversion gain is

$$\frac{\sqrt{2}}{\pi} \left| \frac{Z_L(\omega_{LO})}{Z_L(\omega_{LO}) + Z_C(\omega_{LO}) + R_{SW} + \frac{2}{\pi^2} Z_{BB}(\omega_m)} \right| \approx \frac{\frac{\sqrt{2}}{\pi} X_C}{\frac{X_C^2}{R_p} + R_{SW} + \frac{2}{\pi^2} R_{BB}}.$$

The equation above indicates that there is an optimum value of X_C that the conversion gain is maximized, and is found by taking the derivative of the gain expression versus X_C,

---

[10] Unless there is a parallel resonance at the output, $\left|\frac{R_p}{X_p}\right|$ is always less than the actual inductor Q. Nevertheless, it is reasonable to assume $\left|\frac{R_p}{X_p}\right|$ is large.

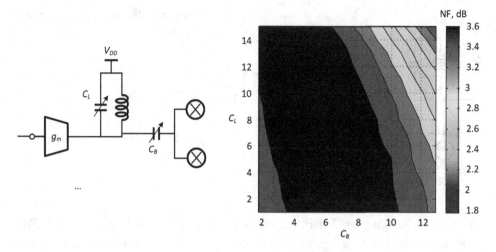

Figure 8.47: Receiver noise figure optimized based on load and series capacitances

$$X_{C,opt} = \sqrt{R_p \left(R_{SW} + \frac{2}{\pi^2} R_{BB} \right)},$$

which leads to a maximum conversion gain of

$$\frac{1}{\sqrt{2}\pi} \sqrt{\frac{R_p}{R_{SW} + \frac{2}{\pi^2} R_{BB}}}.$$

For typical values of switch resistance, inductor Q, and the TIA input impedance, this leads to a substantially higher gain if one were to resonate Z_L and choose a large C. This *passive gain* in front of the TIA helps reduce its noise contribution. To emphasize the importance of this, shown in Figure 8.47 is the measured noise figure of a 40nm 3G receiver[11] versus a sweep of values of the load capacitor (C_L) and the series capacitor between the g_m cell and the mixer input (C_B). The capacitors are programmable, and what is shown on the plot is the programming code applied to each. A suboptimum design may lead to as much as 2dB degradation in the receiver noise figure! It is interesting to note an acceptable noise figure is obtained based on several different combinations of C_L and C_B. For that reason the actual optimization may be performed considering other receiver aspects such as gain or linearity.

Using a very similar procedure, we can show that if the RF current resides at the low side of LO (at $\omega_{LO} - \omega_m$), the current flowing into to the capacitor is a result of a current division between Z_L, Z_C, and $Z_{IN}(\omega_{LO} - \omega_m) \approx R_{SW} + \frac{2}{\pi^2} Z_{BB}(-\omega_m) = R_{SW} + \frac{2}{\pi^2} Z_{BB}^*(\omega_m)$, and thus the output current is

$$I_{BBI} = I_{BBQ} e^{+j\frac{\pi}{2}} = \frac{\sqrt{2}}{\pi} e^{+j\frac{\pi}{4}} \frac{Z_L^*(\omega_{LO})}{Z_L^*(\omega_{LO}) + Z_C^*(\omega_{LO}) + R_{SW} + \frac{2}{\pi^2} Z_{BB}(\omega_m)} I_{RF}.$$

---

[11] The details of the receiver design are presented in Chapter 12 in the practical transceiver design concerns section.

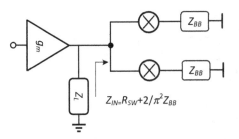

Figure 8.48: Mixer schematic to illustrate different high- and low-side gains

We must note that since the signal is at the negative side, once the capacitor current is downconverted to baseband, the resultant current is conjugated with respect to the case of high side, that is, its imaginary part is 180° out of phase.

Comparing this equation with the one obtained for the signal at $\omega_{LO} + \omega_m$, we conclude that unless Z_{BB} is purely resistive, the mixer exhibits different conversion gains whether the signal is located at the lower or higher side of the LO. That is because in general

$$\left| Z_L(\omega_{LO}) + Z_C(\omega_{LO}) + R_{SW} + \frac{2}{\pi^2} Z_{BB}^*(\omega_m) \right| \neq \left| Z_L(\omega_{LO}) + Z_C(\omega_{LO}) + R_{SW} + \frac{2}{\pi^2} Z_{BB}(\omega_m) \right|.$$

To understand this intuitively, consider Figure 8.48, which shows a simplified block diagram of the mixer, and let us assume for the moment that $X_C \approx 0$, and thus not shown in the figure.

Z_L then naturally resonates at the center frequency, which is ω_{LO}. Let us assume that Z_{BB} consists of a resistor in parallel with a capacitor modeling the TIA input. For high-side, then, when upconverted to the RF, the impedance looking in is *capacitive*, appearing in parallel with Z_L. On the other hand, for low-side, the upconverted impedance is *inductive* (due to the term $\frac{2}{\pi^2} Z_{BB}^*(\omega_m)$), whereas Z_L has not appreciably changed, and thus the g_m stage is loaded differently, leading to different conversion gains. This situation is substantially worse if the mixer LO were 50% due to the interaction of I and Q switches and appearance of the image current [14].

If the high- and low-side conversion gains are not identical, as the signal has *uncorrelated energy* at both sides of the spectrum, once downconverted, it will be distorted, which leads to EVM degradation.

To minimize this, one must ensure that the TIA input (Z_{BB}) is dominantly real at the passband of interest. Alternatively, if Z_L resonates with C, then $Z_L(\omega_{LO}) + Z_C(\omega_{LO})$ is real, and thus the high- and low-side gains are identical (of course within the passband of the resonance) regardless of Z_{BB}. This will also lead to the optimum gain as we established before.

Example: For $Z_{BB} = 100\Omega || 25\text{pF}$, $R_{SW} = 15\Omega$, $L = 2\text{nH}$, $R_p = 250\Omega$, the simulated high- and low-side gains for both the conventional and optimum designs are shown in Figure 8.49. The LO frequency is 2GHz.

Continued

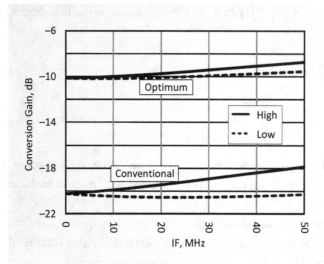

Figure 8.49: Simulated high- and low-side gains of the mixer

Not only the optimum design has a higher gain, but also the difference between high- and low-side gains is very small. Moreover, as IF increases, the capacitive part of the TIA input impedance becomes more dominant, and thus more difference is observed as expected.

8.4.4 Passive Mixer Noise

If the mixer LO signals have no overlap like shown in Figure 8.33, their noise contribution is very small. Since the switches are in the Triode region with zero bias current, their $1/f$ noise contribution is negligible as well. This is however not always the case, as it has been experimentally shown in [20], and analyzed in [21], that depending on the bias conditions, the passive mixers are indeed capable of producing $1/f$ noise. In a well-designed passive mixer however, the $1/f$ noise, if it exists at all, is negligible.

If there are mismatches between the switches, however, the $1/f$ noise of the g_m stage may leak to the output. An analysis very similar to the one performed for the active mixers reveals that the leakage gain is

$$2\frac{V_{OS}}{S \times T},$$

where V_{OS} models the switch mismatches as an input-referred offset voltage [22], and $S \times T$ is the normalized slope of the LO signals. This noise, however, may be easily suppressed if any kind of highpass filtering between the g_m cell and switches is incorporated, the most simple one being the insertion of a blocking capacitor (Figure 8.50).

Using a similar approach as we took to analyze the mixer current gain, and considering that the mixer noise is cyclostationary (see Chapter 5), we can easily show that as far as switches' thermal noise is concerned, they can be modeled as the Thevenin equivalent circuit shown in Figure 8.51 [23].

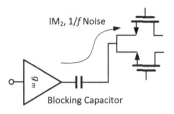

Figure 8.50: Highpass filter to block $1/f$ noise

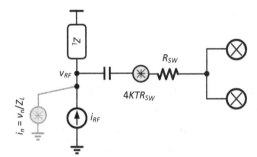

Figure 8.51: Thevenin equivalent circuit to model switch noise

This can be intuitively understood as at any given point of the time, there is one and only one switch on whose resistance and its corresponding noise appears in series with the current flowing from the g_m cell. This noise voltage, whose spectral density is then $4KTR_{SW}$ (R_{SW} is the switch resistance), may be referred to as the g_m cell output as also shown by the shaded circuit source in Figure 8.51. The current noise spectral density is

$$\frac{4KTR_{SW}}{|Z_L|^2}.$$

If the g_m cell is ideal, then Z_L is infinite, and the switches contribute no noise. That is simply because the switch noise appears in *series with an ideal current source*. In practice, they do however, but this noise contribution is quite small compared to the g_m cell itself, as the switch resistance is typically quite small compared to the output impedance of the g_m cell.

8.4.5 Passive Mixer Linearity

The operation of passive mixer in the current domain, along with the impedance transformation properties explained earlier, results in a superior linearity compared to active mixers [24]. Consider Figure 8.52 where a small desired signal is accompanied by a large blocker.

If the TIA ideally presents a very small input impedance to the switches, and assuming the switches' resistance is small, then the g_m cell is going to see a very small impedance at its output. This ensures that the swing at the g_m cell output is small. Moreover, if a large capacitance is present at the TIA input (large enough to present a short circuit at the blocker offset frequency, Δf_B), a bandpass filter is created that will suppress the blocker. Once down-converted to IF, the signal resides at DC, whereas the blocker is several tens of MHz away, and is easily suppressed by a lowpass filter. The only challenge is to ensure that the filtering is such that the desired signal is not affected, while the blocker is suppressed adequately. This is generally the case, as in most applications the ratio of the blocker offset to the signal bandwidth

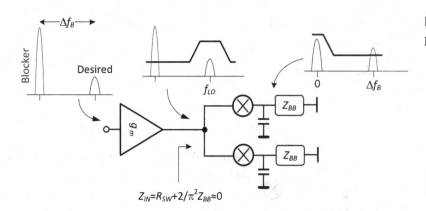

Figure 8.52: CM-mode passive mixer linearity

is usually large (5–10×), allowing room for sufficient filtering even if a simple 1st-order RC filter is exploited. The main trade-offs are the size of the switches, as the larger they are made, the more suppression of blocker is achieved, but more power consumption in the mixer buffer is resulted. Moreover, the design of the TIA is critical, especially to guarantee a low enough input impedance. This will be discussed shortly.

This property of current-mode passive mixers makes them unique and results in what is known as a *current-mode receiver*, where the blocker tolerance is substantially boosted compared to the traditional voltage-mode receivers.

8.4.6 Passive Mixer 2nd-Order Distortion

Passive mixer 2nd-order distortion arises from similar mechanisms as described for active mixers, which can be summarized as follows: (1) Direct leakage of IM_2 components created inside the LNA to the baseband in presence of mismatches among the switches of the passive mixer (this leakage gain was shown to be $2\frac{V_{OS}}{S \times T}$). (2) RF-to-LO coupling in the downconversion mixer. This mechanism may be also analyzed very much like we did for the active mixer [25], resulting in the exact same expression,

$$\text{IIP}_2 = 2\frac{S \times T}{\pi \alpha},$$

where α is the leakage gain and must be optimized by proper layout. We must note, however, that in the case of passive mixer the slope of the LO, which is a square-wave rail-to-rail signal, is somewhat larger than that of the active mixer driven by a sinewave. (3) Nonlinearity of the mixer switches in the presence of β and threshold voltage mismatches.

Unlike the active mixers, the presence of a series capacitor between the LNA or the RF g_m cell and the switches attenuates those IM_2 components generated inside the LNA; therefore, they do not contribute to the IIP_2 of the receiver. The rest of the IM_2 sources mentioned above can contribute to the IIP_2 in two ways: (a) by modulating the resistance of the switches of the passive mixer and (b) through modulating the on/off instants of the mixer switches. Fortuitously, it appears that modulation of the window during which the switches are on does not contribute much to the IM_2 products compared to the resistance modulation of the switches [25]. This is because the switches are clocked by rail-to-rail signals with fast rise and fall times. In conclusion, in a properly designed passive mixer with 25% clocks, we expect the mixer switches IIP_2 contribution to be reasonably small.

Example: We wish to find the impact of RF to LO feedthrough in a passive current-mode mixer.

The simplified mixer along with the LO waveforms are depicted in Figure 8.53. The RF feedthrough is modeled as a single-ended source with the magnitude of αv_{IN} appearing at the gate of one switch.

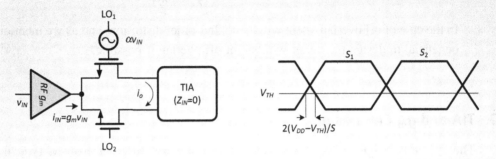

Figure 8.53: Analysis of RF to LO feedthrough in passive mixer

Suppose the on conductance of the switches is g_1, and g_2, respectively. During the on overlap (that is, the period of time when both switches are on, given finite rise and fall times), the current splits as

$$i_O = \frac{g_2}{g_1+g_2}i_{IN} - \frac{g_1}{g_1+g_2}i_{IN} = \frac{g_2-g_1}{g_1+g_2}i_{IN} = \frac{g_2-g_1}{g_1+g_2}g_m v_{IN}.$$

Ideally no current appears at the TIA output if the switches are identical, but in practice there is some current given the RF feedthrough. Assuming square-law characteristics,

$$g_1 = \beta(S_1 - V_{TH} - \alpha v_{IN})$$

$$g_2 = \beta(S_2 - V_{TH}),$$

where

$$\beta = \mu C_{OX} \frac{\omega}{L}.$$

Therefore

$$i_O = \frac{\beta(S_2-S_1) + \beta\alpha v_{IN}}{\beta(S_1+S_2) - 2\beta V_{TH} - \beta\alpha v_{IN}} g_m v_{IN}.$$

Note that the on overlap happens only if the zero-crossings are above the FET switches' threshold voltage as shown in the figure (on overlap).

Assuming symmetric waveforms with linear rise and fall times, and for $\alpha \ll 1$, the reader can show (see also Problem 16) that the normalized 2nd-order term is

Continued

$$\frac{\Delta t}{T/2} \times \frac{\alpha g_m}{2(V_{DD}-V_{TH})} v_{IN}^2 = 2\frac{\alpha g_m}{S \times T} v_{IN}^2,$$

where $\Delta t \approx \frac{2(V_{DD}-V_{TH})}{S}$ is the overlap time as shown in the figure. Since the mixer desired gain is $\frac{2}{\pi} g_m$,

$$\text{IIP}_2 = \frac{\frac{2}{\pi} g_m}{2 \frac{\alpha g_m}{S \times T}} = \frac{S \times T}{\pi \alpha}.$$

In the case of off overlap, there will be no 2nd-order distortion, and as we mentioned, it is beneficial to design the mixer with slight off overlap if possible.

8.4.7 TIA and g_m Cell Design

The g_m cell design as mentioned earlier is not very different from a typical tuned RF amplifier. Particularly, if it is combined with the LNA, as is very commonly done, all the design trade-offs described in Chapter 7 are applicable. Thus we shall describe only the TIA design choices here.

Similar to active mixers, one may consider the TIA (or the I–V converter) comprising simply a pair of resistors R_{BB} where the downconverted current is directly fed to (Figure 8.54). This approach, however, suffers from the fundamental drawback that the mixer overall gain, and the TIA input impedances are set by the same resistor R_{BB}. Thus, the optimum condition of a high gain and a low input impedance may be never satisfied simultaneously.

Consider the equivalent model shown on the right, where, as we established before, the RF amplifier current is divided between its output resistance (R_L) and the input impedance upconverted by the switches. This current is then downconverted by a loss factor of $\frac{\sqrt{2}}{\pi}$, and ultimately appears as a voltage at the output once multiplied by R_{BB}. Thus, with a reasonable approximation ignoring the switch resistance, the mixer total conversion gain is

$$g_m \frac{R_L}{R_L + \frac{2}{\pi^2} R_{BB}} \frac{\sqrt{2}}{\pi} R_{BB}.$$

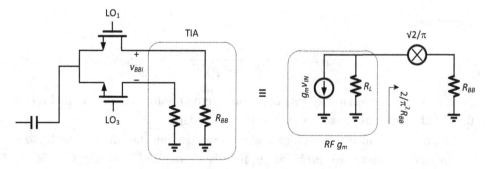

Figure 8.54: A simple TIA comprising only a pair of resistors

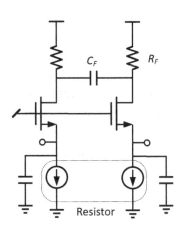 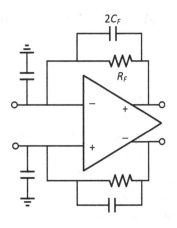

Figure 8.55: Common practical TIA circuits

Common-Gate TIA **Opamp-based TIA**

Unless the RF amplifier output resistance R_L is infinite, which is not the case in practice, and for moderately large values of R_{BB}, to the first order the mixer gain becomes independent of the TIA resistance. There are also severe linearity repercussions. To keep the swing at the amplifier output low, one must choose R_{BB} to be very small, resulting in zero gain for the mixer.

To break this trade-off, we can consider one of the two choices: a common-gate amplifier, or an opamp in feedback, both of which are shown in Figure 8.55. Assuming the input devices have the same size and bias current between the two circuits, then the single-ended input resistance looking into each structure is approximately $\frac{1}{g_{mBB}}$, where g_{mBB} is the input device transconductance. This of course assumes a one-stage opamp. Large capacitance at the input ensure that the TIA input impedance is kept low at the blocker offset frequency.

The gain, however, is set independently by the load or feedback resistor R_F. The mixer gain expression is now

$$g_m \frac{R_L}{R_L + \frac{2}{\pi^2} g_{mBB}} \frac{\sqrt{2}}{\pi} R_F.$$

Consequently, the gain and input impedance may be set independently.

A feedback capacitor in parallel sets the TIA bandwidth, and can be treated as a single-pole RC filter.

Since the load capacitance can be implemented differentially in the case of CG topology, for the same bandwidth it requires four times less capacitance. Thus the CG structure tends to be smaller, while the opamp may be superior in terms of linearity. Another drawback of the CG structure is the noise contribution of the current source. If headroom allows, it may be replaced by a resistor. Otherwise, very large devices may be needed to ensure a low flicker noise.

The opamp itself may be a two-stage design, or simply a complementary single-stage structure (basically an inverter). If a two-stage structure is chosen, it must be ensured that the opamp has sufficiently wide bandwidth to ensure a good linearity across all the downconverted blocker frequencies.

8.5 PASSIVE VOLTAGE-MODE MIXERS

Unlike the current mixers, the voltage-mode passive mixers commutate the RF AC voltage. The circuit driving the mixer, whether it is the LNA or some buffer, will not be much different from that of the current-mode mixers, as RF amplifiers act neither like true voltage amplifiers that have zero output impedance, nor like true current amplifiers that have an infinite output impedance. This is more prominent in CMOS, as a true voltage amplifier with low output impedance is not common. Instead, most CMOS amplifiers are transconductors.

The main distinguishing factor then between the voltage- and current-mode passive mixers is the circuit that follows the switches. Shown in Figure 8.56, a voltage-mode passive mixer drives ideally a voltage buffer whose input impedance is infinite, as opposed to a current-mode mixer, where the TIA input impedance is ideally zero.

The RF amplifier that is preceding the switches, however, whether it is the LNA or a separate buffer, is not much different from a current-mode mixer. In the case of the voltage-mode mixer though, as the switches upconvert a large baseband impedance to RF, then the RF amplifier acts in voltage mode. If the condition $R_{SW} \ll Z_{IN}$ met, then the mixer has a *voltage* conversion loss of $\frac{2}{\pi}$ if the LO is sharp enough. Nonetheless, same as current-mode design, the baseband buffer must have a very low impedance at the blocker offset frequency to ensure that the swings at the output of the RF amplifier are kept low.

The mixer principle operation as well as noise [21], [22] or linearity trade-offs are very similar to those of the current-mode design, and we will not spend any more time in this section describing them. See also [26] for more details on analysis of passive mixers in general. One main distinction between the voltage- and current-mode passive mixers is the fact that in the latter circuit, the switch is in series with a current source. Thus, the switch adds little noise or distortion of its own. However, switch distortion in the desired signal frequency is not very important. At the blocker frequency, assuming the voltage mixer buffer input impedance has

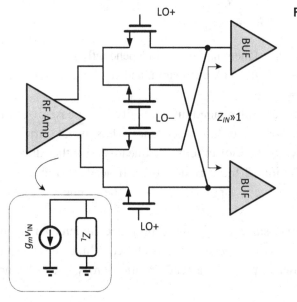

Figure 8.56: A voltage-mode passive mixer

8.5 Passive Voltage-Mode Mixers

Table 8.1: **A performance summary of different mixer structures**

Parameter	Commutation	White NF	1/f	Linearity	Loading	LO Swing
Active	DC current		Poor	Modest	Capacitive	Modest
CM passive	AC current	Comparable	Small	Good	BB impedance transformation	Rail-rail
VM passive	AC voltage		Small	Good		

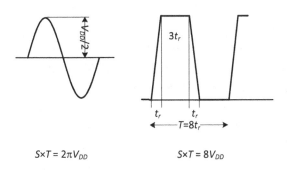

Figure 8.57: Mixer LO drive comparison

been reduced enough, the current- and voltage-mode mixers behave very similarly. Nevertheless, the current-mode mixer appears to be superior from the mixer switch noise point of view.

Shown in Table 8.1 is a comparison between different types of the mixer. We assume the overall power consumption of the mixer and its supporting circuitry is comparable between different structures.

It is instructive to have a closer comparison between the mixer LO drive requirements. Shown in Figure 8.57, let us assume a rail-to-rail sinusoid LO signal is applied to the active mixer. The corresponding normalized slop of the LO is $2\pi V_{DD}$.

Passive mixers on the other hand almost always require nonoverlapping clocks with <50% duty cycle. This prevents them from using sinusoidal LOs. Let us first assume a 50% LO is used. We also assume that the signal has an equal rise and file time of t_r, and is shaped as shown in Figure 8.57, where the LO stays flat for at least $3 \times t_r$. The rationale behind this choice is that it leads to a normalized slope of $8V_{DD}$ for the passive mixer LO, comparable to that of the active mixer.

Using CML type logic along with tuned buffer, generating a sinusoidal signal for the active mixer can be done quite efficiently, and at frequencies of up to one-third or so of the process transit frequency, which is several 100GHz for nanometer CMOS. On the other hand, for a typical 40nm CMOS process, the rise and fall time of logic gates are on the order of 10ps, limiting the 50% LO signal to only 12.5GHz. For a 25% LO design, which is preferred, this value is roughly reduced in half, to only 6.25GHz, that is, the maximum allowable frequency. In an 8-phase mixer, this number reduces to 3.1GHz. In general, the increased number of phases requires a larger normalized slope, and so a sharper rising/falling edge is needed.

It is clear that despite all the advantages, passive mixers may be used only at radio frequencies, although this scales with technology. Moreover, it shows why passive mixers have not been used as extensively until very recently.

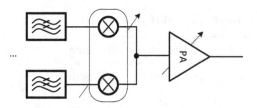

Figure 8.58: A generic Cartesian transmitter

8.6 TRANSMITTER MIXERS

Similar requirements and design trade-offs apply to upconversion mixers more or less. Both linearity and noise are critical, as we established in Chapter 6, to meet the close-in and far-out mask. The second-order distortion is typically not a concern as the IM_2 components are low frequency once upconverted.[12] In addition to the concerns mentioned above, the gain control and LO isolation are more critical in transmitter mixers. As we saw in Chapter 6, the LO feedthrough can potentially cause EVM issues, and must be properly dealt with.

In this section we will present both the active and passive mixer choices for transmitters, and comment on pros and cons of each qualitatively. The exact operation is very similar to that for downconversion mixers, which has been already presented.

8.6.1 Active Upconversion Mixers

Shown in Figure 8.58 is a generic Cartesian transmitter, where the IQ upconversion mixers are followed by a power amplifier (or a power amplifier driver).

Similar to our discussion on downconversion mixers, it is more appropriate to discuss the mixer performance issues in the context of a quadrature upconverter, consisting of I and Q branches, and to include any possible interaction. An active quadrature upconverter is shown in Figure 8.59.

The gain control is achieved through breaking the mixer into several unit cells turned on/off by controlling the bias (GC_1), in addition to a possible current steering on top (GC_2). The addition or subtraction is simply accomplished by tying the differential currents of the I and Q branches. This current is subsequently fed into a tuned circuit to boost the gain, and alleviate the headroom issue by setting the output bias at V_{DD}. The input stage nonlinearity may be also improved, employing a current mirror as shown in Figure 8.60. There are still three devices stacked, and realizing the circuit through a low supply voltage may be a challenge, especially given the large swing at the output.

The *current* generated by the DAC/LPF may be directly fed into the input transistors through a mirror, as opposed to apply a *voltage*, and converted back to a current by the input devices. The channel length modulation may be alleviated if the mirror is implemented through a feedback as shown on the right. This is one advantage of TX mixers, as since the input is at

---

[12] Unless the IM_2 products are created in baseband prior to upconversion.

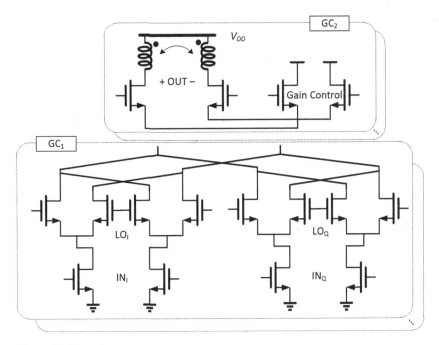

Figure 8.59: Active upconverter

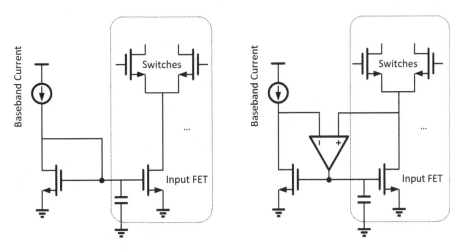

Figure 8.60: Active mixer input-stage linearity enhancement

low frequency, more linearization may be applied. In either case, a capacitor at the gate can provide some light filtering relaxing the preceding lowpass filter order.

In addition to input-stage linearity enhancement, unlike the receiver mixers, $1/f$ noise and second-order distortion are not a big concern anymore. Still, the upconversion active mixer suffers from a few drawbacks:

– The need for several devices stacked could potentially lead to poor linearity despite the mixer output being biased at V_{DD}. It is unlikely for the mixer to function properly at low supply voltages, given the relatively large swing at its output.

- The gain control may cause some challenges. Apart from the complexity in routing and layout, reducing the input signal strength is not desirable. In some applications such as 3G or LTE, a very stringent gain control range of over 70dB is required, a big portion of which is expected to be provided by the active mixer. The transmitter EVM, however, needs to stay unaffected throughout the course of the gain control. Since the absolute level of the LO feedthrough is relatively constant, mostly dominated by the LO signal directly leaking to the output, the transmitter EVM may be compromised at lower gains.

Given the above challenges, let us now consider the passive mixers.

8.6.2 Passive Upconversion Mixers

All the advantages of 25% versus 50% mixers remain applicable to transmitter mixers as well. Also, unlike the receivers that current-mode passive mixers seemed like a more natural choice, they are not as suitable for transmitters. This is shown in Figure 8.61, where the RX and TX current-mode mixers are compared.

In the case of a receive mixer, the RF current is always directed to one of the TIA's inputs, as one of the switches remains on. In the case of transmitter as shown on the left, the relatively large baseband current fed to the switches becomes disconnected over a large portion of the period when its corresponding switch is turned off. In addition to that, designing a TIA with low input impedance is more challenging at high frequency.

With all those options ruled out, let us focus on *25% voltage-mode mixers*, as shown in Figure 8.62.

The input applied to the mixer is relatively strong (perhaps several hundreds of mV), to ensure a good SNR is maintained throughout the TX chain. As long as the output stage of the lowpass filter driving the mixer has a low output resistance, this will not be an issue as the mixer switches are fairly linear. The mixer outputs are tied together (a single-ended design shown for simplicity) and are connected to the input of the PA driver. The PA driver is very similar to the active mixer shown in Figure 8.59, except the switches are removed. It consists

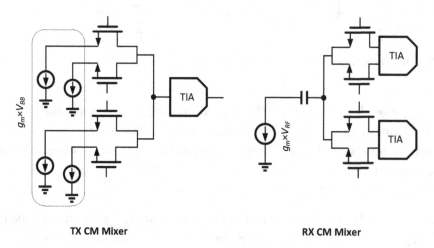

Figure 8.61: Current-mode passive mixers used in transmitters

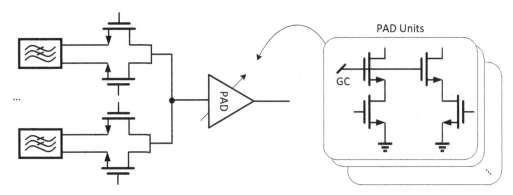

Figure 8.62: A 25% voltage-mode upconversion passive mixer

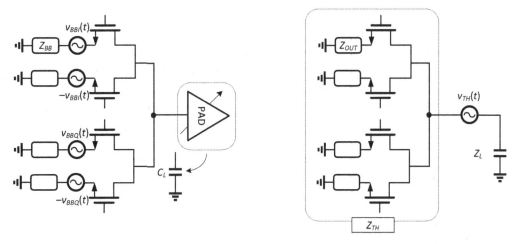

Figure 8.63: Passive mixer Thevenin equivalent circuit

of several unit stages, each turned on/off by the cascode bias, with an additional current steering applied on top (although not shown in Figure 8.62). The elimination of the switches given the large LO swing applied results in better linearity at lower voltage. Moreover, the relative ratio of the signal to LO feedthrough remains constant, and good EVM is expected over the entire gain control.

To analyze the mixer performance, it is best to use a Thevenin equivalent circuit as the passive mixer is *linear* (although time-variant) [27]. As shown in Figure 8.63, taking a very similar approach as the one we presented for the RX mixer (Figure 8.51), we arrive at the Thevenin equivalent illustrated on the right. The PA driver can be simply modeled as a capacitor looking into its input.

A similar analysis to the one performed for the RX reveals that the Thevenin impedance is

$$Z_{TH}(\omega) \approx R_{SW} + \frac{2}{\pi^2} Z_{BB}(\omega - \omega_{LO}),$$

and the equivalent Thevenin voltage is found to be

$$v_{TH}(t) = v_{BBI}(t)[S_1(t) - S_3(t)] + v_{BBQ}(t)[S_2(t) - S_4(t)],$$

indicating the upconversion of baseband IQ signals. $S_k(t)$, $k = 1, \ldots, 4$, is the effective LO signal driving the switches, and is one when the corresponding LO is high, and zero otherwise.

The above analysis ignores the impact of higher order harmonics.

Using the equivalent circuit, we can make an approximation and use a *linear time-invariant* analysis to find the mixer transfer function. It easily follows that the Thevenin voltage is divided between the mixer output equivalent impedance, and the load presented by the PA driver. Thus the RF voltage at PA driver input is

$$V_{RF}(\omega) \approx \frac{\sqrt{2}}{\pi} \left[e^{j\frac{\pi}{4}} V_{BBI}(\omega - \omega_{LO}) + e^{-j\frac{\pi}{4}} V_{BBQ}(\omega - \omega_{LO}) \right] \frac{Z_L(\omega)}{Z_L(\omega) + R_{SW} + \frac{2}{\pi^2} Z_{BB}(\omega - \omega_{LO})},$$

where the first term is a result of upconversion of the baseband quadrature inputs, whereas the second term shows the voltage division.

Example: To verify how accurate the above approximation is, shown in Figure 8.64 is a comparison between the exact linear time-variant analysis, approximated time-invariant analysis, as well as simulations for values: $Z_{BB} = 50\Omega \| 100\text{pF}$, $R_{SW} = 15\Omega$, and $C_L = 1.5\text{pF}$. The LTV model and simulations match very closely. Although the LTI model deviates a little, it still predicts the actual circuit behavior fairly accurately.

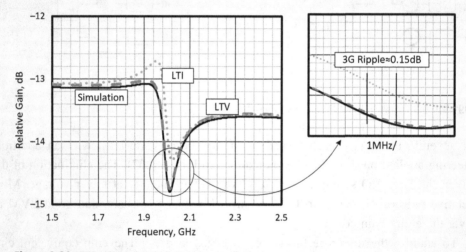

Figure 8.64: TX mixer simulation versus analysis

Similar to downconversion mixers, if Z_{BB} is not dominantly real, and is large enough with respect to the load impedance (Z_L), different high- and low-side gains may be observed. For the example shown in Figure 8.64, the parameters chosen result in very little high- and low-side gain difference over the ±2MHz 3G band that the design was intended for.

Furthermore, to maximize the mixer gain, one must ensure that the circuit driving mixer has sufficiently low output impedance (i.e. low Z_{BB}) over the entire band of interest.

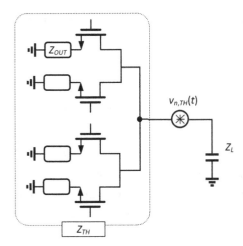

Figure 8.65: Passive mixer noise

Similarly, we can easily show that from a noise point of view, assuming a switch resistance of R_{SW}, the following Thevenin models can be used, where the noise of all switches is replaced by one source at RF whose power spectral density is

$$\overline{v_{n,TH}^2} = 4KTR_{SW},$$

where $v_{n,TH}^2$ represents the combined switch noise (Figure 8.65).

This noise then will approximately (if we were to use the LTI model) appear in series with the input-referred noise of the PA driver.

8.7 HARMONIC FOLDING IN TRANSMITTER MIXERS

While in downconversion mixers the LO component around the 3rd harmonic results in unwanted blocker and noise downconversion, in the case of upconversion mixers it may result in linearity degradation [27]. This applies to both passive and active realization.

Suppose the mixer output consists of a desired component $v_d(t)$ as follows:

$$v_d(t) = A(t)\cos(\omega_{LO}t + \phi(t)).$$

Due to the LO 3rd harmonic, there exists an undesired component around $3f_{LO}$:

$$v_u(t) \approx \frac{1}{3}A(t)\cos(3\omega_{LO}t - \phi(t)).$$

Now suppose the mixer output is passed through a PA (or PA driver) with 3rd-order non-linearity, whose input–output characteristics are

$$y = a_1 x + a_3 x^3,$$

as shown in Figure 8.66. Thus the PA driver output is

$$a_1 v_d + a_1 v_u + a_3 (v_d + v_u)^3,$$

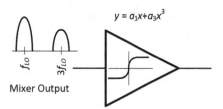

Figure 8.66: Mixer output passing through a nonlinear next stage

where the first term (v_d) is the desired output, while the second term (v_d) is at $3f_{LO}$ and subject to the filtering at the PA driver output. The third term consists of several components that may be problematic due to the mixing and folding of the undesired signal at $3f_{LO}$, combined with the PA driver 3rd-order nonlinearity. Focusing on this last term, after expansion we have

$$a_3(v_d+v_u)^3 = a_3v_d^3 + 3a_3v_d^2v_u + 3a_3v_dv_u^2 + a_3v_u^3.$$

The first three terms are expected to have components around f_{LO}, whereas the term $a_3v_u^3$ gives rise to components only at $3f_{LO}$ and higher frequency. Thus, we shall consider only the first three terms. Upon expansion, focusing only on the components that appear around f_{LO} at the PA driver output (assuming that the rest are subject to filtering), we have the following signals:

$$a_3v_d^3 \to \frac{3}{4}a_3A(t)^3 \cos(\omega_{LO}t + \phi(t))$$

$$3a_3v_d^2v_u \to \frac{1}{4}a_3A(t)^3 \cos(\omega_{LO}t - 3\phi(t))$$

$$3a_3v_dv_u^2 \to \frac{1}{6}a_3A(t)^3 \cos(\omega_{LO}t + \phi(t)).$$

Thus the contribution of the PA driver output nonlinearity may be summarized as

$$a_3(v_d+v_u)^3 \to \frac{11}{12}a_3A(t)^3 \cos(\omega_{LO}t + \phi(t)) + \frac{1}{4}a_3A(t)^3 \cos(\omega_{LO}t - 3\phi(t)).$$

If the 3rd harmonic were not present, the only nonlinear term would have been $\frac{3}{4}a_3A(t)^3 \cos(\omega_{LO}t + \phi(t))$, consistent with our previous definition of $IIP_3 = \sqrt{\frac{4}{3}\left|\frac{a_1}{a_3}\right|}$. However, in practice, the presence of a strong 3rd harmonic results in worse nonlinearity, as one would have expected considering only the PA driver nonlinearity. To gain more perspective, shown in Figure 8.67 is the PA driver output in response to a 2-tone signal with fundamental amplitude of A, and frequencies of $\omega_{LO}+\omega_1$ and $\omega_{LO}+\omega_1$ at the mixer output. We assume ω_1 and $\omega_2 \ll \omega_{LO}$.

This suggests that the IIP_3 is expected to degrade, as the energy of unwanted IM components nearly doubles.

8.7 Harmonic Folding in Transmitter Mixers

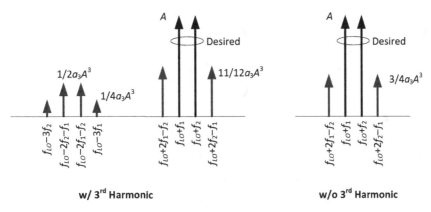

Figure 8.67: PA driver 2-tone output

It must be noted that the two terms $\frac{11}{12}a_3 A(t)^3 \cos(\omega_{LO} t + \phi(t))$ and $\frac{1}{4}a_3 A(t)^3 \cos(\omega_{LO} t - 3\phi(t))$ may have very different power spectral densities, and depending on the signal statistics, their correlation may vary.

Example: Shown in Figure 8.68 is the simulated ACLR (adjacent channel leakage ratio) for a 3G signal fed into an otherwise ideal mixer, but with 3rd-order nonlinearity, passed through a typical 3G PA driver. For some typical nonlinear characteristics of the PA driver used in this simulation, the ACLR at 5MHz degrades from −44dBc to −38dBc once the impact of the 3rd harmonic is included.

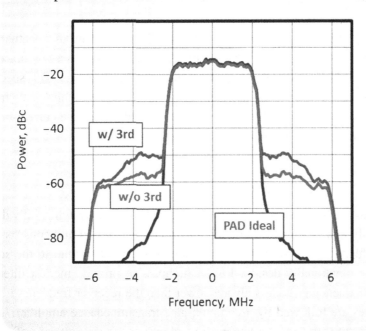

Figure 8.68: Simulated spectrum of a 3G output as a result of mixer 3rd-order harmonic folding

To alleviate this issue, one may consider employing some filtering at the mixer output to attenuate the signal around the 3rd harmonic. In the case of the active mixer shown before

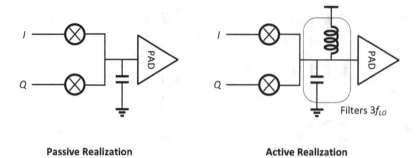

Figure 8.69: Comparison of active and passive mixers in the presence of 3rd harmonic folding

(Figure 8.59), this is readily achieved as the mixer is loaded with a tuned LC circuit resonating at f_{LO}.

On the other hand, implementing a sharp filter at the passive mixer output is not quite feasible, as unlike the active mixer, the output of the passive mixer is relatively low-impedance, and a strong 3rd harmonic is expected. The two designs are compared against each other as shown in Figure 8.69. For the passive design, it may be beneficial to include some kind of notch filtering to trap the 3rd harmonic before entering the PA driver. Alternatively, an 8-phase mixer may be employed at the expense of higher power consumption in the LO generation circuitry.

8.8 LNA/MIXER CASE STUDY

We conclude this chapter by discussing a case study detailing the design of an actual low-noise amplifier and a mixer. Designed for a 2/3/4G receiver, the circuit features a low-noise transconductance amplifier (LNTA), followed by a current-mode 25% passive mixer. This specific choice of architecture arises from the relatively stringent noise and linearity requirements imposed by the full duplex operation of LTE and 3G radios. As we showed earlier, the current-mode nature of the circuit boosts the linearity, yet is capable of achieving a low noise figure.

8.8.1 Circuit Analysis

The schematic of the LNTA and mixer is shown in Figure 8.70. The design is fully differential, but for simplicity only half is shown. The matching circuit is external and consists of a shunt L and a series C, which downconverts the impedance of the LNTA input to the source. The matching network is intentionally designed to be highpass to provide modest filtering at the transmitter leakage frequency (which is always lower than the receiver frequency).

The mixer switches are followed by a common-gate transimpedance amplifier. The choice of a common-gate circuit is to ensure a lower area and more convenient means of gain control (see Section 8.4.7). The main transistor (M_2) is in fact followed by a cascode device, which enables some gain control through switching unit devices on and off (the details are not shown though).

8.8 LNA/Mixer Case Study

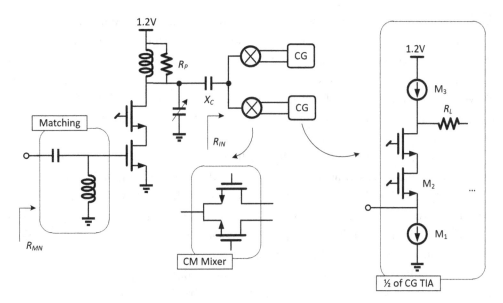

Figure 8.70: Low-noise transconductance amplifier and mixer simplified schematic

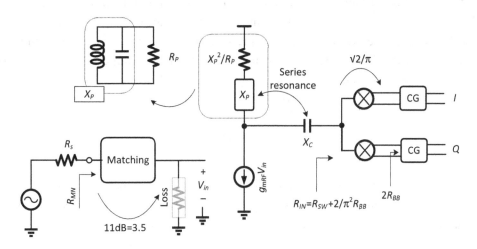

Figure 8.71: Front-end model for noise and gain calculations

A simplified model of the front end used for gain and noise calculations is shown in Figure 8.71.

For high bands, the LNTA features a tuned load using a 1.8nH inductance with a Q of 14 at 2GHz. The total shunt impedance as a result is about 280Ω at 2GHz, including the impact of finite output resistance of the LNTA and other losses. As most cellular receivers need to support various bands of operation, there are several amplifiers connected to the output, in which the parasitic capacitances limit the inductance to a relatively small value.[13] The matching circuit

---

[13] See Chapter 12 detailing the design of a multimode multiband cellular transceiver as a case study.

design and noise trade-offs are for the most part the same as those described for the voltage-mode LNAs in Chapter 7. We have chosen not to implement a degeneration inductor for area reasons, especially knowing that several inductors are needed given the multiband nature of the design. This comes at the expense of somewhat worse noise figure and narrower matching. However, the combination of the matching circuit loss and the shunt feedback resulted by the gate-drain capacitance is sufficient to provide a reasonable matching with sufficiently low noise figure.

The LNTA drains 6mA differentially, leading to a transconductance of about 30mS for each of the core devices biased at about 125mV overdrive. If only limited by the transistor input nonlinearity, this overdrive voltage would result in an IIP_3 of about 1V (or 10dBm referred to 50Ω) for a single device. The LNTA device size is 288μm/60nm. A minimum channel device is not used since it turns out that the impact of low DC gain due to poor r_o on gain and linearity is far more detrimental compared to the parasitic capacitances.

For this application an input return loss of about –10dB is targeted. If the noise contribution of the LNTA transistors ignored (we shall shortly verify that), we showed in Chapter 7 that this leads to a noise factor of 1.5. Given the shunt feedback formed by the gate-drain capacitance, the actual noise figure would be somewhat less, as the real part of the input impedance would be made up of a combination of the matching loss, and a noiseless part created by the shunt feedback. Once all the package and parasitic reactances are included, the matching network provides a net available voltage gain of about 11dB (or 3.5), large enough to suppress the LNTA transistors noise. Given a transconductance of 30mS for the LNTA core device, when referred to the input, the noise will be

$$\approx \frac{\frac{4KT\gamma}{30mS}}{\left(\frac{3.5}{2}\right)^2} = 4KT \times 10\Omega,$$

assuming $\gamma = 1$. The noise degradation is relatively small, and somewhat offset by the fact that the shunt feedback caused by gate-drain capacitance lowers the noise factor below 1.5, that is, if the real of the input impedance were entirely made up of a noisy resistance.

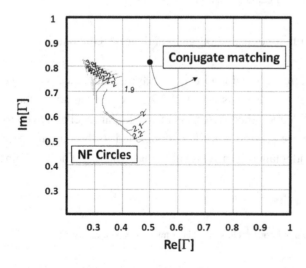

Figure 8.72: Noise figure circles and optimum conjugate matching

As an example, noise figure circles are plotted in Figure 8.72 for band II (2GHz), as well as the optimum impedance for conjugate matching (Figure 8.72). If matched to the source, the noise figure will be around 3dB. In practice, a compromise may be made to improve the noise figure, yet maintaining a reasonable input return loss.

The mixer switch size is 80μm/40nm, corresponding to an on resistance of about 10Ω. Each TIA branch is biased at 0.5mA. The top and bottom current sources are sized to lower their transconductance and thus their noise as much as possible, yet maintain a reasonable headroom. The corresponding transconductances are 3.6mS and 2.8mS for the bottom and top devices. The core device (M_2) on the other hand is biased at weak inversion to maximize g_m/I_D, which happens to be about $22V^{-1}$ for the process used. This leads to a transconductance of 11mS for the main device, and consequently an input resistance of about 90Ω. All the devices are long channel to lower flicker noise. The mixer input resistance around the LO frequency is then equal to

$$R_{IN} = R_{SW} + \frac{2}{\pi^2} R_{BB} \approx 28\Omega,$$

which is close to the simulated value of 32Ω.

To calculate the overall gain, let us consider Figure 8.71. We showed earlier that the total available transconductance gain from the source to the TIA output is

$$G_{MN} g_{mRF} \frac{\sqrt{2}}{\pi} \left| \frac{X_P}{\frac{X_P^2}{R_p} + j(X_P + X_C) + R_{IN}} \right|,$$

where G_{MN} is the matching network gain, g_{mRF} is the LNTA transconductance, and $R_{IN} = R_{SW} + \frac{2}{\pi^2} R_{BB}$ is the mixer input impedance. We also proved that the gain is maximized if $|X_C| = |X_P| = \sqrt{R_p R_{IN}}$. The optimum transconductance gain to each of the I or Q differential outputs will then be

$$\frac{G_{MN} g_{mRF}}{\sqrt{2}\pi} \sqrt{\frac{R_p}{R_{IN}}}.$$

Considering $G_{MN} = 3.5$, $g_{mRF} = 24$mS, $R_p = 280\Omega$, the available transconductance gain comes out to be 60mS. Note that even though the LNTA transistors have a transconductance of 30mS, the effective LNTA transconductance is somewhat less (24mS) due to the poor DC gain of the devices, leading to some signal loss at the cascade node, despite not using minimum channel length.

As we showed in Section 8.4.4, the impact of mixer switches on the overall noise figure is minimal, which leaves us with the TIA. Shown in Figure 8.73 is a simplified schematic of the mixer interface to the TIA input.

As the TIA is a current-mode buffer with low input impedance, its total output noise is a function of the impedance appearing at its input. For the moment, we shall assume a time-invariant single-ended mixer output resistance of R appearing at each branch. The TIA total output noise current when monitored differentially at the output is then equal to

498 | **Mixers**

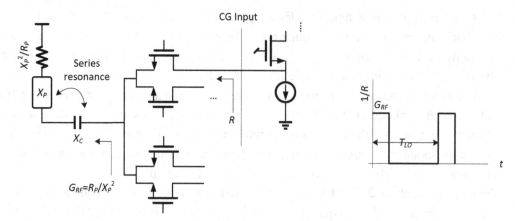

Figure 8.73: CG TIA noise contributions

$$\overline{i_{on}^2} = 2\left[\left|\frac{g_{m2}R}{1+g_{m2}R}\right|^2 \overline{i_{n1}^2} + \left|\frac{1}{1+g_{m2}R}\right|^2 \overline{i_{n2}^2} + \overline{i_{n3}^2}\right],$$

where $\overline{i_{n1}^2}$, $\overline{i_{n2}^2}$, and $\overline{i_{n3}^2}$ are the spectral densities of the bottom current source (M_1), the main transistor (M_2), and the top current source (M_3), respectively, and g_{m2} is the main device transconductance (Figure 8.70).

To find the mixer output resistance, R, consider the right side of Figure 8.73. The resistance seen by the TIA is evidently time-variant, and the exact analysis is quite involved in the math. To gain some perspective though, we resort to the following approximation using an LTI model: For a 25% mixer, at a given branch only at a quarter of a period does the RF conductance (G_{RF}) appear at the output when the corresponding switch is on. For the remainder of the period, the conductance seen at the mixer output is zero. Hence, the effective output conductance could be approximated by taking the time average. Furthermore, since for the optimum gain the series capacitance (X_C) must resonate with the LNTA shunt reactance (X_P), the admittance seen at the mixer input is mostly resistive (ignoring the switch parasitic) and is equal to

$$G_{RF} = \frac{R_P}{X_P^2}.$$

Consequently,

$$R \approx 4\left(R_{SW} + \frac{X_P^2}{R_P}\right),$$

which is found to be about 150Ω for our example. The factor of four is a result of time averaging and the 25% operation of the mixer. In many cases, however, this approximation may not hold, and the actual output impedance is probably smaller than this. It is because all the harmonic folded impedances appear in parallel as well, and the LNTA's output impedance at higher harmonics could be a lot smaller than at the fundamental.

There appears to be an optimum for the mixer output resistance R. A larger R leads to higher noise contribution from the current source, whereas a smaller R increases the TIA core device

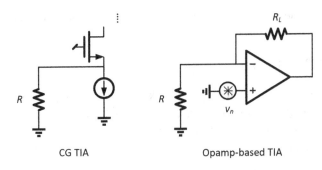

Figure 8.74: Comparison between CG and opamp-based TIA noise

noise. However, if the noise of the bottom current source and the core device are comparable, to the first order there is not a strong dependence on R, and it is best to maximize the gain by providing a series resonance as we showed in Section 8.4.3. Furthermore, given a rather weak dependence on the value of R, we shall continue to use the approximated formula given above to estimate the TIA noise.

The situation may be quite different if an opamp-based TIA is used. As shown in Figure 8.74, assuming a feedback resistance of R_L, the opamp input-referred noise is amplified by $\left(\frac{R_L}{R}\right)^2$. Hence, a low output resistance at the mixer may exacerbates the TIA noise contribution substantially. From a noise point of view, thus, there may exist a different optimum value for the series capacitance (X_C) as the one suggested by series resonance. On the extreme case, if one chooses a more traditional design with large series capacitance ($X_C \approx 0$), and a parallel resonance at the LNTA output, the mixer output resistance will be considerably higher, which may be advantageous from TIA noise point of view.

Returning to our common-gate example, the total differential output noise current was shown to be

$$\overline{i_{on}^2} = 8KT\gamma \left[\left| \frac{g_{m2}R}{1+g_{m2}R} \right|^2 g_{m1} + \left| \frac{1}{1+g_{m2}R} \right|^2 g_{m2} + g_{m3} \right].$$

Assuming $\gamma = \frac{2}{3}$ for long channel devices, using the values provided earlier, the noise will be

$$\overline{i_{on}^2} = 4KT \times 8\text{mS}.$$

When referred to the input, given the total *available* transconductance gain of 60mS, the noise will be

$$4KT \frac{8\text{mS}}{\left(\frac{60\text{mS}}{2}\right)^2} = 4KT \times 10\Omega,$$

which leads to a noise figure of 0.8dB if the LNTA were noiseless. Once the noise of the LNTA included, the total noise figure is

$$F \approx 1 + 0.7 + \frac{10}{50} + \frac{10}{50} = 1.9 = 2.7\text{dB}$$

The first term is the noise of the source, the second is matching network noise, the third is that of the LNTA input device, and the last term is the noise contribution of TSA. The results agree well with the measurements.

As for the linearity, assuming the input transistor is the dominant source, considering the matching network gain of 11dB, the IIP_3 is

$$\text{IIP3}_3 = 10\text{dBm} - 11\text{dB} = -1\text{dBm}.$$

Any filtering on the blocker caused by the matching network helps improving the IIP_3. On the other hand, the impact of finite r_o and output nonlinearity, which has been ignored, may become dominant. For this design, an out-of-band IIP_3 of about +1dBm is measured for band II at an offset of 80MHz, where the TX leakage resides.

Although the circuit ideally operates in current mode, it is beneficial to find the *voltage gain* of the LNTA from the source to its output. Assuming a series resonance for the optimum gain, the LNTA output impedance is

$$R_P \| jX_P \| (-jX_C + R_{IN}) = R_P \| \frac{X_P^2 + jX_P R_{IN}}{R_{IN}}.$$

Since $X_P^2 = R_P R_{IN}$, the impedance will be

$$R_P \| (R_P + jX_P) \approx R_P/2.$$

The voltage gain will be then

$$G_{MN} g_{mRF} \frac{R_P}{2}.$$

If this gain is too high, the large swing at the LNTA output caused by the blocker may become very problematic. Interestingly, for the series resonance case, this gain is not a direct function of the mixer input impedance. However, this is not quite true. To achieve the same overall transconductance gain, a lower R_{IN} leads to a lower R_P, and thus less voltage gain.[14] On the other hand, in a more traditional design where the series capacitance is larger, the voltage gain is directly set by the mixer input impedance: $\approx G_{MN} g_{mRF} R_{IN}$. This clearly indicates a trade-off between the gain and linearity.

8.8.2 Design Methodology

Although in the previous section we provided analytical details of the circuit of Figure 8.70, we are yet to offer a methodical approach to obtain the design parameters. The first step is to choose the right circuit topology. The correct topology along with its pros and cons has been discussed in great detail in the previous sections as well as in Chapter 6.

Once the right architecture is selected (say the one we just discussed), the next step is to find the right combination of the design parameters. This is typically quite application dependent, and there often may not exist a unique answer. In this section, however, we offer a qualitative approach to highlight the trade-offs, and provide some design guidelines:

---

[14] Although once the output inductor is set, the value of R_P is fixed, one may choose a smaller inductance to lower R_P and hence improve the linearity.

- The LNTA transconductance and matching network gain are directly determined by the noise requirements as discussed in Chapter 7. This sets a lower bound for the noise figure. We expect the TIA and the subsequent stages, if designed properly, to not contribute much.
- Once the LNTA device transconductance and the matching are determined, the LNTA bias current is set based on the linearity requirements. A combination of the matching network gain and the input device overdrive sets an upper limit on the IIP$_3$. In most modern processes, it may not be surprising to see a nonlinearity contribution from the output given the poor device gain and low supply voltages.
- The LNTA load is determined based on the overall gain of the front end, needed to suppress the noise of the IF circuits, the estimated parasitic capacitances at the output, and the linearity. The overall transconductance gain was found to be $\frac{G_{MN}g_{mRF}}{\sqrt{2\pi}}\sqrt{\frac{R_p}{R_{IN}}}$. The value of the shunt impedance (R_p) sets the transconductance gain (which impacts the noise), as well as the voltage gain (which impacts the linearity). The linearity improves with lowering R_p, yet the same overall gain may be maintained by lowering R_{IN} proportionally. The latter determines the bias current of the TIA.
- The TIA bias is obtained based on the overall transconductance gain and noise requirements. In either common-gate or opamp-based designs, it is best to use long channel devices biased in weak inversion for the core devices. A higher bias results in lower R_{IN}, and higher gain for the same linearity (or LNTA output swing). Suppose the TIA core device has a transconductance of g_{mTIA}. In the case of the CG design for instance, the output noise current was found to be proportional to $4KT\gamma g_{mTIA}$. On the other hand, the overall transconductance gain is $\frac{G_{MN}g_{mRF}}{\sqrt{2\pi}}\sqrt{\frac{R_p}{R_{IN}}} \approx \frac{G_{MN}g_{mRF}}{\sqrt{2\pi}}\sqrt{R_p g_{mTIA}}$. The TSA noise then, if referred to the input, will be to the first order independent of g_{mTIA}. However, a larger g_{mTIA} does lower the noise contribution of the stages following the TIA such as the ADC or lowpass filter.
- The switch resistance is also a determining factor for the mixer input resistance (R_{IN}). As a rule of thumb, we set the switch size such that the two terms $\frac{2}{\pi^2}R_{BB}$ and R_{SW} are comparable. A smaller switch size may be chosen to lower the LO power consumption, and the overall R_{IN} may be adjusted by burning more current in the TIA, and thus reducing R_{BB}.

Once an initial set of parameters is chosen according to the application, often a few iterations are needed to fine-tune the design.

If the RF transconductance amplifier does not incorporate a tuned load, for instance if it is built of complementary FETs (basically an inverter), the design flow will be much more straightforward. Suppose the RF transductor, shown in Figure 8.75, sees a shunt impedance of $R_p \| \frac{1}{jC_p\omega}$ at its output. In the case of an inverter-like transconductor, $R_p = r_{on} \| r_{op}$, and C_P represents all the parasitic capacitances at the output node.

The RF transconductance and the matching network are designed according to noise tradeoffs as before. The series capacitance is simply large enough to provide an AC short at the frequency of interest. If possible, it may be removed altogether and replaced by a CMFB circuit (see Section 8.4 and Figure 8.29). To minimize the RF loss, clearly $R_{IN} \ll R_p$. The total

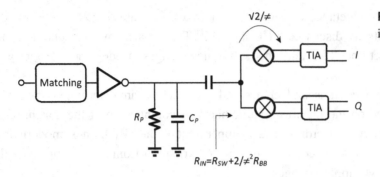

Figure 8.75: Passive mixer with invert-like RF transconductor

bandwidth will then be set by the combination of the output parasitic capacitance C_p and the mixer input impedance R_{IN}. Additionally, R_{IN} may be reduced further to lower the voltage gain (which is now $G_{MN}g_{mRF}R_{IN}$) to minimize the blocker swing at the mixer input. If the conditions above are satisfied, the total transconductance gain will be

$$\sqrt{2}\frac{G_{MN}g_{mRF}}{\pi}.$$

This topology, when designed properly, leads to a wideband and blocker tolerant scheme, commonly employed in the context of software-defined radios (see Chapter 12 for more details). The main challenge with the circuit of Figure 8.75 arises from the dependence of the LNA noise and input impedance on the load resistance seen at the output, which is largely a function of the mixer and the TIA. Thus, it is very critical to codesign the LNTA along with the mixer and TIA very carefully. A cascode inverter may prove to be helpful, although it compromises the headroom and linearity.

8.9 Summary

This chapter discussed the analysis and design of mixers for receivers and transmitters.

- Sections 8.1 and 8.2 covered the fundamental properties of the mixers, the general requirements, and their evolution from vacuum tube era to modern nanometer CMOS.
- Active downconversion mixers were discussed in Section 8.3. Section 8.3 also covered the nonideal effects, including noise and distortion. Several topologies to circumvent noise and 2nd-order distortion were explained as case studies.
- Passive current-mode mixers were presented in Section 8.4, along with their nonideal effects. Multiphase passive mixers and their properties were covered in this section as well.
- Section 8.5 discussed the voltage-mode passive mixers.
- Transmitter mixers were discussed in Section 8.6. The relevant topic of harmonic folding in transmitter mixers and its impact on the transmitter performance was covered in Section 8.7.
- Finally, a 40nm receiver front end as a case study was presented in Section 8.8.

8.10 Problems

1. In the mixer shown below, assume the FET is square-law, and remains in saturation all the time. If $v_{RF}(t) = a\sin\omega_0 t$, and $v_{LO}(t) = A\sin\omega_{LO}t$, find the mixer transconductance conversion gain. How does that compare to the FET transconductance if it were to be used as a linear amplifier? **Hint:** The upper bound of the LO voltage is set by the requirement of the FET staying on. **Answer:** $\frac{G_c}{G_m} = \frac{\beta A/2}{\beta(V_{GS0}-V_{TH})} < \frac{1}{2}$.

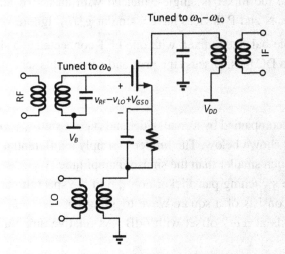

2. The circuit shown below, known as a differential pair mixer, is intended to be used in a zero-IF receiver. Assuming the BJTs remain in forward active mode, analyze the circuit operation and find the mixer transconductance conversion gain. **Answer:** $G_c = \frac{A}{4V_T}\frac{\alpha}{R_E+r_{e1}}$, where A is the LO amplitude, $V_T = \frac{KT}{q}$, and $\alpha = \frac{\beta}{\beta+1}$.

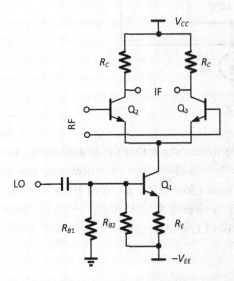

3. In the previous problem, assume the supply voltage is ±5V, and the LO amplitude is 200mV peak. Design the circuit for a bias current of 1mA, and a voltage gain of 6dB.

4. Find the noise figure of a single-balanced active mixer with active load. Do not ignore the load resistor noise.

5. Using the g_m/I_D and IIP_3 curves provided in Chapter 6, design a double-balanced active mixer with passive load with the following specifications: gain: 0dB, NF: 10dB, IIP_3:10dBm. Noise and IIP_3 are referred to 50Ω for convenience. Find the proper LO swing and its DC level. The supply voltage is 1.2V.

6. Repeat Problem 5, but assume the mixer is single-balanced with an active load, and the required gain is 12dB. Assume N and P devices have a similar g_m/I_D. Ignore r_o.

7. Find the noise figure of a double-balanced mixer, with one FET connected to a single-ended input, and the other one left at a DC bias. Discuss the pros and cons of this scheme. **Answer:**
$$F = 1 + \frac{8KT\gamma g_m + 16KT\gamma\frac{1}{\pi A} + \frac{8KT}{R_L}}{4KTR_s\left(\frac{2}{\pi}g_m\right)^2}.$$

8. A sinusoid signal ($A \sin \omega_0 t$) accompanied by a small sideband ($v_1 = a \sin(\omega_0 + \omega_m)t$, $a \ll A$) is applied to an ideal limiter as shown below. The limiter is simply a differential pair whose devices overdrive voltage is much smaller than the sinusoid amplitude ($V_{eff} \ll A$). Using the same analysis provided for the switching pair flicker noise, find the spectrum at the limiter output. **Answer:** The output consists of a square-wave toggling between $\pm I_0$ at the main signal frequency, and sidebands at $\pm \omega_m$ offset with 6dB less relative amplitude than the fundamental at ω_0.

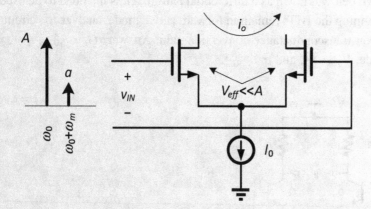

9. For the following single-balanced active mixer, show that the instantaneous transconductance of the switching pair is $G_m(t) = \frac{\partial i_{OUT}}{v_{LO}} = 2\frac{g_{m1}(t)g_{m2}(t)}{g_{m1}(t)+g_{m2}(t)}$. Show that the instantaneous spectral density of the output current noise (due to the noise of both switches) is $S_{i_{OUT}}(f) = 8KT\gamma G_m(t)$, where the noise of each device is $S_{i_{n1/2}}(f) = 4KT\gamma g_{m1/2}(t)$. Find the output noise power spectrum assuming a sinusoid LO with fast switching. **Hint:** The switches add noise only at zero crossings where $v_{LO} \approx 0$.

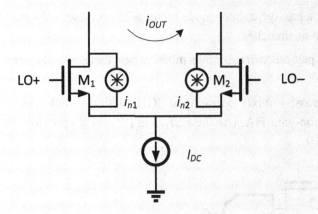

10. To combat flicker noise of switches in the active mixer used in a zero-IF receiver, a student proposes the following active mixer circuit with two transconductances and a large capacitor to store flicker noise and random input-referred DC offset due to mismatches. Explain the flaw with this mixer circuit and why it cannot be used in the zero-IF receiver. **Hint:** Show that the downconverted current has to pass through a capacitor, which is open at DC.

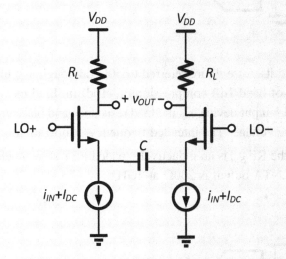

11. Find the IIP_2 of a single-balanced active mixer where the load resistors are mismatched (one is R_L, the other is $R_L(1+\alpha)$). Simplify your answer assuming the input transistors are square-law. **Answer:** $IIP_2 = \frac{16}{\pi}\frac{V_{\text{eff}}}{\alpha}$.

12. Design a 6-phase mixer with ideal LO signals, each with one-sixth of a cycle duration. Find the coefficients of each branch, the effective LO, and the mixer gain. Discuss the pros and cons of the mixer compared to a 4- and an 8-phase design.

13. Repeat Problem 12 for a 3-phase mixer.

14. Show that the minimum number of phases needed to distinguish the signal from its image is 3. Why are 4-phase mixers more popular in zero-IF receivers?

15. Find the leakage gain of a passive current-mode mixer with an offset voltage at the gate of one switch modeling the mismatches.

16. Calculate the IIP_2 of a passive current-mode mixer when there is a nonzero RF to LO feedthrough.

17. Design a current-mode passive mixer operating at 2GHz with a complementary (inverter-like) RF g_m, and a common-gate TIA. Use the g_m/I_D and IIP_3 curves provided in Chapter 6. The device DC gain is 10.

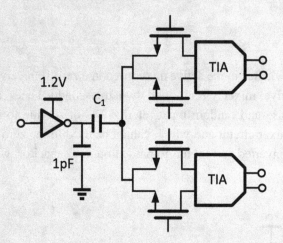

The mixer gain is 30dB, and its noise figure referred to 50Ω (although it is unmatched) is 3dB. The mixer has an out-of-band 1dB compression of −10dBm. Find the switch resistance, the value of C_1, the TIA input device g_m, the load resistance and bias current, and the RF g_m current and overdrive voltage. The intended frequency of operation is 2GHz.

18. Repeat Problem 17, where the RF g_m is an inductively loaded FET shown below. Assume the shunt resistance at the LNTA output is 200Ω at 2GHz.

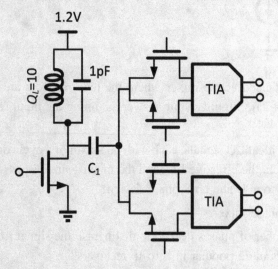

19. Discuss the pros and cons of using an opamp-based TIA for Problems 17 and 18.

20. Consider the fully differential passive current-mode mixer shown below, driven by a single-ended g_m cell.

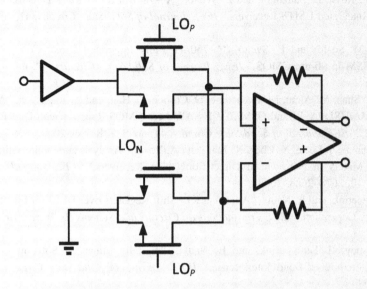

Argue why one may choose this scheme where the second set of switches are connected to ground (as opposed to removing them). Discuss the pros and cons of this scheme, and particularly any degradation in mixer noise figure.

8.11 References

[1] R. Rafuse, "Symmetric MOSFET Mixers of High Dynamic Range," in *IEEE International Solid-State Circuits Conference. Digest of Technical Papers*, 1968.

[2] B. Gilbert, "A Precise Four-Quadrant Multiplier with Subnanosecond Response," *IEEE Journal of Solid-State Circuits*, 3, no. 4, 365–373, 1968.

[3] P. R. Gray and R. G. Meyer, *Analysis and Design of Analog Integrated Circuits*, John Wiley, 1990.

[4] M. Terrovitis and R. Meyer, "Intermodulation Distortion in Current-Commutating CMOS Mixers," *IEEE Journal of Solid-State Circuits*, 35, no. 10, 1461–1473, 2000.

[5] H. Darabi and A. Abidi, "Noise in RF-CMOS Mixers: A Simple Physical Model," *IEEE Journal of Solid-State Circuits*, 35, no. 1, 15–25, 2000.

[6] D. Binkley, C. Hopper, J. Cressler, M. Mojarradi, and B. Blalock, "Noise Performance of 0.35um SOI CMOS Devices and Micropower Preamplifier from 77–400k," *IEEE Transactions on Nuclear Science*, 51, no. 6, 3788–3794, 2004.

[7] J. Chang, A. Abidi, and C. Viswanathan, "Flicker Noise in CMOS Transistors from Subthreshold to Strong Inversion at Various Temperatures," *IEEE Transactions on Electron Devices*, 41, no. 11, 1965–1971, 1994.

[8] M. Terrovitis and R. Meyer, "Noise in Current-Commutating CMOS Mixers," *IEEE Journal of Solid-State Circuits*, 34, no. 6, 772–783, 1999.

[9] H. Darabi and J. Chiu, "A Noise Cancellation Technique in Active RF-CMOS Mixers," *IEEE Journal of Solid-State Circuits*, 40, no. 12, 2628–2632, 2005.

[10] D. Manstretta, M. Brandolini, and F. Svelto, "Second-Order Intermodulation Mechanisms in CMOS Downconverters," *IEEE Journal of Solid-State Circuits*, 38, no. 3, 394–406, 2003.

[11] M. Brandolini, P. Rossi, D. Sanzogni, and F. Svelto, "A +78dBm IIP2 CMOS Direct Downconversion Mixer for Fully Integrated UMTS Receivers," *IEEE Journal of Solid-State Circuits*, 41, no. 3, 552–559, 2006.

[12] M. Brandolini, M. Sosio, and F. Svelto, "A 750mV Fully Integrated Direct Conversion Receiver Front-End for GSM in 90-nm CMOS," *IEEE Journal of Solid-State Circuits*, 42, no. 6, 1310–1317, 2007.

[13] D. Kaczman, M. Shah, M. Alam, M. Rachedine, D. Cashen, L. Han, and A. Raghavan, "A Single-Chip 10-Band WCDMA/HSDPA 4-Band GSM/EDGE SAW-Less CMOS Receiver with DigRF 3G Interface and 90dBm IIP2," *IEEE Journal of Solid-State Circuits*, 44, no. 3, 718–739, 2009.

[14] A. Mirzaei, H. Darabi, J. Leete, X. Chen, K. Juan, and A. Yazdi, "Analysis and Optimization of Current-Driven Passive Mixers in Narrowband Direct-Conversion Receivers," *IEEE Journal of Solid-State Circuits*, 44, no. 10, 2678–2688, 2009.

[15] X. Gao, E. Klumperink, and B. Nauta, "Advantages of Shift Registers over DLLs for Flexible Low Jitter Multiphase Clock Generation," *IEEE Transactions on Circuits and Systems II: Express Briefs*, 55, no. 3, 244–248, 2008.

[16] Z. Ru, N. Moseley, E. Klumperink, and B. Nauta, "Digitally Enhanced Software-Defined Radio Receiver Robust to Out-of-Band Interference," *IEEE Journal of Solid-State Circuits*, 44, no. 12, 3359–3375, 2009.

[17] D. Murphy, H. Darabi, A. Abidi, A. Hafez, A. Mirzaei, M. Mikhemar, and M.-C. Chang, "A Blocker-Tolerant, Noise-Cancelling Receiver Suitable for Wideband Wireless Applications," *IEEE Journal of Solid-State Circuits*, 47, no. 12, 2943–2963, 2012.

[18] J. Weldon, R. Narayanaswami, J. Rudell, L. Lin, M. Otsuka, S. Dedieu, L. Tee, K.-C. Tsai, C.-W. Lee, and P. Gray, "A 1.75-GHz Highly Integrated Narrow-Band CMOS Transmitter with Harmonic-Rejection Mixers," *IEEE Journal of Solid-State Circuits*, 36, no. 12, 2003–2015, 2001.

[19] A. Mirzaei, H. Darabi and D. Murphy, "Architectural Evolution of Integrated M-Phase High-Q Bandpass Filters," *IEEE Transactions on Circuits and Systems I: Regular Papers*, 59, no. 1, 52–65, 2012.

[20] W. Redman-White and D. Leenaerts, "1/f Noise in Passive CMOS Mixers for Low and Zero IF Integrated Receivers," in *Proceedings of the 27th European Solid-State Circuits Conference*, 2001.

[21] S. Chehrazi, R. Bagheri, and A. Abidi, "Noise in Passive FET Mixers: A Simple Physical Model," in *Proceedings of the IEEE Custom Integrated Circuits Conference*, 2004.

[22] S. Chehrazi, A. Mirzaei and A. Abidi, "Noise in Current-Commutating Passive FET Mixers," *IEEE Transactions on Circuits and Systems I: Regular Papers*, 57, no. 2, 332–344, 2010.

[23] A. Mirzaei, H. Darabi, J. Leete, and Y. Chang, "Analysis and Optimization of Direct-Conversion Receivers with 25% Duty-Cycle Current-Driven Passive Mixers," *IEEE Transactions on Circuits and Systems I: Regular Papers*, 57, no. 9, 2353–2366, 2010.

[24] E. Sacchi, I. Bietti, S. Erba, L. Tee, P. Vilmercati, and R. Castello, "A 15mW, 70kHz 1/f Corner Direct Conversion CMOS Receiver," in *Proceedings of the IEEE Custom Integrated Circuits Conference*, 2003.

[25] S. Chehrazi, A. Mirzaei, and A. Abidi, "Second-Order Intermodulation in Current-Commutating Passive FET Mixers," *IEEE Transactions on Circuits and Systems I: Regular Papers*, 56, no. 12, 2556–2568, 2009.

[26] M. Soer, E. Klumperink, P.-T. de Boer, F. van Vliet, and B. Nauta, "Unified Frequency-Domain Analysis of Switched-Series-Passive Mixers and Samplers," *IEEE Transactions on Circuits and Systems I: Regular Papers*, 57, no. 10, 2618–2631, 2010.

[27] A. Mirzaei, D. Murphy, and H. Darabi, "Analysis of Direct-Conversion IQ Transmitters with 25% Duty-Cycle Passive Mixers," *IEEE Transactions on Circuits and Systems I: Regular Papers*, 58, no. 10, 2318–2331, 2011.

[28] R. Brederlow, W. Weber, C. Dahl, D. Schmitt-Landsiedel, and R. Thewes, "Low-Frequency Noise of Integrated Polysilicon Resistors," *IEEE Transactions on Electron Devices*, 48, no. 6, 1180–1187, 2001.

[29] A. van der Wel, E. Klumperink, J. Kolhatkar, E. Hoekstra, M. Snoeij, C. Salm, H. Wallinga, and B. Nauta, "Low-Frequency Noise Phenomena in Switched MOSFETs," *IEEE Journal of Solid-State Circuits*, 42, no. 3, 540–550, 2007.

[30] B. Cook, A. Berny, A. Molnar, S. Lanzisera, and K. Pister, "Low-Power 2.4GHz Transceiver with Passive RX Front-End and 400mV Supply," *IEEE Journal of Solid-State Circuits*, 41, no. 12, 2757–2766, 2006.

9 Oscillators

In this chapter we present a detailed discussion on various types of oscillators, including ring and crystal oscillators. The LC resonators and integrated capacitors and inductors have been already discussed in Chapter 1, and are essential to this chapter. Furthermore, some of the communication concepts that we presented in Chapter 2, such as AM and FM signals, as well as stochastic processes, are frequently used in this chapter.

Given the nonlinear nature of the oscillators, we will offer a somewhat different view on noise compared to much of the material presented on amplifiers and mixers. This is captured in the "Cyclostationary Noise" section (9.3.5), which is a continuation of our discussion in Chapter 5. Even though we introduced the concept of phase noise in Chapter 6, we present a detailed analysis based on the concept of *energy balance*, briefly touched on in Chapter 1.

Although the focus of this chapter is mainly LC oscillators, which are widely used in RF transceivers, we will also cast brief discussions on ring and quadrature oscillators. Both schemes are not as common given the performance limitation that we shall discuss. We conclude the chapter by presenting crystal oscillators basic properties, and discuss several commonly used topologies.

The specific topics covered in this chapter are:

- Linear oscillators and Leeson's model
- Nonlinear oscillators and energy balance model
- AM and PM noise
- Cyclostationary noise
- Bank's general result
- Two-port oscillators
- Examples of common oscillators topologies: NMOS, CMOS, Colpitts, and class C
- Ring oscillators
- Quadrature oscillators
- Crystal oscillators

For class teaching, we recommend the linear oscillator (Section 9.1), and selected oscillator topologies from Section 9.4. If deemed necessary, a summary of Bank's results (Section 9.3.8) may be presented to understand the phase noise derivation of various topologies better. A detailed discussion of phase noise (Sections 9.2 and 9.3) must be deferred to a more advanced course. Sections 9.7, 9.8, 9.9, and 9.10 may be assigned as reading.

9.1 THE LINEAR LC OSCILLATOR

An ideal LC oscillator consists of an inductor and capacitor connected in parallel. As discussed in Chapter 1, any initial energy stored in this resonator (or *tank*) will oscillate back and forth between the two components. Unfortunately, any physical realization of these passive components will have some associated loss, which will result in a decaying oscillation. Therefore, to ensure a constant oscillation amplitude, a practical LC oscillator requires a circuit that periodically injects energy into the resonator in order to balance the energy dissipated by losses in the tank.

Practical (i.e., physically realizable) LC oscillators are by necessity nonlinear circuits. Nevertheless, many basic insights into their operation can be gleaned by studying a simple linear abstraction, and so the theoretical *linear* LC oscillator is first reviewed in this section.

9.1.1 The Feedback Model

Shown in Figure 9.1 is a lossy resonator connected to an energy-restoring circuit. In RF-CMOS the active circuit is typically some form of transconductance, and therefore we have modeled it as such. The noise associated with the lossy resonator and the transconductance are both modeled as current sources (i.e., $i_{nRp}(t)$ is the noise source associated with the tank loss, and $i_{ngm}(t)$ is the noise associated with the transconductance). An aperiodic current source, $i_{SU}(t)$, is also included, which will be used to *start-up* the oscillator. Using straightforward manipulation, the circuit can be recast in the form of a feedback system (also shown in Figure 9.1), where the noise sources and the start-up current become the inputs to an LTI system whose transfer function is given by

$$H(s) = \frac{Z_{tank}(s)}{1 + g_m Z_{tank}(s)} = \frac{s \dfrac{1}{C_p}}{s^2 + s\left(\dfrac{1 + g_m R_p}{R_p C_p}\right) + \dfrac{1}{L_p C_p}}.$$

From feedback theory, it follows that the loop is stable if the poles of $H(s)$ are in the left-half plane ($g_m > -1/R_p$), the loop is unstable if the poles of $H(s)$ are in the right-half plane

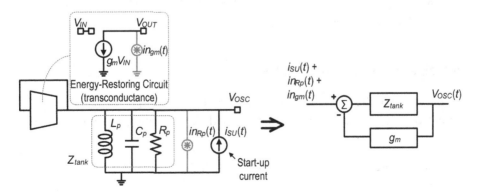

Figure 9.1: The linear LC oscillator

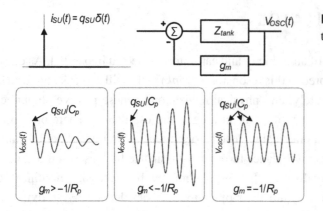

Figure 9.2: Impulse response of the linear LC oscillator

($g_m < -1/R_p$), while the loop is marginally stable if the poles of $H(s)$ reside along the imaginary-axis ($g_m = -1/R_p$).

Now, consider the case where the start-up current is a single impulse, i.e., $i_{SU}(t) = q_{SU}\delta(t)$, where q_{SU} is in units of amperes times second (i.e., coulombs). The response of this linear system to the impulse for the three different cases (assuming complex poles) is shown in Figure 9.2. If the loop is stable, the injected energy will produce an oscillation that will eventually decay, if the loop is unstable the oscillation amplitude will grow without bound (drawing ever more energy from the ideal transconductance), while in the marginally stable case a constant oscillation is produced with an amplitude proportional to the energy injected by the start-up current,[1] i.e., q_{SU}/C_p.

The marginally stable condition results in the only desirable output; however, this condition cannot be achieved in practice because any infinitesimal deviation in g_m will result in a growing or decaying oscillation (hence why a *linear* oscillator is just an abstraction). Instead, oscillators are designed such that $g_m < -1/R_p$, which results in a growing oscillation. This growing oscillation will eventually self-limit because a practical implementation of the transconductance cannot draw infinite power. Accordingly, the amplitude of oscillation in a real oscillator is independent of the start-up current, and instead depends on the nonlinear characteristics of the energy-restoring circuit.

As an oscillator is designed to ensure $g_m < -1/R_p$, an explicit start-up current is generally not required. This is because in this unstable state an arbitrarily small amount of injected energy will excite a growing oscillation. This small amount of energy is typically just the noise sources in the system (i.e., $i_{nRp}(t)$ and/or $i_{ngm}(t)$).

9.1.2 Phase Noise in the Linear Oscillator

In 1966, Leeson used the linear feedback model to derive an expression for phase noise [1]. We will present a simplified restatement of that work here. Taking the linear model (shown in

---

[1] This result can be obtained using the initial value theorem, i.e., $\lim_{t \to 0} v_{osc}(t) = \lim_{s \to \infty} sH(s)q_{SU}$.

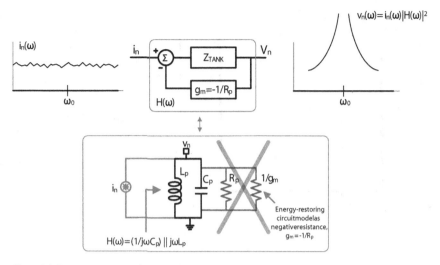

Figure 9.3: Noise transfer function of the linear oscillator

Figure 9.3), we note that the noise transfer function for the marginally stable condition (i.e., $g_m = -1/R_p$) is easily deduced as

$$|H(j\omega)|^2 = \left|\frac{\omega L_p}{1 - \omega^2 L_p C_p}\right|^2,$$

which is simply the squared impedance of a lossless resonator. This is more obvious if we redraw the memoryless transconductance as a negative resistance equal $-1/R_p$ as shown in Figure 9.3. In this case, the negative resistance ($1/g_m$) perfectly cancels out R_p and the noise current doesn't see either the equivalent tank resistance or the negative resistance provide by the energy-restoring circuit. The noise current only sees an ideal inductor and capacitor in parallel.

Because of the strong filtering action of the lossless resonator, the transfer function close to the resonance frequency is the only region of interest in any RF application, i.e., around $\omega = \omega_0 \pm \Delta\omega$, where $\omega_0 = 1/\sqrt{L_p C_p}$ and $\Delta\omega \ll \omega_0$. This allows for a further simplification of the noise transfer function,

$$|H(j(\omega_0 + \Delta\omega))|^2 \approx \left|\frac{1}{2\Delta\omega C_p}\right|^2 = \left|\frac{\omega_0^2 L_p}{2\Delta\omega}\right|^2 = \frac{R_p^2}{4}\frac{1}{Q^2}\left(\frac{\omega_0}{\Delta\omega}\right)^2,$$

where $Q = \dfrac{R_p}{L_p \omega_0}$ is the resonator quality factor. Given that we are still dealing with an LTI system, the power spectral density (PSD) of the voltage noise around the oscillation frequency is therefore simply $\overline{v_n^2} = \overline{i_n^2}|H(j(\omega_0 + \Delta\omega))|^2$. Now, as discussed in Chapter 6 and shown in

514 | **Oscillators**

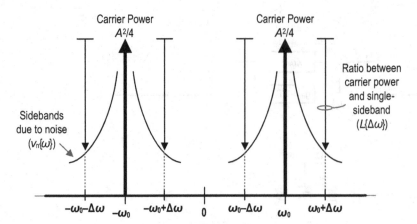

Figure 9.4: Phase noise is the single-sideband noise power normalized to the carrier power. The figure shows the double-sided frequency spectrum, but the phase noise definition is valid whether a single-sided or double-sided frequency definition of power is used

Figure 9.4, the definition of phase noise is this voltage spectral density normalized to carrier power, i.e.,

$$\mathcal{L}\{\Delta\omega\} = \frac{\overline{v_n^2}(\omega_0 + \Delta\omega)}{A^2/4},$$

where A is the peak oscillation amplitude.[2] Generally, this is quoted and/or plotted in unit of decibels below carrier per hertz [dBc/Hz], i.e.,

$$\mathcal{L}\{\Delta\omega\}_{\text{dBc/Hz}} = 10\log_{10}\mathcal{L}\{\Delta\omega\}.$$

In performing this normalization in the context of a linear oscillator, two *fudges* are employed to account for fact we are modeling a real nonlinear oscillator with a linear abstraction:

1. We divide $\overline{i_n^2}$ by 2 before the normalization to account for the fact that in a nonlinear oscillator generally half the noise is rejected because it is amplitude noise, i.e., $\overline{v_n^2} = \left(\overline{i_n^2}/2\right)|H(j(\omega_0 + \Delta\omega))|^2$. The reasoning behind this maneuver will be fully detailed in a later section, but at his point it is sufficient to assume that roughly half of the noise modulates the phase of the carrier, while the other half will modulate the amplitude of the carrier. In practical oscillators, amplitude noise will be heavily suppressed by the nonlinear nature of the circuit, while phase noise will persist.
2. As the large signal dynamics of the nonlinearity are unknown, we cannot (at this stage) quantify the amount of noise that the nonlinearity itself injects. We will therefore represent

---

[2] All spectral density terms used in this chapter (unlike previous chapters) are defined on a double-sided frequency axis, *not* a single-sided frequency axis. For the simple case of the spectral density of the voltage noise of resistor, R, the definition we use is $S_v = S_v^{DS} = 2kTR$ rather than $S_v^{SS} = 4kTR$.

Furthermore, as the noise spectral density, $\overline{v_n^2}$, is defined on the doubled-sided frequency spectrum, it is normalized to $A^2/4$ rather than the RMS power $A^2/2$. See Figure 9.4 for further clarification.

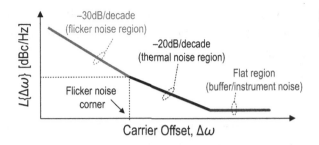

Figure 9.5: Typical phase noise profile (where carrier offset is plotted on a log scale)

the injected total current noise[3] as $\overline{i_n^2} = \overline{i_{Rp}^2} + \overline{i_{gm}^2} = 2kTF/R_p$, where the unknown constant F, which we will term "oscillator noise factor," accounts for the uncertainty in the injected noise. If the energy-restoring circuit is noiseless, $F = 1$ (much like the definition for LNA noise factor).

Given these adjustments, which will be justified when we study a nonlinear oscillator, the phase noise of the linear oscillator can be expressed as

$$\mathcal{L}\{\Delta\omega\} = \left(\frac{\overline{i_n^2}}{2}\right) \frac{|H(j(\omega_0 + \Delta\omega))|^2}{A^2/4} = \frac{kTFR_p}{A^2} \frac{1}{Q^2} \left(\frac{\omega_0}{\Delta\omega}\right)^2,$$

which is Leeson's well-known expression for phase noise.

When $\mathcal{L}\{\Delta\omega\}_{\text{dBc/Hz}}$ is viewed on a spectrum analyzer (see Figure 9.5), three distinct regions will generally be seen: a -20dB/decade region due to thermal noise in the system; a -30dB/decade region due to flicker noise in the system; and a flat region far from the carrier that is not due to the oscillator, but is attributable to either buffer noise or test equipment noise. The above expression qualitatively predicts the -20dB/decade region, but it does not explain the -30dB/decade region. This latter region can be explained only using more sophisticated nonlinear analysis, which will be presented later in this chapter.

9.1.3 Efficiency

In order to deliver real power to the tank, the active circuit must take it from a power source, which is typically the DC supply rail. Especially in wireless applications, it is important to inject energy into the tank in the most efficient manner possible. In the case of an oscillator, this power efficiency is defined as

$$\eta = \frac{\overline{P_{tank}(t)}}{P_{DC}} = \frac{A^2/(2R_p)}{V_{DD}I_{DD}},$$

where $\overline{P_{tank}(t)}$ is the average power burned in tank over one oscillation period, and P_{DC} is the DC power drawn from the supply. Since there are no energy sources in the oscillator circuitry

---

[3] Note that since this chapter assumes a doubled-sided frequency axis, $\overline{i_n^2} = 2kTF/R_p$ and not $\overline{i_n^2} = 4kTF/R_p$.

other than the supply, power efficiency cannot exceed 1 (or a 100%). Ideally all the DC power is burned in the tank (compensating for the losses), and *no* power is burned in the active circuit.

Power efficiency can also be recast in the following form,

$$\eta = \frac{1}{2}\left(\frac{I_{tank@\omega 0}}{I_{DD}}\right)\left(\frac{A}{V_{DD}}\right) = \eta_I \frac{A}{V_{DD}},$$

where η_I is the active circuit's current efficiency [2]. It is a measure of how effectively the supply current is injected into the tank at ω_0 and is defined as

$$\eta_I = \frac{I_{tank@\omega 0}}{2I_{DD}},$$

where $I_{tank@\omega 0} = A/R_p$.

9.1.4 Oscillator Figure of Merit

It is common to benchmark various LC oscillator designs using the following figure of merit [3],

$$\text{FOM} = \frac{\left(\frac{\omega_0}{\Delta\omega}\right)^2}{(P_{DC}/1\text{mW})\,\mathcal{L}\{\Delta\omega\}},$$

which is essentially the *phase noise per unit power normalized to oscillator frequency and carrier offset*. In general, FOM varies with carrier offset (see Figure 9.6), but when reported as a single number it is assumed that FOM was calculated using measurements from the thermal noise region where the FOM plateaus. We will follow this convention. To gain some insight we substitute Leeson's expression in the above definition for FOM and recast in the following form,

$$\text{FOM} = \left(\frac{\eta Q^2}{F}\right)\left(\frac{2}{kT}\right)10^{-3},$$

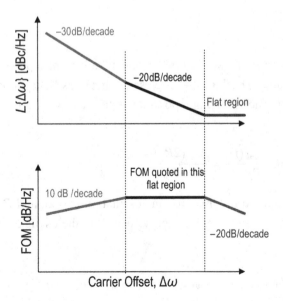

Figure 9.6: The oscillator figure of merit is a function of carrier offset frequency. Typically, the FOM in the thermal noise region is quoted

which demonstrates that FOM depends on only three variables: the quality factor of the LC tank (Q), the efficiency of the oscillator (η), and the oscillator's noise factor (F).

Maximizing Q (or, equivalently minimizing tank losses) is the single best way to improve the performance of any LC oscillator. However, Q is technology dependent and largely outside the control of an RF circuit designer. Moreover, it is independent of a specific oscillator topology, and therefore, when assessing a given architecture we can say that a good architecture is simply one that maximizes efficiency and minimizes noise factor. That is, the *best* topology is one that maximizes η/F. By definition the maximum value of η/F is 1, which would correspond to a noiseless, 100% efficient active element [4].

9.2 THE NONLINEAR LC OSCILLATOR

At this point the astute reader has many unanswered questions: What is the value of the noise factor F? Why can we disregard amplitude noise? How can we predict the oscillation amplitude? How is the −30dB/decade region explained? To answer these and other important questions, we need to analyze the LC oscillator as a nonlinear, time-varying circuit.

9.2.1 Intuitive Understanding

Shown in Figure 9.7 is a nonlinear LC oscillator with a tank loss of $R_p = 100\Omega$. Compared to Figure 9.1, the only difference is that energy-restoring-circuit[4] is drawn as an explicit nonlinear conductance, whose instantaneous conductance is some function of the voltage across its terminals. We assume the energy-restoring-circuit is memoryless, which is to say it contains no reactive components (i.e., no inductors, capacitors, or any other energy-storage devices).

As deduced from the I–V characteristic of the negative resistance (also shown in Figure 9.7), the conductance is −20mS around $v_{nr} = 0$ and, so, given that the tank resistance is 100Ω, the oscillator will start up. However, as also shown in Figure 9.7, the waveform will eventually self-limit. Plotting both $g_{nr}(t) = di_{nr}(t)/dv_{nr}(t)$ and $v_{osc}(t)$, we see that during start-up $g_{nr}(t) < -1/R_p$ and consequently we see a growing oscillation, but when the amplitude stabilizes $g_{nr}(t)$ is less than $-1/R_p$ for only a small portion of the oscillation. This is because of the compressive nature of the negative conductance, since it cannot provide an arbitrarily large amount of current when presented with an arbitrarily large voltage. As all circuits will eventually compress if driven hard enough, this is the mechanism through which the oscillation waveform self-limits.

As a thought experiment, suppose we separately drive the negative resistance and the tank with a sinusoid whose frequency is the same as the tank's resonant frequency (Figure 9.8). In this setup, the amount of power dissipated in the tank and the amount of power provided by the active circuit will be a function of the amplitude of the applied sinusoid. Notably, for amplitudes less than 0.57V, the active resistor will provide more power than the tank dissipates, while for

---

[4] Depending on the context, the oscillator's energy-restoring circuit is referred to as a negative resistance, a negative conductance, a nonlinear resistance, a nonlinear conductance, an active element, and/or an active circuit. These terms are essentially synonymous.

518 Oscillators

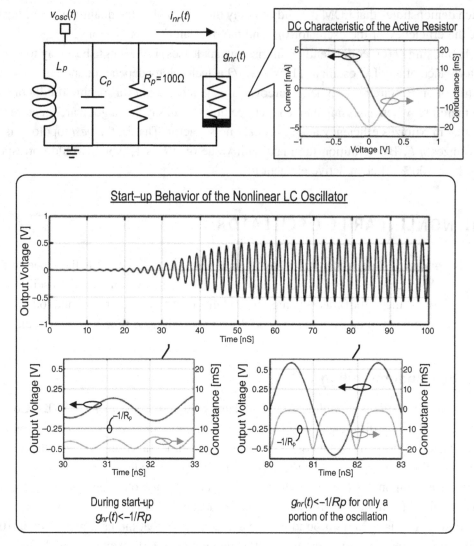

Figure 9.7: The nonlinear LC oscillator

amplitudes greater than 0.57V, the tank will dissipate more power than the circuit can provide.[5] Accordingly, in the oscillator of Figure 9.7, the output waveform will grow if the amplitude is less than 0.57V, while it will decay if the amplitude is greater than 0.57V. As a result, the output amplitude settles to 0.57V because it is the only stable large signal operating point. Even if a disturbance (noise or otherwise) changes the instantaneous amplitude, it will eventually return to this operating point. Therefore, unlike the linear oscillator, the oscillation amplitude of a non-linear oscillator is independent of the initial energy stored in the tank.

In the next section, a mathematical description of this power balance is used to derive the "effective" or "large signal" conductance provided by the active element. This effective conductance, defined as $G_{\textit{eff}}$, is equal to the negative of the tank loss conductance when sustained oscillation occurs (as also shown in Figure 9.8).

---

[5] The balance point (in this case 0.57V) depends on the nonlinearity used and the tank loss.

9.2 The Nonlinear LC Oscillator

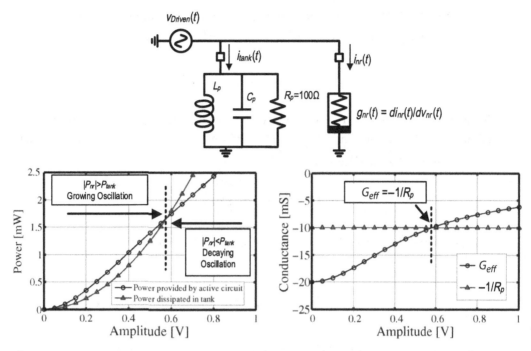

Figure 9.8: The power dissipated and provided in an LC oscillator

9.2.2 Power Conservation Requirements

Assuming the oscillator has started up, the average power delivered to the resonator must be matched by the average power dissipated in the tank, i.e., $<P_{nr}(t)> = - <P_{Rp}(t)>$, or in terms of currents and voltages,

$$\int_0^T v_{osc}(t) i_{nr}(t) dt = - \int_0^T v_{osc}(t) i_{Rp}(t) dt,$$

where $v_{osc}(t)$ is the output voltage, $i_{nr}(t)$ is the current flowing into the active element, $i_{Rp}(t)$ is the current flowing into the tank resistance, and T is the oscillation period. If this power conservation requirement was not satisfied, we would observe a growing or decaying oscillation. The frequency domain representation[6] of this identity is expressed as

$$\sum_{k=-\infty}^{\infty} V_{osc}[k] I_{nr}[-k] = - \sum_{k=-\infty}^{\infty} V_{osc}[k] I_{Rp}[-k],$$

where $V_{osc}[k]$, $I_{nr}[k]$, $I_{Rp}[k]$ respectively define the Fourier coefficients of $v_{osc}(t)$, $i_{nr}(t)$, and $i_{Rp}(t)$. We assume a nearly sinusoidal output,[7] that is, $v_{osc}(t) = A\cos\omega t$, which is a reasonable

---

[6] The Fourier series coefficients used throughout this chapter are defined with respect to the double-sided frequency spectrum.

[7] Throughout this chapter, we assume the output waveform has no initial phase offset, i.e., $v_{osc}(t) = A\cos(\omega t + \phi)$, where $\phi = 0$. This is to simplify the math and to make it more readable. Essentially, we are selecting the time reference $t = 0$ to coincide with an oscillation peak. Moving the time reference will have no effect on the final noise result.

assumption for LC oscillators with moderate Q values. This implies that all values of $V_{osc}[k]$ are filtered when $k \neq \pm 1$, and $V_{osc}[-1] = V_{osc}[1] = A/2$, which allows us to write

$$I_{nr}[-1] + I_{nr}[1] = -\frac{2V_{osc}[1]}{R_p} = -\frac{A}{R_p}.$$

Note that although $v_{osc}(t)$ is assumed to be nearly sinusoidal, $i_{nr}(t)$ can and typically does contain significant harmonic content. Now, if we define the instantaneous conductance of the active resistance as $g_{nr}(t) = di_{nr}(t)/dv_{nr}(t)$, we can write $i_{nr}(t)$ as

$$i_{nr}(t) = I_{nr(DC)} + \int \left(g_{nr}(t) \frac{dv_{nr}(t)}{dt} \right) dt,$$

and, therefore, the Fourier's series of $i_{nr}(t)$ can be calculated as

$$I_{nr}[k] = \begin{cases} I_{nr(DC)}, & |k=0 \\ \sum_{l=-\infty}^{\infty} \left(\frac{l}{k}\right) V_{nr}[l] G_{nr}[k-l], & |k \neq 0 \end{cases}.$$

Again, assuming a nearly sinusoidal output, $v_{osc}(t) = v_{nr}(t) = A \cos \omega t$, this condition implies that for $k \neq 0$,

$$I_{nr}[k] = \left(\frac{A}{2k}\right)(G_{nr}[k-1] - G_{nr}[k+1]),$$

which can be used along with the result of the power conservation requirement (i.e., $I_{nr}[-1] + I_{nr}[1] = -A/R_p$) to define the effective conductance [5], [6] of the active element:

$$G_{eff} = G_{nr}[0] - \frac{1}{2} G_{nr}[-2] - \frac{1}{2} G_{nr}[2] = -\frac{1}{R_p}.$$

This is the same effective conductance that is plotted in Figure 9.8, and must always equal $-1/R_p$ under steady state conditions. Note that the effective transconductance is not a *time-averaged* transconductance, which would simply be $G_{nr}[0]$. If we assume the nonlinear conductance is memoryless, the instantaneous conductance will be symmetric around $t = 0$ (i.e., $g_{nr}(t) = g_{nr}(-t)$). This symmetry is because the output waveform, which excites the memoryless conductance, is defined to be symmetric around $t = 0$ (i.e., $v_{nr}(t) = v_{nr}(-t)$). This means that $G_{nr}[-2] = G_{nr}[2]$ and the effective conductance simplifies to

$$G_{eff} = G_{nr}[0] - G_{nr}[2] = -\frac{1}{R_p}.$$

Note that for a sustained oscillation to occur in a *linear* oscillator, the small signal conductance must equal the tank loss (i.e., $g_m = -1/R_p$ in Figure 9.2), but this is a marginally stable state that is impossible to achieve in practice. By contrast, in a *nonlinear* oscillator, a sustained oscillation implies that the effective or *large signal* conductance balances the tank loss: $G_{eff} = -1/R_p$.

9.2.3 Oscillation Amplitude

Given the analysis in the previous section, the oscillation amplitude can be calculated as the tank resistance times the fundamental of the current provided by the active element, i.e.,

$$A = 2V_{osc}[1] = -(I_{nr}[-1] + I_{nr}[1])R_p.$$

Depending on the I–V characteristic of a given nonlinearity $I_{nr}[\pm 1]$ (and thus A) may or may not be straightforward to calculate.

9.3 PHASE NOISE ANALYSIS OF THE NONLINEAR LC OSCILLATOR

In order to make the noise analysis of a nonlinear oscillator manageable, we are going to analyze the noise mechanisms much like we analyze noise in a mixer,[8] that is:

1. We will initially assume the circuit is completely noiseless, and solve for frequency, amplitude, operating point, etc.
2. We will then inject *small* noise sources into circuit, and note its effect on the output spectrum. We assume that "small" noise sources result in *small* perturbations in the output spectrum.
3. Finally, we will treat the oscillator as an LTV system (with respect to noise) and sum the phase perturbations from each of the noise sources in the system.

Despite these simplifications, such an approach can adequately quantify phase noise in the thermal noise region of $\mathcal{L}\{\Delta\omega\}$ (i.e., the –20dB/decade region). However, as this approach defines a fixed frequency with respect to a noiseless oscillator, it cannot quantify noise in the spectrum that occurs because of frequency modulation effects. The -30dB/decade region in $\mathcal{L}\{\Delta\omega\}$ can be fully explained only using these effects. Therefore, separate analysis methods that account for frequency modulation effects will be discussed in a later section.

We start with some basic definitions of phase, frequency, and amplitude noise and present simple ways of visualizing them.

9.3.1 Defining Phase, Frequency, and Amplitude Noise

Consider an amplitude, frequency, and phase modulated continuous sinusoid,

$$\begin{aligned}v_{osc}(t) &= A(1+m(t))\,\cos\left(\int_0^t (\omega_0 + \omega_\Delta(\tau))d\tau + \phi(t)\right) \\ &= A(1+m(t))\,\cos\left(\omega_0 t + \int_0^t \omega_\Delta(\tau)d\tau + \phi(t)\right),\end{aligned}$$

---

[8] The analysis approach presented in this chapter is based on [5], [6], [8], [9], [32].

where $m(t)$ modulates amplitude, $\omega_\Delta(\tau)$ modulates frequency, and $\phi(t)$ modulates phase. Assuming small modulation indexes, i.e., $m(t) \ll 1$, $\omega_\Delta(\tau) \ll \omega_0$, and $\phi(t) \ll 2\pi$, this can be rewritten using the small angle approximation as

$$v_{osc}(t) \approx A\cos\omega_0 t + m(t) A\cos\omega_0 t - \left(\phi(t) + \int_0^t \omega_\Delta(\tau)d\tau\right) A\sin\omega_0 t,$$

which in the frequency domain is

$$V_{osc}(\omega) \approx A\pi\delta(\omega - \omega_0) + A\pi\delta(\omega + \omega_0) + \left(\frac{A}{2}M(\omega - \omega_0) + \frac{A}{2}M(\omega + \omega_0)\right)$$
$$- \left(\frac{A}{2j}\Phi(\omega - \omega_0) - \frac{A}{2j}\Phi(\omega + \omega_0)\right) - \left(-\frac{A}{2(\omega - \omega_0)}\Omega(\omega - \omega_0) + \frac{A}{2(\omega + \omega_0)}\Omega(\omega + \omega_0)\right),$$

where $M(\omega)$, $\Omega(\omega)$, and $\Phi(\omega)$ are the Fourier transforms of $m(t)$, $\omega_\Delta(\tau)$, and $\phi(t)$, respectively. We assume $m(t)$, $\omega_\Delta(\tau)$, and $\phi(t)$ vary at a rate much less than ω_0. If each of these terms are noise sources, the spectral density of noise around the carrier at an offset of $\Delta\omega$ is

$$S_N(\omega_0 + \Delta\omega) = S_{PM}(\omega_0 + \Delta\omega) + S_{FM}(\omega_0 + \Delta\omega) + S_{AM}(\omega_0 + \Delta\omega),$$

which is the sum of the noise around the carrier that is due to phase fluctuations, frequency fluctuations, and amplitude fluctuation (respectively). These terms are defined as

$$S_{PM}(\omega_0 + \Delta\omega) = \frac{A^2}{4} S_\phi(\Delta\omega)$$

$$S_{FM}(\omega_0 + \Delta\omega) = \frac{A^2}{4} \frac{S_\Omega(\Delta\omega)}{|\Delta\omega|^2}$$

$$S_{AM}(\omega_0 + \Delta\omega) = \frac{A^2}{4} S_M\{\Delta\omega\}.$$

The total noise power, $S_N(\omega_0 + \Delta\omega)$, normalized to the total carrier power, is the definition of phase noise, i.e.,

$$\mathcal{L}\{\Delta\omega\} = \frac{S_N(\omega_0 + \Delta\omega)}{A^2/4} = S_\phi(\Delta\omega) + S_M(\Delta\omega) + \frac{S_\Omega(\Delta\omega)}{|\Delta\omega|^2}.$$

Therefore this definition for phase noise, i.e., single-sideband noise normalized to carrier power, is somewhat of misnomer as it includes noise that causes perturbations in phase, frequency, and amplitude. However, as we will soon discover, in practical LC oscillators, it is generally the case that $S_M(\Delta\omega)$ is small and can be neglected, and so the term can be approximated as

$$\mathcal{L}\{\Delta\omega\} \approx S_\phi(\Delta\omega) + \frac{S_\Omega(\Delta\omega)}{|\Delta\omega|^2}.$$

Moreover, as demonstrated in the next section, we cannot distinguish between frequency noise and phase noise. Therefore, if $S_\Omega(\Delta\omega)/|\Delta\omega|^2$ exists at all, it is considered as a part of $S_\phi(\Delta\omega)$. Consequently, it is common to write[9]

$$\mathcal{L}\{\Delta\omega\} \approx S_\phi(\Delta\omega).$$

9.3.2 Similarity of FM and PM Noise

To understand why we cannot distinguish between FM and PM noise, consider that a stationary noise source can be modeled as an infinite sum of sinusoids that are uncorrelated in phase and separated in frequency by 1Hz [7],

$$n(t) = \sum_{k=0}^{\infty} \sqrt{4S_n(2\pi k)} \cos(2\pi k t + \theta_k),$$

where $S_n(2\pi k)$ is the spectral density per unit hertz of $n(t)$ at frequency of k Hz, and θ_k is the uncorrelated random phase offset of each sinusoid. This was proven in Chapter 2 in the context of randomly phased sinusoid. Also in Chapter 6, it was exploited to calculate the spectral density of band limited white noise squared.

Now consider a sinusoid, whose frequency is modulated by a noise source,

$$v_{FM}(t) \approx A \cos \omega_0 t - \left(\int_0^t \omega_\Delta(\tau)d\tau\right) A \sin \omega_0 t,$$

where

$$\omega_\Delta(\tau) = \sum_{k=0}^{\infty} \sqrt{4S_\Omega(2\pi k)} \cos(2\pi k t + \theta_k).$$

Calculating the integral gives us

$$v_{FM}(t) \approx A \cos \omega_0 t - \left(\sum_{k=0}^{\infty} \frac{\sqrt{4S_\Omega(2\pi k)}}{2\pi k} \cos(2\pi k t - \pi/2 + \theta_k)\right) A \sin \omega_0 t.$$

Now consider a sinusoid, whose phase is modulated by a noise source, i.e.,

$$v_{PM}(t) \approx A \cos \omega_0 t - \left(\sum_{k=0}^{\infty} \sqrt{4S_\phi(2\pi k)} \cos(2\pi k t + \theta_k)\right) A \sin \omega_0 t.$$

As $\theta_{[k=0,1,2,\ldots]}$ is a set of random uncorrelated phases, we note that a noise source with spectral $(2\pi k)^2 S_x(2\pi k)$ that modulates frequency will have exactly the same effect on the output

---

[9] In some books, a factor of ½ relates $S_\phi(\Delta\omega)$ to $\mathcal{L}\{\Delta\omega\}$. However, as we have defined $S_\phi(\Delta\omega)$ on the double-sided frequency axis, that term does not appear, in other words, $\mathcal{L}\{\Delta\omega\} = S_\phi^{DS}(\Delta\omega) = S_\phi^{SS}(\Delta\omega)/2$.
   When viewed/measured on a spectrum analyzer $\mathcal{L}\{\Delta\omega\}$ is single-sideband noise normalized to carrier power. However, the IEEE definition of $\mathcal{L}\{\Delta\omega\}$ is $\mathcal{L}\{\Delta\omega\} = S_\phi^{DS}(\Delta\omega) = S_\phi^{SS}(\Delta\omega)/2$, which ignores amplitude noise even if it is present.

spectrum as a noise of spectral $S_x(2\pi k)$ that modulates phase. Therefore, we cannot distinguish between the two by observing the output, just as was the case for FM and PM signals discussed in Chapter 2. Moreover, a nonlinear element will respond to frequency noise and phase noise in exactly the same way, and therefore we can treat frequency noise and phase noise as one and the same.

9.3.3 Recognizing AM and PM Sidebands

A small source that modulates the amplitude or phase of a carrier generates correlated sidebands around the carrier. If the modulation index is small, it is not possible to know whether the sidebands are the result of amplitude or phase modulation by simply looking at the sideband magnitude. However, by observing the relative phase of the sidebands with to respect to the carrier, we are able to make this distinction [7], [8]. Consider an arbitrary PM source given by

$$\phi(t) = \Delta\phi \sin(\Delta\omega t + \varphi)$$

that modulates an arbitrary carrier, resulting in

$$v(t) = A \cos(\omega_0 t + \phi(t) + \theta).$$

Assuming $\Delta\phi \ll 2\pi$, this can be approximated as

$$v(t) \approx A \cos(\omega_0 t + \theta) - \Delta\phi \sin(\Delta\omega t + \varphi) A \sin(\omega_0 t + \theta),$$

and can be written in the form

$$v(t) \approx v_c(t) + v_{LSB}(t) + v_{USB}(t),$$

where the unmodulated carrier, the lower sideband (LSB) and the upper sideband (USB) are given, respectively, by

$$v_c(t) = A \cos(\omega_0 t + \theta)$$

$$v_{LSB}(t) = \frac{\Delta\phi A}{2} \cos((\omega_0 - \Delta\omega)t + \pi + \theta - \varphi)$$

$$v_{USB}(t) = \frac{\Delta\phi A}{2} \cos((\omega_0 + \Delta\omega)t + \theta + \varphi).$$

Note that the magnitudes LSB and USB are equal (i.e., $\Delta\phi A/2$), while the mean of the phase of LSB and USB is 90° offset from the phase of the carrier. This is visualized in the phasor plot shown in Figure 9.9.

Now consider an arbitrary AM carrier given by

$$m(t) = \Delta m \sin(\Delta\omega t + \varphi)$$

$$v(t) = A(1 + m(t)) \cos(\omega_0 t + \theta),$$

which can be again written in the form

$$v(t) \approx v_c(t) + v_{LSB}(t) + v_{USB}(t).$$

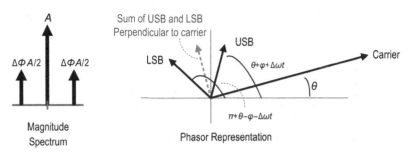

Figure 9.9: The sum of the PM sidebands will always be perpendicular to the carrier

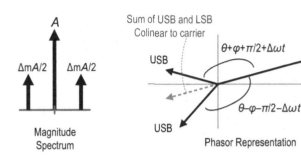

Figure 9.10: The sum of the AM sidebands will always be colinear (i.e., in phase or 180° out of phase) to the carrier

The unmodulated carrier, LSB, and USB are given by

$$v_c(t) = A\cos(\omega_0 t + \theta)$$

$$v_{LSB}(t) = \frac{\Delta m A}{2} \cos\left((\omega_0 - \Delta\omega)t + \frac{\pi}{2} + \theta - \varphi\right)$$

$$v_{USB}(t) = \frac{\Delta m A}{2} \cos\left((\omega_0 + \Delta\omega)t - \frac{\pi}{2} + \theta + \varphi\right).$$

Again the amplitude of the sidebands are the same, but unlike phase modulation, in amplitude modulation the mean of the phase of the USB and LSB is either in phase or 180° out of phase with the carrier. This is visualized in Figure 9.10.

9.3.4 Decomposing an SSB into AM and PM Sidebands

It is possible to represent a small uncorrelated sideband as a set of small correlated AM and PM sidebands each with half the magnitude of the original sideband (as shown in Figure 9.11) [7], [9]. This decomposition is useful when analyzing the response of a nonlinear circuit to a modulated waveform because, as we will see, PM and AM sidebands respond differently when passed through a nonlinear circuit. For example, if we pass a modulated waveform through a hard-limiting nonlinearity the PM sideband-to-carrier will be preserved, but the AM sideband-to-carrier will be removed. Therefore, as shown in Figure 9.12, passing a sinusoid modulated by a single-sideband through a limiter results in a sinusoid with phase modulated sidebands at the

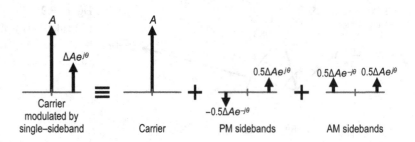

Figure 9.11: Representing a single band as the sum of PM and AM sidebands

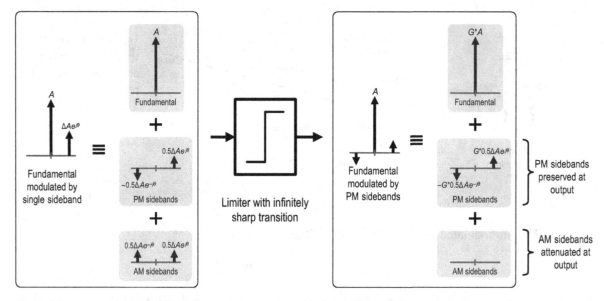

Figure 9.12: PM and AM sidebands response to a nonlinearity. In general, PM sidebands experience the same gain as the carriers, while AM sidebands are attenuated (assuming a compressive nonlinearity).

output, which are half the magnitude of the original single-sideband, while the AM sidebands will be removed.

To show this decomposition, consider the simple case of a continuous sinusoid, $v_c(t) = A \cos \omega_0 t$, modulated by a very small sideband $v_{SSB}(t) = \Delta A \cos((\omega_0 + \Delta \omega)t + \theta)$. This can be written in the form

$$v(t) = v_c(t) + v_{SSB}(t)$$
$$= A \cos \omega_0 t + \Delta A \cos((\omega_0 + \Delta \omega)t + \theta)$$
$$= v_c(t) + v_{PM}(t) + v_{AM}(t),$$

where

$$v_{PM}(t) = \frac{\Delta A}{2} \cos((\omega_0 + \Delta \omega)t + \theta) - \frac{\Delta A}{2} \cos((\omega_0 - \Delta \omega)t - \theta)$$

$$v_{AM}(t) = \frac{\Delta A}{2} \cos((\omega_0 + \Delta \omega)t + \theta) + \frac{\Delta A}{2} \cos((\omega_0 - \Delta \omega)t - \theta).$$

To reconcile the above decomposition with the definition of a phase and amplitude modulated sinusoid described previously, we can write $v(t)$ in the form of

$$v(t) = A\cos(\omega_0 t) + m(t)A\cos\omega_0 t - \phi(t)A\sin\omega_0 t,$$

where the modulation components are given by

$$\phi(t) = \frac{\Delta A}{A}\sin(\Delta\omega t + \theta)$$

$$m(t) = \frac{\Delta A}{A}\cos(\Delta\omega t + \theta).$$

As stated previously, noise can be modeled as an infinite sum of sinusoids (or in this context an infinite sum of small sidebands). Therefore in the case of additive stationary noise where a noise source $n(t)$ with a PSD of $S_n(\omega)$ is simply added to a continuous sinusoid, $v(t) = A\cos\omega_0 t$, the decomposition reveals that half of the additive noise will modulate the phase of the carrier, while the other half will modulate the amplitude:

$$S_{PM}(\omega_0 + \Delta\omega) = S_{AM}(\omega_0 + \Delta\omega) = \frac{1}{2}S_n(\omega_0 + \Delta\omega).$$

Unfortunately not all noise sources in an oscillator are stationary, as discussed next.

9.3.5 Cyclostationary Noise

In an oscillator, the noise injected by the parasitic tank resistance, R_p, is a stationary noise source (with $S_v(\omega) = 2kTR_p$), but this is typically not the case for noise sources associated with the active element. The spectral density of such sources will normally vary periodically because of the changing operating points of the transistors used in the circuit [10]. As stated in Chapter 5, this type of periodically varying noise source is called a cyclostationary noise source and can be defined as

$$n_{CYCLO}(t) = n(t)w(t),$$

where $n(t)$ is a stationary noise source, and $w(t)$ is a periodic waveform, which will be referred to as a *noise-shaping waveform*. The autocorrelation and spectral density of cyclostationary noise were calculated in Chapter 5. Given their importance to oscillator noise discussion, we shall present a summary here.

The autocorrelation of the cyclostationary noise, was found to be

$$R_{n_{CYCLO}}(t,\tau) = R_n(\tau)w(t)w^*(t+\tau),$$

where $R_n(\tau)$ is the autocorrelation of $n(t)$. This can be expanded as

$$R_{n_{CYCLO}}(t,\tau) = R_n(\tau)\sum_{k=-\infty}^{\infty}\sum_{m=-\infty}^{\infty}W[k]W^*[m]e^{j(k-m)\omega_0 t}e^{-jm\omega_0\tau},$$

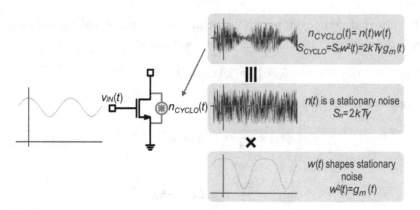

Figure 9.13: The output noise of a transistor is cyclostationary when the gate is driven by a periodic waveform

where $W[k]$ are the Fourier coefficients[10] of the periodic noise shaping waveform $w(t)$. If $R_{n_{CYCLO}}(t, \tau)$ is time-averaged, the time (t) dependence is removed, and the spectral density is found to be

$$\overline{S_{n_{CYCLO}}}(\omega) = \sum_{k=-\infty}^{\infty} |W[k]|^2 S_n(\omega - k\omega_0).$$

For the case of white cyclostationary noise, given the Parseval's energy relation, the above expression was shown to simplify as follows,

$$S_{n_{CYCLO}}(\omega, t) = S_n |w(t)|^2,$$

where S_n is the white noise spectral density and is frequency independent. Figure 9.13 shows a practical example of a transistor with a time-varying gate voltage. In this case the noise current can be modeled as a stationary current source with spectral density $S_n = 2kT\gamma$, modulated by the *noise-shaping waveform* $w(t) = \sqrt{g_m(t)}$. Given this, the PSD is

$$S_{n_{CYCLO}}(\omega, t) = S_n w^2(t) = S_n g_m(t) = 2kT\gamma g_m(t),$$

which, of course, varies periodically with time. When viewed on a spectrum analyzer the PSD displayed will be a time-average of $S_{n_{CYCLO}}(\omega, t)$, i.e.,

$$\overline{S_{n_{CYCLO}}}(\omega) = 2kT\gamma \overline{g_m(t)}.$$

A stationary noise source implies that the noise voltages at two distinct frequencies are completely uncorrelated.[11] However, as a cyclostationary noise is modeled as a stationary source modulated with a periodic waveform, its spectrum can be highly correlated. For example, low-frequency noise when mixed with a square waveform generates correlated noise around harmonics of the square waveform. In the context of oscillator noise analysis, we need to carefully account for this correlation in order to distinguish between noise that modulates the

---

[10] As in this chapter we deal with several variables with complex notations, we use $W[k]$ to represent the Fourier series coefficients, as opposed to w_k used previously.

[11] This is not to be confused with colored noise or noise that is *shaped* in the frequency domain. Colored noise is uncorrelated in the frequency domain, but correlated in the time domain. Stationary noise, which may or may not be colored, is always uncorrelated in the frequency domain.

phase of a carrier, and noise that modulates the amplitude of a carrier. The previous section dealt with the straightforward case of stationary noise. To make this distinction for the more general case of cyclostationary noise, consider again the example of an amplitude and phase modulated continuous sinusoid. Assuming that $m(t) \ll 1$, and $\phi(t) \ll 2\pi$, we can write

$$v_{osc}(t) \approx A\cos\omega_0 t + m(t)A\cos\omega_0 t - \phi(t)A\sin\omega_0 t.$$

We also assume that both $m(t)$ and $\phi(t)$ are varying at rate much less than ω_0. Let us model the modulation terms (the second and third terms in the expression) as a cyclostationary noise source, i.e.,

$$n_{CYCLO}(t) = n(t)w(t) = Am(t)\cos\omega_0 t - A\phi(t)\sin\omega_0 t,$$

where $n(t)$ is a stationary noise source and $w(t)$ is an arbitrary *noise-shaping waveform*. Consequently, the PM/AM components can be retrieved as follows:

$$\phi(t) = \left\langle -\frac{2}{A}n(t)w(t)\sin\omega_0 t \right\rangle$$

$$m(t) = \left\langle \frac{2}{A}n(t)w(t)\cos\omega_0 t \right\rangle.$$

The function $\langle \cdot \rangle$ retains only low-frequency components. To find the spectral density of the AM and PM components, let us take an intermediate step first, and define the following two variables,

$$\tilde{\phi}(t) = -\frac{2}{A}n(t)w(t)\sin\omega_0 t$$

$$\tilde{m}(t) = \frac{2}{A}n(t)w(t)\cos\omega_0 t,$$

which are the *unfiltered* versions of $\phi(t)$ and $m(t)$. This is conceptually shown in Figure 9.14.

Both $\tilde{\phi}(t)$ and $\tilde{m}(t)$ can be understood as cyclostationary noise where the *noise-shaping waveforms* are, respectively, $w_{\tilde{\phi}}(t) = (-2/A)w(t)\sin\omega_0 t$ and $w_{\tilde{m}}(t) = (2/A)w(t)\cos\omega_0 t$. Now, as $\tilde{\phi}(t)$ and $\tilde{m}(t)$ represent noise, we can calculate the PSD of the phase noise and amplitude noise as

$$S_{\tilde{\phi}}(\omega) = \sum_{k=-\infty}^{\infty} \left| W_{\tilde{\phi}}[k] \right|^2 S_n(\omega - k\omega_0) = \frac{1}{A^2}\sum_{k=-\infty}^{\infty} |W[k-1] - W[k+1]|^2 S_n(\omega - k\omega_0)$$

$$S_{\tilde{m}}(\omega) = \sum_{k=-\infty}^{\infty} |W_{\tilde{m}}[k]|^2 S_n(\omega - k\omega_0) = \frac{1}{A^2}\sum_{k=-\infty}^{\infty} |W[k-1] + W[k+1]|^2 S_n(\omega - k\omega_0),$$

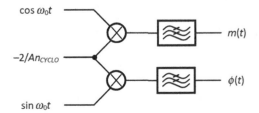

Figure 9.14: Obtaining AM and PM sidebands

where $W[k]$, $W_{\tilde{\phi}}[k]$, and $W_{\tilde{m}}[k]$ respectively define the Fourier coefficients of $w(t)$, $w_{\tilde{\phi}}(t)$, and $w_{\tilde{m}}(t)$, and S_n is the PSD of $n(t)$. As we are concerned only with noise around the carrier, evident from the lowpass filtering shown in Figure 9.14, these terms assuming a small carrier offset lead to

$$S_\phi(\Delta\omega) = \frac{1}{A^2} \sum_{k=-\infty}^{\infty} S_n(\Delta\omega - k\omega_0)|W[k-1] - W[k+1]|^2$$

$$S_M(\Delta\omega) = \frac{1}{A^2} \sum_{k=-\infty}^{\infty} S_n(\Delta\omega - k\omega_0)|W[k-1] + W[k+1]|^2.$$

The folding involved in the generation of phase noise is visualized in Figure 9.15, while the folding involved in the generation of amplitude noise is visualized in Figure 9.16. Note that stationary noise is a special case of the above analysis when the *noise-shaping waveform* is unity, i.e., $w(t) = 1$.

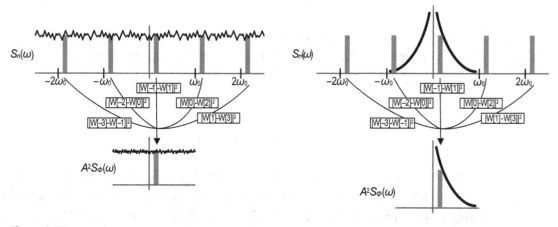

Figure 9.15: A stationary noise modulated by periodic waveform $w(t)$ results in phase noise when reference to a cosine carrier through the weightings shown

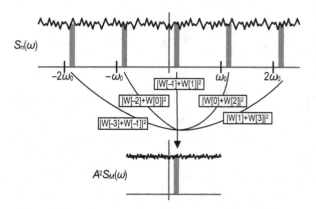

Figure 9.16: A stationary noise modulated by periodic waveform $w(t)$ results in amplitude noise when reference to a cosine carrier through the weightings shown

9.3 Phase Noise Analysis of the Nonlinear LC Oscillator

If $n(t)$ is a white noise source with PSD equal to S_n, this simplifies to

$$S_\phi(\Delta\omega) = S_n \overline{|w_\phi(t)|^2} = \frac{2}{A^2} S_n \overline{p(t)(1 - \cos 2\omega_0 t)} = \frac{2}{A^2} S_n \left(P[0] - \frac{1}{2}P[-2] - \frac{1}{2}P[2]\right)$$

$$S_M(\Delta\omega) = S_n \overline{|w_m(t)|^2} = \frac{2}{A^2} S_n \overline{p(t)(1 + \cos 2\omega_0 t)} = \frac{2}{A^2} S_n \left(P[0] + \frac{1}{2}P[-2] + \frac{1}{2}P[2]\right),$$

where $p(t) = w^2(t)$, and $P[k]$ is the Fourier series of $p(t)$. As discussed previously, these terms can be related back to the PSD of sidebands around the carrier,

$$S_{PM}(\omega_0 + \Delta\omega) = \frac{A^2}{4} S_\phi(\Delta\omega) = S_n \left(\frac{1}{2}P[0] - \frac{1}{4}P[-2] - \frac{1}{4}P[2]\right)$$

$$S_{AM}(\omega_0 + \Delta\omega) = \frac{A^2}{4} S_M(\Delta\omega) = S_n \left(\frac{1}{2}P[0] + \frac{1}{4}P[-2] + \frac{1}{4}P[2]\right).$$

Example: In the case of the transistor with the time-varying input, $S_n = 2kT\gamma$, $w(t) = \sqrt{g_m(t)}$, and $p(t) = w^2(t) = g_m(t)$, the phase noise sidebands will be

$$S_{PM}(\omega_0 + \Delta\omega) = 2kT\gamma \left(\frac{1}{2}G_M[0] - \frac{1}{4}G_M[-2] - \frac{1}{4}G_M[2]\right),$$

where $G_M[k]$ is the Fourier series of $g_m(t)$.

Another important case is when $n(t)$ originates from a low-frequency noise source, for example flicker noise in a CMOS device or low-frequency noise on a supply line. In this case, $S_\phi(\omega)$ is negligible except for frequencies much less than the fundamental, and we can write

$$S_{PM}(\omega_0 + \Delta\omega) = \frac{A^2}{4} S_\phi(\Delta\omega) = \frac{1}{4} S_n(\Delta\omega)|W[-1] - W[1]|^2$$

$$S_{AM}(\omega_0 + \Delta\omega) = \frac{A^2}{4} S_M(\Delta\omega) = \frac{1}{4} S_n(\Delta\omega)|W[-1] + W[1]|^2.$$

If the noise-shaping function is symmetric (i.e., $W[-1] = W[1]$), the phase noise sidebands $S_{PM}(\omega_0 + \Delta\omega)$ are nulled, and the low-frequency noise component does not corrupt the oscillation output's phase. This is the case for all the active elements discussed in this chapter because they are all considered to be memoryless.

Example: In the trivial case of a stationary white source, the *noise shaping waveform* is equal to unity (i.e., $w(t) = 1$), which means that $p(t) = w^2(t)$ is also equal to unity. Accordingly, a stationary noise source is decomposed as follows,

Continued

$$S_{PM}(\omega_0 + \Delta\omega) = \frac{A^2}{4} S_\phi(\Delta\omega) = \frac{S_n}{2}$$

$$S_{AM}(\omega_0 + \Delta\omega) = \frac{A^2}{4} S_M(\Delta\omega) = \frac{S_n}{2},$$

which is consistent with our previous analysis of stationary noise where we found that half of the noise modulates the phase of the carrier and half modulates the amplitude.

Now that we are comfortable decomposing noise into phase and amplitude modulating components around a carrier, it is time to see why this decomposition is so useful in nonlinear circuits.

9.3.6 Noise Passing through a Nonlinearity

Consider the nonlinear transconductance shown in Figure 9.17, where the output current (i_{out}) is some function of the input voltage (v_{in}). Using the Taylor series approximation, we can calculate the output given a small deviation of the input (Δv_{in}), i.e.,

$$i_{out}(v_{in} + \Delta v_{in}) = i_{out}(v_{in}) + \frac{di_{out}(v_{in})}{dv_{in}} \Delta v_{in} + \cdots.$$

Assuming Δv_{in} is very small, we can neglect higher order derivatives, and the output becomes

$$i_{out}(v_{in} + \Delta v_{in}) = i_{out}(v_{in}) + g(t)\Delta v_{in}(t).$$

Therefore the deviation in output current due to $\Delta v_{in}(t)$ is simply $g(t)\Delta v_{in}(t)$, where $g(t)$ is the instantaneous transconductance defined as $g(t) = di_{out}(v_{in})/dv_{in}$. As $g(t)$ is a periodic waveform, we can express the deviation in output current as

$$\Delta i_{out}(t) = \Delta v_{in}(t) g(t) = \Delta v_{in}(t) \left(\sum_{k=-\infty}^{\infty} G[k] e^{jk\omega_0 t} \right),$$

where $G[k]$ defines the Fourier coefficients of $g(t)$. If $\Delta v_{in}(t)$ is a small single-sideband around the carrier, the output is an LTV *not* an LTI response. This is seen in Figure 9.17 where the input to the nonlinearity contains only a USB, while the output contains both an LSB and a USB. However, if $\Delta v_{in}(t)$ is composed of phase modulating sidebands, i.e., $\Delta v_{in}(t) = -A\phi(t)\sin\omega_0 t$, the output in the frequency domain is calculated as

$$\Delta i_{out}(t) = -A\phi(t) g(t) \sin\omega_0 t = -\frac{A\phi(t)}{2j} \left(\sum_{k=-\infty}^{\infty} (G[k-1] - G[k+1]) e^{jk\omega_0 t} \right).$$

Now, if $\phi(t)$ has only low-frequency components, and we are concerned only with $\Delta i_{out}(t)$ around ω_0, the only values of k in the above summation that are relevant are ± 1. Therefore, the expression simplifies to

$$\Delta i_{out}(t) = -\frac{A\phi(t)}{2j} \left((G[0] - G[2]) e^{j\omega_0 t} + (G[-2] - G[0]) e^{-j\omega_0 t} \right).$$

9.3 Phase Noise Analysis of the Nonlinear LC Oscillator

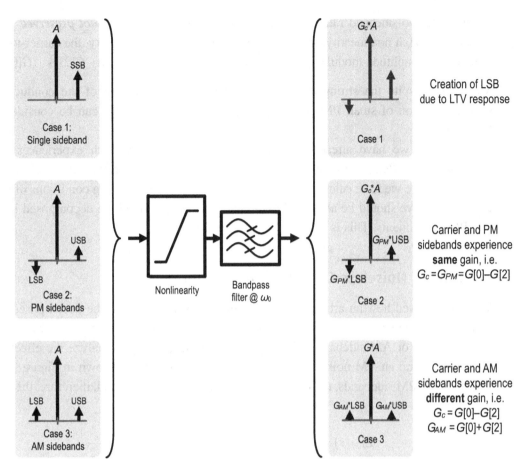

Figure 9.17: Output of a nonlinearity due to SSB-modulated, PM-modulated, and AM-modulated carriers. The bandpass filter is included as we are concerned only with the output around ω_0

Assuming $G[-2] = G[2]$, which is the case for a memoryless active circuit,

$$\Delta i_{out}(t) = -A\phi(t)((G[0] - G[2])\sin\omega_0 t) = (G[0] - G[2])\Delta v_{in}(t),$$

which is an LTI response (i.e., $G_{PM} = G[0] - G[2]$). To reiterate, we have assumed that we can neglect components far from carrier, which will be eventually filtered by the resonator. If $\Delta v_{in}(t)$ is composed of amplitude-modulating sidebands, i.e., $\Delta v_{in}(t) = Am(t)\cos\omega_0 t$, the output around the carrier is calculated using a similar approach,

$$\Delta i_{out}(t) = (G[0] + G[2])\Delta v_{in}(t),$$

which, again, is an LTI response (i.e., $G_{AM} = G[0] + G[2]$). This is also visualized in Figure 9.17.

Note that, as was shown in Section 9.2.2, the carrier gain is also given by $G_{eff} = G_{PM} = G[0] - G[2]$, and so we can state the following assuming a small modulation index:

- The carrier-to-sideband ratio of a ***phase-modulated*** signal is ***preserved*** when it is passed through nonlinearity.

- The carrier-to-sideband ratio of an *amplitude-modulated* signal is *not preserved* when it is passed through nonlinearity. In the case of a compressive nonlinearity, the carrier-to-sideband ratio of an amplitude-modulated signal will be attenuated, since $(G[0] + G[2]) < (G[0] - G[2])$.

Moreover, despite the strongly nonlinear and time-varying action of the conductance, the transfer function of small PM sidebands, and small AM sidebands can be considered as an LTI response.

Of course, we have already seen how a single-sideband (which experiences an LTV-response) can be decomposed in PM and AM (which both experience separate LTI responses). Therefore, once we have calculated the noiseless steady-state operating conditions of the active circuit, $G[k]$, we should be able to able to analyze noise provided we decomposed it into AM and PM components. This is done in the next section.

9.3.7 Reaction of Noiseless Oscillator to an External Noise

We have showed how an arbitrary cyclostationary noise source can be decomposed into AM and PM components. We also showed how the active element exhibits an LTI response with respect to PM or AM sidebands. This allows us to independently analyze the effect of a PM noise source and an AM noise source on a noiseless oscillator, as shown in Figure 9.18. With respect to the PM sidebands, the response for every device is linear and, therefore, the following must hold:

$$Z_{PM}(s) = \frac{v_{PM}(s)}{i_{PM}(s)} = \frac{1}{sC_P} \| sL_P \| R_P \| \frac{1}{G_{nr}[0] - G_{nr}[2]}.$$

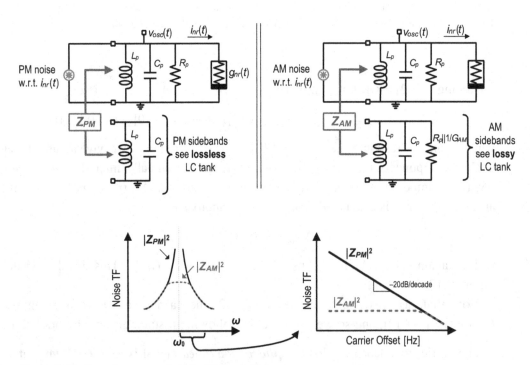

Figure 9.18: Impedance perceived by PM and AM noise sources

Because of the power conservation requirement, the effective conductance is given by $G_{eff} = G_{nr}[0] - G_{nr}[2] = -1/R_p$, which simplifies the above expression to

$$|Z_{PM}(s)|^2 = \left|\frac{\omega L}{1-\omega^2 L_P C_P}\right|^2,$$

and, so, around the carrier the impedance is given by

$$|Z_{PM}(j(\omega_0 + \Delta\omega))|^2 \approx \left|\frac{\omega_0^2 L}{2\Delta\omega}\right|^2 = \frac{R_p^2}{4}\frac{1}{Q^2}\left(\frac{\omega_0}{\Delta\omega}\right)^2.$$

Remarkably, this is the same noise transfer function as the perfectly linear oscillator. Therefore, if the injected current phase modulates the carrier, it sees a lossless impedance, which is the assumption made during the derivation of Leeson's expression (see Figure 9.3). In the case of the AM sidebands, the transfer function is given by

$$Z_{AM}(s) = \frac{v_{PM}(s)}{i_{PM}(s)} = \frac{1}{sC_P}||sL_P||R_P||\frac{1}{G_{nr}[0]+G_{nr}[2]},$$

which is a lossy LC tank. Both transfer functions are plotted versus carrier offset in Figure 9.18. Close to the carrier $Z_{AM}(s) \ll Z_{PM}(s)$, which is why amplitude is generally of little concern in an LC oscillator. Accordingly, once a noise source has been decomposed into AM and PM components, its contribution to $\mathcal{L}\{\Delta\omega\}$ is calculated as

$$\frac{\overline{v_{PM}^2} + \overline{v_{AM}^2}}{A_c^2/4},$$

where

$$\overline{v_{PM}^2} = \overline{i_{PM}^2}|Z_{PM}(j(\omega_0 + \Delta\omega))|^2$$

$$\overline{v_{AM}^2} = \overline{i_{AM}^2}|Z_{AM}(j(\omega_0 + \Delta\omega))|^2.$$

In the case of a stationary white noise source, $\overline{i_n^2}$, the decomposition is $\overline{i_{PM}^2} = \overline{i_{AM}^2} = \overline{i_n^2}/2$. This explains why amplitude noise can generally be disregarded and why we needed to introduce the factor of 2 division when deriving Leeson's expression. In the more general case of a colored cyclostationary noise source, $n_{CYCLO}(t) = n(t)w(t)$, the resultant AM and PM contributions are

$$\overline{v_{PM}^2} = \frac{|Z_{PM}(j(\omega_0 + \Delta\omega))|^2}{4}\sum_{k=-\infty}^{\infty} S_n(\Delta\omega - k\omega_0)|W[k-1] - W[k+1]|^2$$

$$\overline{v_{AM}^2} = \frac{|Z_{AM}(j(\omega_0 + \Delta\omega))|^2}{4}\sum_{k=-\infty}^{\infty} S_n(\Delta\omega - k\omega_0)|W[k-1] + W[k+1]|^2.$$

If the cyclostationary noise is white, it simplifies to

$$\overline{v_{PM}^2} = |Z_{PM}|^2 S_n\left(\frac{1}{2}P[0] - \frac{1}{4}P[-2] - \frac{1}{4}P[2]\right)$$

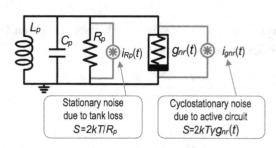

Figure 9.19: Negative g_m LC oscillator satisfying Bank's general equation

$$\overline{v_{AM}^2} = |Z_{AM}|^2 S_n \left(\frac{1}{2}P[0] + \frac{1}{4}P[-2] + \frac{1}{4}P[2] \right).$$

Again, it should be emphasized that both $|Z_{PM}|^2$ and $|Z_{AM}|^2$ are linear transfer functions. Therefore, the noise analysis is greatly simplified provided we refer all noise currents across the resonator and decompose them into AM and PM components.

9.3.8 Bank's General Result

We now possess the tools needed to analyze the negative-g_m LC oscillator model, which is redrawn in Figure 9.19 along with explicit noise sources. However, it is useful to first summarize three important results:

1. Due to energy-balance in an LC oscillator, the active circuit has to precisely balance the tank loss, R_p. Accordingly, if we define the instantaneous conductance of the element as $g_{nr}(t)$, this requirement implies that

$$G_{nr}[0] - G_{nr}[2] = -\frac{1}{R_p},$$

   where $G_{nr}[k]$ are the Fourier coefficients of $g_{nr}(t)$.

2. An arbitrary cyclostationary noise source, ($n_{CYCLO}(t)$), can be modeled as a stationary noise source ($n(t)$) modulated by a periodic waveform ($w(t)$), i.e., $n_{CYCLO}(t) = n(t)w(t)$. This noise can then be decomposed into AM and PM components with respect to some carrier. The PM component of such a noise source with respect to a zero-phase cosine waveform is

$$S_{PM}(\Delta\omega) = \frac{1}{4} \sum_{k=-\infty}^{\infty} S_n(\Delta\omega - k\omega_0)|W[k-1] - W[k+1]|^2,$$

   where $S_n(\omega)$ is the spectral density of $n(t)$, and $W[k]$ are the Fourier coefficient of $w(t)$. If $S_n(\omega)$ is white, this simplifies to

$$S_{PM}(\Delta\omega) = S_n \left(\frac{1}{2}P[0] - \frac{1}{4}P[-2] - \frac{1}{4}P[2] \right),$$

   where $P[k]$ are the Fourier coefficient of $p(t) = w^2(t)$.

3. If a PM noise current is applied to a noiseless LC oscillator, it "sees" a lossless LC tank. In other words, the transfer function of a PM current source to PM voltage noise is given by

$$\overline{v_{PM}^2} = \overline{i_{PM}^2}|Z_{PM}(j(\omega_0+\Delta\omega))|^2 = \overline{i_{PM}^2}\frac{R_p^2}{4}\frac{1}{Q^2}\left(\frac{\omega_0}{\Delta\omega}\right)^2.$$

Returning to Figure 9.19, the current noise due to tank loss is simply $2kT/R_p$, while we assume the noise of the active resistor can be modeled as cyclostationary white noise source with PSD equal to $2kT\gamma g_{nr}(t)$, where $g_{nr}(t)$ is the instantaneous conductance of the active element circuit, and γ is some constant. Note that this assumption has some physical basis as the active element is generally composed of transistors operating in saturation where $\gamma = 2/3$ in long channel devices. Ignoring the contribution of amplitude noise, $\mathcal{L}\{\Delta\omega\}$ is calculated as

$$\mathcal{L}\{\Delta\omega\} = \frac{\overline{v_{Rp(PM)}^2} + \overline{v_{nr(PM)}^2}}{A_c^2/4} = \frac{\overline{i_{Rp(PM)}^2} + \overline{i_{nr(PM)}^2}}{A_c^2/4}|Z_{PM}(j(\omega_0+\Delta\omega))|^2,$$

where the PM components of noise are given by

$$\overline{i_{Rp(PM)}^2} = \frac{\overline{i_{Rp}^2}}{2} = kT/R_p$$

$$\overline{i_{nr(PM)}^2} = \frac{2kT\gamma}{2}(G_{nr}[0] - G_{nr}[2]) = kT\gamma/R_p.$$

Note that because of energy-balance $\overline{i_{nr(PM)}^2}$ does not depend on the specifics of $g_{nr}(t)$. Simplifying the expression for phase noise gives

$$\mathcal{L}\{\Delta\omega\} = \frac{kTR_p(1+\gamma)}{A^2}\frac{1}{Q^2}\left(\frac{\omega_0}{\Delta\omega}\right)^2.$$

Comparing this expression to Leeson's analysis, we see that the unknown *oscillator noise factor* is given by $F = 1 + \gamma$. This is actually quite a powerful result because phase noise was quantified knowing only the relationship between the noise PSD and the instantaneous conductance of the active circuit. In other words,

> If the PSD of a noise source injected into the tank is directly proportional to the time-varying conductance of the active circuit, its contribution to noise factor F depends only on the proportionality constant that relates the noise and the instantaneous conductance.

This is Bank's general result [11], and as we will see, many of the most common LC topologies satisfy this condition (at least partially).

Furthermore, as we are concerned only with the spectrum around the oscillation frequency, a single-tank nearly-sinusoidal LC oscillator can generally be redrawn in the form of the negative-g_m model. Doing this often simplifies the analysis.

Example: Shown in Figure 9.20 is a distributed lossy LC tank and a two terminal negative conductance. Using straightforward analysis, the circuit is recast in the form of the negative-g_m model using the transformation variables α and β. Depending on where

Continued

the terminals of the conductance are connected, α and β can assume any value from 0 to 1. Given this, and assuming the PSD of the active element current noise is equal to $2kT\gamma g(t)$, Bank's general result can be used to quantify the noise factor as

$$F = 1 + \frac{\beta}{\alpha}\gamma.$$

The oscillator *power* efficiency can be defined in terms of the power efficiency of the oscillator if the active circuit is connected across the entire tank (η_g), i.e.,

$$\eta_{osc} = \beta^2 \eta_g.$$

Therefore the FOM is given by

$$\text{FOM} = \frac{\beta^2}{\left(1 + \frac{\beta}{\alpha}\gamma\right)} \eta_g \frac{2Q^2}{kT} 1\text{mW},$$

which is maximized by connecting the negative conductance across the entire tank, which ensures $\alpha = \beta = 1$. Therefore, it is generally best practice to connect the energy-restoring circuit directly across the LC tank, rather than tapping some intermediate node of the tank.

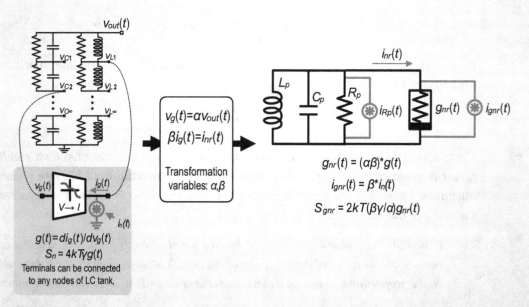

Figure 9.20: Recasting an arbitrary topology in the form of the negative-g_m model

Since the optimum connection arrangement is $\alpha = \beta = 1$, and the maximum efficiency of any active circuit is 1, we can state that the maximum achievable FOM of any single-tank, nearly sinusoidal LC oscillator is given by

$$\text{FOM}_{\text{MAX}} = \frac{Q^2}{1+\gamma}\frac{2}{kT} 1\text{mW}.$$

This maximum achievable FOM will be used to evaluate the various topologies discussed in the next few sections.

9.3.9 Two-Port Oscillators

So far we have primarily focused on one-port oscillators, that is, oscillators where the active circuit has a one (differential or single-ended) port and that can be easily redrawn in the form of the negative-g_m LC model. This is the simplest and mostly widely used arrangement; however, other topologies do exist and have some potential benefits. Before we discuss a few examples of the one-post oscillators in the next section, it is appropriate to say a few words about two-port oscillators.

As described by Mazzanti and Bevilacqua [12], if an oscillator is partitioned into a two-port reactive element and a memoryless conductance, as shown in Figure 9.21, the noise factor is $F = 1 + \gamma/|Av|$, where γ is the constant that relates the noise current of the conductance with its instantaneous conductance, and Av is the voltage gain across the two-port resonator. Accordingly, as Av becomes very large, conductance noise is suppressed, F approaches one, and the optimum FOM, in dBc/Hz, becomes

$$\text{FOM}_{\text{MAX}} = \frac{2Q^2}{kT} 1\text{mW},$$

or in dB,

$$\text{FOM}_{\text{2PORT(MAX)}} = 176.8 + 20\log_{10}Q,$$

which is equal to the FOM of a 100% efficient oscillator with a noiseless conductance quoted in [4].

While excellent results (particularly in terms of flicker noise suppression) have been achieved using this "step-up" technique [2], [13], [14], [15], making $|Av| > 1$ can be problematic. First, in order to maximize oscillator efficiency, the swing at the output of the conductance should always be maximized, which in the case of a CMOS conductance means the swing will be V_{DD} or, in the case of a NMOS-only or PMOS-only conductance, will be $2V_{DD}$. This implies that the swings on gate oxides will be $|Av|V_{DD}$ (or $|Av|2V_{DD}$), which can compromise reliability if a large $|Av|$ is employed. Second, the step-up transformer used in such a design should not result in an effective tank Q degradation, i.e., the Q of the impedance parameter Z_{21} of the two-port tank [4] that is degraded with respect to the Q of a tank that uses a single inductor occupying the same area. Although these issues can be managed, such designs do not currently outperform designs with $|Av| = 1$, which corresponds

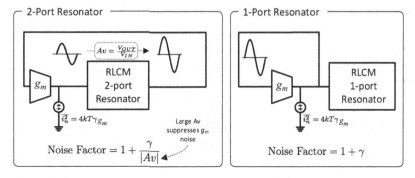

Figure 9.21: The distinction between oscillators using two-port and one-port resonators

to the one-port oscillator also shown in Figure 9.21. In the case of this one-port design, the optimum FOM becomes

$$\text{FOM}_{1\text{PORT(MAX)}} = 176.8 + 20\log_{10}Q - 10\log_{10}(1+\alpha).$$

When a single transistor or a differential pair is used as the negative conductance, the noise coefficient α is approximately equal to $\gamma = 0.67$ and, so, the one-port limit becomes

$$\text{FOM}_{1\text{PORT(MAX)}} = 174.6 + 20\log_{10}Q,$$

which is only 2.2dB less than optimal. As will be shown in the next sections, a number of modern topologies get within 1dB of this practical limit (or 3dB of the absolute limit).

9.4 LC OSCILLATOR TOPOLOGIES

So far we have taken an abstract approach to understanding the operation of an LC oscillator. It is now time to explore some specific transistor-level examples.

9.4.1 The Standard NMOS Topology

The LC oscillator with a negative-conductance consisting of a current-biased, cross-coupled differential pair is, perhaps, the most fabricated of all CMOS LC oscillator. Shown in Figure 9.22, the topology has more recently been referred to as a class B topology [16], but we will simply refer to it as the standard NMOS topology.

In this topology, the differential pair provides the negative conductance, which is shown separately in Figure 9.23 along with its I–V characteristic and transconductance. In order for the oscillator to start-up, it is required that $-g_{nr0} < 1/R_p$, where g_{nr0} is the negative conductance

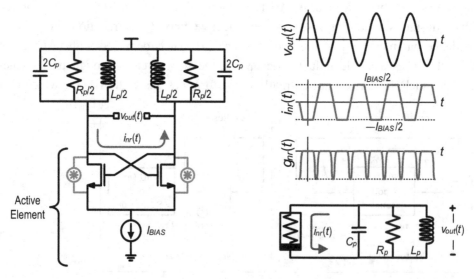

Figure 9.22: Standard LC oscillator topology

9.4 LC Oscillator Topologies

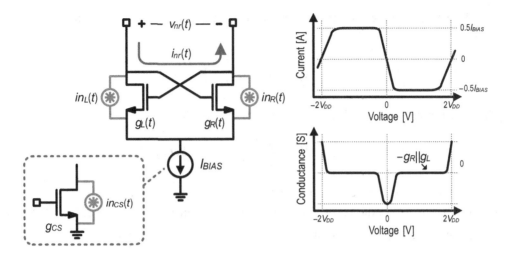

Figure 9.23: Active element used in class B topology, and the associated I–V characteristic and conductance plot

when the differential voltage is zero. Assuming $-g_{nr0} \ll 1/R_p$, the I–V characteristic can be approximated as a hard switching nonlinearity defined as

$$i_{NR}(v_{nr}) \approx -\frac{I_{BIAS}}{2}\operatorname{sgn}(v_{nr}) = \begin{cases} \dfrac{I_{BIAS}}{2}, & v_{nr} < 0 \\ -\dfrac{I_{BIAS}}{2}, & v_{nr} > 0 \end{cases},$$

where sgn(·) is the sign or signum function. Thus, under oscillation conditions $i_{NR}(t)$ will approximate a square wave (see waveforms in Figure 9.22), which implies that the fundamental component of the current injected into the tank is given by $I_{NR}[1] = -(1/\pi)I_{BIAS}$, which in turn implies a differential oscillation amplitude of

$$A = \frac{2}{\pi}I_{BIAS}R_p$$

and a power efficiency of

$$\eta = \frac{1}{\pi}\frac{A}{V_{DD}},$$

which is consistent with the analysis in Chapter 1. Therefore to maximize the topology's power efficiency, the amplitude must be as large as possible. In a practical design, the maximum amplitude will be limited by the headroom requirement of the current source, i.e., $A_{MAX} \approx 2(V_{DD} - V_{eff})$, where $V_{eff} = V_{GS} - V_{TH}$ is the overdrive voltage of the current source. By minimizing V_{eff}, the maximum achievable efficiency of this topology can approach $2/\pi$ (or 63.7%).

In order to analyze the topology's phase noise performance, we redraw the circuit in the form of the negative-g_m model (see the right-hand side of Figure 9.22). We note that the noise of the differential pair is directly proportional to the conductance of the differential pair. That is, assuming the current source stays in saturation, the differential conductance is given by

$$g_{nr}(t) = -\frac{g_R(t)g_L(t)}{g_R(t) + g_L(t)},$$

while the noise due to the differential pair can be quantified as

$$\overline{i_{gnr}^2} = 2kT\gamma\frac{g_R(t)g_L(t)}{g_R(t) + g_L(t)} = 2kT\gamma g_{nr}(t).$$

This satisfies Bank's general result in that the noise due to the differential pair is directly proportional to its conductance. Consequently, the contribution of the differential pair to the oscillator's noise factor is simply the proportional constant γ (or in other words, the PM contribution of the tank loss and differential pair are related as follows: $\overline{i_{gnr(PM)}^2} = \gamma \overline{i_{Rp(PM)}^2}$).

The current source noise $i_{CS}(t)$ – much like noise associated with the input transconductance of a single-balanced active mixer (see Chapter 8) – is mixed with a unity square wave and injected into the tank, and so it appears as a cyclostationary noise source across the tank given by

$$i_{CS}(t)(\tfrac{1}{2})\text{sgn}(v_{OSC}(t)) = \begin{cases} \dfrac{i_{CS}(t)}{2}, & v_{osc}(t) < 0 \\ -\dfrac{i_{CS}(t)}{2}, & v_{osc}(t) > 0 \end{cases}.$$

This is essentially a stationary noise $i_{CS}(t)$ modulated by the *noise-shaping waveform* $w(t) = (\tfrac{1}{2})\text{sgn}(v_{OSC}(t))$. Noting that $i_{CS}(t)$ is a white noise source and $p(t) = w^2(t) = 1/4$, we can use the PM decomposition documented previously to derive the PM contribution as

$$\overline{i_{cs(PM)}^2} = \overline{i_{cs}^2}\left(\frac{1}{2}P[0] - \frac{1}{4}P[-2] - \frac{1}{4}P[2]\right) = \frac{\overline{i_{cs}^2}}{8} = (1/4)kT\gamma g_{CS}.$$

Accordingly, the phase noise of the standard NMOS topology is calculated as

$$\mathcal{L}\{\Delta\omega\} = \frac{\overline{i_{Rp(PM)}^2} + \overline{i_{gnr(PM)}^2} + \overline{i_{cs(PM)}^2}}{A^2/4}|Z_{PM}(j(\omega_0 + \Delta\omega))|^2$$

$$\mathcal{L}\{\Delta\omega\} = \frac{kTFR_p}{A^2}\frac{1}{Q^2}\left(\frac{\omega_0}{\Delta\omega}\right)^2,$$

where the noise factor is

$$F = 1 + \gamma + \frac{\gamma g_{CS} R_p}{4}.$$

Note that the third term (due to the current source) can be minimized by increasing the current source's overdrive voltage. Therefore, we can say that the absolute minimum noise factor of this topology is $F = 1 + \gamma$. However, this will come at the expense of current source headroom, which will in turn limit the maximum amplitude and efficiency. To see this we can rewrite the above expression in terms of the overdrive voltage of the current source device (or equivalently its headroom):

$$F = 1 + \gamma + \gamma \frac{\pi A}{4 V_{eff}}.$$

We need to carefully choose the V_{eff} of the current source in order to optimize FOM. If the current source is on the edge of saturation (which is a good design choice), the FOM is given by

$$\text{FOM} = \frac{A}{F V_{DD}} \frac{2Q^2}{kT} 1\text{mW} = \frac{2\left(1 - \frac{V_{eff}}{V_{DD}}\right)}{\pi \left(1 + \gamma + \frac{\gamma \pi}{2}\left(\frac{V_{DD}}{V_{eff}} - 1\right)\right)} \frac{2Q^2}{kT} 1\text{mW}.$$

If $\gamma = 2/3$, the best choice of effective voltage is around $V_{eff} = (1/2) V_{DD}$, which corresponds to a FOM of roughly 7.5dB less than the maximum achievable figure of merit ($\text{FOM}_{1\text{port(MAX)}}$).

The above analysis assumes the current source transistor always operates in saturation. If the current-source enters triode, both phase noise and FOM begin to degrade, as shown in Figure 9.24. This is because when the differential pair is hard-switched, the LC tank sees a resistance composed of a differential-pair transistor in triode along with the current-source in triode. This is a relatively low impedance (compared to a device in saturation), and so the effective Q of the LC tank will be degraded. Accordingly, a good design will ensure that the current source is just at the edge of saturation, but does not drop into triode for any significant portion of the oscillation period.

This implies that once the tank inductance and Q are set (which are largely a function of the oscillation frequency and the tuning range), the oscillator bias current is set. Thus, all is left is to adjust the device sizes to achieve a reasonably sharp switching. The oscillator bias current may be reduced then only if R_p is increased. This in turn may be accomplished only by increasing the inductance, given a maximum allowable Q for the technology used. Increasing the inductance however typically comes at the expense of lower tuning range.

If a lower current design is desired, one may choose a complementary structure (next section) or a class C topology (7.2).

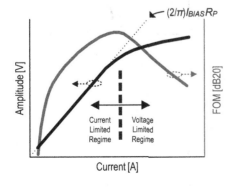

Figure 9.24: Transition between current-limited and voltage-limited regime occurs when current source transistor enters the triode region for some portion of the oscillation

9.4.2 The Standard CMOS Topology

It is possible to realize a negative resistance with cross-coupled PMOS instead of NMOS transistors. It is also possible to employ both, as shown in Figure 9.25. This standard CMOS topology uses the active element, shown separately in Figure 9.26, to provide the required negative conductance. In this case, the fundamental of the current switched into the tank is $I_{NR}[1] = -(2/\pi)I_{BIAS}$, and the differential voltage is therefore

$$A = \frac{4}{\pi} I_{BIAS} R_p,$$

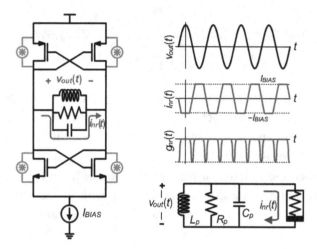

Figure 9.25: Standard CMOS LC topology

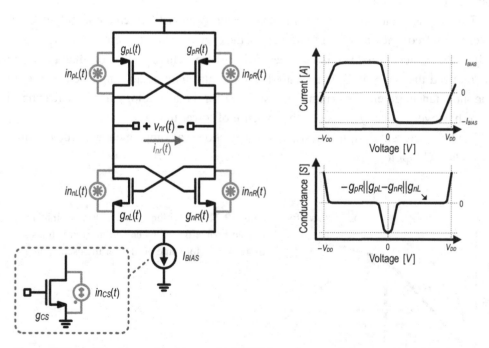

Figure 9.26: Active element used in standard CMOS topology, and the associated I–V characteristic and conductance plots

which is twice as large as the standard NMOS topology (assuming the same bias current). The power efficiency is given by

$$\eta = \frac{2}{\pi}\frac{A}{V_{DD}},$$

which is again twice as large as the standard NMOS topology (assuming the same amplitude). At first glance, this would suggest the CMOS topology is a more attractive topology, however, the maximum differential amplitude of the CMOS is limited to V_{DD}, while the NMOS-only design is limited to $2V_{DD}$. Accordingly, the maximum achievable efficiency of both topologies is comparable when achievable amplitudes are maximized.

In terms of phase noise, the time-varying conductance and noise due to the differential pair transistors are related, as follows,

$$g_{nr}(t) = -\frac{g_{nR}(t)g_{nL}(t)}{g_{nR}(t)+g_{nL}(t)} - \frac{g_{pR}(t)g_{pL}(t)}{g_{pR}(t)+g_{pL}(t)}$$

$$\overline{i^2_{gnr(PM)}} = 2kT\left(\gamma\frac{g_{nR}(t)g_{nL}(t)}{g_{nR}(t)+g_{nL}(t)} + \gamma\frac{g_{pR}(t)g_{pL}(t)}{g_{pR}(t)+g_{pL}(t)}\right) = 2kT\gamma g_{nr}(t),$$

where we assumed the channel noise coefficient, γ, is the same for both the PMOS and NMOS devices [6]. As the noise contribution of the differential pair is proportional to its conductance, Bank's result can again be used to state that $\overline{i^2_{gnr(PM)}} = \gamma \overline{i^2_{Rp(PM)}}$.

As was done in the NMOS-topology, the noise due to the current source can be referred across the tank as a cyclostationary noise source given by $i_{CS}(t)\,\mathrm{sgn}\,(v_{OSC}(t))$, which has a PM component equal to $\overline{i^2_{cs(PM)}} = kT\gamma g_{CS}$. Therefore the complete expression for phase noise is given by

$$\mathcal{L}\{\Delta\omega\} = \frac{\overline{i^2_{Rp(PM)}} + \overline{i^2_{gnr(PM)}} + \overline{i^2_{cs(PM)}}}{A^2/4}|Z_{PM}(j(\omega_0+\Delta\omega))|^2$$

$$\mathcal{L}\{\Delta\omega\} = \frac{kTFR_p}{A^2}\frac{1}{Q^2}\left(\frac{\omega_0}{\Delta\omega}\right)^2,$$

where the noise factor is

$$F = 1+\gamma+\gamma g_{CS}R_p = 1+\gamma+\gamma\frac{\pi A}{V_{eff}}.$$

Compared to the NMOS-only, the CMOS topology can achieve lower phase noise, higher amplitude, and better efficiency for the same current and tank. However, if the amplitude of both topologies is maximized (as is the case in a good design), the NMOS-only achieves better phase noise, while achieving similar FOM and efficiency numbers. This is visualized in Figure 9.27. Note that to gain some insight, the current source is assumed to be noiseless.

It should be noted that the above analysis assumes that the tank capacitance and any parasitic capacitance appears differentially across the tank. If a substantial amount of capacitance is connected from either output terminal to ground, the PMOS transistors have the potential to load the tank and substantially degrade performance [6], [17]. (When the PMOS devices enter the triode region, a single-ended capacitor will present a low impedance path to ground, while a differential cap will not.) Because of this, the NMOS topology generally exhibits better performance when large single-ended capacitors banks are used to provide wideband tuning.

9.4.3 The Colpitts Topology

The Colpitts oscillator, shown in Figure 9.28, is another widely used topology. Assuming the conducting transistor stays in the saturation region when on, the transistor will inject current into tank for a short time during the trough of the oscillation waveform. The conduction time is generally kept small in order to maximize the oscillator's efficiency. If we approximate this current injection as an impulse train equal to $i_{ds}(t) = I_{BIAS} T \sum_{n=-\infty}^{\infty} \delta(t - nT)$, where T is the oscillation period, the fundamental component of $i_{ds}(t)$ is $I_{ds}[1] = I_{BIAS}$, and we can approximate the oscillation amplitude as

$$A = 2I_{BIAS} \frac{C_2}{C_1 + C_2} R_p,$$

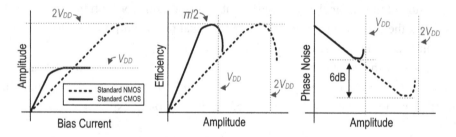

Figure 9.27: Comparison of the standard NMOS and standard CMOS topologies assuming identical tanks and a noiseless current source

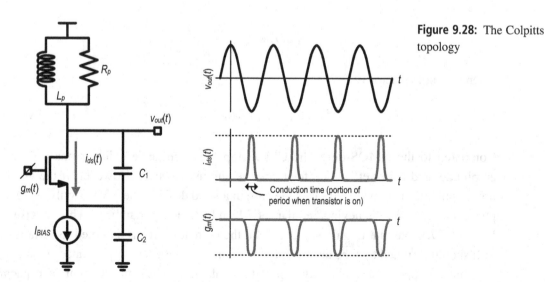

Figure 9.28: The Colpitts topology

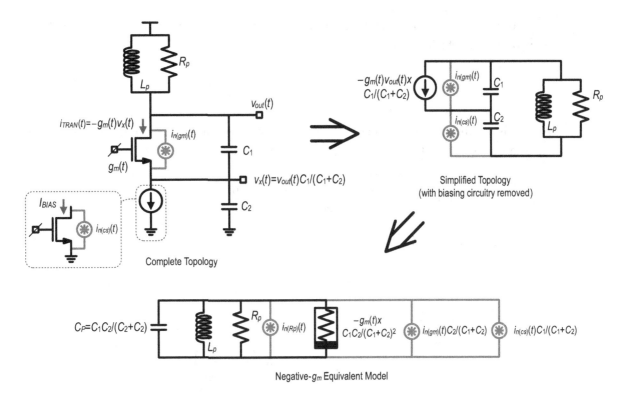

Figure 9.29: Redrawing the Colpitts topology in the form of the negative-g_m model

which results in an efficiency of

$$\eta = \frac{C_2}{C_1+C_2}\frac{A}{V_{DD}}.$$

Naturally, a finite conduction time will result in an efficiency and amplitude of less than that predicted by these expressions. A simple way to calculate the phase noise of the Colpitts oscillator is to redraw the circuit in the form of the negative-g_m model. This is done in Figure 9.29, where the equivalent time-varying resistance is calculated as

$$g_{nr}(t) = -g_m(t)\frac{C_1C_2}{(C_1+C_2)^2},$$

and the cyclostationary noise due to the conducting transistor referred across the tank is

$$\overline{i^2_{gnr(PM)}} = 2kT\gamma g_m(t)\left(\frac{C_2}{C_1+C_2}\right)^2 = 2kT\gamma g_{nr}(t)\frac{C_2}{C_1}.$$

Given that the noise PSD of the conducting transistor is directly proportional to the time-varying conductance $g_{nr}(t)$, we can use Bank's general result to deduce that injected PM noise is related to the tank noise as follows:

$$\overline{i^2_{gnr(PM)}} = \gamma\frac{C_2}{C_1}\overline{i^2_{Rp(PM)}}.$$

The noise of the current source can also be referred across the tank using a simple AC transformation, which results in a PSD of

$$\overline{i^2_{cs-tank}} = 2kT\gamma g_{cs}\left(\frac{C_1}{C_1+C_2}\right)^2.$$

As this is a stationary noise source, its PM component is simply half this value. Accordingly the phase noise of the Colpitts topology is calculated as

$$\mathcal{L}\{\Delta\omega\} = \frac{\overline{i^2_{Rp(PM)}} + \overline{i^2_{gnr(PM)}} + \overline{i^2_{cs-tank(PM)}}}{A^2/4}|Z_{PM}(j(\omega_0+\Delta\omega))|^2,$$

which simplifies to

$$\mathcal{L}\{\Delta\omega\} = \frac{kTFR_p}{A^2}\frac{1}{Q^2}\left(\frac{\omega_0}{\Delta\omega}\right)^2,$$

where the noise factor is

$$F = 1 + \frac{C_2}{C_1}\gamma + \gamma g_{CS}R_p\left(\frac{C_1}{C_1+C_2}\right)^2.$$

If we ignore the third term due to the current source, the noise factor becomes $F=1+(C_2/C_1)\gamma$. Therefore, if C_1 is designed to be greater than C_2, a noise factor less than that of the standard cross-coupled topology (where $F=1+\gamma$) can be attained. However, it must be noted that increasing C_1 also reduces the oscillator's efficiency, and it can be shown that the achievable FOM of a Colpitts oscillator is always less than the standard topology [18].

9.4.4 Oscillator Design Methodology

Table 9.1 summarizes the key features of the three topologies discussed. Note that the Colpitts design is single-ended (although it can be extended to differential), whereas the NMOS and CMOS topologies are differential.

For a given application, typically the nominal frequency of oscillation, the tuning range, and the phase noise at a certain offset are given.

To understand the trade-offs involved, let us say a few words about the tuning range first. As we saw in Chapter 1, most practical oscillators incorporate some kind of switchable array of capacitors to provide coarse frequency tuning. Let us assume that there is a total capacitance of C_P at the output due to the device, inductor, and routing parasitics. We also assume the switched capacitor array exhibits a capacitance of C_F when all the switches are off, resulting from switch parasitic and capacitor top plate, with an additional capacitance of C_V accounting for the variable part. Thus, the total capacitance of the tank varies from $C_{MIN}=C_P+C_F$ to $C_{MAX}=C_P+C_F+C_V$. To achieve a reasonable Q for the tunable array, typically C_F and C_P are comparable. In Chapter 1, we saw an example of a discrete tunable capacitor with 32 units of 40fF capacitors, ranging from 430fF to 1.36P with a Q of better than 45 at 3.5GHz. Thus, for

Table 9.1: **Comparison between various oscillator topologies**

Topology	NMOS	CMOS	Colpitts
Differential amplitude	$\dfrac{2}{\pi} I_{BIAS} R_p$	$\dfrac{4}{\pi} I_{BIAS} R_p$	$2 I_{BIAS} \dfrac{C_2}{C_1+C_2} R_p$
Maximum amplitude	$2V_{DD}$	V_{DD}	V_{DD}
Efficiency	$\dfrac{1}{\pi} \dfrac{A}{V_{DD}}$	$\dfrac{2}{\pi} \dfrac{A}{V_{DD}}$	$\dfrac{C_2}{C_1+C_2} \dfrac{A}{V_{DD}}$
Noise factor	$1+\gamma+\gamma \dfrac{\pi A}{4 V_{eff}}$	$1+\gamma+\gamma \dfrac{\pi A}{V_{eff}}$	$1+\dfrac{C_2}{C_1}\gamma+\gamma \dfrac{\pi A}{V_{eff}} \left(\dfrac{C_1}{C_1+C_2}\right)^2$
Noise factor, excluding the current source[12]	$1+\gamma$	$1+\gamma$	$1+\dfrac{C_2}{C_1}\gamma \approx 1+\gamma$

that particular design, $C_F = 430\text{fF}$. Assuming a tank inductance of L_P, the lowest and highest frequencies of oscillation are

$$\omega_L = \dfrac{1}{\sqrt{L_P C_{MAX}}}$$

$$\omega_H = \dfrac{1}{\sqrt{L_P C_{MIN}}},$$

and the nominal oscillation frequency is $\omega_0 = \dfrac{\omega_L + \omega_H}{2}$. The tuning range may be then obtained as (Problem 11)

$$TR = \dfrac{\omega_H - \omega_L}{\omega_0} = 2 \dfrac{\sqrt{\dfrac{C_{MAX}}{C_{MIN}}} - 1}{\sqrt{\dfrac{C_{MAX}}{C_{MIN}}} + 1}.$$

Consequently, if the total parasitic capacitance C_P is small, the tuning range to the first order is independent of the tank inductance and nominal frequency of oscillation, as it depends only on the ratio of on/off mode capacitance of the array. For a constant Q, the unit capacitor and the switch size may be scaled while keeping the ratio constant. Although this is not quite true in practice, we shall make use of it to find a first-cut estimate of the oscillator design parameters. In our tunable capacitor example of Chapter 1, if C_F and C_P are comparable, the tuning range is calculated to be 36%.

We further assume that considering the technology and area constraints, the tank quality factor, mostly dominated by that of the inductor (at least at GHz frequencies), is given. Finally, we make the observation that the oscillation optimum amplitude is also given for a certain

---

[12] The current source noise may be eliminated using a class C or bias filtering scheme, as we shall discuss in Section 9.7.

technology, as it solely depends on the supply voltage, that is, $A = 2V_{DD}$ for an NMOS design for instance.

With these assumptions in mind, once a certain topology is chosen:

- Given the acceptable phase noise, the tank resistor (R_p) is set. This is based on the assumption that the phase noise given by $\mathcal{L}\{\Delta\omega\} = \frac{kTFR_p}{A^2}\frac{1}{Q^2}\left(\frac{\omega_0}{\Delta\omega}\right)^2$ depends only on F, A, Q, and R_p. The first three parameters are mostly set based on technology, and area constraints.
- Once R_p is set, the oscillator bias current (I_{BIAS}), known as the oscillation amplitude, is known and given by $A = \frac{2}{\pi}I_{BIAS}R_p = 2V_{DD}$ (for an NMOS oscillator for instance).
- Assuming a small signal transconductance of g_m for the NMOS design for instance, the oscillator *small signal gain* is

$$\frac{g_m}{2}R_p \approx \frac{\frac{I_{BIAS}}{2}}{\frac{V_{effc}}{2}}R_p = \frac{\pi A}{8V_{effc}} = \frac{\pi V_{DD}}{4V_{effc}},$$

where V_{effc} is the NMOS core transistors overdrive voltage. Assuming a small signal gain of 5 is needed as a rule of thumb to ensure hard switching and start-up over PVT variations, we obtain $V_{effc} = \frac{\pi V_{DD}}{20}$. The device size is then chosen accordingly, which determines the device parasitic capacitance at the output.

- With R_p and Q known, the tank inductance $L_p = \frac{R_p}{Q\omega_0}$ is obtained.

With this choice of tank inductance, the nominal oscillation frequency may come out to be less than ω_0, depending on how large C_P is. If this is the case, a new value of inductance may need to be chosen, which sets a new R_p and I_{BIAS}. Often one or two iterations may be required to finalize the oscillator parameters and complete the design. Note that the new sets of parameters may lead to a different and possibly better phase noise if L_p and hence R_p need to be lowered. In consequence, while in an optimum design the bias current should be determined based on the required phase noise, the maximum oscillation frequency may invoke a higher current. However, this is usually not the case in most RF oscillators operating at several GHz.

9.4.5 Optimum Tank Q

A useful exercise would be to find the optimum frequency where the net LC tank quality factor is maximum. This is motivated by the fact that the switched capacitor Q usually monotonically reduces as the frequency increases, whereas the inductors' Q tend to increase by frequency, at least up to a certain frequency. Consequently, we set up the following exercise: We start with a 1.2nH differential inductor optimized for the best Q used in a 4GHz LC tank. The choice of 1.2nH is somewhat arbitrary, but a value in this range is quite typical for a 4GHz VCO. We further assume a required tuning range of 20% for the VCO. From the last section, this sets the ratio of the maximum to minimum total capacitance of the tank $\left(\frac{C_{MAX}}{C_{MIN}}\right)$ to about $1.5\times$. For that

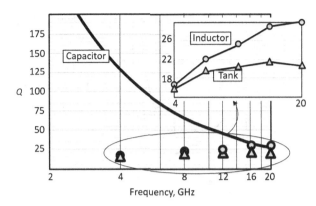

Figure 9.30: Simulated quality factor of the LC tank and individual capacitor and inductor

we design a switched capacitor array similar to the ones showed in Chapter 1. For this exercise, we chose a 16nm process, with roughly a simulated ratio of 5.5× maximum to minimum capacitance for each unit. In reality, the routing and other parasitic elements will degrade this somewhat. Four more cases are considered for frequencies of 8, 12, 16, and 20GHz, where in each case the inductance value is scaled such that $L_p\omega_0$ is kept constant. If the Q remains the same, so will the net tank quality factor, as well as $R_p = L_p\omega_0 Q$, but that is certainly not the case. As the frequency goes up, a smaller number of units is used for the capacitance, but the units as well as the $\frac{C_{MAX}}{C_{MIN}}$ ratio are maintained. For all cases the inductors are designed using only the thick RDL layer, and optimized for the best Q. Naturally, for lower inductances designed for higher frequencies, the physical size of the inductor reduces, though no particular constraint was put on the size. With the exception of 4GHz, all the inductors are single-turn with sizes varying from 160μm to 290μm.

The results are illustrated in Figure 9.30, where shown are the simulated quality factor over frequency for the switched capacitor unit cell, the inductors, and the overall LC tank.

As expected the tank Q raises quickly and somewhat flatten after 8GHz, reaching an optimum value of close to 21 at 16GHz. Beyond that point, the increase in the inductor Q is not much, while the capacitance Q linearly reduces with frequency, leading to a net drop in the tank Q. This result is clearly technology dependent, and is also a function of the assumed tuning range. Furthermore, the simulated quality factor and the maximum to minimum unit capacitance used here are optimistic, as the post-layout routing and other parasitic factors will kick in and degrade them. This overall leads to a shift of the optimum point to the left. Roughly speaking though, it appears that if one ever has an option of choosing an optimum frequency for the VCO, something around 10–15GHz gives the best trade-off between the L and C quality factors. Moreover, this frequency range is reasonably away from the point that the tank capacitance becomes too small, where the parasitic elements dominate.

9.5 Q-DEGRADATION

In the case of the standard NMOS and standard CMOS, the amplitude and the achievable FOM are limited by the headroom requirement of the current source. The question, therefore, arises as to why this current source is required at all.

Oscillators

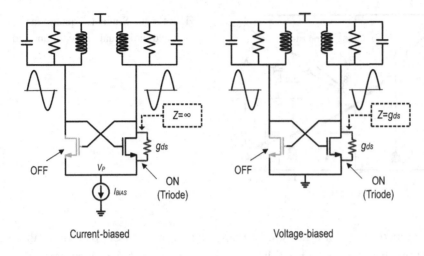

Figure 9.31: The current source in the standard topology prevents the differential pair loading the tank during triode operation

Looking at Figure 9.23, we can see that active element provides a negative conductance only when the input voltage is small. For larger voltages, one of the transistors in the differential pair will be *off*, while the other will be *on*, and the conductance drops to zero. This is because the on transistor will be degenerated by the high impedance of the current source. Even when the on transistor enters the triode region, the current through it will be completely determined by the current source. This is shown in Figure 9.31.

If the current source is removed, when either of the transistors enters the triode region a time-varying positive resistance due to the drain-source resistance (g_{ds}) is seen by the LC tank (see Figure 9.31 and Figure 9.32). In this case, the time-varying conductance is given by

$$g_{nr}(t) = -\left(\frac{g_{mL}(t) + g_{mR}(t)}{2}\right) + \left(\frac{g_{dsL}(t) + g_{dsR}(t)}{2}\right),$$

which consists of the required negative-conductance due to the transconductance of the devices (i.e., g_{mL} and g_{mR}), and the unwanted positive-conductance due to the drain-source resistance (i.e., g_{dsL} and g_{dsR}). This unwanted positive-conductance degrades the tank's Q and, as a result, the oscillator's phase noise and FOM. Accordingly, this *voltage-biased* oscillator should be avoided.

Even if the current source is retained, it is important that its impedance is large enough to degenerate the on transistor when the differential pair is hard-switched. This is straightforward to do at DC, but a large impedance is ideally needed at all even order harmonics with $2\omega_0$ being the most critical (in a perfectly balanced oscillator, only even-order harmonics appear at the source of the differential pair, i.e., node V_P in Figure 9.31). For high-frequency oscillators, parasitic capacitance at the source of the differential pair can reduce the impedance at these even-order harmonics frequencies and allow the differential pairs to load the tank. Therefore, the *current-biased* oscillator can behave like a *voltage-biased* oscillator. Figure 9.33 shows the effect of increasing capacitance on efficiency and FOM.

9.5 Q-Degradation

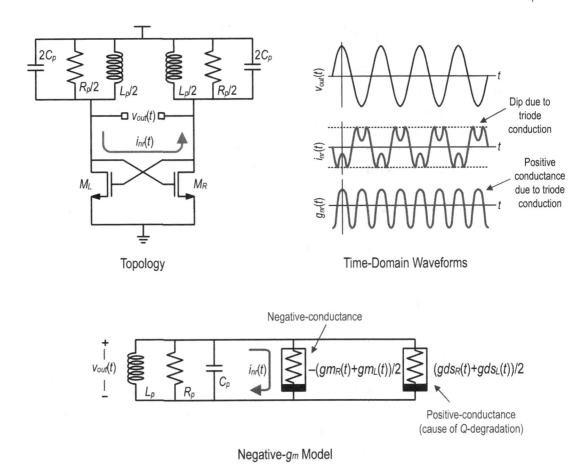

Figure 9.32: The voltage-biased LC oscillator

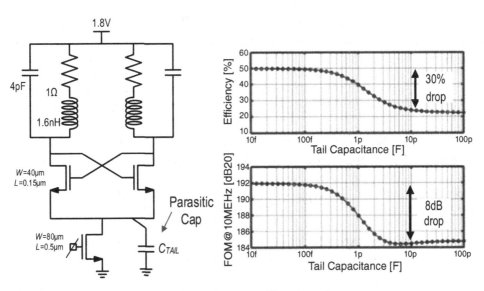

Figure 9.33: Effect of parasitic tail capacitor on oscillator's performance

9.6 FREQUENCY MODULATION EFFECTS

The noise mechanisms documented to this point suggests that flicker-noise only produces amplitude noise, which can be ignored. However, as anyone familiar with CMOS oscillator design can attest, flicker noise does appear in the output spectrum, and can often dominate in a bad design. The shortcoming in the analysis is because flicker noise and other low-frequency noise are up-converted to the output spectrum through frequency-modulation effects that were not captured in the model. The problem being that we assumed a fixed frequency defined by the equivalent noiseless oscillator. Therefore to augment the analysis, this section shows how low-frequency noise sources can cause perturbations in the oscillation frequency, which contributes directly to $\mathcal{L}\{\Delta\omega\}$. Two specific mechanisms are reviewed: one due to nonlinear reactive elements [19] and one due to harmonic currents [9].

9.6.1 Nonlinear Capacitance

We have so far assumed the RLC resonant tank is completely linear, but this is rarely the case in practical circuits. Device capacitance associated with the active circuit can be very nonlinear, and almost all VCOs employ varactors to provide some level of continuous tuning (as discussed in Chapter 1, the capacitance of varactors is a strong nonlinear function of the oscillation amplitude). Therefore, the amplitude of oscillation will generally have some effect on the frequency of oscillation. Assuming a linear inductance, which is reasonable for any integrated or bond-wire inductance, the oscillation frequency is generalized as

$$\omega_0(A) = \frac{1}{\sqrt{L\,C(A)}},$$

where the oscillation frequency is a function of amplitude because of the dependence of the nonlinear capacitance on amplitude. As a result, small changes in amplitude will also produce small changes in the oscillation frequency [19]. The rate of this AM–FM conversion is quantified as

$$K_{AMFM} = \frac{d\omega_0}{dA} = -\frac{1}{2\sqrt{LC(A)}}\frac{1}{C(A)}\frac{dC(A)}{dA}.$$

Because of this AM–FM conversion process, amplitude noise can produce small perturbations in frequency. The spectral density of this frequency modulation is calculated as

$$S_\Omega(\Delta\omega) = |K_{AMFM}|^2 \overline{v_{AM}^2} = \frac{\omega_0^2}{4C^2}\left|\frac{dC}{dA}\right|^2 \overline{v_{AM}^2},$$

where $\overline{v_{AM}^2}$ is the power spectral density of amplitude noise, which although small will never be zero. This can then be related to phase noise as follows:

$$\mathcal{L}\{\Delta\omega\} = \frac{S_\Omega(\Delta\omega)}{|\Delta\omega|^2} = \frac{|K_{AMFM}|^2 \overline{v_{AM}^2}}{|\Delta\omega|^2}.$$

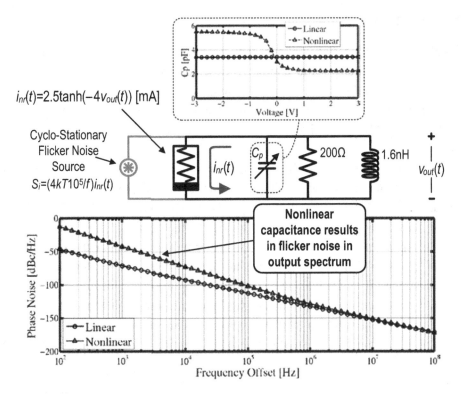

Figure 9.34: Nonlinear reactive elements results in flicker noise appearing in the output spectrum

Therefore if K_{AMFM} is large and the carrier offset $|\Delta\omega|$ is small, even a small amount of residual amplitude noise present in the oscillator can be converted into a significant amount of frequency noise, which as discussed previously is indistinguishable from phase noise.

This AM–FM conversion is generally the dominant mechanism through which flicker-noise and other low-frequency noise sources are up-converted to the output frequency. Consider the negative-g_m model shown in Figure 9.34: If the tank contains a linear capacitor, no flicker noise appears in the phase noise plot. However, if the linear capacitor is replaced with a nonlinear capacitor, flicker noise does appear in the phase noise profile.

Because of this AM–FM mechanism, it is not advisable to design a VCO with one large varactor. Instead, as explained in Chapter 1, a good VCO design will contain a digitally controlled capacitor bank to provide coarse frequency tuning, and a small varactor to provide just enough continuous tuning between digital codes. As a digitally controlled capacitor bank can be much more linear than a varactor, flicker noise upconversion can be minimized [20].

It should be noted that the nonlinear capacitance of the active element can also present significant nonlinear capacitance, as shown in Figure 9.35. In this case, the nonlinear capacitance is a strong function of the tail capacitance, which generally results in a larger value of $|dC/dA|^2$, which is yet another reason to minimize the tail capacitance.

Finally it is important to remember that this AM–FM mechanism is dependent on the value of $|dC/dA|^2$, not on the total capacitance. Therefore for a specific amplitude, A, it is possible that $|dC/dA|^2$ will be nulled and the flicker noise upconversion will be eliminated [19]. However, since nonlinear capacitance is difficult to model, and tight control over the oscillation amplitude is not easy, it is always best to minimize the total amount of nonlinear capacitance rather than seeking to $|dC/dA|^2$ for a specific value of A.

9.6.2 Effective Nonlinear Capacitance

More insight into the impact of nonlinear tank capacitance, and particularly the varactor, may be obtained using the quasi-static approach presented in [19]. Shown in Figure 9.36 is a schematic of a simplified LC oscillator tank, along with typical MOS varactor characteristics. The varactor C–V curve was analyzed in Chapter 1, and was indicated there that an accumulation-mode MOS capacitance is commonly employed, which results in the characteristics shown below on the left.

For a standalone MOS varactor, it was shown that the threshold where the capacitance goes from low to high is around 0V, where the device moves from depletion region to accumulation. In an actual VCO on the other hand, this depends on the control voltage and the VCO output DC bias, which is typically at V_{DD}. As the control voltage is also limited to supply, one may employ an independent DC bias to achieve a wider tuning range for the same varactor size (Figure 9.36 right side). This however comes at the expense of additional parasitic capacitance, and imposes a trade-off on the biasing resistor size to optimize the loading on the tank, and the resistor noise contribution. To generalize, we label the transition threshold as an arbitrary voltage V_0, which is a function of the control voltage, and the varactor biasing (V_{DD} or otherwise), and that can be adjusted as needed. Assuming a nearly sinusoidal VCO output

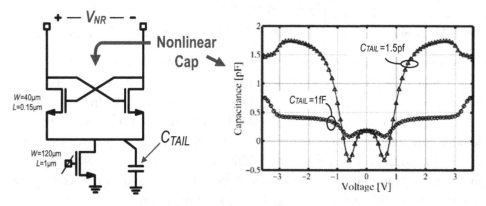

Figure 9.35: The active element can be a source of significant nonlinear capacitance. The presence of a large tail capacitance can significantly affect the nonlinearity

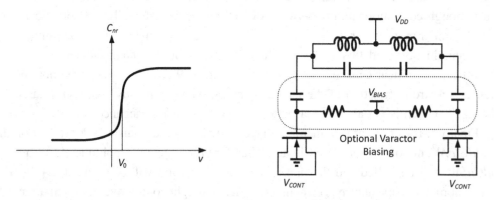

Figure 9.36: A typical CMOS VCO tank along with the varactor characteristics

voltage of $A\cos\omega_0 t$, at steady state, a certain effective capacitance (C_{eff}) is presented to the tank which not only depends on V_0, but also is a function of the VCO output amplitude, A. While this was already shown in Chapter 1, we shall present an approximate mathematical expression for C_{eff} here based on [19].

9.6.2.1 Graphical Interpretation

Consider a memoryless system, such as the nonlinear resistor employed in the VCO to sustain steady oscillation, whose *I–V* curve is shown in Figure 9.37 left side. Since the negative resistance is memoryless, the area described by the instantaneous point of *work* in the (i_{nr}, v_{nr}) coordinate must be zero for a total period of the fundamental frequency. Consequently, the line integral of the *I–V* locus must be equal to zero:

$$\oint i_{nr} dv_{nr} = 0.$$

That is to say, moving along the curve back and forth results in no net work, as it can be clearly seen from the figure. On the contrary, if the system has hysteresis for instance, moving back and forth takes place on two different paths, leading to a nonzero line integral.

Now, let us consider a linear (for the moment) capacitance of C, where a sinusoidal voltage of $A\cos\omega_0 t$ is applied across its terminals. The capacitor has memory, and intuitively we expect a nonzero line integral. We have

$$v_C(t) = A\cos\omega_0 t$$

$$i_C(t) = -AC\omega_0 \sin\omega_0 t.$$

Consequently, we can write

$$\left(\frac{i_C}{AC\omega_0}\right)^2 + \left(\frac{v_C}{A}\right)^2 = 1,$$

which indicates an ellipse in the (i_c, v_c) coordinate (Figure 9.37 right side). The line integral $\oint i_C dv_C$ is clearly no longer zero, and is described by the *area of the ellipse*:

$$\oint i_C dv_C = \pi A^2 C\omega_0.$$

If the capacitor is nonlinear, as in the case of a varactor, the *I–V* locus driven by a periodic waveform consists of several ellipses of different heights but the same width. This will be clarified further next, when a simplified MOS varactor is analyzed.

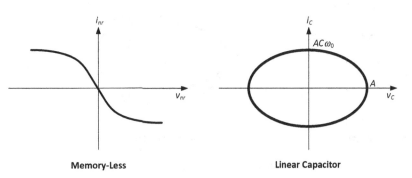

Figure 9.37: Comparison of *I–V* characteristics of a memoryless system with a capacitor

9.6.2.2 Mathematical Expression

Consider Figure 9.38, which shows a MOS varactor driven by a sinusoidal periodic signal of amplitude A.

For simplicity, let us assume the varactor varies between two values of C_L and C_H, with a step transition as the voltage reaches V_0. The corresponding capacitance over time, along with its I–V locus, are shown in Figure 9.39.

Let us define the small signal nonlinear capacitance as $C_{nr}(t) = \frac{dQ}{dV}$, whose value over time is shown above for the simplified MOS varactor. Since the nonlinear capacitance is periodic, we may describe it using a Fourier series:

$$C_{nr}(t) = \sum_{m=-\infty}^{+\infty} C_{nr}[m] e^{jm\omega_0 t}.$$

Note that the fixed (or linear part of the) tank capacitance is inconsequential in this analysis, and will be dealt with shortly. The tank voltage (which is the same as the capacitance voltage) is periodic as well, and may be expressed as

$$v_C(t) = \sum_{n=-\infty}^{+\infty} V_C[n] e^{jn\omega_0 t}.$$

Thus, the inductor current is

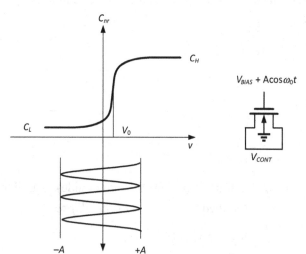

Figure 9.38: Varactor C–V curve swept by the VCO large sinusoidal output

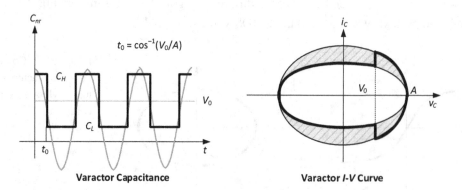

Figure 9.39: Nonlinear capacitance over time and its I–V locus

$$i_L(t) = \frac{1}{L}\int v_C(t)dt = \frac{1}{L}\sum_{n=-\infty}^{+\infty}\frac{V_C[n]}{jn\omega_0}e^{jn\omega_0 t}.$$

The capacitor current is

$$i_C(t) = \frac{dQ}{dt} = \frac{dQ}{dv_C}\frac{dv_C}{dt} = C_{nr}(t)\frac{d}{dt}\sum_{n=-\infty}^{+\infty}V_C[n]e^{jn\omega_0 t}.$$

After expansion,

$$i_C(t) = \sum_n\sum_m jn\omega_0 C_{nr}[m]V_C[n]e^{j(m+n)\omega_0 t}.$$

KCL demands,

$$i_C(t) + i_L(t) = \frac{1}{L}\sum_{n=-\infty}^{+\infty}\frac{V_C[n]}{jn\omega_0}e^{jn\omega_0 t} + \sum_n\sum_m jn\omega_0 C_{nr}[m]V_C[n]e^{j(m+n)\omega_0 t} = 0.$$

The *quasi-static approximation* entails a nearly sinusoidal tank voltage, i.e.,

$$V_C[1] = V_C[-1] = \frac{A}{2}$$

$$V_C[n] \approx 0, \text{ for } n \neq \pm 1.$$

As a consequence, with higher harmonics ignored, we have

$$\frac{\frac{A}{2}}{jL\omega_0}\left(e^{j\omega_0 t} - e^{-j\omega_0 t}\right) + j\omega_0\frac{A}{2}\left[\left(C_{nr}[0] - \frac{1}{2}C_{nr}[2]\right)e^{j\omega_0 t} - \left(C_{nr}[0] - \frac{1}{2}C_{nr}[-2]\right)e^{-j\omega_0 t}\right] = 0,$$

which simplifies to

$$\frac{1}{jL\omega_0} + j\left(C_{nr}[0] - \frac{1}{2}C_{nr}[2] - \frac{1}{2}C_{nr}[-2]\right)\omega_0 = 0.$$

Note that since $C_{nr}(t)$ is an even symmetric function, $C_{nr}[m] = C_{nr}[-m]$ for all m.

Thus, we shall define the effective tank capacitance as

$$C_{eff} = C_{nr}[0] - \frac{1}{2}C_{nr}[2] - \frac{1}{2}C_{nr}[-2].$$

The first term is the *time-average capacitance*, which includes any *fixed* linear capacitance in parallel with the nonlinear tank capacitance, while the other two terms are described by the nonlinear capacitance 2nd-order Fourier coefficients. Clearly, the time-average capacitance alone does not correctly account for the effective capacitance, as it does not properly deal with the balance of the tank voltage and current. Note that a very similar expression was derived in Section 9.2.2 for the *effective conductance* of the nonlinear active element $\left(G_{eff} = G_{nr}[0] - \frac{1}{2}G_{nr}[-2] - \frac{1}{2}G_{nr}[2]\right)$.

The above expression is not entirely accurate, as it neglects the effect of mixing of higher harmonics of the nonlinear capacitance with higher order components of the tank voltage. If the tank quality factor is large, however, this expression is reasonably correct.

Note that, as shown in Problem 14, we can say in general

$$C_{\text{eff}} = \frac{\oint i_C dv_C}{\pi A^2 \omega_0},$$

which is obvious for the linear capacitance, and may be easily extended to a nonlinear capacitance as well.

9.6.3 Groszkowski Effect

In analyzing LC oscillators we have thus far assumed for the most part that the output is sinusoidal, this is to say the harmonics of the output waveform have been neglected entirely. Although small, these harmonics do exist (even in a perfectly linear LC tank), and can affect the frequency of oscillation. In 1933, Groszkowski [21] explained and quantified how this harmonic current produces a *downward shift* in the oscillation frequency.

Shown in Figure 9.40 is a simplified LC oscillator, where the active element is excited by the output waveform $v_{nr}(t)$, and generates an energy-restoring current $i_{nr} = f(v_{nr})$, with significant harmonic content. The fundamental of this current compensates for tank losses, while higher order harmonics (which must go somewhere) flow through the least resistance path, i.e., capacitor. Since the negative resistance is memoryless, as we showed previously, the area described by the instantaneous point of work in the (i_{nr}, v_{nr}) coordinate must be zero for a total period of the fundamental frequency:

$$\oint i_{nr} dv_{nr} = 0.$$

Switching to the frequency domain, this condition leads to the following identity

$$\sum_{k=-\infty}^{\infty} k I_{nr}[k] V_{nr}[-k] = 0,$$

where $I_{nr}[k]$ and $V_{nr}[k]$ are the kth Fourier coefficients of the current and voltage, respectively. This can be further rewritten as

$$\sum_{k=-\infty}^{\infty} \frac{k|V_{nr}[k]|^2}{Z[k]} = 0,$$

where $Z[k]$ is the tank impedance at the kth harmonic. Collecting the fundamental components on one side of the equation, this becomes

$$\frac{|V_{nr}[1]|^2}{|Z[1]|} = -\sum_{k=2}^{\infty} k \frac{|V_{nr}[k]|^2}{|Z[k]|}.$$

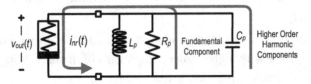

Figure 9.40: Higher-order harmonics of the current from the active element flows through the capacitor. This harmonic content produces a downward shift in the oscillation frequency (compared to that predicted by linear analysis).

Around the fundamental, the tank impedance is given by

$$|Z[1]| \approx \frac{L_p \omega_0^2}{2\Delta\omega},$$

where $\Delta\omega$ is the shift in frequency with respect to the natural resonant frequency, and $\omega_0 = 1/\sqrt{L_p C_p}$. At higher harmonics ($k > 1$), the tank impedance is capacitive, and its magnitude is approximately equal to

$$|Z[k]| \approx \frac{k}{(k^2 - 1)} L_p \omega_0.$$

Thus,

$$\frac{\Delta\omega}{\omega_0} \approx -\frac{1}{2} \sum_{k=2}^{\infty} (k^2 - 1) \frac{|V_{nr}[k]|^2}{|V_{nr}[1]|^2}.$$

Alternatively, we could have expressed the results in terms of harmonic *current* content,

$$\frac{\Delta\omega}{\omega_0} \approx -\frac{1}{2Q^2} \sum_{k=2}^{\infty} \frac{k^2}{k^2 - 1} \frac{|I_{nr}[k]|^2}{|I_{nr}[1]|^2},$$

where $Q = \frac{R_p}{L_p \omega_0}$ is the tank quality factor.

If no harmonic components are present, evidently $\Delta\omega/\omega_0 = 0$, which is satisfied at the natural resonant frequency of the tank. However, if significant harmonic content exists, the identity demands that $\Delta\omega/\omega_0 < 0$, which is satisfied only when oscillation frequency is less that natural resonant frequency of the tank. Naturally a higher Q will result in a smaller frequency shift for the same harmonic current.

This effect may be also qualitatively explained as follows: In the presence of harmonic current, the electrostatic energy stored in the capacitor is larger than if harmonics did not exist. Because the active circuit is memoryless, this extra energy must be balanced by the inductor, which is accomplished by a downward shift in the oscillation frequency, thereby allowing it to store more electromagnetic energy.

We have previously deduced that in the case of an oscillator with linear reactive elements, flicker noise will only modulate the amplitude of the oscillation waveform, and will not produce phase noise. This, however, was based on the nearly sinusoidal approximation. If the harmonic content of the output is not ignored, noise that modulates any harmonic of the output waveform will produce frequency noise, which directly contributes to $\mathcal{L}\{\Delta\omega\}$ [9].

To eliminate this frequency modulation effect, the active element should be made as linear as possible, so as to not introduce any harmonic content. It should be noted, however, that designs with hard-switching nonlinearities are much more power efficient and, so, using a linear (or soft-limiting) active element is not advisable for a CMOS LC oscillator. Moreover, even with a linear active-element, frequency modulation effects due to nonlinear reactive elements would still be present.

9.6.4 Supply Pushing

Low-frequency supply noise and spurs, just like flicker noise, can produce AM sidebands, which are in turn converted to frequency noise through nonlinear capacitance. The magnitude of this "supply pushing" is an important consideration, especially in modern SoCs, which may have relatively noisy supplies. For this reason, the supply of a VCO is typically generated by a separate low-dropout regulator (LDO), which isolates the circuit from noise elsewhere on the chip.

9.7 MORE LC OSCILLATOR TOPOLOGIES

Improved understanding of the operation of LC oscillators has led to the development of topologies with enhanced FOMs. Two important topologies are now discussed, which achieve close to the maximum theoretical FOM of any one-port LC oscillator.

9.7.1 The Standard Topology with Noise Filter

In the standard NMOS topology, the current source ultimately limits the achievable FOM. If a large current source device is used, its transconductance will be large and will inject a large amount of noise into tank. If a small device is used, the headroom requirement will be large, and the oscillation amplitude will be reduced. However, as noted previously, the current source needs only to present a high impedance around the second harmonic, and so it is possible to employ a simple passive network that resonates at $2\omega_0$ [22]. As shown in Figure 9.41, a large device that functions as a current source can be retained, but its noise is filtered off using a large capacitor. A series inductor is then designed to resonate at $2\omega_0$ using the parasitic capacitance at the source node of differential pair transistors. In this way the current source introduces no noise, the differential pair does not load the tank, and the headroom of the current source can be made arbitrarily small. As a result, the noise factor is simply

$$F = 1 + \gamma,$$

while, assuming the output is driven rail to rail, the efficiency can approach (and even exceed) $\eta = 2/\pi = 63.7\%$. Accordingly the FOM is given by

$$\text{FOM} = \frac{4/\pi}{(1+\gamma)} \frac{Q^2}{kT} 1\text{mW},$$

which is only 2dB less than the maximum achievable FOM of a CMOS oscillator. This FOM is preserved even if the tail transistor is removed, and the oscillator is voltage-biased.

To emphasize the power of this common-mode resonance consider the idealized oscillator shown in Figure 9.42. The transistors are thick oxide devices from a 28nm process. The tail inductor is half the size of the tank inductance and has a similar Q for the same frequency. A tail capacitance is then swept and the FOM is plotted. For a tank of quality factor of 12, the maximum achievable FOM is 196. When the tail inductor resonates at $2\omega_0$ (when the tail capacitance is equal 6.25pF), the FOM of this oscillator is within 1dB of this value. Equally the

9.7 More LC Oscillator Topologies

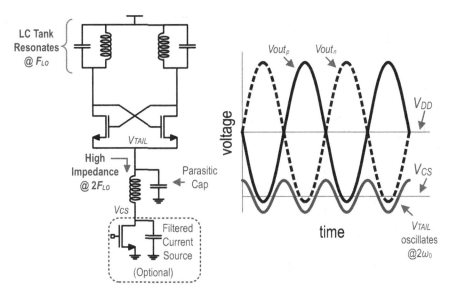

Figure 9.41: The standard NMOS topology with noise filter

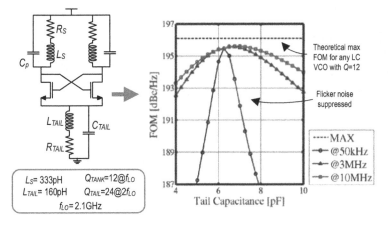

Figure 9.42: Importance of common-mode resonance on FOM

flicker noise is almost entirely eliminated as the FOM is preserved whether it is measured at 50kHz or at a 10MHz offset. Note that the current switched into the tank is not a perfect square, but rather as a waveform with finite rise/falls times, which results in more energy at ω_0 and better efficiency.

The standard CMOS oscillator can also employ this technique. In order to prevent the PMOS transistor loading the tank, it is advisable to include a third inductor series inductor as shown in Figure 9.43.

It is also possible to employ Hegazi's common-mode resonance technique without the use of an additional inductor. This is because in any differential oscillator design the tank has two distinct modes, i.e., a common mode and a differential mode. If we ensure that common-mode resonance frequency occurs at twice the differential resonance frequency, all the benefits of Hegazi's oscillator will be reaped without the need for an additional inductor. These two different modes are shown in Figure 9.44, and can be controlled by either adjusting the

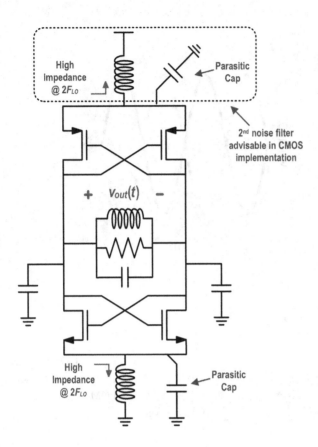

Figure 9.43: The standard CMOS topology with two noise filters

differential inductors magnetic coupling, k, or adjusting the arrangement of capacitors in the tank.

9.7.2 The Class C Topology

As shown in Figure 9.45, if the active element of an LC oscillator converts the DC supply current into a square wave at ω_0, the maximize possible power efficiency is given by

$$\eta = \frac{2}{\pi} \frac{A}{V_{DD}},$$

which reaches a maximum of 63.7% when $A = V_{DD}$. If, instead, the supply current is injected into the tank as series of impulses, the efficiency becomes

$$\eta = \frac{A}{V_{DD}},$$

which can potentially reach 100% when $A = V_{DD}$. This is analogous to class C power amplifiers (see Chapter 11).

9.7 More LC Oscillator Topologies

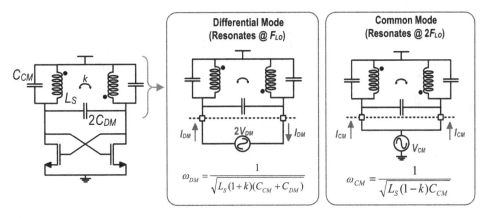

Figure 9.44: An oscillator with implicit common-mode resonance

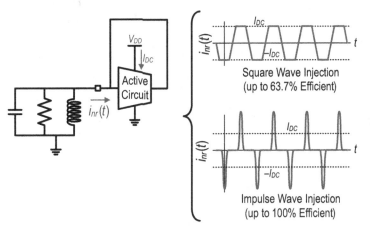

Figure 9.45: Injecting the energy-restoring current as an impulse can improve oscillator efficiency

In [2], an active element was proposed that sought to generate this impulse injection. As shown in Figure 9.46, by introducing a DC offset in the biasing the differential pair transistors, the majority of the current is injected as an impulse at the peak and trough of each oscillation. The offset effectively increases the *turning on instant* of each device, and so a larger voltage swing is required, which results in the current being injected only when the instantaneous voltage is large. The complete circuit, known as a class C topology, is shown in Figure 9.47. If the differential pair transistors don't enter the triode region, and the current source noise is stripped off with a large capacitor,[13] the noise factor is given by $F = 1 + \gamma$ (because of Bank's general result), and the FOM can be calculated as

$$\text{FOM} = \frac{1}{(1+\gamma)} \frac{A}{V_{DD}} \frac{Q^2}{kT} \text{1mW}.$$

---

[13] It is suggested [2] that the size of the capacitor should be limited to two to three times the total tank capacitance in order to avoid an amplitude instability effect known as "squegging."

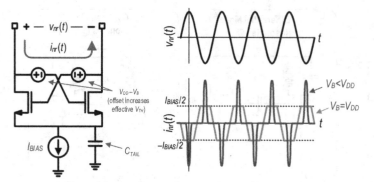

Figure 9.46: The active element of the class C topology. The shift in the DC bias of differential pair transistors is modeled with ideal series voltage sources

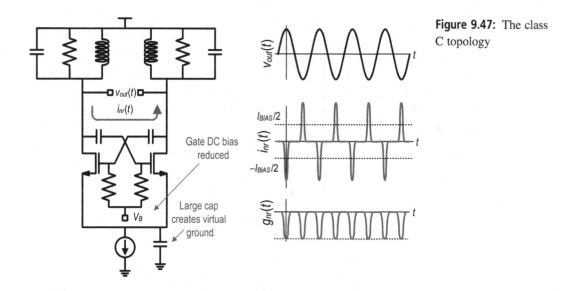

Figure 9.47: The class C topology

Therefore, if the A can be made close to $2V_{DD}$, the topology can theoretically achieve close to the maximum achievable FOM of an LC oscillator. Published performance metrics [2] show that this topology can achieve a similar FOM to that of the standard topology employing a tail filter [22], and has the advantage not requiring an additional inductor. Furthermore, biasing the devices closer to threshold leads to a lower power design compared to the standard NMOS topology. Hence, it is fair to say that in typical applications, the class C topology tends to offer a better figure of merit and a lower bias current, but not necessarily a better phase noise.

9.8 RING OSCILLATORS

Ring oscillators are another well-known type of CMOS oscillator. This section presents a simple overview of their operation and noise performance with a particular focus on the inverter-based design. The primary goal of this section is to show that the phase noise per unit power of a ring oscillator (i.e., FOM) is significantly worse than that of an LC oscillator.

9.8.1 Basic Operation

At its simplest a ring oscillator consists of a chain of an odd number of inverting amplifiers connected in unity feedback. A basic inverter-based oscillator is shown in Figure 9.48. Assuming, the small signal gain of the amplifier chain is greater than 1, and the delay of the chain is nonzero (which is always the case in a real design), the circuit will oscillate. In the case of the inverter-based amplifier, the oscillation period will be equal to the time a particular transition edge takes to propagate twice around the loop, which is calculated as

$$f_0 = \frac{1}{T_0} = \frac{1}{N(\tau_{inv(r)} + \tau_{inv(f)})},$$

where T_0 is oscillation period, N is the number of inverter stages, $\tau_{inv(r)}$ is the propagation delay of the inverter given a rising edge output, and $\tau_{inv(f)}$ is the delay of the inverter given a falling edge output. Assuming the inverters are sized to give approximately equal rising and falling edges (i.e., $\tau_{inv} = \tau_{inv(r)} = \tau_{inv(f)}$), the oscillation frequency simplifies to

$$f_0 = \frac{1}{2N\tau_{inv}}.$$

Therefore, the frequency of a ring oscillator can be increased by reducing the number of stages (ensuring that N is always odd and greater than or equal to 3), or decreasing the inverter delay. Given that the total load capacitance at the output of each inverter is completely charged and discharged every cycle, the power burned by each inverter can be approximated as

$$P_{INV} \approx C_{LOAD} V_{DD}^2 f_0,$$

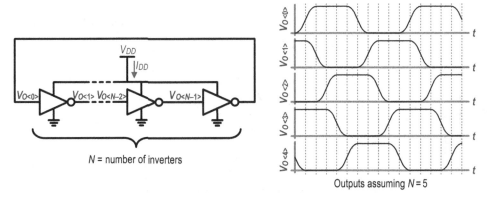

Figure 9.48: A simple inverter-based ring oscillator. N should be odd and greater than 3

where C_{LOAD} is the total load capacitance perceived by the inverter, and V_{DD} is the supply voltage. The power dissipated by the oscillator is simply the sum of the power dissipated in each inverter, i.e.,

$$P_{OSC} = NP_{INV} \approx NC_{LOAD}V_{DD}^2 f_0,$$

which gives a total oscillator current of

$$I_{DD} \approx NC_{LOAD}V_{DD}f_0.$$

Typically it is best to calculate the inverter delay through simulation. Consider the inverter shown in Figure 9.49. If the input changes instantaneously, the propagation delay given a rising edge output (i.e., the time for the output to charge up to $V_{DD}/2$) is calculated as

$$\tau_{inv(r)} = \int_0^{V_{DD}/2} \frac{C_{LOAD}(v)}{I_{PMOS}(v)} dv,$$

where I_{PMOS} is the current through the PMOS device. In general, both the device current and the capacitance, C_{LOAD}, are strong nonlinear functions of the output voltage. Moreover, the equation does not even account for an input with a finite slope. This makes calculation of an accurate close-form expression for the delay of an inverter embedded in a ring oscillator difficult [23].

Nevertheless, some valuable design insight can be attained by making some crude simplifications. Rewriting the expression for oscillator current in terms of frequency gives

$$f_0 \approx \frac{1}{N} \frac{I_{DD}}{C_{LOAD}V_{DD}}.$$

Now, at any given instance one node in the oscillator is being pulled up and one node is being pulled down. Assuming the inverters are sized to give equal rise and fall times, we can say that at all times one PMOS device is conducting (charging up a node) and one NMOS device is conducting (discharging a different). Therefore we can get a rough estimate of the upper bound

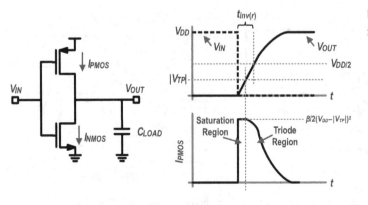

Figure 9.49: Response of a CMOS inverter to a step function

of I_{DD} as the quiescent current of one fully on NMOS (or PMOS) device, which assuming the square law model is given by

$$I_{DD} \approx \frac{1}{2}\beta(V_{DD} - |V_{th}|)^2,$$

where $\beta = \mu(W/L)C_{OX}$. This gives an approximate upper bound on the oscillation frequency of

$$f_0 \approx \frac{1}{2N} \frac{\beta(V_{DD} - |V_{th}|)^2}{C_{LOAD}V_{DD}}.$$

Therefore, as well as adjusting the number of stages, adjusting C_{LOAD}, V_{DD}, and device sizes (via $\beta \propto W/L$) can be used to set the oscillation frequency. Note that if C_{LOAD} arises primarily from the inverter gate capacitance (i.e., $C_{LOAD} \propto WLC_{OX}$), increasing the device width will not have a strong effect on frequency since it increases both β and C_{LOAD} in the same proportion. If, however, C_{LOAD} is dominated by the routing capacitance, the buffer load, or is simply an explicit fixed capacitance, increasing the device width will increase the oscillation frequency (this is typically the case in most practical designs where the oscillator is needed to drive a load). Naturally a more accurate model will provide a more accurate frequency prediction, but the qualitative effect of N, C_{LOAD}, V_{DD}, and β on the oscillation frequency is valid.

9.8.2 Estimating Phase Noise in Hard-Switching Circuits

The output of a ring oscillator generally toggles between supply rails and so resembles a square wave rather than a sinusoid as was the case for an LC oscillator. Therefore it is more natural to describe the frequency instability of such an oscillator in terms of the time deviation of its period from the expected period of a noiseless oscillator as measured at waveform transitions.

Shown in Figure 9.50 is the simple case of an inverter toggled periodically between 0 and V_{DD}. In the presence of device noise, the transition time can be seen to either advance or retard with respect to a noiseless device. By sampling the output noise at the expected (i.e., noiseless) transition time, a good estimate of the phase noise can be calculated. As highlighted in Figure 9.50, the sampled phase noise at the rising and falling edges can be defined as

$$\phi_{\delta r}(t) = \frac{2\pi}{T_0}\frac{1}{\lambda_r}v_{\delta r}(t)$$

$$\phi_{\delta f}(t) = \frac{2\pi}{T_0}\frac{1}{\lambda_f}v_{\delta f}(t),$$

where λ_r and λ_f are the rising and falling edge slopes at the transition point, and the sampled noise voltage is given by

570 Oscillators

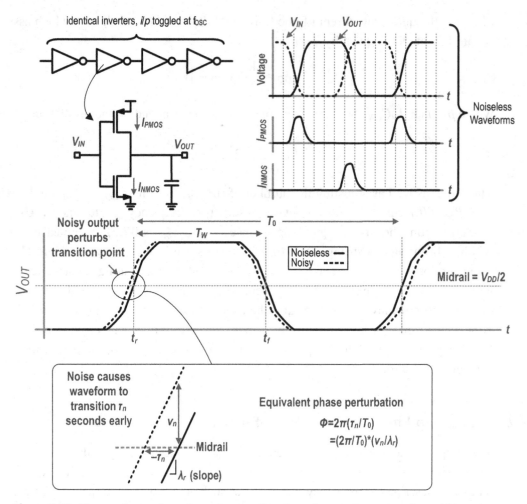

Figure 9.50: Phase noise can be estimate by sampling the voltage at the expected transition point

$$v_{\delta r}(t) = \sum_{n=-\infty}^{\infty} v_n(t)\delta(t - t_r - nT_0)$$

$$v_{\delta f}(t) = \sum_{n=-\infty}^{\infty} v_n(t)\delta(t - t_f - nT_0),$$

where $v_n(t)$ is the noise voltage of output waveform, t_r is the time of the first (noiseless) rising transition through the midrail (i.e., $V_{DD}/2$), and t_f is the time of the first (noiseless) falling transition through the midrail. If we assume this sampled phase noise persists between the measuring instances, we can say that a good estimate of phase instability (i.e., $\phi(t)$) is a sampled and held version of the sampled phase noise, i.e.,

$$\phi_{EST}(t) = \phi_{\delta r}(t) * w_r(t) + \phi_{\delta f}(t) * w_f(t).$$

The symbol $*$ denotes convolution and $w_r(t)$ and $w_f(t)$ are unity waveforms given by

$$w_r(t) = \begin{cases} 1, & 0 \leq x < T_W \\ 0, & \text{otherwise} \end{cases}$$

$$w_f(t) = \begin{cases} 1, & 0 \leq x < T_0 - T_W \\ 0, & \text{otherwise} \end{cases},$$

where T_W is time between a rising and falling transition (i.e., T_W/T_0 is the duty cycle of the waveform; see Figure 9.50). While presented as a theoretical construct, this estimate is essentially the phase deviation perceived by subsequent circuits after the output of the oscillator has passed through hard switching buffers. For simplicity, let's assume the duty cycle is 50% and the rise and falls times are equal (i.e., $T_W = T_0/2$, $|\lambda| = |\lambda_r| = |\lambda_f|$, and $t_f = t_r - T_0/2$), which gives an estimate of phase deviation as

$$\phi_{EST}(t) = (\phi_{\delta r}(t) + \phi_{\delta f}(t)) * w(t),$$

or, in terms of sampled voltage,

$$\phi_{EST}(t) = \frac{2\pi}{T_0} \frac{1}{|\lambda|} (v_{\delta r}(t) - v_{\delta f}(t)) * w(t),$$

where

$$w(t) = \begin{cases} 1, & 0 \leq x < T_0/2 \\ 0, & \text{otherwise} \end{cases}.$$

If the rising and falling edge voltages are uncorrelated (as is generally the case for the noise of a buffer), the spectral density becomes

$$S_{\phi_{EST}}(\omega) \approx \left(S_{\phi_{\delta r}}(\omega) + S_{\phi_{\delta f}}(\omega)\right) \left(\frac{\sin(\frac{\omega T_0}{4})}{\omega/2}\right)^2.$$

In a ring oscillator the noises associated with the rising and falling edges are strongly correlated (i.e., $v_{\delta f}(t) \approx -v_{\delta r}(t - T_{OSC}/2)$), which implies the spectral density is closer to

$$S_{\phi_{EST}}(\omega) \approx 4 S_{\phi_\delta}(\omega) \left(\frac{\sin(\frac{\omega T_0}{4})}{\omega/2}\right)^2,$$

where $S_{\phi_\delta}(\omega)$ is the spectral density of the phase noise sampled at either a rising or falling edge. The reason rising and falling edges are correlated will become obvious in the next section. Generally, we are concerned only with the low-frequency component of ω, which is responsible for noise close to the carrier. Therefore, we can say that a reasonable estimate of phase noise is related to the spectral density of $\phi_\delta(t)$ as follows,

Oscillators

$$\mathcal{L}\{\Delta\omega\} \approx S_{\phi_{EST}}(\Delta\omega) \approx S_{\phi_\delta}(\Delta\omega)T_0^2,$$

where $\Delta\omega \ll \omega_{OSC}$. Alternatively, we can define phase noise in terms of the spectral density of the sampled noise at the transition point, i.e.,

$$\mathcal{L}\{\Delta\omega\} \approx \frac{4\pi^2}{|\lambda|^2} S_{v_\delta}(\Delta\omega).$$

Again, this estimate can be measured at either a rising or falling edge, since, in the context of ring oscillators, the noise on a rising and falling edge will be strongly correlated. In the case of uncorrelated noise on the rising and falling edge noise, this equation would be approximately given by

$$\mathcal{L}\{\Delta\omega\} \approx \frac{\pi^2}{|\lambda|^2} \left(S_{v_{\delta r}}(\Delta\omega) + S_{v_{\delta f}}(\Delta\omega)\right).$$

9.8.3 Simple Ring Oscillator Noise Model

A ring oscillator can be viewed as a delay line with unity feedback, which enables a simplified noise analysis [24]. Figure 9.51 models the inverter chain as a delay line, and models the total noise of inverter chain as an additive noise source, $v_{CHAIN}(t)$. Given this setup, the noise voltage of the oscillator is given by

$$v_{RING}(t) = -v_{RING}\left(t - \frac{T_0}{2}\right) + v_{CHAIN}(t),$$

alternatively,

$$v_{RING}(t) = v_{RING}(t - T_0) + v_{CHAIN}(t) - v_{CHAIN}\left(t - \frac{T_0}{2}\right).$$

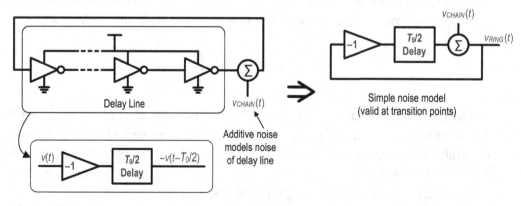

Figure 9.51: Simplified noise model of ring oscillator

This model is only valid around the transition points, where the delay line gain is −1. If we sample the noise voltage at these transition points (i.e., a rising or falling edge), we may write

$$v_{\delta RING(r)}(t) = v_{\delta RING(r)}(t - T_0) + v_{\delta CHAIN(r)}(t) - v_{\delta CHAIN(f)}(t)$$

$$v_{\delta RING(f)}(t) = v_{\delta RING(f)}(t - T_0) + v_{\delta CHAIN(f)}(t) - v_{\delta CHAIN(r)}(t),$$

where

$$v_{\delta RING(r)}(t) = \sum_{n=-\infty}^{\infty} v_{RING}(t)\delta(t - nT_0)$$

$$v_{\delta RING(f)}(t) = \sum_{n=-\infty}^{\infty} v_{RING}(t)\delta\left(t - \frac{T_0}{2} - nT_0\right)$$

$$v_{\delta CHAIN(r)}(t) = \sum_{n=-\infty}^{\infty} v_{CHAIN}(t)\delta(t - nT_0)$$

$$v_{\delta CHAIN(f)}(t) = \sum_{n=-\infty}^{\infty} v_{CHAIN}(t)\delta\left(t - \frac{T_0}{2} - nT_0\right).$$

To simplify matters it is assumed that one of the rising edge transition points coincides with $t = 0$. Moreover, we assume the rising and falling edge transition points are $T_{OSC}/2$ seconds apart (i.e., a 50% duty cycle).

In general, the noise due to the rising and falling edge of a delay line will be uncorrelated (for instance, in a simple inverter the rising edge noise will be dominated by the PMOS, while the falling edge will be dominated by the NMOS device). Therefore, in the frequency domain, the spectral density of the voltage noise of the oscillator measured at either the rising or falling edge is given by

$$S_{v_\delta(RING)}(\omega) = \frac{S_{v_\delta(CHAIN-r)}(\omega) + S_{v_\delta(CHAIN-f)}(\omega)}{|1 - e^{-j\omega T_0}|^2} = \frac{S_{v_\delta(CHAIN-r)}(\omega) + S_{v_\delta(CHAIN-f)}(\omega)}{\frac{1}{4}\sin^2\left(\frac{\omega T_0}{2}\right)},$$

where $S_{v_\delta(CHAIN-r)}$ and $S_{v_\delta(CHAIN-f)}$ are the spectral density of the noise voltage of the chain sampled at the rising and falling edge, respectively. We are concerned only with components of $S_{v_\delta(RING)}$ that lie close to harmonics, since it is only noise close to the harmonics that has the ability to be frequency translated close to the carrier. Moreover, because of impulse sampling, the noises around all the harmonics are identical, and so we need only look at the very low frequency of $S_{v_\delta(RING)}(\omega)$. Therefore, using the small angle approximation, we can write

$$S_{v_\delta(OSC)}(\Delta\omega) \approx \left(S_{v_\delta(CHAIN-r)}(\omega) + S_{v_\delta(CHAIN-f)}(\omega)\right)\frac{1}{4\pi^2}\left(\frac{\omega_{OSC}}{\Delta\omega}\right)^2,$$

where $\Delta\omega \ll \omega_{OSC}$. This expression can be used to generate an estimate of $\mathcal{L}\{\Delta\omega\}$, i.e.,

$$\mathcal{L}\{\Delta\omega\} \approx \left|\frac{1}{\lambda}\right|^2 \left(S_{v_\delta(CHAIN-r)}(\omega) + S_{v_\delta(CHAIN-f)}(\omega)\right)\left(\frac{\omega_{OSC}}{\Delta\omega}\right)^2.$$

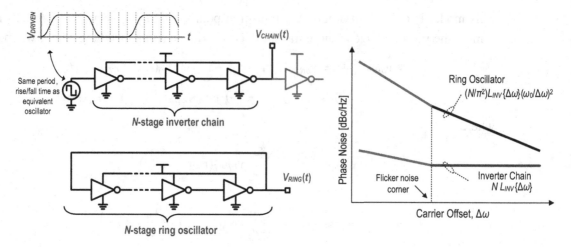

Figure 9.52: Relationship between phase noise of inverter chain and a ring oscillator

In an N-stage inverter chain, the noise contribution of each inverter is typically independent, and the spectral density of the time delay due to noise accumulates, and so

$$S_{v_\delta(CHAIN-r)} = \left(\frac{N+1}{2}\right) S_{v_\delta(INV-r)} + \left(\frac{N-1}{2}\right) S_{v_\delta(INV-f)}$$

$$S_{v_\delta(CHAIN-f)} = \left(\frac{N-1}{2}\right) S_{v_\delta(INV-r)} + \left(\frac{N+1}{2}\right) S_{v_\delta(INV-f)},$$

where $S_{v_\delta(INV-r)}$ and $S_{v_\delta(INV-f)}$ are the spectral densities of noise voltage sampled at rising and falling edge of a single inverter, respectively. In keeping with our notation from the previous section, N is the number of inverter stages, which is assumed to be odd and greater than 3. Accordingly, the phase noise of the ring oscillator can be written as

$$\mathcal{L}\{\Delta\omega\} \approx \frac{N}{|\lambda|^2} \left(S_{v_\delta(INV-r)}(\Delta\omega) + S_{v_\delta(INV-f)}(\Delta\omega)\right) \left(\frac{\omega_0}{\Delta\omega}\right)^2.$$

Therefore, if we know the sampled spectral density of a single inverter, we can calculate a closed form expression for the ring oscillator. Indeed, we can rewrite this expression directly in terms of inverter phase noise as

$$\mathcal{L}\{\Delta\omega\} \approx \frac{N}{\pi^2} \mathcal{L}_{INV}\{\Delta\omega\} \left(\frac{\omega_0}{\Delta\omega}\right)^2,$$

where $\mathcal{L}_{INV}\{\Delta\omega\}$ is the phase noise of a driven inverter assuming the same toggling frequency as the ring oscillator, and the same rise and fall times. This is visualized in Figure 9.52.

9.8.4 Phase Noise of a Single Inverter

In practice, it is best to calculate the noise of an inverter (i.e., $S_{v_\delta(INV-r)}(\Delta\omega)$, $S_{v_\delta(INV-f)}(\Delta\omega)$ or simply $\mathcal{L}_{INV}\{\Delta\omega\}$) through simulation. This is because the device used in the inverter

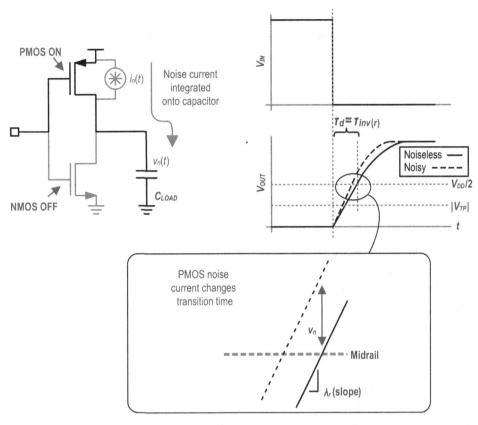

Figure 9.53: Phase noise model of inverter. For simplicity, input is assumed to switch instantaneously, and charging current is from device in saturation

transitions through many different regimes. Nevertheless, simplified models have been proposed that are useful in guiding design and determining achievable performance. Notably, in [23] the inverter noise is portioned into two sources. On the rising edge, the noise that perturbs the zero-crossing is a combination of the PMOS noise being integrated onto the load capacitor (see Figure 9.53), and kT/C noise due to the NMOS device that was in triode before the switching event. Equally, on a falling edge, the noise is modeled as a combination of the NMOS noise integrated onto the load capacitor, and the kT/C noise due to the PMOS device that was in triode before the switching event. Following this approach it is possible to generate closed form expressions for sampled voltage noise both on the rising and falling edge. To simplify matters, however, we note that simulation suggests that the phase noise of an inverter in the thermal noise region (with equal rise and fall times) is proportional to kT/C, i.e.,

$$S_{v_\delta(INV-r)}(\Delta\omega) \propto \frac{kT}{C_{LOAD}} \frac{1}{T_0}$$

$$S_{v_\delta(INV-f)}(\Delta\omega) \propto \frac{kT}{C_{LOAD}} \frac{1}{T_0},$$

where T_0 is the period of the toggling frequency. Simulation suggests that the proportional constant is close to 4 over a wide range of operating conditions. In the preceding section, we found that the phase noise of an N-stage ring oscillator is

$$\mathcal{L}\{\Delta\omega\} \approx \frac{N}{|\lambda|^2} \left(S_{v_\delta(iINV-r)}(\Delta\omega) + S_{v_\delta(INV-f)}(\Delta\omega) \right) \left(\frac{\omega_0}{\Delta\omega} \right)^2.$$

Substituting in the approximations for $S_{v_\delta(iINV-r)}(\Delta\omega)$ and $S_{v_\delta(iINV-f)}(\Delta\omega)$, and making the simplifying assumptions that $T_o = 1/f_o = 2NC_{LOAD}V_{DD}/I_{DD}$ and $\lambda = I_{DD}/C_{LOAD} = 2NV_{DD}/T_0$, this expression becomes

$$\mathcal{L}\{\Delta\omega\} \approx \frac{4kT}{V_{DD}I_{DD}} \left(\frac{\omega_0}{\Delta\omega} \right)^2.$$

Therefore, the only way to improve the phase noise performance of a ring oscillator is to increase its power consumption either through increasing I_{DD} by increasing the device size, or raising V_{DD}. A more sophisticated analysis [23] shows some dependence on the threshold device $|V_{th}|$,

$$\mathcal{L}\{\Delta\omega\} \approx \frac{2kT}{I_{DD}} \left(\frac{2\gamma}{V_{DD} - |V_{th}|} + \frac{1}{V_{DD}} \right) \left(\frac{\omega_0}{\Delta\omega} \right)^2,$$

which suggests subthreshold operation should be avoided. In practice, the performance of ring oscillators close to subthreshold does not degrade as rapidly as the equation suggests (owing to the breakdown of the square law model used in derivation).

9.8.5 Ring Oscillator and LC Oscillator Comparison

In order to draw a comparison between the achievable performance of a ring oscillator and an LC oscillator, we return to the concept of figure of merit (FOM), which is normalized phase noise per unit power. In the case of a ring oscillator where $V_{DD} \gg |V_{th}|$ this is calculated as

$$\text{FOM} = \frac{\left(\frac{\omega_0}{\Delta\omega} \right)^2}{(P_{DC}/1\text{mW}) \mathcal{L}\{\Delta\omega\}} \approx \frac{10^{-3}}{4kT}.$$

Therefore the ratio of FOM of an LC oscillator and a ring oscillator (when measured in the thermal region) is given by approximately

$$\frac{\text{FOM}_{LC}}{\text{FOM}_{RING}} \approx \left(\frac{\eta Q^2}{F} \right) 8.$$

Assuming a well-design LC oscillator with a $F = 2$ and $\eta = 1/2$, this expression becomes

$$\frac{\text{FOM}_{LC}}{\text{FOM}_{RING}} \approx 2Q^2.$$

This is quite a dramatic difference. For instance, in this example, a ring oscillator would need to burn 200 times more power to achieve the same phase noise performance as a well-designed LC oscillator with a Q of 10. (Published results show an upper-limit FOM for ring oscillators of 167dB, while an LC VCO with an FOM in excess of 190dB is not uncommon.) Moreover, the flicker noise corner of a ring oscillator is typically much larger than a well-designed LC oscillator. For these reasons, LC oscillators are favored in RF-front ends.

The poor noise performance of a ring oscillator compared to an LC oscillator can be traced to the time reference inherent in each circuit [25]. In the ring oscillator, the time constant is related to the time required for a device current to charge (or discharge) a capacitor (i.e., $T \propto I/C$). At temperatures above 0 kelvin, a nonzero amount of thermal noise will be associated with such a current, and so I/C is an inherently noisy time reference. By contrast, in an LC oscillator the time reference is related to the inductor and capacitor size, i.e., $T \propto \sqrt{LC}$. An ideal LC tank has no associated thermal noise and so is a noiseless time reference. In other words, any noise in the system is a parasitic effect and not fundamental to the time reference. Q is a measure of the parasitic loss of an LC tank, and so increasing Q naturally improves an LC oscillator's performance.

9.9 QUADRATURE OSCILLATORS

In previous chapters we have shown that most practical transceivers rely on quadrature LO signals for downconversion as well as upconversion. We also showed in Chapter 4 two different methods of quadrature LO generation, one based on the divider, and the other based on polyphase filtering. There is yet another mechanism that relies on quadrature LC VCOs. As we will show shortly, there are several drawbacks with this latter scheme that makes them somewhat uncommon in GHz RF applications. Nevertheless, we shall have a brief qualitative glance on quadrature oscillators (QOSC) in this section, and explain the reasons behind their unpopularity.

The conventional QOSC consists of two cross-coupled LC oscillator cores with an inversion in the feedback path (Figure 9.54). The two cores are ideally identical, and mutually locked to a common frequency.

While a linear model may describe the oscillator performance, we choose to use a more realistic model based on hard-limiting transconductors, as shown in Figure 9.55. The latter

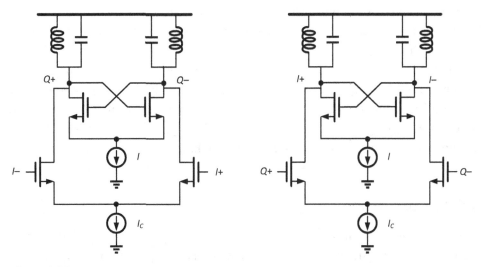

Figure 9.54: Generic quadrature oscillator

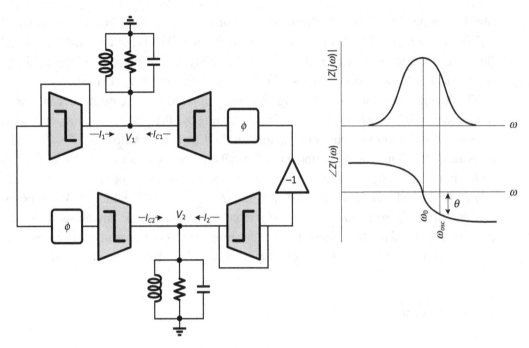

Figure 9.55: Quadrature oscillator model

approach proves to be more accurate, and predicts the oscillation amplitude properly. It also anticipates all possible modes of oscillation, and their stability. We assume that the QOSC is designed differentially and the oscillation amplitude is large enough to completely steer the tail current from one side to the other.

As shown in Figure 9.55, two arbitrary phase-shifters ϕ are also inserted in the two coupling paths prior to the coupling transconductors. For the conventional QOSC that does not have these phase-shifters, ϕ is zero. The transconductors commute their corresponding currents at the zero-crossings of their input voltages. Filtering in the LC circuit attenuates all higher order harmonics of currents leading to a quasi-sinusoidal tank voltage. Therefore, we need to consider only fundamental components of the transconductor output currents. The LC tank resonance frequency is ω_0, and magnitude and phase of its impedance versus frequency are also shown in the figure.

Two coupled oscillator cores are locked and oscillate at a common frequency ω_{osc}, which can be different from ω_0. V_1, and V_2 are phasors of oscillator outputs, and without loss of the generality we may assume V_2 lags V_1 by an unknown angle ψ. Output currents of the transconductors, I_1, I_2, I_{C1}, and I_{C2} are depicted accordingly. The voltage of the first tank, V_1, is obtained by summing the current I_1 and I_{C1} and multiplying that by the tank impedance whose phase and magnitude is shown in Figure 9.55. Similarly, V_2 is obtained by summing the corresponding current, and passing through the tank.

9.9.1 Modes of Oscillation

Performing a vector analysis we can obtain four possible modes of oscillations [26], [27]. Specifically, we can show $|I_1|=|I_2|=I$, and $|I_{C1}|=|I_{C2}|=I_C$ which is expected from the symmetry of the circuit. The four solutions are summarized as follows:

1. $\psi = +\frac{\pi}{2}$, and $\omega_{osc} > \omega_0$.
2. $\psi = -\frac{\pi}{2}$, and $\omega_{osc} < \omega_0$.
3. $\psi > -\frac{\pi}{2}$, and $\omega_{osc} < \omega_0$.
4. $\psi < -\frac{\pi}{2}$, and $\omega_{osc} < \omega_0$.

The first and second modes are quadrature with identical oscillation amplitudes in the two cores. However, the quadrature sequences of these two modes are the opposite. The third and fourth modes are nonquadrature with different oscillation amplitudes in each of the two cores.

Using the perturbation analysis method [27], it can be shown that the third and fourth modes of oscillation are unconditionally unstable. Furthermore, the second mode is stable only for $\phi < \sin^{-1}(I_C/I)$, which is the case for most practical realizations of the quadrature oscillator, as ϕ, is typically equal to zero. If, however, one can insert a phase shift large enough to break the condition $\phi < \sin^{-1}(I_C/I)$, the second mode will be ruled out. However, creating a phase shift, say by using an RC circuit, proves to be a challenge. A small resistor loads the tank, while a large resistor adds noise. For that reason, most practical realizations of the quadrature oscillator do not include any phase shift, and both modes may exist. This could become a major problem, as image rejection mixers would dangerously suffer from the existent ambiguity in the sequence of quadrature signals.

It is worth pointing out that the first mode has higher amplitude than the second one. This emphasizes the tendency of this mode to prevail even if the second mode is stable. The reason for this has to do with the series resistor of the inductor, which makes the tank frequency response somewhat asymmetric around ω_0 (Figure 9.55). Thus, the first mode may predominate during the start-up process, although this is not guaranteed for any given design.

Considering only the first mode with $\phi = 0$, from Figure 9.55 it can be easily shown that

$$\tan\theta = \frac{I_C}{I}.$$

Using the LC tank equation, the oscillation frequency is readily obtained:

$$\omega_{osc} \cong \omega_0 + \frac{\omega_0}{2Q}\frac{I_C}{I}.$$

Evidently, the frequency of oscillation is generally a function of the coupling ratio $m = \frac{I_C}{I}$. In fact, m can be utilized as a means to control the oscillation frequency. Applying KCL on the V_1 node leads to

$$V\left(\frac{1}{R_P} + j\left(C\omega_{osc} - \frac{1}{L\omega_{osc}}\right)\right) = \frac{4}{\pi}\left(I + I_C e^{j\frac{\pi}{2}}\right).$$

Separating the real and imaginary parts results in the amplitude of oscillation as follows,

$$V = \frac{4}{\pi}R_P I,$$

where R_P is the tank resistor, and I is the core bias current.

9.9.2 Quadrature Accuracy Due to Mismatches

If the two oscillator cores perfectly match, their output phases are in precise quadrature. However, mismatches between the two cores would cause the outputs to depart from quadrature. In the following section, we study quadrature inaccuracy caused by mismatches in resonance frequencies of the two tanks. A more complete analysis including other sources of mismatch may be found in [26].

Let us assume that the two tanks have different resonance frequencies, ω_{01} and ω_{02}. It is also assumed that the mismatch is small enough such that both cores oscillate at a common frequency ω_{osc}. Defining the phase deviation of $\Delta\psi$ from quadrature, either geometrically or algebraically, it can be shown that

$$\frac{\Delta\psi}{\Delta\theta} = \frac{1+m^2+2m\sin\phi}{2m(m+\sin\phi)},$$

where $m = \frac{I_C}{I}$ is the coupling factor. Furthermore, by making use of impedance equation of the LC tank, $\Delta\theta$ is found in terms of the mismatch $\Delta\omega = \omega_{01} - \omega_{02}$:

$$\frac{\Delta\theta}{\Delta\omega} \cong -\frac{2Q}{\omega_0}.$$

From Figure 9.55, θ is the phase of each tank at the frequency of oscillation. Given the tanks are assumed to have different center frequencies, we expect different values of θ for each, as shown above. Thus, the resulting quadrature error is obtained as

$$\Delta\psi \cong -Q\frac{1+m^2+2m\sin\phi}{m(m+\sin\phi)}\frac{\Delta\omega}{\omega_0}.$$

It is instructive to compare the resilience of two extreme cases against the mismatch $\Delta\omega$ between the two tank resonance frequencies. If there is no phase-shifter $\phi = 0$, which is the case for most designs, the quadrature error reduces into $\Delta\psi = -Q(1+\frac{1}{m^2})\left(\frac{\Delta\omega}{\omega_0}\right)$. Since the coupling factor m is typically a small number usually between 0.1 and 0.4, the phase inaccuracy due to the mismatch between the tank center frequencies could be quite substantial. A phase shifter helps improve the dependence to $1 + 1/m$ but has implementation issues, as pointed out.

Also, the deviation is proportional to the quality factor Q. This is intuitively expected, as higher Q results in a steeper phase variation of the tank impedance.

9.9.3 Phase Noise Analysis

In this section we present a qualitative description of QOSC white phase noise. The flicker noise analysis and more details on the derivations may be found in [27].

In the noise analysis of a classical LC oscillator it was proven that the noise factor F of the oscillator due to thermal noise sources is $1 + \gamma$. This minimum noise factor happens when the thermal noise of the tail current source is filtered out and does not contribute to the phase noise.

If a similar phase noise analysis is performed on the quadrature oscillator, the oscillator phase noise is found to be

$$\mathcal{L}\{\Delta\omega\} = \frac{kTFR_P}{2A^2} \frac{1}{Q^2} \left(\frac{\omega_0}{\Delta\omega}\right)^2,$$

where F is given by

$$F = 1 + \left(\frac{m \cos \phi}{1 + m \sin \phi}\right)^2 + \frac{\gamma}{1 + m \sin \phi} \left(1 + m\left(\frac{m + \sin \phi}{1 + m \sin \phi}\right)^2\right).$$

For a more common case of no phase shift,

$$F = 1 + m^2 + \gamma(1 + m^3).$$

The first term reflects the contribution of the tank loss on the phase noise, while the second term predicts that of the switches in the coupling and regenerative transconductors. Also, as the coupling factor m approaches zero, F converges to $1 + \gamma$. Assuming m is small, phase noise of the QOSC is 3dB lower compared to the corresponding standalone LC oscillator. Since the power consumption is also doubled, the FOM remains intact. In fact, the ratio of FOM of the QOSC FOM_{QOSC} to that of the corresponding standalone LC oscillator FOM_{core} is found to be

$$\frac{\text{FOM}_{\text{QOSC}}}{\text{FOM}_{\text{core}}} = \frac{1 + \gamma}{(1 + m)[1 + m^2 + (1 + m^3)\gamma]} \approx \frac{1}{1 + m},$$

where in the above derivation the current of the standalone oscillator is adjusted to have the same amplitude as that of the QOSC. Thus, the FOM of a QOSC cannot be better than its standalone core oscillator. Only if the coupling is weak (small m), the QOSC approaches to have a similar phase noise performance (for the same total current), as a single VCO.

Given the quadrature ambiguity, relatively large quadrature inaccuracy due to mismatches, a bigger area, as well as a somewhat worse phase noise in the quadrature oscillators, they are not very commonly used. Furthermore, the use of a divide by two reduces the transmitter sensitivity to pulling, as we disused in Chapter 6, which makes them a more attractive means of providing IQ LO signals, despite running the VCO at twice the frequency of the carrier.

9.10 CRYSTAL AND FBAR OSCILLATORS

A crystal oscillator (XO) is a critical component of every RF system, providing the reference clock to various key building blocks, such as the frequency synthesizer, data converters, calibration circuits, baseband, and peripheral devices (Figure 9.56).

The reason why one employs a *piezoelectric crystal resonator* in place of an LC tank is that the crystal quality factor is substantially higher (several thousand or more), which directly leads to much better frequency stability. Hence crystal oscillators are used as fixed-frequency, highly stable references for frequency or timing.

9.10.1 Crystal Model

To facilitate the derivation of the crystal oscillator principles, we first discuss the crystal electrical model. The main material used as a mechanical resonator for oscillator is the *crystalline quartz*. The property of this material differs in different directions, and in addition, each different *cut* may be mounted and vibrated in several different ways. The crystal is commonly modeled as several parallel RLC circuits, in addition to a fixed shunt capacitance, C_0, as shown in Figure 9.57. As frequency of oscillation is normally very close to the series resonance of only one branch, the model is simplified to only that branch, in parallel with C_0, as shown on the right. The multiple branches are a result of *mechanical vibrations* at approximately odd harmonics, commonly known as *overtones*.

Since the mechanical frequency of the fundamental vibration is proportional to the crystal dimensions, practical considerations limit the crystal frequency to tens of MHz. By operating on an overtone, a higher frequency is possible (usually the third overtone), although overtone crystal oscillators are not very common. This is mainly due to the fact that the crystal does not want to resonate on its overtone frequency voluntarily, and the oscillator circuit needs to force the correct overtone, which creates some overhead, and possibly performance degradation. The most common type of overtone crystal oscillators employ an LC resonance tank tuned to the overtone frequency such that the positive feedback dies out at the main crystal frequency, while remaining effective at the overtone frequency. An example of a practical overtone crystal oscillator will be discussed at the end of this section.

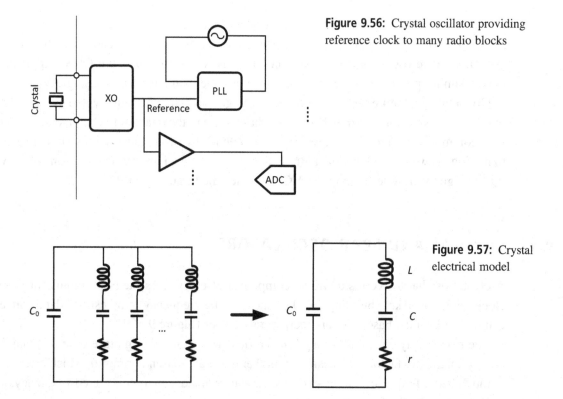

Figure 9.56: Crystal oscillator providing reference clock to many radio blocks

Figure 9.57: Crystal electrical model

9.10 Crystal and FBAR Oscillators

The quality factor of crystal, determined by the *equivalent series resistor* (ESR), is several or tens of thousands. The series branch components consisting of r, L, and C are known as motion components, while C_0 is the total parallel capacitor resulted from mounting. The crystal is intended to produce the exact frequency of oscillation specified when loaded properly with a certain total parallel capacitance. For instance, for a commercially available 8pF 26MHz crystal, the following values are commonly specified: $C_0 = $ 1pF, $C = $ 3.641fF, $L = $ 10.296mH, and $r = $ 26Ω. The crystal produces exactly 26MHz when loaded by a total shunt capacitance of 8pF. As typically the value of the inherent parallel capacitors C_0 is less than that, it is common to load the crystal with extra capacitance, preferably tunable, to obtain the exact desired frequency (Figure 9.58). The corresponding crystal quality factor is over 64000.

The total shunt capacitance is now $C_0 + \frac{C_1 C_2}{C_1 + C_2}$, which must be 8pF. C_1 and C_2 may include the oscillator and other related parasitic.

The motion capacitance, C, is much smaller than C_0, a property that results in crystal stability, as we shall discuss shortly. For simplicity, we assume C_0 consists of the total shunt capacitance (that is, C_1 and C_2 are included).

The total admittance is

$$Y(s) = sC_0 \frac{s^2 + \frac{r}{L}s + \left(1 + \frac{C}{C_0}\right)\omega_0^2}{s^2 + \frac{r}{L}s + \omega_0^2},$$

where $\omega_0 = \frac{1}{\sqrt{LC}}$, and the quality factor is $Q = \frac{L\omega_0}{r}$. Since $C \ll C_0$, we expect the poles and zeros of Y(s) to be very close to each other, as shown in Figure 9.59.

There are two modes of operation where the crystal oscillator is built upon accordingly:

– Series: Crystal behaves as a low impedance element, and resonates near the pole of $Y(s)$. Crystal is typically placed in the feedback path directly, for instance to shunt a large resistor which will otherwise prevent the oscillation.

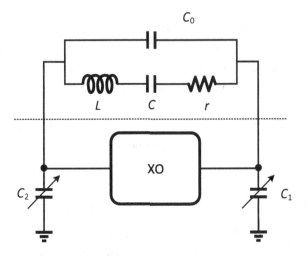

Figure 9.58: Properly loaded crystal

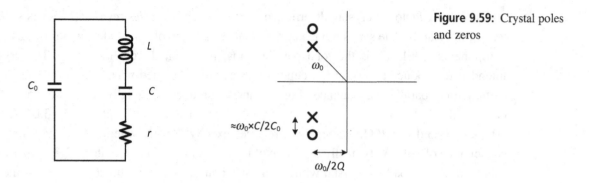

Figure 9.59: Crystal poles and zeros

- Parallel: Crystal exhibits a high impedance, and resonates near the zero of $Y(s)$. Thus, the crystal behaves as a large shunt inductance that is resonating with the circuit capacitances.

Most common practical crystal oscillators rely on the parallel resonance, and thus we shall focus mostly on the parallel mode.

To understand the reason for crystal oscillator stability, consider the parallel resonance as an example, and suppose the total shunt capacitance varies from $C_0 \to C_0 + \Delta C_0$. This can be caused by temperature or process variations in the oscillator, or inaccurate modeling of the capacitances. The oscillation frequency, which is near the zero of $Y(s)$ (see Figure 9.59), then shifts by

$$\Delta \omega \approx \left(\frac{C}{2(C_0 + \Delta C_0)} - \frac{C}{2C_0} \right) \omega_0 \approx -\frac{C}{2C_0} \omega_0 \times \frac{\Delta C_0}{C_0}.$$

Thus, as the frequency of oscillation is roughly $\left(1 + \frac{C}{2C_0}\right)\omega_0$, the crystal sensitivity is

$$S \approx \left| \frac{\frac{\Delta \omega}{\omega_0}}{\frac{\Delta C_0}{C_0}} \right| = \frac{C}{2C_0}.$$

Since $C \ll C_0$, the frequency variation as a result of oscillator process variations is small. For instance, for the values presented earlier for a 26MHz crystal, 10% variation of the total shunt capacitance leads to only 22.75ppm variation in frequency. This in fact is exploited to tune the crystal frequency in the presence of circuit variations or aging to a given desired accuracy. We shall discuss this more shortly.

Intuitively, given our discussion of LC oscillators previously, we expect a substantially higher stability of the oscillation frequency thanks to a much larger resonator quality factor.

9.10.2 Practical Crystal Oscillators

Apart from frequency stability, a crystal oscillator must satisfy several other exacting requirements. It must have low phase noise in order not to degrade the receive SNR or transmit EVM.

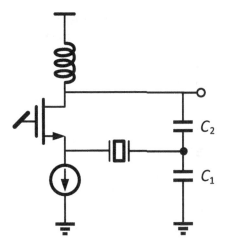

Figure 9.60: Series-mode crystal oscillator

For supplying clocks to external peripheral devices, the phase noise requirements are even more stringent. For instance for the WLAN IEEE 802.11n 5GHz band, the required phase noise is around −145dBc/Hz at a 10kHz frequency offset relative to 26MHz, a commonly used crystal for handsets. The oscillator must also be very low power, as it is the only block that remains on while the system is idle. Moreover, in many applications, such as those for cellular phones, the crystal oscillator must have a wide tuning capability to cover crystal and circuit variations.

A well-designed crystal oscillator must stop oscillating if the crystal is removed, and should minimize the crystal driving current and voltage. The latter is to avoid physical damage caused by *overdriving*, and to minimize self-heating. This is typically accomplished by proper circuit design, and often by employing some kind of amplitude control loop to avoid crystal overdriving.

An example of a series-mode crystal oscillator is shown in Figure 9.60. It is based on a Colpitts oscillator topology (Section 9.4.3), except for the crystal is placed in the feedback path. As a result, the positive feedback is only enforced around the zero of the crystal impedance (or the pole of $Y(s)$). The drain LC tank is tuned around the intended oscillation frequency.

Two examples of parallel-mode crystal oscillators are shown in Figure 9.61 (bias details not shown). In both cases the crystal behaves like a series RL circuit, and the drain inductors are simply placed for biasing purposes.

A more practical version of the Pierce XO is shown in Figure 9.62, which is based on a complementary structure (simply a self-biased inverter) to avoid the biasing inductor. When the inverter is sized properly, it results in a stable and robust design, and for that reason this structure is commonly adopted in many radios today. The capacitor arrays are added for tuning purposes (see the next section).

An example of an overtone Pierce crystal oscillator is shown in Figure 9.63. To understand the operation of the oscillator, we first note that in the fundamental oscillator on the left, the inverter creates 180° of phase shift, and that along with the combination of the crystal and the capacitors (C_1 and C_2) give a total phase shift of 360°, establishing the intended positive feedback at the frequency of the interest. Now, compared to the fundamental oscillator, in the

586 Oscillators

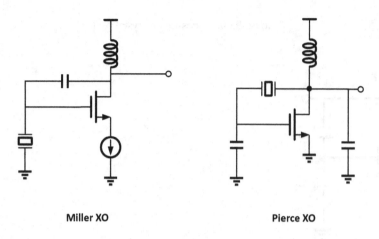

Figure 9.61: Parallel-mode crystal oscillator examples

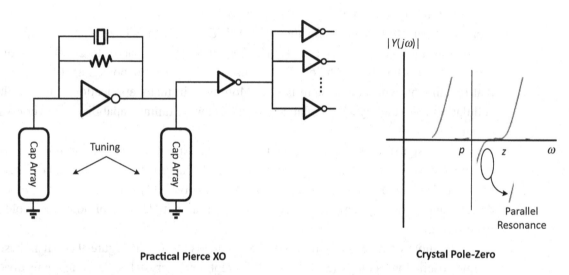

Figure 9.62: Practical Pierce crystal oscillator

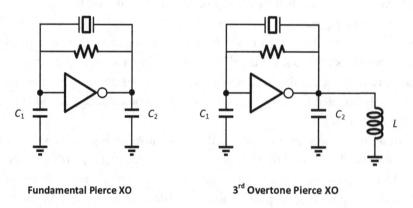

Figure 9.63: A fundamental and an overtone Pierce crystal oscillator

overtone oscillator the inductor L has been added in parallel to C_2, such that the impedance of the parallel L–C_2 is inductive at the fundamental frequency, but remains capacitive at the overtone frequency. The inductive impedance prevents the oscillation at the fundamental frequency, as the desired 360° will not be established. See also Problems 18 and 19 for a more analytical justification.

Alternatively, one could add a highpass filter at the output of the active circuit to reduce the loop gain at the fundamental frequency, effectively *starving* the gain of the oscillator at that frequency.

Apart from the need for additional hardware, there are several other disadvantages associated with the overtone crystal oscillators. The overtone series resistance is typically higher, leading to a worse Q, and a higher power consumption in the active circuitry to establish the required loop gain. Also the motion capacitance is smaller (nine times less in a 3rd overtone oscillator, for instance), which makes the tuning or calibration harder (see the next section for tuning requirements). On the other hand, as fundamental crystals over tens of MHz are not common (the thickness of the crystal which is inversely proportional to its frequency becomes too small to handle properly in volume production), the overtone oscillators are a viable solution to extend the frequency.

9.10.3 Tuning Requirements

Cellular standards require that the average frequency deviation of the transmitted carrier from handsets to be better than 0.1ppm with respect to the receiving carrier frequency in base stations. To achieve such stringent accuracy, an automatic frequency control (AFC) loop in the handset is employed to synchronize the handset crystal oscillator with the base station. The handset generates the corresponding digital codes based on the frequency difference to vary the digitally controlled crystal oscillator (DCXO) frequency until the desired oscillation frequency is achieved. In practice, after taking into account the crystal and DCXO variations as well as AFC loop imperfection, the frequency deviation must be much smaller than 0.1 ppm, which poses a design challenge for the DCXO tuning circuitry.

The tuning is typically achieved in two steps: a coarse calibration for a one-time frequency correction at the factory, and a fine calibration for time-varying frequency errors in the phone. The coarse calibration covers the crystal frequency tolerance of ± 10ppm due to crystal process and part-to-part variations, and the DCXO circuit process variations. The remaining residual errors due to aging, temperature variations, and voltage drift are corrected through the fine calibration continuously running as a part of an AFC loop in the transceiver. To ensure that little of the total frequency budget of 0.1ppm is consumed, a fairly wide range with a resolution of better than 0.01ppm (about one-tenth of the specification) is targeted. Moreover, achieving a small frequency error required for proper AFC operation mandates monotonic tuning characteristics.

These requirement are typically satisfied by incorporating switchable array of capacitances with fine resolution (Figure 9.62), properly laid out to ensure monotonicity and the required accuracy [28]. An alternative approach is to use a varactor driven by a high-resolution DAC. This may not be as desirable, however. Although this approach makes the capacitor array design easier, as the varactor inevitably requires a large gain to cover the range, it makes the

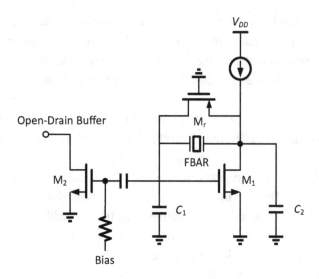

Figure 9.64: A low-power 2GHz FBAR oscillator

oscillator very sensitive to supply pushing and control-voltage noise. In actual implementations, this condition might lead to a much larger frequency error. Moreover, for other sensitive circuits, the DAC noise and clock harmonics become problematic.

9.10.4 FBAR Oscillators

In Chapter 4 we presented the circuit model of an FBAR resonator, which was very similar to the crystal electrical model discussed in Figure 9.57. As such, we expect to see large similarity between the crystal and FBAR oscillators, with the main difference being the frequency of operation. Shown in Figure 9.64 is an example of a Pierce FBAR oscillator operating at 2GHz [29]. The P transistor M_r is in the triode region and self-biases the core device M_1. M_2 is an open-drain buffer for testing purposes, and could be replaced by an on-chip self-biased inverter for on-chip distribution. Otherwise the oscillator operates very similarly to the parallel resonance Pierce crystal oscillator shown in Figure 9.61.

The oscillator dissipates 25µW at 2GHz (using a forward-bias substrate technique explained in [29]), and has an excellent figure of merit of over 220dB at 100kHz offset.

While FBAR resonators operate at much higher frequencies than the crystal, their temperature drift is substantially worse. They also have generally lower quality factors. More examples of FBAR oscillators can be found in [30], [31].

9.11 Summary

This chapter covered various oscillators and VCOs topologies, and presented a detailed and rigorous discussion of the oscillator phase noise.

- In Section 9.1 a linear feedback model of the oscillators was presented.
- A more realistic and practical nonlinear model of the oscillators was presented in Section 9.2.

- Oscillator phase noise analysis based on the nonlinear feedback model was discussed in Section 9.3. Cyclostationary noise as well as AM and PM noise were covered in this section as well.
- Various commonly used LC oscillator topologies such as NMOS, CMOS, and Colpitts oscillators were discussed in Section 9.4. A general design and optimization methodology was presented as well. More VCO topologies were discussed in Section 9.7.
- LC tank Q degradation due to oscillator nonlinearity was discussed in Section 9.5.
- Frequency modulation effects due to the tank nonlinearities were discussed in Section 9.6.
- Section 9.8 discussed the design and phase noise properties of ring oscillators.
- Quadrature oscillators were covered in Section 9.9.
- Section 9.10 presented an overview of crystal and FBAR oscillators.

9.12 Problems

1. Prove that the linear oscillator shown in Figure 9.1 can be redrawn in the form of a linear feedback system (also shown in Figure 9.1). Derive the resultant transfer function $H(s)$.
2. Prove that the amplitude of oscillation in a noiseless linear LC oscillator is equal to q_{SU}/C_p if the startup current is an impulse defined as $i_{SU}(t) = q_{SU}\delta(t)$. Assume the circuit is marginally stable. **Hint:** Use the initial value theorem, $f(t=0^+) = \lim_{s \to \infty} sF(s)$.
3. Given the generic negative-g_m topology shown below, which active element will ensure start-up? Which active element can never sustain an oscillation? Which element could sustain an oscillation, but would require some separate startup mechanism? Suggest such a mechanism.

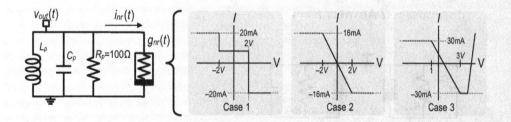

4. Given the generic negative-g_m shown below, what is the expected voltage spectral density noise around the carrier of the oscillator for the two of the active elements shown? Assume the noise injected by each active element is proportional to its instantaneous conductance. Further, assume the oscillator has started up. **Hint:** Use Bank's general result.

Answer: $S_V(\Delta\omega) = \dfrac{A^2}{4}\mathcal{L}\{\Delta\omega\} = \dfrac{\theta\omega kT(1+\alpha)R_p}{4}\dfrac{1}{Q^2}\left(\dfrac{\omega_0}{\Delta\omega}\right)^2.$

Oscillators

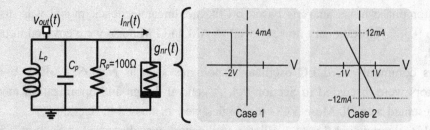

5. Prove that in limit as the active element of the negative-g_m LC oscillator becomes linear, amplitude and phase noise will be equal in magnitude, i.e., $S_\phi\{\Delta\omega\} = S_M\{\Delta\omega\}$.

6. In regard to Figure 9.20, prove that the optimum arrangement is to connect the input and output terminals of the negative transconductance to $v_{out}(t)$.

7. Derive the noise factor for the Hartley oscillator shown below. Assume the transistor never operates in the triode region. **Hint:** For simplicity you may assume both resonators have the same Q. Use the narrowband transformer approximation introduced in Chapter 3.

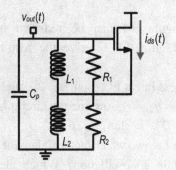

8. Derive the noise factor for the standard topology shown below where the current source has been replaced by a simple resistor. What are the advantages of this topology? What are the disadvantages? **Answer:** $F = 1 + \gamma + \dfrac{R_p}{4R_{BIAS}}$.

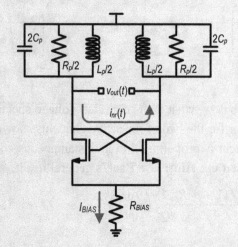

9. For a set bias current and inductor size, which is a better choice of topology: the standard NMOS topology or the standard CMOS topology? Assume the current is low enough that both operate in the current-limited regime (i.e., the current source doesn't enter the triode region).

10. The Colpitts oscillator can provide a noise factor arbitrarily close to 1. Under what limiting condition is this possible? And explain why it is generally not a good idea to design at this point. (**Hint:** Think in terms of power efficiency.) Derive an expression for the FOM of a Colpitts oscillator.

11. Show that the oscillator tuning can expressed in terms of tank maximum and minimum capacitance as follows:

$$TR = 2 \frac{\sqrt{\frac{C_{MAX}}{C_{MIN}}} - 1}{\sqrt{\frac{C_{MAX}}{C_{MIN}}} + 1}.$$

12. Derive the maximum amplitude of the class C topology that will ensure startup, but will not drive the differential pair transistors into triode. If startup is ensured through some other means, what is the maximum amplitude? **Hint:** Both answers are a function of the DC gate bias.

13. It is often stated that amplitude noise is of little concern because it will be stripped out by any subsequent hard-switching circuit (e.g., a hard-limiting buffer or mixing). Under what conditions might a circuit designer be concerned about amplitude noise?

14. Assuming a near sinusoidal tank voltage, show that

$$C_{eff} = \frac{\oint i_C dv_C}{\pi A^2 \omega_0}.$$

15. Prove that square wave injection of bias current can result in a maximum power efficiency of $\eta = 2/\pi$, while impulse-injection can result in a maximum power efficiency of $\eta = 1$.

16. Design an NMOS LC VCO intended for cellular transmitter applications. The VCO must cover a range of 3GHz to 4GHz, and have a phase noise of better than −152dBc/Hz at 20MHz offset. Assume the supply is 1.2V, and the tank Q is 10. Ignore $1/f$ noise, and assume $\gamma = 1$. Also assume the current source has an overdrive voltage of 200mV.

17. Repeat the previous problem for a CMOS topology, assuming identical N and P transistors.

18. In the two circuits shown below, the transistor is in saturation (bias details are not shown). Ignore r_o and the internal capacitances of the FET.
    a. Find the impedances (Z_1 and Z_2) looking into the nodes 1-1'.
    b. If a crystal is connected to the nodes 1-1', which circuit is capable of potentially oscillating?

    Answer: $Z_1 = \frac{1}{jC_1\omega} + \frac{1}{jC_2\omega} - \frac{g_m}{C_1 C_2 \omega^2}$ and $Z_2 = \frac{1}{jC_1\omega} + jL_2\omega + \frac{g_m L_2}{C_1}.$

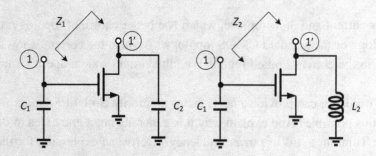

19. The simplified AC model of a Pierce crystal oscillator is shown below, where the crystal in parallel resonance is approximated by a series RL circuit (r and L in the figure).
   a. Using the findings of the previous problem, determine the required transistor g_m such that the circuit is on the verge of oscillating.
   b. What is the frequency of oscillation?
   c. By tracing the transistor input voltage (v_{gs}) through the feedback loop, argue how the 360° phase shift needed for oscillation is established.

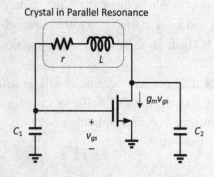

9.13 References

[1] D. Leeson, "A Simple Model of Feedback Oscillator Noise Spectrum," *Proceedings of the IEEE*, 54, no. 2, 329–330, 1966.
[2] A. Mazzanti and P. Andreani, "Class-C Harmonic CMOS VCOs, with a General Result on Phase Noise," *IEEE Journal of Solid-State Circuits*, 43, no. 12, 2716–2729, 2008.
[3] P. Kinget, "Integrated GHz Voltage Controlled Oscillators," in *Analog Circuit Design*, Kluwer, 1999, 353–381.
[4] M. Garampazzi, S. D. Toso, A. Liscidini, D. Manstretta, P. Mendez, L. Romanò, and R. Castello, "An Intuitive Analysis of Phase Noise Fundamental Limits Suitable for Benchmarking LC Oscillators," *IEEE Journal of Solid-State Circuits*, 49, no. 3, 635–645, 2014.
[5] C. Samori, A. L. Lacaita, F. Villa, and F. Zappa, "Spectrum Folding and Phase Noise in LC Tuned Oscillators," *IEEE Transactions on Circuits and Systems II: Analog and Digital Signal Process*, 45, no. 7, 781–790, 1998.

[6] D. Murphy, J. Rael, and A. Abidi, "Phase Noise in LC Oscillators: A Phasor-Based Analysis of a General Result and of Loaded Q," *IEEE Transactions on Circuits and Systems I: Fundamental Theory and Application*, 57, no. 6, 1187–1203, 2010.

[7] W. P. Robins, *Phase Noise in Signal Sources: Theory and Applications*, Institution of Electrical Engineers, 1984.

[8] J. J. Rael and A. A. Abidi, "Physical Processes of Phase Noise in Differential LC Oscillators," in *IEEE Custom Integrated Circuits Conference (CICC)*, 2000.

[9] E. Hegazi, J. J. Rael, and A. A. Abidi, *The Designer's Guide to High-Purity Oscillators*, Springer, 2004.

[10] J. Phillips and K. Kundert, "Noise in Mixers, Oscillators, Samplers, and Logic: An Introduction to Cyclostationary Noise," in *IEEE Custom Integrated Circuits Conference (CICC)*, 2000.

[11] J. Bank, *A Harmonic-Oscillator Design Methodology Based on Describing Functions*, Chalmers University of Technology, 2006.

[12] A. Mazzanti and A. Bevilacqua, "On the Phase Noise Performance of Transformer-Based CMOS Differential-Pair Harmonic Oscillators," *IEEE Transactions on Circuits and Systems I: Regular Papers*, 62, no. 9, 2334–2341, 2015.

[13] M. Babaie and R. B. Staszewski, "An Ultra-Low Phase Noise Class-F 2 CMOS Oscillator with 191 dBc/Hz FoM and Long-Term Reliability," *IEEE Journal of Solid-State Circuits*, vol. 50, no. 3, 679–692, 2015.

[14] M. Babaie, A. Visweswaran, Z. He, and R. B. Staszewski, "Ultra-Low Phase Noise 7.2–8.7 Ghz Clip-and-Restore Oscillator with 191 dBc/Hz FoM," in *Proceedings of the IEEE Radio Frequency Integrated Circuits Symposium (RFIC)*, 2013.

[15] M. Shahmohammadi, M. Babaie, and R. B. Staszewski, "A 1/f Noise Upconversion Reduction Technique for Voltage-Biased RF CMOS Oscillators," *IEEE Journal of Solid-State Circuits*, 51, no. 11, 2610–2624, 2016.

[16] L. Fanori and P. Andreani, "Class-D CMOS Oscillators," *IEEE Journal of Solid-State Circuits*, 48, no. 12, 3105–3119, 2013.

[17] P. Andreani and A. Fard, "More on the $1/f^2$ Phase Noise Performance of CMOS Differential-Pair LC-Tank Oscillators," *IEEE Journal of Solid-State Circuits*, 41, no. 12, 2703–2712, 2006.

[18] P. Andreani, X. Wang, L. Vandi, and A. Fard, "A Study of Phase Noise in Colpitts and LC-Tank CMOS Oscillators," *IEEE Journal of Solid-State Circuits*, 40, no. 5, 1107–1118, 2005.

[19] E. Hegazi and A. A. Abidi, "Varactor Characteristics, Oscillator Tuning Curves, and AM-FM Conversion," *IEEE Journal of Solid-State Circuits*, 36, no. 12, 1033–1039, June 2003.

[20] A. Kral, F. Behbahani, and A. Abidi, "RF-CMOS Oscillators with Switched Tuning," in *IEEE Custom Integrated Circuits Conference (CICC)*, 1998.

[21] J. Groszkowski, "The Interdependence of Frequency Variation and Harmonic Content, and the Problem of Constant-Frequency Oscillators," *Proceedings of the Institute of Radio Engineers*, 21, no. 7, 958–981, 1933.

[22] E. Hegazi, H. Sjoland, and A. A. Abidi, "A Filtering Technique to Lower LC Oscillator Phase Noise," *IEEE Journal of Solid-State Circuits*, 36, no. 12, 1921–1930, 2001.

[23] A. A. Abidi, "Phase Noise and Jitter in CMOS Ring Oscillators," *IEEE Journal of Solid-State Circuits*, 41, no. 8, 1803–1816, 2006.

[24] A. Homayoun and B. Razavi, "Relation between Delay Line Phase Noise and Ring Oscillator Phase Noise," *IEEE Journal of Solid-State Circuits*, 49, no. 2, 384–391, 2014.

[25] R. Navid, T. Lee, and R. Dutton, "Minimum Achievable Phase Noise of RC Oscillators," *IEEE Journal of Solid-State Circuits*, 40, no. 3, 630–637, 2005.

[26] A. Mirzaei, M. Heidari, R. Bagheri, S. Chehrazi, and A. Abidi, "The Quadrature LC Oscillator: A Complete Portrait Based on Injection Locking," *IEEE Journal of Solid-State Circuits*, 42, no. 9, 1916–1932, 2007.

[27] A. Mirzaei, "Clock Programmable IF Circuits for CMOS Software Defined Radio Receiver and Precise Quadrature Oscillators," Doctoral dissertation, University of California, Los Angeles, 2006.

[28] Y. Chang, J. Leete, Z. Zhou, M. Vadipour, Y.-T. Chang, and H. Darabi, "A Differential Digitally Controlled Crystal Oscillator with a 14-Bit Tuning Resolution and Sine Wave Outputs for Cellular Applications," *IEEE Journal of Solid-State Circuits*, 47, no. 2, 421–434, 2012.

[29] A. Nelson, J. Hu, J. Kaitila, R. Ruby, and B. Otis, "A 22μW, 2.0GHz FBAR oscillator," in *Proceedings of the IEEE Radio Frequency Integrated Circuits Symposium*, 2011.

[30] W. Pang, R. C. Ruby, R. Parker, P. W. Fisher, M. A. Unkrich, and J. D. Larson, "A Temperature-Stable Film Bulk Acoustic Wave Oscillator," *IEEE Electron Device Letters*, 29, no. 4, 315–318, 2008.

[31] K. A. Sankaragomathi, J. Koo, R. Ruby and B. P. Otis, "25.9 A ±3ppm 1.1mW FBAR Frequency Reference with 750MHz Output and 750mV Supply," in *Proceedings of the IEEE International Solid-State Circuits Conference (ISSCC) – Digest of Technical Papers*, 2015.

[32] Q. Huang, "Phase Noise to Carrier Ratio in LC Oscillators," *IEEE Transactions on Circuits and Systems I: Fundamental Theory and Application*, 47, no. 7, 965–980, July 2000.

10 PLLs and Synthesizers

In this chapter we present the phase-locked loops and synthesizers, built upon the discussions of the previous chapter, VCOs and crystal oscillators. It is a new chapter in this edition, although some pieces of it existed in the previous edition under Chapter 8. A detailed discussion of PLLs and synthesizers would perhaps require an entire book of its own, and our objective here is only to establish enough background to allow design and analysis of synthesizers for typical radio applications.

A phase-locked loop is a mixed-mode nonlinear feedback system. As we have already established, most modern radios require precise tunable LO signals typically created through a frequency synthesizer, which relies on the concept of phase locking. We therefore begin our discussion with basics of phase-locked loops and their building blocks, followed by presenting both fractional and integer synthesizers. A frequency synthesizer generates a range of frequencies from a single reference frequency, and relies on frequency multiplication or division using phase-locked loops. Through an approximate linear model commonly used among PLL designers, noise and transient properties of PLLs and synthesizers are discussed, and design guidelines are offered.

We also present a detailed discussion on latches and multi-modulus frequency dividers as they are key building blocks of every frequency synthesizer. We conclude this chapter by offering an introductory discussion of digital PLLs and theirs pros and cons.

The analysis of PLL dynamics and its noise performance offered in this chapter is particularly important as it sets the foundation for our discussion on polar and PLL-based transmitters in the next chapter.

The specific topics covered in this chapter are:

- Phase-locked loops building blocks
- Type I and II PLLs
- PLL linear analysis
- Integer and fractional synthesizers
- PLL noise sources
- Frequency dividers
- Introduction to digital PLLs

For class teaching, we recommend spending no more than two lectures to cover the basics of PLLs (Sections 10.1, 10.2, 10.3), and an introduction to integer and fractional synthesizers (selected parts of Sections 10.4 and 10.5). If time permits though, a lecture may also be spent on the divider topologies (Section 10.6).

10.1 PHASE-LOCKED LOOPS BASICS

Like operational amplifiers, RF oscillators are seldom used open-loop. As in most applications a precise carrier frequency is needed, it is common to stabilize the oscillator within a feedback system, namely a phase-locked loop or PLL. Like any feedback system, a PLL requires a sense and a return mechanism to correct for VCO frequency drift. Unlike amplifier feedback loops, though, the variable of interest here is neither voltage nor current, but is rather phase (or frequency). With that in mind, one can build the PLL as follows:

- A phase detector (PD) that takes the phase difference between the two inputs and produces an output voltage or current accordingly. Thus, the PD output may be ideally expressed as $K_{PD}\Delta\phi$, where $\Delta\phi = \phi_{IN1} - \phi_{IN2}$ denotes the input signals phase difference (Figure 10.1), and K_{PD} is the phase detector gain.
- A loop filter, whose role is to remove the unwanted frequency components produced at the phase detector output. We will find out later that this is not the only purpose of the loop filter in a PLL. In fact the loop filter plays a critical role in the PLL stability and its transient response.
- And of course the VCO itself, which from the PLL point of view takes a voltage (sometimes current) as the input and produces an output frequency proportional to that. In practice, the VCO output may be applied to subsequent blocks through a buffer, which is often an inverter. Since it is not going to affect the PLL functionality, for simplicity it is omitted here.

So ideally, one may build a PLL as shown in Figure 10.2, where the phase detector provides the means of sense, and what is sensed is, of course, the VCO phase (or frequency). What is returned is the phase detector output after being lowpass filtered, which is a voltage and is applied to the VCO control voltage input.

Of the three blocks involved, the VCO has been covered in details already in the previous chapter. The loop filter could be as simple as an RC circuit whose role will be more clear once we get to the PLL transfer function and its dynamics in the next section. So for the remainder of this section, we will spend some time on the phase detector.

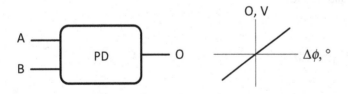

Figure 10.1: An ideal phase detector output–input characteristic

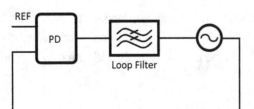

Figure 10.2: A simple phase-locked loop consisting of a VCO, a phase detector, and a loop filter

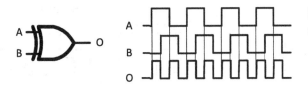

Figure 10.3: Phase detector realization using an XOR gate

10.1.1 Phase Detectors

A phase detector may be simply realized through multiplying the two inputs. For instance if the A and B inputs are

$$v_A(t) = A_0 \cos(\omega_1 t + \phi_1)$$

$$v_B(t) = B_0 \cos(\omega_2 t + \phi_2),$$

the product once lowpass filtered will be

$$\frac{A_0 B_0}{2} \cos((\omega_1 - \omega_2)t + \phi_1 - \phi_2),$$

and if the frequencies are the same, the output becomes proportional to $\cos(\phi_1 - \phi_2)$. Since the phase detector input signals are often rail-to-rail digital-type waveforms, in practice the multiplication may be accomplished by an XOR gate as shown in Figure 10.3.

If the inputs are perfectly aligned, the XOR output is zero. As one starts to lag (say B in our example above), the XOR produces short pulses at twice the frequency, whose DC component tracks the input phase difference linearly. The XOR output reaches a maximum if the inputs are $180°$ out of phase.

Example: The XOR circuit can be implemented by PMOS (or NMOS) switches as shown in Figure 10.4. Also shown is the XOR gate truth table. The output will be equal to V_{DD} or logic one, if either one of the inputs is high with the other one low, as desired.

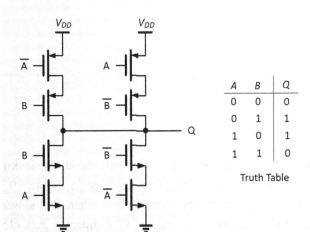

Figure 10.4: CMOS XOR gate

A	B	Q
0	0	0
0	1	1
1	0	1
1	1	0

Truth Table

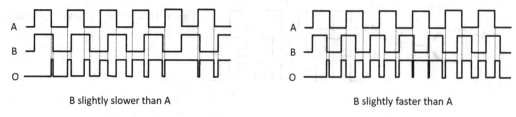

Figure 10.5: Phase detector waveforms when the two inputs have different frequencies

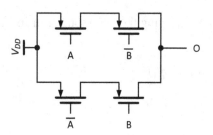

Figure 10.6: Current-mode XOR circuit realization using PMOS switches

In practice, the inputs of PD could have different frequencies prior to the PLL locking. Figure 10.5 shows two examples of the XOR PD output, where on the left the B signal is slower than A, while in the right B is faster. In either case, the PD creates a proper output, which once filtered and applied to the VCO results in its frequency to increase or decrease as desired.

An example of a current-mode XOR gate using only PMOS switches is shown in Figure 10.6.

Interestingly, even though the main objective of the PLL is to stabilize the VCO *frequency*, a *frequency detector* alone is not enough. This is because, like any feedback system, the output tracks only *the changes* in the input around a certain operating point, which is to say in the case of a frequency detector, the frequencies track but are not necessarily the same. On the other hand, if the loop phase locks, the phases track, and the frequencies that are the derivative of the phases are then identical. As we will see in Section 10.3.1, it is very common though to use a *phase-frequency detector* that accomplishes both tasks.

10.2 TYPE I PLLs

With the background earlier, we are now ready to take a closer look at the PLL of Figure 10.2.

10.2.1 PLL Qualitative Analysis

Let us first try to understand the PD and VCO behavior qualitatively during the locking process. Suppose initially the VCO is slower than the reference as depicted in Figure 10.7. The PD produces a sequence of pulses that, once passed through the loop filter, will slowly raise the control voltage, and thus the VCO frequency. As the VCO frequency gets closer to the reference, the PD pulses become smaller, and eventually the loop locks, that is, the VCO and

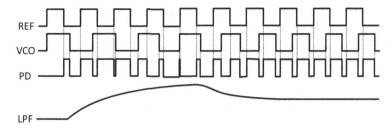

Figure 10.7: Type I PLL time domain behavior

reference have the same frequency, but there is a nonzero phase difference to set the required control voltage for the VCO to operate.

Depending on the loop dynamics, there may be an overshoot (like the example) or even some ringing, though obviously not desirable. This has to do with the specific choice of the loop filter bandwidth and the VCO and PD gains. We will discuss the transient behavior of the PLL quantitatively in the next section.

Example: As a practical circuit implementation of type I PLL, consider the circuit shown in Figure 10.8, consisting of a PMOS XOR (Figure 10.6) as a phase detector, an RC loop filter, and a ring oscillator whose frequency is adjusted through changing its supply voltage, effectively varying the inverter's transconductance or delay.

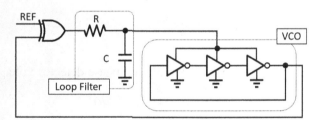

Figure 10.8: Example of a simple type I PLL circuit realization

The XOR output will sit below V_{DD}, such that when the loop is locked the proper control voltage (i.e., supply) is fed to the ring oscillator. The PD could be thought of a mixer (or multiplier) creating a DC output (once filtered), and is capable of providing the DC current required by the oscillator.

10.2.2 PLL Linear Model

Since the PLL transient response is a nonlinear phenomenon, it is not easy to derive a simple formula to capture all the mechanisms involved. Nevertheless, it is very common to use an approximate linear model to obtain the PLL transfer function, and consequently to gain insight into PLL dynamics and the trade-offs involved.

While the PLL building blocks such as VCO or phase detector produce voltage outputs, it is only meaningful to view the PLL as a hybrid system that deals with phase (or frequency) as well as voltage. Accordingly, a linear model can be derived, as shown in Figure 10.9.

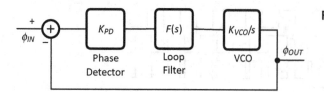

Figure 10.9: Type I PLL linear phase model

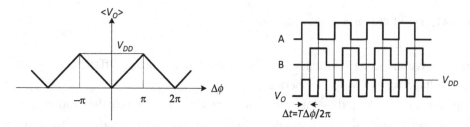

Figure 10.10: Voltage–phase characteristics of the XOR phase detector of Figure 10.6

The phase detector compares the phase difference between the reference and the VCO, and creates a voltage with the appropriate gain K_{PD}, measured in V/rad. For example, for the XOR phase detector of Figure 10.6, the voltage–phase characteristics are depicted in Figure 10.10, where the *y*-axis represents the *average* output voltage. This average voltage is effectively created by the loop filter whose transfer function is generally denoted by $F(s)$. Hence, for the XOR,

$$K_{PD} = \frac{V_{DD}}{\pi}.$$

The VCO produces a frequency proportional to the filtered phase detector voltage,

$$v_{VCO}(t) = A\cos(\omega_0 t + K_{VCO}\int v_{CTRL}(t)dt),$$

where $v_{CTRL}(t)$ is the loop filter output (or the VCO input). Since the output phase is the integral of frequency, in our linear model the VCO may be generally described by $\frac{K_{VCO}}{s}$, where K_{VCO} is the VCO gain measured in rad/s/V. The factor *s* represents the integration in Laplace domain.

From the Figure 10.9 block diagram, assuming the loop filter is a simple RC circuit like that of Figure 10.8, then the PLL is modeled as a linear unity feedback loop, with an open-loop gain of

$$\frac{K_{VCO}K_{PD}}{s}F(s) = \frac{K_{VCO}K_{PD}}{s(1+RCs)}.$$

Since the open-loop gain has *one* zero at the origin (due to the VCO), this type of PLL is referred to as type I. Consequently, the closed-loop transfer function, $H(s)$, becomes

$$H(s) = \frac{\Delta\Phi_{OUT}}{\Delta\Phi_{IN}} = \frac{\frac{K_{VCO}K_{PD}}{RC}}{s^2 + \frac{1}{RC}s + \frac{K_{VCO}K_{PD}}{RC}}.$$

This is a 2nd-order response, and using familiar control theory notations, it can be expressed as

$$H(s) = \frac{\omega_n^2}{s^2 + 2\zeta\omega_n s + \omega_n^2},$$

where $\omega_n = \sqrt{\frac{K_{PD}K_{VCO}}{RC}}$, and $\zeta = \frac{1}{2RC\omega_n} = \frac{1}{2\sqrt{RCK_{PD}K_{VCO}}}$. Now, we can justify the transient response of the PLL shown in Figure 10.7. For a critically damped response, which is often desirable, $\zeta = 1$, whereas the example depicted in Figure 10.7 has a small overshoot, and corresponds to an underdamped response, or $\zeta < 1$.

A few observations are in order:

- The PLL transfer function is lowpass, i.e., the PLL tracks only the low-frequency variations of the reference. This is expected, given the presence of a lowpass filter in the feedforward path of the feedback system. For a given loop filter cutoff, the PLL simply will not be fast enough to track the reference high-frequency perturbations.
- When locked, there will be a nonzero static phase difference between the input and the VCO. In other words, the PD has a nonzero output that maintains the right frequency of oscillation. This means that in general the phase difference between the reference and VCO ($\phi_{OUT} - \phi_{IN}$) will be a constant. Since frequency is the derivative of phase, however, $f_{OUT} - f_{IN} = 0$ as desired. Furthermore, despite the nonzero phase difference, any *change* of the input phase will be exactly followed by the VCO, which is to say $\Delta\phi_{OUT} - \Delta\phi_{IN} = 0$.
- To have a favorable transient response, and to achieve a sufficient phase margin, it is desirable to keep the damping factor, ζ, close to one. This presents a trade-off between the loop stability, and the lowpass filter bandwidth. A lower loop filter cutoff enjoys more filtering, but reduces the damping factor, and hence produces less stability.

The linear model predicts the PLL dynamics only when there is a small phase (or frequency) perturbation around the desired lock point. It fails to give us any information on the VCO frequency transient up to the vicinity of the loop locking. Nonetheless, it provides useful insights into designing and optimizing the PLL building blocks and the entire system, and is widely used among RF designers.

10.3 TYPE II PLLs

Despite their simplicity, type I PLLs are seldom used in most modern radios due to their limited lock range, the bandwidth–stability trade-off, and, to a lesser extent, the nonzero phase difference between the reference and the VCO during the lock. *Lock range* is a measure of how different the VCO and reference frequency can be from one another, and yet the PLL still acquires the lock. In the case of a type I PLL, the limited lock range mainly arises from the fact that the phase detector simply has no sense of its input *frequency* difference.[1] This issue is

---

[1] More information on the lock range and the PLL dynamics during the lock acquisition may be found in [21], [22].

addressed if a *phase-frequency detector* is exploited rather than a simple phase detector, and is discussed next.

10.3.1 Phase-Frequency Detection

Ideally, the phase-frequency detector (or PFD) must:

- Be insensitive to the inputs duty cycles or their shape (only zero crossings matter)
- Create the appropriate output(s) if the inputs frequencies are different, say positive if one is higher, negative if lower
- Create the appropriate output(s) if the inputs phases are different

A very commonly used circuit that satisfies all the above properties is shown in Figure 10.11. The edge-triggered D flip-flops (in this case positive edge-triggered) eliminate any dependence on the duty cycle, and only respond to the positive edge of either input. Both flip-flops inputs are connected to V_{DD} (or logic 1), and the inputs (A and B) are applied to the clock port. If the input A is leading B (as shown as an example on the right), the positive edge of A arrives first, which results in the upper flop output (U or up) going high. The U output stays high until the positive edge of B input arrives, which will result in the D or down output to go high as well. The U and D outputs will stay high only shortly after, until the AND gate resets the two D flip-flops. So the U output consists of pulses tracking the phase lag between A and B, while the D output consists of narrow pulses whose width is equal to the delay of the D flip-flop plus the AND gate, and are known as *reset delay* pulses. Reset delay pulses are often useful to avoid the *dead-zone* in the charge pump, that is, a region where neither output is sufficiently large to turn on the subsequent stages. This usually occurs when the phase difference between the two inputs is so small that the PFD cannot generate wide enough up and down pulses to turn on the next stage.

The circuit implementation of the D flip-flops is very straightforward, and is covered in Section 10.6.1. Since the B input is lagging, the PFD is effectively commanding the B signal (being represented by the VCO in the context of a PLL) to speed up, and hence a series of *up* pulses are issued. The scenario of course will be reversed if A is lagging B.

It is evident that the PFD outputs create the desired signal whether the inputs have phase or frequency difference.

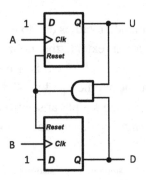
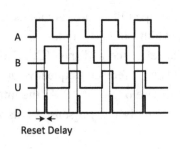

Figure 10.11: Phase-frequency detector

Example: Consider Figure 10.12, where on the left A and B have the same frequency but A is lagging, while on the right B has a higher frequency.

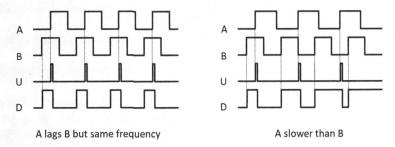

Figure 10.12: PFD timing diagram with one input lagging or slower than the other

In either case, a series of down pulses are created to effectively slow down the B signal.

Generally, one can utilize the PFD in the PLL by simply subtracting the outputs and apply to the loop filter, as illustrated in Figure 10.13.

We will show next that this task is performed more elegantly by employing a charge pump circuit.

10.3.2 Charge Pump

A more effective way of taking advantage of the up and down pulses created by the PFD is through applying them to two switched current sources known as a charge pump (or CP), shown in Figure 10.14. The current sources are always accompanied by a capacitor, whose need will be clear shortly. The timing diagram for the case of A leading B as an example is shown on the right.

Given the timing diagram on the right, since A is leading (or A is faster), U output consists of a series of up pulses, while D output has only the reset delay pulses. Once U is high, the top current source is on, pushing a current of I_{CP} to the capacitor C at the output. Consequently the output voltage rises as the capacitor charges up. Once the reset delay comes, both current sources are on, and no net current flows to the output. We will discuss shortly what happens if

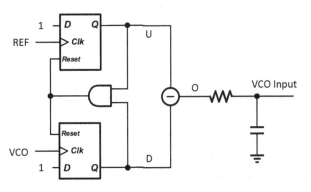

Figure 10.13: Phase-frequency detector conceptual use in a PLL

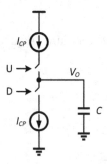 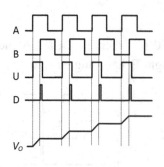

Figure 10.14: Schematic of a typical charge pump and the corresponding waveforms

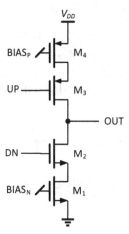 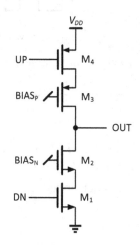

Figure 10.15: Examples of simple charge pump circuits

the two current sources are not identical. So overall the output voltage linearly goes up in discrete steps. We can postulate that if this voltage is applied to the VCO (after some appropriate filtering), it will raise the VCO frequency as it should. Obviously if A is slower than B, now the down pulses are effective, pushing the same current of I_{CP} to the output, but in a reverse direction, causing the capacitor to discharge.

10.3.2.1 Charge Pump Circuit Implementation

Two simple candidates to implement the charge pump circuit are shown in Figure 10.15. The circuit on the left is exactly how one would implement the conceptual design shown in Figure 10.14: transistors M_1 and M_4 are biased to create the current, while M_2 and M_3 act as switches.

Alternatively, one could use the circuit on the right, where the switches are moved to the source and away from the output. This could be advantageous, as clock feedthrough or charge sharing caused by the up and down signals' sharp transition are alleviated.

Figure 10.16 shows a differential implementation, where the P and N differential pairs create *current-mode switches*. Unlike the two examples of Figure 10.15, in the differential implementation the current sources always stay on, and the differential pair steer the current only either to the output or to a dummy path, whose voltage is kept the same as the output through a unity gain buffer.

This has a few advantages: It avoids any undesirable charge to be deposited to the output due to the current sources turning on or off. Furthermore, it will be a faster design, though in many

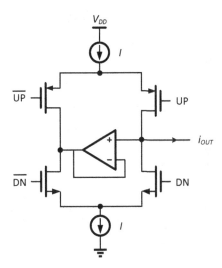

Figure 10.16: Charge pump implementation using a differential pair as a current-mode switch

cases the PFD/CP circuits run at relatively low frequencies (e.g., tens of MHz). The disadvantages are the need for a differential PFD, and a unity gain opamp with rail-to-rail input as the charge pump output usually is biased at the mid-rail, and should ideally swing as large as the headroom allows. The opamp does not add any noise, however, as its output is connected to the dummy path.

Apart from charge sharing and charge feedthrough, which are typically not a major concern, one potential issue in the charge pump arises from inevitably unequal up and down currents. This leads to nonlinearity and unwanted spurs at the PLL output, which we will address in details in Sections 10.3.3 and 10.5.3.2. To alleviate this, in all the three circuits shown, cascode current sources may be used to achieve more accuracy at the expense of voltage headroom. It is also possible to exploit feedback to equalize the up and down currents. An example of the latter is shown in Figure 10.23, and will be discussed in Section 10.3.3.

10.3.2.2 Charge Pump Modeling

In the case of type I PLLs, it was rather straightforward to quantify the phase detector gain or K_{PD} (Figure 10.10 for instance), used subsequently to derive the PLL transfer function. To do the same for the PFD/CP circuit, let us take a closer look at the timing diagram of Figure 10.14. First we observe that the charge pump output is not linear; e.g., if the input's phase skew doubles, the output waveform is not stretched two times. Second, the charge pump output is really a discrete signal, and thus must be treated as such. On the other hand, one can think of the charge pump output current as pulse-width modulated (PWM) bursts, or effectively as a continuous waveform sampled at the reference frequency as depicted in Figure 10.17.

The width of the pulses, ΔT, is proportional to the input's phase difference,

$$\Delta T = \frac{\phi_{REF} - \phi_{VCO}}{\omega_{REF}},$$

where $\omega_{REF} = \frac{2\pi}{T_{REF}}$ is the reference frequency, and ϕ_{REF} and ϕ_{VCO} are the reference (that the VCO must lock into) and VCO phases, respectively. The PWM currents sequence may be assumed to be a continuous current sampled by a train of impulses at the reference frequency,

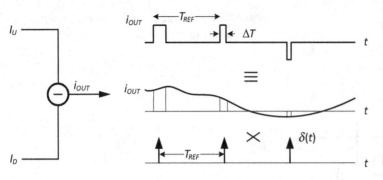

Figure 10.17: Charge pump output viewed as PWM currents sampled by the reference

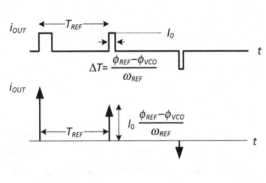

Figure 10.18: Charge pump current approximated by a train of impulses

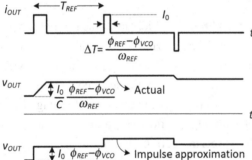

Figure 10.19: Charge pump output voltage as a result of current sequence integrated on the capacitor

provided that the rate of variation of the current pulses going into the capacitor is much less than the reference frequency. That is usually the case, as the loop filter bandwidth that directly determines this rate is much smaller than the reference frequency. The approximated current impulses as well as the original burst are depicted in Figure 10.18 as a comparison. We have assumed that the up and down currents are the same and equal to I_0.

Either waveform in Figure 10.18 will eventually lead to the correct charge pump voltage once integrated by the charge pump capacitor, the only difference being the abrupt change in the capacitor voltage if the impulse approximation is utilized (Figure 10.19).

In the frequency domain, the train of impulses becomes

$$\frac{1}{T_{REF}} \sum_k \delta(f - kf_{REF}),$$

which is convolved by the continuous current whose height at each sampling point is $I_0 \frac{\phi_{REF} - \phi_{VCO}}{\omega_{REF}}$. The principal spectrum of current appears around DC, with all the other images at the harmonics of the reference removed by the loop filter. The height of the current impulses at each point is

$$\frac{1}{T_{REF}} I_0 \frac{\phi_{REF} - \phi_{VCO}}{\omega_{REF}} = \frac{I_0}{2\pi}(\phi_{REF} - \phi_{VCO}),$$

or in the Laplace domain,

$$I_{OUT}(s) = \frac{I_0}{2\pi}(\phi_{REF}(s) - \phi_{VCO}(s)).$$

Thus, the PFD/CP gain is equal to $\frac{I_0}{2\pi}$, with I_0 being the charge pump current. With the capacitor at the output included, the charge pump output voltage will be

$$V_{OUT}(s) = \frac{1}{sC} \frac{I_0}{2\pi}(\phi_{REF}(s) - \phi_{VCO}(s)).$$

That is, it behaves like an *integrator*, and therefore has infinite gain at DC. Intuitively this is justified by noting that, when locked, a phase difference between the VCO and reference, however small, will result in a small charge to be deposited on the capacitor at every cycle, which goes on forever. So in a type II PLL, at least with an ideal charge pump, the VCO and reference are not only frequency locked, but also phase aligned. We will show, however, that in a more realistic charge pump with unequal up and down currents, there will be a small phase difference, causing unwanted reference spurs at the output.

10.3.3 Type II PLL Analysis and Nonideal Effects

The combination of the PFD, CP, and output capacitor provides, for the most part, the same functionality as the PD and loop filter in the case of a type I PLL. Consequently, one can surmise the PLL may be built like shown in Figure 10.20, only that there is an additional resistor in series to the capacitor as well, whose role will be clear in a moment.

Let us assume first there is no resistance, as has been the case thus far, that is $R=0$ in the figure. Then the open-loop gain is

$$G(s) = \frac{I_0}{2\pi} \frac{1}{sC} \frac{K_{VCO}}{s}.$$

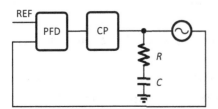

Figure 10.20: A simple type II PLL

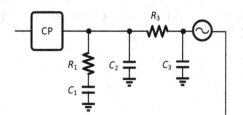

Figure 10.21: More practical loop filter implementation of type II PLLs

The open-loop gain has two zeros at origin, and hence the terminology type II PLL, but once put in the unity feedback it creates an oscillatory system. To avoid this, a resistor is added in series, which makes the open-loop gain

$$G(s) = \frac{I_0}{2\pi}\left(R + \frac{1}{sC}\right)\frac{K_{VCO}}{s}.$$

There are still two zeros in the origin, but now a left-half plain zero is added, which improves the phase margin. The closed-loop PLL transfer function is now

$$H(s) = \frac{\omega_n^2(1 + RCs)}{s^2 + 2\zeta\omega_n s + \omega_n^2},$$

where $\omega_n = \sqrt{\frac{\frac{I_0}{2\pi}K_{VCO}}{C}}$, and $\zeta = \frac{RC\omega_n}{2} = \frac{R}{2}\sqrt{\frac{I_0}{2\pi}CK_{VCO}}$. Unlike type I PLL, here reducing the loop filter bandwidth (or PLL bandwidth) will not necessarily reduce the damping factor. Furthermore, the addition of the LHP zero helps tremendously with the phase margin and stability, a feature not present in a type I PLL.

In practice, to achieve a sharper roll off to suppress far-out noise and spurs, additional poles are often added, and the actual loop filter looks more like the one shown in Figure 10.21. As a rule of thumb, usually $C_2 = \frac{C_1}{10}$, and the third pole set by R_3/C_3 is chosen to be about 5–10 times higher than the PLL bandwidth. So apart from minor degradation in the phase margin, they have little impact on PLL close-in response or bandwidth, but certainly provide much-needed rejection at higher frequencies. This is especially important in fractional PLLs, to be discussed in Section 10.5.3.3.

It must be emphasized that the linear analysis presented is based on the assumption that the rate of variation of the VCO phase is small compared to the reference frequency, i.e., the loop bandwidth is sufficiently small with respect to the reference. If this assumption does not hold, the charge pump can no longer be approximated as a continuous circuit, and the PLL must be analyzed in the discrete domain. Given the stability concerns as well as the general desire of having sufficient filtering from the loop filter, this assumption is usually satisfied, and thus the simpler linear model works fine for most purposes.

More on the PLL transfer function and its noise sources can be found in Sections 10.4.1 and 10.4.2.

Apart from noise concerns which will be discussed in the next section, the main issue in the type II PLLs arises from the unwanted spurs located around the VCO at the reference

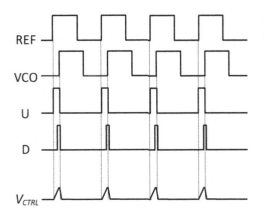

Figure 10.22: Demonstration of reference spur in PLLs in the presence of mismatch in up and down currents

frequency and its harmonics, known as the *reference spurs*. This primarily (there may be also direct feedthrough due to parasitic paths from the reference to the VCO) arises from unequal up and down currents in charge pump. Consider Figure 10.22, where we have assumed the up current is slightly less than the down current. If the currents are equal, the VCO and reference will be exactly phase aligned, and the charge pump only turns on shortly during the reset delay, but since the equal up and down currents offset each other, it will have no impact on the control voltage or VCO. On the other hand, if the up current is less, as shown in the figure, the up pulse must be wider such that the net charge deposited on the capacitor during one cycle is zero. This will cause the control voltage to go up first, after which point is reduced back to the steady-state level due to a larger down current, which is effective only during the reset delay phase. Naturally, the VCO and reference are not phased aligned anymore,[2] but the feedback loop of course ensures that the frequencies are the same.

Consequently, the VCO control voltage will not be flat, rather it has a small perturbation on it happening at the reference frequency. Once applied to the VCO, it will frequency modulate its output, creating sidebands at the reference frequency and all its harmonics. Since this waveform is subject to the lowpass loop filtering, usually the main harmonic is the most troublesome, though it ultimately depends on where exactly the blockers are located.

The unequal up and down currents are inevitable, and are caused mainly by the mismatches and/or the channel length modulation in the current sources. The former is often not a dominant factor, and can be improved by sizing the current sources properly. Since the charge pump output is often biased at mid-rail, and ideally should entertain as large a swing as possible,[3] the finite output resistance of the current sources due to the channel length modulation is the most problematic factor. Typical remedies are to use cascode current sources at the expense of precious headroom, or use feedback to equalize the currents.

---

[2] This is usually unimportant.

[3] This is primarily to reduce the VCO gain as much as possible for a given desired locking range, which in turn is to avoid low-frequency noise modulating the VCO through a large varactor as discussed in the previous chapter.

Example: Shown in Figure 10.23 is a charge pump circuit utilizing feedback to alleviate the channel length modulation [1].

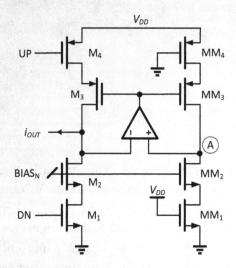

Figure 10.23: Charge pump with feedback loop to improve current matching [1]

A similar idea is used in low-voltage bandgaps as well [2], and works as follows: The high gain opamp ensures that the charge pump output and node A are at the same potential. Thus, between the main charge pump devices (M_1–M_4), and the replica path (MM_1–MM_4), all have the same gate, source, and drain voltage. Since KCL forces the N and P currents to be the same in the replica path (MM_2 and MM_3), so should be the main charge pump up and down currents. In other words, the opamp sets the bias of the P transistor (M_3) to offset the channel length modulation, as opposed to the traditional implementation (Figure 10.15), where the P bias is created externally. The trade-offs here are the need for a rail-to-rail opamp, which adds noise and power consumption. It may be possible though to at least partially filter the opamp noise by placing a capacitor at its output.

Example: Another possible source of reference spurs in PLLs, shown in Figure 10.24, is the leakage current caused by the varactor or the loop filter capacitors (especially if MOS capacitors are used for the lower silicon area).

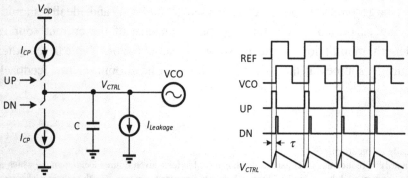

Figure 10.24: Leakage casing unequal up and down currents, leading to reference spurs

The leakage current results in the control voltage continuously decreasing, and hence to make up for that, the UP signal must turn on a little bit earlier, causing the reference to lead the VCO. Assuming a reference lead time of τ, we should have

$$(I_{CP} - I_{Leakage})\tau = I_{Leakage}(T_{REF} - \tau),$$

where I_{CP} is the charge pump current, $I_{Leakage}$ is the leakage current, and T_{REF} is the reference period. Therefore,

$$\tau = \frac{I_{Leakage}}{I_{CP}} T_{REF}.$$

The strength of the spur of course depends on how large the leakage current, or the VCO and reference phase lead, are.

10.4 INTEGER-*N* FREQUENCY SYNTHESIZERS

Although PLLs as described so far are used in radios for purposes of LO generation, what actually is needed is a frequency synthesizer. Two simple reasons: First, the VCO frequency is usually in GHz range to supply the mixers, while the practical references have frequencies on the tens of MHz range due to the limitation of crystals. Second, the VCO frequency, which provides the RX or TX LOs, must be tunable, usually in a wide range, while the crystals are not.

The issues are resolved by simply inserting a programmable divider in the feedback path, known as the *feedback divider*, as shown in Figure 10.25 (the divide-by-*N* circuit). Furthermore, the reference must be also divided by a proper value (K in the figure, known as the *reference divider*) so that the PFD input frequency becomes equal to the required channel spacing such that by varying *N* in integer steps, the desired RF channel is selected.

The feedback loop in the PLL naturally forces the output frequency to be

$$f_{OUT} = N\frac{f_{REF}}{K} = Nf_{Channel},$$

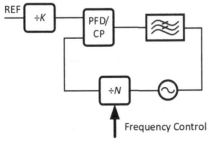

Figure 10.25: Integer-*N* frequency synthesizers in wireless radios

where $f_{Channel}$ is the required channel spacing (Figure 10.25). Since the synthesizer produces frequencies only in integer multiples, it is often referred to as an *integer-N synthesizer*, or simply an integer synthesizer, as opposed to the fractional synthesizers discussed in Section 10.5.

> **Example:** In Bluetooth, the desired channel located in the ISM band can assume any frequency between 2402MHz to 2480MHz in steps of 1MHz. Thus, $f_{Channel} = 1$MHz (the reference divide ratio of course depends on the crystal frequency), and a programmable divider with a divide ratio of 2402 to 2480 is needed. The dividers will be discussed at the end, in Section 10.6.

In the remainder of this section, we will extend our PLL linear analysis to include the feedback divider, and also discuss the noise sources in the synthesizer. For the most part, the synthesizer of Figure 10.25 behaves much the same as the PLL (Figure 10.20) discussed already in detail, and suffers from very similar issues (e.g., the reference spur, lock range, etc.).

10.4.1 Signal Transfer Functions

The synthesizer transfer function and its phase noise may be obtained by using the linear model shown in Figure 10.26. The noise of each block is injected accordingly as shown in the figure.

The feedback divider simply divides the VCO phase (or frequency) by N, and is represented by $\frac{1}{N}$. The open-loop gain is given by

$$G(s) = K_{PFD} F(s) \frac{K_{VCO}}{s} \frac{1}{N},$$

where $F(s)$ is the loop filter transfer function, and K_{PD} is the PD or PFD/CP gain, whichever is used. In the more common case of using a type II PLL, K_{PD} is $\frac{I_0}{2\pi}$ as derived in Section 10.3.2.2. Accordingly, the PLL output (including the noise sources) may be expressed as

$$\phi_{OUT} = \frac{NG}{1+G}\phi_{IN} + \frac{NG}{1+G}v_{nDIV} + \frac{1}{1+G}v_{nVCO} + \frac{N}{K_{PD}}\frac{G}{1+G}(v_{nPD} + v_{nLF}).$$

Ignoring the noise sources for the moment, the PLL closed-loop transfer function is

$$H(s) = \frac{\phi_{OUT}}{\phi_{IN}} = N\frac{G(s)}{1+G(s)}.$$

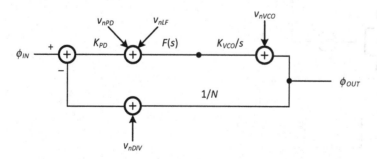

Figure 10.26: PLL linear model

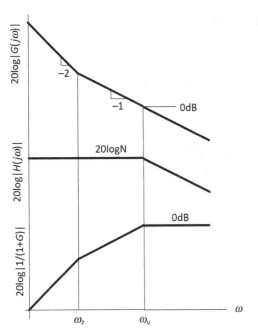

Figure 10.27: Key synthesizer transfer functions

Shown in Figure 10.27 is the Bode plot of the open-loop gain ($G(s)$) as well as the PLL closed-loop transfer function ($H(s)$). Also shown is the transfer function from the VCO to the output, which is $\frac{1}{1+G(s)}$.

In the case of a type I synthesizer, there is one pole at origin in the loop gain resulted from the frequency integration inherent in the VCO ($\frac{K_{VCO}}{s}$). In the case of a type II PLL, the loop filter has an additional pole at origin. This justifies why the loop gain reduces by a slope of -40dB/Dec initially, as shown in Figure 10.27. Furthermore, as we discussed in Section 10.3.3, to ensure stability, the loop filter may employ a left-half plane zero (ω_z), at which point the loop gain drops by -20dB/Dec slope, reaching one at the unity gain frequency, ω_u. The exact location of the unity gain frequency depends on the loop filter transfer function.

At frequencies well below ω_u, $|G(j\omega)| \gg 1$, and thus $|H(j\omega)| \approx N$. This implies that the output frequency must be exactly N times that of the input (or reference). On the other hand, at frequencies well above ω_u, $|G(j\omega)| \approx 0$, and $|H(j\omega)|$ follows $|G(j\omega)|$, hence reducing with a 20dB/Dec slope. The PLL closed-loop bandwidth is then expected to be roughly equal to ω_u. Right around the unity gain frequency, the exact shape of the PLL closed-loop transfer function will depend on the loop filter characteristics. It may peak or smoothly drop according to the loop filter components values.

The VCO transfer function (the third plot in Figure 10.27) may be justified similarly.

To derive some numerical values for the PLL bandwidth, let us assume the charge pump output drives a simple RC circuit (Figure 10.20) only, and ignore the second and third poles that may exist in a more realistic loop filter (Figure 10.21). As indicated, the resistors and capacitors are often chosen such that the additional poles or zeros created are far from the PLL bandwidth. This is to ensure the stability, and a good phase margin for the PLL. Consequently, the additional filtering may help only at far-out frequencies.

Therefore, the loop filter transfer function may be simply expressed as

$$F(s) = R + \frac{1}{Cs} = \frac{1+RCs}{Cs}.$$

Accordingly, the synthesizer closed-loop transfer function is

$$H(s) = N \frac{\omega_n^2(1+RCs)}{s^2 + 2\zeta\omega_n s + \omega_n^2},$$

where $\omega_n = \sqrt{\frac{K_{PD}K_{VCO}}{CN}}$, and $\zeta = \frac{RC\omega_n}{2}$. For a CP PII, $K_{PD} = \frac{I_o}{2\pi}$ to ensure a critically damped response, $\zeta = 1$, which follows $RC\omega_n = 2$. Under these circumstances, the synthesizer 3dB bandwidth is calculated to be

$$\omega_{3dB} = \frac{2\sqrt{3+\sqrt{10}}}{RC} \approx \frac{5}{RC}.$$

Or expressed in terms of the synthesizer parameters,

$$\omega_{3dB} \approx 2.5\sqrt{\frac{K_{PD}K_{VCO}}{CN}},$$

and the unity gain frequency is

$$\omega_u = \frac{2\sqrt{2+\sqrt{5}}}{RC} \approx \frac{4}{RC}.$$

Example: We wish to calculate the PLL phase margin for the critically damped case. The open-loop transfer function was found to be

$$G(s) = \frac{\omega_n^2}{s^2}\left(1 + \frac{s}{\omega_z}\right),$$

where $\omega_z = \frac{1}{RC}$. The corresponding Bode plot is shown in Figure 10.28.

Since $\omega_u \approx 4\omega_z$, the phase margin is found to be

$$PM = \tan^{-1}(4) = 76°.$$

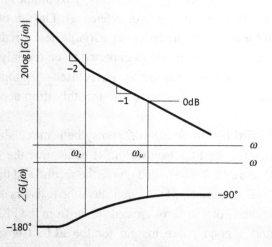

Figure 10.28: Open-loop transfer function of the PLL Bode plot for the critically damped case

10.4 Integer-N Frequency Synthesizers

Example: Consider a Bluetooth transmitter. The synthesizer reference is 1MHz, same as channel spacing, and suppose we set the PLL bandwidth to be one-tenth of the reference, 100kHz. Assume K_{VCO} is equal to $2\pi \times 15\text{MHz/V}$. With $\zeta = 1$, we have

$$RC = \frac{5}{\omega_{3dB}} = \frac{5}{2\pi \times 100\text{kHz}} = 8\mu S.$$

For noise concerns (see the next section), we choose $R = 20k$, which leads to $C = 400p$, a fairly large capacitance, though it can be partially implemented with MOSCAPs. Since $\sqrt{\frac{K_{PD}K_{VCO}}{CN}} = \omega_n = \frac{2}{RC} = 0.4\omega_{3dB}$, and with $N = 2440$, we find $K_{PD} = 0.0013\text{V/rad}$, which leads to charge pump current of 4.1mA. This is rather large, and can be reduced by either reducing the loop filter capacitor or by increasing the VCO gain, both of which have noise implications of course. As the divide ratio changes to cover the 80MHz ISM band, the PLL characteristics change slightly. Since this is predictable, one may choose to compensate by varying the charge pump current accordingly, though since its impact is small, it is often left alone.

Interestingly, if we could have used a smaller divide ratio, a lower charge pump current would have been obtained. This is not possible in integer PLLs, but as we will see shortly, fractional PLLs allow for much higher reference frequencies, which in turn imposes a much more reasonable trade-off on the charge pump value and the loop filter size.

10.4.2 Noise Sources in Synthesizer

As we already have the transfer function of various noise sources to the output, we may now explain the PLL phase noise profile as was shown previously in Chapter 5, and depicted again in Figure 10.29.

The reference, phase detector, and divider go through a lowpass transfer function given by $\frac{G(s)}{1+G(s)}$ (with different scaling factors, however). Shown in Figure 10.30 is the reference noise transfer function to the output.

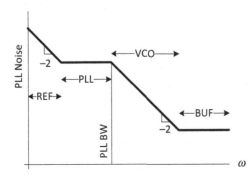

Figure 10.29: PLL general phase noise profile, and various noise contributors

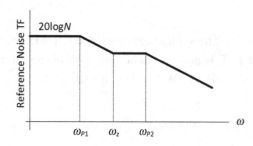

Figure 10.30: Reference noise transfer function to the output that is lowpass

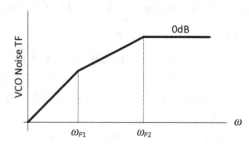

Figure 10.31: VCO noise transfer function to the output that is bandpass

Assuming a simple RC loop filter, the transfer function follows as

$$N\frac{\omega_n^2(1+RCs)}{s^2+2\zeta\omega_n s+\omega_n^2},$$

and the two poles of the transfer function (assuming overdamped function) are given by

$$\omega_{P1,2}=\omega_n\left(\zeta\pm\sqrt{\zeta^2-1}\right).$$

The VCO noise, on the other hand, enjoys a highpass transfer function determined by $\frac{1}{1+G(s)}=\frac{s^2}{s^2+2\zeta\omega_n s+\omega_n^2}$, and is shown in Figure 10.31.

Consequently, for frequencies below the PLL bandwidth, we expect the PLL components such as the phase detector or divider to dominate. As the reference is typically a crystal oscillator, at very low frequencies, where its noise rises with 20dB/Dec slope, we expect it to dominate. Beyond the PLL bandwidth, the noise of the PLL is suppressed given the lowpass nature of the response. On the other hand, the VCO noise appears directly at the output, and dominates. Finally at very far-out frequencies, the noise flattens as the VCO buffer and following stages start to dominate.

Intuitively, we may explain the PLL noise profile as follows: At low frequencies within the PLL bandwidth, any fluctuation at the VCO frequency resulted from its noise is suppressed by the PLL due to the presence of a strong feedback. As the VCO frequency is expected to follow the reference precisely, the reference noise (as well as that of the divider or PD) is expected to appear at the output. On the other hand, at high frequencies, the feedback is essentially ineffective, and the VCO noise appears at the output unfiltered.

10.4 Integer-N Frequency Synthesizers

Example: We shall find the synthesizer phase noise due to the loop filter resistor, as shown in Figure 10.32.

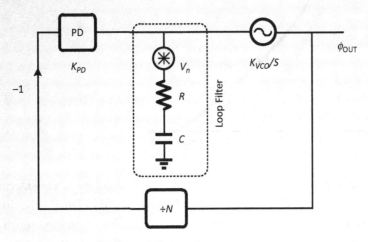

Figure 10.32: Loop filter resistor noise

We can show that the resistor noise transfer function to the output is (see Problem 2)

$$\frac{\phi_{OUT}}{v_n}(s) = K_{VCO}\frac{s}{s^2 + 2\zeta\omega_n s + \omega_n^2}.$$

The transfer function is *bandpass*, and peaks at the PLL natural frequency, ω_n, located at $0.4\omega_{3dB}$, assuming critically damped condition (ω_{3dB} is the PLL bandwidth). Thus, the peak phase noise caused by the loop filter resistor is

$$\mathcal{L}\{\omega_n\} = 2KTR\left|\frac{\phi_{OUT}}{v_n}(j\omega_n)\right|^2 = \left|\frac{K_{VCO}}{2\omega_n}\right|^2 2KTR.$$

Note that what is shown above is the *SSB phase noise*, and thus the resistor noise is considered: $2KTR$. For a given PLL bandwidth, the required phase noise determines the resistor size, and consequently, the capacitor value.

As an example, Figure 10.33 shows the noise contribution of the loop filter resistor, the reference, and the VCO one at a time with everything else noiseless, as well as the composite phase noise with all the sources acting simultaneously. Given the linear model, the superposition applies, and therefore we can find the noise contribution of each building block separately, and then add them all up:

- The reference and the VCO can be simulated free running and then translated to the output using the proper transfer function.

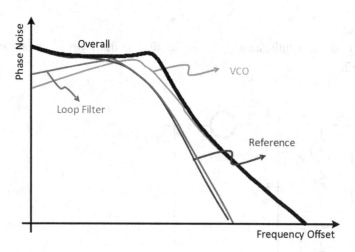

Figure 10.33: Synthesizer typical noise contributors

- The PFD/CP noise can be estimated running AC noise simulation, considering the time charge pump is on, or run a PSS[4] simulation based on a realistic PFD/CP behavior. In integer PLLs, the charge pump is on only shortly and is not expected to contribute much noise. This is, however, not the case in fractional PLLs, as we will find out soon.
- Loop filter noise can be estimated as $\mathcal{L}\{\omega_n\} = \left|\frac{K_{VCO}}{2\omega_n}\right|^2 2KTR$, as was just described, or can be simply simulated using an AC analysis of the loop filter through the linearized PLL model.
- The divider noise can be estimated using a PSS simulation with an ideal input at the proper frequency.

Then every noise source is translated to the output using the proper transfer function and added up. That is the thick black curve on the plot of Figure 10.33. At very far-out offsets the noise profile follows the VCO noise, and at very low offsets it follows the reference. In between, all the PLL blocks contribute, and it often follows the PLL transfer function. However, it is not correct to estimate the PLL bandwidth or transfer function based on solely its output noise profile in general.

The linear model proves to be very convenient and fast and is widely used. Alternatively, one could run transient noise simulation, which is more realistic but is a very slow simulation. It may be helpful to use an ideal VCO, and run transient only to capture the PLL noise.

10.5 FRACTIONAL-*N* FREQUENCY SYNTHESIZERS

Despite their simplicity, the integer-*N* synthesizers have a major drawback arising from the fact that the reference frequency cannot be higher than the channel spacing, which is typically limited to a few hundred kHz, or a few MHz. This in turn sets the upper bound of the loop bandwidth, which causes several implications on:

---

[4] Periodic steady state simulation in Cadence Spectre RF.

10.5 Fractional-N Frequency Synthesizers

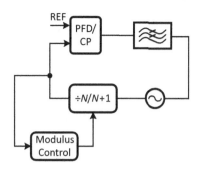

Figure 10.34: Fractional-N synthesizer high-level schematic

- The size of the loop filter capacitors and resistors
- Synthesizer settling time
- Close-in and far-out phase noise
- Ability to work with any arbitrary crystal frequency
- Reference spurs, and the amount of filtering they could be subject to

The above issues can be alleviated if a fractional divider could be implemented. This in practice can be devised if the fix divider by N is replaced with a dual-modulus divider by $N/N+1$, in such a way that for several cycles it divides by N, and for several other cycles divides by $N+1$. Then, *on average,* over a long period of time, a *fractional* divide ratio of $N+\alpha$ is achieved, where $0 \leq \alpha \leq 1$. The high level concept is illustrated in Figure 10.34, where the modulus control block is responsible for choosing the divider modulus properly to accomplish the required fractional divide ratio.

For instance, as a simplest form of modulus control, imagine the divider is programmed to divide by N for $K-1$ cycles, and $N+1$ for one last cycle, and this repeats periodically. Since the PLL negative feedback forces the PFD inputs to have the same frequency, then on average for K cycles of the reference, a divide ratio of $\frac{(K-1)N+(N+1)}{(K-1)+1} = N + \frac{1}{K}$ is achieved, where $0 \leq \frac{1}{K} \leq 1$ represents the fraction ($\alpha = 1/K$).

Example: In the Bluetooth synthesizer of the previous example, we chose K_{VCO} to be $2\pi \times 15$MHz/V. Assuming a fractional synthesizer with reference frequency of 40MHz, we set the loop bandwidth to 200kHz now as the channel spacing constraint is removed, and with $\zeta = 1$, we have $RC = 4\mu S$. This leads to a smaller loop filter capacitance of $C = 200p$ for the same resistance of $R = 20k$. Furthermore, with a reference frequency of 40MHz, $N = 61$, and since $\sqrt{\frac{\frac{I_{CP}}{2\pi}K_{VCO}}{CN}} = 0.4\omega_{3dB}$, we find $I_{CP} = 205\mu A$.

The main drawback of the simple approach outlined in Figure 10.34 is the creation of fractional spurs. To demonstrate this, shown in Figure 10.35 is a hypothetical example of dividing by N for three cycles, and by $N+1$ for the fourth cycle. Now, assume the PLL feedback loop forces the output steady-state frequency to be

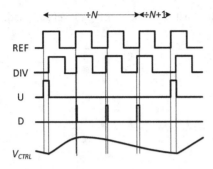

Figure 10.35: Demonstration of fractional spurs due to the periodic variation of the control voltage

$$f_{OUT} = f_{REF}\left(N + \frac{1}{K}\right),$$

as desired, where f_{REF} is the reference frequency, and for this example $K = 4$.

For the first three cycles of dividing by N, the divider frequency is

$$f_{DIV} = \frac{f_{REF}\left(N + \frac{1}{K}\right)}{N} > f_{REF}.$$

Thus, the PFD starts to generate *down* pulses to slow down the VCO, which in turn causes the control voltage to reduce (the reset delay pulses are not shown for simplicity in Figure 10.35). On the fourth cycle, however, the VCO is divided by $N+1$, resulting in divider output to be slower than the reference, and hence an *up* signal is created that raises the control voltage. This pattern repeats every $K = 4$ cycles, leading to the VCO being modulated by a periodic signal with a period of $\frac{f_{REF}}{K} = \alpha.f_{REF}$, and hence creating sidebands at $\pm \alpha. f_{REF}$, and its harmonics. On the other hand, to accommodate the desired channel selection, $\alpha. f_{REF}$ could be as small as the channel spacing, and hence the same limitation as the integer-N synthesizer exists, that is, the presence of low-frequency spurs as low as the channel spacing.

10.5.1 Noise Shaping

One very common and elegant technique to deal with fractional spurs is through exploiting the noise shaping concept widely used in data converters [3], [4].

First, we realize that the *periodicity* of the control voltage could be broken if the divider modulus is *randomly* selected between N and $N+1$, but still maintaining the desired average value. By doing so, the deterministic fractional spurs appearing as tones will be converted to noise spreading around the VCO signal, with the total energy remaining the same of course.

We can therefore write the relation between the divider and output frequencies as

$$f_{DIV} = \frac{f_{OUT}}{N + b(t)},$$

where $b(t)$ is a random stream of 0s and 1s with an average value of α. The bit stream can be then broken up into the desired value, α, and an additive quantization noise, $q(t)$ [5], [6]:

$$b(t) = \alpha + q(t).$$

Thus,

$$f_{DIV} = \frac{f_{OUT}}{N+\alpha+q(t)} = \frac{\frac{f_{OUT}}{N+\alpha}}{1+\frac{q(t)}{N+\alpha}} \cong f_{DIV,nom}\left(1 - \frac{q(t)}{N_{nom}}\right),$$

where $f_{DIV,nom} = \frac{f_{OUT}}{N+\alpha}$ is the divider nominal frequency, and $N_{nom} = N+\alpha$ is the nominal divide ratio. Thus, when referred to the divider output, the *normalized* instantaneous frequency departure, $y(t)$, is

$$y(t) = -\frac{q(t)}{N_{nom}}.$$

If the spectral density of the quantization noise is $S_q(f)$, the spectral density of the normalized frequency departure is

$$S_y(f) = \frac{S_q(f)}{N_{nom}^2}.$$

Since phase is the integral of frequency, in terms of phase noise then

$$S_{\phi,DIV}(f) = \frac{f_{DIV,nom}^2}{f^2} S_y(f) = \left(\frac{f_{DIV,nom}}{N_{nom}f}\right)^2 S_q(f).$$

Note that $S_y(f)$ is the normalized frequency departure, hence in our derivation above there is the additional multiplication by $f_{DIV,nom}$. When referred to the VCO output,

$$S_{\phi,OUT}(f) = \left(\frac{f_{OUT}}{N_{nom}f}\right)^2 S_q(f) = \left(\frac{f_{REF}}{f}\right)^2 S_q(f).$$

Despite the fact that the randomization process eliminates the fractional spurs, it could still rise to a substantial amount of phase noise, which is typically unacceptable for most applications. The latter issue is dealt with by shaping the noise through $\Delta-\Sigma$ modulation, as used in the realization of oversampled data converters [6], [7].

A simplified block diagram of a 1-bit $\Delta-\Sigma$ analog to digital converter (ADC) is shown in Figure 10.36 to illustrate the concept. It consists of an adder (Σ), an integrator (Δ), a 1-bit ADC (comparator), and a 1-bit DAC.

Ignoring the ADC quantization noise for the moment, the signal transfer function is

$$\frac{Y}{X}(s) = \frac{\frac{1}{s}}{1+\frac{1}{s}} = \frac{1}{s+1},$$

which is lowpass. For simplicity it is assumed that the integrator transfer function is $\frac{1}{s}$. If the ADC quantization noise is modeled as an additive source, $q(t)$, at its output, the noise transfer function is highpass and given by

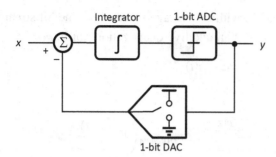

Figure 10.36: A 1-bit analog $\Delta-\Sigma$ ADC

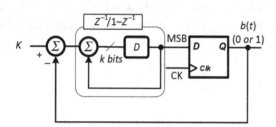

Figure 10.37: Block diagram of a 1-bit digital $\Delta-\Sigma$ modulator

$$\frac{Y}{Q}(s) = \frac{s}{s+1}.$$

This means that if the signal of interest is lowpass, it passes through ADC, whereas the ADC quantization noise around low frequencies where the signal resides is suppressed, a very desirable property.

A digital implementation of the 1-bit $\Delta-\Sigma$ modulator above is shown in Figure 10.37. The analog integrator is replaced with a delay cell in feedback loop. Given that a delay cell has a discrete-domain transfer function of z^{-1}, the discrete integrator transfer function (shown in the dashed box) is

$$H(z) = \frac{z^{-1}}{1-z^{-1}}.$$

Commonly used in switched-capacitor filters [8], this in fact represents the forward Euler $s-$ to $z-$ domain conversion, denoted by $s \leftrightarrow \frac{1}{T}(z-1)$. Since $z = e^{j\omega T}$ (T is the clock frequency),

$$H(j\omega T) = \frac{1}{e^{j\omega T}-1} \cong \frac{1}{j\omega T},$$

as expected.

Since the modulator is implemented digitally, the 1-bit D/A is not required, and finally the comparator is replaced by simply a D flip-flop fed by the integrator MSB, discarding all the other $k-1$ bits of the integrator. Thus, for the k-bit integrator, the input (which is the digital representation of the desired fractional offset) has $k-1$ bits, and the output is simply 1-bit.

This clearly results in large levels of quantization noise, which is modeled by the additive noise source shown in Figure 10.38. Similar conclusions to that of the analog modulator of Figure 10.36 can be made here.

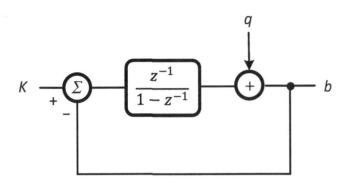

Figure 10.38: Simple equivalent model of the 1-bit digital $\Delta-\Sigma$ modulator

The signal transfer function is

$$\frac{B}{K}(z) = \frac{\frac{z^{-1}}{1-z^{-1}}}{1 + \frac{z^{-1}}{1-z^{-1}}} = z^{-1},$$

i.e., the output simply tracks the input with a delay, while the noise transfer function becomes

$$\frac{B}{Q}(z) = 1 - z^{-1},$$

which is highpass again.

To find the corresponding phase noise, we first express the noise in terms of the divider,

$$N(z) = N_{nom} + (1 - z^{-1})Q(z),$$

where $N_{nom} = N + \alpha$ is the fractional divider value as defined earlier, and $N(z)$ is the discrete-domain divide ratio.[5] The output frequency can be expressed in the z-domain as well:

$$F_{OUT}(z) = N(z)f_{REF} = N_{nom}f_{REF} + f_{REF}(1 - z^{-1})Q(z).$$

The first term is the desired output frequency, whereas the second term is the impact of the shaped quantization noise. What we are interested in is ultimately the phase noise, which is the integral of frequency. In continuous form, the phase noise is $\phi_{OUT} = \int \omega_{OUT}(t)dt = 2\pi \int f_{OUT}(t)dt$, which in the discrete domain translates to

$$\Phi_{OUT}(z) = 2\pi \frac{T}{z-1} f_{REF}(1 - z^{-1})Q(z).$$

Or, the noise spectral density,

$$S_{\phi,OUT}(f) = \left|\frac{2\pi}{z-1}\right|^2 |1 - z^{-1}|^2 S_q(f),$$

where $z = e^{j\omega T} = e^{j2\pi fT}$.

The term $|1 - z^{-1}|^2 = 4|\sin(\pi fT)|^2$ is the *noise shaping function*, which is highpass as expected. Consequently, the quantization noise is suppressed around DC as desired. The noise suppression function $(4|\sin(\pi fT)|^2)$ has been plotted in Figure 10.39.

---

[5] To distinguish from the dual-modulus divide ratio N, we are using a bold notation for $N(z)$.

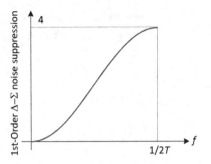

Figure 10.39: Noise suppression in a 1st-order $\Delta-\Sigma$ modulator

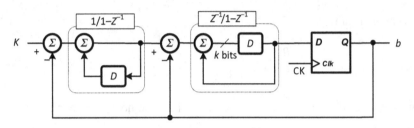

Figure 10.40: An example of a 2-bit $\Delta-\Sigma$ modulator

Finally, assuming a 1-bit quantizer has a uniform quantization error, the error power is then $\frac{\delta^2}{12}$, where δ is the minimum step size of the quantizer [4]. Because we are quantizing to integers, $\delta = 1$, and the quantization error power is $\frac{1}{12}$. Since this power is spread over a bandwidth of f_{REF}, the power spectral density of noise is

$$S_q(f) = \frac{1}{12 f_{REF}}.$$

10.5.2 Higher Order $\Delta-\Sigma$ Modulators

The integrator stage of the $\Delta-\Sigma$ modulator introduces a zero at the origin, thus noise shaping the quantization noise in the frequency domain. For the 1st-order $\Delta-\Sigma$ modulation illustrated in Figure 10.37, the integration applied when converting from frequency to phase removes the zero from the noise-shaping function. That is evident from the phase noise expression derived earlier:

$$S_{\phi,OUT}(f) = \left|\frac{2\pi}{z-1}\right|^2 |1 - z^{-1}|^2 S_q(f) = (2\pi)^2 S_q(f) = \frac{(2\pi)^2}{12 f_{REF}}.$$

Furthermore, the 1st-order $\Delta-\Sigma$ modulation also fails to randomize the quantization error [9], and consequently spurious frequency components become a problem. However, a second or higher order of integration can be used to reduce the practical impact of this periodicity as demonstrated by numerous $\Delta-\Sigma$ modulator structures used in oversampled data converters [7]. An example of a 2nd-order $\Delta-\Sigma$ modulator implementation is shown in Figure 10.40.

The noise shaping function is derived the same way as the 1st-order modulator (see Problem 7). Ignoring the input, with only the quantization noise present, the output will be

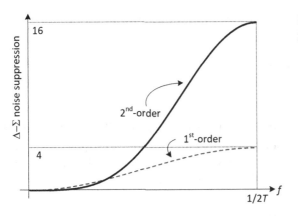

Figure 10.41: A comparison between the 1st- and 2nd-order noise shaping in $\Delta-\Sigma$ modulators

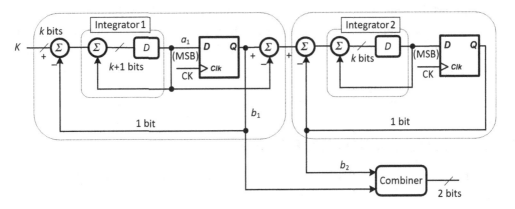

Figure 10.42: A 1-1 cascaded $\Delta-\Sigma$ modulator

$$\frac{B}{Q}(z) = \left(1 - z^{-1}\right)^2.$$

Consequently, the noise shaping function will be $|2\sin(\pi fT)|^4$. Shown in Figure 10.41 is a comparison between the 1st- and a 2nd-order noise shaping functions.

A 2nd-order noise shaping enjoys lower in-band noise levels, but the far-out noise rises more rapidly. This is potentially an issue for many applications where the far-out noise is important and is dealt with in Section 10.5.3.3.

It is certainly possible to add more integrator stages to obtain sharper noise suppression. However, feedback loops containing more than two integrators are potentially unstable, requiring various stabilization techniques [7]. In general, for an mth-order modulator, it can be shown that the phase noise spectral density is

$$S_{\phi,OUT}(f) = \frac{(2\pi)^2}{12 f_{REF}} |2\sin(\pi fT)|^{2(m-1)} \cong \frac{(2\pi)^{2m}}{12 f_{REF}} \left(\frac{f}{f_{REF}}\right)^{2(m-1)}.$$

An alternative approach to avoid stability concerns is to take advantage of cascade of stages as shown in Figure 10.42. Here, a subtractor finds the difference between the input (a_1) and the output (b_1) of the first quantizer. If this is applied to another 1st-order $\Delta-\Sigma$ modulator (shown in the dashed box), we expect the quantization error to be less.

Modeling the quantization noise of each stage as an additive source at the output of each D flip-flop as we did before, we can write

$$B_1 = Kz^{-1} + (1 - z^{-1})Q_1$$
$$B_2 = Q_1 z^{-1} + (1 - z^{-1})Q_2,$$

where Q_1 and Q_2 are the quantization noise of each stage. Note that the input to the second stage, $B_1 - A_1$, is actually the quantization noise of the first stage. The quantization noise of the first stage can be eliminated if B_1 is multiplied by z^{-1}, and B_2 is multiplied by $(1 - z^{-1})$ and the two are subtracted inside the combiner. Then the 2-bit combiner output ($B_{combiner}$) will be

$$B_{combiner} = Kz^{-2} - (1 - z^{-1})^2 Q_2,$$

which has the same noise shaping as a 1-bit 2nd-order $\Delta-\Sigma$ modulator has.

The cascaded structures (also known as MASH) can achieve higher order of noise shaping without the risk of instability. They need multi-modulus dividers, however, as the output has more than one bit. For instance, the *MASH 1-1* structure of Figure 10.42 requires division by $N-1$, N, $N+1$, and $N+2$.

10.5.3 $\Delta-\Sigma$ Modulator Nonideal Effects

In this section we discuss several issues associated with fractional synthesizers and $\Delta-\Sigma$ modulators.

10.5.3.1 Low-Frequency Tones and Dithering

Consider the 1st-order $\Delta-\Sigma$ modulator of Figure 10.37. The input K is a $(k-1)$-bit digital signal fed to the digital integrator. If this input is constant, it is easy to see that the modulator output is not really random (though the quantization noise is still shaped). Depending on the value of K, the repetitive pattern that exists in the output (also known as limit cycle) ultimately leads to spurious tones, which if lying within the PLL bandwidth, will not enjoy PLL filtering.

This is shown in Figure 10.43, where for a small K, the signal at the integrator output starts to build up slowly in small increments, producing a low-frequency pulse at the $\Delta-\Sigma$ modulator output. The average value of the pulse naturally represents the small fraction set by K, but creates a low-frequency spur in the VCO output.

The simple fix is to break the periodicity, say toggle the LSB of K randomly between 0 and 1. Known as *dithering*, this will eliminate the spur but introduces some noise whose level is usually low and acceptable in many cases.

10.5.3.2 Charge Pump Nonlinearity

As we discussed in Section 10.3, the PFD/CP charge-phase characteristic is not perfectly linear, due to a number of reasons, such as up and down current mismatches or channel length modulations. This nonlinearity typically leads to excess in-band noise and spurious tones [10], [11].

10.5 Fractional-N Frequency Synthesizers

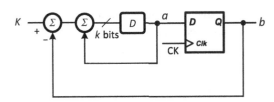

Figure 10.43: Limit cycles in $\Delta-\Sigma$ modulator

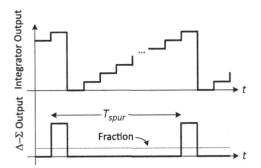

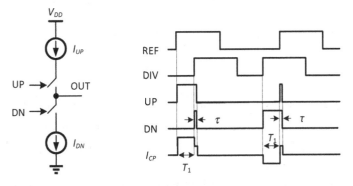

Figure 10.44: Charge pump nonlinearity due to unequal up and down currents

To understand the sources of nonlinearity better, consider a simple charge pump with different up and down currents of I_{UP} and I_{DN} (Figure 10.44). The difference arises from several factors, such as random mismatches or channel length modulation, though the latter is often dominant.

During the first pulse, since the reference is leading the feedback divider, an up signal is created, leading to a positive net charge pump current going to the loop filter. There is a small period of time (τ in Figure 10.44) when both up and down current sources are on due to the PFD reset delay. The total charge going to the loop filter is then

$$Q_{OUT} = I_{UP}T_1 + (I_{UP} - I_{DN})\tau,$$

where T_1 is the reference lead time. The second term above will not exist if the up and down currents are equal, while in the figure, exaggeratedly an up current of twice the down has been assumed. Regardless, this term leads to only a small phase difference between the reference and the divider outputs once the loop is locked.

In the next pulse, if the divider is now leading (say by the same amount T_1), and the total output charge delivered becomes

$$Q_{OUT} = -I_{DN}T_1 + (I_{UP} - I_{DN})\tau.$$

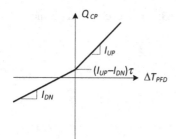

Figure 10.45: The charge pump characteristic is nonlinear in the presence of different up and down currents

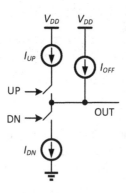
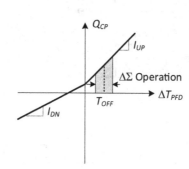

Figure 10.46: Adding an offset current moves the charge pump output away from the nonlinear region

It is clear from the equations and the figure that the charge delivered to the loop filter is exhibiting nonlinearity, unless $I_{UP} = I_{DN}$. The charge pump total charge versus the PFD inputs delay time is plotted in Figure 10.45. The slope of the curve is obviously changing depending on whether the reference is leading or lagging.

Recall that the main advantage of using a $\Delta-\Sigma$ modulator is its ability to shape the noise favorably, that is, the noise is moved to higher frequencies (Figure 10.39), which will be subsequently attenuated by the loop filter. However, this is true only in the context of a PLL linear model as has been previously developed. The presence of any nonlinearity results in noise folding and could increase the PLL in-band noise significantly.

To improve this, one can use the feedback charge pump circuit shown in Figure 10.23, although still the random mismatches could cause unequal up and down currents, which is of a lesser importance and can be alleviated by increasing the transistors sizes. As we said, the main drawback of the feedback circuit is the need for a rail-to-rail input opamp, which adds noise and power consumption.

Another common way to avoid the nonlinearity is to shift the charge pump characteristic away from the origin by adding a constant offset current (in either direction) as shown in Figure 10.46.

Example: To find the amount of shift (T_{OFF} in Figure 10.46), let us examine the PFD input signals in the presence of such an offset, but let us assume that the PLL is locked. As shown in Figure 10.47, given that the offset current ($I_{OFF} < I_0$ in Figure 10.46) is positive, the

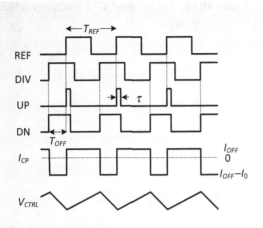

Figure 10.47: PFD input signals in locked mode when there is an offset current in charge pump (Figure 10.46)

divider leads the reference, and the corresponding down pulse will result in a current of $I_{OFF} - I_0$ lasting T_{OFF} seconds, which will offset the continuous current of I_{OFF} for the remainder of the cycle. We have assumed for simplicity $I_{UP} = I_{DN} = I_0$, but the same conclusion may be drawn for the more realistic case of unequal currents. Therefore, $I_{OFF}T_{REF} = I_0 T_{OFF}$, or

$$T_{OFF} = \frac{I_{OFF}}{I_0} T_{REF},$$

where T_{REF} is the reference signal period.

Note that in practice, unlike the integer PLLs, the inputs of the PFD are not perfectly aligned at the same point in every cycle even in the locked condition. Due to this phase (or time) skew, the offset current must be sufficiently large to stay away from the origin given the largest skew possible (Figure 10.46). Fortunately, for a given $\Delta-\Sigma$ modulator design, the time skew is predictable and can be planned for accordingly. Increasing the number of bits or the order of the $\Delta-\Sigma$ modulator generally causes more vulnerability to charge pump nonlinearity given a larger skew. See Problem 11 for an example.

The main drawback of the offset current is creation of more noise from the charge up as it stays on longer, and more ripple on the control voltage. Interestingly, in integer PLLs unequal up and down currents do not result in excess noise, but only lead into reference spurs as demonstrated earlier in Figure 10.22.

10.5.3.3 Out-of-Band Noise

It must be emphasized that the $\Delta-\Sigma$ modulation does not change the total noise energy, rather the noise shaping just shifts the noise from low frequencies to high frequencies. This is certainly problematic for many applications where far-out phase noise is important. However, since the $\Delta-\Sigma$ modulator enjoys the same lowpass loop suppression as the divider or reference do, the high-frequency noise is subject to the inherent filtering provided by the PLL (Figure 10.48). So overall, the $\Delta-\Sigma$ output noise will be

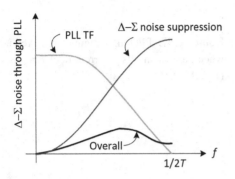

Figure 10.48: The noise of Δ−Σ modulator subject to the PLL lowpass filtering

$$S_{\phi,OUT,PLL}(f) = S_{\phi,OUT}(f)|H_{PLL}(f)|^2,$$

where $H_{PLL}(f)$ is the PLL lowpass transfer function, and $S_{\phi,OUT}(f)$ is the Δ−Σ output referred noise calculated earlier.

There is likely some peaking outside the PLL 3dB bandwidth, where the PLL suppression is not sufficient to balance the sharp rise of the Δ−Σ modulator noise. This alone may not be much of a problem, depending on the other PLL noise contributors in that region, especially the VCO, which could be dominant. Nonetheless, the noise shaping must be carefully evaluated to ensure that the Δ−Σ modulator noise contribution is within the requirement at all the offset frequencies.

Higher order Δ−Σ modulators that enjoy a lower in-band noise experience a potentially higher far-out noise considering a more steep increase of the noise profile.

As an example, assuming a simple RC loop filter like the one in Figure 10.20, the PLL closed-loop transfer function was shown to be

$$|H_{PLL}(f)|^2 = \omega_n^4 \frac{1+(2\zeta\omega\omega_n)^2}{(\omega_n^2-\omega^2)^2+4\zeta^2\omega_n^2\omega^2},$$

where $\omega = 2\pi f$. Note that the divide ratio (N) has been dropped from the equation as the Δ−Σ modulator noise is already referred to the output. So the overall phase noise at the PLL output assuming a 2nd-order Δ−Σ modulator is

$$S_{\phi,OUT,PLL}(f) = \frac{(2\pi)^2}{12 f_{REF}} |2\sin(\pi f T)|^2 \omega_n^4 \frac{1+(2\zeta\omega\omega_n)^2}{(\omega_n^2-\omega^2)^2+4\zeta^2\omega_n^2\omega^2}.$$

While at far-out frequencies the PLL rolls off with a slope of f^2, the 2nd-order modulator noise rises with the same slope, and the noise is expected to plateau. In practice, the loop filter looks more like the one shown Figure 10.21, which results in ample noise suppression at very high frequencies.

10.6 FREQUENCY DIVIDERS

The need for programmable divider in synthesizers has been already established. In this section, we will discuss a few common circuit ideas to realize these dividers.

10.6.1 Latches and D Flip-Flops

A latch is the basic building block of any frequency divider. The latch may be implemented using standard CMOS logic. Shown in Figure 10.49 is an example of such implementation. It takes an input (D), a complementary clock signal (CK, $\overline{CK}$), and has an output (Q). The latch consists of an input inverter enabled by the clock, followed by two inverters back to back creating the latch function.

When the clock is on, the inverter is enabled, and the latch is transparent, that is, $Q=D$. When the clock turns off, the input inverter is disabled, and the value it had at that instant of the clock going low is stored in the back-to-back inverters. The positive feedback enforced in the back-to-back inverters makes sure that the output holds its state despite the input being disconnected. The latch of Figure 10.49, given its simplicity, works quite well up to several tens of GHZ in current nanometer CMOS processes, and is widely adopted.

Shown in Figure 10.50 is the dynamic realization of the latch, where the back-to-back inverters are removed. The (parasitic) capacitance at the latch output is in principle sufficient to hold the output when the clock is disabled. This would naturally lead to somewhat higher speed, but the output may be easily disturbed by any noise or interference (such as one on the supply), which may not desirable.

An analog version of the CMOS latch of Figure 10.49 is shown in Figure 10.51, often referred to as a current-mode logic (CML) latch. There are several attributes that potentially make the CML latch faster:

- It comprises only NMOS transistors, which are traditionally faster. This is however not true for the more recent CMOS processes (28nm and beyond, including Fin-FET), where N and P FETs are almost equal.
- The resistive load leads to less parasitic at the output, hence a higher speed. This again is not necessarily the case in the recent CMOS nodes where the PMOS transistors are as good as NMOS.
- The current-mode differential pairs are faster to turn on and off, and require less voltage swing. This is analogous to the active versus passive mixer speed trade-offs discussed in Chapter 8.

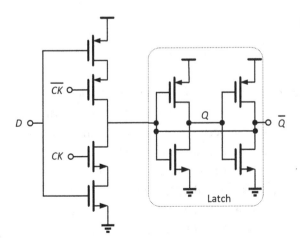

Figure 10.49: Schematic of a static CMOS latch

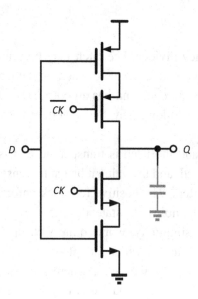

Figure 10.50: Dynamic CMOS latch

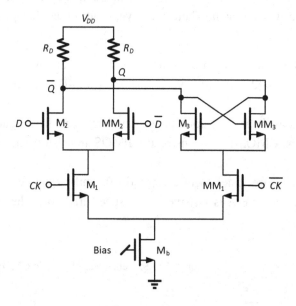

Figure 10.51: An analog current-mode latch

Apart from design complexity, the main drawback of the analog topology is headroom and consumption of static power. The former especially is important considering the trend of reducing supply voltage at lower nodes of process. It is not uncommon to remove the tail current source (M_b in the figure) to alleviate the headroom concerns.

The latch functionality is very similar to that of the CMOS version shown in Figure 10.49. When the clock is high, the pair M_2–MM_2 is enabled, acting like an analog inverter along with the load resistors, and the latch is transparent. When clock goes low, the input inverter is disabled, and the cross-coupled pair (M_3–MM_3) hold the output through enforcing the positive feedback. The size of the cross-coupled transistors must be chosen such that the negative resistance created due to the positive feedback dominates the load resistance.

Two latches put back to back with opposite clocks form a D flip-flop (DFF), as shown in Figure 10.52. Shown also in the figure is the timing diagram of the output (Q) versus the input and clock. In this example, the output is latched at the positive edge of the clock.

A circuit level example of the D flip-flop using the dynamic latches is shown in Figure 10.53.

Once put in a feedback loop, the D flip-flops becomes a divide-by-two circuit as shown in Figure 10.54, where the $\bar{Q}$ output is fed back to the D input. The corresponding timing diagram is shown on the right.

When the clock (input) goes high, the first latch is transparent, and sees the output of the second latch, which is low, and since $\bar{Q}$ of the second latch is fed back, it goes high. When the

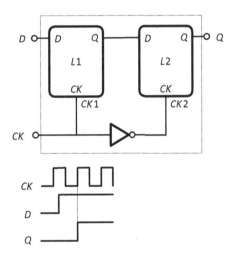

Figure 10.52: D flip-flop formed with two back-to-back latches

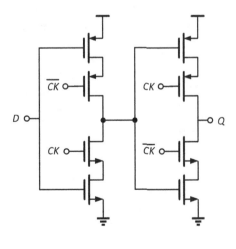

Figure 10.53: A dynamic CMOS flip-flop

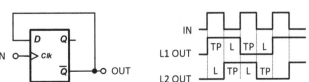

Figure 10.54: Divide-by-2 using a toggle flip-flop (flip-flop in a feedback loop)

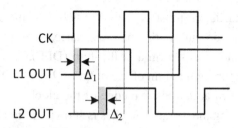

Figure 10.55: Speed concerns in a divide-by-2 circuit

input goes low, the high logic is latched in the first, and the second latch output, which is now transparent, goes high. Once the clock goes back up the output, which is high, is latched, and a low is fed back to the input latch, which is now transparent. Consequently, the latch outputs are at half the frequency of the input as desired.

For the divider to function properly, the delay of one latch must be less than half the input signal period as depicted in Figure 10.55.

Once the input goes high, the first latch is transparent, and its input is expected to go up, but with some delay in practice (Δ_1 in the figure). This signal must be *ready* before the second latch becomes transparent, which is when the input goes back low, or else the divider fails. In 16nm CMOS for instance, the CMOS latch (Figure 10.49) delay is of the order of tens of pS, and thus the divider functions up to tens of GHz comfortably.

10.6.2 Dual-Modulus Dividers

With the D flip-flops used as the main building block, and with the aid of a few simple gates, we can construct the dual- or multi-modulus dividers that are needed in the synthesizer design. A few examples will be covered in this section.

Example: Shown in Figure 10.56 is a programmable divider by 2 or 3 widely used in synthesizers. Also shown is the divider timing diagram for the case of dividing by 3. The

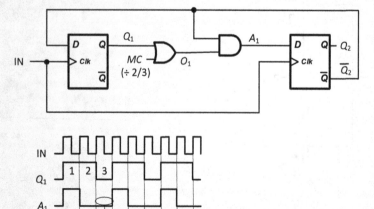

Figure 10.56: An example of a divide-by-2/3 circuit and the corresponding timing diagram

signal MC sets the divide ratio. If $MC=1$, the OR output, O_1, is high all the time, so the first DFF is invisible from the second DFF, which is configured to act like a divide-by-2 as $A_1 = \overline{Q_2}$.

With $MC=0$, $O_1 = Q_1$, so the OR is effectively disabled, and the back-to-back D flip-flops create a divide-by-3. Without the AND gate, the reader can easily show that the back-to-back flip-flops put in a feedback loop ($\overline{Q_2}$ connected to the first DFF input) create a divide-by-4 circuit. The AND, however, causes the output to go high one cycle earlier, effectively realizing a divide-by-3 as shown in the timing diagram. Assuming positive edge D flip-flops with some nonzero delay, we start with all the outputs at zero ($Q_1 = Q_2 = 0$, $\overline{Q_2}=1$). When the first clock edge arrives, Q_1 goes high, since the first DFF input, $\overline{Q_2}$, is high. On the other hand, the AND output (A_1) has been zero, and given the gate delays, the second DFF remains to see zero at the clock edge, and thus Q_2 remains zero. Now, both of the AND input are high, and thus at the second clock edge, Q_2 goes high ($\overline{Q_2}$ goes low after some delay). Since right at the second clock edge $\overline{Q_2}$ is still high, Q_1 remains high. At the third clock edge, since $\overline{Q_2}$ is low, so is the AND output, which results in Q_2 going low or $\overline{Q_2}$ going up. This means that at the fourth clock edge, Q_1 will go back high, resulting in a divide-by-3. Without the AND, Q_2 would have remained high ($\overline{Q_2}$ low) at the third clock edge, resulting in Q_1 to remain low at the fourth clock edge. It is interesting to point out that at the first two clock cycles, since Q_1 is high, the state of MC is unimportant. It is only at the third input cycle that MC must be zero to realize a divide-by-3. This becomes important in cases that the divide-by-2/3 will be used to create more complex dual- or multi-modulus dividers. We will discuss a few examples of this shortly.

As a simple rule of thumb, to create a certain divide ratio, n, the number of latches needed is usually chosen such that 2^n is the closest integer larger than the desired divide ratio, e.g., two latches for a divide-by-3, or three latches for a divide-by-5, or 6.

Example: As a second case study, shown in Figure 10.57 is a divide-by-4/5 circuit [12]. If $MC=0$, N_2 is high, and the third DFF output is always high. Thus, $N_1 = \overline{Q_2}$, and the output (Q_2) divides the input by four.

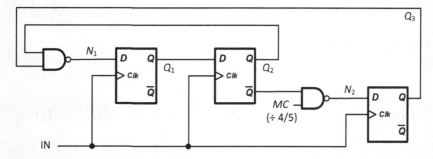

Figure 10.57: Divide-by-4/5 circuit

Continued

On the other hand, if $MC=1$, $N_2=Q_2$, and we have the first two D flip-flops in feedback, with the addition of $\overline{Q_3}$ appearing at the first DFF through the N_1 AND gate. The timing diagram (starting from reset point) is shown below in Figure 10.58, and is self-explanatory.

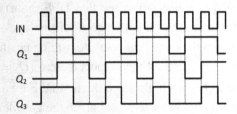

Figure 10.58: Waveforms of the divide-by-5 circuit of Figure 10.57

Note that Q_2 is following Q_1 with one input cycle delay, and Q_3 is following $\overline{Q_2}$ with one input cycle delay.

Example: Figure 10.59 shows a divide-by-8/9 circuit [13]. The divider operation can be analyzed much the same way as the examples before. With $MC=0$, the divide-by-2/3 operates in divide-by-2 mode, followed by two other divide-by-2s (the second and third D flip-flops in feedback), resulting in a cascaded divide-by-8 circuit. If $MC=1$, the circuit will divide by 9, and the analysis is left to the reader.

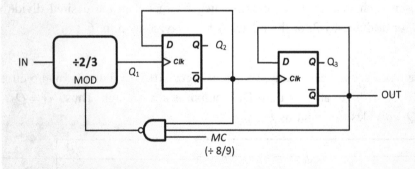

Figure 10.59: Divide-by-8/9

Several other examples of divider circuits may be found in [12], [14], [15].

Example: As the final example for this section, shown in Figure 10.60 is a divide-by-4 circuit that creates an 8-phase LO for an 8-phase blocker tolerant receiver [16], [17], employing a shift register-based (or Johnson) divider. Note that the divider is not a part of

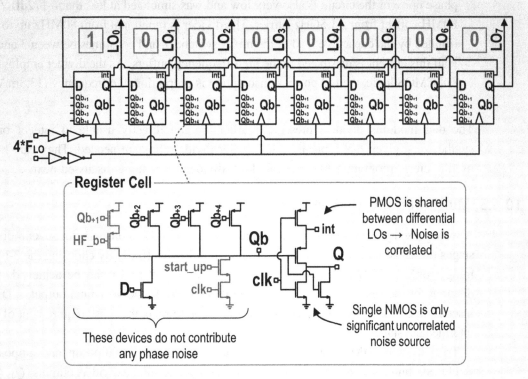

Figure 10.60: An 8-phase divide-by-4 to generate receiver LO

the frequency synthesizer, rather it is used to create the proper LO signals for the receiver mixer directly from the VCO.

One register cell stores a logic high, while all other registers store a low. The input, at four times the output frequency, moves this logic high along the register to generate the required 8-phase nonoverlapping clocks. The register cell is designed with the following in mind: a negative clock transition should propagate a high present at the D input, while a positive clock transition should always pull the output low. The logic high is propagated via the internal node. This node is pulled low by input when the previous stage output is high, while inputs precharge the internal node and enforce the condition that only one register outputs a high at any given time. Importantly, this internal node enables the pull-up PMOS transistor only in the output stage, and so the transistors to the left of ideally contribute no phase noise. The output of each cell is triggered by one of the high-frequency clocks, and the same clock triggers output clocks offset by 180° (in fact, every other cell is triggered by the

Continued

same high-frequency clock). This retiming limits the source of uncorrelated noise between clocks offset by 180° to the single highlighted NMOS device, thereby limiting the deterioration in noise figure from LO-to-RF coupling. Because a master clock strobes every stage, the phase noise of the circuit is also very low and was simulated at less than −172dBc/Hz at an 80MHz offset from a 1.5GHz carrier. The divider is functional from 80MHz up to 2.7GHz (limited by the capacitive load of the mixer switches), and consumes between 3 and 36mA. Half this current is dissipated in the high-frequency buffers. As the divider employs rail-to-rail CMOS logic, the LO power consumption is proportional to frequency (13.3mA/GHz).

The dual-modulus dividers presented earlier are not directly usable in integer or fractional synthesizers, given that typically much larger divide ratios are needed. They can, however, be used to create programmable multi-modulus dividers, which are discussed next.

10.6.3 Multi-Modulus Dividers

A widely used programmable multi-modulus divider (MMD) consisting of several divide-by-2/3 stages presented earlier (Figure 10.56) is shown in Figure 10.61. By choosing the 3-bit programming word, $M_2M_1M_0$, a programmable divide ratio of 8 to 15 can be achieved. The timing diagram for the case of divide-by-15 is shown as well. At the very final output, a D flip-flop is inserted to retime to the input, a common technique to relax the noise requirement of the divider intermediate stages.

To analyze the MMD, for the moment let us ignore the final D flip-flop, and suppose $M_2M_1M_0 = 111$. Starting from zero state, with the input rising edge, all the dividers outputs (Q_1, Q_2, and Q_3) go high. With both Q_2 and Q_3 as well as M_0 high, the first divider is going to divide-by-3 (the MOD port is zero, see Figure 10.56), and the input remains low for the third input cycles as shown in Figure 10.61 timing plot. At the fourth input edge, Q_1 which is effectively the clock of the second divider goes high, which causes Q_2 to go low. The third divider output remains high, until the second rising edge of Q_2 which acts as its input. With Q_3 and M_1 high, the second divider divides

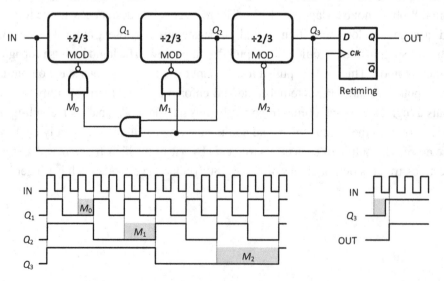

Figure 10.61: Multi-modulus divider using a cascade of divide-by-2/3s

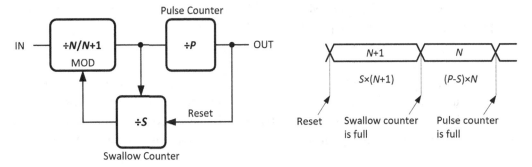

Figure 10.62: Pulse-swallow multi-modulus divider

by 3 as well, staying low for two cycles of Q_1, which is equivalent to four cycles of the input, as the first stage is now dividing by 2, given that Q_2 is low. Once Q_2 goes high, Q_3 goes low and will divide by 3 as M_2 is high, skipping two cycles of Q_2, or four cycles of Q_1, or eight input cycles. Note that with Q_3 going low and staying low for the remainder of the output, the second stage will divide by 2. So overall one output cycle is worth $3+4+8=15$ input cycles.

With any of modulus control bits zero, the MOD input of the corresponding divide-by-2/3 will be high, and thus it will act like a divide-by-2. The reader can easily work out the other combinations, and show that the modulus can be programmed as follows,

$$N = 8 + 2^0 M_0 + 2^1 M_1 + 2^2 M_2,$$

providing a divide range of 8~15. The design can be extended to create any arbitrary divide ratio through cascading more stages and with the aid of the necessary logic. Since each subsequent divider operates at at least half the frequency, its size and power may be scaled, leading to a very efficient design at GHz frequencies. The circuit or its variations are commonly used in fractional synthesizers.

The purposes of the last D flip-flop is to *retime* the output with the clean input, and reduce the accumulated noise through several stages of the divider. The corresponding timing diagram is shown on the bottom right. When the clock edge arrives, the output of the last divider (Q_3) goes high after some delay. Assuming this delay is less than one input cycle, the actual output will be latched at the rising edge of the next input cycle, and whose jitter will only depend on the last flop noise and the input itself, thus effectively bypassing the noise of the rest of the divider.

Another widely used multi-modulus divider, especially in integer synthesizers that need very large divide ratios, is shown in Figure 10.62, commonly known as a pulse-swallow divider.

It consists of a dual-modulus divide-by-$N/N+1$ (any of the circuits of Figure 10.56, Figure 10.57, or Figure 10.59 for example), or a programmable swallow counter (S in the figure, it essentially divides its input [the dual-modulus divider output] by S)), and a pulse counter (P in the figure, which divides the dual-modulus output by P). Usually N and P are fixed. After the reset, the swallow counter output is zero, and the dual-modulus divider divides by $N+1$ until the swallow counter is full, that is, after $S \times (N+1)$ cycles of the input. Once full, it flags high, resulting in the dual-modulus divider (also known as prescaler) to divide by N, while the swallow counter starts from 0. Thus far, the program counter has counted S cycles of the prescaler, and has another $P-S$ cycles to fill up (naturally the design requires $P > S$). After $(P-S) \times N$ additional cycles of the input (recall that the prescaler is dividing by N), the program counter fills up, after which point resets everything. So the output has counted overall $[S \times (N+1)] + [(P-S) \times N] = NP + S$ cycles.

> **Example:** For the example of the Bluetooth integer-N synthesizer, recall that the divide ratio required must be between 2402 to 2480. This can be accomplished for instance by using a divide-by-4/5, a program counter of $P = 480$, and a programmable swallow counter of $S = 2\sim80$. The counter is realized by cascading of several shift registers (flip-flops basically).

10.7 INTRODUCTION TO DIGITAL PLLs

We have dedicated this section to present a brief overview on digital PLLs. More information on the digital PLL fundamental design and trade-offs may be found in [18].

Before discussing digital PLLs (DPLLs) details, let us review several important limitations of the analog PLLs first.

- Depending on the noise requirements and the frequency of the reference, the size of the loop filter can become large. While the capacitors do scale with technology, still their size does not reduce as favorably as the digital gates.
- Often a large charge pump current is required to achieve a low in-band phase noise.
- PFD/CP linearity could be a concern for certain applications.
- There is an important trade-off between the noise contribution of the PLL building blocks, and particularly the VCO and the $\Delta-\Sigma$ modulator. As explained earlier, the $\Delta-\Sigma$ modulator noise follows the PLL lowpass transfer function, while the VCO noise transfer function to the output is highpass. As often the VCO is the most power consuming part of the synthesizer, it may be quite beneficial to increase the loop bandwidth of the PLL to suppress the VCO in-band phase noise contribution, and hence reduce its power consumption. For a given integration bandwidth, this in turn will result in the noise contribution of the reference and particularly the $\Delta-\Sigma$ modulator to increase. This trade-off is illustrated in Figure 10.63, where the same PLL noise contributors, and most notably the VCO and $\Delta-\Sigma$ modulator, are highlighted for two loop bandwidths of 350kHz and 800kHz, with everything else the same. While increasing the loop

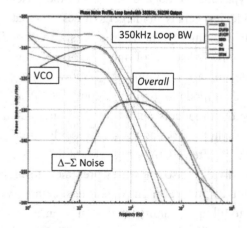
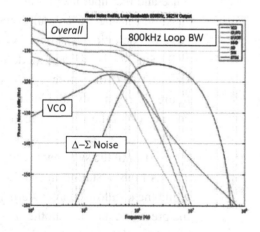

Figure 10.63: Analog PLL noise for two different loop bandwidths and $\Delta-\Sigma$ noise impact

10.7 Introduction to Digital PLLs

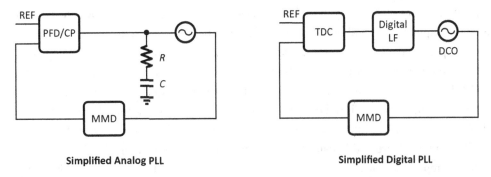

Figure 10.64: A simplified block diagram of a digital PLL, and comparison to its analog counterpart

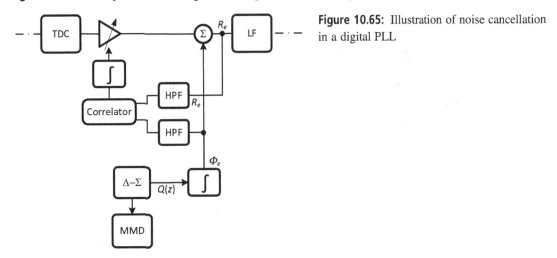

Figure 10.65: Illustration of noise cancellation in a digital PLL

bandwidth does reduce the VCO contribution, the $\Delta-\Sigma$ modulator noise especially at far-out frequencies substantially go up. On the other hand, if somehow the modulator noise could be cancelled, a wider loop leads to considerable power saving in VCO and the overall PLL.

Considering that many of the analog PLL building blocks are pseudo-digital (e.g., phase detector), one may consider devising a digital version as depicted in Figure 10.64.

Instead of a phase detector, one could use a time-to-digital converter (TDC) which effectively measures the inputs (reference and MMD output) time difference, and creates a proportional digital output. This signal now can be processed digitally and fed back to the VCO. The VCO however can only take a digital control line (hence often called digitally controlled oscillator), which could be accomplished either through a switchable cap array, or by inserting a DAC and feeding the corresponding analog signal to the varactor. This arrangement offers several advantages:

- The loop filter and phase detection (TDC) are all digital, which could potentially occupy a smaller area in general, but more importantly their area and performance scales with technology.
- The noise (and spurs) of the $\Delta-\Sigma$ modulator can be digitally corrected.

The $\Delta-\Sigma$ modulator noise cancellation is conceptually depicted in Figure 10.65. The phase error Φ_e is estimated by integrating the quantization error at the $\Delta-\Sigma$ modulator output ($Q(z)$), which is known beforehand. This phase error can be subtracted from the TDC output to cancel. The quantization error Φ_e will pass through a highpass filter (HPF) to filter out DC, and is correlated with the residual error R_e. If there is any residual error, the correlator output will have nonzero average output, which will be integrated to control the gain correction block after the TDC. Note that the perfect cancellation of the modulator noise requires the exact knowledge of the phase detector (or TDC) gain, which can be effectively corrected through the calibration loop as shown.

While this can be done at least conceptually in the analog PLL as well, it proves to be a lot more convenient in the digital domain, and in the context of a digital PLL. The main limitation of analog implementation is the need for a very linear and low-noise DAC to inject the error to the loop filter, whereas this can be done much more efficiently in the digital domain.

In the remainder of the section, we will briefly discuss the digital PLL building blocks in more details.

10.7.1 Time-to-Digital Converters

The TDC may be thought of an ADC except for its input is time rather than voltage. As such, the TDC creates some quantization noise that will ultimately limit the overall PLL phase noise. Similar to an ADC, the TDC quantization noise is shown to be $\frac{\Delta T_{res}^2}{12}$, where ΔT_{res} is the TDC resolution. This noise is usually uniformly spread between $\pm \frac{f_{REF}}{2}$, and thus the TDC quantization noise spectral density is

$$S_q(f) = \frac{\Delta T_{res}^2}{12 f_{REF}}.$$

Similar to the noise of the Δ–Σ modulator, the noise spectral density of the TDC when referred to the PLL output is shown to be

$$S_{\phi, OUT}(f) = \frac{\Delta T_{res}^2}{12 f_{REF}} (2\pi f_{REF})^2 |H_{PLL}(f)|^2 \approx \frac{(2\pi \Delta T_{res})^2}{12} f_{REF} N^2,$$

considering that for in-band noise, $|H_{PLL}(f)|^2 \approx N^2$.

> **Example:** Consider a 4GHz digital PLL for LTE applications, where the reference frequency is 40MHz. Assuming an in-band phase noise of −120dBc/Hz is budgeted for the TDC, we can write
>
> $$-120 = 10 \log \left[\frac{(2\pi \Delta T_{res})^2}{12} f_{REF} N^2 \right].$$
>
> With $N = 100$ and $f_{REF} = 40$MHz, the TDC resolution is found to be $\Delta T_{res} = 0.87$pS.

10.7.1.1 TDC Circuit Realization

Below we will discuss a few examples of TDC circuit implementation.

Shown in Figure 10.66 is a delay line TDC circuit. Using the same notation as before, we label the TDC input signals as A and B (typically representing the reference and MMD output)

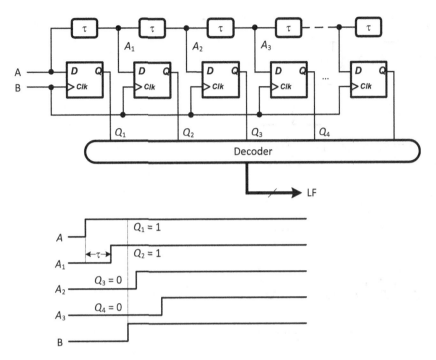

Figure 10.66: A delay line TDC

with the corresponding timing diagram. Once the input A arrives, a sequence of the signal along with its delayed replica are fed to an array of D flip-flops. When the B input arrives, only the ones that have received a high signal at their input are triggered high. Naturally, when the outputs of DFFs are decoded, they will represent the delay between the rising edges of the A and B inputs, which is precisely what we are looking for.

Obviously, the TDC resolution is limited by the minimum delay of each delay cell. Furthermore, the mismatch between various delay cells could create nonlinearity.

Another TDC realization based on an oscillator and a counter is shown in Figure 10.67. Shown also in the figure is the corresponding timing diagram. For the moment, let us assume that the enable signal turning the ring oscillator on is always high. The counter start and stop point are determined by the rising edges of the input signals, and what is counted is the number of the cycles of the ring oscillator. Note that in either implementation the exact value of the phase or time difference is not going to affect the PLL locking functionality; however, it will affect the loop dynamics, as will be discussed in Section 10.7.4. Same as the previous TDC, the resolution is determined by the minimum delay of the ring oscillator cells. Furthermore, the ring oscillator is continuously running, which could be power consuming.

To improve the oscillator-based TDC power consumption, the ring oscillator may be *enabled* by the reference input, and disabled by the MMD, with its initial state being stored and used as initial condition for the next cycle. Known as a gated ring oscillator TDC [19], the gating effectively realizes a 1st-order noise shaping, which will also help reduce the phase noise contribution of the TDC for a given achievable delay. For instance, the raw delay achievable in a 16nm CMOS process is on the order of 4–5pS, whereas with gating sub-pS effective delays are achievable.

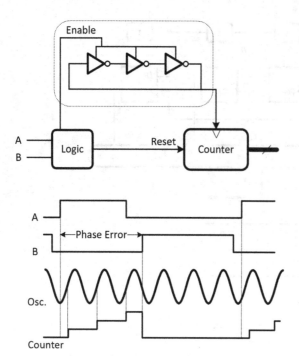

Figure 10.67: Oscillator-based TDC

Finally, an alternative approach using an ADC-based scheme utilizing a slop detector is discussed in [20], which we leave to the interested reader.

10.7.2 Digital Loop Filters

As will be discussed in Section 10.7.4, the same analysis performed to capture the analog PLL transfer function is directly applicable to DPLLs. Consequently, one may estimate the required values of the loop filter components (namely R and C for say the type II PLL of Figure 10.20) based on the required DPLL transfer function. Once the analog loop filter is determined in the s-domain, forward Euler s- to z-domain approximation, denoted by $s \leftrightarrow \frac{1}{T}(z-1)$, may be utilized to determine the digital loop filter. This is similar to the design procedure often used in the switched capacitor filters, as explained in Section 10.5.1. The final loop filter transfer function can then be simulated and optimized in the z-domain. This approach works well if the loop bandwidth is well below the reference frequency, which is often the case.

Example: Consider Figure 10.68, where a type II RC filter is shown on the top.
In the analog domain we can write

$$H_{LF}(s) = \frac{V_{OUT}}{I_{IN}}(s) = R + \frac{1}{Cs}.$$

Replacing s with $\frac{1}{T}(z-1)$, we have

$$H_{LF}(z) = \frac{V_{OUT}}{I_{IN}}(z) = R + \frac{T}{C}\frac{z^{-1}}{1-z^{-1}}.$$

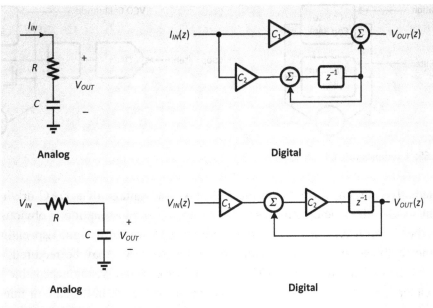

Figure 10.68: Digital loop filter examples

The corresponding z-domain implementation is shown on the right. The transfer function is readily found to be

$$\frac{V_{OUT}}{I_{IN}}(z) = C_1 + \frac{C_2 z^{-1}}{1 - z^{-1}}.$$

Thus, $C_1 = R$, and $C_2 = \frac{T}{C}$.

The type I filter may be designed similarly shown on the bottom of the figure (see Problem 17). It can be shown that

$$\frac{V_{OUT}}{I_{IN}}(z) = \frac{\frac{T}{RC}}{1 - \frac{T}{RC}} \frac{\left(1 - \frac{T}{RC}\right)z^{-1}}{1 - \left(1 - \frac{T}{RC}\right)z^{-1}} = \frac{C_1 C_2 z^{-1}}{1 - C_2 z^{-1}},$$

where $C_1 = \frac{\frac{T}{RC}}{1 - \frac{T}{RC}}$, and $C_2 = 1 - \frac{T}{RC}$.

10.7.3 Digitally Controlled Oscillators

Two variations of a DCO are depicted in Figure 10.69. As explained in Chapter 9, along with a varactor, almost all VCOs utilize a digitally controlled array of capacitors to coarse tune the VCO and effectively reduce the VCO gain for a given range.

Shown on the left, the varactor may be replaced with another array of switchable capacitors to create the required *fine-tuning*. Depending on the acceptable frequency error set by the application, realizing such fine steps may lead to impractical values, and often proves to be a difficult task.

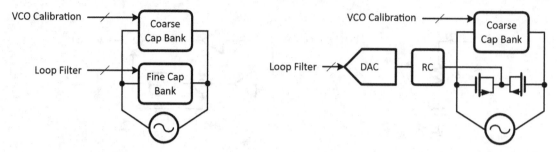

Figure 10.69: Examples of DCO realization

An alternative approach shown on the right takes advantage of a DAC driving a more traditional VCO with varactor. The DAC noise and power consumption is obviously adding some overhead, but it typically leads to a more straightforward design. Especially in high-performance applications, a high dynamic range for the DAC may be required. The DAC LSB is set based on the required DCO resolution, or effectively, the acceptable frequency error. In either approach, Δ−Σ modulation may be incorporated. In the case of fine array, the Δ−Σ helps achieving finer resolutions at the cost of extra noise or possibly spurs, nonetheless it is often used; while in the DAC approach, Δ−Σ modulation helps boost the DAC dynamic range.

> **Example:** Consider a continuous VCO with $K_{VCO} = 15\text{MHz/V}$ to be used in a DPLL along with a DAC. Assuming a 10-bit DAC with a full-scale voltage of 0.5V, the DAC LSB will be about 0.5mV. The frequency resolution will then be $0.5\text{mV} \times 15\text{MHz/V} = 7.5\text{kHz}$. If used for a 2.4GHz WLAN application, the frequency resolution will be about 3ppm.

10.7.4 DPLL Linear Analysis

The DPLL linear transfer function and noise analysis may be carried out much the same way as the analog PLLs discussed in Sections 10.4.1 and 10.4.2. Replacing K_{PFD} with K_{TDC}, and K_{VCO} with K_{DCO}, same equations as before may be used with the PLL closed-loop transfer function expressed as

$$H(s) = N\frac{\omega_n^2(1+RCs)}{s^2 + 2\zeta\omega_n s + \omega_n^2},$$

where $\omega_n = \sqrt{\frac{K_{TDC}K_{DCO}}{CN}}$ and $\zeta = \frac{RC\omega_n}{2}$.

As before, noise from reference, TDC, and MMD goes through a lowpass transfer function, whereas the noise transfer function from the DCO to the output is highpass. The noise of each block may be estimated and simulated in block level, and eventually added up by multiplying by the proper transfer function.

10.8 Summary

Phase-locked loops, frequency synthesizers, and frequency dividers were covered in this chapter.

- Section 10.1 discussed basic properties and building blocks of a phase-locked loop.
- Type I PLLs were presented in Section 10.2. A linear model of the PLL commonly used in PLL phase noise analysis was also discussed.
- Type II PLLs were discussed in Section 10.3. Key building blocks such as phase frequency detectors and charge pumps were discussed.
- Section 10.4 discussed the integer frequency synthesizers.
- Fractional synthesizer were presented in Section 10.5. The concepts of noise shaping and $\Delta-\Sigma$ modulators were covered in this section.
- Latches, frequency dividers, as well as dual- and multi-modulus dividers were discussed in Section 10.6.
- A brief overview of digital PLLs was presented in Section 10.7.

10.9 Problems

1. In the type I PLL of Figure 10.8, the reference is 5GHz, and $K_{VCO} = 1$GHz/V. Design the PLL for a bandwidth of 500MHz, assuming critically damped condition. What is the PLL phase margin?

2. In the following type II PLL, the loop filter resistor and PD noise are modeled by a series voltage source and a parallel current source. Find the noise transfer function for each.

   Answer: $S_{\phi_{OUT}} = \left|\frac{N}{K_{PD}}\right|^2 \left|\frac{G}{1+G}\right|^2 \left(S_{i_n} + \frac{S_{v_n}}{|F|^2}\right).$

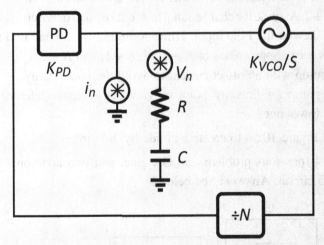

3. Assuming the charge pump noise is given by $S_{i_n} \approx 2KTI_{CP}$, show that the phase noise at the PLL natural frequency due to the charge pump noise alone is given by $S_{\phi_{OUT}}(\omega_n) = \frac{5\pi^2 N^2}{I_{CP}} 2KT.$

4. A charge pump PLL intended for LTE applications uses a 52MHz reference frequency, and creates an output at 1960MHz. The VCO gain is 40MHz/V. Find the charge pump current,

and the loop filter components such that the PLL 3dB bandwidth is 100kHz, and the resistor phase noise is less than −100dBc/Hz (the loop is critically damped). Assuming a VCO phase noise of −90dBc/Hz at PLL natural frequency (ω_n), what is the total phase noise at ω_n (including the charge pump noise a stated in the previous problem)? **Answer:** $I_{CP} = 9.5\mu A$, $R = 50k\Omega$, $C = 160pF$, −93.8dBc/Hz.

5. Consider a fractional synthesizer where the VCO is first divided by 2 before being applied to the MMD. Discuss the pros and cons of this topology compared to the one where the VCO is directly fed to the MMD. **Hint:** Consider the Δ−Σ modulator quantization noise contribution in each case.

6. Analyze the divide-by-8/9 circuit in Figure 10.59. Draw the timing diagram for either divide ratio.

7. Modeling the quantizer as an additive noise source, find the noise shaping function and the output–input characteristic of the 2nd-order Δ−Σ modulator of Figure 10.40.

8. Show that for an mth-order Δ−Σ modulator, the SSB phase noise is given by
$S_{\phi,OUT}(f) = \frac{(2\pi)^2}{12 f_{REF}} |2\sin(\pi f T)|^{2(m-1)}$.

9. Consider a type-II PLL with a reference frequency of 40MHz, creating an output frequency of 2GHz. Assume the VCO gain is 100MHz/V.
    a. Find the charge pump current and the loop filter components if the PLL 3dB bandwidth is to be 1MHz.
    b. Assuming there is an up offset current of 25% of the nominal charge pump current, sketch the PFD input and output signals, and the charge pump current over one cycle.

10. Redo Problem 9, part b for the case where the up current is 10% lower than the down current.

11. Consider a fractional synthesizer with a reference frequency of 50MHz, and an output frequency of 1GHz. The synthesizer uses a MASH 1-1 Δ−Σ modulator, dividing the output by $N − 1, N, N + 1$, and $N + 2$. Assume the charge pump has equal up and down currents of 200µA.
    a. Find out the time skew at the PFD input. **Hint:** Assume $N \approx 20$ and find the divider output frequencies for various values of $N − 1, N, N + 1$, and $N + 2$.
    b. Design a charge pump with an offset current to avoid the nonlinearity.
    c. From the charge pump nonlinearity point of view, is a higher reference frequency more desired or a lower one?

12. Modify the circuit of Figure 10.56 to create a divide-by-3/4 circuit.

13. Using a divide-by-3/4 (previous problem), an OR gate, and two additional D flip-flops, design a divide-by-15 circuit. **Answer:** See below!

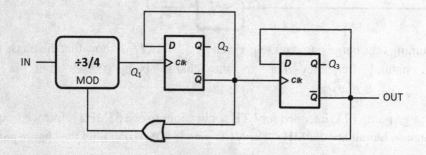

14. Shown below is a divide-by-3 circuit creating a 6-phase output. Analyze the circuit, and plot the corresponding timing diagram.

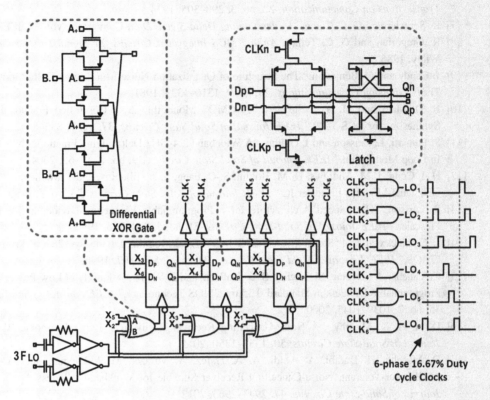

15. Using four divide-by-2/3 circuits (Figure 10.56), and the appropriate logic, design a divide-by-8~31 multi-modulus divider.

16. Prove that for the MMD of Figure 10.61, $N = 8 + 2^0 M_0 + 2^1 M_1 + 2^2 M_2$. Work out the timing diagram of two cases of $M_2 M_1 M_0 = 100$, and $M_2 M_1 M_0 = 011$.

17. Using the same approach as Section 10.7.2, synthesize the digital equivalent of the type I RC filter of Figure 10.68.

18. Considering that the TDC quantization noise spectral density is $S_q(f) = \frac{T_{res}^2}{12 f_{REF}}$, calculate the PLL in-band noise due to the TDC quantization noise.

10.10 References

[1] J.-S. Lee, M.-S. Keel, S.-I. Lim, and S. Kim, "Charge Pump with Perfect Current Matching Characteristics in Phase-Locked Loops," *Electronics Letters*, 36, no. 11, 1907–1908, 2000.

[2] P. R. Gray and R. G. Meyer, *Analysis and Design of Analog Integrated Circuits*, John Wiley, 1990.

[3] T. Riley, M. Copeland, and T. Kwasniewski, "Delta-Sigma Modulation in Fractional-N Frequency Synthesis," *IEEE Journal of Solid-State Circuits*, 28, no. 5, 553–559, 1993.

[4] B. Miller and R. J. Conley, "A Multiple Modulator Fractional Divider," *IEEE Transactions on Instrumentation and Measurement*, 40, no. 6, 578–583, 1991.

[5] J. Candy, "Use of Double Integration in Sigma Delta Modulation," *IEEE Transactions on Communication*, 33, no. 3, 249–258, 1985.

[6] J. Candy, "Use of Limit Cycle Oscillations to Obtain Robust Analog-to-Digital Converters," *IEEE Transactions on Communication*, 22, no. 3, 298–305, 1974.

[7] R. Schreier and G. C. Temes, *Understanding Delta-Sigma Data Converters*, vol. 74, IEEE Press, 2005.

[8] R. Gregorian and G. C. Temes, *Analog MOS Integrated Circuits for Signal Processing*, vol. 1, John Wiley, 1986.

[9] J. Candy and O. Benjamin, "The Structure of Quantization Noise from Sigma-Delta Modulation," *IEEE Transactions on Communications*, 29, no. 9, 1316–1323, 1981.

[10] B. D. Muer and M. S. J. Steyaert, "A CMOS Monolithic $\Delta;\Sigma$-Controlled Fractional-N Frequency Synthesizer for DCS-1800," *IEEE Journal of Solid-State Circuits*, 37, no. 7, 835–844, 2002.

[11] S. Pamarti, L. Jansson and I. Galton, "A Wideband 2.4-GHz Delta-Sigma Fractional-NPLL with 1-Mb/s In-Loop Modulation," *IEEE Journal of Solid-State Circuits*, 39, no. 1, 49–62, 2004.

[12] H.-I. Cong, J. M. Andrews, D. M. Boulin, S.-C. Fang, S. J. Hillenius, and J. Michejda, "Multigigahertz GHz Dual-Modulus Prescaler IC," *IEEE Journal of Solid-State Circuits*, 23, no. 5, 1189–1194, 1988.

[13] P. Larsson, "High-Speed Architecture for a Programmable Frequency Divider and a Dual-Modulus Prescaler," *IEEE Journal of Solid-State Circuits*, 31, no. 5, 744–748, 1996.

[14] J. Craninckx and M. Steyaert, "A 1.75GHz 3V Dual-Modulus Divide-by-128/129 Prescaler in 0.7µm CMOS," *IEEE Journal of Solid-State Circuits*, 31, no. 7, 890–897, 1996.

[15] C. Vaucher, I. Ferencic, M. Locher, S. Sedvallson, and Z. Wang, "A Family of Low-Power Truly Modular Programmable Dividers in Standard 0.35µm CMOS Technology," *IEEE Journal of Solid-State Circuits*, 35, no. 7, 1039–1045, 2000.

[16] D. H. H. X. Murphy, "A Noise-Cancelling Receiver Resilient to Large Harmonic Blockers," *IEEE Journal of Solid-State Circuits*, 50, 1336–1350, 2015.

[17] D. Murphy, H. Darabi, A. Abidi, A. A. Hafez, A. Mirzaei, M. Mikhemar, and M.-C. F. Chang, "A Blocker-Tolerant, Noise-Cancelling Receiver Suitable for Wideband Wireless Applications," *IEEE Journal of Solid-State Circuits*, 47, 2943–2963, 2012.

[18] R. B. Staszewski and P. T. Balsara, *All-Digital Frequency Synthesizer in Deep-Submicron CMOS*, John Wiley, 2006.

[19] C. Hsu, M. Z. Straayer, and M. H. Perrott, "A Low-Noise Wide-BW 3.6-GHz Digital $\Delta\Sigma$ Fractional-N Frequency Synthesizer with a Noise-Shaping Time-to-Digital Converter and Quantization Noise Cancellation," *IEEE Journal of Solid-State Circuits*, 43, no. 12, 2776–2786, 2008.

[20] X. Gao, L. Tee, W. Wu, K. Lee, A. A. Paramanandam, A. Jha, N. Liu, E. Chan, and L. Lin, "9.4 A 28nm CMOS Digital Fractional-N PLL with −245.5dB FOM and a Frequency Tripler for 802.11abgn/ac radio," in *Proceedings of the IEEE International Solid-State Circuits Conference Digest of Technical Papers*, 2015.

[21] F. M. Gardner, *Phaselock Techniques*, John Wiley, 2005.

[22] D. H. Wolaver, *Phase-Locked Loop Circuit Design*, vol. 177, Prentice Hall, 1991.

11 Power Amplifiers

In this chapter we study the challenging problem of delivering power to antenna *efficiently*. The linear amplifier topologies that we have discussed thus far are fundamentally incapable of achieving high efficiency. Considering the high demand for improving battery life, this shortcoming becomes very critical when delivering hundreds of mW or several watts of power into the antenna. This issue is exacerbated in most modern radios that employ complex modulation schemes to improve the throughput without raising the bandwidth. As we discussed in Chapter 6, such systems demand more linearity on the power amplifier, and hence achieving a respectable efficiency becomes more of a challenge.

We start this chapter by general description of challenges and concerns, as well as linearity–efficiency trade-offs. We then have a detailed description of different classes of PAs. We conclude the chapter by discussing various techniques employed to linearize the PA, or to improve the efficiency in an attempt to ameliorate the challenging linearity–efficiency issues.

The specific topics covered in this chapter are:

- Power amplifier basic operation
- PAs classes A, B, and C
- Switching power amplifiers (classes D, E, and F)
- Digital transmitters and digital PAs
- Power combining techniques
- Predistortion
- Envelope elimination and restoration
- Envelope tracking
- Dynamic biasing
- Doherty power amplifiers

For class teaching, we recommend covering Section 11.1, and selected parts of Sections 11.2–11.8 (for instance, only classes A, B, and E or F). The discussion on Doherty PAs and linearization techniques may be deferred to a more advanced course, although most of the material is easy to follow, if assigned as reading.

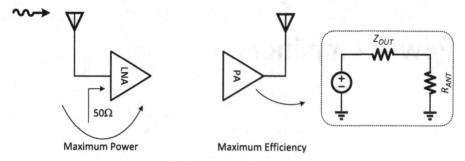

Figure 11.1: Maximum power transfer in receivers and transmitters

11.1 GENERAL CONSIDERATIONS

When dealing with power amplifiers, there are several important factors that distinguish them from the small signal amplifiers we have shown so far:

- The notion of maximum power transfer accomplished by *conjugate matching* is not quite applicable to PAs (Figure 11.1).

As we showed in Chapter 3, in a receiver, conjugate matching to the *source* results in maximum power transfer from the antenna to the LNA input, but also causes 50% power loss. For instance, in the case of a GSM PA that needs to deliver 2W of power to the antenna, the 50% efficiency results in dissipating 2W of power in the PA itself, which is not acceptable, considering that there are many other practical limitations that lower PA efficiency further. In fact from Figure 11.1, the output power at the antenna is proportional to

$$P_{OUT} \propto \frac{1}{2} R_{ANT} \left| \frac{1}{R_{ANT} + Z_{OUT}} \right|^2,$$

and for a given antenna impedance R_{ANT}, the output power is maximum if the PA output impedance (Z_{OUT}) is zero. For that reason, most practical PAs are designed to deliver a certain power set by the application with the *highest possible efficiency*. As we saw earlier, even for the receivers, apart from certain impedance requirements imposed by the external components, matching to source does not always result in optimum performance.[1]

- Similar to the LNAs, the PAs rely on matching networks quite often, but for an entirely different reason. Consider the example of the GSM PA that needs to deliver 2W to a 50Ω antenna. This results in a swing of about 14V peak at the antenna, well exceeding the 3.7V nominal supply available to most handsets. As shown in Figure 11.2, a lossless transformer may be used to alter the swing at the transistor output favorably, yet delivering the intended power to the antenna.

---

[1] In the case of a receiver that would be mainly the noise figure.

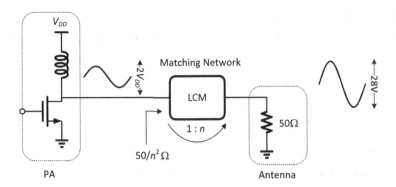

Figure 11.2: Role of the matching network in a PA

The matching network may be realized by any of the circuit topologies that we discussed in Chapter 3, but in almost all cases it tends to *downconvert* the load impedance to the device output for obvious reasons (thus n is always ≥ 1). While in the case of LNA the matching loss proved to be critical due to noise figure concerns, in the case of a PA, it is important to be minimized so as not to degrade the efficiency. That, along with the need to deal with large swings, limits the choice of matching networks components to high-Q inductors, capacitors, and transformers sometimes even realized externally.[2] Similar to LNAs, we simply assume there is a load resistance of R_L at the PA output, and the lossless matching network transfers the power and impedance as desired.

There are several other features associated with the matching network: they are often used to absorb the parasitic capacitances (which could be very large given the size of the transistors), and they filter the *undesired harmonics*.

– Another concern arises from the fact that the PA devices, especially the ones at the later stages, are constantly under stress given the large voltage swing that they must undergo. Reliability and aging are common design concerns, which may sometimes limit the swing at the PA output. Although one may argue that regardless of the swing the required power is always delivered through the matching network, in practice a larger impedance transformation results in more loss given the finite Q of the elements (see Problem 1 that explores the impact of transformer ratio on loss).

Finally, let us present two common definitions for the efficiency that prove to be the biggest design metric for the PAs: The *drain efficiency* (in the case of a MOS PA) is defined as

$$\eta = \frac{P_{OUT}}{P_{DC}},$$

where P_{OUT} is the power delivered to the load, and P_{DC} is the power dissipated in the PA. On the other hand, the *power added efficiency* is defined as

$$\eta = \frac{P_{OUT} - P_{IN}}{P_{DC}},$$

---

[2] It is not uncommon to employ a package substrate or PCB for matching purposes.

where P_{IN} is the power delivered to the PA. If the PA has sufficiently large gain (power gain to be exact), the two metrics will be very close. Throughout this chapter we use the former definition unless otherwise stated.

11.2 CLASS A PAs

The main distinguishing feature between the classes A, B, and C is how the device is biased, and its impact on efficiency. Otherwise, the PA schematics look almost identical.

A simplified class A power amplifier is shown in Figure 11.3. The load inductance, L, primarily sets the output at V_{DD}, which is important to maximize the swing, and could also absorb some of the parasitic. We assume it is lumped with the rest of the matching network, which provides a net resistance or R_L to the transistor output. The class A device is biased such that it always stays in a linear (or semilinear) mode of operation. Thus the input is biased such that the signal never goes below the device threshold voltage, and the output swing is such that the transistor stays in saturation. Hence, the device current never reaches zero. The corresponding transistor drain current and voltage are shown in Figure 11.3.

Assuming a drain swing of A, the drain voltage is

$$V_{DS} = V_{DD} + A \sin \omega_0 t,$$

and given the load resistance R_L, the drain current is

$$I_{DS} = I_{DC} + \frac{A}{R_L} \sin \omega_0 t,$$

where I_{DC} is the device quiescent current. Clearly, $I_{DC} > \frac{A}{R_L}$ to ensure operation in class A. Assuming a lossless matching network, the efficiency then is

$$\eta = \frac{P_{OUT}}{P_{DC}} = \frac{\frac{A^2}{2R_L}}{I_{DC} V_{DD}}.$$

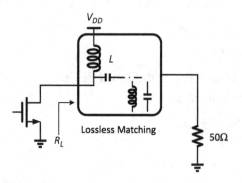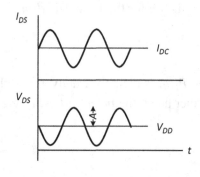

Figure 11.3: A simplified class A PA.

To maximize the efficiency, then, one must maximize the swing and minimize the bias current. Given the constraints imposed by linearity and supply, we have

$$\eta_{MAX} = \frac{\frac{A^2}{2R_L}}{\frac{A}{R_L}V_{DD}} = \frac{A}{2V_{DD}} = \frac{1}{2} = 50\%.$$

This condition is never reached though, as the drain voltage cannot go below $V_{DS,SAT}$, and the device current must not reach zero. In addition, the loss of the matching network degrades the efficiency further.

Apart from using much larger devices to deliver the required power, we can see that the main difference between a class A PA and the small signal amplifiers we have discussed before is that to maintain a reasonable efficiency, in a class A PA, the device AC current is comparable to its quiescent current.

Assuming the PA DC current is fixed at $I_{DC} = \frac{V_{DD}}{R_L}$, the efficiency is

$$\eta = \frac{A^2}{2V_{DD}^2},$$

which is plotted in Figure 11.4 as a function of drain swing.

It is often desirable for the PA to provide some kind of gain control in case less power is needed to be delivered. This, however, leads to degradation in efficiency, which *quadratically* depends on the swing at the PA output. This issue may be alleviated in two ways:

- The matching network may be altered to present a larger impedance to the device output, thus while keeping the swing at V_{DD}, less power is delivered to the load.
- The PA supply may be reduced proportionally.

Both options, however, still result in lower efficiency at lower powers, simply because the efficiency is a function of A^2. Considering that

$$\eta = \frac{A^2}{2R_L I_{DC} V_{DD}},$$

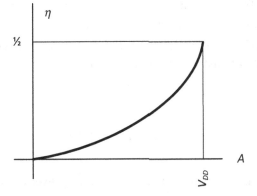

Figure 11.4: Class A PA efficiency versus output swing

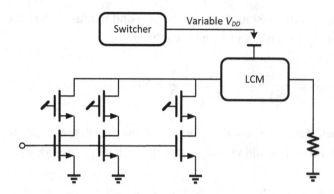

Figure 11.5: Class A PA with gain control

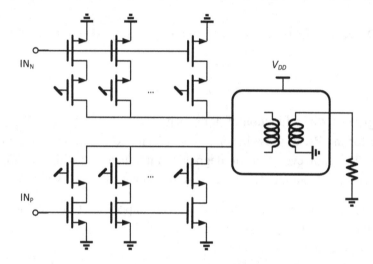

Figure 11.6: A fully differential class A PA

another option would be to reduce the *supply and PA DC bias* simultaneously as the swing is reduced, assuming R_L is fixed. A more practical class A PA based on this idea is shown in Figure 11.5.

The cascode devices may incorporate *thick oxide* devices that are more tolerant of larger voltage swings, while the core devices use thin oxide to provide more gain. Moreover, by selectively turning the cascode bias on or off, the PA bias current may be varied to maintain a better efficiency at lower output powers. Similarly, a variable V_{DD} is supplied using a switching regulator. Changing the bias current may affect matching, and thus it may be needed to employ some kind of switchable capacitor in the matching network to accommodate for that. Finally, the PA may be entirely designed differentially, and only a single-ended output is created at the last stage of the matching network through a transformer (Figure 11.6).

11.3 CLASS B PAs

The somewhat poor efficiency of class A PAs is a direct result of leaving a large DC current *unused* in the device, that is to say, a current that is not contributing to the power delivered to

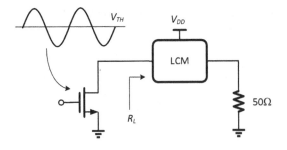 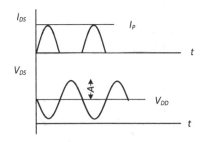

Figure 11.7: A simplified class B PA

the load. It is clear from Figure 11.3 that if the conduction angle is reduced (pushing I_{DC} lower) such that only at a *fraction of cycle* both drain voltage and current are nonzero, a better efficiency is expected. In a class B PA this is accomplished by biasing the PA input right at the threshold, and thus only conducting at half the cycle (Figure 11.7).

Although it may seem that this kind of biasing results in substantial nonlinearity, one must note that the nonlinear drain current resulting from the 50% conduction angle is filtered at the output given the narrowband matching network, and still a sinusoid drain voltage is expected. Clearly, in the case of class B PAs, and in fact other high-efficiency topologies that we will discuss later, a high-Q matching network is mandatory to lower the *far-out emission* caused by the nonlinear PA.

The output current is approximately a half-wave rectified sinusoid. In each half cycle that the device is off, the load current is directly sourced from the supply, pointing to a better efficiency. For the other half cycle, the on transistor sinks current from the load. To determine the peak current, I_P, let us first express the drain current (Figure 11.7) in terms of its Fourier series:

$$I_{DS} = \frac{I_P}{\pi} + \frac{I_P}{2} \sin \omega_0 t - \frac{2I_P}{3\pi} \cos 2\omega_0 t + \cdots.$$

The DC term, $\frac{I_P}{\pi}$, determines the supply average current, whereas the fundamental leads to the output voltage assuming the higher order harmonics are filtered by the tank. When the fundamental component of the current reaches its peak, $\frac{I_P}{2}$, the total current sank from the load is $\frac{A}{R_L}$, assuming a peak swing of A at the drain. Hence,

$$\frac{I_P}{2} = \frac{A}{R_L}.$$

The efficiency can be readily found to be

$$\eta = \frac{P_{OUT}}{P_{DC}} = \frac{\frac{A^2}{2R_L}}{\frac{I_P}{\pi} V_{DD}} = \frac{\pi}{4} \frac{A}{V_{DD}},$$

which reaches a maximum of $\frac{\pi}{4} = 78\%$ when the output swing approaches V_{DD}. Interestingly, the efficiency not only is higher, but now drops linearly as the output swing is lowered (Figure 11.8).

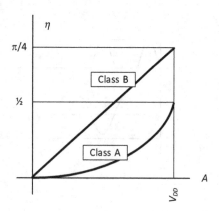

Figure 11.8: Efficiency of class A and B PAs

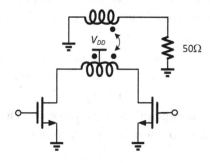

Figure 11.9: A differential push–pull class B PA

A more practical class B PA employs a push–pull [1] scheme to supply current at both half cycles. Although PMOS devices used in more conventional push–pull amplifiers are not as suitable for RF PAs, in a fully differential design an NMOS-only PA may be easily realized by employing a transformer (Figure 11.9).

Assuming the transformer is lossless, this configuration is not going to affect the efficiency: each device drains the same supply current, and delivers the same power to the load as calculated before. The output powers are then simply combined at the other end of the transformer. The advantage, however, is that the same output power is achieved with half the swing at the drain of each transistor, which is important with the new process nodes as lower supply voltages are used. Alternatively, with the same swing, a higher impedance at the device output results in the same total output power, leading to lower loss in the matching network.

The idea of *power combining* in general appears as a very attractive option in PAs to overcome the issues associated with low supply voltages, devices stress, and inefficiency of on-chip transformers with more than 1:1 turn ratio. One realization especially in discrete implementation is to use transmission-line based combiners, such as the Wilkinson power combiner [2]. Shown in Figure 11.10, when connected to a matched load, it results in a low-loss 2:1 power combining. This structure can be readily extended to an N-way combiner/divider. However, in GHz RF applications, transmission lines are not practical to be realized on-chip, given their large physical geometries and relatively high losses associated with Si substrate and routing.

A more practical approach is accomplished by combining the outputs of several push–pull stages using a *distributed active transformer* [3], [4]. A simplified schematic of such structure is

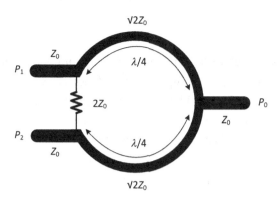

Figure 11.10: Wilkinson power combiner

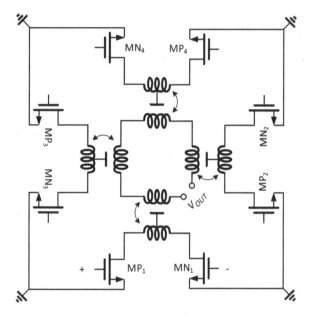

Figure 11.11: Four push–pull stages combined by an distributed active transformer

shown in Figure 11.11, where four push–pull stages are combined. For simplicity, the tuning capacitors are not shown.

The advantage is that for each core, the impedance transformation ratio is less (ideally 1), yet the overall desired output power is achieved through combining the powers of four stages. This minimizes the loss of the matching network, especially if on-chip components are used. There are several other advantages, offered by the *circular* geometry. For instance, the AC grounds are conveniently available between the devices of adjacent push–pull cores (for instance, MN_1 and MP_2), minimizing the ground loss and sensitivity to parasitics. Finally the power combining is accomplished by introducing a 1-turn metal strip inside the circular geometry to act as a magnetic pickup of the output power.

One challenge is the input signal distribution, which must be carefully done through a similar distributed transformer. It must be noted that this topology is not unique to a class A or B design, and could be extended to any other class of PA.

11.4 CLASS C PAs

The higher efficiency of class B PAs as a result of the smaller conduction angle may inspire us to further reduce the conduction angle, and possibly enjoy better efficiency. Known as class C PAs, the typical waveforms are shown in Figure 11.12, where the input is pushed further lower to cause conduction duration of less than 50%.

Assuming a conduction duration of $\tau < \frac{T}{2}$, where T is one cycle, we can see that the device turns on and off at the two points:

$$t_{ON/OFF} = \frac{T}{4} \mp \frac{\tau}{2}.$$

Given the input waveform as shown in Figure 11.12, $\tau = T\frac{\cos^{-1}\frac{V_{TH}}{V_P}}{\pi}$. Accordingly, we define the conduction angle as $\theta = \frac{\pi}{T}\tau = \cos^{-1}\frac{V_{TH}}{V_P}$. A class B PA has a conduction angle of $\frac{\pi}{2}$, or a duration of half a cycle. Although the calculations are somewhat more complex, same as class B, all that is needed is to express the drain current in terms of its Fourier series, and find the average and the fundamental components.

Assuming that as soon as the device turns on, that is, when the input reaches V_{TH}, it starts conducting linearly, or that it behaves like an ideal current source, the drain current then consists of narrow pulses that may be approximated by top pieces of sinusoids (Figure 11.12). During the conduction, we can express the drain current as

$$I_{DS} = I_P(-\cos\theta + \sin\omega_0 t) \qquad t_{ON} \leq t \leq t_{OFF},$$

where the term $-\cos\theta$ accounts for the fact that the current pulses exist for less than half the cycle. One can verify that for $t = t_{ON/OFF}$, $I_{DS} = 0$. Calculating the Fourier series, we have

$$I_{DS} = \frac{I_P}{\pi}\left[(\sin\theta - \theta\cos\theta) + \left(\theta - \frac{\sin 2\theta}{2}\right)\sin\omega_0 t + \cdots\right],$$

where higher order harmonics are not shown. Assuming the tank Q is high enough, the second term results in a sinusoidal voltage with an amplitude of A, as shown in Figure 11.12, whereas

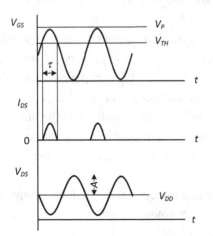

Figure 11.12: Typical waveforms of a class C PA

the first term represents the average power dissipated. Therefore, to establish a swing of A at the output load R_L, during the conduction we must have

$$I_P = \pi \frac{\frac{A}{R_L}}{\left(\theta - \frac{\sin 2\theta}{2}\right)}.$$

The efficiency is

$$\eta = \frac{P_{OUT}}{P_{DC}} = \frac{\frac{A^2}{2R_L}}{\frac{I_P}{\pi}(\sin\theta - \theta\cos\theta)V_{DD}} = \frac{\theta - \frac{\sin 2\theta}{2}}{\sin\theta - \theta\cos\theta} \frac{A}{2V_{DD}},$$

and allowing a drain swing of V_{DD},

$$\eta = \frac{\theta - \frac{\sin 2\theta}{2}}{2(\sin\theta - \theta\cos\theta)},$$

which indicates that if the conduction angle approaches zero, the efficiency reaches 100%. Interestingly, setting the conduction angle to π or $\frac{\pi}{2}$, we arrive at the exact same expression we had for class A and B PAs.

By setting $t = \frac{T}{4}$ in the drain current expression, $I_{DS} = I_P(-\cos\theta + \sin\omega_0 t)$, we can find the peak current,

$$I_{DS,Peak} = I_P(1 - \cos\theta) = \frac{A}{R_L} \pi \frac{(1 - \cos\theta)}{\left(\theta - \frac{\sin 2\theta}{2}\right)}.$$

If we let $\theta \to 0$,

$$I_{DS,Peak} \approx \frac{3}{4} \frac{A}{R_L} \frac{\pi}{\theta},$$

which indicates that to maintain the same output swing (to deliver the same desired power), the drain current peaks to infinity. This makes sense, as the pulse width approaches zero when the conduction angle approaches zero, and to deliver the same fundamental current to the load, the peak reaches infinity. The total area of each pulse is then expected to remain constant and is calculated to be $\frac{A}{R_L}$.

A plot of efficiency versus the conduction angle is shown in Figure 11.13. Also shown in the plot is the drain peak current normalized to the load peak current, $\frac{A}{R_L}$. Note that for class A the drain current has an AC peak equal to the load peak current, but also carries a constant DC current, and thus its net peak is twice the load peak current, same as class B.

We can see that class A and B PAs could have been treated as special cases of the general class C that we discussed. Accordingly, one may consider a conduction angle of between $\left(\frac{\pi}{2}, \pi\right)$

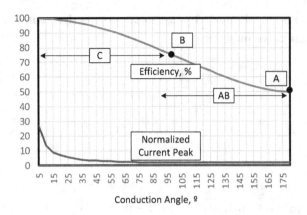

Figure 11.13: Class A/B/C efficiency

to compromise the efficiency for linearity. This type of PA is known as *class AB* PA, and is a common choice for high-linearity applications such as WLAN.

11.5 CLASS D PAs

The classes of PA discussed so far treat the device as a current source to deliver the required power to the load. Class D PAs as well as a few other topologies that we will present shortly, rely on the active device used as a *switch*. The notion of switch as being a passive component implies zero (ideally) power dissipation, and thus 100% efficiency. Naturally, the switching action involved makes these classes of PAs suitable only for constant envelope applications such as GSM or FM radio, as the amplifier is inevitably nonresponsive to the input signal AM content.

The class D PA is traced back to as early as 1959 [5]. A simplified push–pull class D PA is shown in Figure 11.14. Similar to class B, the design can be extended to an NMOS-only push-pull topology using center-tapped transformers (Figure 11.9).

Apart from the switching nature of the design, guaranteed by the high-gain preamplifier, the circuit differs from previous classes as a *series RLC tank* is incorporated, consistent with the voltage switching-mode nature of the circuit, as opposed to controlled current source designs we saw before. However, the class D amplifier may also be implemented in a *current-mode* switching scheme (Figure 11.15), in which case a parallel tank is used (known as inverse[3] class D, or class D^{-1}). Nonetheless, in either case, the tank is expected to filter the undesired harmonic, and the voltage across the load resembles a sinusoid.

Assuming fast switching, an equivalent circuit for the class D PA (Figure 11.14), along with the corresponding waveforms (for NMOS switch) are shown in Figure 11.16. The switch as well as the tank inductor are assumed to have a small resistance. In the half cycle that the N switch is on, the current is sunk from the load. Since the series RLC filters the undesired harmonics, the load voltage as well as the drain current are expected to be sinusoidal. In the

---

[3] The inverse classes, derived based on the *duality* principle, apply to other switching power amplifiers too such as class F.

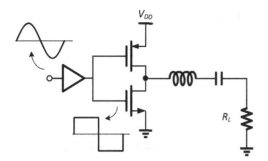

Figure 11.14: A simplified class D PA

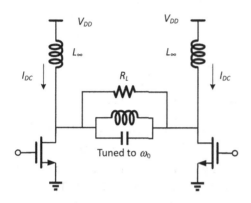

Figure 11.15: Class D^{-1} power amplifier

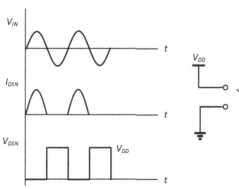

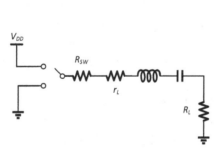

Figure 11.16: Class D PA waveforms

next half cycle, the N switch turns off, and the drain voltage is pulled to V_{DD} by the P switch, which sources current to the load through the supply.

The load voltage peak amplitude is

$$V_{Load} = \frac{2}{\pi} V_{DD} \frac{R_L}{R_L + r_L + R_{SW}},$$

and the load current amplitude is

$$I_{Load} = \frac{2}{\pi} V_{DD} \frac{1}{R_L + r_L + R_{SW}}.$$

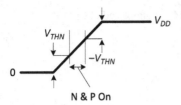

Figure 11.17: Demonstration of shoot-through in class D power amplifiers

The DC current drained from the supply corresponds to the half-sinusoid shown in Figure 11.16, and is equal to

$$I_{DC} = \frac{I_{Load}}{\pi}.$$

Thus the efficiency is

$$\eta = \frac{P_{OUT}}{P_{DC}} = \frac{\frac{V_{Load} \times I_{Load}}{2}}{I_{DC} V_{DD}} = \frac{R_L}{R_L + r_L + R_{SW}},$$

which can reach 100% if the switch resistance is small, and inductor Q is large. This of course relies on the fast switching of the transistor, such that it always sees either a zero current, or a zero drain voltage. In practice, nonideal switching guarantees some nonzero power dissipation, especially at the higher frequencies that the switching speed degrades at. The main trade-off arises from the fact that increasing the switch size reduces R_{SW}, which helps improve the efficiency, but in turn results in an increase in the dynamic power dissipation ($CV_{DD}^2 f$) due to the increase in the switch parasitic capacitance.

Apart from nonzero switch resistance and capacitance, another potential issue arises from the finite rise and fall times of the switching signal. Consequently, there will be a certain period of time during which both the N and P transistors turn on as shown in Figure 11.17.

Known as *shoot-through*, since the switch resistances are small, this could lead to large current drain during the overlap time. As such, often nonoverlapping clocks are devised to minimize that, but it could lead into some distortion.

Since the output power depends only on V_{DD}, the only viable choice for gain control is to vary the supply voltage, but the efficiency is expected to remain high even at lower output powers.

One common application of class D PAs is in audio. Evidently, amplitude modulation cannot be supported given the switching nature of the amplifier. Instead, it is common to utilize what is known as pulse-width modulation[4] (PWM) [6], [7]. The basic concept is illustrated in Figure 11.18.

A sawtooth signal when compared to the input ($x(t)$) results in the waveform $x_P(t)$ shown, which has a constant amplitude of A, but whose width linearly varies with the input signal amplitude at the time location t_k.

---

[4] Also known as pulse-duration modulation (PDM).

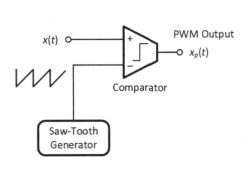 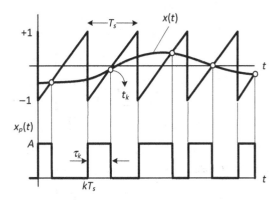

Figure 11.18: Basic concept of PWM and the corresponding waveforms

Examination of waveforms in Figure 11.18 reveals that the modulation duration depends on the message value at the time location t_k of the pulse edge, rather than the apparent sample time kT_s. As shown in Chapter 2, the duration of each pulse, τ_k, may be defined as

$$\tau_k = \frac{T_s}{2}(1+x(t)),$$

where T_s is the sawtooth signal period. To prevent missing pulses or negative durations, let us assume that $x(t)$ is normalized such that $|x(t)| < 1$. Since the rate of the input signal variation is assumed to be much less than the sampling frequency, we may assume uniform sampling, which is to say τ_k could be treated as nearly a constant. Thus, we can write the Fourier series coefficients as

$$a_n = \frac{1}{T_s}\int_{T_s} x_P(t)e^{-j2\pi nf_s t}dt = \frac{A}{T_s}\int_{-\tau_k/2}^{\tau_k/2} e^{-j2\pi nf_s t}dt = \frac{A}{\pi n}\sin\left(\frac{n\pi}{2}(1+x(t))\right),$$

which leads to

$$x_P(t) = \frac{A}{2}(1+x(t)) + \sum_{n=1}^{\infty} \frac{2A}{n\pi}\sin n\phi(t)\cos n\omega_s t,$$

where $\phi(t) = \frac{\pi}{2}(1+x(t))$. The PWM signal contains the input $x(t)$ plus a DC component as well as phase-modulated waves at the harmonics of f_s. As the phase modulation has negligible overlap in the input signal band provided that the sampling frequency is high, $x(t)$ can be fully recovered by lowpass filtering with a DC block.

11.6 CLASS D DIGITAL PAs

Most modern wireless communication standards employ modulation with a time-varying envelope to improve the spectral efficiency. As mentioned in the last section, conventional class D PA achieves the high efficiency by switching rail to rail, thus eliminating amplitude information from the input signal. For narrowband signals, PWM is used to change the output

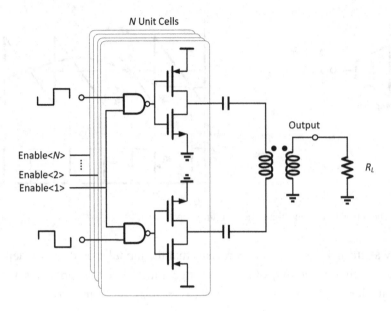

Figure 11.19: Segmented differential class D PA

signal amplitude by changing the duty cycle of the input signal as shown in Figure 11.18. For wider bandwidth signals, however, that require high dynamic range, PWM is not as suitable, and often digital class D PAs are used.

In a digital PA (or DPA) the total size of the PA is designed to deliver the target maximum power, then the total size is divided into N unit cells that can be individually turned on or off. Each cell is driven by a rail-to-rail input, which passes through only if the cell is selected with an enable selection bit (Figure 11.19). Similarly a segmented inverse class D PA can be constructed as shown in Figure 11.20, where each cell has a differential switchable current source and two cascode devices. The differential currents from all units are summed up in the load balun and converted to single-ended voltage on the load resistance [8], [9].

The DPA is essentially a high-speed digital-to-analog converter (DAC). Like any DAC, the nonlinearity mechanisms can be divided into two groups:

1. Analog nonlinearity caused by the dependence of each cell output on the value of the combined output. For example, the switch resistance of an on MOS device depends on the drain to source voltage, which is the total output voltage. For instance, as was proved in Section 11.5, the output voltage is function of the switch resistance.
2. Unit cell mismatch, where random and systematic mismatch mechanisms cause the output of each cell to be different. For example, the random mismatch of the threshold voltages of the MOS devices would cause a mismatch in the switch resistance and therefore the output voltage. Mismatches could result in differential nonlinearity (DNL) or integral nonlinearity (INL).

A poor DNL results in high quantization noise, which degrades the noise floor, typically important for out-of-band requirements. On the other hand, high INL degrades the in-band linearity. In the context of DPA, the INL is related to OIP_3, OIP_5, and so forth. Therefore, a poor INL would degrade EVM and results in spectral regrowth that might violate the transmission mask. To achieve good linearity, the accumulated knowledge of high performance DACs

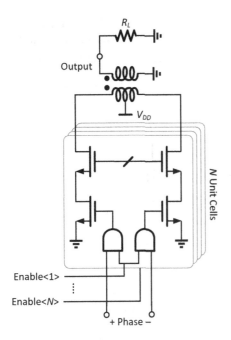

Figure 11.20: Segmented inverse class D PA

is leveraged. For instance, to optimize the DNL, the DPA cells are divided into two sections: unary and binary. The unary bank of cells are all identical and controlled with thermometer code, while the binary bank has regular binary weighted cells. The segmentation is done as explained in [10], as a trade-off between DNL and the area and power overhead of segmentation.

To improve the linearity over the segmented class D PA of Figure 11.19 a switched capacitor (SC) architecture may be sought [11]. In the switched capacitor PA, the matching capacitor is segmented as well, such that each cell has a small unit cap with an impedance higher than that of the driving inverter, as depicted in Figure 11.21. This topology improves the linearity substantially, because the gain of each cell is determined by an extremely linear passive capacitor in comparison to the nonlinear MOS switch resistance. Capacitors also can handle larger swing, scale well with the technology (same as switch), and have excellent matching properties.

Moreover, the power efficiency of SC-PA at back-off is better than class D because the gain setting element is reactive and does not consume power in the off state.

The DPA can be designed with polar or quadrature architectures. In a polar arrangement, the input clock to the DPA is modulated with the phase information, while the amplitude information controls how many unit are turned on. As will be discussed in the next chapter, polar architecture requires CORDIC module to generate the phase and amplitude data from the I and Q data. It also requires delay matching between the amplitude and phase paths, which makes it more suitable for narrowband modulation schemes where the delay requirement can be practically met.

On the other hand, in a quadrature DPA, the I and Q data can be combined in digital or analog. In analog combining, I and Q data are applied to separate DPA and the output current or voltage is combined in one balun. The more convenient digital combining is shown in

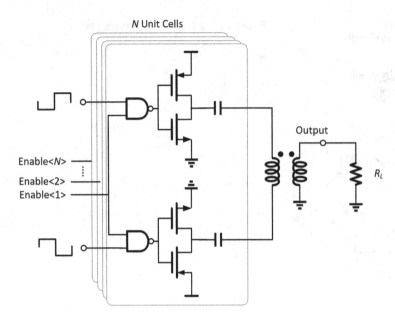

Figure 11.21: Switched capacitor digital PA with improved linearity

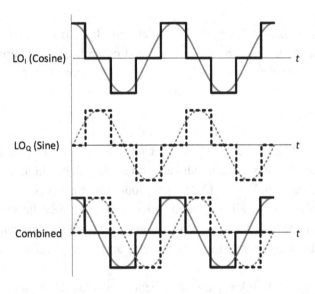

Figure 11.22: Digital mixing of I and Q data

Figure 11.22, where 25% duty-cycle clocks are used to multiplex the I and Q data for each cell. These clocks are globally multiplexed using the sign bit of the I and Q data [12].

11.6.1 Practical Limitations of DPA

The high efficiency of the DPA and the continuous improvement in its performance with technology scaling have made it an attractive choice for integrated CMOS PAs. Yet DPAs suffer from practical limitations that make the design challenging. The main limitation of the DPA is its code-dependent nonlinearity. For example, in the current-mode segmented inverse class D DPA, the output impedance of the current sources changes based on the code, resulting

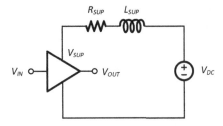

Figure 11.23: Supply impedance model of DPA

in a code-dependent gain, which degrades the linearity substantially. Similarly, in SC-PA, the inevitable supply routing resistance and inductance will cause unintended modulation of the supply voltage with the signal envelope (Figure 11.23). This is called remodulation and results in degraded linearity in the form of EVM degradation and spectral regrowth.

It has been demonstrated that digital predistortion DPD can be used to correct for various types of nonlinearities to a level suitable for commercial products [13] (see also Section 11.9.1). It should be noted, however, that DPD, especially the two-dimensional [14], adds complexity to the system and comes with an area and power overhead. However, with technology downscaling, it is expected that the DPD overhead will be more manageable and will make the DPA suitable for more applications.

11.7 CLASS E PAs

The main challenge of class D PAs is to minimize: the voltage across the switch when it carries current, and the current through it when there exists a voltage across. These conditions imply fast switching, which may not be feasible at high frequencies. If the switch voltage and current overlap, as they do in practice due to finite rise and fall times, the efficiency degradation may be substantial. This may be avoided by proper design of the matching network,[5] which adds enough degrees of freedom to ensure that there is no appreciable time duration that voltage and current simultaneously exist [15].

A simplified class E power amplifier is shown in Figure 11.24.

The drain inductor is large and is used for biasing purposes.[6] The load resistance R_L and the supply voltage are chosen a priori, based on the maximum available or allowable (given the device stress concerns) supply, and the power delivery requirements.

The matching network consists of the series RLC circuit (L, C_2, R_L), as well as the parallel capacitance C_1 that includes the device and the drain inductor parasitic. This capacitance ensures that when the device turns off, the drain voltage stays low until the drain current reaches zero. This is especially important at high frequencies where the switch does not respond as quickly as needed. The matching network components (L, C_1, and C_2) are chosen as follows:

---

[5] This type of network is often referred to as a ZVS (zero voltage switching) network, for the reasons we will describe shortly.
[6] A class E PA may be designed with a modest drain inductor, absorbed in the rest of the matching network.

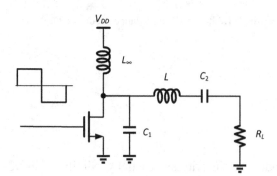

Figure 11.24: Class E power amplifier

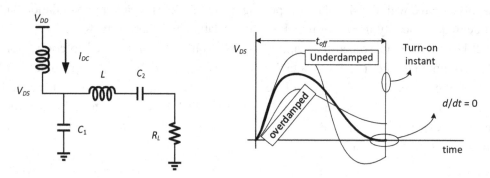

Figure 11.25: Off-mode class E PA

When the switch turns off, the circuit simplifies to a damped RLC network (Figure 11.25). The drain inductor current which may be assumed constant if the inductance is large, charges up C_1. If at the instant that switch turns on the capacitor voltage is nonzero, this voltage is dumped to ground, and causes power loss. Furthermore, the nonzero capacitor voltage violates the condition that the voltage across the switch must be zero when it carries current.

Thus, the quality factor of the RLC circuit, $Q_L = \frac{L\omega_0}{R_L}$, must be chosen such that at the instant of turning on, the voltage across C_1 (V_{DS} in Figure 11.25) has reached zero. If the circuit is too damped, it will never reach zero, and if it is underdamped, it may become negative which could ultimately turn on the switch in the reverse mode. Furthermore, it is desirable for this voltage to reach zero, with a zero slope to reduce the *sensitivity* to process variations. There is also a trade-off between the quality factor of series RLC circuit (Q_L), and the amount of filtering provided to remove the undesirable harmonic.

The exact analysis of the circuit requires solving nonlinear differential equations and is given in [16]. To gain some perspective, we offer a more simplified approach here when the optimal conditions are satisfied. Let us first focus on the off half cycle ($\frac{T}{2} < t < T$). Assume that the current flowing to the load contains only the fundamental, and thus can be generally expressed as

$$i_{OUT} = I_0 \cos(\omega_0 t + \theta_0).$$

During the off mode ($\frac{T}{2} < t < T$), the drain inductor current (I_{DC}), which can be assumed constant if the inductance is large enough, flows through C_1, and the capacitor current according to KCL is (Figure 11.25)

$$i_{C_1} = I_{DC} - I_0 \cos(\omega_0 t + \theta_0).$$

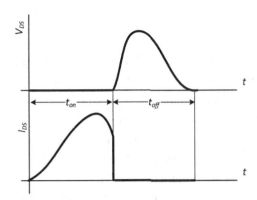

Figure 11.26: Class E PA optimum waveforms

The capacitor voltage, which is equal to the drain voltage, is the integral of the current,

$$v_{C_1} = \frac{I_{DC}t}{C_1} - \frac{I_0}{\omega_0 C_1}\sin(\omega_0 t + \theta_0) + K,$$

where K is a constant resulted from integration. Now these conditions must be met: First, $v_{C_1}|_{t=\frac{T}{2}} = 0$, as that is the instant of the switch turning off, until which time the capacitor has been shorted by the switch. Second, $v_{C_1}|_{t=T} = 0$ to ensure that when turning the switch back on, no power is lost due to nonzero capacitance voltage. Finally, $\frac{dv_{C_1}}{dt}\big|_{t=T} = 0$ to ensure low sensitivity to variation. With these conditions met, we obtain

$$v_{C_1} = \frac{I_{DC}}{C_1\omega_0}\left(\omega_0 t - \frac{\pi}{2}\frac{\sin(\omega_0 t + \theta_0)}{\sin\theta_0} - \frac{3\pi}{2}\right),$$

where $I_0 = I_{DC}\sqrt{1 + \frac{\pi^2}{4}}$, and $\theta_0 = \tan^{-1}\frac{\pi}{2} = 1$ rad.

At $t = \left(1 - \frac{\theta_0}{\pi}\right)T = 0.68T$, the drain voltage peaks to $\frac{I_{DC}}{C_1\omega_0}(\pi - 2\theta_0) = 1.13\frac{I_{DC}}{C_1\omega_0}$, and eventually falls to zero as desired at the end of the cycle.

By obtaining the capacitor voltage Fourier series, and satisfying the KVL and energy conservation (see problem set) we can show that when the proper design conditions are met, the drain current and voltage resemble the ones shown in Figure 11.26. The matching network as stated before is expected to extract the main harmonic of the signal when appearing at the load, although the inductor need not necessarily resonate with C_2 or C_2 in series with C_1.

The drain voltage is expected to reach a peak value of $\pi \sin^{-1}\frac{\pi}{1+\frac{\pi^2}{4}}V_{DD} = 3.562V_{DD}$, and fall to zero with zero slope at the end of the cycle. The drain current peaks to $I_{DC} + I_0 = I_{DC}\left(1 + \sqrt{1 + \frac{\pi^2}{4}}\right)$, and eventually falls to $2I_{DC}$ at the instant of the switch turning off $\left(t = \frac{T}{2}\right)$ [16].

11.8 CLASS F PAs

A class E PA has a somewhat poor power capability as it results in a large peak drain current and voltage. Also the large current drain may degrade the efficiency as practical switches have a nonzero on resistance. These issues may be avoided in a class F design that relies on the matching network to shape the switch voltage and current to advantage. Specifically, if the even order contents of the drain voltage are removed, while its odd harmonics are amplified, one expects the drain voltage to resemble an ideal square-wave that results in perfect efficiency. Shown in Figure 11.27 is the circuit diagram of such implementation.

A parallel tank tuned to fundamental frequency ensures that the impedance seen on the right side of the transmission line is equal to R_L. If the line characteristics impedance Z_0 is equal to R_L, on the left side the impedance seen is also R_L, as desired. At other frequencies and particularly at the harmonics, the parallel tank is a short circuit. As we saw in Chapter 3, a quarter wavelength line reverses the impedance. Thus the short appears as an open circuit at all odd harmonics, but remains as an open circuit at all even harmonics as the effective length is equal to half-wavelength or its integers. This is exactly what is needed to produce a square-like waveform at the output.

As the device sees open circuit at all the odd harmonics, only the fundamental component of the current flows to the load, and thus the drain current, is expected to be a sinusoid. When the device turns off, the load current is directly sourced by the supply, similar to class B PA. The corresponding drain voltage and current waveforms are shown in Figure 11.28.

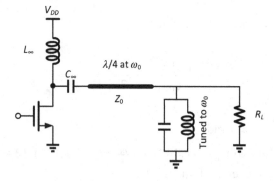

Figure 11.27: Class F power amplifier using a $\lambda/4$ transmission line

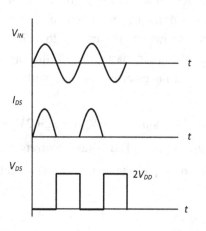

Figure 11.28: Class F device waveforms

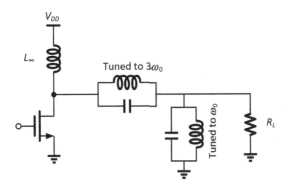

Figure 11.29: Class F power amplifier using lumped elements

Assuming all the harmonics are subject to the filtering, the peak voltage at the load is $\frac{4}{\pi}V_{DD}$, and the drain current peak is $\frac{8}{\pi}\frac{V_{DD}}{R_L}$, twice as large as the output peak current.

An alternative class F PA using lumped elements is shown in Figure 11.29. Instead of the transmission line, a parallel tank tuned to $3\omega_0$ removes the second harmonic.

Although this is not going to produce a perfect square-wave, in practice it may be sufficient to achieve a respectable efficiency. Adding more resonance branches is an option, although the loss created due to their finite Q may offset the efficiency improvement caused by a more favorable waveform.

A more analytical description of switching power amplifiers in general, and classes E and F in particular, is found in [17].

11.9 PA LINEARIZATION TECHNIQUES

The efficiency-linearity trade-offs pointed out thus far may result in considerably poor efficiency in many applications with complex modulation schemes that demand high linearity. One idea to break this trade-off is to start with an efficient yet nonlinear power amplifier, and compensate for the linearity through some kind of feedback or feedforward mechanism. In this section we discuss some of these schemes and discuss their pros and cons.

11.9.1 Predistortion

Widely used in wireless transmitters, the predistortion (PD) technique compensates the power amplifier's nonlinear gain or phase responses by shaping the input signal such that the overall cascade response becomes linear. The predistorter realizes the inverse of the power amplifier's response to counterbalance its nonlinearity. This way the PA can be biased in high-efficiency regions such as class AB while maintaining a good linearity. Since power amplifiers are typically compressive, the predistorter has an expansive response.

The predistortion can be performed directly at the RF in the analog domain using diodes and other nonlinear devices. An example of the RF predistortion making use of analog devices is shown in Figure 11.30.

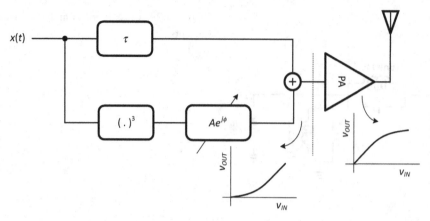

Figure 11.30: An RF cubic predistorter

The intention is to partially cancel the nonlinearity of the power amplifier by adding a cubic component of the signal with proper phase and magnitude at the input of the PA. The PA input is divided into two paths. In the main path the transmitted signal is applied after experiencing a delay of τ, to make up for the delay created in the second path. The second path is responsible for providing the cubic nonlinearity followed by a scaling using a variable gain and phase amplifier. The two signals are added together to construct the predistorted signal, and are applied to the power amplifier that is weakly nonlinear.

In modern wireless transmitters the predistortion is realized at the baseband digital processor. A digital predistortion can lead to a better correction and is the most suitable structure for wideband applications. Through a feedback, the predistortion can adaptively follow the process, voltage, and temperature variations (PVT) of the PA, making it a robust and reliable linearizer. Figure 11.31 shows a simplified block diagram of an IQ transmitter with an adaptive predistortion performed in the digital baseband.

The PA output is attenuated and downconverted to the baseband by a feedback receiver. The downconverted signal is digitized and fed to the DSP to evaluate the transmitted signal both within the channel (for EVM) and out of it (for ACLR). An inverse predistortion function needed to linearize the PA's nonlinear characteristic is calculated and implemented on the digital IQ signals. The distorted baseband signals occupy a wider bandwidth than that of the original undistorted ones. For example, the presence of the cubic terms in the modified baseband signals would triple the occupied bandwidth. Ignoring the scaling factor, assume that the ideal RF input to a linear PA is supposed to be $x_{RF}(t) = x_{BB,I}(t)\cos\omega_c t - x_{BB,Q}(t)\sin\omega_c t$. However, to compensate for the PA nonlinearity, its input must be $x_{RF}(t) + \beta x_{RF}^3(t)$. It can be readily proven that the baseband signals must be modified to the following,

$$x_{BB,I,mod} = x_{BB,I}(t) + \frac{3}{4}\beta\left\{x_{BB,I}^2(t) + x_{BB,Q}^2(t)\right\}x_{BB,I}(t)$$

$$x_{BB,Q,mod} = x_{BB,Q}(t) + \frac{3}{4}\beta\left\{x_{BB,I}^2(t) + x_{BB,Q}^2(t)\right\}x_{BB,Q}(t),$$

indicating that the bandwidth of the modified baseband signals widens by a factor of three.

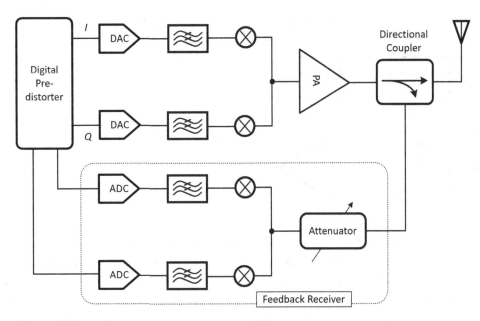

Figure 11.31: Transmitter with adaptive IQ predistorter

Consequently, the bandwidth of the lowpass filters following the DACs must be adequately high. The increased bandwidth of the lowpass filter may increase the transmitter noise and degrade the emission mask, and thus may lead to some design trade-offs.

The distortion cancellation architecture is generic and is applicable to PAs with smooth nonlinearity such as class A or class AB. However, the nonlinearity of PAs with steep slopes in the characteristic curve may not be corrected.

A major issue in the described IQ transmitter with adaptive predistortion is possible nonlinear distortion caused by the feedback receiver. This nonlinear distortion can adversely impact the open loop linearity, and produce unwanted post-predistortion nonlinear components. Another drawback is related to the power consumption associated to the DSP processing and bandwidth limitations enforced by the DSP clock. The feedback can be activated with a duty cycle to save in the power consumption. The correction coefficients must be refreshed with an adequate rate to cover and correct effects of variations caused by PVT as well as changes on the antenna impedance as the mobile device is moved.

The adaptive predistortion technique described for IQ transmitters, can also be applied to any other transmitter architectures such as polar and outphasing (see Chapter 12 for more details on polar and outphasing architectures). The operation principle remains the same, meaning that the transmitted signal is detected by the feedback receiver and its quality is evaluated and necessary corrections are applied to the underlying signal components.

11.9.2 Envelope Elimination and Restoration

The concept of envelope elimination and restoration (EER) is presented in Figure 11.32, where the envelope and the phase information of the modulated transmitted signal are decoupled across two separate paths. The modulated signal's envelope is extracted with the help of an

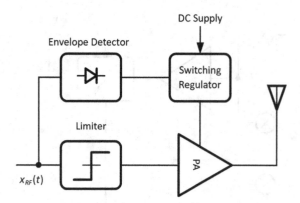

Figure 11.32: Operation principle of EER

envelope detector and the phase-only-modulated signal is constructed by passing the input signal through a *limiter*, which is typically a hard-limiting buffer. The phase-modulated RF signal has a constant amplitude, therefore can be amplified using a highly efficient nonlinear PA. The PA supply voltage is now modulated by the extracted envelope signal. A switching regulator with a high efficiency is responsible to provide this supply voltage to the PA, and typically is the main building block that limits the bandwidth of the transmitted signal using the EER architecture.

Linearity of the EER-based transmitters is independent of the linearity of the constituent power transistors of the PA and is only a function of the switching regulator's performance. The EER architecture generally presents a good linearity for applications with narrow or modest envelope bandwidths. Besides the envelope bandwidth, the linearity and accuracy performance of the transmitter is greatly affected by the timing alignment of envelope and phase components.

The EER technique was proposed by Kahn in 1952 [18] and has been successfully utilized in high-power transmitters with good efficiencies. With the advent of the digital CMOS signal processing technology, the envelope and phase information can be prepared in the baseband with the minimum hardware overhead eliminating the need for the limiter and the envelope detector. Of course, the transmitter must perform phase-modulation into the carrier. This way, not only the analog imperfections of the limiter and the envelope detector are eliminated, but also the delay between the envelop and the phase paths can be controlled with a greater accuracy.

The concept of EER has evolved into polar architecture, which was covered in more detail in Chapter 12.

11.9.3 Envelope Tracking

The maximum theoretical efficiency of a linear power amplifier happens at the onset of the saturation level. This maximum efficiency is 50% and 78.5% ($\pi/4$) for class A and class B PAs, respectively (Figure 11.8). When the swing of the PA input is reduced from its peak at the onset of saturation, the efficiency drops too. As the input level departs further from the peak the drop in the efficiency would be substantial. This problem especially becomes severe

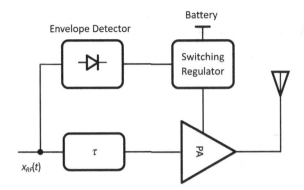

Figure 11.33: Concept of envelope tracking

when the input signal is modulated and has a large peak-to-average power ratio (PAPR). Modulations with large PAPRs are now common in modern wireless communication systems such as quadrature amplitude modulation (QAM). These modulation signals with variable envelopes require linear power amplifiers where the PA is forced to operate away from its saturation level for most of the times, leading to a low average efficiency. In other words, a large PAPR mandates a large back-off from the saturation level causing degradation in efficiency. Typical values of efficiency as low as 10% or less are reported for some high data-rate modulation standards.

To boost the efficiency of linear power amplifiers, the use of envelope tracking (ET) technique has become an attractive and viable option. According to the ET technique, through a switching regulator the supply voltage of a linear PA tracks the envelope of the RF signal to be amplified. With the ET architecture assuming the efficiency of the switching DC–DC converter is 100%, the maximum efficiency of a linear PA is theoretically achievable with any modulated signal, for instance, 50% for a class A PA. The block diagram of an ET transmitter is shown in Figure 11.33, where the linear PA is supplied by a switching DC–DC converter with a bandwidth wide enough to faithfully track the envelope of the input RF signal. This DC–DC converter is called the envelope amplifier (EA).

In the ET architecture, the intention is to boost the overall efficiency by using a linear amplifier in the vicinity of its saturation level all the time. The input to the PA is the actual and the final form of the intended modulated signal to be transmitted. Thus, typically a linear IQ transmitter is responsible to generate the modulated RF signal to the PA input.

Assuming a class A PA is used, the bias current remains almost constant at all time over the entire range of the input swing. Note that the efficiency would not be improved if the envelope amplifier is a linear regulator, because only location of the power loss would be shifted from the PA to the regulator. Consequently, the envelope amplifier has to be in the switching-mode with high efficiency. Usually the switching DC–DC converter is some derivation of the basic buck converter [19]. This form of DC–DC converter is very attractive for its simplicity, high efficiency, and relatively fast dynamic response. A simplified diagram of the switcher is shown in Figure 11.34.

The main idea arises from the fact that a switched inductor can ideally produce an output voltage whose value may be controlled through switching rate or *duty cycle*. A simplified block

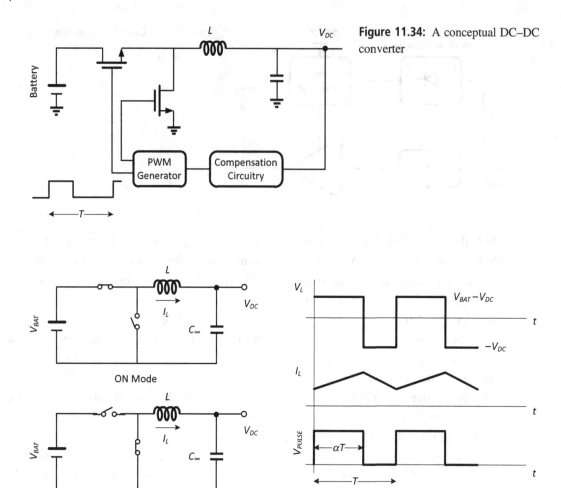

Figure 11.34: A conceptual DC–DC converter

Figure 11.35: Basic operation of a buck converter

diagram describing the buck converter basic operation is shown in Figure 11.35. The steady-state waveform corresponding to the *continuous mode* (the inductor current never goes to zero) is also shown on the right. The converter may be analyzed through volt-second and charge balance principles [21], and is briefly outlined below.

During the *on* mode, the inductor voltage is positive and equal to

$$V_L = V_{BAT} - V_{DC}.$$

Thus, the inductor current increases, and the inductor stores magnetic energy. The net increase in the current is

$$\Delta I_{L+} = \frac{V_{BAT} - V_{DC}}{L} \alpha T,$$

where αT is the on mode duration. During the *off* mode, the inductor current decreases, and similarly, we can show that the net decrease in the current is

$$\Delta I_{L-} = \frac{-V_{DC}}{L}(1-\alpha)T.$$

In the steady-state, we must have

$$\Delta I_{L+} + \Delta I_{L-} = 0,$$

which results in

$$V_{DC} = \alpha V_{BAT}.$$

So the output voltage is simply controlled by the duty cycle (α) of the pulse applied. Clearly the output voltage cannot exceed the input, and thus this kind of converter is known as a *step-down (or buck) converter*. If the inductor loss and switch resistance are negligible, unlike a linear regulator, the buck converter of Figure 11.34 is capable of reducing the output voltage efficiently. That is because the load current is drained from the battery only during the on mode. Consequently, the battery current is scaled by the duty cycle, as is the output voltage. See also Problems 5 and 6 to gain more insight into switched-inductor circuits, and DC–DC converters. Problem 6 presents an example of a *boost or step-up* converter.

The envelope amplifier and the DC–DC converter must have a sufficient bandwidth to generate an acceptable version of the envelope signal as the PA supply. A switcher with a feedback system is slower than one with no feedback (open-loop). As we showed, the output voltage is proportional to the duty cycle of the input to the converter (Figure 11.34), which can be controlled in an open-loop manner. The duty-cycle must then be proportional to the envelope of the RF signal to ensure the regulated output voltage of the converter follows the RF signal's envelope. Note that this linear relationship between the regulated output voltage and the input is valid only when the switcher operates in the *continuous conduction mode*; otherwise, the relationship would be heavily nonlinear [21].

In the ET the PA supply roughly tracks the instantaneous signal envelope without introducing compression within the power amplifier. Since the envelope is a nonlinear function of the quadrature baseband signals (Figure 11.36), the envelope bandwidth is much larger than those of the constituent quadrature signals.

To ensure the switching regulator is fast enough to track the wideband envelope, stringent requirements on the loop bandwidth and slew rate are imposed. Under such circumstances the

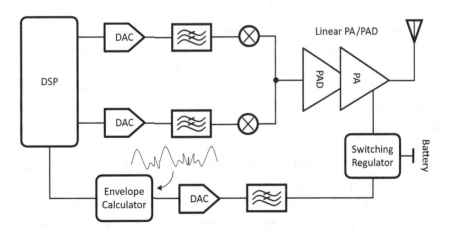

Figure 11.36: Simplified IQ transmitter with envelope tracking

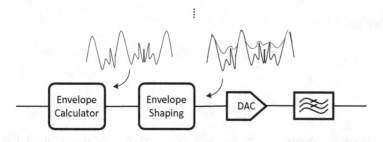

Figure 11.37: Envelope tracking with modified envelope signal

power consumption overhead caused by the extra signal processing of the envelope calculator and the following DAC, LPF, and the DC–DC converter may defeat the intended power saving in the PA. Furthermore, the DSP, DAC, and switcher need to work at higher clock speeds, further exacerbating the situation. For modern high data-rate applications, the clock speed and power consumption of the DC–DC converter may prove to be impractical. One technique to relax the bandwidth requirements of the envelope tracker is shown in Figure 11.37, where a modified signal replaces the RF signal's envelope as the input to the DC–DC converter. This modified signal is derived from the original envelope such that it has smoother transitions with less dynamic range than the envelope itself by limiting its minimum (beyond a certain low value the envelope waveform is clipped).

To obtain a good efficiency, the PA and the switching regulator must be codesigned. As the combined PA and switching regulator is a nonlinear system, extensive digital predistortion and calibration can be utilized to improve the system performance. Also, the PA gain typically falls off with reduced supply voltage, which could cause AM–AM nonlinearity, and must be compensated. The sensitivity of the PA gain to its supply variations would also reduce the overall power supply rejection ratio (PSRR) from the main DC supply of the switching converter to the transmitter output. A thorough system modeling of DSP/DAC/LPF/converter/PA is critical for a digital predistortion and nonlinear signal processing. Also, the design of output matching network is critical. Specifically, the matching must be preserved while the PA supply is modulated across its entire range. Since harmonics of the transmitted signal degrade abruptly as the amplifier enters the compression region, the matching network must suppress harmonic spurious growth. Finally, the interface between the envelope tracker (the switching regulator) and the PA is critical for both ET and EER-based transmitter architectures. This matter entails design trade-offs between the interface capacitance, the stability, the system bandwidth, and the overall PSRR.

As we discussed, the transmitter based on the envelope elimination and restoration (EER) architecture also employs a switching regulator as an envelope amplifier. In fact, linearity and noise requirement of the envelop amplifier in an EER architecture is much more stringent than that of an ET transmitter. This is due to the fact that in the ET architecture the envelope tracking does not have to be exact and small deviations are tolerable, as in the ET the PA input is the RF signal in its final format and the PA is linear. However, unlike the ET transmitter, in the EER transmitter the PA input signal is only phase-modulated and the PA is nonlinear operating deep in compression for maximum efficiency. Consequently, the envelope amplifier is responsible to reconstruct the envelope of the transmitted signal, which is why it must provide the envelope information with a greater accuracy to the PA's supply voltage. This justifies why in ET, the DC–DC converter or the envelope amplifier has much more relaxed requirements.

11.9.4 Dynamic Biasing

We have seen that bandwidth efficient modulations such as 64QAM embed a great portion of the modulation information in envelop of a signal. Transmitting such signals with varying amplitudes needs a highly linear power amplifier to drive the transmitter antenna with an acceptable EVM and emission mask. Dynamic biasing of the PA is another simple approach to improve efficiency of linear PAs employed for these applications. The philosophy of the dynamic biasing technique is to improve the PA efficiency by dynamically lowering the bias current drawn from the main supply at times when the output power level is low. The dynamic biasing of the PA leads to a reliable, low-power, and low-cost solution for improving the PA efficiency.

Consider a commonly used CMOS PA based on a common-source structure. The dynamic biasing technique applicable to this PA can be categorized into two main forms: (1) dynamic biasing of the gate and (2) dynamic biasing of the drain. Let us briefly discuss the latter first. With the dynamic bias of the drain, the drain voltage of the PA is appropriately reduced whenever the transmitter signal level is low. The reason for this reduction of the PA supply voltage is to ensure that the PA operates close to the saturation level in order to maintain a good efficiency. In other words, this dynamic reduction of the PA supply voltage adjusts the load line such that the RF signal swings across almost the entire load line all the time. The *envelope tracking* scheme we discussed earlier operates based on this principle. Note that for PAs using dynamic biasing of the drain, almost the entire current drawn by the PA is provided by the envelope amplifier, which is a switching regulator with good efficiency.

In the dynamic biasing of the gate, however, the gate voltage, which defines the quiescent current of the PA, is adjusted. Under the power back-off situation, the gate voltage is reduced to increase the efficiency. A simple PA circuit operating based on this power saving idea is shown in Figure 11.38. Since modifying the gate voltage can alter the characteristic curve of the PA considerably, the AM–PM and AM–AM nonlinearities can be substantial, and the system may require a predistortion to correct them.

The nonlinearity effect could become severe as the gate bias voltage approaches the transistor threshold voltage. The AM-to-PM and AM-to-AM characteristics of PAs utilizing dynamic biasing are typically lower when the drain is dynamically biased instead of the gate. Another

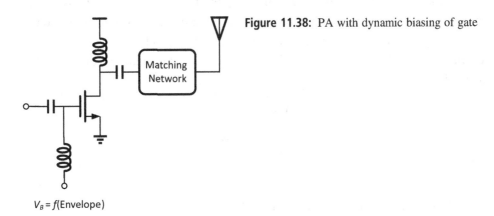

Figure 11.38: PA with dynamic biasing of gate

$V_B = f(\text{Envelope})$

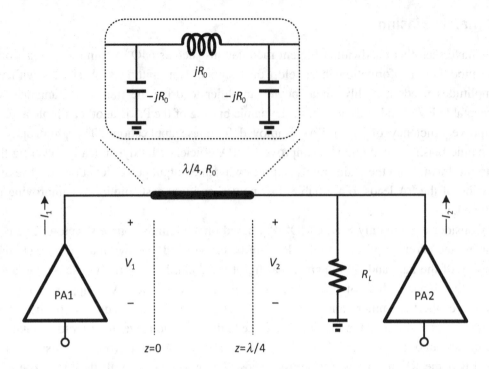

Figure 11.39: Form of high-efficiency power amplifier

potential issue with this approach is its susceptibility to any noise sources capable of effectively modifying the gate voltage.

The major benefit of the dynamic biasing of the gate is its relatively small and compact size. Since the dynamic biasing circuit does not need to be routed beyond the gate of the input device, the circuit size is typically very small and has a negligible die area.

11.9.5 Doherty Power Amplifier

Another topology to improve the efficiency yet maintaining acceptable linearity is to use multiple power amplifiers, each responsible for some subset of the power range. The earliest realization of such scheme is traced back to 1936 by Doherty [20].

First let us consider the circuit shown in Figure 11.39, consisting of two arbitrary power amplifiers[7] separated by a quarter wavelength transmission line. The transmission line may be approximated by a π lumped LC circuit as shown in Figure 11.39.

From Chapter 3, the voltage and current in the transmission line can be generally expressed as

$$V(z) = V^+ e^{-j\beta z} + V^- e^{+j\beta z}$$

$$I(z) = \frac{1}{R_0}\left(V^+ e^{-j\beta z} - V^- e^{+j\beta z}\right),$$

---

[7] The original design proposed by Doherty uses tubes.

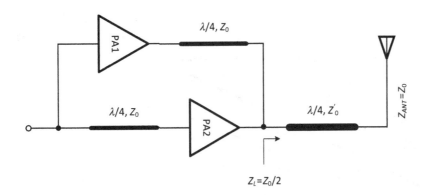

Figure 11.40: A simplified Doherty PA

where z is an arbitrary point in the line, and $\beta = \frac{2\pi}{\lambda}$.

Assuming the current of the second PA is related[8] to the first one by $I_2 = Ae^{-j\phi}I_1$, solving for V_1/I_1 and V_2/I_2 at boundaries ($z=0$ and $\frac{\lambda}{4}$), yields

$$Z_1 = \frac{V_1}{-I_1} = R_0\left(\frac{R_0}{R_L} - jAe^{-j\phi}\right),$$

which is the impedance seen by PA1. Assuming that the input voltages of the PAs are 90° out of phase ($\phi = \frac{\pi}{2}$), and for the specific choice of $R_L = \frac{R_0}{2}$,

$$Z_1 = R_0(2 - A).$$

As in the quarter wavelength lines the input impedance is *inversely proportional* to the terminating impedance, the network of Figure 11.39 presents to the first PA an impedance of R_0, when the effective terminating impedance is also R_0, that is, when the second PA is contributing to half the power to the load ($A = 1$). However, should the second PA be removed from the circuit, or prevented from contributing to the output ($A = 0$), the terminating impedance is reduced to $\frac{R_0}{2}$. Consequently, the impedance seen by the first PA increases to $2R_0$.

Devised according to this principle, the block diagram of a Doherty PA is shown in Figure 11.40. It is composed of a main PA (PA1) operating in the class B region, and an auxiliary PA (PA2) operating in the class C region. The 90° phase shift is generated by an additional transmission line in front of the second PA. Furthermore, to create a load impedance of half the antenna impedance, another transmission line with characteristic impedance of $Z'_0 = \frac{Z_0}{\sqrt{2}}$ is placed between the antenna (whose impedance is Z_0) and the output.

The main PA is active when the power level is low, and the input swing is not large enough to activate the auxiliary PA. The total output power is entirely obtained from the first PA, which is working at $2R_0$ ohms, twice the impedance it is working at when delivering its peak power. The corresponding output voltage and DC current of each of the amplifiers versus the

---

[8] This is accomplished by feeding input voltages that are related accordingly.

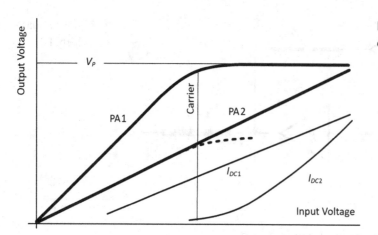

Figure 11.41: Voltage and currents of the two amplifiers in Doherty PA

input are shown in Figure 11.41. Note that the output voltage is the same as the second PA voltage.

At a certain input power, denoted as the *carrier power*, PA1 is saturated. Beyond this point, the output characteristics, *unassisted*, would flatten (the dashed line in Figure 11.41). The second PA, however, is permitted to come into play, and not only delivers power of its own, but through the action of impedance inversion causes an effective reduction in the impedance of the first PA. Thus, PA1 may increase its power without increasing its output voltage, which has already reached a maximum. At the peak of the input, the voltages of each PA are the same (V_P), and they each deliver a power of $\frac{V_P^2}{2R_0}$, twice the carrier power. So the total instantaneous power is four times the carrier power.

Prior to carrier, the voltage of PA1 is twice as that of the output, and 90° out of phase, a property that can be shown using the transmission line characteristics (see the problem sets). The current of the second PA is almost zero up to the carrier, at which point it starts to raise twice as rapidly as the PA1 current. At the peak input, the currents are equal, indicating that the two PAs contribute to the output *equally*.

One can recognize that the Doherty power amplifier resembles a class B push–pull PA, where half the power is handled by half the circuit. In fact, it can be shown that the instantaneous efficiency increases linearly with the output power, reaching a maximum efficiency of 78.5%, similar to an ideal class B PA efficiency.

The main application of Doherty PAs is for base station transmitters for the superior efficiency they offer. Also, for applications with very high peak-to-average ratio, where the amplifier is delivering a modest average power, but occasionally large powers due to the peaks, the Doherty PA may be suitable. This is largely due to the fact that the power delivery is mostly done by the main PA, which is low-power and efficient, and the large peaks are assisted by the high-power auxiliary PA. The design of the two PAs, and the threshold that the second PA must conduct, then may largely depend on the statistical properties of the modulation used.

The use of a Doherty PA for wireless devices in GHz range is very limited, mainly due to the bulky $\lambda/4$ transmission lines on the PCB, which can be several centimeters in length. Efforts to replace these transmission lines with equivalent lumped elements were unsuccessful due to narrowband operation and loss of efficiency. Furthermore, the distortion may be a concern, and often feedback is required to enhance that.

11.10 Summary

The power amplifier and digital transmitters are presented in this chapter:

- In Section 11.1 general properties of power amplifiers and the challenges involved were discussed.
- Various classes of PAs were presented in Sections 11.2 to 11.8.
- Digital transmitters were discussed in Section 11.6.
- An overview of PA linearization techniques such as predistortion and envelope tracking was given in Section 11.9.

11.11 Problems

1. An LC matching circuit with a fixed inductor Q downconverts the load impedance in a PA as illustrated in the figure. Show that the larger the impedance transformation (n), the more power is lost in the matching. **Answer:** Relative power loss is $\approx \frac{n}{Q}$.

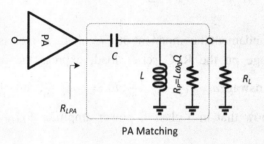

PA Matching

2. Calculate class C efficiency and output power. Show all the steps.
3. Explain why a class B topology cannot be used for the 3rd harmonic peaking.
4. Plot the N and P switches' waveforms in a class D push–pull PA. Expand the Fourier series up to 3rd harmonic, and find the output waveform, and the load and supply powers.
5. Find the output waveforms of the following circuit where a periodic input is applied ($\alpha < 1$). Assume ideal switching in the transistor. Consider the two cases of $T \gg L/R$, and $T \ll L/R$.

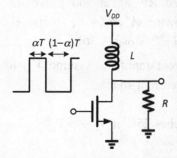

6. Find the output voltage of the following circuit, known as a *boost* switching regulator. Plot the inductor current and voltage in steady-state conditions. Assume the load impedance is very large, and that the diode is ideal. **Answer:** $V_{DC} = \frac{V_{BAT}}{1-\alpha}$.

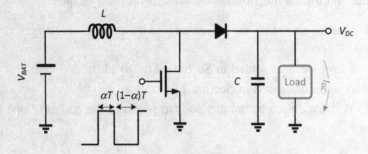

7. For the previous problem, suppose the load drains a constant current of I_{DC}. Plot the inductor and capacitor voltage and current waveforms. What is the average inductor current? Assume $\frac{T}{\sqrt{LC}} \ll 1$. **Answer:** The capacitor voltage discharges/charges almost linearly between $\frac{V_{BAT}}{1-\alpha} + \frac{I_{DC}}{C}\alpha T$ and $\frac{V_{BAT}}{1-\alpha}$. The inductor current is $i_L(t) \approx \begin{cases} \frac{I_{DC}}{1-\alpha} + \frac{V_{BAT}}{L}(t-\alpha T) \\ \frac{I_{DC}}{1-\alpha} - \frac{\alpha}{1-\alpha}\frac{V_{BAT}}{L}(t-\alpha T) \end{cases}$ for on/off modes. The inductor average current is $\frac{I_{DC}}{1-\alpha} - \frac{V_{BAT}}{2L}\alpha T \approx \frac{I_{DC}}{1-\alpha}$, as expected from power conservation.

8. Using Fourier series, find the fundamental component of C_1 voltage in a class E amplifier. By equating that to the voltage on the RLC series load, express the impedance of C_1 based on the load resistor. **Answer:** $a_1 = \left(\frac{2}{\pi} - \frac{\pi}{4}\right)\frac{I_{DC}}{C_1\omega_0}$, $b_1 = -\frac{1}{2}\frac{I_{DC}}{C_1\omega_0}$, and $\frac{1}{C_1\omega_0} = R_L\frac{1+\frac{\pi^2}{4}}{\frac{\pi}{2}}$.

9. Using energy conservation, show that in a class E power amplifier $R_L I_{DC} = \frac{2V_{DD}}{1+\frac{\pi^2}{4}}$, and $\frac{I_{DC}}{C_1\omega_0} = \pi V_{DD}$.

10. Show that in class E design, under optimum conditions we have $\frac{1}{C_2\omega_0} = L\omega_0\left(1 + \frac{\frac{\pi}{4}\left(1-\frac{\pi^2}{4}\right)}{Q_L}\right)$. Clearly C_2 and L do not resonate at ω_0. $Q_L = \frac{L\omega_0}{R_L}$ is the series circuit quality factor.

11. Calculate the peak voltage at the drain of a class E amplifier. **Answer:** $(\pi - 2\theta_0)\pi V_{DD} = 3.6 V_{DD}$.

12. Find the peak drain current of a class E amplifier.

13. In an IQ transmitter assume that prior to the up-conversion the quadrature baseband signals experience a memoryless nonlinearity given by $y = x + \beta x^3$. If the ideal output RF signal was to be $A\cos(\omega_0 t + \theta)$, derive the modified RF signal and prove the existence of undesired components proportional to the following: $A^3\cos(\omega_0 t + \theta)$ and $A^3\cos(\omega_0 t - 3\theta)$. **Answer:** $\hat{y}(t) = \left(A + \frac{3}{4}\beta A^3\right)\cos(\omega_0 t + \theta) + \frac{1}{4}\beta A^3\cos(\omega_0 t - 3\theta)$.

14. Consider the following source follower power amplifier. Assume V_{eff} to be effective voltages for the input transistor as well as the current source.
    a. Derive equation for the drain efficiency.
    b. Prove that the maximum efficiency approaches 25% if $V_{DD} \gg V_{eff}$.

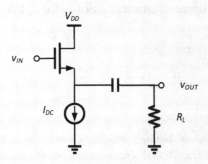

15. Consider the following common-source power amplifier with a resistive load, where V_{eff} is the FET overdrive voltage. Derive an equation for the drain efficiency, and show that the maximum efficiency approaches 25% if $V_{DD} \gg V_{eff}$.

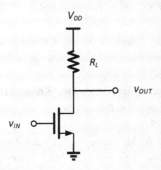

16. Consider the PA with dynamic gate biasing. Assume that gate is biased at $V_{TH} + f(A)$, where A is the envelope of the RF signal. To preserve the linearity the envelope of the input RF signal is predistorted given by $g(A)$, i.e. $g(A)\cos(\omega_{LO}t + \theta)$ is applied to the PA input. Assume that the NMOS transistor is modeled by the long channel equations. Find $g(A)$ such that the output remains the same as the case where the gate bias voltage is fixed at $V_{GS, Q}$. **Answer:** $\dfrac{A(V_{GS,Q} - V_{th})}{f(A)}$.

17. For the previous problem, assume $f(A)$ is chosen to be $\sqrt{(V_{GS,Q} - V_{th})A}$. Prove that ratio of the average currents drawn from the supply voltage between two dynamic and fixed biases for a give amplitude value A is $\dfrac{3A/2(V_{GS,Q} - V_{th})}{(V_{GS,Q} - V_{th})^2 + 1/2A^2}$. For a given probability density function $p(A)$ for the envelope of a modulated signal, derive equation for ratio of the average currents drawn from the supply voltage. **Answer:** $\dfrac{3/2(V_{GS,Q} - V_{th})E[A]}{(V_{GS,Q} - V_{th})^2 + 1/2E[A^2]}$.

18. The envelope of a white bandpass signal has a Rayleigh distribution $p(A) = (A/\sigma^2)e^{-A^2/2\sigma^2}$, where $p(A)$ is the probability density function, and A is the envelope. If this signal is passed once through a class A PA with an efficiency of $\eta = A^2/2V_{DD}^2$ and another time through a class B PA with an efficiency of $(\pi/4)A/V_{DD}$, calculate the average efficiency in both cases. Assume $\sigma \ll A_{max} = V_{DD}$, such that the limit of integrals can be taken from 0 to $+\infty$. **Answer:** $\overline{P_{out}} = \int_0^{+\infty} P_{out}(A)p(A)dA$, $\overline{P_{DC}} = \int_0^{+\infty} P_{DC}(A)p(A)dA$ and $\eta = \overline{P_{out}}/\overline{P_{DC}}$.

19. Find the voltage of PA1 as a function of the output voltage in a Doherty power amplifier. Assume a piecewise linear characteristics for PA1. **Answer:** $V_1 = -j\frac{Z_1}{R_0}V_2 = -j(2-A)V_2$.

11.12 References

[1] P. R. Gray and R. G. Meyer, *Analysis and Design of Analog Integrated Circuits*, John Wiley, 1990.

[2] E. Wilkinson, "An N-Way Hybrid Power Divider," *IRE Transactions on Microwave Theory and Techniques*, 8, no. 1, 116–118, 1960.

[3] I. Aoki, S. Kee, D. Rutledge, and A. Hajimiri, "Fully Integrated CMOS Power Amplifier Design Using the Distributed Active-Transformer Architecture," *IEEE Journal of Solid-State Circuits*, 37, no. 3, 371–383, 2002.

[4] I. Aoki, S. Kee, R. Magoon, R. Aparicio, F. Bohn, J. Zachan, G. Hatcher, D. McClymont, and A. Hajimiri, "A Fully-Integrated Quad-Band GSM/GPRS CMOS Power Amplifier," *IEEE Journal of Solid-State Circuits*, 43, no. 12, 2747–2758, 2008.

[5] P. Baxandall, "Transistor Sine-Wave LC Oscillators. Some General Considerations and New Developments," *Proceedings of the IEEE – Part B: Electronic and Communication Engineering*, 106, no. 16, 748–758, 1959.

[6] K. K. Clarke and D. T. Hess, *Communication Circuits: Analysis and Design*, Krieger, 1994.

[7] A. B. Carlson and P. B. Crilly, *Communication Systems: An Introduction to Signals and Noise in Electrical Communication*, vol. 1221, McGraw-Hill, 1975.

[8] D. Chowdhury, S. V. Thyagarajan, L. Ye, E. Alon, and A. M. Niknejad, "A Fully-Integrated Efficient CMOS Inverse Class-D Power Amplifier for Digital Polar Transmitters," *IEEE Journal of Solid-State Circuits*, 47, no. 5, 1113–1122, 2012.

[9] D. Chowdhury, L. Ye, E. Alon, and A. M. Niknejad, "An Efficient Mixed-Signal 2.4-GHz Polar Power Amplifier in 65-nm CMOS Technology," *IEEE Journal of Solid-State Circuits*, 46, no. 8, 1796–1809, 2011.

[10] C.-H. Lin and K. Bult, "A 10-b, 500-MSample/s CMOS DAC in 0.6 mm²," *IEEE Journal of Solid-State Circuits*, 33, no. 12, 1948–1958, 1998.

[11] S. Yoo, J. S. Walling, E. C. Woo, B. Jann, and D. J. Allstot, "A Switched-Capacitor RF Power Amplifier," *IEEE Journal of Solid-State Circuits*, 46, no. 12, 2977–2987, 2011.

[12] Z. Deng, E. Lu, E. Rostami, D. Sieh, D. Papadopoulos, B. Huang, R. Chen, H. Wang, W. Hsu, C. Wu, and O. Shanaa, "9.5 A Dual-Band Digital-WiFi 802.11a/b/g/n Transmitter SoC with Digital I/Q Combining and Diamond Profile Mapping for Compact Die Area and Improved Efficiency in 40nm CMOS," in *Proceedings of the IEEE International Solid-State Circuits Conference (ISSCC)*, 2016.

[13] R. Winoto, A. Olyaei, M. Hajirostam, W. Lau, X. Gao, A. Mitra, O. Carnu, P. Godoy, L. Tee, H. Li, E. Erdogan, A. Wong, Q. Zhu, T. Loo, F. Zhang, L. Sheng, D. Cui, A. Jha, X. Li, W. Wu, K. Lee, D. Cheung, K. W. Pang, H. Wang, J. Liu, X. Zhao, D. Gangopadhyay, D. Cousinard, A. A. Paramanandam, X. Li, N. Liu, W. Xu, Y. Fang, X. Wang, R. Tsang, and L. Lin, "9.4 A 2×2 WLAN and Bluetooth Combo SoC in 28nm CMOS with On-Chip WLAN Digital Power Amplifier, Integrated 2G/BT SP3T Switch and BT Pulling Cancelation," in *Proceedings of the IEEE International Solid-State Circuits Conference (ISSCC)*, 2016.

[14] H. Wang, C. Peng, Y. Chang, R. Z. Huang, C. Chang, X. Shih, C. Hsu, P. C. P. Liang, A. M. Niknejad, G. Chien, C. L. Tsai, and H. C. Hwang, "A Highly-Efficient Multi-band Multi-mode All-Digital Quadrature Transmitter," *IEEE Transactions on Circuits and Systems I: Regular Papers*, 61, no. 5 1321–1330, 2014.

[15] N. Sokal and A. Sokal, "Class E-A New Class of High-Efficiency Tuned Single-Ended Switching Power Amplifiers," *IEEE Journal of Solid-State Circuits*, 10, no. 3, 168–176, 1975.

[16] R. Zulinski and J. W. Steadman, "Class E Power Amplifiers and Frequency Multipliers with Finite DC-Feed Inductance," *IEEE Transactions on Circuits and Systems*, 34, no. 9, 1074–1087, 1987.

[17] S. Kee, I. Aoki, A. Hajimiri and D. Rutledge, "The Class-E/F Family of ZVS Switching Amplifiers," *IEEE Transactions on Microwave Theory and Techniques*, 51, no. 6, 1677–1690, 2003.

[18] L. Kahn, "Single-Sideband Transmission by Envelope Elimination and Restoration," *Proceedings of the IRE*, 40, no. 7, 803–806, 1952.

[19] B. Sahu and G. Rincon-Mora, "System-Level Requirements of DC-DC Converters for Dynamic Power Supplies of Power Amplifiers," in *ASIC 2002 Proceedings*, 2002.

[20] W. Doherty, "A New High Efficiency Power Amplifier for Modulated Waves," *Proceedings of the Institute of Radio Engineers*, 24, no. 9, 1163–1182, 1936.

[21] R. Erickson and D. Maksimovic, *Fundamentals of Power Electronics*, Springer, 2001.

12 Transceiver Architectures

By now we have a general idea of what the receivers and transmitter look like. Nonetheless, the exact arrangement of the blocks, the frequency planning involved, the capabilities of digital signal processing, and other related concerns result in several different choices that are mostly application dependent. While our general goal is to ultimately meet the standard requirements, arriving at the proper architecture is largely determined by the cost and power consumption concerns. The goal of this chapter is to highlight these trade-offs, and present the right architecture for a given application, considering the noise, linearity, and cost trade-offs, which were generally described in Chapters 5 and 6. Moreover, we will see that the proper arrangement of the building blocks is a direct function of the circuit capabilities that we presented in Chapters 7–11.

We start this chapter with general descriptions of the challenges and concerns when realizing RF transceivers. We then offer a detailed description of both receivers and transmitter architectures, and present several case studies. At the end of the chapter, a practical transceiver design example is presented, along with some discussion on production-related issues, packaging requirements, and integration challenges.

The specific topics covered in this chapter are:

- Transceiver general considerations
- Super-heterodyne receivers
- Zero- and low-IF receivers
- Quadrature downconversion and image reject receivers
- Dual-conversion receivers
- Blocker tolerant receivers
- ADC, filtering, and gain control in receivers
- Linear transmitters
- Direct-modulated and polar transmitters
- Out-phasing transmitters
- Transceiver case study
- Packaging and product qualification
- Production-related concerns

For class teaching, we recommend covering only Sections 12.1, 12.2.1, 12.2.2, 12.2.3, and 12.6.1. A brief introduction to nonlinear transmitters (Sections 12.6.3 and 12.6.4) may be also very helpful, if time permits. The remaining sections, and particularly Section 12.7, are easy to follow and may be appealing to practicing RF engineers.

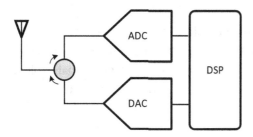

Figure 12.1: An ideal transceiver comprising a pair of data converters

12.1 GENERAL CONSIDERATIONS

Since almost all the radios today rely on digital modulation, the transceivers inevitably require a very sophisticated digital processing unit. On the other hand, the electromagnetic waves transmitted and received are analog in nature. Hence we expect our transceivers to at least include a pair of data converters, as shown in Figure 12.1.

This architecture in fact has been proposed recently in the context of *software-defined radios*, that is, radios that are capable of receiving or transmitting any standard at any desired frequency [1]. Despite being wideband in nature and very flexible, the problem with this idealistic approach is the stringent blocker and mask requirements for the receiver and transmitter, respectively.

Example: Consider Figure 12.2, where it shows the desired receiver signal at −99dBm is accompanied by in- and out-of-band blockers that are as large as −23/0dBm, respectively, as specified in GSM.

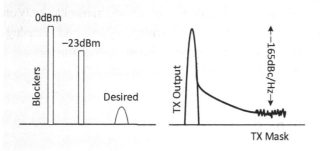

Figure 12.2: Receiver blocker and transmitter mask requirements

To receive the signal successfully at −99dBm, let us assume that the ADC quantization noise needs to be at least 10dB below the signal. Assuming that the ADC needs to receive up to 2GHz to cover various bands (and thus it requires a minimum clock frequency of 4GHz), then the ADC noise level needs to be at

$$-99\text{dBm} - 10 + 10\log\left(\frac{2\text{GHz}}{200\text{kHz}}\right) = -69\text{dBm},$$

given that the GSM signal is 200kHz wide. On the other hand, to be able to tolerate a 0dBm blocker with say 3dB of margin, the ADC full scale comes out to be +3dBm. Thus

Continued

> the dynamic range needed is 72dB, or roughly 12bits at 4GHz sampling frequency. This kind of requirement leads to a substantially large power consumption in the ADC alone.
>
> Even if a front-end filter is incorporated to attenuate the out-of-band blocker, the receiver still needs to take −23dBm in-band blockers, and an 8-bit 4GHz ADC is needed that is quite power hungry.

Similarly, relatively stringent requirements are needed to satisfy the far-out mask requirements if one chooses to use a DAC to interface the output directly.

> **Example:** Assume an out-of-band noise level of −165dBc/Hz is required for the transmitter to avoid desensitizing the nearby receivers. Assuming a clock frequency of 4GHz, the noise floor is
>
> $$-165\text{dBc/Hz} + 10\log(2\text{GHz}) = -72\text{dBc}.$$
>
> Depending on the application, the DAC dynamic range would be
>
> $$\text{Dynamic range} = 72 + \text{PAPR},$$
>
> where PAPR is the transmitted signal peak-to-average power. For a 64QAM WLAN transmitter, PAPR ≈ 10dB, and roughly a 14-bit DAC is needed. The DAC dynamic range is far better than what is needed to guarantee the EVM requirements, and is mostly limited by the OOB noise spec. For instance, to achieve an EVM of −50dBc, a 9-bit DAC suffices.

Given these challenges, as we pointed out before, both receivers and transmitters rely on some kind of frequency translation, often provided by the means of a mixer to perform the analog and digital signal processing much more efficiently at lower power consumption. With this in mind, let us take a closer look at various choices to realize the RX and TX architecture.

12.2 RECEIVER ARCHITECTURES

From our discussion before, we expect a given receiver to look something like what is illustrated in Figure 12.3, where the need for the building blocks shown is justified as follows:

- An optional front-end filter to suppress the out-of-band blockers
- The LNA to suppress the noise of following stages
- The mixer[1] to provide the frequency translation, and to ease the burden on the following blocks

---

[1] Although not explicitly shown, throughout this chapter we assume the downconversion mixer output employs some kind of modest lowpass filtering (say an RC stage) to remove the component at the sum frequency.

12.2 Receiver Architectures

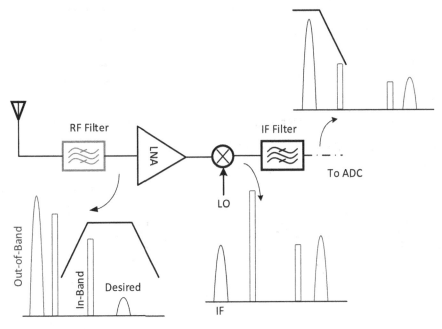

Figure 12.3: A basic receiver topology

- Partial or full filtering (or channel selection) at IF to attenuate the downconverted blockers, mostly in-band if an RF filter is used, and thus relax the subsequent stages linearity

After proper amplification and filtering, the signal may be passed to a relatively low-frequency and low-power ADC for digitization.

Figure 12.3 also shows how the out- and in-band blockers are progressively filtered throughout the receive chain, while the desired signal is subject to amplification.

Invented by Armstrong,[2] this architecture is known as a super-heterodyne[3] receiver [2], and has been around since as early as 1918. We shall have a closer look in the next section and discuss its pros and cons.

12.2.1 Super-Heterodyne Receiver

While downconversion to IF eases the power consumption requirement of the following stages, it introduces several issues of its own. As we showed in Chapter 6, there are several blockers that are subject to downconversion as well, and given the harmonically rich nature of the mixer LO, as well as nonlinearities present in the chain, this leads to SNR degradation [3]. Consider the image and half-IF blockers in a GSM[4] receiver shown in Figure 12.4. The GSM signal may be anywhere from 925 to 960MHz. In the extreme case of the signal at the edge of the band, say at 960MHz with low-side injection, the image resides at the lower end as shown in Figure 12.4. The SAW filter stopband typically starts at 20MHz away from the edge of the passband, which

---

[2] Edwin Armstrong (1890–1954) was an American electrical engineer and inventor, best known for developing FM radio and the super-heterodyne receiver system.
[3] Heterodyning is a radio signal processing technique invented by Canadian engineer R. Fessenden in which a new frequency is generated by combining (or mixing) two frequencies. The word *heterodyne* is Greek and means *different-power*.
[4] The reason to choose GSM in many of our examples has to do with its very stringent blocking requirements.

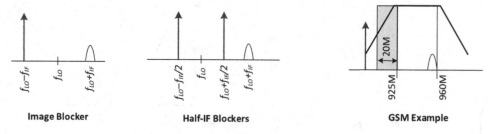

Figure 12.4: Image and half-IF blockers in a GSM receiver

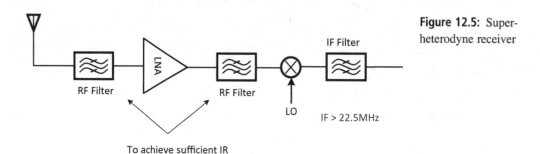

Figure 12.5: Super-heterodyne receiver

is at 925MHz. Thus, to achieve an appreciable rejection on the image, it must be at least 55MHz (960 − 925 + 20) away, and thus the IF must be at least 55/2 = 22.5MHz. Depending on the receiver 2nd-order nonlinearity and LO harmonics, if half-IF blockers are dominant, a much higher IF is needed. For instance, for the one closer to the signal, the IF must be at least 55×2 = 110MHz to force the blocker to reside in the filter stopband.

In the example of GSM, out-of-band blockers are at 0dBm, whereas the desired signal is at −99dBm, and thus to achieve a reasonable SNR, a total rejection of over 105dB is needed. As the external filters provide a typical rejection of 40–50dB, often two filters are needed, which along with the LNA modest filtering, an adequate rejection may be provided. As a result, a typical super-heterodyne receiver looks like what is shown in Figure 12.5, where the IF is typically tens or hundreds of MHz.

Of the two RF filters needed, one is usually placed *before* the LNA to adequately attenuate the 0dBm blockers and thus relax the LNA linearity, whereas the other one is placed *after* the LNA. Otherwise, the *noise figure degradation* due to the loss of two filters may be unacceptable. The two RF filters are inevitably external, but so is the IF filter, which is centered at several tens of MHz. Since the GSM signal is 200kHz wide, a >22.5MHz IF filter with such a narrow bandwidth cannot be realized on-chip, as we discussed in Chapter 4. Thus a super-heterodyne receiver, despite its robustness, is too expensive to implement as it requires several external filters. Moreover, since these filters are generally 50Ω matched, driving them externally requires somewhat power hungry amplifiers.

12.2.2 Zero-IF Receivers

An alternative approach to relax the filtering and consequently to reduce the cost of a super-heterodyne receiver is to set the IF at zero, in which case the image or other problematic

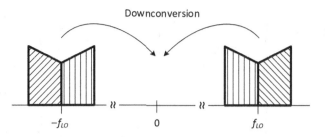

Figure 12.6: Image problem in a zero-IF receiver

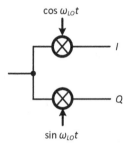

Figure 12.7: Quadrature receiver to reject one sideband

blockers will be of a less concern.[5] This, however, creates a new problem as the signal is now its own image. As we established before, almost all RF signals rely on single-sideband spectrum generation. Thus, at either sideband of the carrier there exists uncorrelated spectrums, that even though they may look the same, they are not. As shown in Figure 12.6, when such signal is downconverted to zero IF, the negative and positive sides fold and fall on each other.

A similar scheme as a transmitter, where quadrature LO signals are used to upconvert the baseband components, is applicable to receivers. Shown in Figure 12.7, a quadrature downconverter consisting of I and Q branches clocked with LO signals 90 out of phase is capable of distinguishing high or low sidebands.

Suppose the signal of interest is a tone residing on the high side of the LO, as shown in Figure 12.8. The frequency spectrum of the I and Q LO signals is shown as well. Multiplication in the time domain results in *convolution* in the frequency domain, and leads to the downconverted spectrum shown on the right side.

If the signal resides on the low-side, the resultant spectrum is depicted in Figure 12.9.

Comparing Figure 12.8 and Figure 12.9, it is evident that even though the downconverted spectrums of both high- and low-side signals reside at f_m, the phase contents are not identical. This could be taken advantage of to discriminate the two sides.

In Chapter 2 we showed that applying Hilbert transform to a signal results in multiplication of the positive frequency components by $-j$, and those of negative frequencies by $+j$. We also showed that while in principle Hilbert transform is a noncasual operation, it may be well approximated over a reasonable frequency range. Shown in Figure 12.10, by applying Hilbert transform on the Q output, and adding or subtracting that from the I channel, only one sideband

---

[5] Zero-IF receivers are also called homodyne or direct-conversion.

Transceiver Architectures

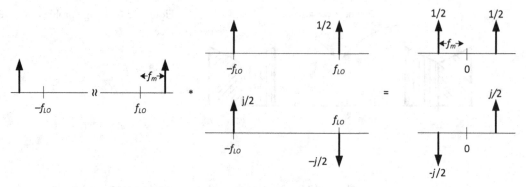

Figure 12.8: High-side signal downconverted by quadrature LOs

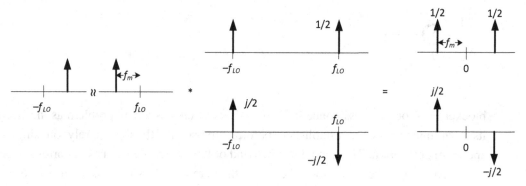

Figure 12.9: Low-side signal downconverted by quadrature LOs

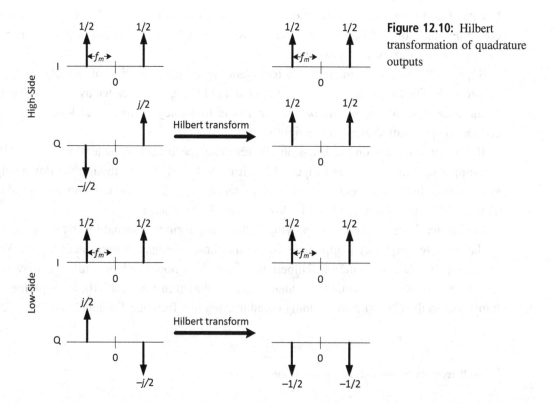

Figure 12.10: Hilbert transformation of quadrature outputs

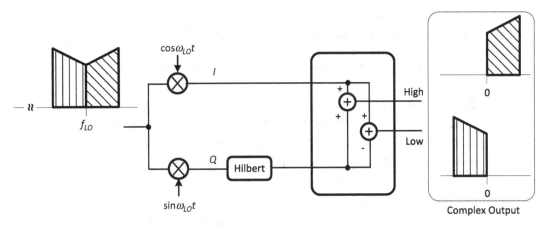

Figure 12.11: Complex output of a quadrature zero-IF receiver

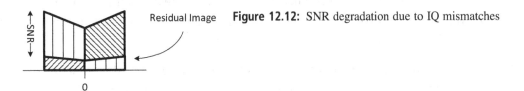

Figure 12.12: SNR degradation due to IQ mismatches

may be selected. This is exactly the opposite of single-sideband selection in transmitters we showed in Chapter 2.

A more complete block diagram of the quadrature receiver used to detect the zero-IF signal is shown in Figure 12.11. The Hilbert transform and other processing involved is typically performed in digital domain where the I and Q downconverted spectrums are passed to the modem after the ADC.

It must be noted that the quadrature receivers as shown in Figure 12.11 are not unique to zero-IF applications. Such a scheme may be used to reject the image if the IF is not zero. However, while a necessity in a zero-IF receiver, quadrature downconversion may be replaced by RF filtering in a super-heterodyne receiver as long as the IF is high enough.

One drawback of the quadrature receiver may seem to be the need for duplicating the entire IF path. This, however, is not true, as the signals on the I and Q outputs eventually add constructively, while the noise of the IF chain blocks are uncorrelated. Thus, for the same total power and area as a single path (that is, half the power and area for each of the building blocks of the I and Q paths), the overall SNR remains the same. Moreover, as the image is rejected *after* the downconversion, DSB noise figure applies to the mixer when calculating the overall receiver noise. However, quadrature LO generation does come at a cost, as typically the techniques mentioned in Chapter 2 result in some power consumption overhead.

If there is some I and Q imbalance either in the LO or in the receive chain, effectively one undesired sideband appears at the other and vice versa. This is shown in Figure 12.12, and is a source of SNR degradation.

In high throughput applications such as WLAN or LTE, as a large SNR is required, this may impose some challenges. For instance, in 802.11ac application, an SNR of better than 40dB over a wide bandwidth of ±80MHz is required, which is typically not feasible given the mismatches present in integrated circuits. To circumvent this, often an IQ calibration scheme is needed.

In addition to quadrature generation and IQ accuracy concerns, there are several other issues that must be dealt with when using zero-IF receivers [4]:

- As the signal is now located at zero IF, any low-frequency noise or interference directly competes with the desired signal. This includes the flicker noise, as well as any 2nd-order distortion. As we discussed in Chapter 5, $1/f$ noise may not as problematic in wider band applications (such as 3/4G or WLAN), where the noise is integrated over a wide bandwidth. Still, 2nd-order distortion is a concern (for instance in a full-duplex radio such as an LTE transceiver), and must be dealt with using proper circuits or through calibration (see [5] for an example of IP_2 calibration).
- LO self-mixing: We showed in Chapter 8 that RF self-mixing is a source of 2nd-order distortion. Similarly, LO self-mixing results in an unwanted DC offset at the zero-IF mixer output. The value of the offset is channel dependent, as the amount of LO leakage varies with frequency. This along with the other static DC offsets present in the chain due to mismatches as well as the DC offset created due to the 2nd-order nonlinearities must be removed using one of the known DC offset calibration schemes.

Fortuitously, most of the aforementioned problems may be avoided by employing calibration circuitry that leverages on the strong digital signal processing available in CMOS radios today. A complete zero-IF receiver with several calibration circuitry involved is shown in Figure 12.13.

Since the signal is located at DC, only low-Q lowpass filters are needed for channel selection, which leads to a lower cost and power receiver. The DC offset as well as the 2nd-order distortion may be monitored in the digital domain, and corrected in the radio. The quadrature inaccuracy may be also compensated for in the digital domain, leading to a fairly robust performance at a low power consumption. Still the $1/f$ noise may be a concern, and thus this architecture may not be as suitable for narrowband applications such as GSM where the signal is only $\pm 100\text{kHz}$ wide.

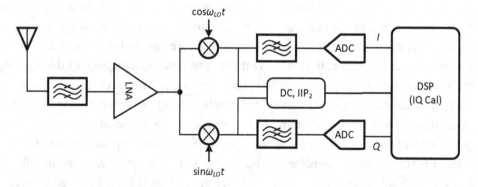

Figure 12.13: Zero-IF receiver with enhanced performance using calibration

12.2 Receiver Architectures

Example: Consider a zero-IF receiver where a desired signal located at ω_0 accompanied by a modulated blocker at an offset frequency of $\Delta\omega_B$ is down-converted to zero IF by a noisy LO. We wish to find the receiver noise due to the reciprocal mixing of the modulated blocker with the LO phase noise.

We assume the blocker is stationary, which is true for most practical cases. Furthermore, the blocker offset frequency is assumed to be large compared to its bandwidth, which again is a fair assumption for any typical out-of-band blocker. The blocker modulation is however assumed to be arbitrary.

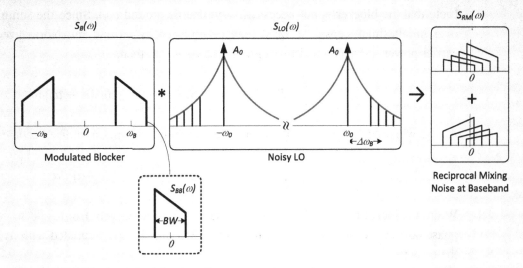

Figure 12.14: Reciprocal mixing in a direct-conversion receiver

From chapter 9 the LO may be expressed as an infinite sum of noise sinusoids that are uncorrelated in phase and separated in frequency by 1Hz:

$$x_{LO}(t) = A_0 \cos \omega_0 t - \left(\sum_{k=0}^{\infty} \sqrt{4S_\phi(2\pi k)} \cos(2\pi kt + \theta_k) \right) A_0 \sin \omega_0 t$$

where $S_\phi(2\pi k)$ is the phase noise per unit Hertz at an offset frequency of k Hz from the carrier, and θ_k is the uncorrelated random phase with a uniform distribution.

The reciprocal mixing noise component may be then expressed as:

$$x_{RM}(t) = -\frac{A_0}{2} \left(\sum_{k=0}^{\infty} \sqrt{4S_\phi(2\pi k)} [\sin(\omega_0 t + 2\pi kt + \theta_k) + \sin(\omega_0 t - 2\pi kt - \theta_k)] \right) x_B(t)$$

Continued

where $x_B(t)$ is the modulated blocker. Therefore, as depicted in Figure 12.14, in frequency domain replicas of the blocker are *convolved* (or downconverted to baseband) with each of the noise impulses and folded on top of each other. Knowing that the sum frequency components are lowpass filtered at the mixer output, the reciprocal mixing noise spectral density around baseband is:

$$S_{RM}(\omega) = \frac{A_0^2}{4} \sum_{k=-\infty}^{\infty} S_\phi(2\pi|k|)[S_{BB}(\omega - \Delta\omega_B + 2\pi k) + S_{BB}(-\omega - \Delta\omega_B + 2\pi k)]$$

where $S_{BB}(\omega)$ is the blocker *baseband* spectral density replica as shown in Figure 12.14 (note that the blocker is not necessarily symmetric around ω_B). Since the summation is over infinitesimal steps of 1Hz, it may be replaced by an integral. Normalized to the carrier power, the reciprocal mixing noise ($\mathcal{L}_{RM}(\omega)$) is then:

$$\mathcal{L}_{RM}(\omega) = \int_{-\infty}^{\infty} S_\phi(|\alpha + \Delta\omega_B|)[S_{BB}(\omega + \alpha) + S_{BB}(\omega - \alpha)]d\alpha$$

At DC where the desired signal is centered, the reciprocal mixing noise becomes:

$$\mathcal{L}_{RM}(0) = \int_{-\infty}^{\infty} S_\phi(|\alpha + \Delta\omega_B|)[S_{BB}(\alpha) + S_{BB}(-\alpha)]d\alpha$$

With the blocker bandwidth of BW, the integration is effectively from $-\frac{BW}{2}$ to $\frac{BW}{2}$. If the phase noise is assumed to stay flat in this range, it may be approximated with its value at the center:

$$S_\phi(|\alpha + \Delta\omega_B|) \approx S_\phi(\Delta\omega_B)$$

Thus:

$$\mathcal{L}_{RM}(0) \approx S_\phi(\Delta\omega_B) \int_{-\infty}^{\infty} [S_{BB}(\alpha) + S_{BB}(-\alpha)]d\alpha$$

On the other hand, by definition, the integral represents the blocker average power:

$$P_B = \int_{-\infty}^{\infty} [S_{BB}(\alpha) + S_{BB}(-\alpha)]d\alpha$$

Therefore:

$$\mathcal{L}_{RM}(0) = S_\phi(\Delta\omega_B)P_B$$

which is in agreement with the result obtained in chapter 6. To the 1st order, the noise is only a function of the blocker average power, and the LO phase noise at the blocker offset frequency, but independent of the blocker modulation.

12.2.3 Low-IF Receivers

For narrower band applications where the low-frequency noise or distortion may be too difficult to cope with, a compromise may be made to set the IF to some higher value, high enough to push the signal well above the problematic noise or distortion. This may then sound exactly like the super-heterodyne receiver, yet there is fundamental difference between the low-IF receiver as we will describe shortly, and a heterodyne radio.

In a super-heterodyne receiver, to attenuate the image and other problematic blockers the IF is intentionally high enough to locate those blockers outside the passband of the RF filter. This results in (1) a very demanding requirement for image rejection as the out-of-band blockers are very strong; (2) higher cost due to the need for external filters, particularly for IF channel selection. In contrast, in a low-IF receiver, the IF is low enough to ensure the image is in-band, and thus not only an integrated IF filter becomes feasible, but the image rejection requirement is also more manageable. However, the image may now be removed only through quadrature downconversion, whose principles where described previously. Thus, we expect a low-IF receiver to conceptually look as shown in Figure 12.15.

This type of receiver architecture is generally known as a *Hartley image-reject receiver* [6], and is obviously suitable for low-IF applications. It may also be incorporated in a more conventional heterodyne receiver to improve the image rejection, and thus relax the external filtering, and in consequence, the cost.

As we showed in Figure 12.10, the Hilbert transform and the subsequent addition/subtraction results in only one sideband to be selected, and thus the image, located on the other sideband, is entirely canceled in the absence of mismatches. In practice the mismatches set an upper limit on the image rejection feasible. It is instructive to quantify the amount of image rejection in terms of the I and Q accuracy. Let us assume the I channel LO is $\cos \omega_{LO} t$, whereas the Q channel LO is $(1-\alpha)\sin(\omega_{LO}t + \beta)$. Ideally both α and β are zero, but in practice they are not due to mismatches. Although there may be mismatches in any of the other blocks of the receiver, for simplicity we have assumed they are all lumped in the LO. This is a good approximation, especially for the gain mismatches. Note that α and β may be a function of both RF (due to LO), and IF (due to post-mixer blocks) frequencies. Assuming the signal applied to the input is $\cos(\omega_{LO} \pm \omega_m)t$, we can easily show that the following signals appear at the output of receiver of Figure 12.15:

$$\text{Desired sideband}: \cos(\omega_m t) + (1-\alpha)\cos(\omega_m t - \beta)$$
$$\text{Image sideband}: \cos(\omega_m t) - (1-\alpha)\cos(\omega_m t + \beta)$$

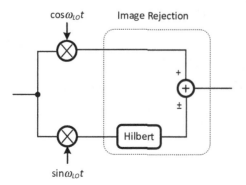

Figure 12.15: Low-IF receiver basic operation

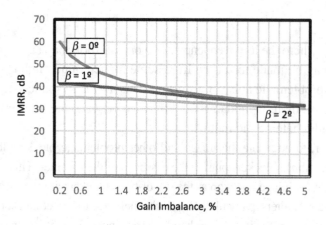

Figure 12.16: Image rejection in a low-IF receiver versus gain and phase imbalance

In the absence of mismatches, $\alpha=\beta=0$, and the second terms representing the unwanted sideband disappears. For nonzero values of α and β, the signal at the image side is not entirely canceled, and its relative amplitude with respect to the main component is

$$\sqrt{\frac{\alpha^2 + 4(1-\alpha)\left(\sin\frac{\beta}{2}\right)^2}{\alpha^2 + 4(1-\alpha)\left(\cos\frac{\beta}{2}\right)^2}}.$$

A plot of image rejection ratio in dB versus gain imbalance (α) is shown in Figure 12.16 for three different values of phase imbalance (β).

Typical values of gain and phase imbalance achievable in modern CMOS processes limit the *uncalibrated* image rejection to around 30–40dB.

The amount of achievable image rejection sets an upper limit on the acceptable IF. This has to do with adjacent blockers profile, and the fact that they progressively become stronger as the frequency offset from the desired signal increases.

Example: To understand this trade-off better, consider a GSM receiver where the image is one of the adjacent blockers located at an integer multiple of 200kHz away from the desired signal (Figure 12.17).

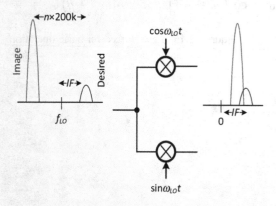

Figure 12.17: Image blocker in a GSM low-IF receiver

After downconversion, the blocker folds on top of the desired signal and is located at

$$n \times 200\text{kHz} - IF,$$

away from the desired signal. The GSM adjacent blockers for $n = 1, 2$, and 3 are shown in Figure 12.18. Shown on the right is the amount of image reception ratio (IMRR) as a function of IF.

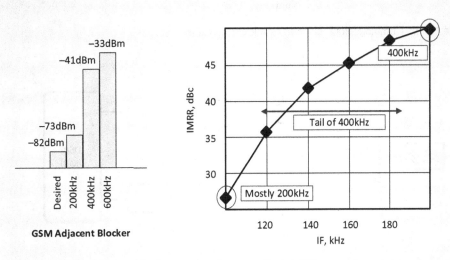

Figure 12.18: Adjacent blockers setting the upper limit of IF

Consider the case where IF is 100kHz. Then the image is the first adjacent blocker, which is only 9dB stronger, leading to a relaxed image rejection requirement. On the other hand, if the IF is 200kHz, the image is the second adjacent blocker, which is 41dB stronger, leading to a stringent requirement of about 50dB. For values in between, the tail of the second adjacent blocker folding over the signal is still large enough to cause a steep increase in image rejection. Most practical GSM receivers use an IF of about low 100kHz, which is still advantageous with respect to noise or 2nd-order distortion, but not too high to demand an impractical image rejection. From the two graphs shown in Figure 12.16 and Figure 12.18, we conclude that unless a digital enhancement is employed, the IF may not exceed 130–140kHz.

Rather than using Hilbert transform, a practical low-IF receiver may be realized by polyphase filtering as was discussed in Chapter 4 [7]. The polyphase filter rejects the unwanted sideband, and may be combined with the following active IF filter. Accordingly, an *active polyphase* filter makes a popular choice for the low-IF receivers. A complete schematic of a low-IF receiver is shown in Figure 12.19. Similar to zero-IF receivers, the RF filter, though it does not provide any image rejection, may be still needed to attenuate some of the large out-of-band blockers.

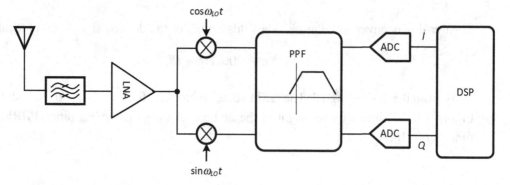

Figure 12.19: Complete block diagram of a low-IF receiver

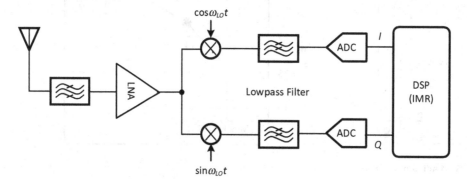

Figure 12.20: An alternative low-IF receiver

There may be yet a better way of realizing the low-IF receiver, if the ADC has high enough dynamic range to tolerate the unfiltered image blocker. The main disadvantage of the low-IF receiver shown in Figure 12.19 is that after passing through the polyphase filter, the image and signal cannot be distinguished anymore. Hence, applying any digital correction afterward is not feasible, and the amount of image rejection is set by what is achievable given the RF and analog inaccuracies. Shown in Figure 12.20, the receiver may employ only *real lowpass filters*, and thus keeping the I and Q paths separate.

The advantage is that the image is entirely canceled in the digital domain, and thus image rejection may be enhanced digitally. However, the image blocker is subject to very little filtering. For instance, suppose the IF is set at 135kHz where an image rejection of about 40dB is needed. Since the GSM signal is ± 100kHz, the lowpass filter passband must be at least 235kHz not to affect the desired signal. The 400kHz adjacent blocker, the one residing on the image sideband, once downconverted, appears at $400 - 135 = 265$kHz, which is subject to almost no filtering. This puts some burden on the ADC design. We shall discuss the trade-offs involved with ADC and IF filtering shortly.

By comparison, the low-IF receiver shown in Figure 12.20 is almost identical to the zero-IF receiver of Figure 12.13. All digital enhancements for IQ correction, DC offset removal, and

IIP$_2$ improvement are applicable to both architectures. The only difference lies in the choice of LO, and whether the desired signal is placed directly at DC, or at a slightly higher frequency. The latter may be critical for narrower band applications such as GSM, GPS, or Bluetooth. The receiver is completely reconfigurable, however, and can support zero-IF architecture if the signal is wide enough, for instance if the radio switches to 3G or LTE modes.

12.2.4 Weaver Receiver

An alternative approach to provide on-chip image rejection is known as the Weaver image-reject receiver [8], and is shown in Figure 12.21.

If the input is located at ω_0, assuming low-side injection for both stages of downconversion, we must have $\omega_0 = \omega_{LO1} + \omega_{IF1} = \omega_{LO1} + (\omega_{LO2} + \omega_{IF2}) = \omega_{LO1} + \omega_{LO2} + \omega_{IF2}$. The other combinations of high- and low-side injection may be treated similarly and are covered in Problem 9.

From our previous discussion, we can imagine that the second quadrature downconversion is effectively providing the Hilbert transform (see Problem 12). It is then easy to conceive that the image is subject to cancellation at the second IF when the I and Q sides are added (or subtracted for high-side injection). The signal and image progression along the receive chain are shown in Figure 12.22.

Unfortunately, unless the second IF is zero,[6] the Weaver receiver suffers from the issue of the second image. We can see that (Figure 12.22) the second image after the *first down-conversion* is located at

$$\omega_{Image2} = \omega_{IF1} - 2\omega_{IF2} = -\omega_0 + \omega_{LO1} + 2\omega_{LO2}.$$

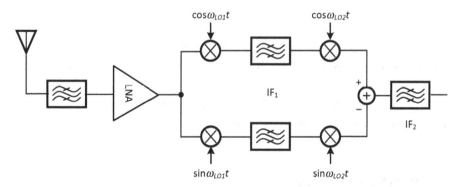

Figure 12.21: Weaver image-reject receiver

---

[6] Setting the second IF to zero requires a quadrature second downconversion to create I and Q signals in the second IF. See also Problem 10.

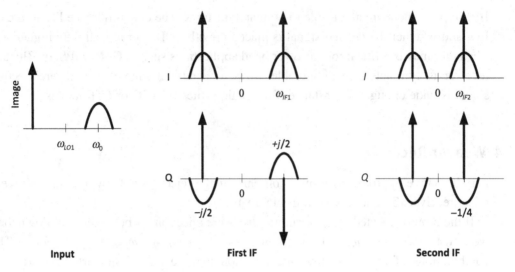

Figure 12.22: Signal and image progression in Weaver receiver

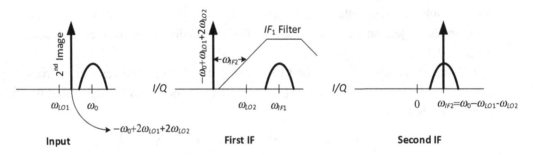

Figure 12.23: Weaver receiver second image problem

Or equivalently, it resides at the input $-\omega_0 + 2\omega_{LO1} + 2\omega_{LO2}$. As shown in Figure 12.23, the second image may be removed only by filtering at the first IF (or at the receiver input if it is far enough).

Due to this issue, along with the need for two stages of downconversion in the signal path, the Weaver receiver is not as commonly used as the low-zero IF receivers.

12.2.5 Dual-Conversion Receivers

Another compromise to the zero-IF receiver may be made if the receiver employs two steps of downconversion [9], as shown in Figure 12.24.

The main advantage is that as we discussed in Chapter 8, the mixer related issues such as $1/f$ noise or 2nd-order distortion directly improve if the relative slope of the LO is increased, or equivalently, if a lower frequency LO is used. At the first IF, where the LO frequency is high, the low-frequency noise and distortion are not important, whereas at the second IF, which may be chosen to be zero or very low, the aforementioned problems are expected to improve thanks

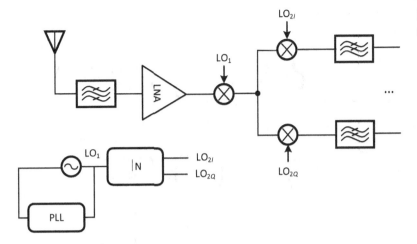

Figure 12.24: A dual-conversion receiver

to lower frequency LO. Moreover, only the second LO needs to be quadrature, and is readily created by dividing down the first single-phase LO signal. Consequently, the receiver requires only one VCO/PLL. Using a divide by $N = 2^n$ (to produce quadrature signal), we have

$$\text{IF}_1 = \frac{f_{RF}}{N \pm 1}.$$

The appearance of $\pm$ in the denominator is whether we use low- or high-side LO. This kind of frequency planning results in the first IF to vary with the RF channel, and hence is known as a *sliding* IF receiver.

The main disadvantages of this scheme are introducing an additional mixer in the main receiver path, as well as the image issue for the first IF. For this reason the first IF is typically very large, limiting the choice of divider usually to $N = 2$ or 4. The second IF image is not critical if the second IF is very low or zero.

12.3 BLOCKER-TOLERANT RECEIVERS

In almost all the architectures we have discussed so far, the existence of RF filter seems inevitable, particularly due to the very large out-of-band blockers. For instance, in the case of GSM, blockers as large as 0dBm could easily compress the receiver, and consequently desensitize it. In recent years tremendous work [10], [11], [12], [13], [14] has been done to relax or remove the RF filtering, and we will briefly discuss some of these topologies. In all the presented architectures, the choice of IF, whether zero or not, is arbitrary, but they all rely on quadrature downconversion as the RF filter is intended to be removed.

12.3.1 Current-Mode Receivers

In Chapter 8, when discussing the passive mixer IIP_3, we briefly introduced the concept of current-mode receivers. A current-mode receiver consists of a low-noise transconductance

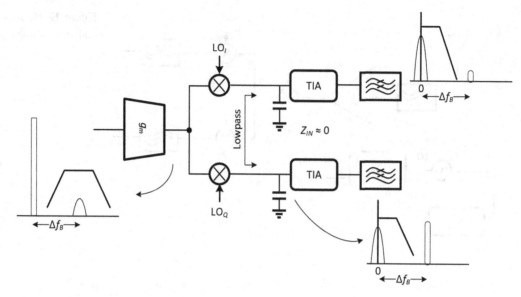

Figure 12.25: Current-mode receivers

amplifier (LNTA), followed by passive mixers that operate in current domain, as shown in Figure 12.25 [15].

Relying on impedance transformation properties of the passive mixers, this scheme provides two critical features:

- If the TIA input impedance that the mixer switches drive is low, and that the mixer switch resistance is low as well, the LNTA is loaded by a very low-impedance circuit, and hence the swing at its output is very low.
- A lowpass roll-off of the TIA translates into a high-Q bandpass filter appearing at the LNTA output.

As a result, we expect good linearity at the RF front end, and progressive filtering of the blockers along the receive chain. Once downconverted, the blocker is several tens of MHz away, and is filtered by the TIA or the subsequent IF filter if one is needed. A large capacitor at the TIA input helps attenuate the blocker given that the TIA input impedance tend to increase at higher frequencies.

Example: As a case study, let us consider a multimode, multiband receiver designed for 3G applications [16] as shown in Figure 12.26. The receiver consists of two sets of LNTAs, one for low bands (869–960MHz), and one for high bands (1805–2170MHz). The LNTA outputs are combined and drive a quadrature current mode passive mixer, whose LO is created by dividing the master clock at around 4GHz by either 2 or 4. The mixer output consist of a common-gate TIA which provides a 1st-order roll-off (set at 270kHz or 2.2MHz, for 2/3G modes), followed by a biquad resulting in a 3rd-order lowpass response.

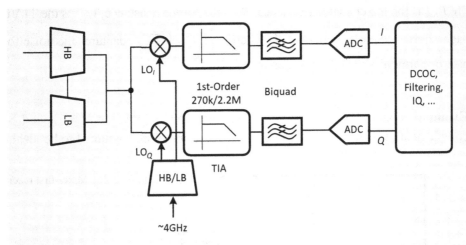

Figure 12.26: An example of a 3G receiver

The receiver achieves a noise figure of better than 2.5dB for all the bands, and the out-of-band IIP_3 is about −2dBm, sufficient for 3G applications.

12.3.2 Mixer-First Receivers

Even though the current mode receivers improve the linearity substantially, external filtering (directly or through the duplexer) may still be required to tolerate the 0dBm GSM blocker. Another proposal to boost the receiver linearity and consequently remove the RF filter is to dispense with the low-noise amplifier entirely [10], [14]. A simplified schematic of such a receiver using an *M*-phase mixer is shown in Figure 12.27. Such a receiver would clearly result in substantial improvement of linearity, but the noise figure is expected to suffer.

Based on the analysis we offered for *M*-phase passive mixers earlier in Chapter 8 (also see Problem 13), the receiver noise figure may be expressed as

$$F \approx \left[1 + \frac{R_{SW}}{R_s} + \frac{\overline{v_{bb}^2}}{4MKTR_s}\right]K,$$

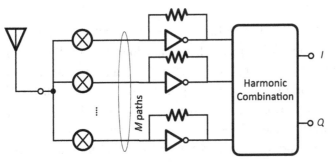

Figure 12.27: *M*-phase mixer-first receiver

where R_{SW} is the mixer switch resistance, R_s is the source resistance, $\overline{v_{bb}^2}$ is the TIA input noise voltage referred to each single-ended switch, and $K = \left(\frac{\frac{\pi}{M}}{\sin\frac{\pi}{M}}\right)^2$ captures the noise folding. The impact of harmonics have been ignored.

Example: Figure 12.28 shows the mixer-first noise figure for $M = 4$, and 8, and a complementary TIA biased at 1 or 2mA. Everything else is assumed to be ideal.

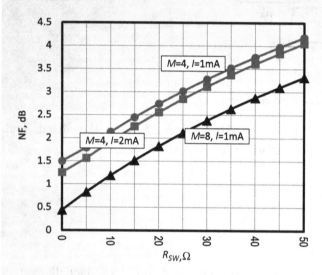

Figure 12.28: Mixer-first receiver noise figure

Evidently, even for the 8-phase design that enjoys less aliasing but comes at the expense of more LO power consumption, the receiver noise figure reaches 3dB and higher once the input matching condition is satisfied ($R_{SW} \approx 50\Omega$). In practice, the receiver noise figure is even worse, once the impact of capacitances and other parasitic elements included.

There are certain cases where a somewhat worse noise figure may be acceptable, especially for low-power applications such as Bluetooth or low-power WLAN. In those cases the mixer-first receiver may be an appropriate choice. In more demanding applications such as cellular, the higher noise figure is typically not acceptable.

12.3.3 Noise-Canceling Receivers

The mixer first receiver may be thought of an LNA using explicit 50Ω resistance for matching. Although linear and wideband, the noise figure of such amplifier is poor. If, however, one can alleviate the noise issue of this topology by providing both voltage and current measurements and taking the advantage of noise cancellation [17] (as we discussed in Chapter 7), then an arbitrarily low noise figure may be achieved, regardless of matching [12]. The concept is shown in Figure 12.29, which is similar to the noise-canceling LNAs, but in the context of a receiver. The mixer first path provides a current measurement, and the 50Ω matching is accomplished

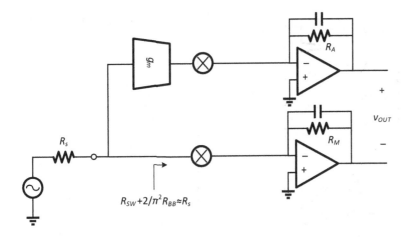

Figure 12.29: The concept of noise-canceling receiver

by adjusting the mixer switch and TIA input resistance. Ignoring the mixer's loss, the total available current gain to this path (we call it the *main* path) is $\frac{1}{R_s}$, and the available voltage gain to the TIA output is $\frac{-R_M}{R_s}$. A second path, which we call the *auxiliary* path, is added to provide the voltage measurement by employing a linear current-mode receiver consisting of a low-noise transconductance stage and current-mode mixers. The total gain of this path is $g_m \times R_A$.

Thus, for noise cancellation to be satisfied, as we showed in Chapter 7,

$$g_m R_A = \frac{R_M}{R_s}.$$

This criterion is simply met by adjusting the feedback resistors of each TIA. Once the noise cancellation condition is satisfied, an arbitrarily low noise figure with sufficiently high linearity is achievable.

As a case study, a complete noise-canceling receiver [12] based on the concept shown in Figure 12.29 is illustrated in Figure 12.30. The receiver is intended to be used in wideband applications, and thus must be very resilient to blockers. To achieve harmonic rejection, 8-phase mixers are used in both main and auxiliary paths, and hence all the harmonics up to the 7th are rejected. The low-noise transconductor linearity is critical, and thus a complementary structure is chosen, which as we showed in Chapter 7 can tolerate a large input as long as its output is loaded with a low resistance. In this design, the mixers' resistance is adjusted such that the RF transconductor sees a low impedance of 10Ω. On the other hand, the mixer first path is designed to provide 50Ω. The eight outputs of main and auxiliary TIAs are fed into a harmonic rejection as well as noise cancellation block.

With the noise of the main path (the mixer-first portion) canceled, and as the noise aliasing for an 8-phase mixer is very close to unity, the receiver noise figure is

$$F \approx 1 + \frac{\gamma}{g_m R_s},$$

where g_m is the transconductance of the RF cell. In the above equation, we have ignored the noise contribution of TIAs in the auxiliary path as they are suppressed by the large gain of the RF g_m cell.

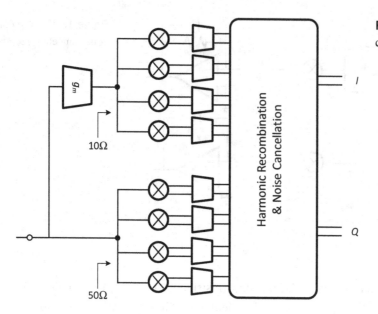

Figure 12.30: An 8-phase noise-canceling receiver

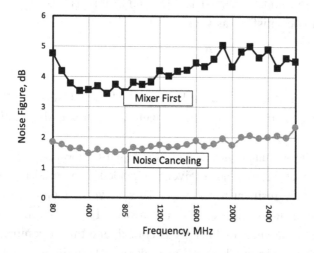

Figure 12.31: Noise-canceling receiver noise figure

The receiver measured noise figure over an input frequency range of 0.1–3GHz is shown in Figure 12.31.

Without noise cancellation (auxiliary path off), the receiver is simply a mixer-first, and as expected a poor noise figure is measured. Once the noise cancellation is enabled, a sub-2dB noise figure for the entire receiver is achieved over a wide input range. The receiver can tolerate blockers as large as 0dBm without much gain compression or noise degradation. The measured 0dBm blocker noise figure is 4dB.

This receiver combines the three ideas presented before to satisfy opposing requirements of good noise figure and linearity simultaneously: current-mode receiver concept, noise cancellation, as well as mixer-first receiver idea based on low-noise passive mixers.

Example: Since the noise cancellation is a function of the source impedance R_s, a legitimate question arises as to how sensitive the noise-canceling receiver (or the noise-canceling LNA for that matter) is to source impedance variation. We are going to study that in the context of an example shown in Figure 12.32. The mixers are eliminated for simplicity creating a linear time-invariant noise model of the receiver, though the main path mixer is replaced by its equivalent resistance R_s providing the matching. Furthermore, the noise of the TIAs is ignored, which is a reasonable assumption. Other relevant noise sources, particularly those of the RF g_m and main path mixers, are explicitly shown. The source impedance is assumed to be arbitrary, and equal to Z_s.

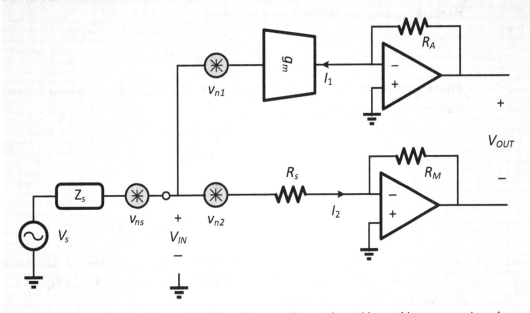

Figure 12.32: A simple noise model of the noise-canceling receiver with an arbitrary source impedance

Using superposition, the receiver input voltage is

$$V_{IN} = \frac{R_s}{R_s+Z_s}(V_s+v_{ns}) - \frac{Z_s}{R_s+Z_s}v_{n2}.$$

Finding the current of each path (I_1 and I_2), the output voltage is

$$V_{OUT} = \frac{g_m R_A R_s + R_M}{R_s+Z_s}(V_s+v_{ns}) + g_m R_A v_{n1} + \frac{R_M - g_m R_A Z_s}{R_s+Z_s}v_{n2},$$

which, considering that $g_m R_A = \frac{R_M}{R_s}$, simplifies to

$$V_{OUT} = -g_m R_A \left[\frac{2R_s}{R_s+Z_s}(V_s+v_{ns}) + v_{n1} + \frac{R_s-Z_s}{R_s+Z_s}v_{n2}\right].$$

If the source impedance Z_s is equal to the nominal value of R_s, we arrive at the same conclusions as before. In general though, the receiver noise factor is

Continued

$$F = 1 + \frac{\left|\frac{R_s+Z_s}{2R_s}\right|^2 \overline{v_{n1}^2} + \left|\frac{R_s-Z_s}{2R_s}\right|^2 \overline{v_{n2}^2}}{4KT\text{Re}[Z_s]} = 1 + \frac{\gamma}{g_m \text{Re}[Z_s]} \left|\frac{R_s+Z_s}{2R_s}\right|^2 + \frac{R_s}{\text{Re}[Z_s]} \left|\frac{R_s-Z_s}{2R_s}\right|^2.$$

The above equation tells us that in the presence of a nonideal source[7] not only is the noise of the RF g_m increased (the first term), but only partial noise cancellation is achieved (the second term). The former is common in any other receiver as well, whereas the latter is unique to noise-canceling receivers. Note that the output noise is normalized to $4KT\text{Re}[Z_s]$, and not $4KTR_s$, as, by the definition, the signal-to-noise ratio at the input is determined by $\frac{|V_s|^2}{4KT\text{Re}[Z_s]}$.

It is often customary to express the noise figure degradation in terms of the source standing wave ratio (*VSWR*). So let us define

$$\Gamma_s = \frac{Z_s - R_s}{Z_s + R_s} = \rho e^{j\phi}.$$

The magnitude of the source reflection coefficient ρ may be expressed in terms of its standing wave ratio (*VSWR*):

$$\rho = \frac{VSWR - 1}{VSWR + 1}.$$

Rearranging the noise equation, we arrive at

$$F = 1 + \frac{F_{nom} - 1 + \rho^2}{1 - \rho^2} = \frac{F_{nom}}{1 - \rho^2},$$

where $F_{nom} = 1 + \frac{\gamma}{g_m R_s}$ is the receiver noise figure under nominal source impedance of R_s.

The receiver noise figure over various *VSWR* values is illustrated in Figure 12.32. The nominal noise figure is assumed to be 3dB.

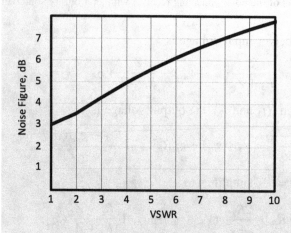

Figure 12.33: The noise figure of the noise-canceling receiver under mismatch

---

[7] This is often characterized by performing *source pull* measurements in the receiver.

If one chooses to calibrate the noise cancellation for every given source (Z_s), we must have $g_m R_A = \frac{R_M}{Z_s}$, and the noise factor simply becomes

$$F = 1 + \frac{\gamma}{g_m \text{Re}[Z_s]} = 1 + (F_{nom} - 1)\frac{1 + \rho^2 + 2\rho \cos\phi}{1 - \rho^2}.$$

That is essentially how a regular receiver behaves over mismatch. In general, the difference in noise figure is fairly small for moderate values of *VSWR* when the noise-canceling receiver is used. Furthermore, if need be, the noise cancellation can indeed be recalibrated under source mismatch as described in [18]. The actual source pull measurements confirm this, indicating that the noise cancellation is fairly insensitive to source mismatches.

12.4 RECEIVER FILTERING AND ADC DESIGN

In this section we will discuss the receiver ADC requirements. The ADC dynamic range is set based on the combination of three concerns:

- The receiver noise figure, which ideally must not be affected by the ADC quantization noise.
- The receiver available filtering, and particularly the IF filtering.
- The strength of the signals appearing at the ADC input, whether desired or blocker, or any other signal due to receiver nonidealities such as uncorrected DC offsets. If the desired signal is the main concern, the receiver gain control is then a factor as well.

As the desired signal is expected to receive plenty of amplification by the time entering the ADC, the unfiltered blockers perhaps are the most critical factor. There often exists a subtle trade-off between the IF filter order, and the ADC effective number of bits. For a given bandwidth, once the acceptable noise is set, the higher the order, the bigger the filter, and thus more cost. As shown in Figure 12.34, if a blocker is close to the desired signal, and hence

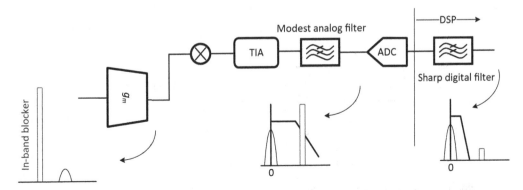

Figure 12.34: Receiver filtering and ADC requirements

subject to little *analog filtering*, it appears as a strong signal at the ADC input. As long as the ADC has high enough dynamic range so as not to be compressed, the strong blocker is subject to subsequent more aggressive *digital filtering*, which is much cheaper.

It is fair to assume that the out-of-band blockers are subject to enough filtering by the time they appear at the ADC input. Thus, our main focus is mostly on the in-band blockers, which are subject only to post-downconversion filtering. Let us consider a few examples:

Example: *GSM ADC Requirements* — As we discussed in previous section, in the case of the low-IF receiver shown in Figure 12.20, we expect almost no filtering applied to the strong 400kHz blocker residing at the image side. We expect this blocker, which is 41dB stronger, to set the bottleneck, and as such we calculate the ADC dynamic range as follows:

- We leave 10dB of margin for down and up *fading*. The fading arises from the fact that the mobile user may be subject to a sudden drop of signal (for example the vehicle entering a tunnel), and the receiver gain control is not fast enough to respond. Since the signal and blocker are at different frequencies, the fading may happen in opposite directions for each.
- A margin of 5dB is considered for inaccuracy in gain control and other nonidealities caused by process or temperature variations.
- We would like the signal to be 20dB above the ADC noise. This may be too generous of a margin, but ensures that the receiver sensitivity is mainly dominated by the receiver RF blocks that are generally more difficult to be managed.

When everything is added up, this leads to an ADC dynamic range of 86dB as shown in Figure 12.35. As a result of this lineup, the desired signal lies 56dB below the ADC full scale, which is to leave room for the blocker, up-fading, and nonidealities.

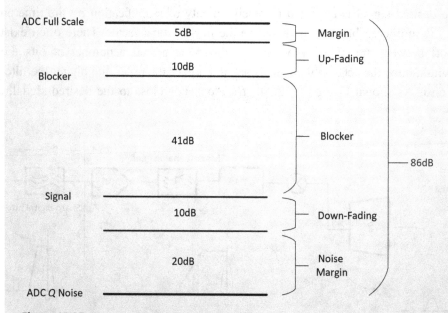

Figure 12.35: GSM ADC requirements

One can go through other blocker scenarios and show that indeed the 400kHz blocker is the most stringent as far as the ADC is concerned. For instance, even though the 600kHz blocker is 8dB stronger (Figure 12.17), once downconverted it will reside at 600 − 135 = 465kHz. Assuming a filter passband of 235kHz (for a 135kHz IF), even a 1st-order roll-off is adequate to make up for the 8dB difference.

Example: *3G ADC Requirements* — In the case of 3G, the ADC dynamic range is mainly set by the need for higher SNR. As we mentioned in Chapter 6, as the desired signal increases, the system supports higher throughput by switching into more complex modulation schemes. This not only results in a higher SNR needed, but also the processing gain reduces to support higher bandwidth. In the case of 3G, a data rate of up to 21Mbps may be supported when using a 64QAM modulation scheme, which requires a net SNR of 30dB or higher. Moreover, as we showed in Chapter 5, the more complex modulated spectrums cause a higher peak-to average power ration (PAR).

Once all these factor taken into account, we arrive at a total dynamic range of 72dB for the 3G ADC (Figure 12.36).

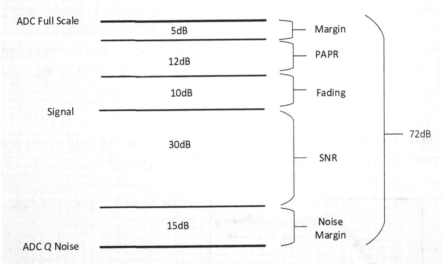

Figure 12.36: 3G ADC requirements

It turns out that this dynamic range comfortably covers the unfiltered in-band blockers even if a simple low-cost filter is used.

In many modern receivers, higher dynamic range ADCs are enabled by providing them a high clock frequency. Since the noise requirements of such clock signal is relatively stringent, it is common to feed a divide-down version of the RF LO. This results in the ADC clock to vary as the RF channel changes, but can be compensated for using a *rate adapter* in digital domain, which is relatively cheap in nanometer CMOS processes (Figure 12.36).

718 Transceiver Architectures

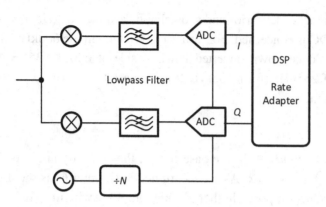

Figure 12.37: ADC clocked by RF VCO

It is not uncommon to see ADCs with sampling frequencies of several GHz in many modern radios.

12.5 RECEIVER GAIN CONTROL

As the desired signal increases, the receiver gain is expected to reduce to accommodate for the larger signal. This, however, must be done carefully, as it is desirable to improve the receiver SNR to support higher data rates when the signal is strong. To accomplish this goal, the receiver front-end gain (such as the LNA gain) is ideally expected not to change unless the desired signal is very large, strong enough to compress the front end. Ideally we expect a 1dB increase in SNR for a 1dB increase in the signal level, which can be accomplished only if the receiver noise figure stays constant.

Example: A practical GSM receiver [19] SNR is shown in Figure 12.38. Also shown is the ideal SNR curve which is a straight line with a slope of one.

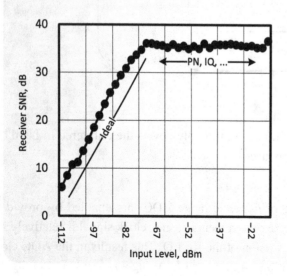

Figure 12.38: The measured SNR of a GSM receiver

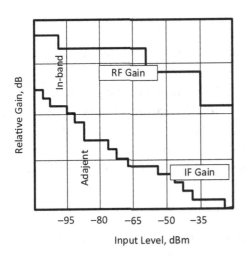

Figure 12.39: Receiver RF and IF gain

There are several factors that the actual SNR could deviate from the ideal one:

- It may be desirable to reduce the front-end gain slightly to accommodate for the large in-band blockers. In the case of GSM, the 3MHz blocker at −23dBm is strong and could potentially saturate the receiver. As the desired signal is specified to be at −99dBm, around this region the receiver front-end gain is slightly reduced, and thus there is a slight deviation from the ideal curve. If a *blocker detection mechanism* is incorporated, this may be done more efficiently, only when a large blocker is present.
- At very large inputs, the receiver SNR is not determined by the receiver chain thermal noise any more. Other contributors such as the LO in-band phase noise, or IQ mismatches, which are independent of the signal power, limit the SNR. At this point (around −65dBm for the example of Figure 12.38) it may be appropriate to reduce the front-end gain.
- At around −82dBm, there are several adjacent blocker tests. These blockers are typically not strong enough to saturate the front end, but the IF gain must be adjusted properly. A good example of this is the 400kHz image blocker that was discussed previously. Fortuitously, in most cases, the reduction of the IF gain has a negligible impact on the SNR.

The receiver relative RF and IF gains are hypothetically depicted in Figure 12.39. The IF gain is expected to provide more resolution and track the signal more closely.

As we showed in the previous section, the desired signal is well below the ADC full scale. Thus, at early stages the gain control is not as critical, and it may be deemed necessary only to accommodate the blockers.

12.6 TRANSMITTER ARCHITECTURES

As we saw in Chapter 6, from a requirements point of view, the transmitter was more or less dual of the receiver. We shall see in this section that from the architectural point of view, this is also somewhat true.

12.6.1 Direct-Conversion Transmitters

We showed in Chapter 2 that a single-sideband modulator consisting of a quadrature LO has been evolved to be the basis of a Cartesian transmitter, also known as a direct-conversion transmitter (Figure 12.40). The I and Q signals are created digitally,[8] and are fed into IQ DACs that provide the analog equivalent input to the quadrature upconverter. A lowpass filter after the DAC is typically needed to remove the DAC image signal and possibly improve the far-out noise. In addition, an RF filter before the antenna may be necessary to clean the output spectrum to ensure that the far-out modulation mask is met. In the case of FDD application, the *duplexer* provides the filtering automatically.

Despite versatility, direct-conversion transmitters suffer from several important drawbacks:

- As we pointed out in Chapter 6, the feed-through of the LO signal to the output as well as the I and Q imbalances lead to EVM degradation. Similar to direct-conversion receiver, IQ calibration and LO feed-through cancellation may be employed to enhance the performance.
- The far-out noise of the transmitter is somewhat poor as all the building blocks in the TX chain, as well as the LO generation circuitry contribute. This problem is dealt with using alternative topologies that we will discuss shortly.
- As the transmitter output signal coincides in frequency with the VCO, this strong signal disturbs the VCO operation once leaked to the VCO output due to finite isolation.

The latter issue, as we discussed in Chapter 6, is known as *pulling*, and in certain applications it may prohibit the use of the direct-conversion architecture. To understand the impact of the pulling issue better, consider Figure 12.41, which shows a direct-conversion TX transmitting a tone at a frequency offset of f_m from the carrier. As a result, at the power amplifier output a strong signal at $f_{LO} + f_m$ appears. Given the finite isolation between the PA output and the VCO, this component leaks back to the VCO, and creates unwanted sideband at $\pm f_m$ away from

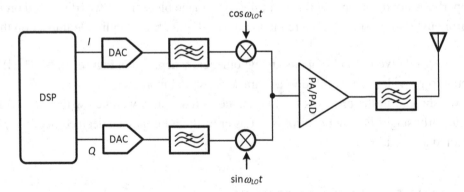

Figure 12.40: Direct-conversion transmitters

---

[8] As we said in Chapter 2, the I and Q baseband signals are statistically independent in most modern radios.

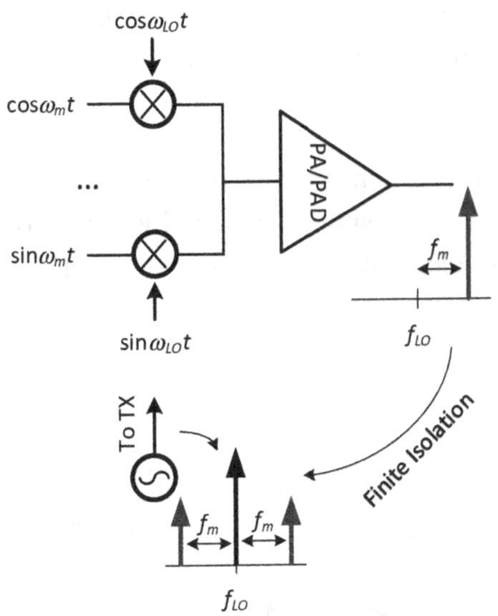

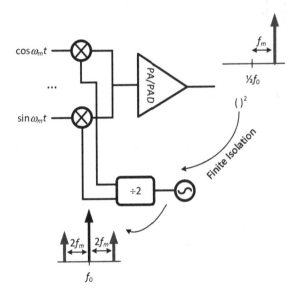

Figure 12.41: Pulling in direct-conversion transmitters

Figure 12.42: Using divider to alleviate the pulling

the VCO output signal. Consequently, once this corrupted spectrum is fed to the mixers to upconvert the baseband input, the transmitter EVM and modulation mask are affected adversely [20]. See Chapter 6 for more details.

To alleviate this issue, one may consider using a VCO at twice the frequency along with a divider by two. This has the added advantage of creating quadrature LO signals readily. By employing a divider, the VCO and PA output frequency are spaced far from each other, and the pulling is expected to improve, as the pulling strength is shown to be inversely proportional to the frequency offset between the signals [20], [21], [22]. This, however, is not quite true as the pulling could be still significant due to the 2nd-order nonlinearity of the PA, which is often very strong (Figure 12.42).

Alternatively, one may choose a dual-conversion architecture, where the VCO and TX frequencies are not harmonically related anymore. We shall discuss this next.

> **Example:** As a case study, an example of an LTE transmitter using the direct-conversion architecture is shown in Figure 12.43. The pulling is simply avoided by using a divide by 2/4 to produce high/low band LO signals, along with a careful layout, and package design.
>
>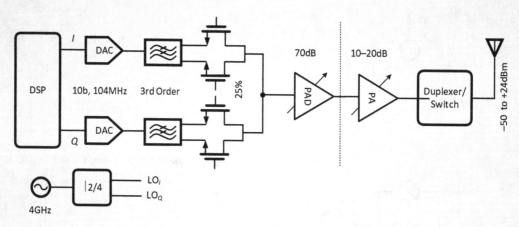
>
> Figure 12.43: A typical 4G direct-conversion transmitter
>
> In the 3G mode, a 10b DAC followed by a 3rd-order lowpass filter to clean that DAC output is used. The filter is about 2MHz wide. The mixers are 25% passive, followed by a PA driver that provides over 70dB of gain control. That along with additional 10–20dB gain control provided by the PA are sufficient to cover the 74dB gain range needed in 3G standard. The LTE mode is also very similar except for the filter that is wider. We shall discuss more details of the transmitter later in this chapter (Section 12.7.2).

12.6.2 Dual-Conversion Transmitters

As shown in Figure 12.44, similar to receivers, a dual-conversion transmitter employs two steps of upconversion [9]. The main advantage is considerably less sensitivity to pulling.

In order to use only one VCO, similar to receiver, the 2nd quadrature LO is obtained through a divider. With the VCO frequency at f_{LO1}, the TX output frequency is

$$f_{TX} = f_{LO1}\left(1 + \frac{1}{N}\right).$$

Clearly the two frequencies are not harmonically related anymore, and thus the pulling is expected to be of little concern. If this architecture is employed for the transmitter for obvious

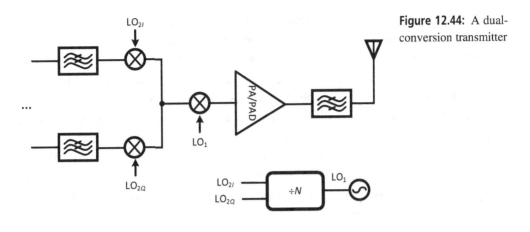

Figure 12.44: A dual-conversion transmitter

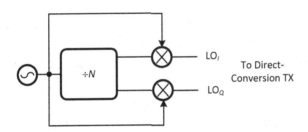

Figure 12.45: LO generation circuitry to avoid pulling

pulling concerns, the choice of the VCO frequency and the frequency planning involved inevitably dictates a dual-conversion architecture for the receiver as well, despite the fact that there may not be as much advantage for the RX.

An alternative architecture that alleviates the pulling issue similar to dual-conversion architecture but still enjoys the simplicity of direct-conversion transceiver is to exploit a VCO frequency shifting circuitry shown in Figure 12.45 [23], [24].

Once the VCO is mixed with its own divided version, it creates LO outputs that are related to the original VCO frequency by

$$f_{LO} = f_{VCO}\left(1 \pm \frac{1}{N}\right),$$

which is similar to the dual-conversion TX. In addition, the quadrature LOs are readily available due to the divider (as long as $N = 2^n$). The drawback of this scheme as well as the dual-conversion TX is the spurious signals produced as a result of mixing. Proper filtering, often with the help of tuned buffers, must be employed, which leads to larger area and cost. In the latter approach of VCO shifting, the mixing is not in the signal path, which is advantageous especially for the receiver.

12.6.3 Direct-Modulation Transmitters

Despite the fact that the direct conversion architecture is very versatile and benefits from relatively low complexity, some important repercussions apply when used in certain applications such as cellular:

1. In Chapter 6 we saw that in order not to desensitize a nearby receiving mobile handset, a very stringent transmitter noise level of −79dBm at the corresponding receive band is specified. This translates to a phase noise of −165dBc/Hz at an offset frequency of 20MHz at the RF IC output. In the case of a direct-conversion TX, in addition to the voltage controlled oscillator (VCO) and the local oscillator (LO) chain, the entire TX path such as the DACs, reconstruction filters, and active or passive mixers contribute to this noise.
2. Even though the GSM signal is only phase modulated, in the case of a direct-conversion architecture the linearity of the modulator and the PA is a concern. This may sound counterintuitive at first, but can be understood by noting that in the switching mixers used in most modulators a replica of the modulated input of opposite phase exists at the LO third harmonic. When mixed down due to the third order nonlinearity of the PA or the PA driver, this signal replica could cause degradation of the modulation spectrum. This was detailed in Chapter 8.
3. As we discussed earlier, despite operating the TX VCO at two or four times the output frequency in most common radios, a direct-conversion scheme suffers from the potential pulling caused by the PA running at 2W, and drawing more than 1A of current from the battery.
4. As we pointed out, direct-conversion transmitters generally require good $I - Q$ imbalance in the signal and LO path. Although the requirements can be met by adopting some kind of calibration, it adds to the complexity. Moreover, the DC offset of the baseband section including the DAC and the following reconstruction filters needs to be corrected.

For these reasons GSM transmitters have traditionally adopted an alternative architecture, using *translational loops* [25] to eliminate the external front-end filter at reasonably low power consumptions as shown in Figure 12.46.

The PLL feedback forces the VCO output to be a replica of the modulated input spectrum fed to the phase frequency detector (PFD), thus *translating* the IF spectrum to RF. Unlike the direct-conversion case, here only the VCO and LO chain contribute to the far-out noise, as usually the PLL bandwidth is narrow enough to suppress the noise of the rest of the loop, including the mixer and its LO. It is clear that

$$f_{VCO} = f_{LO} + f_{IF},$$

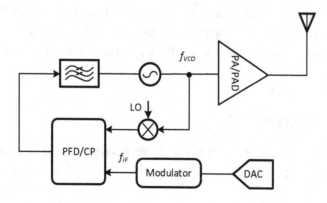

Figure 12.46: Translational loops used as GSM transmitters

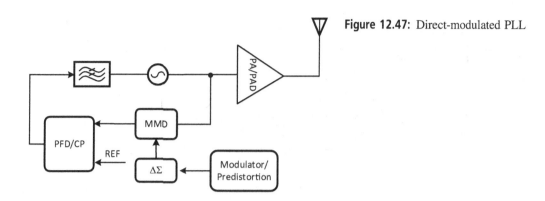

Figure 12.47: Direct-modulated PLL

where LO is a tone created by a second PLL, but the IF signal is frequency modulated, and thus so will be the output signal fed to the antenna.

The translational loop can be simplified to a direct-modulated PLL [19], [26], [27], [28], where modulation is injected via a $\Delta\Sigma$ modulator by changing the fractional divider modulus over time (Figure 12.47).

The main advantage is the elimination of the baseband modulator needed in the translational loop, which could be a potential noise and nonlinearity contributor. The LO needed for the mixing in the loop can be eliminated as well, which reduces the area and power consumption further. This typically requires the PLL to run off the available 26MHz crystal oscillator, although one may choose to adopt an auxiliary PLL to ease some of the typical problems of the fractional PLLs caused by low-frequency reference signal [19]. Overall, if designed properly, the direct-modulated PLL leads to substantial area and power savings, yet meeting the GSM stringent far-out noise requirements.

As we showed in Chapter 6, to meet GSM close-in mask, a stringent phase noise requirement of better than −118dBc at 400kHz for the entire PLL is required. To meet this challenge, the PLL bandwidth is typically chosen to be 100 to 200kHz to minimize the noise contribution of the reference, the loop filter, and the charge pump (CP). Even with such a narrow bandwidth, the rest of the PLL may contribute as much as 1~2dB, leading to a stringent noise requirement of <−120dBc/Hz for the TX VCO. Since the loop bandwidth is relatively narrow, the frequency modulated (FM) equivalent spectrum will be subject to distortion. Such distortion can be compensated with a digital predistortion block whose frequency response is the inverse of the PLL, hence providing an overall all-pass shape for the injected modulation (Figure 12.47). However, as the PLL characteristics tend to vary over process the overall response will generally not be flat, causing phase error degradation in the transmitted signal. It is often critical to employ some kind of calibration to stabilize the PLL bandwidth, particularly due the VCO gain variations [19], [30]. Alternatively a digital PLL may be chosen [27], [28] at the expense of more power consumption, and increase in the spurious level of the output spectrum.

There are also some concerns with the nonlinearity of the PLL components such as the charge-pump or the VCO as that the VCO control voltage is not a DC signal anymore, rather carries modulation. This is not a major concern for GMSK, but as we will discuss shortly, it may become a problem for higher data rate applications such as EDGE or 3G.

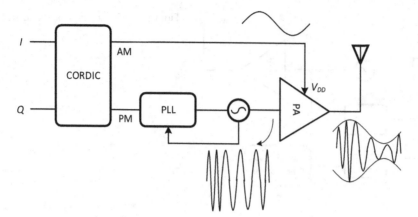

Figure 12.48: A conceptual polar transmitter

12.6.4 Polar Transmitters

The main disadvantage of the PLL-based transmitter is its lack of support for amplitude modulated standards as the VCO is capable of producing only a phase modulated spectrum. A time-varying AM component, as needed for EDGE modulation for instance, can be superimposed on the PM signal, as shown in Figure 12.48. Known as a *polar* transmitter, it has been originally devised to boost the *power amplifier efficiency*. As more complex modulations that support higher data rates come at the expense of higher PAPR, a better linearity from the PA is demanded, leading to a poor efficiency.

To ameliorate this fundamental trade-off, one could break the IQ signals into phase (θ) and amplitude (r), or from *Cartesian* into *polar* domain:

$$I + jQ \leftrightarrow re^{j\theta}.$$

The phase modulated signal is created by a VCO/PLL, which essentially form a direct-modulated transmitter as was discussed. Accordingly, the phase signal is directly fed to the PA, and ideally often drives the amplifier into near saturation, hence not compromising the efficiency. The amplitude content is applied to the PA commonly through modulating its supply voltage (Figure 12.48). Once aligned properly, the composite PA output signal carries the intended phase and amplitude modulation.

A modern polar transmitter based on the concept just described is shown in Figure 12.49 [29], [30], [31]. The AM signal is typically produced by employing another $\Delta\Sigma$ modulator,[9] as shown in Figure 12.49.

The modulator produces the phase (PM) and amplitude (AM) components digitally. The PM is fed to the PLL, which is expected to create a frequency (or phase) modulated spectrum at the VCO output. The AM component, once passed through the $\Delta\Sigma$ DAC, modulates the PA driver, which can be thought of a single-balanced mixer. It is not uncommon to apply the AM into the *PA driver*, rather than the PA, and use a linear power amplifier, often known as *small signal*

---

[9] This is merely to enhance the DAC performance. Otherwise, the use of a $\Delta\Sigma$ modulator for the AM path is not a requirement of polar architecture.

12.6 Transmitter Architectures 727

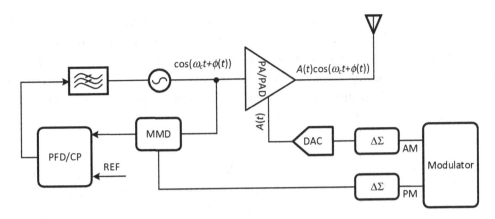

Figure 12.49: Modern polar transmitter

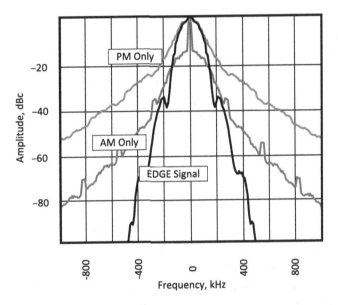

Figure 12.50: PM and AM signals corresponding to an EDGE polar transmitter

polar architecture. This is done merely to avoid extra complexity when the PA nonideal effects are considered. Although this does not take full advantage of the efficiency improvement idea, it still is advantageous compared to the alternative choice of using a direct-conversion transmitter for the reasons already discussed. The PA driver still can operate closer to saturation compared to a linear TX, further improving the power consumption. Additionally, the polar TX enjoys a simpler 50%, single-phase LO path, as opposed to the linear transmitters that typically require quadrature 25% LO, exacerbating the power efficiency problem.

When using a polar transmitter, in addition to the PM path, the AM path is a key factor [30]. The PM issues remain the same as the ones described for the case of direct-modulated TX, however, with a greater impact on performance. To understand this better, it is beneficial to study the phase and amplitude spectrums individually before being combined at the PA driver output. An example of an EDGE spectrum is shown in Figure 12.50.

Compared to the ideal output, both the PM and AM spectrums are considerably wider. Particularly the PM signal is very wide, drooping by only about 30dB at 400kHz offset, compared to 68dB for the ideal EDGE spectrum. Although this spectral growth may be mathematically proven by examining the modulation properties of the I and Q signals, and their nonlinear transformation into polar domain, we shall present a more intuitive perspective here in the context of the following example.

Example: A signal that carries both amplitude and phase modulation is passed through an *ideal limiter*, leading to the amplitude component being stripped away. Thus, we are left with only a phase modulated signal (Figure 12.51). That is because the limiter does not affect the zero-crossing dictated by the phase (or frequency) modulation. In other words, by applying the EDGE signal into an ideal limiter, one expects to produce the PM contents only.

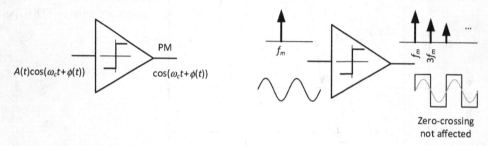

Figure 12.51: Producing the PM signal employing an ideal limiter

Now suppose only a tone at a frequency of f_m representing the highest frequency content of the modulated signal is considered. Clearly this signal has a bandwidth of f_m. Once passed through the limiter, however, strong components at all odd harmonics are produced, and hence the signal at the limiter output has an infinite bandwidth. Particularly the tone at $3 \times f_m$ is only 10dB lower, resulting in substantially wider signal for the PM only contents.

The PM spectral growth demands larger swings at the VCO input, thus raising the sensitivity to nonlinearities. The swing at the VCO input of an EDGE transmitter assuming a VCO gain of about $2\phi \times 40\text{MHz/V}$ is shown in Figure 12.52. Representing the PM signal as

$$\cos\left(\omega_C t + K_{VCO}\int f(\tau)d\tau\right),$$

the swing ($f(t)$) depends only on the EDGE PM characteristics, described by $\phi(t) = K_{VCO}\int f(\tau)d\tau$, which is known, and the VCO gain (K_{VCO}). Although increasing the VCO gain helps, it results in phase noise degradation. If the PLL bandwidth is set to around 200kHz or less as demanded by stringent phase noise requirements at the key 400kHz offset, the loop gain is expected to drop at 400kHz and beyond. Thus, the PLL feedback becomes almost ineffective, pronouncing the negative impact of loop nonlinearities in the presence of larger swings.

Moreover, a narrow bandwidth increases the sensitivity to variations. This is illustrated in Figure 12.53, showing that the desired all-pass response set by the combination of the PLL and a digital predistortion is subject to distortion if the *analog PLL* characteristics vary. As mentioned, either the PLL must be well calibrated, or a digital PLL may be employed.

This issue is dramatically worse if a wider application such as 3G is considered. Now, the composite 3G signal is about ± 1.9MHz wide, whereas the PM component is about 10MHz [31]. Such a wide signal not only suffers substantially more from narrow PLL variations, but also becomes much more sensitive to the VCO nonlinearity. The latter may be resolved either by using a digital VCO/PLL [29], or by incorporating some kind of feedback to linearize the VCO [31], [32]. The former issue may be alleviated by employing a two-point PLL as shown in Figure 12.54 [29], [31].

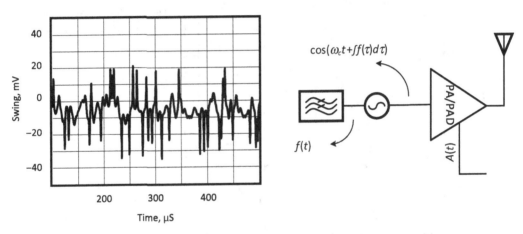

Figure 12.52: Swing at the VCO input for a polar EDGE transmitter

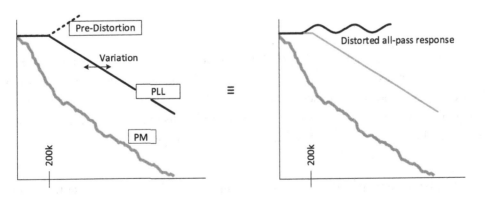

Figure 12.53: Impact of variations of PLL characteristics

730 Transceiver Architectures

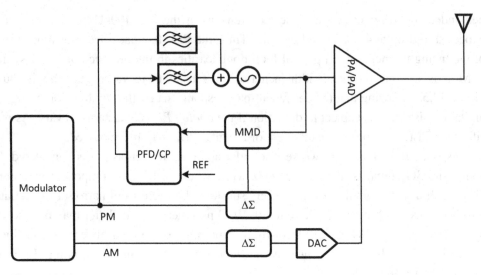

Figure 12.54: A two-point PLL to extend the useful bandwidth

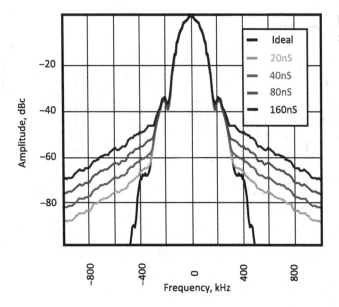

Figure 12.55: AM–PM alignment inaccuracy causing mask violation

Applying a second signal to the VCO control voltage directly creates a *highpass* response, and once the two responses are combined, the net PLL characteristics becomes *all-pass*. For the two frequency responses to match perfectly, still the VCO gain must be very well behaved.

In addition to the nonidealities of the AM and PM paths individually, the alignment between the AM and PM signals becomes critical, as shown in Figure 12.55.

Example: To understand the AM–PM alignment requirement, suppose there is a small delay of τ between the two paths; thus when combined, the output will be

$$A(t-\tau)\cos(\omega_C t + \phi(t)).$$

Using the Taylor expansion of the amplitude component, $A(t)$, around τ we have

$$A(t-\tau) \approx A(t) - \tau\frac{\partial A}{\partial t},$$

which results in the original signal as well as *an error term*:

$$A(t)\cos(\omega_C t + \phi(t)) - \tau\frac{\partial A}{\partial t}\cos(\omega_C t + \phi(t)).$$

The error waveform, whose amplitude is proportional to the small delay, is formed by the *time-derivative* of the desired envelope modulated to the carrier frequency. Differentiation in time will accentuate high-frequency content in the envelope, leading to an error spectral density that could encroach into adjacent channels and possibly violate the transmit mask.

From Figure 12.55, in order not to degrade the EDGE spectrum significantly, an alignment error of less than 20ns is needed. Thus, the PLL bandwidth as well as that of the AM path reconstruction filter must be tightly controlled. For the 3G case, the requirement is considerably more stringent; only a delay of a few nS may be acceptable. Thus a much tighter control of the PLL characteristics is needed.

As for the AM path, both the AM–AM and AM–PM nonlinearities are important factors. Especially AM–PM generated in the PA or the PA driver (in the case of small signal polar TX) is a key contributor. As we discussed before, presence of a strong AM–PM not only causes modulation mask degradation, but also results in an asymmetric spectrum. Shown in Figure 12.56 is the measured +400kHz and −400kHz modulation mask of a polar EDGE transmitter in the presence of AM–PM nonlinearity in the PA driver. With the correct alignment, the modulation mask improves, but remains asymmetric when AM–PM distortion is present. Once the AM–PM is

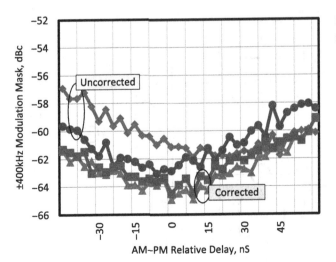

Figure 12.56: Impact of AM–PM nonlinearity on an EDGE polar transmitter

corrected through digital calibration, the two sides become symmetric, and in addition a 2dB improvement in the modulation mask is observed.

Using a nonlinear power amplifier, the AM–PM and AM–AM distortion are expected to be considerably worse. As we said earlier, it may be more practical to consider a small signal polar transmitter, and still benefit from a number of advantages mentioned. This results in a subpar efficiency, but more robust performance for the overall transmitter, especially when production issues considered.

Finally, similar to direct-conversion architecture, the *PM feed-through* (as opposed to LO feed-through as LO is modulated) causes both EVM and mask degradation. The latter has to do with the wide nature of the PM signal, whose magnitude does not drop enough at the key offset frequencies of 400 and 600kHz.

12.6.5 Outphasing Transmitters

Another attractive transmitter architecture that takes advantage of a high-efficiency PA, yet is capable of transmitting nonconstant envelope signals is outphasing, invented by Chireix [33]. Later rediscovered, and named by Cox in 1975 [34], the outphasing transmitter is also known as LINC, which stands for linear amplification using nonlinear components. The basic idea is described in Figure 12.57. The modulated phasor whose magnitude and phase are both time-varying is decomposed as a sum of two constant amplitude but phase-varying phasors. The resultant two phasors are mirrors of each other with respect to the original one. Such decomposition is also called *signal component separation*.

For a general amplitude and phase modulated phasor of $A(t)e^{j\theta(t)}$, defining the mirrored phase component $\phi(t) = \cos^{-1}\frac{A(t)}{2A_0}$, we have

$$A(t)e^{j\theta(t)} = (2A_0 \cos \phi(t))e^{j\theta(t)} = A_0 e^{j(\theta(t)+\phi(t))} + A_0 e^{j(\theta(t)-\phi(t))}.$$

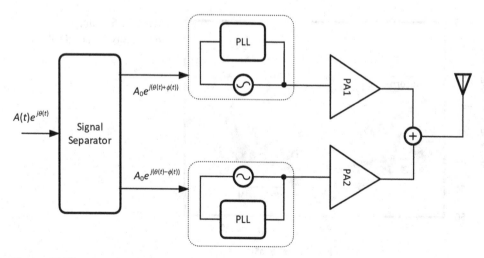

Figure 12.57: Concept of outphasing

The corresponding phase-only modulated RF signals ($A_0 e^{j(\theta(t) \pm \phi(t))}$) can be amplified by two highly efficient nonlinear power amplifiers. The amplified signals can conceptually be combined to reconstruct a high-power version of the intended modulated signal. The gains and phases that the two outphasing signals experience must be well-matched. As the two paths are identical, performing a careful and symmetric layout is a necessity in order to achieve a good phase/gain matching between the two transmitter paths. Any mismatch between the two transmitter paths along with other imperfections such as mutual interactions between the two PAs can be compensated to some extent through digital predistortion of phases in the two signal paths [35].

The major challenge of the outphasing transmitter is combining the outputs of the two high-power PAs. A low-loss power combiner is needed to efficiently add the two PA outputs. It also has to be passive in order to deal with large voltage swings. The combiner should also provide a good isolation between the two PA outputs, while combining their signals and delivering the net power to the antenna. Furthermore, the two PAs must experience a fixed load while the RF signal's envelope varies, in order to maintain their good efficiencies. The latter condition can be met if the two ports of the combiner are impedance-matched. However, it can be shown that a lossless three-port passive network cannot satisfy the matching conditions [36]. To satisfy the matching requirements, a fourth port terminated to a lossy resistor is needed. The signal delivered to the fourth terminal would be the difference of the two PA outputs while their sum would be delivered to the antenna. With this scheme, for a given PAPR at the output, (PAPR-1) times of the transmitted power is wasted in the resistor of the fourth terminal. Therefore, even with a 3dB PAPR (or 2 in linear scale), the wasted power is roughly the same as the transmitted power, imposing an upper limit of 50% for the transmitter efficiency. Including the ohmic loss of the four-port combiner, the overall loss can be as large as 4dB. With a PAPR of 10dB, the upper limit for the efficiency would be 10%, which is no longer acceptable.

Due to its poor efficiency, the four-port combiner is not an attractive solution. Focusing on the three-terminal combiner (also called the *Chireix combiner*), consider Figure 12.58, where the two PAs are to put out two constant-envelope signals, given by $A_0 e^{j(\theta - \phi)}$ and $A_0 e^{j(\theta + \phi)}$. Consequently, the combined output signal delivered to the antenna would be $2NA_0 \cos\phi(t) \times \cos(\omega_{LO} t + \theta(t)) = A(t)\cos(\omega_{LO} t + \theta(t))$, where $A(t) = 2NA_0 \cos\phi(t)$, and N is the transformer's

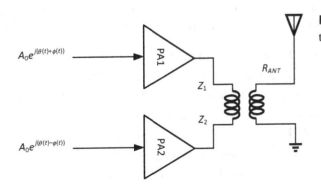

Figure 12.58: Outphasing transmitter with three-point combiner

secondary to primary turn ratio. It can be shown that the impedances seen by two PAs vary over time, and are given by the following two expressions [36],

$$Z_1 = \frac{R_{ant}}{2N^2}(1+j\tan\phi)$$

$$Z_2 = \frac{R_{ant}}{2N^2}(1-j\tan\phi),$$

where R_{ant} is the input impedance looking into the antenna. Therefore, the PA loads are modulated by the outphasing angle, ϕ, indicating that their outputs would no longer be constant envelope, defeating the original purpose of outphasing. This would lead to a large magnitude and phase distortions in the final RF signal, and must be somehow corrected. One approach to correct the PA's load modulation is proposed in [36], which uses switched capacitors at the PAs outputs. By switching them in or out at the outputs of the two PAs based on the instantaneous values of ϕ, the reactive portions of the load modulation are compensated. The real part may be compensated through scaling of the current sources used to deliver the power in the two PAs. There are still several challenges in the proposed scheme: The design is open-loop, and in practice the parasitic effects may limit the performance, especially at high frequencies. Also, the three-port combiner prohibits the differential implementation.

Besides the problems associated with the power combining, the outphasing transmitter architecture is not suitable for applications where a large range of power control is needed. Given all the aforementioned problems, the outphasing transmitter architecture is seldom used in modern wireless devices.

12.7 TRANSCEIVER PRACTICAL DESIGN CONCERNS

We conclude this chapter, and the book, by discussing several practical issues associated with real-life transceivers and provide a few case studies. Although the majority of the designs have been discussed in detail, we mostly focus on practical aspects of them.

Shown in Figure 12.59 is a general description of RF design flow, from inception of a product up to volume production. We have already discussed in Chapters 5 and 6 how to derive the circuit and system parameters for a given standard. We have also shown in Chapters 9–11 how to select the right circuits for a given application, and design them accordingly. Once the design is complete and the radio has been fabricated, there are two levels of verification involved:

- *Device verification and testing* (DVT): This involves radio verification and optimization. If the first silicon (often labeled as A0) does not meet all the requirements, revisions are required. In a metal-only tape-out, the revision number changes (e.g., A1), whereas if an all-layer tape-out is deemed necessary, the letter changes (e.g., B0).
- *Automatic testing equipment* (ATE): This is a fully automated process, and lasts as long as the product is being shipped. Unless the requirements are guaranteed to be met by design, the automatic testing is performed on every part that is to be shipped.

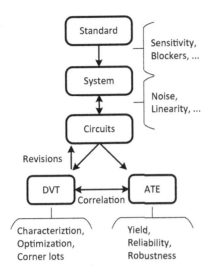

Figure 12.59: RF product design flow

As the ATE and DVT environments are inherently different, often a careful correlation process between the two is needed to ensure that the ATE results are accurate. This is often known as bench-ATE correlation process.

In this section, we will briefly discuss the production-related concerns describe above. We shall start with a few case studies, and then discuss the challenges of volume production.

12.7.1 Receiver Case Study

As our first case study, let us start with a cellular receiver designed in 40nm intended for 2/3G applications. The design can be extended to support LTE by adjusting the IF bandwidth. This radio is currently in volume production, and its die microphotograph is shown in Figure 12.60.

A simplified block diagram of the receiver and the corresponding gain distribution is shown in Figure 12.61. It uses a current-mode passive mixer with 25% duty cycle LO signals. The TIA is a common-gate design with a 1st-order RC roll-off at its output (280kHz/1.7MHz for 2/3G), followed by an active RC biquad.

The 2G receiver uses a low-IF architecture with an intermediate frequency of 135kHz.

The receiver front end is shown in Figure 12.62. The LNA is an inductively degenerated design that uses two stages of current steering to provide three gain steps.

The general performance and key challenges are highlighted below:

- A noise figure of less than 3.5dB over temperature at all frequencies. There is about a typical noise contribution of 25% from the matching, 25% from the LNA devices, and 20% from the TIA.
- An IIP_2 of better than 50dBm at TX frequency.
- In-band IIP_3 of better than −6dBm, and out-of-band IIP_3 of better than −1dBm.
- A blocker noise figure of better than 10dB for the GSM (especially 600kHz and 3MHz blockers).

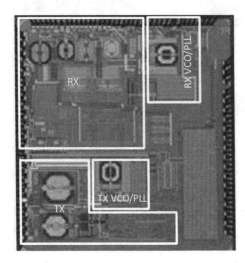

Figure 12.60: 3G transceiver die photo in 40nm CMOS

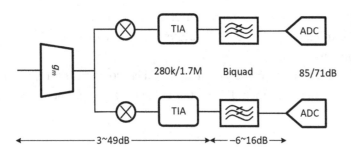

Figure 12.61: 2/3G receiver architecture

- An EVM of less than 2% corresponding to better than 35dB SNR. This ensures that the receiver can support 64QAM modulation, intended to be used in HSPA+ mode, which achieves a data rate of over 20Mbps.

As an example, the 3G receiver sensitivity for band I (2110–2170MHz) for two values of PA power is shown in Figure 12.63.

The receiver sensitivity degrades by only a fraction of a dB when the transmitter operates at its full 23dBm power at the antenna (as opposed to −5dBm for the other case). This shows that the transmitter noise as well as the receiver second-order distortion are adequately met.

12.7.2 Transmitter Case Study

As our second case study, we describe design details of a reconfigurable multiband multimode cellular transmitter covering all four bands of EDGE/GSM 2.5G and five bands of WCDMA 3G. The block diagram of the transmitter is presented in Figure 12.64. The transmitter has four single-ended RF outputs; of those, two cover high bands and the other two cover low bands. Furthermore, for each of the high or low bands, one of the two dedicated RF outputs is assigned for the 3G mode and the other one for the 2G mode. Except for the constant envelope GSM mode where the TX is configured to operate as a direct-modulation transmitter, for all other modes of operation the transmitter architecture is linear IQ direct-conversion. The transmitter

12.7 Transceiver Practical Design Concerns

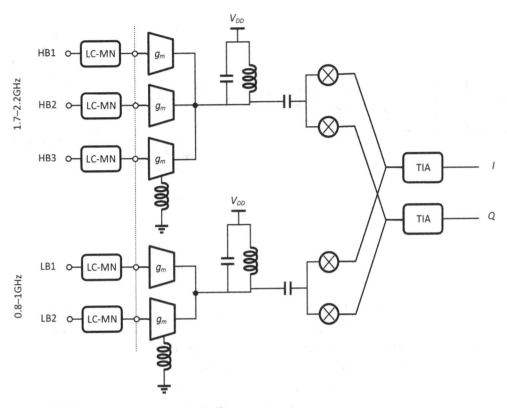

Figure 12.62: 3G receiver front-end details

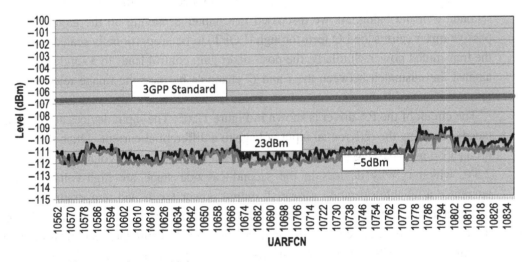

Figure 12.63: 3G receiver sensitivity

features two on-chip baluns as loads of the on-chip PA drivers, one for the high-band RF outputs and one for the low-band ones (Figure 12.64). Besides the differential-to-single-ended conversion responsibility, these baluns also provide on-chip 50 Ω matching for the single-ended RF outputs for all bands and modes with no need to any external matching components. All building blocks of the TX-RF front end prior to the baluns, such as DACs, LPFs, upconversion

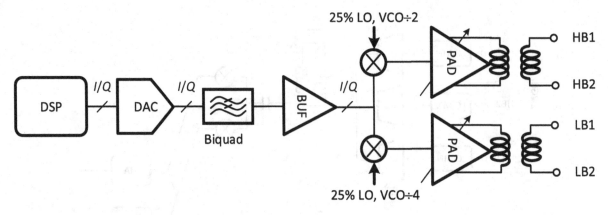

Figure 12.64: 2/3G transmitter block diagram

mixers, and PA drivers, have been designed differentially, although for simplicity only one path is shown.

In the GSM mode, the DACs, LPFs, and upconversion mixers are all powered down, and the corresponding PA driver is biased in the class B region. The PA driver differential inputs are driven directly by phase-modulated LO clocks derived from the VCO through a frequency division of two or four for the high and low bands, respectively.

In EDGE or 3G modes, the upconversion mixer is passive operating in the voltage mode, whose switches are driven by 25% LO clocks. In the linear mode, the PA driver operates in the class A region for a superior linearity. All the gain controls are implemented at the RF side after the upconversion mixer with 72dB gain control embedded in the PA driver and 12dB in the secondary of the balun. As we discussed in Chapter 8, having all the gain control after the passive mixer causes the LO feed-through (LOFT) to be proportionally scaled up or down with the transmitted power. Similarly, the post-mixer gain control leads to a single point calibration against the mismatch between the I and Q channels because the image rejection ratio (IRR) remains intact with the gain control.

The circuit of the PA driver is shown in Figure 12.65. The input devices act as transconductance converting the differential input voltage to a differential current, which flows into the PA driver load after passing through two cascode stages. The supply voltage for the PA driver is 2.5V, which is why devices in the second cascode stage are high-voltage transistors to guarantee a reliable operation across all output power levels and temperature conditions. As mentioned earlier, the PA driver is loaded to a differential-to-single-ended balun. The balun and the PA driver have been codesigned in such a way that the simulated and measured return loss looking into the RFIC is better than −10dB. The capacitor arrays in the primary of the balun are tuned for maximum output power for various bands of operation.

Imposed by the 3G standard, an overall range of 84dB or 14 bits of power control was implemented in the transmitter. Six bits of this power control is embedded in the transconductance stage of the PA driver. As shown in Figure 12.65, this stage is composed of 63 identical unit cells, in which by disconnecting gates of the cascode devices from the bias voltage and connecting them to the ground the unit is disabled or enabled otherwise. The 63 unit cells are laid out as an array of 7 × 9 cells surrounded by extra dummy cells around the boundary for a

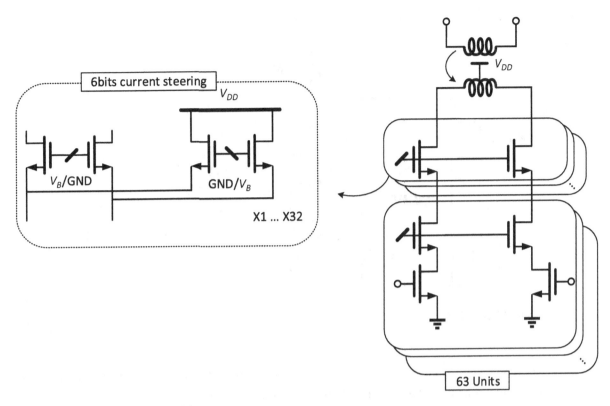

Figure 12.65: Circuit of the PA driver

better matching among the cells. Reducing the power level by disabling the transconductance units lowers the power consumption of the transmitter proportionally, which is very desirable in 3G systems.

Another 36dB or 6 bits of the power control is embedded in the second cascode stage of the PA driver, where a portion of the signal current generated by the transconductance stage can be diverted away from its main path into 2.5V supply voltage. For this purpose this stage is composed of five parallel pairs of cascode devices with their sizes increased in binary format and their drains supplying the primary of the balun. Each pair is accompanied with another pair of transistors with similar sizes but with drains shorted to the PA driver's supply voltage. Gate voltages of transistors in the two pairs are complementary to each other. Another 12dB or 2 bits of the power control is implemented in the secondary of the balun by stealing a portion of the signal current away from the 50Ω load of the external PA into the AC ground.

The lowpass filter is a reconfigurable two-stage third-order Butterworth with complex poles implemented in the first stage, and the real pole in the following buffer (Figure 12.64). The first stage is an active-RC filter. The bandwidth of the filter is adjustable from 100KHz for the 2.5G EDGE mode to 2MHz for the 3G WCDMA mode. Since the passive filter does not have the driving capability, the interstage buffer is placed between the LPF and the passive mixer. This buffer must have the following key stringent performance requirements: (1) a good linearity while handling a differential 1V p–p swing at its output; (2) low output impedance (<5Ω) over the desired signal band in order to accommodate a large conversion gain for the mixer; (3) low output noise at the RX band, i.e., 20MHz for the EDGE mode and 45MHz and beyond for the

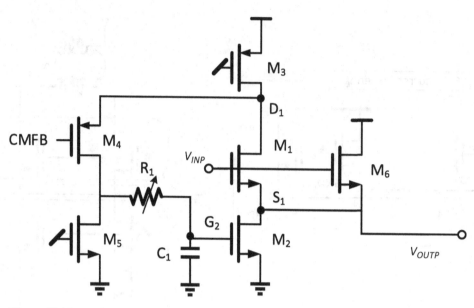

Figure 12.66: Circuit of the buffer driving the TX mixers

3G mode; and (4) large output impedance at the RX band to lower the conversion gain for the noise components of the buffer residing at these frequencies.

While the use of passive mixer with 25% duty cycle provides power and area advantages, it imposes challenging requirements for the previous driving circuit in addition to the aforementioned difficult requirements. The switching action of the passive mixer makes the PA driver input capacitance appear as a resistor seen from the baseband side of the mixer. This resistor is inversely proportional to the product of the input capacitance seen from the PA driver and the TX-LO frequency. For the high-band mode, this resistor is the lowest equal to 500Ω differentially for the designed PA driver. Using a simple source follower as the buffer is not a good option as it would suffer from a moderate linearity, and a poor driving capability due to its large output impedance.

To alleviate these drawbacks without jeopardizing the noise performance, the source-follower buffer was modified to the one shown in Figure 12.66, where feedback has been added. Only one half of the differential circuit is shown for simplicity. The basic idea behind this circuit is that the feedback loop forces the current passing through M_1 and therefore g_{m1} to be constant. As a result, nonlinearity due to the transconductance variations is minimized, and node S_1 follows V_{in} with a fixed gate-source voltage drop. Moreover, this results in the buffer output impedance seen by the upconversion mixer to be very low.

As node D_1 is low impedance (due to M_4), it causes the current source M_3 to remain almost unaltered across the entire range of input and output swings. The only high impedance node in the buffer circuit is the drain of M_4 (and current source M_5), which makes compensation of the feedback loop very convenient. At the RX frequency offset the loop-gain diminishes thanks to the pole created by C_1 and R_1, which increases the output impedance looking into the buffer as a result. The increased output impedance of the buffer lowers the conversion gain for the noise components of the buffer compared to that of the desired signal improving the out-of-band noise performance of the transmitter. For the EDGE mode where the RX band noise is located at 20MHz offset, the resistor R_1 is included in the feedback system, while it is bypassed for the 3G mode to widen the feedback bandwidth.

The transmitter delivers an output power of 6.3dBm for 3G mode, with better than −40dBc ACLR at 5MHz. The measured EVM is 3.4/4.3% for LB/HB outputs. A receive-band noise of better than −160dBc/Hz is achieved.

12.7.3 SoC Concerns

To lower cost and size, it is common to integrate many functions in the same silicon, including RF, analog, mixed-mode, and digital building blocks. Known as system-on-a-chip (SoC), there are several practical concerns arising from the coupling of the noisy baseband blocks to the critical radio components, such as the VCO or LNA. Especially in cellular applications, very stringent noise and spur requirements are targeted, and must be satisfied with careful layout and architectural techniques.

The principles of power and signal integrity were detailed in Chapter 7. We will present some highlights here. There are several coupling sources, such as substrate, supplies, package, and the board. In order to minimize those we may take a two-step approach. One is to minimize the coupling itself, while the second scheme is to improve the radio resilience to coupling, which is inevitable no matter how good of an isolation is achieved. As for the former, shown in Figure 12.67, there are several well-known techniques often employed to minimize the coupling:

1. On-chip linear regulators (LDO or low-drop output) help isolate the supply of the radio blocks, and particularly the sensitive ones such as the VCO. Multiple bypass capacitors

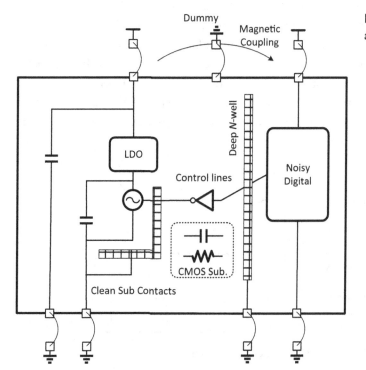

Figure 12.67: Sources of coupling in an SoC

comprising high-density MOS capacitors with linear fringe on top may be used to provide a low-resistance path to the local clean ground.
2. A wide Deep N-well guard band may be placed between the radio and the rest of the baseband, with proper ground connection to a dedicated clean V_{SS} or preferably V_{DD}, to reduce the CMOS substrate coupling.
3. It is common to place the radio on the top corner of the die (see Figure 12.69 as our case study example) to provide two sides for the pads so that critical supplies as well as RF IOs can be accessed with low-inductance routing on packages and PCB. Moreover, less noisy baseband blocks could be placed around the radio to avoid the coupling through substrate and package bondwires.
4. To further reduce the magnetic coupling through the package and bondwires, dummy bondwires connected to ground may be placed between the radio and baseband sections. Moreover, the digital and RF ground planes could be separated on package. To ensure that the charge device model ESD (electrostatic discharge) targets are met, secondary ESD protection between RF and digital domains could be used. Most low-cost handsets do not allow any bypass components at the back side of the phone, where the key pad is placed. Thus the sensitive radio supplies that require low-inductance bypass must be located as close as possible to the periphery of the package.

As it is not possible to entirely eliminate the coupling, especially in a low-cost SoC with a cheap package option, the radio tolerance to noise sources must be enhanced. To achieve that, several architectural techniques can be employed:

1. A *clock shifting* scheme may be implemented in the baseband where a different frequency created by a fractional PLL, is assigned near the harmonics of troublesome baseband clocks, thus shifting the spurs away from the desired channel. Illustrated in Figure 12.68, except for a few blocks that require a clock frequency at the exact integer of 26MHz, such as the modem, the majority of the digital circuits such as the multimedia, the memory controller, and the periphery run at an optional clock frequency close to an integer ratio of 26MHz, but far enough from the critical RF channel of interest susceptible to coupling (such as EGSM channel 5 at 936MHz, which is the 36th harmonic of 26MHz). This puts the spur created by the aforementioned blocks, which compose the majority of the noisy baseband, far away from the desirable channel.

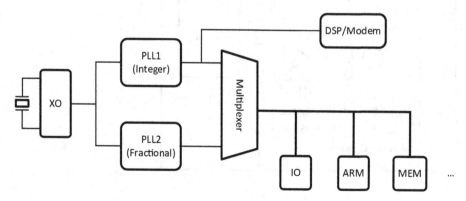

Figure 12.68: Clock shifting architecture to alleviate coupling

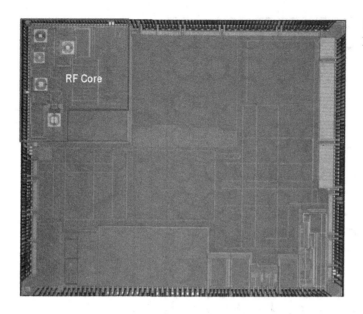

Figure 12.69: An example of a GSM/EDGE SoC

2. A *spur cancellation* block at the receiver output ensures that the downconverted spur resulting from a 26MHz integer harmonic coupling, whose location is well predictable for a given channel, is removed from the band of interest. This, however, is based on two assumptions: First, the spur must be a single tone, and not a wideband modulated spectrum. Second, the digital activity causing the spur should remain constant. Both assumptions hold well for most cases in a receiver.

An example of a 65nm CMOS SoC intended for GSM/EDGE applications [19] is shown in Figure 12.69.

The die is housed in a low-cost 12mm × 12mm 407-pin FBGA package (see next section for package design details). The radio analog and RF portion including the pads occupy an area of 3.95mm^2, a small portion of the 30mm^2 die, which is mostly taken by the baseband. Despite the high level of integration, almost no performance degradation is observed in the sensitivity of the receiver thanks to the techniques mentioned.

12.7.4 Packaging Concerns

The package design is often as important as the radio design itself. If not done properly, despite the best circuits used, it may lead to poor performance, particularly RF signal loss, noise figure degradation, coupling and spurs, and pulling.

There are several types of packages used in modern radios today, the most common ones being FBGA (fine ball grid array) and WCSP (wafer level chip scale). The latter uses *bumps* inside the chip (as opposed to *bonding pads*), which are routed and redistributed by an RDL[10]

---

[10] RDL is an extra level of routing (typically copper) available on chip. Although originally intended for package routing, given its relatively small sheet resistance, it may be used for general signal and supply routing, or inductor design as we saw in Chapter 1. RDL layer is often called AP layer as well.

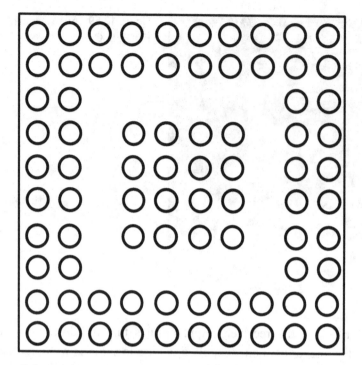
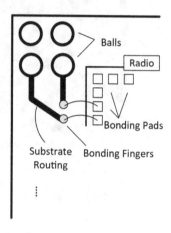

Figure 12.70: An example of an FBGA package

layer (redistribution layer). The package balls may be directly mounted on the die, which is cheaper. On the other hand, in an FBGA package, majority of the routing happens on the package substrate. Two- or four-level substrates are common, although very complex SoCs may use higher number of layers, which leads to higher cost.

Shown in Figure 12.70 is an example of an FBGA package and the design details. On the left is the package back side consisting of 80 balls that are soldered on the PCB. Shown on the right is some details of the package design.

The pads are first bonded to *bonding fingers* inside the package, which are subsequently routed to the balls on different layers on package substrate. Below several important common design practices are summarized:

- The supply and ground routings must be wide and short to minimize inductance. The V_{SS} balls may be distributed evenly across the package, to provide convenient low-inductance connection to different parts of the radio. These V_{SS} balls need to be connected by an extremely low-inductance plane on the substrate (V_{SS} *islands*).
- The RF routings must be as short as possible. For differential IOs, the length of the two paths must be well matched.
- The package substrate may be isolated between sensitive RF section and noisy digital portion. This is clearly a bigger concern for complex SoCs. This however may lead to poor interdomain ESD protection, and is often resolved by using secondary ESD protection.

Compared to WCSP design, FBGA packages offer more flexibility but are typically more expensive. One other advantage of WCSP is smaller inductance due to the elimination of band wires.

12.7.5 Variations

There are several sources of variations that may impact the radio performance:

- Devices: including threshold voltage, mobility, and oxide variations, caused by finite accuracy of the fabrication process. They affect the device transconductance, noise, headroom, gain or linearity. The foundry guarantees a certain window where the device parameters (such as threshold voltage) must fall within. This is typically gauged by *wafer acceptance test* (WAT) done at the foundry. An example of WAT data for NMOS threshold voltage in 65nm process is shown in Figure 12.70. The threshold voltage is typically measured to be 372mV, but can vary to as large as 444mV in slow corner and as low as 320mV in fast corner. This corresponds to $\pm 4.5\sigma$ variations, with the data collected over a large number of wafers.

In addition to process corners, there are temperature and voltage variations that must be included. The circuit performance must be guaranteed within these limits, and for that reason it is crucial to run corner simulations (and eventually split lot measurements).

- Passives: caused by metal width or oxide thickness variations. They affect the inductance and capacitances, and consequently the bandwidth or resonance frequency. As the on-chip inductance is typically a function of the number of turns, metal width, and other parameters discussed in Chapter 1, it is expected to vary over process, but not as strongly as devices or capacitors.
- External components: This includes the supply (or battery), the antenna impedance, external filter passband loss or rejection, as well as the crystal. The circuit parameters such as noise or linearity are typically derived with such variations included. We showed examples of how to deal with crystal variation in Chapter 9.

As we will discuss shortly, as a part of product qualification it is crucial to ensure that the radio performance is within the limits set by the data sheet when measured across a large number of

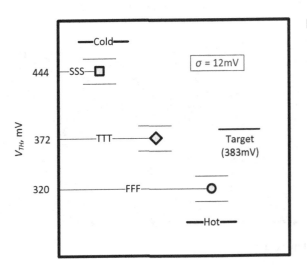

Figure 12.71: WAT data for 65nm process

samples including various corner lots, with temperature, voltage, and external component variations considered.

12.7.6 Product Qualification

Prior to entering mass production, the radio must undergo a rigorous set of tests to ensure its quality. This includes the following steps:

- Characterization over extreme supply, temperature, and corners on a large number of parts.
- ESD (electrostatic discharge) and latch-up test. The ESD test includes HBM (human body model, which typically covers a single voltage domain), CDM (charged device model that is between various domains), and MM (machine model).
- Burn-in test at extreme temperatures to ensure reliability. The circuits that undergo large swings such as VCO or PA are generally of a more concern. To perform the test, a large number of samples are measured under normal condition, and are left to remain at high temperature under extreme conditions (for example maximum transmitter power to ensure PA reliability) for a specified period of time (say 1000 hours). The devices are subsequently tested, and are expected to achieve a similar performance. If there is any degradation, that may be added to the production test limits.

It is not uncommon for RF circuit designers to perform *aging* simulations on critical blocks. It anticipates the device degradation when operating under extreme voltage, and produces new sets of models to reflect the *aged* performance. The design may be modified accordingly if the degradation is unacceptable.

- Package qualification, including thermal shock and temperature cycles
- Achieving acceptable yield

An example of the test setup for product qualification is shown in Figure 12.72.

Many of these tests such as reliability or ESD must be thought of at the very early stages of the design. Underestimating these requirements may result in extra revisions at later stages, despite the fact that the part may be perceived to pass the technical specifications.

 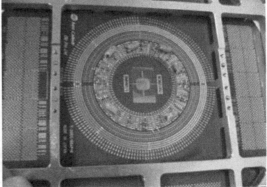

Figure 12.72: Product qualification setup example

12.7.7 Production Issues

As we mentioned earlier, the production testing is fully automated and is performed to ensure the quality of the product is within the limits promised in the *data sheet*. An example of the automated test equipment is shown in Figure 12.73.

There are several challenges associated with the ATE:

- The ATE environment is very different, often less controllable and more unpredictable, compared to the DVT environment. This mostly has to do with the automated process, and the cost concerns associated with it.
- The ATE board is not as optimized and as *RF friendly* as one would desire. Shown in Figure 12.73 is an example of the ATE test board.

To stand the mechanical pressure, the board is usually a few inches thick, and has over 30 layers. This results in longer traces, which leads to more inductance on traces. This in turn causes a few potential issues, such as larger noise figure, or lower output power. In addition, as shown on the right, it is much harder to access the pins, for instance to place bypass capacitor

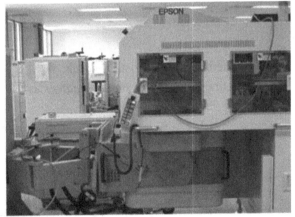

Figure 12.73: An example of ATE

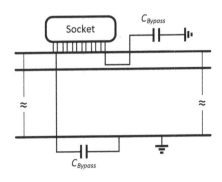

Figure 12.74: ATE test board example

on the critical supplies, or to provide low inductance ground. Moreover, a socket must be inevitably used, as opposed to soldering down the parts as is the case in an actual product, leading to further performance degradation.

- For cost reasons, it is desirable to perform the ATE testing as quickly as possible. *Test times on the order of a few seconds or less are common.* This is another factor that could result in less predictably as we will see shortly.

We shall discuss how these factors may be quantified for a product, and present various ways of dealing with the aforementioned challenges.

ATE-Bench Correlation — The poor performance of the ATE board, and the environment less predictability may not be as bad if one can show that the ATE results are *consistently* worse than the DVT (or *bench*). This requires a subtle and often time-consuming process of *correlating* the test data between the DVT environment and ATE.

The correlation is typically gauged by the *coefficient of correlation* or R^2, indicating how well data points fit into a statistical model – sometimes simply a line or curve. Suppose a data set has values y_i, each of which has an associated modeled value f_i, where f may be simply a linear fit. The variability of the data is defined by the *total sum of squares* as

$$SS_{Tot} = \sum_{i=1}^{n} (y_i - \bar{y})^2,$$

where $\bar{y}$ is the statistical mean, $\bar{y} = \frac{1}{n}\sum_{i=1}^{n} y_i$, and n is the number of observations. We also define the *residual sum of squares* as

$$SS_{Res} = \sum_{i=1}^{n} (y_i - f_i)^2.$$

The coefficient of determination is defined as

$$R^2 = 1 - \frac{SS_{Res}}{SS_{Tot}} = 1 - \frac{\sum_{i=1}^{n}(y_i - f_i)^2}{\sum_{i=1}^{n}(y_i - \bar{y})^2}.$$

A perfect fit results in $R^2 = 1$, but in general: $R^2 \leq 1$. An example of a linear fit is shown in Figure 12.75. The gray squares represent the residual sum of squares (SS_{Res}). The bigger the squares, the worse the correlation, and the smaller R^2 is. As a rule of thumb, we would like R^2 to be greater than 0.8.

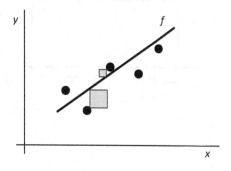

Figure 12.75: Description of a linear fit and the corresponding R^2

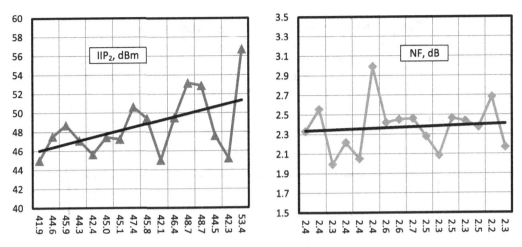

Figure 12.76: Examples of the coefficient of determination and the noise figure

Examples of the coefficient of determination for an EDGE receiver IIP_2 and the noise figure are shown in Figure 12.76.

Using a linear fit, for the IIP_2 an R^2 of 0.976 is calculated, indicating a strong correlation, evident from the plot, whereas the R^2 for the noise figure test is 0.035. The poor correlation for the noise figure is attributed to two factors: First, accurate noise measurements require long averaging, which is not feasible given the test time concerns; and second, as the noise levels are generally very low, the measurement is much more sensitive compared to gain or linearity for instance. Even though the ATE results (shown after *de-embedding* the losses) are within tenths of a dB of the bench data, the correlation is poor. As a result, in this case the noise figure of the radio must be *guaranteed by design* to be within the requirements. This is something that may often need to be thought of at the early stages of the design.

Gauge R&R — The uncertainty of the ATE environment, often exacerbated by the test time concerns, is typically determined by *gauge R&R* (reproducibility and repeatability), indicating how reliable the measured data are.

Let us consider the 400kHz modulation mask of a GSM transmitter as an example. The requirement set by the standard is –60dBc. To leave room for the power amplifier contribution and other production-related concerns, suppose a modulation mask of better than –63dBc is intended for the radio. This means that every radio shipped, under the extreme conditions indicated in the data sheet (temperature, frequency, or supply) must have a ±400kHz modulation mask of no higher than –63dBc.

By performing more averaging, a more reliable measurement is obtained, and thus a lower gauge R&R is observed, but at the expense of test time and ultimately cost. Table 12.1 shows a summary.

As it is common to perform the ATE measurements at room temperature, guard band must be added to account for the extreme temperature and other possible conditions. In this case of 400kHz mask, the high temperature performance is consistently worse by no more than 0.8dB, which must be added to the ATE test limit. Furthermore, performing statistical analysis on the date, it is shown that performing a 50-busrt average results in as much as 0.9dB of variation in the measured data, for the same part at exact same conditions. This is simply a result of ATE

Table 12.1: Gauge R&R for GSM 400kHz modulation mask

	Data Sheet Limit	Guard Band	GRR, 10AVG	GRR, 50AVG	ATE Limit, 10AVG	ATE Limit, 50AVG
400kHz modulation	−63dBc	0.8dB	2.1dB	0.9dB	−65.9dBc	−64.7dBc

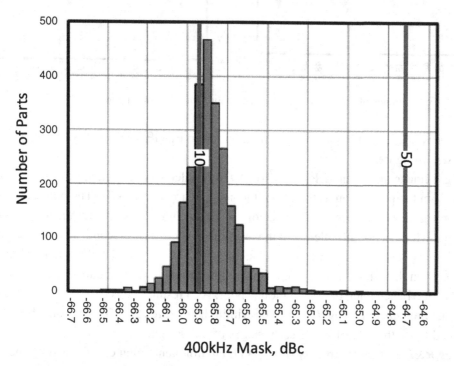

Figure 12.77: Measured 400kHz modulation in the ATE

measurement uncertainty, which must also be added to the test limit. A lower number of averages causes a higher gauge R&R. Thus, the actual ATE limit is determined to be −64.7/−65.9dBc for 50-/10-burst averaging.

Shown in Figure 12.77 is the measured 400kHz mask over a large number of parts (over 10000) in the ATE. Evidently, setting a limit of −65.9dBc corresponding to 10 average results in too big a fallout, and consequently a poor *yield*. Thus, in this case 50 averages may be preferred, which results in almost no yield hit, but comes at the expense of longer test time.

This is why *higher margins* are often desirable.

To reduce test time, if one can find a frequency where the 400kHz mask is consistently the worst, only that frequency may be tested. Otherwise, a range of frequencies (often low, mid, and high) may be considered. For instance, if the 400kHz modulation is mostly affected by the phase noise, it may not be unreasonable to assume that the highest frequency that the radio is intended to operate is the worst case. This must however be statistically established by performing sufficient testing on a large number of samples prior to reaching mass production.

12.8 Summary

In this last chapter of the book we covered the transceiver architectures and some practical aspects of the radio design and related production issues.

- Section 12.1 discussed the general architectural concerns such as noise or blocking requirements.
- In Section 12.2 we presented various commonly used receiver architectures, including super-heterodyne, direct-conversion, and image-reject receivers.
- Section 12.3 discussed some of the more recent receiver topologies such as mixer-first or noise-canceling receivers generally known as blocker-tolerant receivers.
- The receiver gain control, filtering requirements, and the ADC specs were discussed in Sections 12.4 and 12.5.
- Transmitter architectures such direct-conversion, dual-conversion, polar, and out-phasing topologies were presented in Section 12.6.
- The chapter concluded in Section 12.7, where a case study was presented. Furthermore, in this section we discussed practical aspects of radio design such as layout, packaging, and production testing.

12.9 Problems

1. Assuming a 3G transmitter consists of a 2GHz DAC directly connected to the PA, find the DAC effective number of bits. The DAC must put out a power of 0dBm, and the receive band noise must be better than −160dBc/Hz. The 3G signal has a PAPR of 3dB.
2. An RF DAC serves as a WLAN transmitter with 10dB PAPR. Find the DAC effective number of bits such that it desensitizes a nearby LTE receiver with 3dB noise figure by no more than 1dB. The WLAN transmitter and LTE receiver have 20dB of isolation.
3. Redo the previous problem if an ADC is used as the receiver. Assume the receiver needs to have an effective noise figure of 3dB, and is subject to a nearby WLAN blocker as large as −20dBm.
4. In a GSM receiver tuned to 950MHz, the IF is at 110MHz (high-side injection). Assuming a hard-switching differential mixer, identify all the problematic blockers and the filtering needed. The LNA has an IIP_2 of 35dBm, and IIP_3 of 0dBm. Everything else is ideal.
5. Redo Problem 4 with low-side injection. Which one is better?
6. Assume a 2.4GHz WLAN receiver is set up to capture the entire 80MHz ISM band, whereas each channel is only 20MHz wide. Assuming a PAPR of 10dB for 64QAM, and a required SNR of 22dB to demodulate, calculate the ADC dynamic range for two cases:
    a. Only one 20MHz channel is being received
    b. The entire 80MHz band is captured

Assume the signal is accompanied by a −20dBm nearby WLAN blocker, and the receiver has a noise figure of 3dB. We do not want the ADC to add more than 1dB of its own noise to the 3dB noise figure.

7. Using basic sine-cosine trigonometric properties, show how one sideband is rejected in the low-IF receiver below.

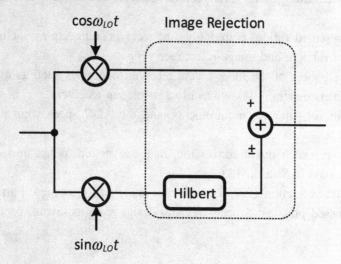

8. Using basic trigonometric properties, show how the first image is canceled in a Weaver receiver.

9. Find the second image location in a Weaver receiver for all the combinations of low- and high-side injection for the two LOs.

10. Argue why in the original Weaver architecture the second IF cannot be zero. Propose a modified Weaver architecture with the second IF at zero. Discuss the advantages and disadvantages.

11. The receiver shown below is represented with an input-referred current noise source of i_{nRX}, and has a nominal noise factor of F_{RX} when matched to the nominal source impedance R_s.
    a. Derive and expression for F_{RX} based on the receiver input-referred noise.
    b. Find the receiver noise figure under mismatch, that is, when the source has an arbitrary input impedance of Z_s. Express your result based on source reflection coefficient.

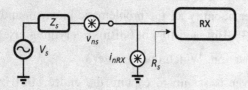

12. Consider the following system where multipliers are ideal and α is an arbitrary phase-shift.
    a. Prove that the overall system is LTI with the following impulse response:
    $h(t)\cos(\omega_0 t + \alpha)$.

b. Prove that this transfer function is the cascade of a system with the impulse response of $h(t)\cos\omega_0 t$ and a Hilbert-like transfer function with magnitude of unity and phase-shift of $+\alpha$ and $-\alpha$ for positive and negative frequencies, respectively.

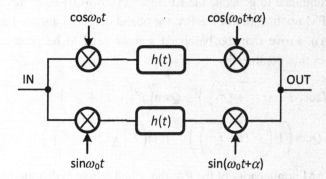

13. Prove the noise figure equation of an M-phase mixer-first receiver: $F \approx \left[1 + \frac{R_{SW}}{R_s} + \frac{\overline{v_{bb}^2}}{4MKTR_s}\right] K$. **Hint:** Use the simplified model shown below and work out the noise contributors based on the analysis we provided for M-phase mixers in Chapter 7.

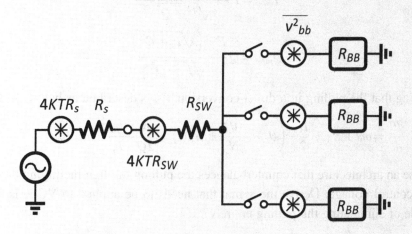

14. In a dual-conversion receiver with sliding IF, a divide-by-2 is employed to create the second LO, intended for use in Bluetooth applications with RF input at the ISM band (2402–2480MHz). Find two possible choices for the LO range, and the IF. Discuss the pros and cons of each.

15. Repeat Problem 14 if the receiver employs a divide-by-4, and is intended for use in 802.11a applications (5180–5825MHz).

16. Find the GSM ADC requirements for 600kHz and 3MHz blockers, assuming a 1st-order RC lowpass filter with 270kHz cut-off. The IF is 135kHz. Which case is more stringent, and how do they compare to 400kHz adjacent blocker case?

17. Calculate ADC requirements for a WLAN receiver as follows: The required SNR to demodulate a 20MHz 64QAM OFDM signal with 10dB PAPR is 22dB. The receiver noise figure is 3dB in the absence of the ADC noise, and must not degrade by more than

0.1dB when ADC is present. A desired signal 3dB above the sensitivity may be accompanied by a nearby blocker of −20dBm, with 10dB of IF filtering.

18. Assume that in an IQ transmitter the quadrature baseband signals $I = A\cos\theta$ and $Q = A\sin\theta$ are upconverted and combined to generate the RF signal $A\cos(\omega_0 t + \theta)$ as the PA input.

    a. Due to the AM-to-PM nonlinearity of the PA, the actual output is distorted and becomes $A\cos(\omega_0 t + \theta + F(A))$. Prove that the baseband signals need to be predistorted to the followings to correct this nonlinearity:

    $$I_{mod} = I\cos\left(F\left(\sqrt{I^2 + Q^2}\right)\right) + Q\sin\left(F\left(\sqrt{I^2 + Q^2}\right)\right)$$

    $$Q_{mod} = Q\cos\left(F\left(\sqrt{I^2 + Q^2}\right)\right) - I\sin\left(F\left(\sqrt{I^2 + Q^2}\right)\right)$$

    b. Due to the AM-to-AM nonlinearity of the PA, the actual output is distorted and becomes $F(A)\cos(\omega_0 t + \theta)$. Prove that the baseband signals need to be predistorted to the following to correct this nonlinearity:

    $$I_{mod} = I\frac{F^{-1}\left(\sqrt{I^2 + Q^2}\right)}{\sqrt{I^2 + Q^2}}$$

    $$Q_{mod} = Q\frac{F^{-1}\left(\sqrt{I^2 + Q^2}\right)}{\sqrt{I^2 + Q^2}}$$

19. Knowing that the pulling in a direct-conversion TX is described as below,

    $$\frac{d\theta}{dt} = \omega_0 + K_{VCO}\frac{I_{CP}}{2\pi}\left(\theta_{ref} - \frac{\theta}{N}\right) * h_{LF}(t) + \frac{\omega_0}{2Q}\frac{\gamma A_{BB}^2}{I_S}\sin(2\theta_{BB} - \Psi),$$

    propose an architecture that counterbalances the pulling through modifying the transmitter VCO control voltage. Derive the signal that needs to be applied to VCO. Is this scheme capable of eliminating the pulling entirely?

12.10 References

[1] J. Mitola, "The Software Radio Architecture," *Communications Magazine, IEEE*, 33, no. 5, 26–38, 1995.
[2] E. H. Armstrong, "The Super-heterodyne – Its Origin, Development, and Some Recent Improvements," *Proceedings of the Institute of Radio Engineers*, 12, no. 5, 539–552, 1924.
[3] E. Cijvat, S. Tadjpour, and A. Abidi, "Spurious Mixing of Off-Channel Signals in a Wireless Receiver and the Choice of IF," *IEEE Transactions on Circuits and Systems II: Analog and Digital Signal Processing*, 49, no. 8, 539–544, 2002.
[4] A. Abidi, "Direct-Conversion Radio Transceivers for Digital Communications," *IEEE Journal of Solid-State Circuits*, 30, no. 12, 1399–1410, 1995.
[5] D. Kaczman et al., "A single-chip 10-bond WCDMA/HSDPA L_1-band GSM/EDGE SAW-less CMOS receiver with digRF 3G interface and 90dBm IIP2," *IEEE Journal of Solid-State Circuits*, 44, no. 3, 718–739, 2009.
[6] R. Hartley, "Single-Sideband Modulator." U.S. Patent No. 1666206, April 1928.

[7] J. Crols and M. Steyaert, "An Analog Integrated Polyphase Filter for a High Performance Low-IF Receiver," in *Symposium on VLSI Circuits, Digest of Technical Papers*, 1995.

[8] D. Weaver, "A Third Method of Generation and Detection of Single-Sideband Signals," *Proceedings of the IRE*, 44, no. 12, 1703–1705, 1956.

[9] M. Zargari, M. Terrovitis, S.-M. Jen, B. J. Kaczynski, M. Lee, M. P. Mack, S. S. Mehta, S. Mendis, K. Onodera, H. Samavati, et al., "A Single-Chip Dual-Band Tri-Mode CMOS Transceiver for IEEE 802.11 a/b/g Wireless LAN," *IEEE Journal of Solid-State Circuits*, 39, no. 12, 2239–2249, 2004.

[10] C. Andrews and A. Molnar, "A Passive Mixer-First Receiver with Digitally Controlled and Widely Tunable RF Interface," *IEEE Journal of Solid-State Circuits*, 45, no. 12, 2696–2708, 2010.

[11] S. Blaakmeer, E. Klumperink, D. Leenaerts, and B. Nauta, "The Blixer, a Wideband Balun-LNA-I/Q-Mixer Topology," *IEEE Journal of Solid-State Circuits*, 43, no. 12, 2706–2715, 2008.

[12] D. Murphy, H. Darabi, A. Abidi, A. Hafez, A. Mirzaei, M. Mikhemar, and M.-C. Chang, "A Blocker-Tolerant, Noise-Cancelling Receiver Suitable for Wideband Wireless Applications," *IEEE Journal of Solid-State Circuits*, 47, no. 12, 2943–2963, 2012.

[13] Z. Ru, N. Moseley, E. Klumperink, and B. Nauta, "Digitally Enhanced Software-Defined Radio Receiver Robust to Out-of-Band Interference," *IEEE Journal of Solid-State Circuits*, 44, no. 12, 3359–3375, 2009.

[14] M. Soer, E. Klumperink, Z. Ru, F. van Vliet, and B. Nauta, "A 0.2-to-2.0GHz 65nm CMOS Receiver Without LNA Achieving $\gg$11dBm IIP3 and $\ll$6.5dB NF," in *Solid-State Circuits Conference – Digest of Technical Papers*, 2009.

[15] E. Sacchi, I. Bietti, S. Erba, L. Tee, P. Vilmercati, and R. Castello, "A 15mW, 70kHz 1/f Corner Direct Conversion CMOS Receiver," in *Proceedings of the IEEE Custom Integrated Circuits Conference*, 2003.

[16] M. Mikhemar, A. Mirzaei, A. Hadji-Abdolhamid, J. Chiu, and H. Darabi, "A 13.5mA Sub-2.5dB NF Multi-band Receiver," in *Symposium on VLSI Circuits*, 2012.

[17] F. Bruccoleri, E. Klumperink, and B. Nauta, "Wide-Band CMOS Low-Noise Amplifier Exploiting Thermal Noise Canceling," *IEEE Journal of Solid-State Circuits*, 39, no. 2, 275–282, 2004.

[18] D. Murphy et al., "A Noise-Cancelling Receiver Resilient to Large Harmonic Blockers," *IEEE Journal of Solid-State Circuits*, 50, 1336–1350, 2015.

[19] H. Darabi, P. Chang, H. Jensen, A. Zolfaghari, P. Lettieri, J. Leete, B. Mohammadi, J. Chiu, Q. Li, S.-L. Chen, Z. Zhou, M. Vadipour, C. Chen, Y. Chang, A. Mirzaei, A. Yazdi, M. Nariman, A. Hadji-Abdolhamid, E. Chang, B. Zhao, K. Juan, P. Suri, C. Guan, L. Serrano, J. Leung, J. Shin, J. Kim, H. Tran, P. Kilcoyne, H. Vinh, E. Raith, M. Koscal, A. Hukkoo, C. Hayek, V. Rakhshani, C. Wilcoxson, M. Rofougaran, and A. Rofougaran, "A Quad-Band GSM/GPRS/EDGE SoC in 65nm CMOS," *IEEE Journal of Solid-State Circuits*, 46, no. 4, 870–882, 2011.

[20] A. Mirzaei and H. Darabi, "Mutual Pulling between Two Oscillators," *IEEE Journal of Solid-State Circuits*, 49, no. 2, 360–372, 2014.

[21] R. Adler, "A Study of Locking Phenomena in Oscillators," *Proceedings of the IEEE*, 61, no. 10, 1380–1385, 1973.

[22] A. Mirzaei and H. Darabi, "Pulling Mitigation in Wireless Transmitters," *IEEE Journal of Solid-State Circuits*, 49, no. 9, 1958–1970, 2014.

[23] H. Darabi, S. Khorram, H.-M. Chien, M.-A. Pan, S. Wu, S. Moloudi, J. Leete, J. Rael, M. Syed, R. Lee, B. Ibrahim, M. Rofougaran, and A. Rofougaran, "A 2.4-GHz CMOS Transceiver for Bluetooth," *IEEE Journal of Solid-State Circuits*, 36, no. 12, 2016–2024, 2001.

[24] H. Darabi, J. Chiu, S. Khorram, H. J. Kim, Z. Zhou, Hung-Ming Chien, B. Ibrahim, E. Geronaga, L. Tran, and A. Rofougaran, "A Dual-Mode 802.11b/Bluetooth Radio in 0.35-um CMOS," *IEEE Journal of Solid-State Circuits*, 40, no. 3, 698–706, 2005.

[25] O. Erdogan, R. Gupta, D. Yee, J. Rudell, J.-S. Ko, R. Brockenbrough, S.-O. Lee, E. Lei, J. L. Tham, H. Wu, C. Conroy, and B. Kim, "A Single-Chip Quad-Band GSM/GPRS Transceiver in 0.18um Standard CMOS," in *Solid-State Circuits Conference, 2005. Digest of Technical Papers*, 2005.

[26] P.-H. Bonnaud, M. Hammes, A. Hanke, J. Kissing, R. Koch, E. Labarre, and C. Schwoerer, "A Fully Integrated SoC for GSM/GPRS in 0.13um CMOS," in *Solid-State Circuits Conference, 2006. Digest of Technical Papers*, 2006.

[27] R. Staszewski, J. Wallberg, S. Rezeq, C.-M. Hung, O. Eliezer, S. Vemulapalli, C. Fernando, K. Maggio, R. Staszewski, N. Barton, M.-C. Lee, P. Cruise, M. Entezari, K. Muhammad, and D. Leipold, "All-Digital PLL and Transmitter for Mobile Phones," *IEEE Journal of Solid-State Circuits*, 40, no. 12, 2469–2482, 2005.

[28] R. Staszewski, D. Leipold, O. Eliezer, M. Entezari, K. Muhammad, I. Bashir, C.-M. Hung, J. Wallberg, R. Staszewski, P. Cruise, S. Rezeq, S. Vemulapalli, K. Waheed, N. Barton, M.-C. Lee, C. Fernando, K. Maggio, T. Jung, I. Elahi, S. Larson, T. Murphy, G. Feygin, I. Deng, T. Mayhugh, Y.-C. Ho, K.-M. Low, C. Lin, J. Jaehnig, J. Kerr, J. Mehta, S. Glock, T. Almholt, and S. Bhatara, "A 24mm^2 Quad-Band Single-Chip GSM Radio with Transmitter Calibration in 90nm Digital CMOS," in *Solid-State Circuits Conference, 2008. Digest of Technical Papers*, 2008.

[29] Z. Boos, A. Menkhoff, F. Kuttner, M. Schimper, J. Moreira, H. Geltinger, T. Gossmann, P. Pfann, A. Belitzer, and T. Bauernfeind, "A Fully Digital Multimode Polar Transmitter Employing 17b RF DAC in 3G Mode," in *Solid-State Circuits Conference Digest of Technical Papers*, 2011.

[30] H. Darabi, H. Jensen, and A. Zolfaghari, "Analysis and Design of Small-Signal Polar Transmitters for Cellular Applications," *IEEE Journal of Solid-State Circuits*, 46, no. 6, 1237–1249, 2011.

[31] M. Youssef, A. Zolfaghari, B. Mohammadi, H. Darabi, and A. Abidi, "A Low-Power GSM/EDGE/WCDMA Polar Transmitter in 65-nm CMOS," *IEEE Journal of Solid-State Circuits*, 46, no. 12, 3061–3074, 2011.

[32] M. Wakayama and A. Abidi, "A 30-MHz Low-Jitter High-Linearity CMOS Voltage-Controlled Oscillator," *IEEE Journal of Solid-State Circuits*, 22, no. 6, 1074–1081, 1987.

[33] H. Chireix, "High Power Outphasing Modulation," *Proceedings of the Institute of Radio Engineers*, 23, no. 11, 1370–1392, 1935.

[34] D. Cox and R. Leck, "Component Signal Separation and Recombination for Linear Amplification with Nonlinear Components," *IEEE Transactions on Communications*, 23, no. 11, 1281–1287, 1975.

[35] J. Qureshi, M. Pelk, M. Marchetti, W. Neo, J. Gajadharsing, M. van der Heijden, and L. de Vreede, "A 90-W Peak Power GaN Outphasing Amplifier with Optimum Input Signal Conditioning," *IEEE Transactions on Microwave Theory and Techniques*, 57, no. 8, 1925–1935, 2009.

[36] S. Moloudi and A. Abidi, "The Outphasing RF Power Amplifier: A Comprehensive Analysis and a Class-B CMOS Realization," *IEEE Journal of Solid-State Circuits*, 48, no. 6, 1357–1369, 2013.

INDEX

Δ–Σ modulator, 621, 624
Δ–Σ modulator noise, 622, 630
Δ–Σ modulator nonideal effects, 626
$1/f$ noise, 292, 449
$1/f$ noise in passive mixers, 478, 603
1dB compression, 357
25% LO, 466
25% mixer, 467
25% upconversion mixer, 489
50Ω, 174
8-phase divide by 4, 637

accumulation-mode MOS capacitor, 44
active filters, 238
active filters ladder design, 238
active filters nonideal effects, 245
active mixer, 461
active mixer 2nd-order distortion, 457
active mixers, 445
active mixers $1/f$ noise, 479
active mixers linearity, 448
active mixers white, 452
active polyphase filters, 263
active upconversion mixer, 486
active upconverter, 487
active-RC integrator, 240
adjacent channel leakage ratio, 369
adler's differential equation, 379
all-pole transfer function, 230
AM sideband, 524–525
AM–AM, 375
Ampere's circuital law, 5
amplitude modulation, 110
amplitude noise, 521
AM–PM, 375
AM–PM nonlinearity, 731
analog multiplier, 444
antenna, 36
antenna characteristics, 36
antenna directivity, 42
antenna effective area, 177
antenna efficiency, 43
antenna gain, 43
antenna radiation resistance, 41
antennas as two-port circuits, 176
approximation theory, 221

ATE, 747
attenuator, 188
autocorrelation, 96
automatic testing equipment (ATE), 734
available power, 139
available power gain, 191, 191
available power gain, reciprocal networks, 142

backed off, 370
balanced AM modulator, 113
band-limited white noise, 346
bandpass filter, 236
bandpass LC filter, 234
Bank's general result, 536
baseband signal, 110
basic receiver topology, 693
Bessel functions, 118
biconjugate matching, 148, 163
Biot–Savart, law, 5
biquad, 243
blocker, 333
blocker-tolerant receiver, 707
bonding pads, 743
Brownian motion, 279
buck converter, 678
bulk acoustic wave, 248
bumps, 743
burn-in test, 746
Butterworth, 223
Butterworth filter, 213, 234

capacitance, 4
capacitive coupling, 427
capacitor lumped model, 45
capacitor noise, 285
carrier, 439
carrier frequency, 110
cartesian transmitter, 486
cascode LNA, 405
cellular network, 332
central limit theorem, 98
characteristic function, 213
characteristics impedance, 15, 31
charge, 2
charge pump, 602

charge pump, 604
charge pump modeling, 605
charge pump nonlinearity, 627
Chebyshev filter, 227
Chebyshev polynomial, 227
Chireix combiner, 733
circulators, 192
circulators in full-duplex radios, 194
class A PA, 654
class A PA with gain control, 656
class B PA, 656
class C oscillator, 564
class C PA, 660
class C PA efficiency, 661
class D digital PA, 665
class D PA, 662
class D^{-1} PA, 663
class E PA, 669
class F PA, 672
class-A PA efficiency, 655
CMOS oscillator, 544
CMOS oscillator noise factor, 521
coaxial cable, 3
coaxial cable capacitance, 5
coaxial shield, 428
coexistence, 336
Colpitts oscillator, 546
Colpitts oscillator noise factor, 521
common-source LNA, 398
common-gate LNA, 400
complementary amplifier, 394
complementary shunt feedback LNA, 403
compression, 357
compressive, 338
conjugate matching, 141, 652
Coulomb's law, 2
coupling, 427
coupling and shielding, 427
cross-modulation, 352
crystal electrical model, 582
crystal model, 582
crystal oscillator, 581
crystal overdriving, 585
crystal poles and zeros, 584
cubic pre-distorter, 674

current switching, 442, 445
current-mode mixer, 471
current-mode passive mixer, 462
current-mode receiver, 707
cyclostationary noise, 293, 527
cyclostationary process, 108

D flip flop, 633
DC gain, 394
DC–DC converter, 678
dead-zone, 602
decoupling capacitors, 425
demodulation, 110
desensitization, 358
device verification and testing (DVT), 734
differential class A PA, 656
differential inductors, 56
differential two-ports, 168
digital loop filter, 644
digital PA, 666
digital PLL, 640
digital PLL linear analysis, 646
digitally controlled oscillator (DCO), 645
diode-ring mixer, 442
dipole, 38
direct-conversion, 379
direct-conversion transmitter, 720
direct-modulation transmitter, 723
discrete tuning, 47
distributed circuits, 13
divide-by-2, 633
divide-by-2/3, 634
divide-by-4/5, 635
Doherty PA, 682–683
double side-band noise, 454
doublet, 84
doubly terminated, 208
downconversion mixer, 441
drain efficiency, 653
dual-conversion receiver, 706
dual-conversion transmitter, 722
dual-modulus divider, 634
duplexer, 253
dynamic CMOS flip flop, 633
dynamic CMOS latch, 632

efficiency, 652
efficiency of class A and B PAs, 658
electric energy, 3
electric field, 2
electric field intensity, 2
electric flux density, 2
electrical balance duplexer, 254
electromechanical, 249
electromotive force (emf), 9
electrostatic discharge (ESD), 746
electrostatic energy, 15

emitter-coupled pair, 446
EMX, 54
energy conservation, 22
envelope elimination and restoration, 675
envelope tracking, 676, 681
equal ripple, 223, 226
equipartition law, 279, 281
ergodicity, 97
error vector magnitude, 371
error vector magnitude (EVM), 371
excess noise ratio, 323
expected value, 96

Faraday cage, 4, 427
Faraday's law, 9
FBAR oscillator, 581, 588
FBAR resonator, 251
FBGA package, 743
feedback, 511
feedback divider, 611
feedback oscillator model, 511
feedback system nonlinearity, 354
feedforward LNA, 410
feedforward noise canceling LNA, 411
ferrites, 132, 136
FET-based mixer, 443
FET-equivalent input noise, 297
FET IIP_3, 340–341
FET thermal noise, 289
fifth-order distortion, 350
filter scaling, 233
filtered random process, 106
flicker noise, 292
FM bandwidth, 116
FM noise, 523
Fourier series, 81
Fourier transform, 80
Fourier transform of periodic signals, 85
Fourier transform properties, 82
four-phase LO, 467
fractional divider, 619
fractional spurs, 620
fractional-N synthesizer, 611, 619
frequency deviation, 116
frequency divider, 630
frequency modulation, 115
frequency noise, 521
Friis transmission formula, 179
fringe capacitance density, 46
fringe capacitor, 45
fringe capacitors, 44
full-duplex, 335
full-duplex division, 335

gain compression, 357
gain method noise measurement, 322
gate resistance, 413
Gauss's law, 2
Gaussian process, 98
Gilbert mixer, 445
Gilbert multiplier, 444
g_m-C integrator, 240
Groszkowski effect, 560
GSM, 333
gyration ratio, 137
gyrator, 136

half-IF blocker, 364
hard limiter, 102
harmonic distortion, 336
harmonic folding, 491
harmonic mixing, 363
harmonic rejection mixer, 364
Hartley image-reject receiver, 701
Helmholtz equation, 28
Hertzian dipole, 36
high-side injection, 363
Hilbert transform, 93, 440

ideal filter, 207
ideal transformer, 58
IIP_2, 345
IIP_3, 337
IIP_3 of cascade of stages, 343
image, 438
image and half-IF blockers, 694
image blocker, 364
image in zero-IF receiver, 695
image rejection, 702
image-reject receiver, 440
impedance transformation, 149
impulse, 83
impulse response, 86
in-band blockers, 333
inductance, 5
inductively degenerated CS LNA, 432, 409
inductor lumped model, 61
inductor noise, 285
inductor Q definitions, 66
inductor shield, 429
initial value theorem, 288, 512
input intercept point, 337
instantaneous frequency, 116
instantaneous spectral density, 293
integer-N synthesizer, 612
integrated capacitors, 43
integrated inductors, 47
interferer, 333

Index

intermediate frequency, 438
inverter phase noise, 574
IQ mismatch, 697
IQ pre-distorter, 675

latch, 631
LC filter design, 229
LC filters, 208
LC oscillator, 511
Leeson, 512
Leeson's phase noise expression, 515
Lenz's law, 10
linear LC oscillator, 511
Linvil stability factor, 146
LNA biasing, 416
LNA case study, 431
LNA gain control, 420
LNA linearity, 417
LNA practical concerns, 413
LNA/mixer case study, 494
LO duty cycle, 462
LO feed-through, 373
local shunt feedback, 401
loop filter, 596
Lorentz theorem, 132
Lorentzian, 292
loss poles, 220
lossless LC circuit, 17
lossless matching, 388
lossless matching network, 388
lossless reciprocal network, 144
lossless transmission, 166
lossy LC circuit energy, 19
lossy transmission line, 173
low-IF receiver, 701
low-noise transconductance amplifier, 495
low-side injection, 363
lumped circuits, 13

magnetic coupling, 419, 428
magnetic energy, 15
magnetic field, 5
magnetic field intensity, 5
magnetic flux, 7
magnetic flux density, 7
MASH, 626
matching, 150, 387
matching network, 150
matching network in PA, 653
maximally flat, 223
maximum power gain, 147
Maxwell's equations, 10
Maxwell–Boltzmann statistics, 280
metal shield, 55

minimum NF, 303
mixer 2nd-order distortion, 480
mixer basic operation, 438
mixer conversion gain, 465, 470
mixer design methodology, 500
mixer impedance transformation, 473
mixer noise spectrum, 452
mixer operation, 440
mixer with active load, 456
mixer-first receiver, 709
modulated spectrum, 110
modulation, 110
modulation index, 111, 117
modulation mask, 366
MOS gate capacitance, 43
MOS g_m/I_D, 396
M-phase mixer, 467
multi-modulus divider, 638
multi-path fading, 332
multi-turn inductor, 51
multi-turn inductors, 51

narrowband FM, 117
narrowband transformers, 165
native layer, 55
network function, 88
NMOS oscillator, 540
NMOS oscillator noise factor, 542
noise bandwidth, 283
noise-canceling receiver, 710
noise circles, 304
noise figure, 299
noise figure, double or single sideband, 454
noise figure measurement, 322
noise figure of a FET, 301
noise figure of a passive lossy network, 302
noise figure of cascade of stages, 312
noise figure versus return loss, 399
noise figure, impact of feedback, 309
noise shaping, 620
noise-canceling LNA, 412
noise-shaping waveform, 527
nonlinear capacitance, 554, 556, 558
nonlinear LC oscillator, 517
nonlinear oscillator, 520
normal process, 98
N-path filters, 255
Nyquist theorem, 285

one-port, 130
one-port oscillator, 539
optimum noise impedance, 303
oscillation amplitude, 521

oscillator efficiency, 515
oscillator feedback model, 511
oscillator figure of merit, 516
oscillator frequency modulation, 554
oscillator noise factor, 537
oscillator power conservation, 519
oscillator Q degradation, 551
oscillator supply pushing, 561
out-of-band blockers, 333
outphasing transmitter, 732

PA dynamic biasing, 681
PA linearization, 673
packaging concerns, 743
Paley–Wiener criterion, 207
parallel resonance circuit, 87
parallel to series impedance transformation, 160
parallel-mode crystal oscillator, 586
parallel–series circuit, 159
Parseval's energy theorem, 81
passive lossless feedback, 310
passive lossy network noise, 285
passive mixer, 460
passive mixer linearity, 479
passive mixer noise, 478
passive polyphase filters, 260
passive upconversion mixer, 488
patterned shield, 55
peak-to-average ratio, 370
Pierce crystal oscillator, 586
permeability, 7
permittivity, 2
phase constant, 167
phase detector, 596–597
phase modulation, 115
phase noise, 316, 359, 521
phase noise definition, 514
phase noise in the linear oscillator, 512
phase noise of nonlinear LC oscillator, 521
phase velocity, 167
phase-frequency detector, 598, 602
piece of wire inductance, 47
PLL transfer function, 600, 608
PM noise, 523
PM sideband, 524–525
polar transmitter, 726
poles, 88
poles and zeros physical interpretation, 220
polyphase filter, 262
positive real, 217
potential difference, 3
power added efficiency, 653

Index

power integrity, 425
power spectral density, 103
power spectrum, 103
practical LC resonator, 18
pre-distortion, 673
probability density function, 95
product design flow, 735
product qualification, 746
propagation velocity, 14
pulling in direct-conversion transmitters, 721
pulling, 379
pulse amplitude modulated, 108
pulse-width modulation, 664
push–pull class B PA, 658
Poynting vector, 34

quadrature downconversion, 439, 463
quadrature filters, 93, 260
quadrature generation, 266
quadrature mixer, 464
quadrature oscillator, 577
quadrature receiver, 695
quadrature signals, 93
quality factor, 18–19, 66, 550

realizability conditions, 219
realizable impedance, 217
receiver ADC design, 715
receiver ADC requirements, 716
receiver architectures, 692
receiver filtering, 715
receiver gain control, 718
reciprocal, 132
reciprocal mixing, 360–361
reciprocal two-port, 132
reciprocity criteria, 133
reciprocity proof, 135
reciprocity theorem, 134
rectangular pulse, 81
reference divider, 611
reference noise transfer function, 616
reflected power, 153, 212
reflection coefficient, 152, 168
reflection factor, 152, 211
resistive feedback LNA, 403
RF amplifier, 392
RF common-gate amplifier, 397
RF common-source amplifier, 397
RF to LO feedthrough, 481
RF tuned amplifier, 392
ring oscillator, 566
ring oscillator noise, 572

SAW resonator, 251
scattering parameters, 185
second-order distortion, 344, 698
self-mixing, 458, 698
self-resonance frequency, 51
sensitivity, 317
series feedback, 406
series feedback LNA, 405
series-mode crystal oscillator, 585
shield, 425
shielding, 427
shifted process, 109
shoot-through, 664
shunt feedback, 402
shunt feedback LNA, 401
signal integrity, 425
sinc function, 80
single sideband (SSB), 525
single sideband (SSB) noise, 454
single-balanced mixer, 448
single-sideband AM, 113
single-sideband receivers, 121
skin depth, 32
skin effect, 31, 48
small signal nonlinearity, 336
Smith chart, 180
software-defined radio, 691
solenoid, 8
spatial frequency, 167
spectral regrowth, 368
spectrum, 80
spiral inductors, 50
spot noise figure, 322
spurious free dynamic range, 356
square-wave-like LO, 455
stability of two-port amplifiers, 145
stacked inductors, 52
standing wave, 169
static CMOS latch, 631
stationary noise, 294, 523, 527
stationary processes, 97
Stieltjes continued fraction, 230
stochastic processes, 95
stopband filter, 229
substrate loss in inductors, 52
substrate noise, 415
superheterodyne receiver, 693
surface acoustic wave, 248
switched capacitor digital PA, 668
switching spectrum, 374
synthesizer, 612

synthesizer noise sources, 615
system-on-a-chip (SoC), 741

tank, 511
tank Q, 550
Tellegen's theorem, 135, 138
terminated transmission lines, 168
thermal noise, 299
thermal noise spectral density, 281
thermal noise variance, 282
third-order nonlinearity, 338
time division multiple access, 373
time-duplex division, 335
time-to-digital converter (TDC), 642
time-varying fields, 9
transceiver practical design concerns, 734
transducer factor, 212
transducer loss, 221
transducer parameters, 214, 211
transducer parameters properties, 216
transformer modeling, 69
transformers, 58, 156
transformers, integrated, 58
transimpedance amplifier (TIA), 482, 497
translational loop, 724
transverse electromagnetic (TEM), 28
transmission line, 13, 166
transmission lines transient response, 171
transmission zeros, 220
transmitter architectures, 719
transmitter mask, 365
transmitter mixer, 486
transmitter nonlinearity, 364
traveling wave, 169
twisted pair, 429
two-phase LO, 467
two-port, 130
two-port equivalent noise, 296
two-port oscillator, 539
two-port stability using, 194
type I PLL, 598, 599
type I PLL linear model, 599
type II PLL, 601
type II PLL analysis, 607
types of noise, 279

uniform plane wave, 26
unilateral two-port, 140
upconversion mixer, 442
uppressed-carrier double-sideband modulation, 110

varactor, 556
VCO drift, 596
VCO noise transfer function, 616
vector magnetic potential, 37
V–I converter, 448
voltage standing wave ratio, 169
voltage switching, 442
voltage-mode passive mixer, 484

wave propagation, 14, 26
wavelength, 13, 29
WCSP package, 743
Weaver receiver, 705
white noise, 283
wideband impedance transformation, 156
Wien bridge oscillator, 90
Wiener–Khinchin theorem, 104

Wilkinson power combiner, 659

XOR gate, 597
XOR phase detector, 597

Y-factor noise measurement, 323

zero-IF receiver, 694
zeros, 88

Printed in the United States
by Baker & Taylor Publisher Services